# Lecture Notes in Physics

Edited by J. Ehlers, München, K. Hepp, Zürich
R. Kippenhahn, München, H. A. Weidenmüller, Heidelberg
and J. Zittartz, Köln
Managing Editor: W. Beiglböck, Heidelberg

## 141

# Seventh International Conference on Numerical Methods in Fluid Dynamics

Proceedings of the Conference, Stanford University,
Stanford, California and NASA/Ames (U.S.A.)
June 23–27, 1980

Edited by W. C. Reynolds and R. W. MacCormack

Springer-Verlag
Berlin Heidelberg New York 1981

Editors

William Craig Reynolds
Mechanical Engineering Dept. Stanford University
Stanford, CA 94305, USA

Robert William MacCormack
Mail Stop 202A-1, NASA-Ames Research Center
Moffett Field, CA 94035, USA

ISBN 3-540-10694-4  Springer-Verlag Berlin Heidelberg New York
ISBN 0-387-10694-4  Springer-Verlag New York Heidelberg Berlin

Contents

V

## Editors' Preface

This volume of Lecture Notes in Physics contains papers presented at the Seventh International Conference on Numerical Methods in Fluid Dynamics, held at Stanford University and the NASA Ames Research Center in the U.S.A., June 23-27, 1980. The papers were selected from abstracts submitted from all over the world by three papers selection groups, one based in the USA, another in Europe, and the third in the USSR. Briefs of the papers were distributed at the Conference. This volume provides the full paper.

The book includes invited papers by D.R. Chapman, M.S. Longuet-Higgins, V.V. Rusanov and H. Viviand, plus 68 contributed papers. The invited papers appear first, followed by the contributed papers in alphabetical order by first author.

The Conference was attended by over 280 scientists. In addition to the strong representation from the USA, there were several delegations from the USSR, France, Germany, China, and many other countries. A list of the participants is given at the end of the volume.

The editors served as the general Conference Co-chairmen. We are indebted to our many colleagues who helped with the details of the meeting, but especially to Mamoru Inouye of the Ames Research Center, who supervised all of the local arrangements, and to Dianne Sinn, the Conference Secretary.

Financial support for the Conference was provided by the National Science Foundation, Air Force Office of Scientific Research and the Office of Naval Research through a grant arranged by the last. In addition, the National Aeronautics and Space Administration's Ames Research Center contributed through provision of facilities and transportation.

We are indebted to Dr. W. Beiglböck and the editorial staff of Springer-Verlag for valuable assistance in preparing these proceedings.


December 11, 1980

W.C. Reynolds
R.W. MacCormack
(Editors)


Papers Selection Committee Chairmen

U.S.A.    R.W. MacCormack
U.S.S.R.  O.M. Belotserkovskii
European  R. Temam

INTERNATIONAL CONFERENCE ON

NUMERICAL METHODS IN FLUID DYNAMICS


First Conference:   Novosibirsk, USSR, 1969
Second Conference:  Berkeley, California, USA, 1970
Third Conference:   Paris, France, 1972
Fourth Conference:  Boulder, Colorado, USA, 1974
Fifth Conference:   Enschede, the Netherlands, 1976
Sixth Conference:   Tbilisi, USSR, 1978
Seventh Conference: Stanford University and NASA/Ames, USA, 1980

TRENDS AND PACING ITEMS IN COMPUTATIONAL AERODYNAMICS

Dean R. Chapman
Stanford University


ABSTRACT

A perspective is presented of trends in computational aerodynamics, and of
important technology development items that pace future advanced applications.
From a survey of AIAA Journal papers published during the past two decades, the
growth trends and the progressively increasing emphasis on code development for
viscous, compressible, turbulent flow are illustrated.  These trends are reflected
in the chronology of introduction by the aerospace industry of new computational
methods in aircraft design.  Key pacing items outlined are:  automatic grid generation for nonlinear inviscid computations; advanced computers, improved efficiency
of numerical methods, and improved turbulence models for Reynolds-averaged Navier-
Stokes computations; advanced computers, time-dependent three-dimensional law-of-
the-wall, code development, improved efficiency of numerical methods, and improved
subgrid-scale turbulence modeling for large eddy simulations.


I.  INTRODUCTION

The evolution of computational aerodynamics in recent years has provided a
major new technological capability of recognized practical importance to the aircraft
industry.  This new capability can substantially increase airplane performance while
reducing risk, design time, and testing requirements (Dillner and Koper, 1979).
Moreover, it is evident that much more evolution of capability still lies ahead.

Of central concern to this paper is the circumstance that some quite diverse
items of technology pace the overall advances in computational aerodynamics.  These
"pacing items" represent key technology developments which will primarily determine
when a new and advanced level of computational capability may become feasible to use
in future aerodynamic applications.  The objectives of this paper are to provide a
synoptic look at underlying trends in computational aerodynamics and at various pacing
items relevant to further advances.  These pacing items refer to developments required
to reach a level at which industry could begin to apply advanced codes.  The actual extent of use in engineering design thereafter can involve other important factors not
considered herein.

II.  TRENDS

To obtain a perspective of the growth trends in computational aerodynamics, a
survey was made of the papers published in the AIAA Journal during the past 20 years.
Over this period the annual number of papers and synoptics did not vary widely,
usually ranging from 200 to 300 per year.  These were classified into the following
categories:  "linearized inviscid," mainly papers on panel methods; "nonlinear inviscid," e.g., transonic flow, supersonic blunt body flow, etc.; "boundary layers,"
including viscous shock-layer papers and papers coupling external flow codes to
boundary layer codes; "Navier-Stokes," including the parabolized and thin-layer
versions; "vortex dynamics;" and "large eddy simulation."  Results of the survey
are presented in Fig. 1.  The fraction of yearly AIAA papers involving computational
aerodynamics has grown from about 1% in the early 1960's to 22% in 1979.  The small
peak in 1966 reflects blunt-body papers relevant to the Apollo program, whereas the
strong growth trend beginning in 1971 reflects papers relevant to aircraft design.
While linearized inviscid computations have remained a small part of the total and
nonlinear inviscid methods have enjoyed a modest growth, the papers on viscous-flow
computation have contributed most of the pronounced growth.  This is to be expected
for papers relevant to aircraft aerodynamics.  Since practical computations of

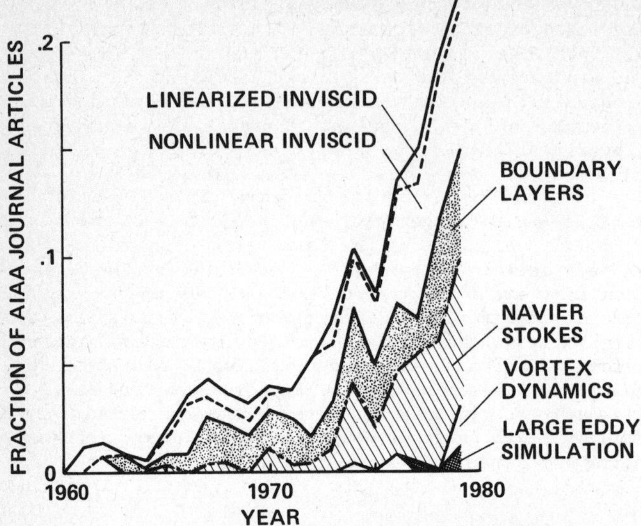



Figure 1.  Growth of computational aerodynamics

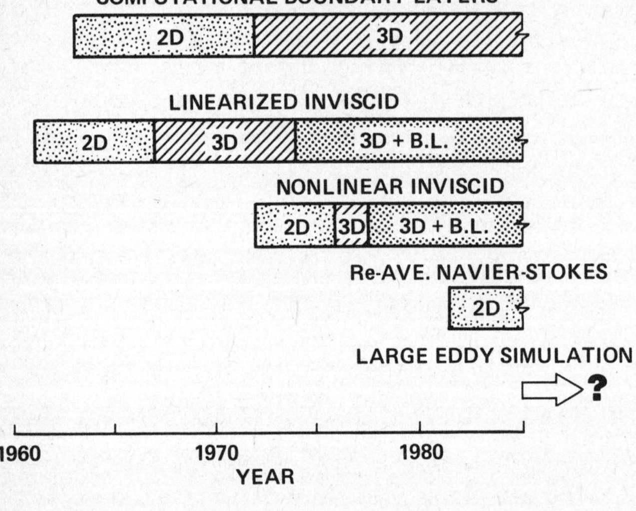



Figure 2.  Chronology of introduction of computational aerodynamic techniques in aircraft
design.

viscous turbulent flow necessarily involve some type of turbulence modeling, e.g. Reynolds-averaged or subgrid scale, it is anticipated that future turbulence modeling for compressible flow will become a pervasive part of computational aerodynamics.

A perspective of this growth trend can be grasped by reference to the other disciplines covered in the AIAA Journal.  Roughly one-half of the publications are devoted to aerodynamics, and one-half to the combined disciplines of propulsion, structural mechanics, thermophysics, and aircraft technology.  Thus, computational aerodynamics, in growing to nearly one-fourth of the AIAA articles, has grown to about one-half of the overall aerodynamics papers.

The growth trend over the past two decades has been reflected at various times throughout this period by the introduction in aircraft design of new and advanced computational aerodynamics methods.  The years during which successively advanced computational techniques were introduced are depicted in the bar graph of Fig. 2.  These data are based partly on publication dates of key papers, and partly on information obtained from colleagues in the aircraft industry.  In each case, the use of a new advanced stage begins with limited application to 2D airfoils, and later evolves to 3D as both computers and codes become further advanced.  It is noted that for both the linearized and the nonlinear inviscid stages, the 3D external flow codes were coupled to 3D boundary-layer codes shortly after both types of code were developed.

Although the Reynolds-averaged Navier-Stokes codes apparently have not yet been applied in practical design, the bar graph indicates that this is anticipated in the early 1980's.  Navier-Stokes codes developed for transonic aileron buzz, afterbody drag, and airfoil buffet onset now seem sufficiently advanced to expect initial application in the near future.  Practical applications of large eddy simulation, however, are many years away.  Both Figs. 1 and 2 are indicative of the main trend in computational aerodynamics, a progressive movement toward the eventual capability to compute compressible, viscous, turbulent flows over practical aerodynamics shapes. Since aircraft geometries are complex and the turbulent flow fields so intricate, this progression will extend over many years.

## III.  MAJOR PACING ITEMS

In Table I, a summary is presented of the author's view on the principal pacing items for four stages of approximation to the full Navier-Stokes equations:  linearized inviscid, nonlinear inviscid (including coupled boundary layer), Reynolds-averaged, and large eddy simulation.  This table refers to external aerodynamics of practical aircraft configurations.  Inasmuch as linearized inviscid panel methods are essentially mature engineering tools already used widely by industry, no major pacing item is listed for this particular stage.  Where more than one item is indicated, the list is in estimated order of importance.  For example, items believed to require the longest lead time and/or offer the greatest degree of potential advancement are listed first.  Reasons for the particular ordering are discussed later in this paper.

TABLE I

| Approximation level to equations of motion | Pacing items |
|---|---|
| Linearized inviscid | None |
| Nonlinear inviscid with coupled boundary layer | • Grid generation for complex geometries |
| Reynolds-averaged Navier-Stokes | • Advanced computers<br>• Improved efficiency of numerical methods<br>• Improved turbulence models<br>  – Large separated regions<br>  – Transitional type separations<br>  – Hypersonic flow |
| Turbulent eddy-simulation | • Advanced computers<br>{ • Time-dependent three-dimensional law of the wall (high Re applications)<br>{ • Code development (low Re applications)<br>• Improved efficiency of numerical methods<br>• Improved subgrid scale turbulence models |

(A) *Nonlinear Inviscid, Including Coupled Boundary Layer*.- For this stage of approximation only one major pacing item is listed — grid generation for complex geometry. More precisely, this item refers to the development of automatically-generated surface-adaptive grids required for the full potential transonic flow equations, the Euler equations, and the boundary layer equations.  The development of future advanced computers is not regarded as a major pacing item for this stage of approximation.  For example, with current algorithms (e.g., Holst, 1979) and current advanced computers, a transonic external flow computation for a practical wing-body with 200,000 grid points would require only about two minutes.  Inclusion of a coupled boundary layer code would add a comparable increment of time. Current computers and algorithms could handle more complex aircraft geometries with nacelles, pylons, winglets, etc., but the appropriate surface-adaptive grid generation techniques have not yet been developed.

A schematic illustration of the rate of past progress in computing 3D transonic flows over successively more complex configurations is shown in Fig. 3. For each degree of geometric complexity, the year that applicable codes were initially developed is indicated in ( ) for the small-disturbance approximation which uses mean-surface boundary conditions, and in [ ] for the full potential approximation which uses surface boundary conditions. The dates correspond respectively to the publication of: Ballhaus and Bailey (1972), and Jameson (1974) for a swept wing; Bailey and Ballhaus (1975), and Caughey and James (1977) for a simple tapered wing on a body with circular cross section; Boppe (1978), and Caughey and Jameson (1979) for a practical wing-fuselage geometry; and Boppe and Stern (1980) for the wing-fuselage-nacelle-winglet configuration. Six years were required to progress from the simple swept wing to a realistic wing-fuselage configuration. In each progressive step the full potential methods evolved one to two years after the corresponding small-disturbance methods. It is anticipated that new techniques, such as using embedded surface-adaptive grid systems in various flow zones, together with appropriate algorithms for joining embedded zones, will enable the transonic flow to be computed over a realistic aircraft configuration with nacelles, pylons, inlets, etc. In any event, grid generation with its associated algorithm development is currently the principal pacing item for nonlinear inviscid computations with or without coupled boundary-layer computations.

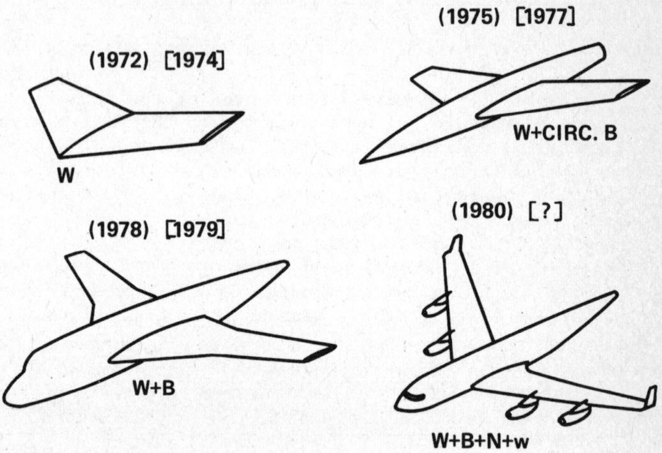


Figure 3. Growth in geometric complexity for three-dimensional transonic flow.

(B) *Reynolds-Averaged Navier-Stokes*.- As indicated in Table I, the major item pacing aerodynamic applications based on this stage of approximation is the development of advanced computers. This can be illustrated by the following tabulation of estimated computer memory and speed requirements (FLOPS = floating point operations per second) for some representative examples:

|  | Re | Grid megapoints | Memory megawords | Speed, megaFLOPS for 1-hr run |
|---|---|---|---|---|
| Compressor blade | $10^6$ | 0.5 | 15 | 140 |
| Wing body | $10^7$ | 2 | 60 | 600 |
| Wing body | $10^8$ | 4 | 100 | 1200 |

These estimated requirements (Chapman, 1979) correspond to an average grid point density per cubic boundary-layer thickness ($\delta$) of $10/\delta^3$, for example, or to $1/\delta$ streamwise, $20/\delta$ across the layer, and $1/2\,\delta$ spanwise. For the case of relatively low Reynolds numbers and simple geometries, such as applicable to a compressor blade, the requirements can be reasonably met by the most advanced new computers (e.g., Cyber 205). For the case of high Reynolds numbers and complex geometries of aircraft, however, the corresponding memory and speed requirements are well beyond commercially available computers of the near future. Thus the development of advanced computers is listed as the principal pacing item for computational aerodynamics based on the Reynolds-averaged Navier-Stokes equations.

A second major pacing item for this stage of approximation, as indicated in Table I, is the development of numerical methods with greatly improved computational efficiency. During the past 15 years, improvements in numerical algorithm efficiency have been roughly comparable to improvements in computer hardware efficiency. In the future, however, it is anticipated that algorithm improvements will fall behind hardware improvements because the former is now relatively closer to fundamental limits than the latter. Estimates have been made of the lower bound on the number of operations per grid point below which finite difference algorithms are not expected to fall. The results of such estimates are presented in Fig. 4. They were made by first counting the number of derivatives and other operations required to approximate the partial differential equations by finite differences, and then taking as the lower bounds one operation per derivative, and five iterations or time steps to reach a converged steady state (two iterations would be required just to know that convergence is attained). It was further assumed that a lower bound for operations involved in the iterative process would be a small fraction (10%) of the operations involved in approximating the governing partial differential equations by finite differences. Such estimates are believed to represent realistic lower bounds which may be approached in a decade or two, but likely not transgressed significantly. Current algorithm efficiency for nonlinear inviscid transonic flows ($2 \times 10^4$ ops/GP) is two orders of magnitude above the corresponding estimated lower bound; and for Reynolds-averaged Navier-Stokes computations of turbulent flows ($10^6$ ops/GP), three orders of magnitude above. Consequently, there appears to be at least one and perhaps two more orders of magnitude of potential improvement in the efficiency of numerical algorithms for computing viscous flows. Improvements of this magnitude are highly significant, although considerably less than the potential improvements in future computer hardware.

The third pacing item for Reynolds-averaged Navier-Stokes computations is that of improved turbulence models. Present models work reasonably well for flows with small amounts of separation.

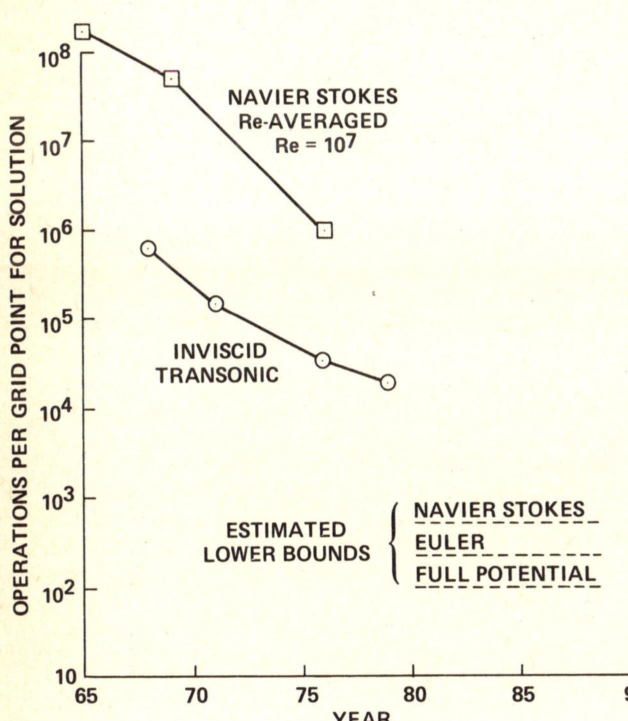


Figure 4. Operations per grid point and estimated lower bounds.

Improvements are needed to make the modeling more realistic for large separated regions, for transitional-type separated flows involving transition between separation and reattachment, and for hypersonic flows. The current lack of such improvements does not preclude computations of these types of flow from being made, but does prevent highly realistic results from being obtained. Such improvements would considerably broaden the present domain of applicability of computational aerodynamics.

(C) *Turbulent Eddy Simulation.*- For wall-bounded flows, a distinction in terminology is made between "large eddy simulation" and "transport eddy simulation." In the former the viscous sublayer is modeled, whereas in the latter it is computed. Subgrid-scale turbulence is modeled in both cases. Computations of turbulent channel flow have provided the first examples for both large eddy simulation (Deardorff, 1970) and transport eddy simulation (Moin, Reynolds and Ferziger, 1978). Since the eddies which transport the principal momentum and energy within the viscous sublayer are extremely small at high Reynolds numbers, transport eddy simulation imposes much greater demands on computer time and memory than large eddy simulation.

Practical applications of large-eddy simulation would require much more powerful computers than are presently available. Consequently, the most important pacing item for turbulent eddy simulation is that of developing advanced computers. The severity of this pacing item may be perceived from the table which follows. Some estimated requirements (Chapman, 1979) on memory and speed for large eddy simulation are listed. The same representative examples are considered as above, but the numbers are now expressed in giga ($10^9$) rather than megaunits.

|  | Re | Grid gigapoints | Memory gigawords | Speed, giga-FLOPS for 1-hr run |
|---|---|---|---|---|
| Compressor blade | $10^6$ | 0.2 | 2 | 30 |
| Wing-body | $10^7$ | 0.4 | 5 | 100 |
| Wing-body | $10^8$ | 1 | 15 | 300 |

Such requirements are well beyond the capacity of near-future computers.

Rough projections can be made of the approximate time when computers of the requisite capability for practical application of large eddy simulation might become feasible. Consideration will be given first to future memory, which is not anticipated to involve radically new technology, but mainly progressive refinements of the scale of integration of silicon chip technology. A plot is presented in Fig. 5 of the past history and projected future of computer central memory. Two sets of data are shown, one representing the envelope of maximum memory for scientific computers, and one for business computers (IBM). The discontinuities about 1971 correspond to the change from magnetic core memory to the much faster semiconductor memory. Greater memory

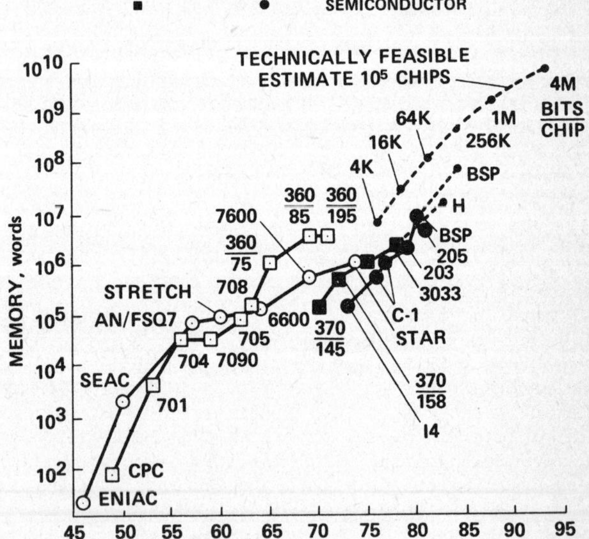


Figure 5. Growth in main memory for scientific and business computers.

was installed in business computers than in scientific computers during the period
1962-1979, although the corresponding difference now is becoming relatively small.
The dashed line represents an estimate of what appears to be technically feasible
without consideration of such things as economics, procurement times, etc. It corresponds to $10^5$ memory chips, a rough upper limit for acceptable mean time between
failures, and is drawn through successive points corresponding to the dates when
4K, 16K, and 64K bits/chip of random access memory first became available, and when
256K, 1M, and 4M chips are estimated to become available. New chips are relatively
expensive when first produced, and become much more cost effective a few years later.
Consequently, the actual maximum memory likely to be realized would be represented
by a curve roughly parallel to the curve of technical feasibility, but delayed in
time by perhaps 3 or 4 years. Hence, semiconductor memories in the range of 1- to
10-gigawords, sufficient for extensive practical aerodynamic applications of large
eddy simulation without I/O buffering, may be available in the 1990's.

Projections of the speed of future supercomputers are more uncertain than memory projections for three reasons: (1) several basically new microelectronic logic technologies are on the horizon which exceed the per-formance of silicon technology, (2) a new DOD program (VHSIC) is underway to accelerate the development of fast integrated logic cir-cuits (Sumney, 1980), and (3) there is not a driving mass market incentive for devel-oping faster logic circuits as there is for developing greater memory. In Fig. 6, a plot is presented of the past history and future projections of computer speed. As in Fig. 5, two curves are shown, one rep-resenting the envelope of maximum speed for scientific computers and one for business computers. It is evident that the

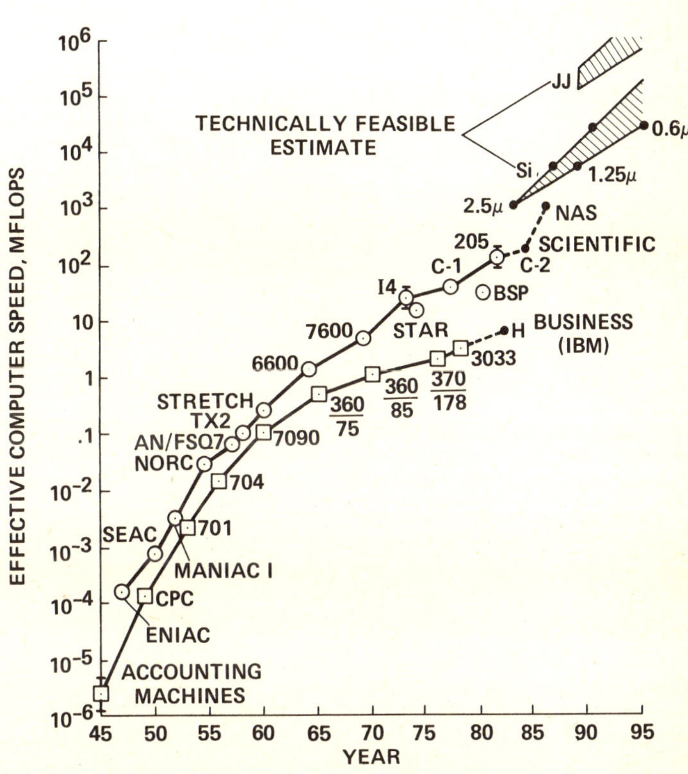


Figure 6. Growth in throughput speed for scientific and business computers.

demand for speed in large-scale scientific computation has always been greater than
in business information processing; and that the magnitude of this difference in
demand is diverging with time. Two zones representing future projections are shown,
one for conventional silicon technology, and one for the cryogenic Josephson Junction
(JJ) technology (a recent compilation of papers on this new technology has been
published in the March 1980 issue of the IBM Journal). Another new technology,
GaAs, which also shows promise, falls in between the zones shown for Si and JJ technology. For each of these two zones the lower boundary corresponds to the normal
pace of logic speed improvement, while the upper boundary corresponds to the estimated
pace if the VHSIC program goals are met. Again, it is noted that the projections

correspond to what appear technically feasible, not to what may be economically realizable. As in the case of memory projections, the anticipated speed performance would be roughly parallel to what is technically feasible but delayed several years in time. Hence, with silicon technology, the estimated requirement for practical large eddy simulation — computer speeds the order of $10^2$ gigaFLOPS — may not be met until the turn of the century, or perhaps the mid-1990's if the goals of the VHSIC program are met. With JJ or other comparable technology, however, such computer speeds might be feasible in the early 1990's. In any event, the requirements for practical large eddy simulation of aerodynamic flows are clearly paced primarily by the development of advanced computers.

The essential measure relevant to turbulent eddy simulation is grid-point resolution capability, i.e., the number of grid points for which flow computations can be made in a practical amount of time and cost. Depending upon the design of a particular computer system, this can be limited by either memory or speed. The historical progress in grid point resolution capability is shown in Fig. 7. To date, turbulent eddy simulations for research purposes have been made with 1- to 2-million grid points (10- to 20-hr runs on ILLIAC IV). In the near future this resolution capability should extend to about 5-million grid points (e.g., 10-hr run on Cyber 205), and, in the latter half of this decade, to about 50 million (e.g., 10-hr run on NAS). The pace of bringing such computers on-line will be the primary factor ·in determining when some limited applications of large eddy simulation could begin.

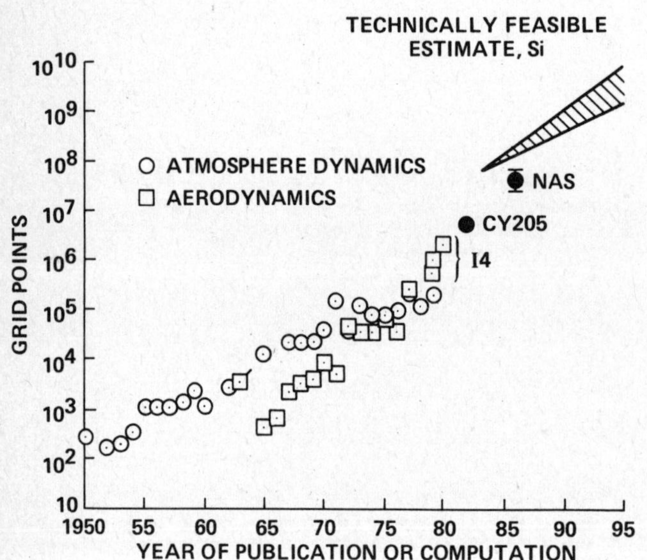


Figure 7.  Growth in grid point resolution capability.

The second major pacing item in advancing turbulent eddy simulation is that of viscous sublayer modeling. At aircraft Reynolds number, the grid point requirements for large eddy simulation (viscous sublayer dynamics modeled) are several orders of magnitude less than for transport-eddy simulation (viscous sublayer dynamics computed). Consequently, the development of an appropriate three-dimensional time-dependent model, to apply along the outer edge of the viscous sublayer as a lower-boundary condition for large-eddy simulation of the outer turbulent flow, would greatly reduce the computations, and therefore advance the time at which practical applications could be made. The two-dimensional time-average law-of-the-wall has long been known, but the appropriate generalization to three-dimensional time-dependent conditions for large-eddy-simulation purposes is not yet known.

The reason for the extremely large grid-point requirements in transport eddy simulation stems from the extremely small scale of the eddies that transport the principal amount of momentum and energy within the viscous layer.' A number of experimental investigations have indicated that the characteristic eddy structures are counter-rotating vortex pairs, highly elongate streamwise, that occur sporadically within the viscous sublayer. The "ejection" of uplifted fluid between the vortex pair accounts for a large fraction of the momentum transport and turbulence production. A sketch of the characteristic structure, adapted from Blackwelder and

Eckelmann (1979), is shown in Fig. 8. The mean spanwise spacing between vortex centers is 50 dimensionless wall units. The corresponding streamwise length is over 1000 wall units. To adequately resolve the thin sheet of fluid ejected between the vortices may require a spanwise grid spacing $\Delta z_+$ considerably less than 10 wall units, and a streamwise grid spacing $\Delta x_+$ less than 100. Thus an estimate of the value of the product $(\Delta z_+)(\Delta x_+)$ required for adequate resolution of such a structure would be the order of $10^2$ to $10^3$. Since the observations of Corino and Brodkey (1969) indicate that substantial Reynolds stress is produced by uplifted fluid elements having streamwise dimensions $\Delta x_+$ of 20 to 40, the product $(\Delta z_+)(\Delta x_+)$ for adequate resolution may be even lower than $10^2$. Resolution in the y-direction is not believed to be critical, since the number of points across the viscous sublayer can be maintained constant through coordinate stretching with changes in Reynolds number.

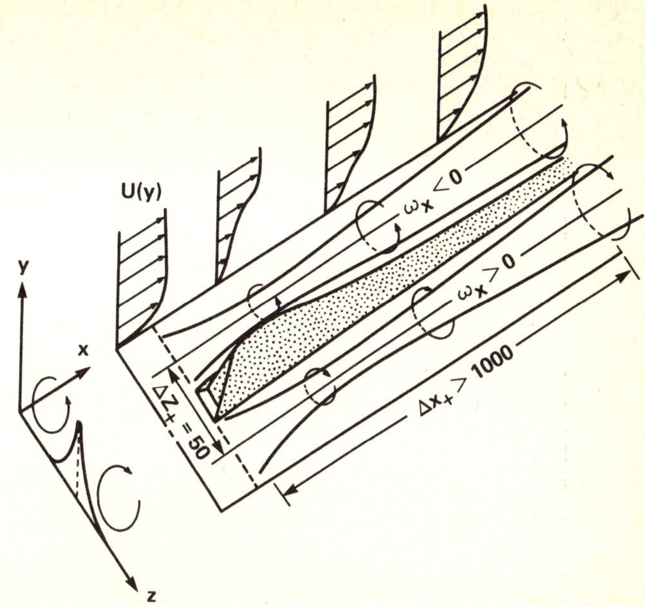


Figure 8. Model of counter-rotating streamwise vortex pair as characteristic structure in viscous sublayer, adapted from Blackwelder and Eckelmann (1979).

The number of grid points required in transport eddy simulations can be extraordinarily large at high Reynolds numbers. This may be seen from Fig. 9 showing some example requirements for turbulent eddy simulation of the flow over an airfoil. In the lower right of this figure, the approximate grid-point resolution capability in various years with various computers is shown for purposes of reference. At the relatively low Re of compressor and turbine blade operation, the grid point requirements are not excessively greater for transport than for large eddy

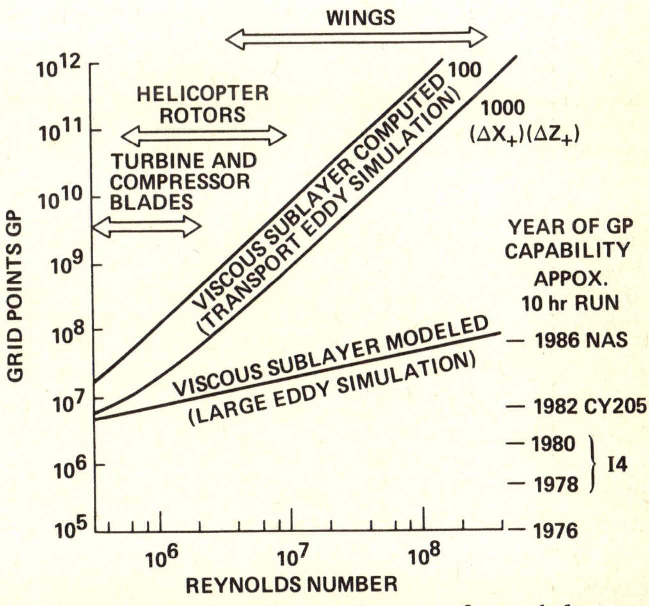


Figure 9. Grid point requirements for turbulent eddy simulation of the flow over an airfoil.

simulation; but at the high Re of aircraft wings and fuselages, the requirement for transport-eddy simulation can be $10^3$ to $10^4$ times greater than for large-eddy simulation. Resolution of the viscous sublayer dynamics at high Re would require that over 99.9% of the grid points be placed in less than 1% of the boundary layer volume. The development of an appropriate three-dimensional time-dependent law-of-the-wall for use in large-eddy simulation is regarded as a major pacing item, second in importance only to the development of advanced computers.

It is significant that with the introduction next year of the Cyber 205 computer, research calculations of airfoil flow using large eddy simulation will become feasible for some low Re applications. Since computer codes for such simulations now are far from available, code development is also a pacing item in the overall advancement of such applications as indicated in Table I.

Two other pacing items for turbulent eddy simulation also listed in Table I are improved efficiency of numerical methods and improved models of subgrid scale turbulence. As noted above, the estimated factor of potential improvement in future numerical algorithm efficiency is believed to be between about 10 and 100. Such improvements would be equivalent to corresponding factors of improvement in computer speed. Improvements in the realism of subgrid scale turbulence models would enable larger grid spacing to be used for a given overall accuracy, and hence also would reduce the computer power required for practical computations. It is difficult to estimate, however, the potential magnitude of future improvements in this particular area.

Acknowledgement

The Ames Research Center is thanked for support in the preparation of this paper, portions of which were developed while the author was an employee of NASA.

<h3 style="text-align:center">REFERENCES</h3>


Bailey, F. R. & Ballhaus, W. F. 1975 Comparisons of Computed and Experimental Pressures for Transonic Flows About Isolated Wings and Wing-Fuselage Combinations, NASA SP-347, Part II, pp. 1213-1231.

Ballhaus, W. F. & Bailey, F. R. 1972 Numerical Calculation of Transonic Flow About Swept Wings. AIAA Paper 72-677.

Blackwelder, R. F. & Eckelmann, H. 1979 Streamwise Vortices Associated with the Bursting Phenomenon. *J. Fluid Mech*. vol. 94, pt. 3, pp. 577-594.

Boppe, C. W. 1978 Computational Transonic Flow About Realistic Aircraft Configurations. AIAA Paper 78-104.

Boppe, C. W. & Stern, M. A. 1980 Simulated Transonic Flows for Aircraft with Nacelles, Pylons, and Winglets. AIAA Paper 80-0130.

Caughey, D. A. & Jameson, A. 1977 Numerical Calculation of Transonic Potential Flow About Wing Fuselage Combinations. AIAA Paper 77-677.

Caughey, D. A. & Jameson, A. 1979 Recent Progress in Finite-Volume Calculations for Wing-Fuselage Combinations. AIAA Paper 79-1513.

Chapman, D. R. 1979 Combutational Aerodynamics Development and Outlook. *AIAA J.*, vol. 17, no. 12, pp. 1293-1313.

Corino, E. R. & Brodkey, R. S. 1969 A Visual Investigation of the Wall Region in Turbulent Flow. *J. Fluid Mech*. vol. 37, pp. 1-30.

Deardorff, J. W. 1970 A Numerical Study of Three-Dimensional Turbulent Channel Flow at Large Reynolds Numbers. *J. Fluid Mech*. vol. 41, pp. 453-480.

Dillner, B. & Koper, C. A. 1979 The Role of Computational Aerodynamics in Airplane Configuration Development. Paper no. 15, AGARD Flight Mechanics Panel Symposium — Munich, Germany.

Holst, T. L. 1979 A Fast, Conservative Algorithm for Solving the Transonic Full-Potential Equation.  AIAA Paper 79-1456, Williamsburg, VA July 23-25, 1979.

Jameson, A. 1974 Iterative Solution of Transonic Flows Over Airfoils and Wings Including Flows at Mach 1.  *Comm. Pure and Appl. Math* vol. 27, pp. 283-309.

Moin, P., Reynolds, W. C., & Ferziger, J. H. 1978 Large Eddy Simulation of Incompressible Turbulent Channel Flow.  Stanford Univ. Thermosciences Div. Rept. TF-12.

Sumney, L. W. 1980 VLSI With a Vengeance.  *IEEE Spectrum,* April, pp. 24-27.

# POLYGON TRANSFORMATIONS IN FLUID MECHANICS

by Michael S. Longuet-Higgins

Department of Applied Mathematics and Theoretical Physics,
Silver Street, Cambridge, England and Institute of
Oceanographic Sciences, Wormley, Surrey.

## I.  INTRODUCTION

In many boundary problems in fluid mechanics where the departures of the boundary from a plane or smooth surface are sufficiently small, a linear perturbation theory can be used, for example in slender ship theory or in the linearised theory of water waves. In other problems such as those to be considered here, we cannot make this assumption.  The boundary, though infinite in extent, may have sharp corners which strongly affect the flow. For example, in gravity waves of limiting steepness there can be sharp corners at the wave crests, as was first shown by Stokes. Similarly in ripples on sandy beaches quite sharp crests are also observed.  To calculate the flow in such cases requires essentially nonlinear methods.

In the present paper we shall describe a technique which is especially suited to cases when the boundary of the fluid has infinitely many sharp-pointed crests, spaced periodically in a horizontal direction. The method involves a very simple conformal transformation of the surface onto the sides of a regular polygon. From there, by a Schwartz-Christoffel transformation, it is easy to map the flow onto the interior of a unit circle.

We shall give three applications. In the first of these the motion in a steep, irrotational gravity wave on deep water is accurately mapped onto the interior of a regular hexagon. This enables the Lagrangian trajectories of the fluid particles to be traced very simply, with the corresponding mean-drift velocities.

Secondly, the problem of the oscillating flow over steep sand ripples is solved by mapping the fluid flow onto the exterior of a regular pentagon. The separation of the boundary layer at the crest of each ripple is represented by the creation of point vortices, with strengths determined by Prandtl's rule. The calculation then proceeds by time-stepping on the set of vortices. It is found that at the end of each stroke of the fluid motion the vortices tend to aggregate into a vortex-pair, which then escapes into the interior. The method enables the total stress on the boundary to be evaluated very simply in terms of the vortex elements.

Though the above model is essentially inviscid, some allowance can be made for finite Reynolds number by assuming each vortex to have an inner core, whose radius increases like $(time)^{\frac{1}{2}}$. Coalescence of vortices is assumed to occur when the cores of adjacent vortices of the same sign overlap.

In the third application, the method is used to study the development of a unidirectional flow over a rough boundary. The shear stress is calculated as a function of the time. The calculation may be relevant to the input of momentum to steep ocean waves by the wind.

## 2. THE POLYGON TRANSFORMATION

Consider a fluid boundary with angled crests (such as one of those shown in Figure 1) and let the position of the crests be given by

$$z = x + iy = mL \qquad (2.1)$$

where $L$ is the wavelength and $m$ is any integer. Then the transformation

$$\zeta = e^{-ikz}, \qquad kPL = 2\pi , \qquad (2.2)$$

takes any $P$ successive crests into the $P$ vertices of a regular polygon on the unit circle $|\zeta| = 1$ .

If we take polar coordinates $(r,\theta)$ in the $\zeta$-plane, so that $\zeta = r\,e^{i\theta}$ , then the equation of each side has the form

$$r\cos(\theta - \theta_j) = \cos(\pi/P) \qquad (2.3)$$

where $\theta_j$ corresponds to the mid-point of that side. Since

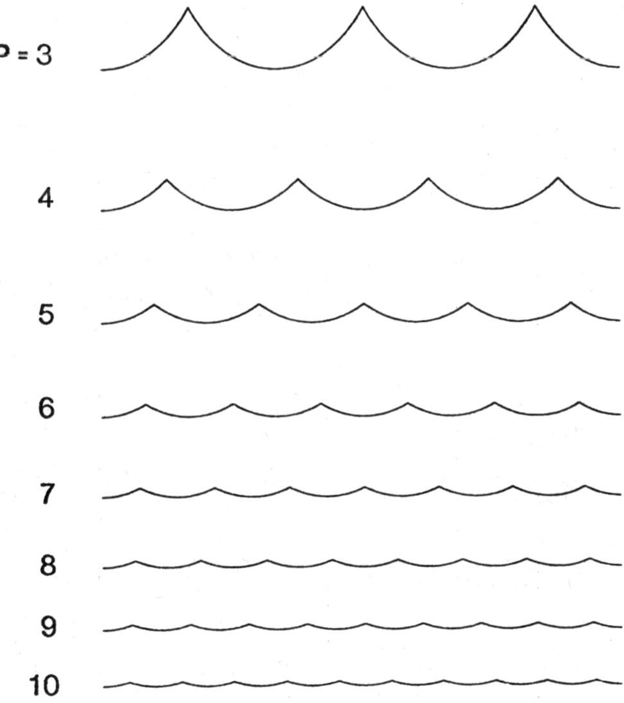


Figure 1.  Periodic wave-forms derived from a regular
P-sided polygon:  P = 2,...10 .

$$k(x + iy) = i \ln \zeta = -\theta + i \ln r \qquad (2.4)$$

it follows that the equation of the boundary in the original (x,y) plane must be

$$k(y - y_j) = \ln \sec \; k(x - x_j) \qquad (2.5)$$

where $x_j$ and $y_j$ are constants, and

$$|k(x - x_j)| \lessapprox \pi/P \; . \qquad (2.6)$$

The wave steepness, that is the ratio of the crest-to-trough height $H$ to the wavelength $L$ , is given by

$$\frac{H}{L} = \frac{\ln \sec (\pi/P)}{2 \, \pi/P} \; . \qquad (2.7)$$

Because the transformation is conformal, the interior angle at the wave crest is equal to the interior angle of the polygon, that is to say $\pi(1 - 2/P)$ .

To handle the problem of the flow, either above or below the wave profile, we transform the exterior or interior of the polygon onto the <u>interior</u> of a unit circle, by means of a Schwartz-Christoffel transformation (see Nehari 1952). Thus the interior of the polygon in the $\zeta$-plane is mapped onto the interior of the unit circle $|W| = 1$ by the transformation

$$\zeta = K \int_0^W \frac{dW}{(1 - W^P)^{2/P}} \qquad (2.8)$$

where $K$ is a constant giving the scale of the transformation. In (2.8) the origins of $\zeta$ and $W$ correspond, and $\zeta = 1$ corresponds to $W = 1$ provided that

$$\frac{1}{K} = \int_0^1 \frac{d\zeta}{(1 - \zeta^P)^{2/P}} = \frac{(-2/P)! \; (1/P)!}{(-1/P)!} \qquad (2.9)$$

where $(\;)!$ denotes the factorial function.

Similarly to map the <u>exterior</u> of the polygon onto the interior of the unit circle we write

$$\zeta = \zeta_o - K' \int_{W_o}^W \frac{(1 - W^P)^{2/P}}{W^2} \, dW \qquad (2.10)$$

where $\zeta_o$ and $W_o$ are suitably chosen constants (neither equal to 0) and

$$\frac{1}{K'} = \frac{(2/P)! \; (-1/P)!}{(1/P)!} \; . \qquad (2.11)$$

We shall see below that the above transformations are not necessarily confined to integral values of $P$ .

### 3.  MASS TRANSPORT IN STEEP GRAVITY WAVES

It is well known that in a progressive, irrotational water wave of small amplitude
the orbits of the particles are not quite closed; superposed on the orbital motion is a
small horizontal motion known as the mass-transport, or Stokes drift velocity. In the
linearised theory of surface waves, where the orbital velocity is small compared with
the phase speed  $c$ , this mass-transport velocity is only a second-order quantity.
But in waves of finite amplitude, and particularly in limiting waves with a sharp
angle at the crest, the orbital velocity becomes comparable to  $c$ . What is then the
mass-transport velocity is this case?

The interior angle at the crest of a progressive, irrotational gravity wave was
shown by Stokes (1880) to be equal to $120^{\circ}$. Taking  $P = 6$  in equation (2.7) we obtain  $H/L = 0.1374$, which is close to the steepness 0.1411 of limiting waves in deep
water. In fact the profile of the free surface as given by equation (2.5) is graphically indistinguishable from that calculated by Michell(1893), Yamada (1957) and
others. (For a recent review see Longuet-Higgins 1980.)

Now in a frame of reference moving with the phase-speed  $c$  the flow appears
steady, and the velocity potential  $\chi = \phi + i\psi$  must be given by

$$\chi = -i(c/k)\ln W \tag{3.1}$$

since this has the correct asymptotic form $(\chi \sim -cz)$  as  $y \to -\infty$ , and also satisfies
the boundary condition  $\psi = $ constant on  $|W| = 1$ . The pressure at the upper surface
is also nearly, but not quite, constant (see Longuet-Higgins 1973). According to (3.1)
the streamlines in the W-plane are concentric circles. In fact if we write $W = \rho e^{i\alpha}$ ,
then

$$\phi = (c/k)\,\alpha, \qquad \psi = -(c/k)\ln\rho \;. \tag{3.2}$$

In the physical z-plane, the streamlines are as shown in Figure 2.

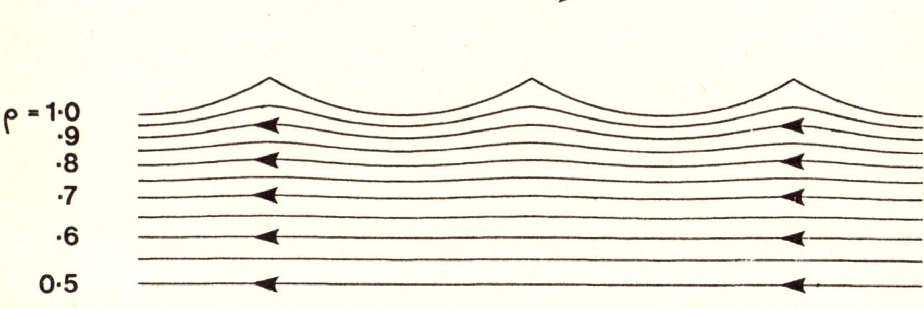


Figure 2.   Streamlines in the irrotational flow given
by (3.1) when P = 6 .

From the above expressions we can now calculate quite accurately the orbital motion in a deep-water wave of maximum steepness. For the coordinates (X,Y) of a particle in a stationary frame of reference are related to those in the moving frame by

$$X + iY = (x + iy) + ct + const .$$  (3.3)

where $t$ is the orbital time, given in general by

$$t = \int \left| \frac{dz}{d\chi} \right|^2 d\phi$$  (3.4)

where $\chi = \phi + i\psi$ is the potential of the steady flow (see Longuet-Higgins 1979). In the present case since

$$\frac{dz}{d\chi} = \frac{dz}{d\zeta} \frac{d\zeta}{dW} \frac{dW}{d\chi} = \frac{i}{k\zeta} \frac{K}{(1-W^6)} \frac{ikW}{c}$$  (3.5)

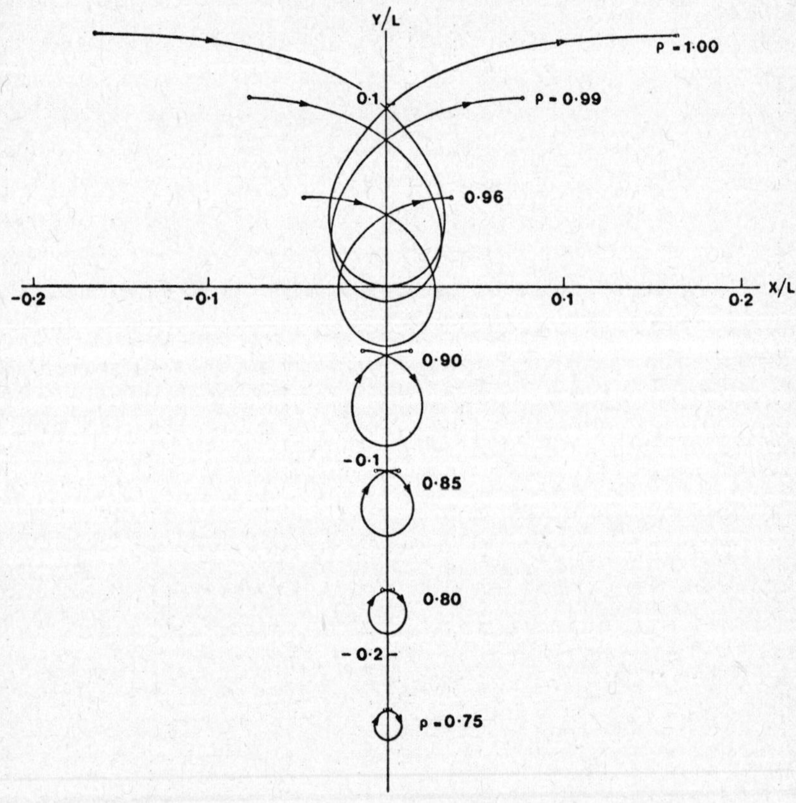


Figure 3.   (from Longuet-Higgins 1979). Particle orbits
in a steep, progressive, irrotational wave.

it follows that

$$ ct = \int \left| \frac{KW}{\zeta(1-W^6)^{1/3}} \right|^2 d\alpha \qquad (3.6) $$

The particle orbits calculated in this way are shown in Figure 3, from Longuet-Higgins (1979). (The origin of (X,Y) was there taken in the wave trough.)

From Figure 3 it can be seen how rapidly the mean displacement of a particle decreases as its depth below the surface is increased. At the free surface itself the mean speed of advance  U  of the particles is given by

$$ U/c \doteq 0.248 \quad . \qquad (3.7) $$

In other words, the drift velocity at the surface is about one-quarter of the phase speed. Further details are given in the paper just quoted.

## 4.  FLOW OVER SAND RIPPLES

Turning now from the upper to the lower boundary of the ocean, we note that the "vortex-ripples" commonly found on the bed of sandy coastlines, just seawards of the swash zone, often have a sharp-crested form such as shown in Figure 26 of Lofquist (1978). It was observed by Bagnold (1946) and others that during each stroke of the oscillating flow associated with sea waves, a vortex tends to form in the lee of each ripple crest. This produces a component of velocity _up_ the lee slope, tending to sweep up sand grains from the troughs of the ripples, and restore them to the ripple crests.

It was also observed by Bagnold, and confirmed by Carstens et al. (1969), that the main features of the flow were largely independent of the _frequency_ of the oscillation, with certain limits, but depended chiefly on the total horizontal excursion of the fluid particles in relation to the length  L  of the ripples. The size and density of the sand grains appears less important, except in so far as these affect the form of the ripples, presumably through the effect of grain shape on the "angle of repose" of the sand grains (see Nielsen 1979).

Further the drag exerted by the ripple bed on the fluid above was found to be remarkably high, with mean-drag coefficients  $\overline{C}_D$  of order $10^{-1}$, some two orders of magnitude  greater than for steady flows.

The observations point to an inviscid mechanism for momentum transfer. So far theoretical attempts to calculate the flow (e.g. Sleath 1975) have been confined to low Reynolds numbers. Now the transformation described above in Section 2 provides an opportunity to try an inviscid approach.

Comparison of Lofquists' "natural ripple" profile with the theoretical profiles of Figure 1 suggests that the case  P = 5  is typical. The angle at the ripple crest in that case is $108^{\circ}$, the angle in a pentagon. Accordingly we may transform five successive ripple crests first into a pentagon and then by transformation (2.10) onto the interior of a unit circle in the W-plane.

Suppose the fluid above the "ripples" is started from rest, in such a way that far above the surface the flow becomes a uniform horizontal stream of strength $U(t)$. Initially the flow will be irrotational, and so, by the same argument as in Section 3, the velocity potential will be given by

$$\chi = -i(U/k) \ln W \ . \tag{4.1}$$

This represents a circular vortex in the W-plane. In the z-plane the streamlines are as shown in Figure 4.

At subsequent times we assume the flow to separate at the ripple crests. To study the overall features of the resulting flow (though not the fine details) we use the method of discrete vortices as reviewed, for instance, by Clements and Maull (1975); see also Saffman and Baker (1979).

In the present analysis the representation of the vortices is particularly simple. The velocity potential $\chi$ associated with a single vortex of strength $\Gamma_n$ situated at the point $W_n$ , together with its image in the circle $|W| = 1$ is

$$\chi = \frac{\Gamma_n}{2\pi i} \ln \frac{W - W_n}{W - \overline{W}_n} \tag{4.2}$$

where $\overline{W}_n = 1/W_n^*$ is the image point. (A star denotes the complex conjugate.) To these we must add, by symmetry, similar vortices at the points $W_n e^{2ij\pi/P}$ , $j = 1,\ldots(P-1)$ giving altogether a system of $2P$ vortices with potential

$$\chi = \frac{\Gamma_n}{2\pi i} \ln \frac{W^P - W_n^P}{W^P - \overline{W}_n^P} \tag{4.3}$$

(see Figure 5). For the complete velocity potential we have in general

$$\chi = -i(U/k) \ln W + \sum_n \frac{\Gamma_n}{2\pi i} \ln \frac{W^P - W_n^P}{W^P - \overline{W}_n^P} \tag{4.4}$$

where the summation is over all the vortices in one typical ripple length, say $0 < x < L$ .

The motion of the vortices themselves is given by

$$\frac{dW_n^*}{dt} = \frac{|W_n|^2}{W_n} \left[ \frac{iU}{k} + \sum_{m \neq n} \frac{\Gamma_m}{2\pi i} \left( \frac{PW_n^P}{W_n^P - W_m^P} - \frac{PW_n^P}{W_n^P - \overline{W}_m^P} \right) \right.$$

$$\left. + \frac{\Gamma_n}{2\pi i} \left( \frac{P-1}{2} - \frac{PW_n^P}{W_n^P - \overline{W}_n^P} + \frac{kW_n}{2iW_n} + \frac{1}{1 - W_n^P} \right) \right] \tag{4.5}$$

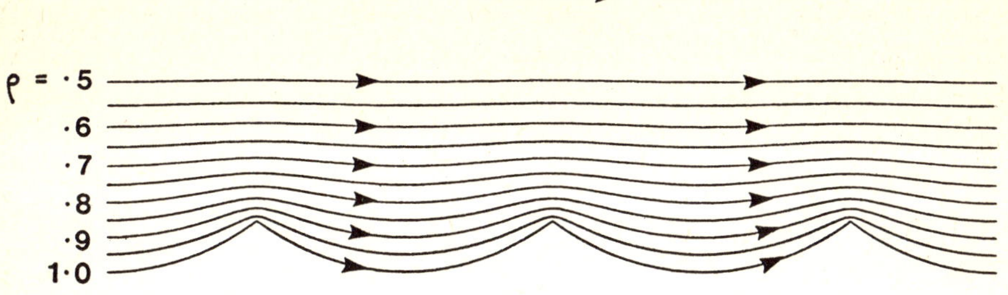



Figure 4.  Streamlines in the irrotational flow given by (4.1) when P = 5 .

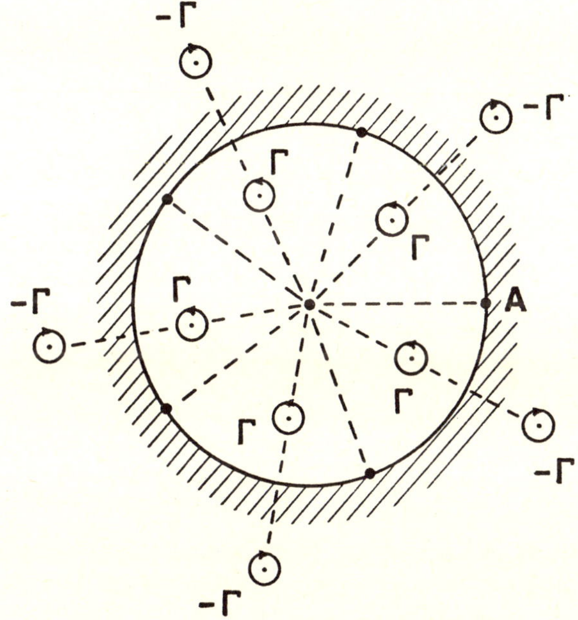



Figure 5.  A P-fold system of vortices and their images in the unit circle  $|W| = 1$ , when P = 5 .

where
$$W_n = \left(\frac{dW}{dz}\right)_{W=W_n} = \frac{i\,k\,\zeta_n W_n^2}{K(1 - W_n^P)^{2/P}} \tag{4.6}$$

(see Longuet-Higgins 1981). The last two terms in (4.5) arise from the well-known "Routh correction".

Following Clements (1973) we assume that vortices are shed from a fixed reference point $W_o = 1 - \delta$ in the neighbourhood of the crest $W = 1$ , at a rate corresponding to Prandtl's rule for the flux of circulation $\Gamma$ :

$$\frac{d\Gamma}{dt} = \tfrac{1}{2}|Q|^2 \tag{4.7}$$

where $Q$ is the (vector) velocity at $W_o$ .

In the present computation, one new vortex was created every three time-steps $\Delta t$, the circulation $\Gamma_n$ being given by

$$\Gamma_n = \sum_j \tfrac{1}{2}|Q_j|^2 \Delta t \tag{4.8}$$

where a suffix $j$ denotes the value of the $j$th time-step and summation is over the steps since creation of the previous·vortex. The initial position $z_n$ of the new vortex was taken as

$$z_n = z_o + \sum_j Q_j \Delta t \tag{4.9}$$

$z_o$ being the fixed position corresponding to $W_o$ . At the $j$th time-step each existing vortex was advanced by an amount

$$W_n(t_j) - W_n(t_{j-1}) = \begin{cases} \dot{W}_n(t_j)\,\Delta t & j = 1 , \\[2mm] \left[\tfrac{3}{2}\dot{W}_n(t_j) - \tfrac{1}{2}\dot{W}_n(t_{j-1})\right]\Delta t , & j = 2 ,3,.. \end{cases} \tag{4.10}$$

where $\dot{W}_n = dW_n/dt$ is given by (4.5).

When programmed in FORTRAN on the IBM 370-95 at Cambridge University the computation took about 3 minutes for 250 time-steps. Thereafter each time-step took a time nearly proportional to the square of the number of vortices.

The results were found to be fairly insensitive to variations in the value of $\delta$ (c.f. Clements 1973) provided $\delta$ was of order 0.1, corresponding to $y/L \simeq 0.038$. They were independent of the time-step $\Delta$ , provided this was not greater than 0.025. Otherwise, some vortices could be lost by crossing the boundary $\rho = 1$ .

Figure 6 shows some results corresponding to the sinusoidal oscillation

$$U = \begin{cases} 0, & t < 0, \\[2mm] a\omega \sin \omega t, & t > 0, \end{cases} \tag{4.11}$$

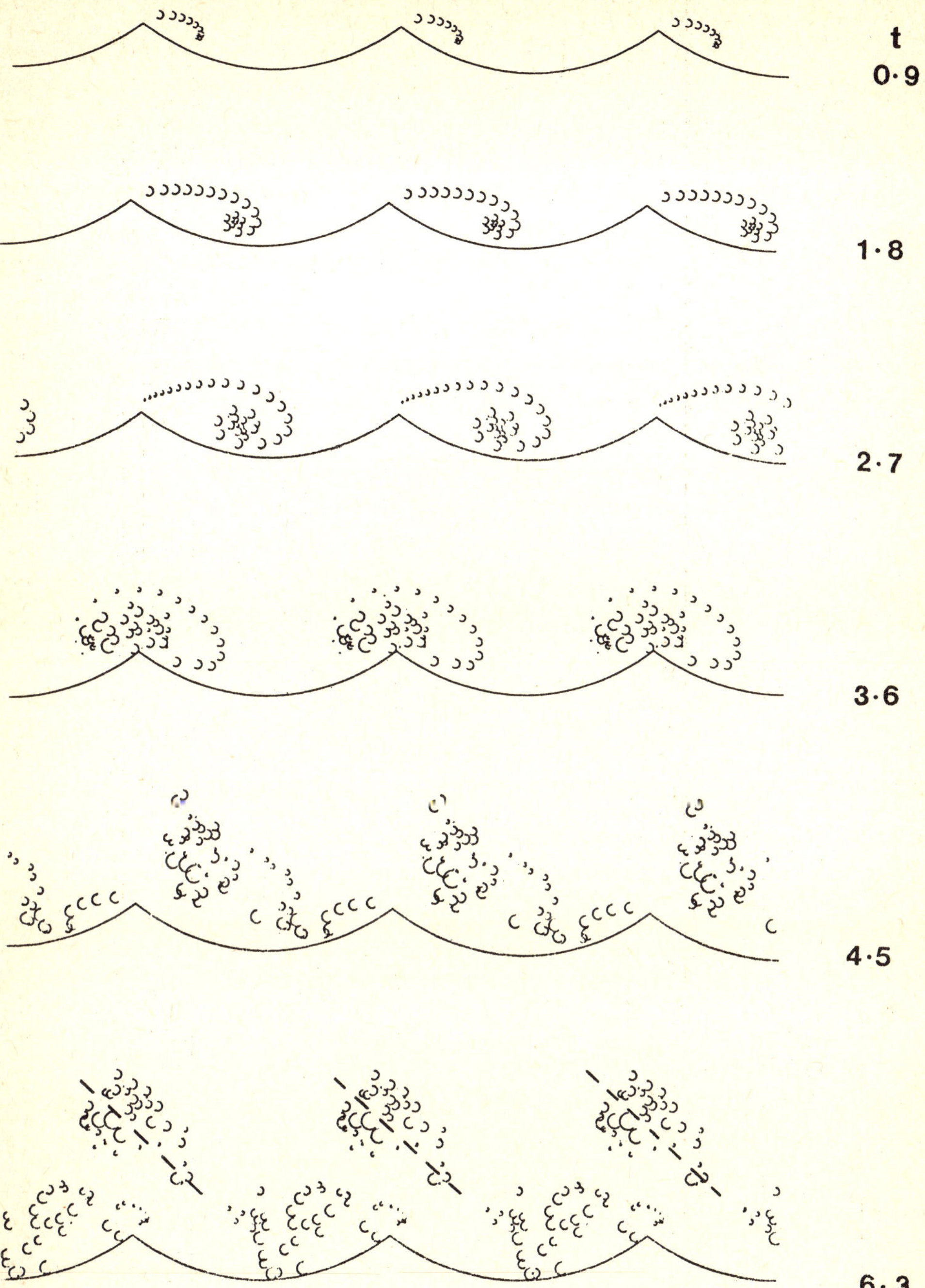



Figure 6. Point-vortices in the sinusoidal flow (4.11),
when P = 5 , k = 1 .

The greatest excursion of the vortices to the right occurs when $\omega t \doteq 2.7$. By this time the stream velocity $U$ is slowing down and the strength of the vortices shed at the crest is quite small. At $\omega t = 3.15$ the velocity at the crest is already reversed. On the return stroke ($\pi < \omega t < 2\pi$) the original cloud of vorticity passes back over the ripple crest, dragging with it some shed vortices of the opposite sign. At $\omega t = 6.30$ the initial vortex comes almost to rest above the crest two ripple-lengths to the left of its origin, having risen to a height of about $\frac{1}{2}L$ .

When $\omega t = 6.3$ , the clouds of vorticity above the crests are detached from the fluid below, and it is possible to draw a line at about $y = 0.2L$ separating these from the vorticity remaining near the surface. On examining the clouds more closely it can be seen that each of them is divided roughly into two halves, by an axis at $45^\circ$ to the horizontal. Above the axis the vorticity has mainly the same sign as the original vortex, i.e. it is negative. Below the axis the vorticity is mainly positive. If we now sum the total positive circulation $\sum_{\Gamma>0} \Gamma_n$ within the cloud it is found to be 0.710, compared with the total negative vorticity, which is $-0.712$. The total circulation in the cloud is therefore almost exactly zero, showing that the cloud is, in effect, a vortex-pair. It is this property that enables the cloud to escape from the neighbourhood of the boundary.


## 5.  MOMENTUM AND DRAG

Expressions for the total momentum and drag on the boundary may easily be derived. By contour integration it can be shown that the total horizontal momentum $I$ contained in one ripple length between $-\frac{1}{2} < x < \frac{1}{2}L$ , say, and between the surface $y = y_s$ and a distant level $y = y_\infty$ is given by

$$I = LU(y_\infty + h) + I_u + I_v \tag{5.1}$$

where

$$\left.\begin{aligned} kI_u &= LU\,(kh + \ln K) \\ kI_v &= -\sum_n \Gamma_n \ln |W_n| . \end{aligned}\right\} \tag{5.2}$$

The two components $I_u$ and $I_v$ arise from the uniform stream and from the shed vortices respectively. Then by considering the rate of change of the momentum, which comes partly from the horizontal force $F$ on the boundary and partly from the horizontal pressure gradient at infinite, we obtain

$$F = F_u + F_v \tag{5.3}$$

where

$$F_u = -\frac{dI_u}{dt}, \qquad F_v = -\frac{dI_v}{dt} . \tag{5.4}$$

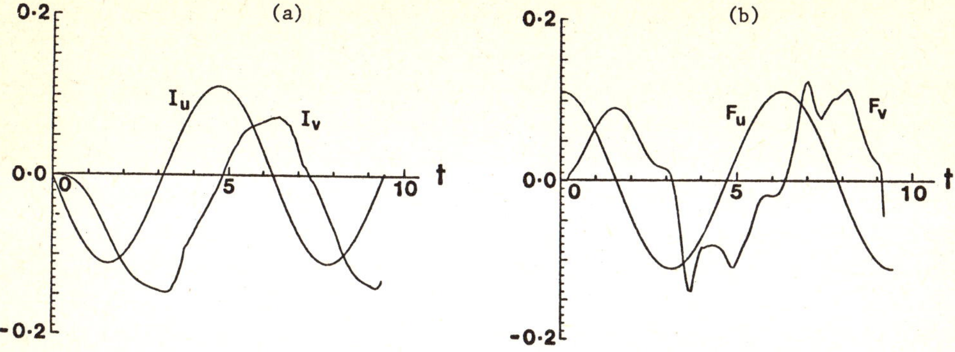


Figure 7. Components of the horizontal momentum I and
of the horizontal force F on the bottom in the
sinusoidal flow (4.11).

Clearly $F_u$ is associated with an added inertia due to the form of the ripples, whereas $F_v$ arises from the shed vortices.

Figure 7(a) shows a time-history of $I_u$ and $I_v$ in the sinusoidal motion (4.11). The consolidation of the vortices at $\omega t = 2\pi$ makes almost no difference to the subsequent graphs. Clearly $I_u$ and $I_v$ are almost $90^o$ out of phase, and like-wise so are $F_u$ and $F_v$. Whereas the inertial term $F_u$ dissipates on the average, no energy, the term $F_v$ is equivalent to a mean drag coefficient

$$\overline{C}_D = \frac{3\pi}{4LU_o^{\,2}} \ \overline{F_v \sin t} \tag{5.5}$$

where an overbar denotes the time-average.

## 6. ARTIFICIAL VISCOSITY

As the computation proceeds, the number of point-vortices below a given level, say $y = 0.25L$, is found to increase indefinitely, and the close approach of random pairs of vortices can give rise to unrealisticly high values of $dW_n/dt$. As a result the computation tends to become "noisy". To avoid this effect we may exclude the neighbourhood of each vortex by a small circle of radius $\delta_n$, say, inside which the velocity is bounded. In fact we assume a quasi-solid core, such that when $|W - W_n| < \delta_n$ the contribution to the velocity field from the vortex at $W_n$ is mul-tiplied by a factor $|W - W_n|^2/\delta_n^2$, the contributions from the other vortices remain-ing unaltered.

This has the desired consequence that the kinetic energy of the flow remains finite. In fact the contribution $E_n$ to the total energy from the neighbourhood of the nth vortex is given by

$$E_n \sim \frac{\Gamma_n^2}{4\pi}\left[\ln(1/\delta n) + \frac{1}{8\pi}\right] \tag{6.1}$$

which is infinite only when $\delta_n \to 0$.

In addition we may assume that the radius $\delta_n$ of the core increases with the time, according to

$$\delta_n = [\varepsilon(t - t_n)]^{\frac{1}{2}} \tag{6.2}$$

where $\varepsilon$ is a small constant, resembling a kinematic viscosity, and $t_n$ denotes the time of generation of the nth vortex. This expansion of the core is similar to that produced within the isolated vortex given by

$$\psi_r = \frac{\Gamma}{2\pi r}(1 - e^{-r^2/4\nu t}) \tag{6.3}$$

$\nu$ being the kinematic viscosity (see Lamb 1932, p.595). The transverse velocity $\psi_r$ is a maximum when $r = 2.24(\nu t)^{\frac{1}{2}}$. We may therefore take $\varepsilon = 5.0\nu$ very roughly.

It is interesting that the rate of loss of energy due to the expansion of the core radius $\delta_n$ is, from (6.1) and (6.2),

$$-\frac{dE_n}{dt} = \frac{\Gamma_n^2}{8\pi}\frac{1}{t - t_n} \tag{6.4}$$

independently of $\varepsilon$; whereas the viscous dissipation in the vortex (6.3) is

$$\int_0^\infty \int_0^{2\pi} \nu(\psi_{rr} - \psi_r/r)^2 \; r \, dr \, d\theta = \frac{\Gamma^2}{8\pi t} \tag{6.5}$$

independently of $\nu$.

Whenever the cores of two neighbouring vortices with circulations $\Gamma_n$ and $\Gamma_m$ of the same sign overlap, that is when

$$|W_n - W_m| < \delta_n + \delta_m, \qquad \Gamma_n \Gamma_m > 0 \tag{6.6}$$

it is natural to assume that they merge into a single vortex with circulation

$$\Gamma_{nm} = \Gamma_n + \Gamma_m, \tag{6.7}$$

such that the cross-sectional area of the new core is the sum of the areas of the two original cores. This implies an equivalent birth-time $t_{nm}$ for the new vortex such that

$$\varepsilon(t - t_{nm}) = \varepsilon(t - t_n) + \varepsilon(t - t_m) \tag{6.8}$$

and so

$$t_{nm} = t_n + t_m - t \, . \qquad\qquad (6.9)$$

We may further assume that when a vortex approaches close enough to the boundary $|W| = 1$ that $|W_n| > (1 - \delta_n)$ , the vortex is absorbed into its image and disappears (cf. Clements 1973).

One advantage of this coalescence of vortices is that the total number of vortices and hence the computation time is greatly reduced. On the other hand the vortices may become so large that the time-stepping interval $\Delta t$ has also to be reduced, so increasing the number of time-steps. Also, when a vortex merges with its image, there can be an appreciable jump in the horizontal momentum.

Some calculations of $\overline{C}_D$ are shown in Table 1 , corresponding to the sinusoidal flow (4.11) with $U = a\omega = 1$ . The drag coefficient is evaluated over successive complete cycles. As expected, the results show some scatter. Probably the onset of "chaotic" behaviour of the vortices could be delayed by employing a rediscretisation technique such as introduced by Fink and Soh (1974). The smoothness might also be improved by applying the Kutta condition at a variable point in the neighbourhood of the point of separation (Sarpkaya 1975). Nevertheless some of the scatter in the calculated values may reflect a real unsteadiness in the flow. The last column of Table 1 gives the mean drag coefficient deduced from experiments by Bagnold (1946). The agreement is reasonable. For further details see Longuet-Higgins (1981).

Table 1.  Values of the mean drag coefficient $\overline{C}_D$ , when $P = 5$, $k = 1$ and $U = \sin \omega t$ , over various ranges of $\omega t$ .

| $2a/L$ | $\varepsilon$ | $(0, 2\pi)$ | $(2\pi, 4\pi)$ | $(4\pi, 6\pi)$ | $(6\pi, 8\pi)$ | $(8\pi, 10\pi)$ | $(10\pi, 12\pi)$ | Mean | Obs. |
|---|---|---|---|---|---|---|---|---|---|
| 1.0 | $10^{-5}$ | .092 | .091 | .098 | .109 | .084 | .123 | .099 | .122 |
| 1.0 | $10^{-4}$ | .093 | .095 | .117 | .135 | .137 | .084 | .110 | |
| 1.5 | $10^{-5}$ | .101 | .107 | .122 | .155 | | | .121 | .120 |
| 1.5 | $10^{-4}$ | .097 | .121 | .169 | .096 | | | .121 | |
| 2.0 | $10^{-5}$ | .091 | .037 | .139 | | | | .089 | .107 |
| 2.0 | $10^{-4}$ | .095 | .098 | .092 | | | | .095 | |
| 3.0 | $10^{-5}$ | .086 | .092 | | | | | .089 | .075 |
| 3.0 | $10^{-4}$ | .099 | .097 | | | | | .098 | |
| 4.0 | $10^{-5}$ | .076 | | | | | | .076 | .063 |
| 4.0 | $10^{-4}$ | .068 | | | | | | .068 | |

## 7.  TRANSIENT FLOW AND WAVE GENERATION BY WIND

Transient flow over rough surfaces with periodically-spaced points of flow separation may be studied by the same method as outlined in Section 4, with interesting results.   Figure 8  shows the development of the flow

$$U = \left\{ \begin{array}{ll} 0, & t < 0 \\ 1, & t > 0, \end{array} \right\} \tag{7.1}$$

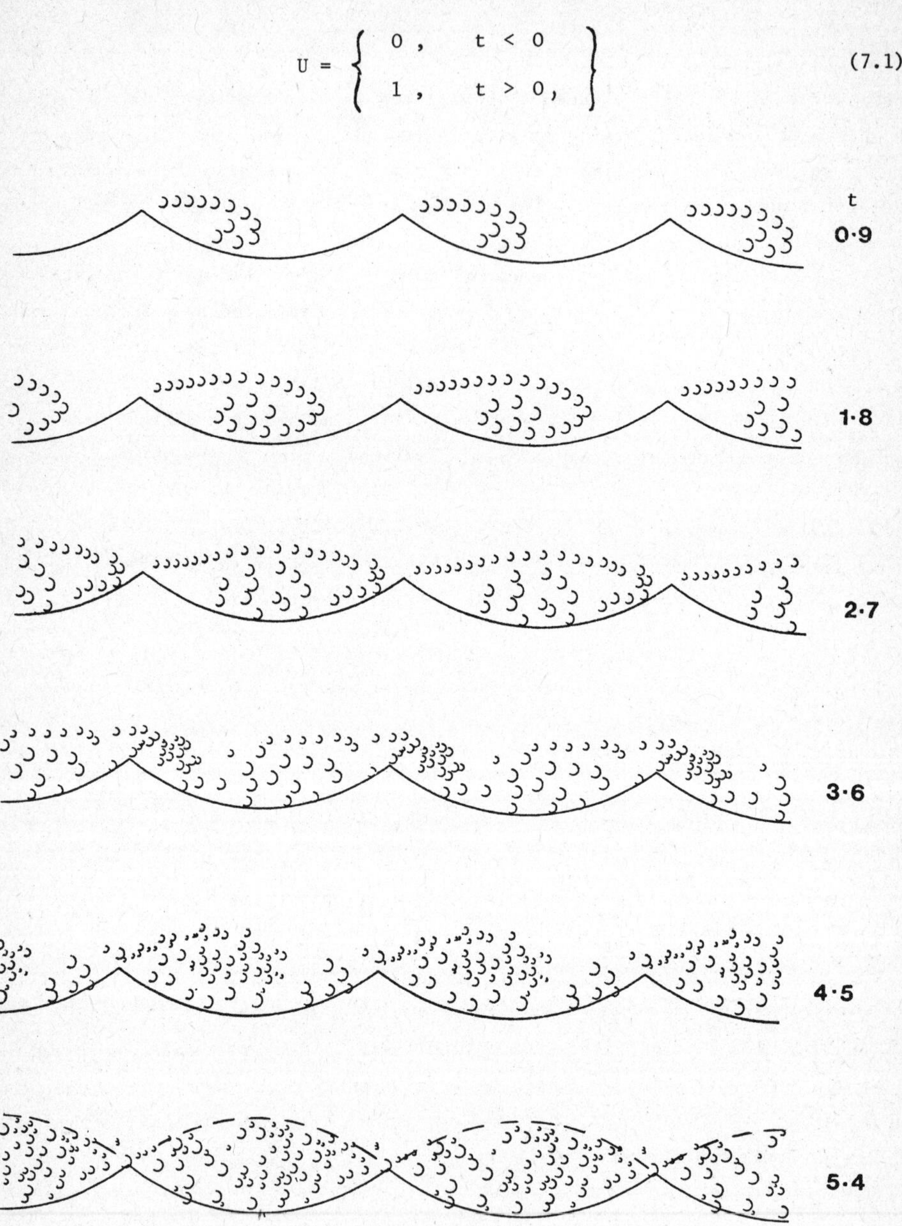

Figure 8.   Development of the uniform flow (7.1) started
from rest at time  $t = 0$ .

which is started impulsively from rest at time  t = 0 . At first the vortices tend to
fill up the wave troughs, just as in the oscillatory flows described earlier. However,
after about  t = 3.15 (Ut/L = 2.5) the troughs, or hollows, are full, and the flow
reaches an almost steady state in which little further vorticity is shed from the
ripple crests. The only noticeable fluctuation is an occasional spilling of vortices
from the trough of each ripple over into the next. The period of this oscillation is
about 2.5, in units of (Ut/L).

The corresponding drag coefficient  $C_D$ , defined as

$$C_D = F/LU^2 \qquad (7.2)$$

where  F  is the total horizontal force on the bottom (see equation (5.3)), is shown
in Figure  9 . The quasi-periodic oscillation is clearly visible.

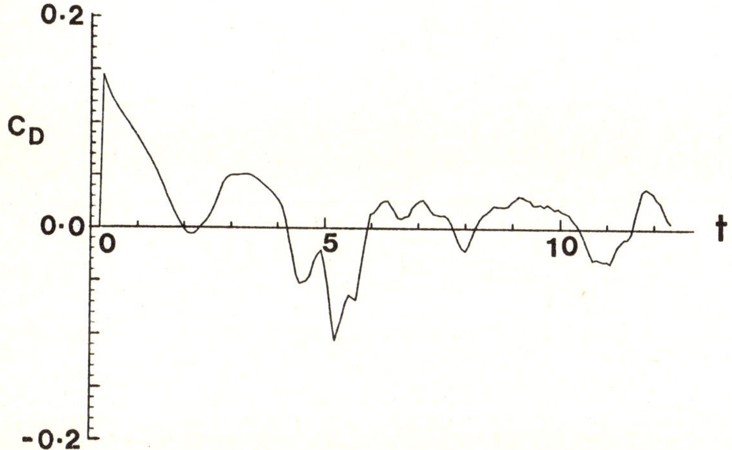


Figure 9.  Time-history of the drag coefficient  $C_D$  in
the flow (7.1).

Especially striking is the swift decrease in the drag coefficient from its value
of about 0.15 when  t << 1 to values one order of magnitude smaller. Indeed after
t ≃ 6  the mean drag coefficient is some two orders of magnitude smaller, in agreement with most observations of drag coefficients in steady flows.

In the quasi-steady flow of Figure 8 , the vorticity fills a lens-shaped region
that is almost symmetric about a horizontal line drawn through the ripple crests, as
indicated by the broken curve. Above this curve the flow is irrotational, and so similar to that shown in Figure 2, but upside down.

These results have suggestive implications for the generation of surface waves
in water by the wind. It was shown experimentally by Banner and Melville (1976) that
separation of the air flow takes place predominantly near the crests of breaking or
sharp-crested waves, and that the flux of momentum from the air to the water is locally enhanced by a factor of 50 or more.

Now on account of the dispersive nature of gravity waves on deep water, the occurrence of whitecaps and breaking waves is intermittent; see Donelan _et al._(1972). The group velocity being only half the phase-speed, individual waves start at the rear of a group as fairly low waves, then advance through the group to their maximum height, and finally die away towards the front of the group. Near the centre of the group they may briefly be steep enough to form whitecaps. However the formation of whitecaps is relatively sudden, as also is the disappearance of the sharp-pointed crests, though foam and bubbles linger for a few seconds afterwards.

Clearly, then, the airflow near a breaking wave must be forced to adjust very rapidly to the forming of the whitecap. In such a situation we may expect the sudden onset of flow separation, and a transient rise in the drag coefficient, similar to that shown in Figure 9.

This argument goes to support the conclusion of Banner and Melville (1976) that the input of momentum from wind to waves is an essentially sporadic phenomemon, localised in the neighbourhood of steep or breaking waves.

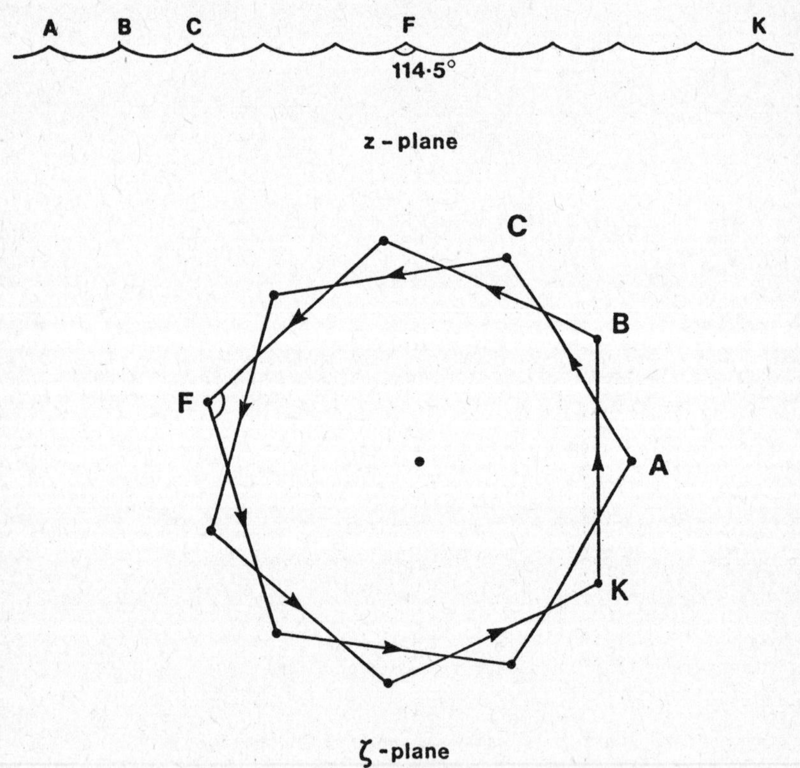


Figure 10.  The polygon transformation corresponding to P = 11/2

## 8.   CONCLUSION

We have used the conformal transformations described in Section 2 in the particular cases  $P = 6$   (progressive waves in deep water) and  $P = 5$  (sand ripples). The
case  $P = 4$  has also been applied to standing waves of limiting steepness, where the
crest angle is nearly, if not exactly, $90^{\circ}$ (see Longuet-Higgins 1973).

However we are by no means confined to integral values of  $P$ . A similar transformation can be used whenever  $P$  is a rational number, that is  $P = N/M \geqslant 2$ , where
$N$  and  $M$  are positive integers. Then equation (2.2) becomes

$$k\,N\,L = 2\,M\,\pi \qquad\qquad (8.1)$$

and we map  $N$  consecutive ripple-lengths onto the sides of a "star-polygon" enclosing
its centre  $M$  times. Figure 10 illustrates the case  $P = 11/2$ . In both the $\zeta$-plane
and the W-plane the flow is mapped onto the  $M$  sheets of a Riemann surface.

Lastly, the case  $P = 2$  corresponds to a <u>digon</u> in the $\zeta$-plane; in the z-plane
we have an infinite sequence of equally-spaced semi-infinite barriers. The corresponding transformation into a circular cylinder in the W-plane is

$$\zeta = \zeta_{0} - K'\int_{W_{0}}^{W} \frac{1 - W^{2}}{W^{2}}\ dW \quad . \qquad\qquad (8.2)$$

Taking  $\zeta_{0} = W_{0} = 1$   as before and  $K' = \tfrac{1}{2}$  from (2.11)   we get the well-known transformation

$$\zeta = \tfrac{1}{2}(W + W^{-1}) \quad . \qquad\qquad (8.3)$$

## REFERENCES


Bagnold, R.A. 1946 Motion of waves in shallow water. Interaction between waves and
sand bottoms. <u>Proc. R. Soc. Lond. A.</u> <u>187</u>, 1-15.

Banner, M.L. and Melville, W.K. 1976 On the separation of air-flow over water waves.
<u>J. Fluid Mech.</u> <u>77</u>, 825-842.

Carstens, M.R., Nielson, F.M. and Altinbilek, H.D. 1969 Bed forms generated in the
laboratory under oscillatory flow. <u>Coastal Eng. Res. Center, Fort Belvoir, Va.,
Tech. Mem.</u> <u>28</u>, 78 pp.

Clements, R.R. 1973 An inviscid model of two-dimensional vortex shedding.
<u>J. Fluid Mech.</u> <u>57</u>, 321-336.

Clements, R.R. and Maull, D.J. 1975 The representation of sheets of vorticity by
discrete vortices. <u>Progr. Aerospace Sci.</u> <u>16</u>, 129-146.

Donelan, M.E., Longuet-Higgins, M.S. and Turner, J.S. 1972 Periodicity in whitecaps.
<u>Nature, Lond.</u> <u>239</u>, 449-457.

Fink, P.T. and Soh, W.K. 1974 Calculation of vortex sheets in unsteady flow and applications in ship hydrodynamics. Proc. 10th Symp. Naval Hydrodynamics, Cambridge, Mass. pp.463-488.

Lamb, H. 1932 Hydrodynamics, 6th ed. Cambridge Univ. Press. 738 pp.

Lofquist, K.E.B. 1978 Sand ripple growth in an oscillatory-flow water tunnel. Coastal Eng. Res. Center, Fort Belvoir, Va., Tech. Pap. 78-5. 101 pp.

Longuet-Higgins, M.S. 1973 On the form of the highest progressive and standing waves in deep water. Proc. R. Soc. Lond. A. 331, 445-456.

Longuet-Higgins, M.S. 1979 The trajectories of particles in steep, symmetric gravity waves. J. Fluid Mech. 94, 497-517.

Longuet-Higgins, M.S. 1980 The unsolved problem of breaking waves. Proc. 17th Int. Conf. Coastal Eng., Sydney, Australia, March 1980.

Longuet-Higgins, M.S. 1981 Oscillating flow over steep sand ripples. J. Fluid Mech. (in press).

Michell, J.H. 1893 The highest waves on water. Phil. Mag. 36, 430-437.

Nehari, L. 1952 Conformal Mapping. New York, McGraw-Hill. 396 pp.

Nielsen, P. 1979 Some basic concepts of wave sediment transport. Inst. Hydraulics and Hydraulic Eng., Tech. Univ. of Denmark, Ser. Pap.20, 160 pp.

Saffman, P.G. and Baker, G.R. 1979 Vortex interactions. Ann. Rev. Fluid Mech. 11, 95-122.

Sarpkaya, T. 1975 An inviscid model of two-dimensional vortex shedding for transient and asymptotically steady flow over an inclined flat plate. J. Fluid Mech. 68, 109-128.

Stokes, G.G. 1880 On the theory of oscillatory waves. Appendix B. Math. and Phys. Pap. 1, 225-228.

Yamada, H. Highest waves of permanent type on the surface of deep water. Rep. Res. Inst. Appl. Mech. Kyushu Univ. 5, 37-52.

## ON THE COMPUTATION OF DISCONTINUOUS
## MULTI-DIMENSIONAL GAS FLOWS

V.V. Rusanov

Keldysh Institute of Applied Mathematics Moscow, USSR

### I. Introduction

In this paper some questions connected with the calculation of
inviscid discontinuous gas flows are considered. The paper is not a
review, and our presentation is by no means a complete study of the
field. Despite the recent advances in the development of the numerical methods for multi-dimensional discontinuous gas flows the number
of unsolved problems is rather great. Here we discuss some basic
questions as well as the difficulties of their solution and the prospects for overcoming these difficulties.

### 2. Statement of the Problem

An inviscid gas flow in a three-dimensional space is determined
by the density $\rho$ , the pressure $p$ and the velocity components $u_x$ ,
$u_y$ , $u_z$ depending on the space coordinates $x$ , $y$ , $z$ and the
time $t$ . We shall consider the $\rho$ , $p$ , $u_x$ , $u_y$ , $u_z$ as the components of a formal vector $W$ . In the gas filled region the surfaces could exist on which the components of $W$ are non-continuous.
Thus the vector $W$ may be considered as the function differentiable
over the whole gas filled region except for the finite (though possibly great) number of the discontinuity surfaces.

Though such representation of the solution makes the theoretical analysis more complicated as compared with the introduction of
the generalized functions, it is more appropriate from the physical
and practical viewpoints. For the practical purposes, the method
yielding the result which is close to the desired solution in the
"weak" sense is not so efficient as the method using the smoothness
of the functions in the domains where it exists and providing there
high accuracy.

While developing such a method, however, one has to overcome
a number of difficulties. The basic of them is that the number of
discontinuities and their topology are not known a priori. Since the
gas dynamic equations are nonlinear, new discontinuities may arising
and interacting discontinuities may be so great that the considera-

tion of each of them is out of question (especially, in the multi-dimensional case). Due to this, the methods of computation "through" the discontinuities were developed, which do not require any information about the presence or the absence of discontinuities. The value of $w$ near the discontinuity surface is then computed with more or less distortion, and there is some difficulty in finding the actual location of the discontinuity in the flow field.

Sometimes it is possible to determine the topology some of the discontinuities beforehand and to select the discontinuities most important in the problem under consideration. They may be considered as the boundaries of the smoothness regions with the appropriate boundary conditions. Then the method has to be developed for the computation of a flow in the region with an unknown boundaries whose location must be computed simultaneously with flow. This method is free from the flaws of the through computation technique but its applicability is limited by the number of boundaries.

All the existing methods of computation of the gas flows with discontinuities are based on one of the two approaches discussed above, or on their combination. In this paper we consider some aspects of these approaches based on the finite-difference approximations.

3. <u>The Through Computation Methods</u>

Neuman and Richtmyer [1] were the first to propose a through computation method for the shock wave. The main idea was to introduce an artificial viscosity that would prevent the "tilting" of the wave profile. Now it is used, in one or other form, in all the through computation methods. The viscosity may be either introduced in the differential equations beforehand or appear as the "scheme" viscosity due to the dissipative terms in the difference scheme. Due to its presence a stretched discontinuity profile of $w$ with the large (but not infinite) gradients is generated instead of the abrupt "jump" in the gas dynamic functions. In some cases the parasitic oscillations may occur near the discontinuity, which amplitude may reach 20-30 percent of the jump. The main task in constructing the through computation schemes is to minimize the profile stretching and oscillation.

The localness of the boundary conditions at the discontinuity surface and their independence on the surface curvature allow to consider as the first approximation the properties of the difference schemes in the one-dimensional case. The numerical investigation shows that the results thus obtained are basically valid for the

multi-dimensional problems too. We shall at first briefly consider
these results.

### 3.1. The Structure of the Linear Discontinuity Profile
### in the Numerical Solution

Let us consider the one-dimensional linear problem:

$$(3.1) \quad u_t + au_x = 0 \qquad u(x,0) = u^0(x) = \begin{cases} 0, & x < 0 \\ 1, & x > 0 \end{cases}$$

and the explicit scheme on the floating grid:

$$(3.2) \quad u^{n+1}(x) = \sum_{\nu=-\nu_1}^{\nu_2} a_\nu u^n(x+\nu h); \quad -\infty < x < +\infty$$

where $h = \Delta x$, $\tau = \Delta t$.

Let us study the asymptotics of the function $u^n(x)$ at $n \to \infty$
and $\tau$, $h$ fixed. Let the scheme (3.2) have the order $p$. Then if it
is stable the theorem [2] is valid, that in any finite interval
$[-X, X]$ for $n$ great enough the function $u^n(x)$ is close
to the function

$$(3.3) \quad u^n(x) \approx F_p\left(\frac{x - na\tau}{(b_{p+1}\,n)^{1/(p+1)}\,h}\right), \quad b_{p+1} > 0, \quad F_p(\infty) = 1$$

where

$$(3.4) \quad F_p(\xi) = \begin{cases} \dfrac{1}{\pi}\displaystyle\int_{-\infty}^{\xi}\int_0^{\infty} e^{-\frac{s^{p+1}}{p+1}} \cos(zs)\,ds\,dz, & p \ \ odd \\[20pt] \dfrac{1}{\pi}\displaystyle\int_{-\infty}^{\xi}\int_0^{\infty} \cos\left(\frac{s^{p+1}}{p+1} + zs\right)ds\,dz, & p \ \ even \end{cases}$$

The Fig. 1 shows the profiles of the functions $F_p(\xi)$ for
$p = 1, 2, 3$. It follows from Fig. 1 and (3.3), (3.4) that

1) When calculated with the scheme of order $p$, the discontinuity profile not only shifts by the value $na\tau$ in accordance with the
equation (3.1), but also  stretches with the velocity proportional
to $n^{1/(p+1)}$.

2) The qualitative nature of the "stretched" profile near the
discontinuity and at $\xi \to \pm\infty$ essentially depends on the scheme
parity. At $\xi < 0$ the even-order schemes have much stronger oscillations damping very slowly at $\xi \to -\infty$. The schemes of order
greater than unity generate nonmonotonic profiles. In other words,
all the schemes with $p > 1$ are asymptotically nonmonotonic.

These results are general enough and are valid for the wide classes of linear schemes including the implicit ones.

Thus, in the framework of the linear uniform schemes the only way to diminish the "stretching" of the linear discontinuity is to increase the order of the scheme accuracy. Since the even-order schemes yield much stronger oscillations and the schemes of order five and above can not be implemented in practice, the third-order schemes are optimal. Their various versions are constructed now by many authors.

The use of the uniform schemes, however, does not solve satisfactory the problem of calculation of the linear (contact or tangential) discontinuities. Even with the third-order scheme, the profile stretching with the velocity $n^{1/4} \sim t^{1/4}$ becomes unacceptable when $n$ is great enough.

Many authors tried to improve the computation of the linear discontinuity. Some success seemed to be achieved in [3] . The method proposed there has a clear theoretical justification and yields good results in the one-dimensional case. Its generalization for the multi-dimensional cases, however, meets some difficulties which are inevitable for such complex problems. It seems to be interesting to study the capabilities of the method [3] in application to the third-order schemes in order to increase the computation accuracy in the regions of smoothness.

### 3.2. The Structure of the Nonlinear Discontinuity Profile in the Numerical Solution

The investigation of the profile structure of the nonlinear discontinuity generated by the difference scheme is a much more complicated problem. We present here some preliminary results which are, however, of fundamental significance.

Let us consider at first the one nonlinear equation

$$(3.5) \quad u_t + [F(u)]_x = 0$$

with the initial data corresponding to the shock wave moving with the velocity $D$ , such as

$$(3.6) \quad u(x, 0) = u^0(x) = \begin{cases} u_1, & x < 0 \\ u_2, & x > 0 \end{cases} \qquad D = \frac{F(u_2) - F(u_1)}{u_2 - u_1} \quad .$$

The exact solution has the form $u(x, t) = u^0(x - Dt)$ .

To solve the problem (3.5)-(3.7) let us use the difference

scheme

$$(3.7) \quad u^{n+1}(x) = \Phi\left\{u^n(x+\nu h)\right\}$$

where $\Phi$ is the nonlinear function of the quantities $u^n(x+\nu h)$, $-\nu_1 \leq \nu \leq \nu_2$, $\nu_1$ and $\nu_2$ are positive integers.

Now we raise the following questions:

1) Has equation (3.7) the stationary (to the accuracy of the shift by $nD\tau$ ) solution $V(\xi)$ satisfying the equation

$$(3.8) \quad V(\xi - D\tau) = \Phi\left\{V(\xi + \nu h)\right\}$$

and the boundary conditions $V(-\infty) = u_1$, $V(+\infty) = u_2$ ?

2) What are the properties of the function $V(\xi)$ if it exists?

3) Does there exist $\lim\limits_{n\to\infty} u^n(\xi + nD\tau)$ equal to $V(\xi)$ ? If it exists, then in what sense?

These questions were implicity formulated by Lax [4] for the equation with $F(u) = u^2/2$ and for the first-order scheme proposed by him. In [5] the necessary existence conditions for $V(\xi)$ were established for an arbitrary three-point difference scheme. In [6] the existence of the limit of $u^n(\xi + nD\tau)$ for the monotonic schemes was proved.

The numerical analysis shows that the function $V(\xi)$ is not only continuous but has the high degree of smoothness. In contrast to the linear discontinuity the shock profile stretches not infinitely but for a finite length depending on the discontinuity and difference scheme parameters. The functions $V(\xi)$ computed for the equation (3.5), $F(u) = 0.5u - 0.1u^2$ , with schemes of accuracy $p$ = 1, 2, 3 are plotted in Figs. 2, $(u_1 = 0, u_2 = 1, D = 0.4)$ .

Note that though $V(\xi)$ is a smooth function on the floating grid, it does not look the same in the calculations on the fixed grid even if $n$ is large enough.

The analysis of the numerical results shows that the scheme order basically has the same influence on the shock profile as in the linear case. The higher the scheme order the less stretching of the profile is in the vicinity of the shock. The oscillations near the shock are much stronger for the even-order schemes, but the first-order scheme does not always generate the monotonous function $V(\xi)$ . Thus, for the nonlinear equation the first-order scheme is not necessarily asymptotically monotonous.

The analogous results were obtained numerically in the case of the nonlinear system of equations. The investigation of second-order schemes shows that there are strong oscillations in all the components of $W$ as in the case of one equation. Therefore the second-order schemes are of no interest for through computation methods. But it is very important that the shock profile structure in the nonlinear third-order scheme is only slightly nonmonotonic in the vicinity of the shock.

### 3.3. On the Shock Profile Structure in the Multi-Dimensional Problems

In the multi-dimensional case two additional factors arise that influence the discontinuity profile structure. These are a curvature of the discontinuity surface and its orientation with respect to the mesh. If the mesh is fine enough one may consider the discontinuity surface to be plane as the first approximation. Then we get a problem to calculate the plane discontinuity which moves in the direction making the arbitrary angle $\theta$ to the axis $x$ .

The numerical experiments show that for the linear discontinuity the speed of the profile stretching is also of the order $O(n^{1/(p+1)})$ , where $p$ is the scheme order, and the difference between the discontinuity profiles for different $\theta$ is very small.

The similar results were obtained for the one nonlinear equation, but in this case the dependence of the profile on the angle can be stronger.

For the system of equations, the structure of the discontinuity profile is much more complicated     In this case the distorsions can appear not only in the transversal direction ( stretching and oscillations) but along the shock too (deflection from rectilinearity). The last distorsions are usually periodic due to the periodic structure of the mesh. It is  an interesting and important problem to construct the difference schemes free from this fault, i.e. "invariant" with respect to the rotation of the coordinate axes.

### 3.4. The Third-Order Schemes

As it was mentioned above the third-order schemes have some advantages compared to the first- and second-order schemes. To-day we have a number of methods of construction of the third-order schemes, mainly for the one-dimensional problems. The development of the third-order schemes and their application to the multi-dimensional problems meet some difficulties, which are common for all methods when the number of the problem dimensions increases. In this section we shall consider the third-order difference scheme for the two-dimensional problem and discuss some of its properties [7, 8] .

The scheme is based on the iterational procedure similar to the Runge-Kutta method for the ordinary equations.

We consider the system

$$(3.9) \quad W_t = F_x + G_y + f$$

where $F = F(w, x, t),\ G = G(w, x, t),\ f = f(w, x, t)$ and introduce the mesh with the steps $h_1$, $h_2$, $\tau$. Let $W_{m,\ell}^n$ be the mesh function at the point $(x_m, y_\ell, t^m)$, $x = mh_1$, $y = \ell h_2$, $t^n = n\tau$. Let the values of $W_{m,\ell}^n$ be given, and necessary to find $W^{n+1}$.

The following sequence of iterationcs is defined: (Fig. 3):

$$W_{m+1/2,\ \ell+1/2}^{(1)} = \Phi^{(1)}(W_{m,\ell}^n,\ W_{m+1,\ell}^n,\ W_{m,\ell+1}^n,\ W_{m+1,\ell+1}^n)$$

$$(3.10) \quad W_{m,\ell}^{(2)} = \Phi^{(2)}(W_{m+\mu,\ \ell+\lambda}^n,\ W_{m+\mu/2,\ \ell+\lambda/2}^{(1)})$$

$$W_{m,\ell}^{(3)} = \Phi(W_{m+\mu,\ \ell+\lambda}^n,\ W_{m+2\mu,\ \ell+2\lambda}^n,\ W_{m+\mu/2,\ \ell+\lambda/2}^{(1)},\ W_{m+\mu,\ \ell+\lambda}^{(2)})$$

where $\lambda$, $\mu$ independently takes the values of $-1$, $0$, $1$. The functions $\Phi^{(j)}$ are the composite functions of their arguments through the functions F, G and $f$. They contain also a number of scalar coefficients as the scheme parameters.

Letting $W_{m,\ell}^{n+1} = W_{m,\ell}^{(2)}$ one can obtain the first- or second-order scheme, depending on the choice of parameters, and letting $W_{m,\ell}^{n+1} = W_{m,\ell}^{(3)}$ the third-order scheme is obtained for some choice of parameters. The $\Phi^{(3)}$ contains also the coefficient $\gamma$ which does not influence the approximation but ensures the stability of the scheme.

The stability of the third-order scheme can be investigated only numerically, but the result can be presented in a rather simple form. As an example, let us consider the case of one equation with the constant coefficients

$$(3.11) \quad u_t = a u_x + b u_y$$

Denote $\sigma = \tau \sqrt{(a/h_1)^2 + (b/h_2)^2}$ and $tg\chi = (bh_1)/(ah_2)$. Figure 4 shows the stability boundaries in the plane $(\sigma, \gamma)$ for different $\chi$. The lines $\gamma = 1$ and the line $\gamma = 2(7 - 2\sigma^2)\sigma^2/3$ corresponding to $\chi = \pi/4$ are the boundaries for all $\chi$. The curve $\gamma = 2(4 - \sigma^2)\sigma^2/3$ corresponds to one dimensional scheme similar to (3.10).

The computations with the use of scheme (3.10) which were

mentioned before show that the discontinuity profiles in the multi-dimensional case differ only by little from the profiles generated by the one-dimensional third-order scheme. But there is two difficulties deterring the wide use of the thrid-order schemes and we shall discuss them now.

The first difficulty is the nonmonotonicity of the discontinuity profile. It may be very inconvenient when the task is to find the values of $W$ strictly at discontinuity surface. The only possible way to eliminate this inconvenience is to change the order of approximation (to the first one) in vicinity of the discontinuity. It may be done only if the discontinuity location can be detected in the course of computation. Thus the problem splits into the detection of the discontinuity and the change of the scheme near it.

Considering the scheme (3.10) we see that the change of its order is of no difficulty and does not require additional computation time. The detection and localization discontinuity is much more difficult. In the one-dimensional case there exist rather simple sufficient criteria for shock waves detection based on the structure of stretched profile. But for the multidimensional flows the satisfactory sufficient criteria has not been found yet. This fundamental problem is common for all through computation methods.

The second inconvenience of the third-order scheme is the necessity to have a regular orthogonal mesh. It is practically impossible to construct even the second-order scheme on an irregular mesh, not saying of the third-order one. At the same time, the domains with curvilinear or even unknown boundaries occur in abundance in the gas dynamic problems. To find his way out one should transform the real flow domain onto another so that the curvilinear mesh will be transformed into the rectangular one. If such a transformation can be realized with the aid of the functions with desired smoothness the problem is solved.

As an example, we consider equation (3.9) which is to be solved in the domain ABCD having the form of a curvilinear quadrangle. Let the transformation

$$(3.12) \quad x = X(\xi, \eta), \quad y = Y(\xi, \eta), \quad \xi = \varphi(x, y), \quad \eta = \psi(x, y)$$

transfer ABCD to the rectangle A'B'C'D' in the plane $(\xi, \eta)$ . It is easy to see that for the preserving of the third order approximation in the plane $(x, y)$ , the functions $X$ , $Y$ must have the continuous second derivatives with respect to $\xi$ , $\eta$ . If the transformation (3.12) can not be written in an explicit form or in the form of simple analytical expressions, one may use local splines of appropriate smoothness. It means that first the grid functions $x_{m,\ell}$ and $y_{m,\ell}$ are given on a set of points $\xi_m$ , $\eta_\ell$ , and then for these functions the local splines of desired smoothness $X(\xi, \eta)$ and $Y(\xi, \eta)$ are constructed, allowing the (numerical) inversion. We are speaking of local splines because the construction of nonlocal ones is impracticable in the two- and multi-dimensional cases.

As the first step in the scheme construction with transformation of variables we consider the case of a stationary supersonic axisymmetric flow about a body (Fig. 5). The initial coordinates are $\eta$ and $z$ ; one of the new variables $(\xi, \eta)$ namely, $\eta = z$ plays the role of time. As a matter of fact, the problem turns out to be one-dimensional. The initial data are given for $\eta = \eta_0$ . It is required to find the solution in the domain $0 \leq \xi \leq 1$ , $\eta > \eta_0$ . The derivatives $\xi_x$ , $\xi_y$ etc depend on an unknown function determining the configuration of the bow shock wave.

The calculation of points in the bow wave and near it was performed by means of the second approach mentioned in Section 2, when the discontinuity is considered as the exact boundary. Such method for approximation of the boundary conditions in mixed problems for the general hyperbolic systems was proposed and described in [9, 10]. It preserves the third-order approximation up to the boundary and requires no additional stability restrictions besides restrictions conditioned by the difference scheme at the internal mesh points. The imbedded shock arising in the flow was calculated with the thrid order through computation method. The computation results are given in Fig. 6 and 7, where the field of characteristics with the shock and the gas dynamic functions of $\xi$ for different $\eta$ are plotted. For comparison in **Fig.** 7 the gas dynamic function are plotted with the use of the second-order scheme. In this case the oscillations are strongly revealed even in the domain where the shock has not yet formed.

4. The Testing of Computation Methods

The "step" type discontinuity is usually applied for testing the computational methods for non-continuous flows. It has an advantage of simple and convenient comparison with the exact solution, but unfortunately it does not allow to estimate the influence of the discontinuity upon accuracy of calculation in the smooth domain. Because the "smooth" region in this test is a region of constant functions all the schemes behave similarly in this region and yield absolute accuracy.

For actual estimation of accuracy in the region where the functions are smooth but not constant, special tests are needed. In the paper [11] the principles for construction such tests were proposed for one-dimensional problems. By using these principles, tests with the arising of shocks, their interaction, reflection, etc. can be developed. In all cases, for computing the exact solution it is sufficient to solve numerically a few (depending on the test complexity) ordinary differential equations with lagging arguments, which is of no problem for to-day's computers.

The results of the schemes investigations by means of these tests were given in [9,10]. An important fact was established that as a distance from the discontinuity increases an error of the computation having the zeroth order near the discontinuity approaches the error determined by the order of approximation.

## 5. Conclusion

The results presented allow to conclude that in calculating the discontinuous solutions the third-order difference schemes have some advantages as compared with the first- and second-order schemes. They provide the higher accuracy in the smooth region and they produce a less stretched profile in the neighborhood of the discontinuity, and much weaker oscillations as compared with the second-order schemes.

However, some technical and fundamental difficulties prevent the wide use of the third-order schemes in computation of multidimensional problems. The main technical difficulties are associated with the development of the schemes on irregular meshes with the high order of approximation. The fundamental difficulties are associatied with an infinite stretching of the linear discontinuity profile, and with the occurrence of non monotonicity near the discontinuity.

The use of splines and other numerical tools for smooth transformation seems to be the good means to overcome the technical difficulties.

In order to improve considerably the quality of the stretched profile without violating the accuracy in the smooth regions, one has to pass over from uniform schemes to non-uniform ones with variable order of accuracy. The main problem here consists in detecting the discontinuity location. The problem is not solved yet in two dimensional case, but it does not appear to be hopeless. As in the one-dimensional case the key to the solution is the detailed study of the stretched profile structure. If the  position of the discontinuity could be determined at any time during the computation the construction of the scheme for achieving the best quality of the stretched profile in the discontinuity zone will be not very hard task.

## References


1. Neuman, von,J., Richtmyer R.D. A Method for Numerical Calculation of Hydrodynamic Shocks. J.Appl. Phys. Vol. 2I, pp.232-237 (I950).
2. Kuznetzov, N.N. The Asymptotic of the Solutions of Finite-Difference Problem (Russian), Zhurnal Vychislit. Matem. i Matemat. Phiziki, vol. 12, N 2, pp.334-35I (I972).
3. Boris, J.P. and Brook, D.L. Flux-Corrected Transport. I. SHASTA, A Fluid Transport Algorithm that Works. J.Comp.Phys., vol. 11, N I, pp.38-69 (I973).
4. Lax, P.D. Weak Solutions of Non-Linear Hyperbolic Equations of Hydrodynamics. Comm. Pure Appl.Math. Vol. 7, N I, pp.I59-I93 (I954).
5. Rusanov, V.V. Non-Linear Analysis of the Shock Profile in Difference Schemes. Proc. Sec. Int.Conf. on Numer.Meth. in Fluid Dynamics. In "Lecture Notes in Physics", vol. 8, Springer Verlag (I970).
6. Rusanov, V.V. and Bezmenov I.V. On the Limiting Profile of Non-Linear Discontinuity in One-Dimensional difference Schemes. Keldysh Inst. Appl. Math. Preprint N 69 (I980).
7. Rusanov, V.V. On Difference Schemes of Third Order Accuracy for Non-Linear Hyperbolic Systems. J.Comp.Phys. Vol. 5, pp.507-5I6 (I970).
8. Izvolskii, V.A. and Rusanov, V.V. The Construction and Investigation of the Third-Order Difference Schemes. (Russian). Keldysh Inst. Appl. Math.Akad. Nauk, Preprint N 3 (I979).
9. Rusanov, V.V. Advanced Techniques for Computation Supersonic Flows. Proc. I5th AIAA Aero. Space Sci. Meeting AIAA Paper, pp.77-I73 (I977).
10. Rusanov, V.V. and Nazhestkina, E.I. Boundary Conditions in Difference Schemes for Hyperbolic Equations. Proc. Third GAMM - Conference on Numerical Method in Fluid Mechanics. Köln (I979).
11. Rusanov, V.V., A Test Case for Checking Computational Methods for Gas Flows with Discontinuities. In GAMM-Workshop "Boundary Algorithms for Multidimensional Inviscid Hyperbolic Flows" K.Förster (Ed.), Verl.F. Vieveg & Sohn Branschweig Wiesbaden, pp.I00-I25 (I978).

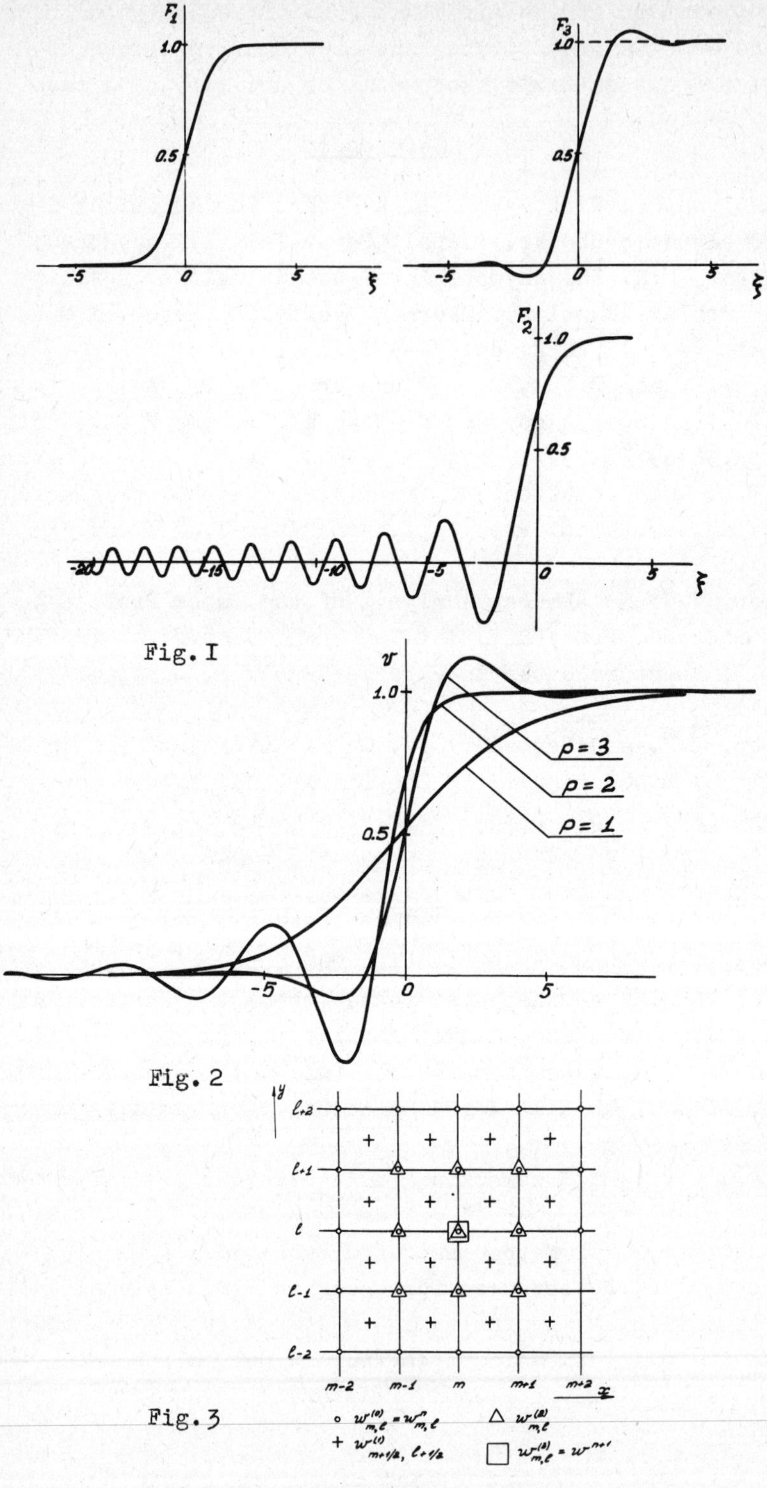



Fig. I



Fig. 2



Fig. 3

$\circ \ w_{m,l}^{(0)} = w_{m,l}^{n}$
$\triangle \ w_{m,l}^{(2)}$
$+ \ w_{m+1/2,\, l+1/2}^{(1)}$
$\square \ w_{m,l}^{(3)} = w^{n+1}$

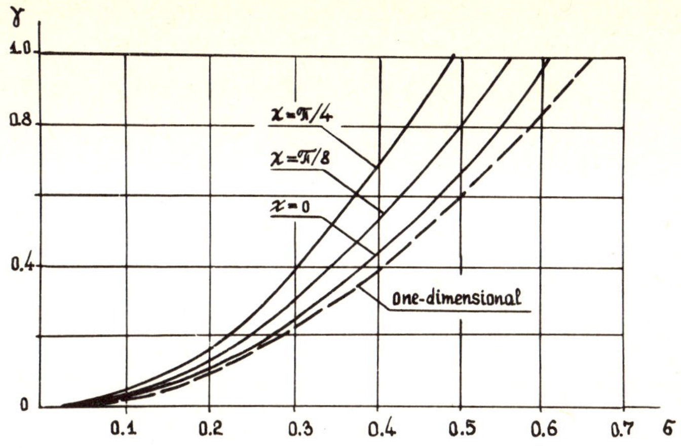


Fig. 4

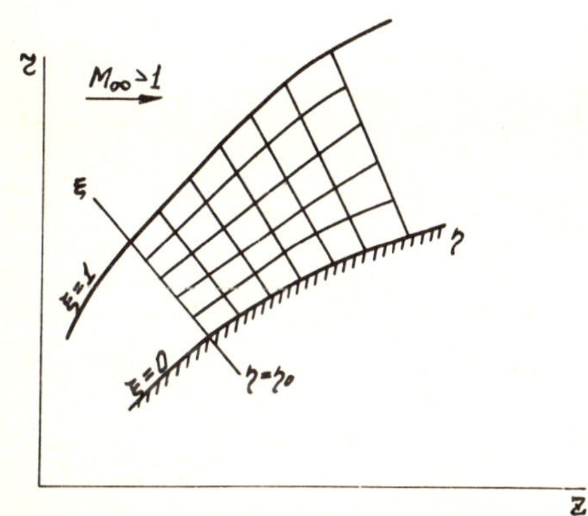


Fig. 5

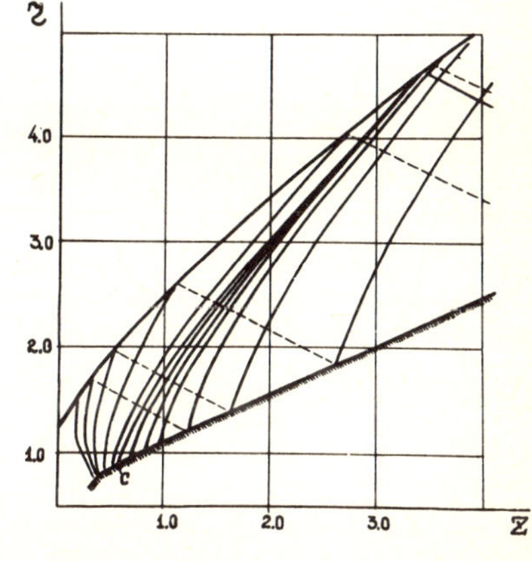


Fig. 6

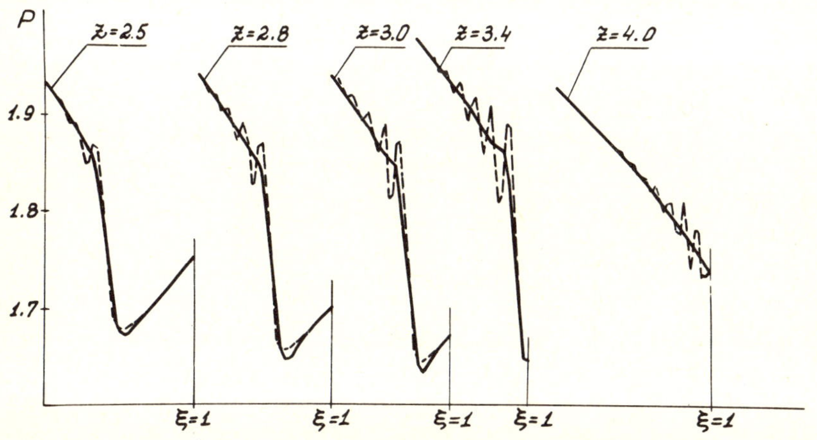


Fig. 7

# PSEUDO-UNSTEADY METHODS FOR TRANSONIC FLOW COMPUTATIONS


H. Viviand

*Office National d'Etudes et de Recherches Aérospatiales (ONERA)*
*92320 Châtillon (France)*


## I. INTRODUCTION

The "natural" approach to compute flow fields is to follow the physics of real flows by solving the unsteady equations of motion step by step in time, possibly reaching an asymptotic steady state. When only the steady state solution (assuming its existence and uniqueness in the framework of a chosen mathematical model) is of interest, the initial and boundary conditions may be chosen so that the transient solution does not correspond to a well-defined physical experiment. Moreover, the unsteady equations themselves can be modified in view of simplifying the calculations and accelerating the convergence towards a steady state ; similarly the numerical scheme need not be consistent in time with the unsteady equations. All these possibilities lead to the so-called false transient or pseudo-unsteady (hereafter abbreviated as P.U.) approach.

As it is well known and has been discussed for example by Lomax and Steger (1975), Jameson (1974, 1975), Wirz (1977), a close connection exists between the relaxation methods used for solving time-independent problems and the P.U. methods in which the time variable plays the role of the iteration parameter. However, from a practical point of view, these two approaches constitute different ways of constructing iterative methods, especially in the case of the non-linear mixed-type equations governing the steady inviscid transonic flows with which we are concerned here. Relaxation methods in this domain have made important and rapid progress in the last ten years and, due to their greater efficiency, have largely dominated over unsteady type methods. Nevertheless, the P.U. approach presents some practical advantages and is open to developments which should allow it to compete with relaxation methods.

In this paper we present a general review of the P.U. approach for steady inviscid transonic flows based on systems of first-order partial differential equations. These P.U. methods are made of two distinct components, namely the system of P.U. equations supplemented with appropriate boundary conditions, and the numerical scheme used to discretize and integrate the equations step by step in time. These two components are uncoupled, hence a large variety of possible methods. Also greater efficiency can be searched for independently in each of these two components, and all P.U. methods can benefit by advances in the numerical solution of general non-linear hyperbolic systems. This review will focus attention on the P.U. formulation itself, which is the characteristic feature of the P.U. approach.

## II. BASIC CONSIDERATIONS ON PSEUDO-UNSTEADY SYSTEMS

The basic features of first-order hyperbolic systems which are relevant to the construction of P.U. systems are briefly reviewed in this section. Detailed mathematical expositions can be found in text books such as Courant and Hilbert (1962) or Jeffrey (1976), and in review papers, e.g. Jeffrey (1978).

### II.1. Hyperbolicity with respect to time

The P.U. systems that we shall consider are first-order quasi-linear systems which can be written in the following developed form :

$$N(f) \frac{\partial f}{\partial t} + \sum_{j=1}^{n} A^j(f) \frac{\partial f}{\partial x_j} = g(f) \tag{1}$$

where $f = (f_1, f_2, \dots f_m)^T$ is the column-vector of the $m$ dependent variables, and

$g = (g_1, g_2, \dots g_m)^T$ is a function of $f$ ; $N$ and $A^j$ ( $j = 1$ to $n$ ) are ( $m \times m$ ) matrices with elements functions of $f$, and $N$ is assumed to be regular. In general, the steady equations to be solved appear in conservative form, and the $A^j$ are Jacobian matrices $\partial F^j(f)/\partial f$ .

The basic condition which the system (1) must satisfy is to be hyperbolic with respect to time. This is a necessary condition for the initial-value problem to be well-posed. Consider the characteristic equation in the unknown $\omega$ for given $f$ :

$$\det [ N\omega - k_j A^j ] = 0 \tag{2}$$

where $k = ( k_1 , k_2 , \dots k_m )$ is the wave vector, and "det [ ]" means the determinant of the matrix inside the brackets. The condition of hyperbolicity in the t-direction is that the $m$ roots of Eq. (2),

$\omega_\ell (k) \; (\ell = 1, 2, \dots m)$, must be real for all real $k \neq 0$, and that the $m$ corresponding left or right eigenvectors, noted respectively $\alpha^\ell$ and $\beta^\ell$, must be linearly independent.

A further requirement on system (1), besides hyperbolicity, is that perturbations corresponding to a system of expansion waves should not tend to coalesce and form and expansion shock. It is known that this condition is satisfied by the exact Euler equations, but it may not be satisfied with P.U. systems. This question has been discussed by Essers (1978, 1979).

### II.2. Asymptotic large-time behaviour.

The question of the convergence of the solution of the initial value problem for system (1) towards a steady state cannot be answered in general. In the case when an energy equation can be derived from the system, the convergence can be discussed in function of the boundary conditions and of the source terms. Useful indications are given in the general case by the study of the linearized system governing the perturbation $\tilde{f} = f - f^0$ about a steady state solution $f^0$ which is regarded as constant compared to the variations of $\tilde{f}$ :

$$N_0 \frac{\partial \tilde{f}}{\partial t} + A_0^j \frac{\partial \tilde{f}}{\partial x_j} = B_0 \tilde{f} \tag{3}$$

where $B = \partial g / \partial f$ , and the subscript 0 means that the matrix is calculated for $f = f^0$ and considered as constant. The temporal behaviour of spatial harmonic modes of the type : $\tilde{f} = \phi \, exp \, (-i \, k_j \, x_j )$ results from the solution of the differential system :

$$\frac{d\phi}{dt} = i \, C \, \phi \tag{4} \qquad\qquad C = N_0^{-1} ( k_j \, A_0^j - i \, B_0 ) \tag{5}$$

The solutions for $\phi$ are of the form :

$$\phi_\ell (t) = exp \, (i \, \tilde{\omega}_\ell \, t) \sum_{p=0}^{r-1} \tilde{\phi}_p \, t^p \tag{6}$$

where $\tilde{\omega}_\ell$ is an eigenvalue of $C$ of order $r$ , and $\tilde{\phi}_p$ are constant column-vectors.

For the perturbation $\tilde{f}$ to remain bounded, it is necessary that, for all admissible values $f_0$ and for all $k$ , all the eigenvalues $\tilde{\omega}_\ell$ have non-negative imaginary parts and all the real multiple eigenvalues are such that $\tilde{\phi}_p = 0$ for $p = 1$ to $r - 1$ .

If $g \equiv 0$ , as it is the case in general for the simple P.U. systems to be discussed in section IV, then the $\tilde{\omega}_\ell$ coincide with the $\omega_\ell$ which are all real. Moreover, since the eigenvectors $\beta^\ell$ are linearly independent, all modes (6) are of the form $exp(i \, \omega_\ell \, t ) \, \beta^\ell$ , even if $\omega_\ell$ is a multiple eigenvalue of order $r$ (then there exist $r$ independent eigenvectors $\beta^\ell$ ). In this case ( $g \equiv 0$ ), the perturbations are neither damped nor amplified. Nevertheless the convergence to a steady state for the initial-value problem is still possible in a bounded domain, depending essentially on the boundary conditions, but it will be rather slow. In practice, in the numerical solution, some local damping is produced by the numerical dissipation due to truncation errors and to artificial viscosity terms.

Techniques for accelerating the convergence and systems especially devised for giving a strong local damping are discussed in sections V and VI.

### II.3. Stability condition

For the numerical solution of system (1) by means of explicit schemes, it is necessary to have an evaluation of the maximum allowable time-step. A simplified C.F.L. stability criterion is easily obtained by writing that the numerical domain of dependence contains the characteristic cone. Assuming for simplicity that the numerical domain of dependence in space is spherical (or circular) with radius $R_N$, we get a simple and general stability condition :

$$\Delta t \, \underset{\ell,k}{Max} \left\{ \, / \, V^{\ell}(k) \, / \, \right\} \leq R_N \tag{7}$$

where
$$V^{\ell} = ( V_1^{\ell} , V_2^{\ell} , \dots V_n^{\ell} ) \quad , \quad V_j^{\ell} = \partial \omega_{\ell}(k) / \partial k_j \, ,$$

is the group velocity associated with the eigenvalue $\omega_{\ell}$.

### II.4. Boundary conditions

For the discussion and the numerical treatment of the boundary conditions, it is convenient to write the system (1) in the form of the $m$ compatibility relations associated with the direction normal to the boundary. Let $\nu = (\nu_1, \nu_2, \dots \nu_n)$ be the unit normal vector to the boundary, directed outwardly, and consider the eigenvalues $\omega_{\ell}$ and the left-eigenvectors $\alpha^{\ell}$ corresponding to the wave vectors $k$ which are proportional to $\nu$, $k = /k/\nu$, for a given value of $f$. The compatibility relation associated with $\alpha^{\ell}$ results simply from the multiplication of (1) with $\alpha^{\ell}$ :

$$\alpha^{\ell} \left( N \frac{\partial f}{\partial t} + A^i \frac{\partial f}{\partial x_j} \right) = \alpha^{\ell} g \tag{8}$$

This relation has the characteristic property that it does not involve the derivative of $f$ along the direction defined by the vector $( -w^{\ell} , \nu )$ in the $(n+1)$ dimensional time-space continuum. It can be written as a combination of transport equations for the components of $f$, with the transport velocity $w^{\ell}\nu$, where $w^{\ell} = \omega_{\ell} / /k/$ is the phase velocity, the source terms depending only on $f$ and its derivatives in directions tangent to the boundary. If $w^{\ell} > 0$ this transport takes place from inside the domain towards the boundary and the compatibility relation (8) can be used at the boundary. If $w^{\ell} < 0$, the compatibility relation cannot be used at the boundary since, from a mathematical point of view, no information is available outside the domain. The missing relations, i.e. those corresponding to $w^{\ell} < 0$ must be replaced by boundary conditions in equal number. These conditions, which are partly dictated by the physics of the problem, must be such that, together with the compatibility relations corresponding to $w^{\ell} > 0$, they allow the determination of $f$ at the boundary.

In practice, the compatibility relations can be obtained directly by combining the equations in the form actually used, and it is also possible to carry out these combinations directly on the discretized equations as indicated by Veuillot and Viviand (1978, 1979).

## III. FIRST-ORDER CONSERVATIVE SYSTEMS FOR STEADY TRANSONIC FLOWS

### III.1. Isoenergetic flows

The basic system for steady isoenergetic flows is the system of the steady Euler equations in conservative form, the energy equation being used in integrated form as the Bernoulli relation :

a) $\quad div \left( \rho \vec{U} \right) = 0 \quad , \qquad$ b) $div \left( \rho \vec{U} \otimes \vec{U} + p \vec{I} \right) = 0$

c) $\quad H = h \left( p , \rho \right) + \frac{1}{2} \vec{U}^2 = H_0$

$$\tag{9}$$

where $\rho$ is the density, $\vec{U}$ the velocity, $p$ the pressure, $h$ the specific enthalpy, and $\vec{I}$ the unit tensor. For later use, we note here that the left-hand-side of Eq. (9-b) can be transformed into :

$$\text{L.H.S. (9-b)} \equiv \rho\, rot\,\vec{U}_\wedge\vec{U} + \vec{U}\, div(\rho\,\vec{U}) + \rho\,(grad\,H - T\,grad\,S) = 0 \tag{10}$$

where $S$ is the specific entropy.

The weak solutions of the conservative system (9) are exact solutions of the steady Euler equations including the proper jumps across discontinuity surfaces (shock waves and slip surfaces).

### III.2. Isoenergetic and homentropic flows

An important class of solutions of system (9) is that of flows in which the entropy is uniform (homentropic flows) : $S(p,\rho) = S_0$ , and therefore shock waves are not permitted (for exact solutions). From Eq. (9-c) and $S = S_0$ , $p$ and $\rho$ are determined as functions of the modulus $U$ of the velocity, and the only remaining unknown is the velocity vector $\vec{U}$ . From Eq. (10), we see that $rot\,\vec{U}_\wedge\vec{U} = 0$ (such flows are known as Beltrami flows). Thus the system of first-order equations to be solved reduces to :

$$\text{a) } div(\rho\,\vec{U}) = 0 \qquad\qquad \text{b) } rot\,\vec{U} = \begin{cases} 0 & in\ 2D\ flow \\ \lambda\,\rho\,\vec{U} & in\ 3D\ flow \end{cases} \tag{11}$$

where, for 3D flows, $\lambda$ is a new unknown which is conserved along streamlines ( $\vec{U}.\,grad\,\lambda = 0$ is a consequence of Eqs. (11). If the upstream flow conditions imply $rot\,\vec{U} = 0$ , then $\lambda \equiv 0$ and the entire flow is irrotational.

### III.3. Homentropic approximation for flows with shocks

It is a common approximation in transonic problems involving only weak shocks to neglect the entropy jumps across these shocks and therefore, assuming that the entropy is uniform upstream, to consider the flow as being homentropic. The flow fields thus defined satisfy all the equations of system (9), point-wise i.e. in the sense of classical solutions, but they satisfy approximate shock relations which are determined by the conservative form in which the equations are actually solved. Thus system (11), which is a conservative system, yields the classical isentropic approximation (to be denoted HS1) for shock jumps :

$$\text{(HS1)} \qquad\qquad \text{a) } [\rho\,U_N] = 0 \qquad\qquad \text{b) } [\vec{U}_T] = 0 \tag{12}$$

expressing the conservation of mass and of tangential velocity ( $\vec{N}$ being a unit vector normal to the discontinuity surface, we note $U_N = \vec{U}.\vec{N}$ , $\vec{U}_T = \vec{U} - U_N\,\vec{N}$ , and $[A] = A^+ - A^-$ is the jump of $A$ in the direction of $\vec{N}$ ). Now Eq. (12-b) implies $[rot\,\vec{U}.\vec{N}] = 0$ , hence from Eq. (11-b) for the 3D case $[\lambda\,\rho\,U_N] = 0$. So, if $\lambda$ is zero upstream, since $(\rho\,U_N)^\pm \neq 0$ we get $\lambda = 0$ downstream, i.e. $rot\,\vec{U} \equiv 0$ everywhere. Thus the system (11) is completely equivalent to the potential equation in the classical conservative form of the continuity equation. We recall that all the conservative relaxation methods currently used for steady transonic flows are based on this form of the potential equation.

Other conservative systems, leading to different isentropic approximations for shock jumps, can be used. For example, we can replace the continuity equation in system (11) by the component of the momentum equation (9-b) in any fixed direction. The jump relation (12-b) applies for all the conservative systems which include Eq. (11-b), and therefore the irrotationality of the flow across shocks is preserved for all these systems. On the other hand, the jump relation (12-b) excludes the slip surfaces or vortex sheets (this is true also of all conservative forms of the potential equation) which cannot be captured automatically, but must be introduced explicitly in the problem as inner boundaries, despite the fact that their equilibrium positions are unknown.

An interesting (though not much used yet) conservative system for homentropic flows, which allows the capture of vortex sheets, is simply the momentum equation (9-b). The corresponding jump relations (to be denoted HS2) are :

$$\text{(HS2)} \qquad a) \; \left[ \rho \, U_N^2 + p \right] = 0 \quad , \quad b) \; \left[ \rho \, U_N \, \vec{U}_T \right] = 0 \tag{13}$$

These relations, which express the conservation of momentum, include isentropic shocks which are different from the HS1 shocks, and also (for $U_N^2 = 0$ ) vortex sheets across which only the direction of the velocity jumps. In 3D the HS2 shocks do not preserve the irrotationality of the flow, and in 2D, where rot $\vec{U} = 0$ , they imply a jump of the potential function because the jump of $/\vec{U}_T/$ is not zero.

To the best of our knowledge, all the calculations of homentropic transonic flows by P.U. methods up to now have been based on systems (11) or (9-b). The system (9-b) seems attractive for 3D flows since it allows the capture of vortex sheets.

Shock polars for HS1 shocks and for HS2 shocks are shown in Fig. 1, and compared to the exact Rankine-Hugoniot shock polar, in the case of a perfect gas with $\gamma = 1.4$ ( $\theta$ is the deflexion angle, $p_2$ the downstream pressure, and $p_i$ the upstream stagnation pressure).

## IV. SIMPLE PSEUDO-UNSTEADY SYSTEMS

In this section we discuss P.U. systems obtained by simply adding time-derivative terms to the steady systems considered in section III. Such systems do not give an internal damping, and means for accelerating the convergence are discussed in section V.

### IV.1. Isoenergetic flows

The exact unsteady continuity and momentum equations constitute an obvious P.U. system for system (9), which differs from the exact unsteady Euler equations only by the use of the steady Bernoulli relation (9-c). This system is now commonly used in place of the exact unsteady Euler equations ; its properties have been studied by Gottlieb and Gustafsson (1976), by Viviand and Veuillot (1978), and more recently by Brochet (1980) in particular on some aspects of 3D boundary conditions. The characteristic cone is made of the trajectory of the particle and of a second-order cone. The elliptic sections of these cones in a $t$ = const. plane for 2D flows of a perfect gas ( $\gamma$ = 1.4) are shown in Fig. 2 for different Mach numbers. The simplified C.F.L. stability condition (7) is shown in Fig. 4 (curve b) and compared with the similar condition for the exact Euler equations (curve a).

A slightly simpler P.U. system, which has been studied by Laffont (1979), is obtained by using $\delta = \sqrt{\rho}$ and $\vec{W} = \delta \vec{U}$ as basic variables, and $\partial \delta / \partial t$ and $\partial \vec{W} / \partial t$ as time-derivative terms for Eqs. (9-a) and (9-b) respectively.

### IV.2. Isoenergetic and homentropic flows

Various P.U. systems have been proposed for 2D homentropic flows, and here they are given a unified presentation based on the use of $U$ (modulus of velocity) and $\theta$ (angle of velocity vector with a reference axis noted $Ox$ ) as basic dependent variables.

Independently of the actual formulation used, all simple P.U. systems for 2D homentropic flows can be written in the following form :

$$N \, \frac{\partial}{\partial t} \begin{pmatrix} U \\ \theta \end{pmatrix} + \begin{pmatrix} \rho^{-1} \, div \, (\rho \vec{U}) \\ \omega \end{pmatrix} = 0 \tag{14}$$

where $\omega$ is the vorticity (rot $\vec{U} = \omega \vec{k}$ , if $\vec{k}$ is the unit vector normal to the plane of the flow), and $N$ a (2 x 2) matrix, function of $U$ and $\theta$ , regular except possibly for $U = 0$ depending on the actual formulation of the equations. For a study of local properties, we write the system (14) in local intrinsic cartesian coordinates, X, Y, with the X-axis in the direction of $\vec{U}$ at the point which is considered, and the Y-axis directly normal to $\vec{U}$ :

$$N \frac{\partial f}{\partial t} + A \frac{\partial f}{\partial X} + B \frac{\partial f}{\partial Y} + \mathcal{E} \frac{U \sin \theta}{y} E = 0 \tag{15}$$

$$f = \begin{pmatrix} U \\ \theta \end{pmatrix} , \quad A = \begin{pmatrix} 1-M^2 & 0 \\ 0 & U \end{pmatrix} , \quad B = \begin{pmatrix} 0 & U \\ -1 & 0 \end{pmatrix} , \quad E = \begin{pmatrix} 1 \\ 0 \end{pmatrix}$$

where $\mathcal{E} = 0$ for plane flows, $\mathcal{E} = 1$ for axisymmetric flows (then $Ox$ is the symmetry axis and $y$ the radial distance) and $M$ is the local Mach number. Let us see examples of these P.U. systems, giving the actual formulation used and the corresponding matrix $N$ .

— Magnus and Yoshihara (1973), Yoshihara (1973) :

$$\frac{\partial}{\partial t} \begin{pmatrix} u \\ v \end{pmatrix} = \begin{pmatrix} 1/\rho \; \operatorname{div} \rho \vec{U} \\ -\omega \end{pmatrix} , \quad N = \begin{pmatrix} -\cos\theta & \sigma \\ \sin\theta & u \end{pmatrix} \tag{16}$$

where $u = U \cos \theta$ , $v = U \sin \theta$ are the velocity components on the absolute cartesian axes $Ox, Oy$.

— Essers (1978, 1979) :

$$\frac{\partial}{\partial t} \begin{pmatrix} u \\ v \end{pmatrix} = \frac{1}{\rho} \operatorname{div} \rho \vec{U} \begin{pmatrix} \cos\theta \\ \sin\theta \end{pmatrix} + b\omega \begin{pmatrix} \sin\theta \\ -\cos\theta \end{pmatrix} , \quad N = \begin{pmatrix} -1 & 0 \\ 0 & U/b \end{pmatrix} \tag{17}$$

with $b = Max \left\{ |1 - M^2| , b_0 \right\}$ , $b_0$ being a small positive constant.

— Veuillot and Viviand (1978, 1979) :

$$\frac{\partial}{\partial t} \begin{pmatrix} \zeta(U) \\ \theta \end{pmatrix} + \begin{pmatrix} \operatorname{div} \rho \vec{U} \\ \omega \end{pmatrix} = 0 , \quad N = \begin{pmatrix} \frac{1}{\rho} \frac{d\zeta}{dU} & 0 \\ 0 & 1 \end{pmatrix} \tag{18}$$

with $d\zeta/dU < 0$ . In practice, $\zeta(U) = - C \rho_* U/a_*$ , where $C$ is a positive constant, $a$ the sound speed, and starred quantities are sonic values.

— Veuillot and Viviand (unpublished) :

$$\frac{\partial}{\partial t} \left( \mu(U) \vec{U} \right) + \operatorname{div} \left( \rho \vec{U} \otimes \vec{U} + p \vec{I} \right) = 0 , \quad N = \begin{pmatrix} \frac{1}{\rho U} \frac{d\mu U}{dU} & 0 \\ 0 & \frac{\mu}{\rho} \end{pmatrix} \tag{19}$$

with $\mu \, d(\mu U)/dU < 0$ . In practice, $\mu(U) = C a_* \rho/U$ , $C > 0$ .

The work of Magnus and Yoshihara (1973) seems to be the first one concerning P.U. methods for homentropic flows ; these authors first proposed a transonic small perturbation (TSP) version of (16) in which $\rho^{-1} \operatorname{div} \rho \vec{U}$ is replaced by its classical TSP approximation, and generalized it in the form (16). However, as shown by Essers (1978, 1979), the system (16) is not always hyperbolic with respect to time in supersonic flow, although its TSP version is.

The systems (16), (17) and (18) yield HS1 shocks (Eq. (12)) at steady state and do not extend to 3D flows, whereas the system (19) yields HS2 shocks (Eq. (13)) and is equally suitable to 2D or 3D flows.

It can be noted that for all these systems except (16) the matrix $N$ is diagonal and is function of $U$ only. The first property means that the variation of $U$ is driven by the residue on $\operatorname{div} (\rho \vec{U})$ whereas the variation of $\theta$ is driven by the residue on the vorticity. The second property is the condition for the system not to depend on the choice of the origin for $\theta$, i.e. of the reference axis $Ox$. In the remaining part of this section, we examine some general properties of systems (14) built from such matrices $N$ of the form $N = \begin{pmatrix} m(U) & 0 \\ 0 & n(U) \end{pmatrix}$.

Such systems are symmetric, as can be seen by multiplying the second equation by $-U$, and the condition of hyperbolicity with respect to time is :

a) $\qquad U/(mn) < 0$ , $\qquad$ or $\qquad$ b) $\quad U/m = 0$ and $1/n = 0$ $\qquad\qquad$ (20)

the system being strictly hyperbolic in case a). The case b) may correspond to regular systems despite the fact that $n$ is infinite ; for example the system (19) with $\mu U = Ca_* \rho$ corresponds to $m = -Ca_* /a^2$ , $n = Ca_* /U$ , and the case b) is obtained for $U = 0$. A "natural" choice of $\mu$ in system (19) would seem to be $\mu = \rho$ , but it does not satisfy (20) in subsonic flow since $d(\rho U)/dU = \rho(1-M^2)$. It can be noted also that the system (18) with $\zeta = -C\rho_* U/a_*$ does not satisfy (20) for $U = 0$ ; however this does not seem to cause any difficulty in practice, probably because $U = 0$ occurs at isolated points. The "natural" choice $\zeta = \rho$ for the system (18) satisfies (20), but it leads to very restrictive stability limits when $U$ is small.

The equation of the characteristic cone is found to be :

$$\frac{4}{(p-q)^2}\left(\frac{X}{\Delta t} - \frac{p+q}{2}\right)^2 - \frac{1-M^2}{pq}\left(\frac{Y}{\Delta t}\right)^2 = 1 \qquad\qquad (21)$$

where $\quad p = (1-M^2)/m$ , $\quad q = U/n$ $\qquad\qquad$ . The sections by $\Delta t =$ const. planes are ellipses centered at $\quad X = (p+q)\Delta t/2$ $\qquad$ , and the vertices on the X-axis are located at $X = p\,\Delta t$ and $X = q\,\Delta t$. These ellipses are shown in Fig. 3a for different Mach numbers in the case of the system (18) with $C = 1$. For the system (17), these ellipses are all centered at $X = 0$ in subsonic flow, and they degenerate into the two points $X/\Delta t = M^2 - 1$ , $\quad Y/\Delta t = \pm\sqrt{M^2-1}$ $\qquad$ in supersonic flow.

The simplified C.F.L. stability criterion (7) allows a general study of the stability of explicit schemes. In particular we require that the ellipses remain bounded for all finite values of $M$, in order that the time step does not got to zero. A case of vanishing time-step occurs with the system (18) for $U = 0$ if we take $\zeta = \rho$ (then $m = -U/a^2$). Stability curves are shown in Fig. 4 for the systems (17), (18) and (19).

When $n$ is a constant, as in the case of system (18), Viviand (1980) has shown that the system admits an additional conservation equation, the new conserved quantity being a convex function of $\zeta$ and $\theta$, and that the associated irreversibility condition, derived by Friedrichs and Lax (1971) for such systems, allows to eliminate expansion shocks.


## V. CONVERGENCE ACCELERATION TECHNIQUES

For simple P.U. systems no local damping mechanism exists and the convergence towards a steady state depends on the boundary conditions of the problem. For the discretized solution however, some local damping occurs due to the artificial viscosity introduced implicily through the truncation errors and also in general through explicit artificial viscosity terms. Both effects can be represented locally in a simplified way by a term of the form $\mu\,\partial^2 f/\partial x_j^2$ ( $\mu > 0$ ) to be added to the RHS of Eq. (1) in which $q = 0$ (taking $N$ to be the unit matrix). The Fourier modes associated with the eigenvalues $\omega_\rho(k)$ are modified by a damping factor of the form : $\exp\left(-\mu\,k_j^2\,t\right)$. However $\mu$ should remain small outside shocks (say of the order of $\Delta x^2\,\Omega$, where $\Delta x$ is a measure of mesh size and $\Omega$ a characteristic frequency, for schemes of second order accuracy), in order that the steady solution be not appreciably perturbed, and, with explicit schemes, the convergence requires a large number of time iterations.

In the "corrected viscosity" method discussed by Couston et al. (1975), Couston (1976), Essers (1978), the perturbation on the steady state solution is much reduced by correcting the artificial viscosity term by means of a time-independent term which is periodically re-evaluated, in the form : $\mu\,\partial^2 f^n/\partial x_j^2 - \mu'\,\partial^2 f^g/\partial x_j^2$

for $q+1 < n < q+Q$, where the superscript $n$, or $q$, on $f$ is the time index ( $t_n = n\,\Delta t$ ), and where ( $\mu - \mu'$ ) is a small positive quantity, of the order of $\mu\,\Delta x$ for example. The optimal rate of convergence depends on the integer $Q$ which should be of the order of $L/\Delta x$ where $L$ is a characteristic length of the computation domain. The value of $\mu$ cannot be taken much larger than the value used without correction because it would make the time step for explicit schemes much smaller.

A similar technique is used in the "corrected damping" method where a damping term of the form : $K\,(f^n - f^q)$ $(K > 0)$, is introduced in the LHS of Eq. (1). The effect of this term on the rate of convergence has been studied by Essers (1978-a) who concludes that in practice this rate is highly sensitive to the values of the parameters $K$ and $Q$, so that the practical interest of this technique seems doubtful.

A different approach, presently studied by Veuillot (unpublished), is based on the modified system (written here in conservative form) :

$$\frac{\partial f}{\partial t} + \frac{\partial}{\partial x_j}\, F_j\left(f + \varepsilon\,\Delta t\, \frac{\partial f}{\partial t}\right) = 0 \tag{22}$$

where $\varepsilon$ is a small positive parameter, and $\Delta t$ is the C.F.L. time step given for example by (7).
The damping given by this method is of the form $\exp(-g t)$ with $g/(\mu k_j^2) \simeq \varepsilon/(\varepsilon' k\,\Delta x)$
where we have written $\mu = \varepsilon'\,\Delta x^2\,\Omega$ (in practice $\varepsilon' < 1$ ) and we have used $k\,\Delta x \simeq \Omega\,\Delta t$.
Therefore this method will be more efficient than the classical artificial viscosity for damping perturbations with wave lengths $\lambda\,(= 2\pi/k)$ such that $\lambda/\Delta x > 2\pi\,\varepsilon'/\varepsilon$. The gain over the artificial viscosity increases with $\lambda$ and may be very important when using fine meshes. In practice the two methods will be combined. Recent calculations of transonic flow past a profile using this technique with the P.U. system (19) indicate a reduction of the number of iterations by a factor of about 2.

Lastly but not least, mention should be made of the well known local time-step technique which applies directly to the discretized equations, and which consists in using at each mesh point the local value of the time-step as given by the stability criterion. From Eq. (7), this value can be written : $\Delta t_{(\ell)} = \delta_0\, R(x)/(w_0\, W(f))$
where $\delta_0$ is a characteristic mesh size and $w_0$ a reference velocity. The scheme is no more consistent in time with the system (1) but with a modified system deduced from (1) by changing $N$ into $WN/R$, and for which the stability condition is simply $\Delta t \leqslant \delta_0/w_0$, a constant. The sections of the characteristic cones for this system are all tangent to the circle of radius $R_N$, as shown on Fig. 3b in the case of the P.U. system (18). This technique is simple to use and quite efficient especially when the mesh is highly non uniform, the variations of $\Delta t_\ell$ being due for a large part to the variations of $R(x)$. Naturally, these variations should be smooth enough, otherwise the calculation may diverge. The values of $\Delta t_\ell$ need not be calculated at each time step, and they can be frozen as the convergence is approached.

## VI. PSEUDO-UNSTEADY SYSTEMS WITH STRONG DAMPING

The techniques described in section V can appreciably reduce the computing costs, but they do not drastically change the efficiency of simple P.U. methods. The difficulty arises from the fact that, for first order systems, a strong damping requires non-derived terms which alter the steady state solutions. With second order equations, a similar damping is obtained with time-derivative terms which do not change the steady state.

A general approach to solve this difficulty is to introduce new artificial unknowns and new equations which contain non-derived damping terms, so that the original steady equations are not perturbed. This approach has been discussed by Wirz (1977) and by Essers (1978) who has presented applications to transonic flows. A brief account of the "dashpot" methods proposed by Essers is given below.

Let the steady-state equations to be solved be $A_j\,\partial f/\partial x_j = 0$, where $f$ is the column-vector of the physical unknowns, as in system (1), and introduce $m$ new artificial unknowns : $q = (q_1, q_2, \dots q_m)$. The "dashpot" methods are based on P.U. systems of the general following form :

$$\frac{\partial q}{\partial t} = A_j \frac{\partial f}{\partial x_j} \quad , \quad \frac{\partial f}{\partial t} + K_1 f + K_2 q = B_j \frac{\partial f}{\partial x_j} + C_j \frac{\partial q}{\partial x_j} \tag{23}$$

Except particular cases, the use of $m$ new unknowns and $m$ new equations can be shown to be necessary in general, i.e. for arbitrary matrices $A_j$ . The supplementary equations must be such that the complete system (23) is hyperbolic with respect to time and that the non-derived terms give an internal damping as strong as possible. This damping will be much more efficient than the one obtained with artificial viscosity type techniques.

Two methods of this type are presented by Essers. The first one (named Dash 1) gives a very fast convergence, but it is restricted to subcritical flows because it produces expansion shocks in mixed subsonic-supersonic flows. The reasons for this behaviour are discussed by Essers (1978). The second method (Dash 2) is less rapid but allows the calculation of transonic flows. Results presented by Essers (1978) for transonic nozzle flows, using the same steady system as in (17) or its TSP version, show that the Dash 2 method can be appreciably faster than line-overrelaxation methods for the potential equation. In the case of 1D transonic nozzle flow with shock calculated by the Dash 2 method using the full Euler equations, Essers reports a reduction of the number of iterations by a factor of 7, and of the computing time by a factor of 4.4, compared to a time accurate method. Of course part of the benefit derived from the strong internal damping is consumed by the increase in computing time due to the doubling of the equations.

## VII. CONCLUSIONS

For applications of the P.U. methods reviewed in this paper we refer to the publications cited and also, as further examples, to the works of Veuillot (1975, 1976), Denton (1976) and Brochet (1980) concerning flows in turbomachines and of Meauzé (1980) concerning inverse type calculations. Applications to external flows past profiles will be found in the Proceedings of a recent GAMM-Workshop (Rizzi and Viviand, Ed. 1980)). Explicit schemes are used in all these applications.

Concerning the efficiency of the P.U. methods, generally speaking, the simple P.U. methods, using explicit schemes with the local time-step technique, appear to be comparable to the classical conservative relaxation methods for the potential equation, which means that they remain appreciably slower than the newest relaxation methods based on approximate factorization or multiple grid techniques (e.g., Holst (1979), Jameson (1979). The "dashpot" methods of Essers may approach the latter on convergence speed measured in number of iterations, but the efficiency in computing time suffers from the increase in equations.

However the P.U. approach certainly presents practical advantages and is open to continued improvements. Practical advantages are the simplicity of use and of programmation, robustness regarding the convergence which is not sensitive to mesh topology and to variations of mesh size, flexibility of use for varied types of boundary conditions including inverse problems with unknown boundaries and time-varying meshes, and also economy of metric coefficients compared to second-order equations in the case of arbitrary meshes.

Progress can be expected from systematic studies of P.U. systems, in particular looking for possibilities of producing a strong damping with a minimum increase in computational cost. Greater efficiency, to be paid for by greater complexity, should also result from the use of implicit schemes, and advances in this domain would probably give a new impetus to P.U. methods.

*Acknowledgments* *The author wishes to thank Mr. J.P. VEUILLOT for many fruitful discussions and for his help in the preparation of this paper. This work has been performed with financial support from DRET.*

**References**

Brochet, J. 1980 La Rech. Aérosp. (to appear).

Courant, R. and Hilbert, D. 1962 Methods of Mathematical Physics, Vol. II Interscience Publishers.

Couston, M. 1976 VKI Lecture Series 84, Rhode St Genèse, Belgium.

Couston, M., McDonald, P.W. and Smolderen, J.J. 1975 VKI Tech. Note 109.

Denton, J.D. 1976 VKI Lecture Series 84, Rhode St Genèse, Belgium.

Essers, J.A. 1978-a VKI Lecture Series 1978-4, Rhode St Genèse, Belgium.

Essers, J.A. 1978-b Coll. Publ. Fac. Sc. Appl., Univ. Liège, No 73.

Essers, J.A. 1979 Appl. Math. Modelling Vol. 3, 55-66.

Friedrichs, K.O. and Lax, P.D. 1971 Proc. Nat. Acad. Sci. U.S.A. Vol. 68, No 8, 1686-1688.

Gottlieb, D. and Gustafsson, B. 1976 Studies in Appl. Math. Vol. 55, 167-185.

Holst, T.L. 1979 AIAA Paper No 79-1456, AIAA Comp. Fluid Dyn. Conf., Williamsburg.

Jameson, A. 1974 Comm. Pure Appl. Math. Vol. 27, 283-309.

Jameson, A. 1975 Proceed.of AIAA 2nd Comp. Fluid. Dyn. Conf., Hartford, 148-161.

Jameson, A. 1979 AIAA Paper No 79-1458, AIAA Comp. Fluid Dyn. Conf., Williamsburg.

Jeffrey, A. 1976 Quasilinear hyperbolic systems and waves, Pitman Publishing.

Jeffrey, A. 1978 ZAMM Vol. 58, T56-T65.

Lafont, P. 1979 ONERA Internal Report

Lomax, H. and Steger, J.L. 1975 Ann. Rev. of Fluid Mech. Vol. 7, 63-88.

Magnus, R. and Yoshihara, H. 1973 NASA CR-2186.

Meauzé, G. 1980 La Rech. Aérosp. No 1980-1, 23-30.

Rizzi, A. and Viviand, H. 1980 (Ed.) Proceed. of a GAMM Workshop, Notes on Num. Fluid Mech. Vol. 3, Vieweg and Sohn.

Veuillot, J.P. 1975 La Rech. Aérosp. No 1975-6, 327-338.

Veuillot, J.P. 1976 ASME Paper No 76-GT-56, ASME Gas Turbine Conf., New Orleans.

Veuillot, J.P. and Viviand, H. 1979 AIAA J. Vol. 17, No 7, 691-692, also AIAA paper 78-1150.

Viviand, H. and Veuillot, J.P. 1978 Publication ONERA No 1978-4.

Viviand, H. 1980 La Rech. Aérosp. (to appear).

Wirz, H.J. 1977 AGARD-VKI Lecture Series 86, Rhode St Genèse, Belgium

Yoshihara, H. 1973 AGARD-VKI Lecture Series 64, Rhode St Genèse, Belgium.

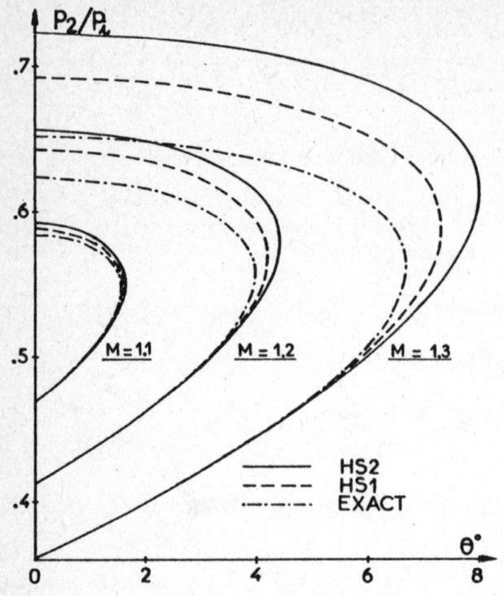


Fig. 1 — Shock polars for the exact shock relations and for the HS1 and HS2 isentropic approximations.

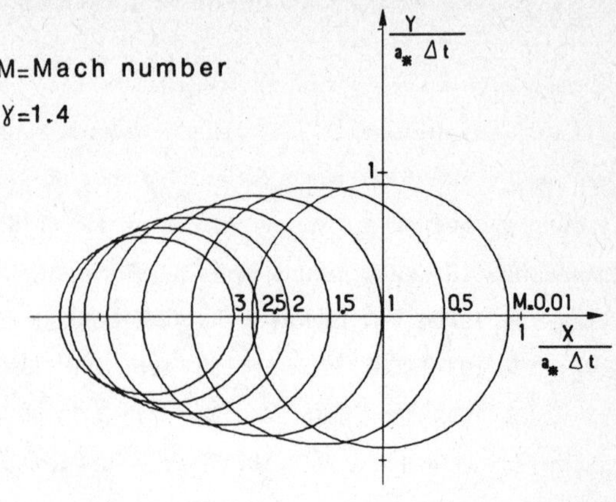


Fig. 2 — Characteristic cones when the total enthalpy H is constant.

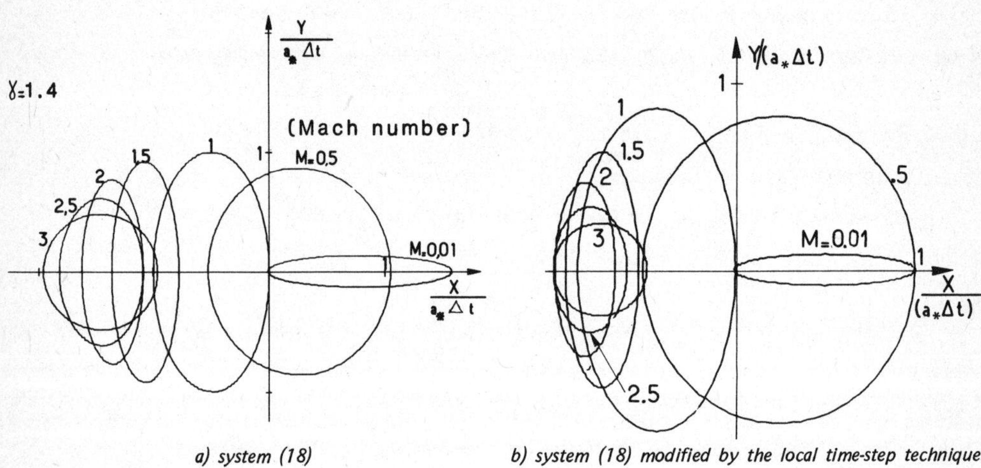


a) system (18)

b) system (18) modified by the local time-step technique.

Fig. 3 — Characteristic cones for the P.U. system (18) (C = 1) and effect of the local time step technique.

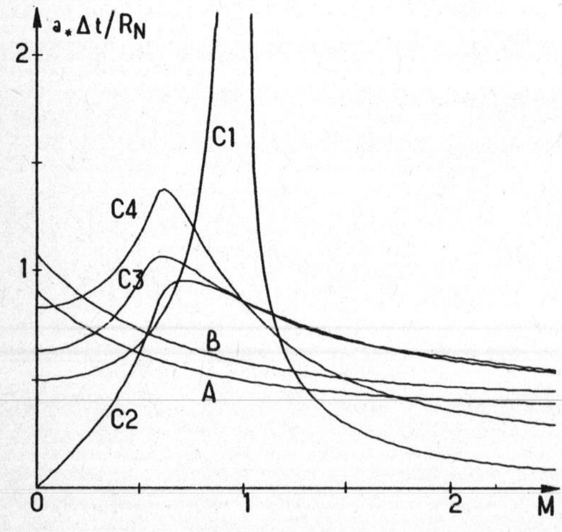


Fig. 4 — Stability curves

a) time-accurate methods
b) methods with H constant
c) methods with H and S constant.
1 - system (17)
2 - system (18) with $\zeta = \rho$
3 - system (18) with $\zeta = -\rho_* U/a_*$
4 - system (19) with $\mu = a_* \rho/U$

AXISYMMETRIC TRANSONIC FLOW COMPUTATIONS USING A MULTIGRID METHOD

B. Arlinger

SAAB-SCANIA AB, Linköping, Sweden

The axisymmetric full potential equation is solved for transonic flow over blunt or
pointed bodies of revolution using an orthogonal coordinate system obtained by conformal mapping. Both the quasilinear and the conservative form of the equation are
solved, applying the concept of artificial compressibility in the conservative case
to get the proper numerical viscosity in supersonic regions. The iterative technique
which is a multigrid method with successive line over-relaxation is shown to yield
very good convergence rate.

## I    Introduction

The present work is motivated by the need of a fast and accurate method for calculation of pressure distributions around blunt or pointed axisymmetric noses of general
shapes. It also illustrates the application of two rather new concepts in numerical
transonics, namely artificial compressibility and multigrid technique.

A difference between the present method and the work by South and Jameson(1973) is
that the full potential equation here is solved in both quasilinear and conservative
formulation. It also uses a strictly orthogonal grid generated by conformal mapping,
a method that has been applied to various two-dimensional configurations examplified
by Jameson(1974), Arlinger(1976).

In discretizing the conservative equation the concept of artificial compressibility,
studied recently by Hafez, Murman and South(1978), is applied to get the proper numerical viscosity in supersonic regions. A case is shown where its application led to non-
unique solutions, one being physically unacceptable.

The multigrid technique used to speed up relaxation processes is not new, but has only
fairly recently been applied to transonic cases, for instance by South and Brandt
(1976), Fuchs(1977), Arlinger(1978), Jameson(1979). In the present application to
both the quasilinear and the conservative equation its effectiveness compared to
single grid SLOR is clearly demonstrated.

## II    Grid generation

A body oriented coordinate system is generated by conformally mapping the physical
flow field in the x,r-plane on a unit square where a simple rectangular grid is applied. The mapping sequence for a pointed body, from the physical plane $z=x+ir$ to
the square plane $\zeta'=\xi'+i\eta'$ comprises the steps: a smooth removal of downstream body

thickness ($z \to z_1$), blunting the nose ($z_1 \to z_2$), square root mapping ($z_2 \to z_3$), mapping to near semi-circle, to exact semi-circle and to quarter plane ($z_3 \to z_4 \to z_5 \to \zeta$) and a final stretching to a square ($\zeta \to \zeta'$). The equations are, using symbols of Fig 1

$$z = z_1 - \frac{R}{\pi}\left[\ln(z_1 - s) + \left(\frac{z_1 - t}{t - s}\right)^2 \ln\frac{z_1 - t}{z_1 - s} + \frac{z_1}{t - s}\right] + Q + iR \tag{1}$$

with t and Q determined by the conditions: $\quad z=0 \to z_1 = 0, \quad z = L + iR \to z_1 = t$

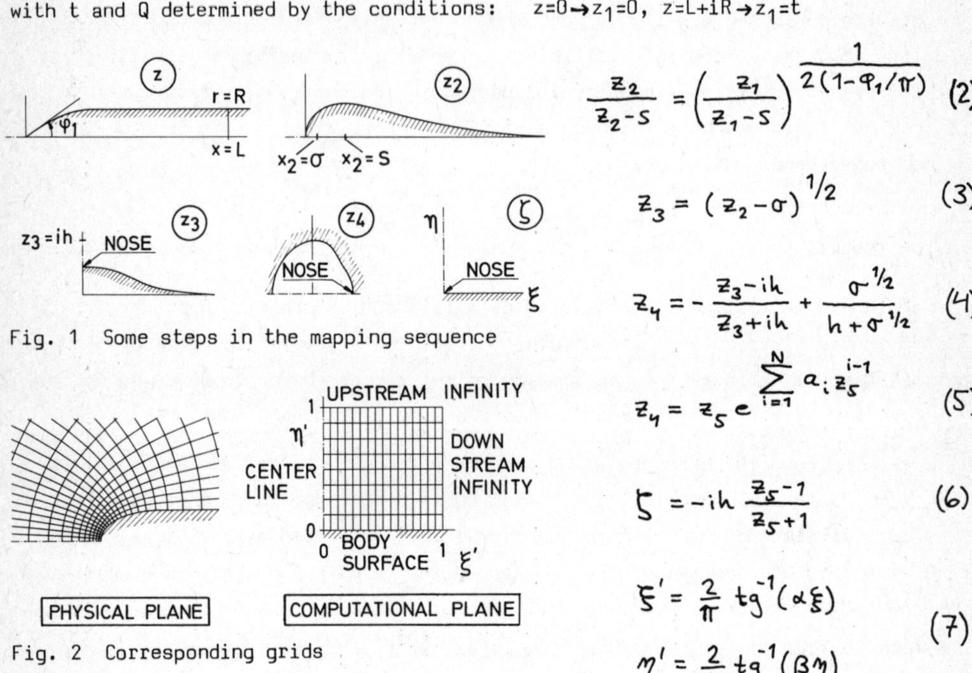



Fig. 1   Some steps in the mapping sequence



Fig. 2   Corresponding grids

$$\frac{z_2}{z_2 - s} = \left(\frac{z_1}{z_1 - s}\right)^{\frac{1}{2(1 - \varphi_1/\pi)}} \tag{2}$$

$$z_3 = (z_2 - \sigma)^{1/2} \tag{3}$$

$$z_4 = -\frac{z_3 - ih}{z_3 + ih} + \frac{\sigma^{1/2}}{h + \sigma^{1/2}} \tag{4}$$

$$z_4 = z_5 \, e^{\sum_{i=1}^{N} a_i z_5^{i-1}} \tag{5}$$

$$\zeta = -ih \frac{z_5 - 1}{z_5 + 1} \tag{6}$$

$$\xi' = \frac{2}{\pi} tg^{-1}(\alpha \xi)$$
$$\eta' = \frac{2}{\pi} tg^{-1}(\beta \eta) \tag{7}$$

The use of Fast Fourier Transforms in the iterative solution of Eq.(5) makes the mapping very fast. The type of grid thus generated is illustrated in Fig 2 for a blunt body.


### III   Governing equations

The mass conservation equation expressed in the unstretched coordinates $\xi, \eta$ of the computational plane may be written in conservative form and developed into the quasi-linear potential equation as

$$\left(r \varrho \phi_\xi\right)_\xi + \left(r \varrho \phi_\eta\right)_\eta = 0 \tag{8}$$

$$(a^2 - u^2)\phi_{\xi\xi} - 2uv\,\phi_{\xi\eta} + (a^2 - v^2)\phi_{\eta\eta}$$
$$+ a^2\left(\phi_\xi \frac{r_\xi}{r} + \phi_\eta \frac{r_\eta}{r}\right) + q^2\left(\phi_\xi \frac{B_\xi}{B} + \phi_\eta \frac{B_\eta}{B}\right) = 0 \tag{9}$$

with $\phi$ as the velocity potential and $\varrho$ as the density. The velocities are defined by

$$u = \phi_\xi / B , \qquad v = \phi_\eta / B , \qquad q^2 = u^2 + v^2 , \qquad B = \left| \frac{dz}{d\xi} \right| \tag{10}$$

where B is the mapping modulus. The local speed of sound a and the density are obtained from the relations

$$a^2 = \frac{1}{M_\infty^2} + \frac{\gamma-1}{2}(1-q^2) , \qquad \varrho^{\gamma-1} = a^2 M_\infty^2 \tag{11}$$

with $M_\infty$ as the free stream Mach number. The density and all velocities are normalized with their free stream values. At the centerline ($\xi = 0$) and along the body contour ($\eta = 0$), the boundary conditions are of homogeneous Neumann type, i.e. $\phi_\xi = 0$ and $\phi_\eta = 0$ respectively.


### IV   Numerical Procedures

The discretizing of the quasilinear equation (9) is made in the standard way, using a rotated scheme for the upwind differencing in supersonic points (Jameson 1974) to get the proper amount of numerical viscosity needed to avoid unphysical solutions.


### A.   Conservative Equation with Artificial Compressibility

The artificial compressibility formulation was applied to the conservative equation (8) to generate the numerical viscosity. It implies that the density in Eq.(8) is replaced by the modified density $\tilde{\varrho}$ which, based on the work by Hafez, Murman and South(1978), here is defined as

$$\tilde{\varrho} = \varrho - \mu V_m \left( \frac{u}{q} \varrho_\xi \Delta\xi + \frac{v}{q} \varrho_\eta \Delta\eta \right) \qquad \text{with} \qquad \mu = max\left[ 0, \left(1 - V_s \frac{a^2}{q^2}\right)\right] \tag{14}$$

The two viscosity parameters $V_m$ and $V_s$ control the magnitude and the switch on Mach number of the density modification. This term in Eq.(14) is equivalent to an artificial viscosity, and the highest derivative term of that (with $V_m$ and $V_s$ equal to 1) is close to the viscosity term in the quasilinear case.

The discretization of Eq.(8) is made according to

$$\left( r\tilde{\varrho}\phi_\xi \right)_\xi = \left[ (r\tilde{\varrho})_{i+\frac{1}{2},j} \left( \phi_{i+1,j} - \phi_{ij} \right) - (r\tilde{\varrho})_{i-\frac{1}{2},j} \left( \phi_{ij} - \phi_{i-1,j} \right) \right] / (\Delta\xi)^2 \tag{15}$$

with an analogue expression for $\left( r\tilde{\varrho}\phi_\eta \right)_\eta$ . The terms $\varrho_\xi \Delta\xi$ and $\varrho_\eta \Delta\eta$ in Eq.(14) are upwind differenced, while each $\varrho_{i+\frac{1}{2},j}$, $\varrho_{i,j+\frac{1}{2}}$ etc is based on either 6 or 8 points of potential values, examplified by the relations

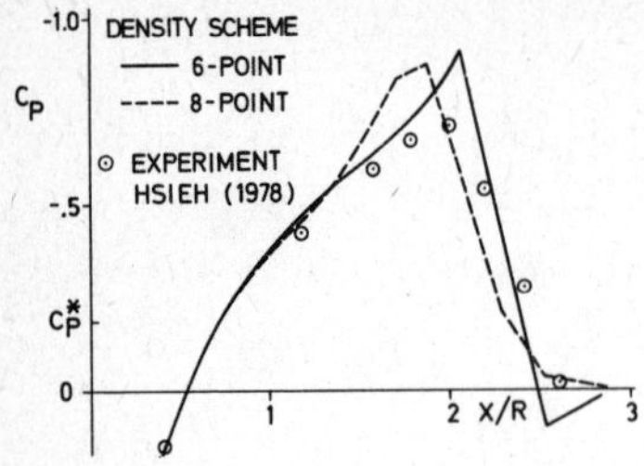

a) Effect of density calculation scheme. Expansion type
solution

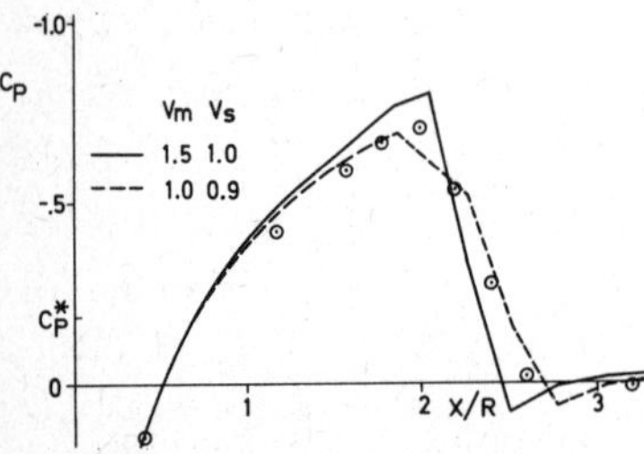

b) Effect of numerical viscosity parameters. 6-point
scheme for density

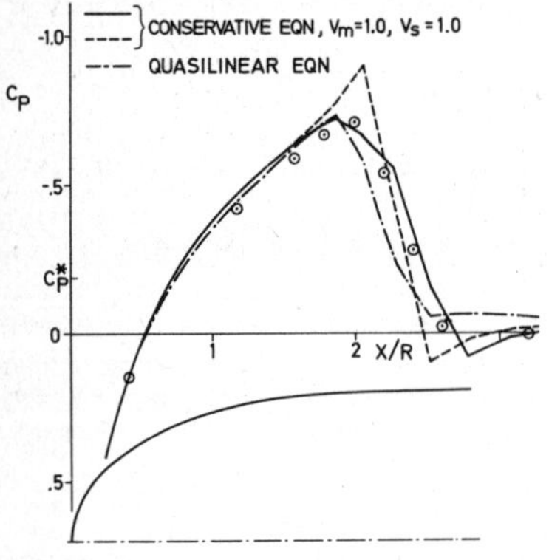

c) Example of non-uniqueness of
conservative eqn solution

Fig. 3 (2:1)-ellipse-nose-cylin-
der at $M_\infty$ =0.9. Mesh 72x32

8-point scheme: $S_{i+\frac{1}{2},j} = \frac{1}{2}\left(S_{ij}+S_{i+1,j}\right), \quad S_{ij} = S(u_{ij},v_{ij})$

$$u_{ij} = \left(\phi_{i+1,j}-\phi_{i-1,j}\right)/\left(2\,\Delta\xi\,B_{ij}\right) \quad \text{etc.}$$

6-point scheme: $S_{i+\frac{1}{2},j} = S\left(u_{i+\frac{1}{2},j}, v_{i+\frac{1}{2},j}\right)$

$$u_{i+\frac{1}{2},j} = \left(\phi_{i+1,j}-\phi_{ij}\right)/\left(\Delta\xi\,B_{i+\frac{1}{2},j}\right)$$

$$v_{i+\frac{1}{2},j} = \left(\phi_{i,j+1}-\phi_{i,j-1}+\phi_{i+1,j+1}-\phi_{i+1,j-1}\right)/\left(4\Delta\eta\,B_{i+\frac{1}{2},j}\right)$$

$$B_{i+\frac{1}{2},j} = \frac{1}{2}\left(B_{i+1,j}+B_{ij}\right), \quad \text{etc.}$$

For simplicity all formulas are here written in
the unstretched coordinates $\xi, \eta$ , and disregarding that a reduced potential $\chi$ , bounded in the
whole flow field was used instead of $\phi$, satisfying
$\chi = \phi - \text{const} \cdot (\xi^2 - \eta^2)$

Applying the two density schemes to the case of
an ellipse-nose-cylinder at $M_\infty$ =0.9 yields the
pressure curves shown in Fig. 3a for $V_m$=1, $V_s$=1.
The physically unacceptable expansions before the
shock wave were sensitive to the viscosity parameters as shown in Fig. 3b, and when $V_s$ was gradually increased again from 0.9 to 1.0 the solution
did not switch back to the expansion type. Fig. 3c
thus illustrates a case of non-uniqueness of the
solutions to the conservative equation.

Generally it seems that the artificial compressibility formulation with the parameter $V_m$ somewhat
larger than 1 and $V_s$ slightly lower than 1 works
well in transonic flow.

## B. Multigrid method

Writing the equations for the potential $\phi$ discre-
tized in a mesh with spacing proportional to h as $L^h\phi$ +F=0, the multigrid principle
is to solve the following correction problem, here written for a grid with doubled
spacing

$$L^{2h}\Delta\phi + I_h^{2h}\left(L^h\phi+F\right) = 0 \tag{16}$$

$\Delta\phi$ is the correction to the potential approximation $\phi$, $L^{2h}$ is the same difference
operator as $L^h$, but written for the coarser grid. The operator $I_h^{2h}$ transfers to each

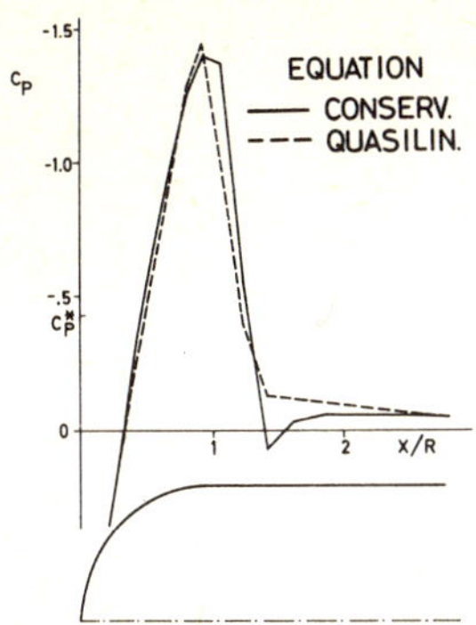



Fig. 4 Pressure about a hemisphere-cylinder, $M_\infty =0.8$

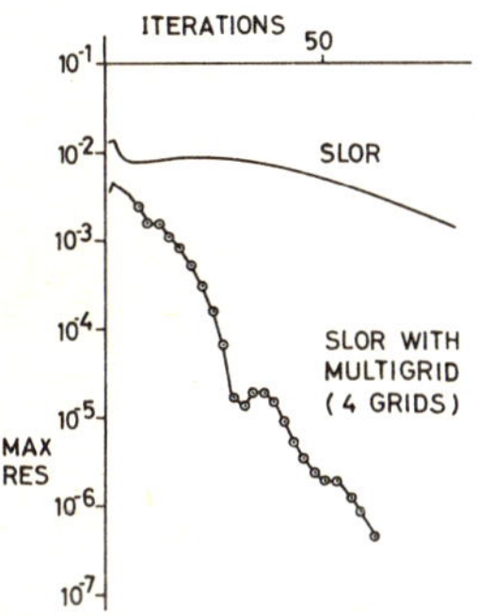



Fig. 5 Convergence for quasilinear equation

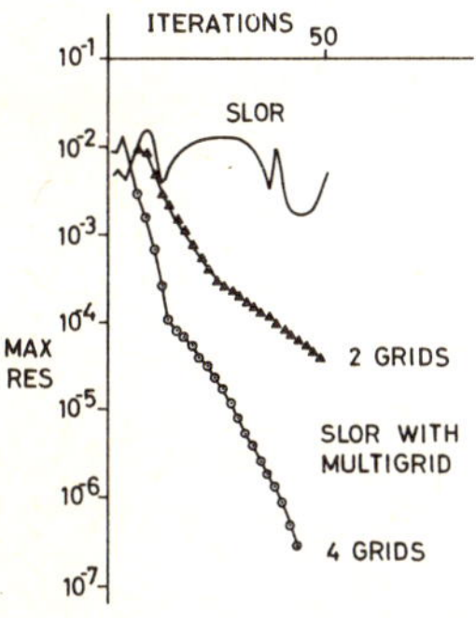



Fig. 6 Convergence for conservative equation

grid point in the coarse grid the value at the corresponding point in the fine grid.

In a typical multigrid sequence Eq.(16) is solved on a series of grids, in the present application starting with a very coarse grid (e.g. 8x4) where a small number of iterations ("smoothings") is performed (e.g. 3-4). Then the $\Delta\phi$-field is interpolated to the next finer grid (mesh spacing halved), again a few iterations are made, etc up to the finest grid. Typical for the present application is that such a multigrid sequence solves a linear problem, because the operators are kept frozen. They are only updated after each multigrid sequence when usually only one nonlinear iteration is performed on the fine grid before the next multigrid sequence. The iteration technique used throughout the present work is successive line over-relaxation (SLOR).

The calculations of the flow field around a hemisphere-cylinder at $M_\infty =0.8$ (Fig. 4) are used to illustrate the effectiveness of the multigrid method as shown in Figs 5 and 6. (One iteration is here equivalent to the work required for one nonlinear iteration on the fine grid). In solving the conservative equation the simple everywhere elliptic operator was used also in the updating algoritm. The ordinary SLOR is then slightly unstable (Fig. 6) as confirmed by a von Neumann analysis, while the multigrid technique yields a very rapid and stable convergence, working for subsonic free stream flow.

## V A Few Computational Examples

Fig. 7 compares results of the present method with an Euler equation solution by Chaussee(1978) who also provides the experimental points. Another comparison with the same author is shown in Fig. 8, illustrating the sensitivity of shock location in a supersonic case to equation type. Fig. 9 shows very good agreement between calculated and measured wave drag for a hemisphere-cylinder throughout the transonic range. Finally, Fig. 10 illustrates an example of supersonic flow at $M_\infty =1.2$ around an ogive-cylinder, defined by $r = -2.4383 + [2.6354^2-(x-1)^2]^{\frac{1}{2}}$ for $x \leq 1$ and $r = 0.1971$ for $x \geq 1$.

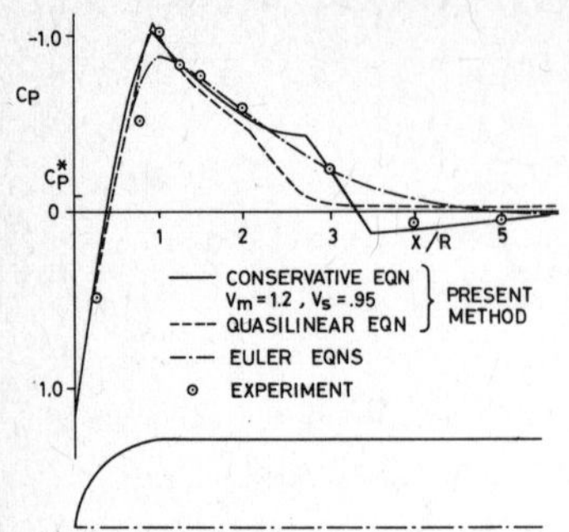



Fig. 7 Pressure distributions about a hemisphere-cylinder at $M_\infty$ =0.95, mesh 64x32

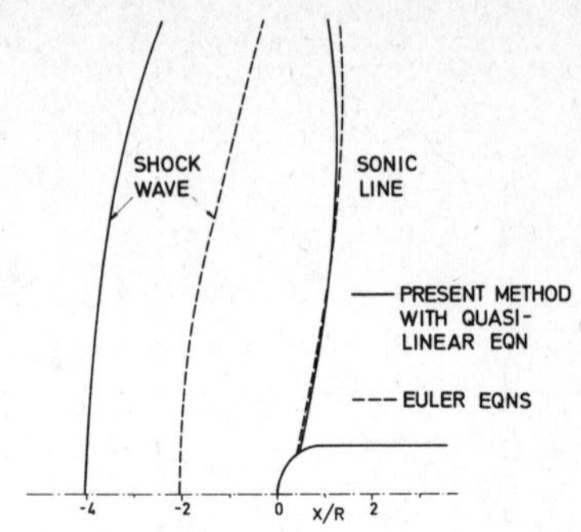



Fig. 8 Supersonic region for flow at $M_\infty$ =0.95 about a hemisphere-cylinder

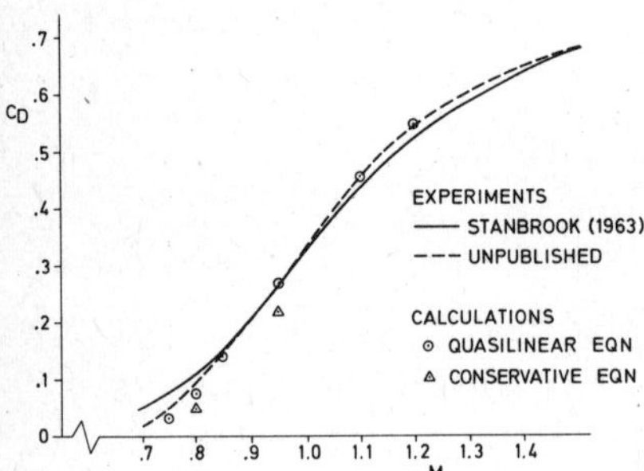



Fig. 9 Calculated and experimental wave drag for a hemisphere-cylinder

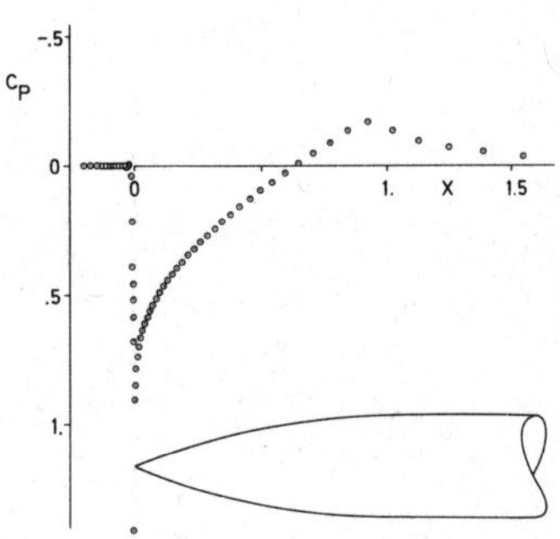



Fig. 10 Quasilinear solution for ogive-cylinder at $M_\infty$ =1.2

References


Arlinger,B.,"Analysis of two-element high lift systems in transonic flow", ICAS Paper 76-13, 1976.
Arlinger,B.,"Multigrid technique applied to lifting transonic flow using full potential equation", SAAB-SCANIA Rep. L-0-1 B439, 1978.
Chaussee,D.,"On the transonic flow field surrounding tangent ogive bodies with emphasis on nose drag calculations", AIAA Paper 78-212, 1978.
Fuchs,L.,"Finite difference methods for plane steady inviscid transonic flows", Techn. Dr. Thesis, KTH, Stockholm, 1977.
Hafez,M., Murman,E. and South,J.,"Artificial compressibility methods for numerical solution of transonic full potential equation", AIAA Paper 78-1148, 1978.
Hsieh,T.,"Unsteady transonic flow over blunt and pointed bodies of revolution",AIAA Paper 78-211, 1978.
Jameson,A.,"Iterative solution of transonic flows over airfoils and wings, including flows at Mach 1", Comm. Pure Appl. Math. 27, p 283-309, 1974.
Jameson,A.,"Acceleration of transonic potential flow calculations on arbitrary meshes by the multiple grid method", AIAA Paper 79-1458, 1979.
South,J. and Jameson,A.,"Relaxation solutions for inviscid axisymmetric flow over blunt or pointed bodies", Proc. AIAA Conf. on Comp.Fluid Dynamics, Palm Springs, p 8-17, 1973.
South,J. and Brandt,A.,"Application of a multi-level grid method to transonic flow calculations", ICASE Report Nr 76-8, Langley Research Center, 1976.
Stanbrook,A.,"A correlation of the forebody drag of cylinders with plane and hemispherical noses at Mach numbers from zero to 2.5", RAE TN Aero 2875, 1963.

Numerical Techniques for Free Surface Motion with Application to Drop

Motion   Caused by Variable Surface Tension

G. R. Baker, Math.  Dept., MIT, Cambridge, MA
M. Israeli, Comp. Sc. Dept., Technion, Haifa, Israel

## I.  INTRODUCTION

Classical potential theory may be used to reduce the calculation of the motion of a free surface of an inviscid, incompressible fluid to one which involves only the determination of quantities at the surface.  In particular, we consider two techniques based on the evaluation of integrals along the surface; Green's formula determines the normal velocity at the surface given the potential there, and the vortex sheet representation determines the velocity via the Biot Savart integral if the vortex sheet strength is known.  The vortex sheet strength obeys a Fredholm integral equation of the second kind.  The vortex sheet representation offers several advantages.

A different approach is based on a truncated series (spectral) least squares filter approximation for the potential in the interior. The surface velocity is  obtained by analytical differentiation.  We apply these techniques to the deformation of a drop which is driven by surface tension whose strength depends on the concentration of a chemical surfactant.

## II.  BOUNDARY INTEGRAL TECHNIQUES

### A.  Green's Formula

Consider an incompressible, inviscid, irrotational fluid enclosed by a surface $\partial B$ which has a continuous tangent plane and normal  n  (the surface may extend to $\infty$ ).  If the velocity potential $\phi$ is known at the surface at some time  t,  Green's formula

$$\int_{\partial B} \frac{1}{|p-q|} \frac{\partial \phi}{\partial n} \, dq = -2\pi\phi(p) + \int_{\partial B} \phi \frac{\partial}{\partial n} \frac{1}{|p-q|} \, dq \; ; \quad p,q \in \partial B$$

is an integral equation for the normal velocity $\frac{\partial \phi}{\partial n}$ of the surface (the principal values of the integrals must be taken).  The surface tangential velocities $\frac{\partial \phi}{\partial s_1}$ , $\frac{\partial \phi}{\partial s_2}$  are determined directly by differentiation, thus the change in the surface's location may be calculated and Bernoulli's equation may be used to calculate the change in $\phi$ at the surface, provided the pressure at the surface is given.  In the simplest case, the exterior of the fluid is vacuum  and hence the pressure is constant at the surface. Longuet-Higgins and Cokelet (1978) have applied this technique to the motion of breaking surface waves.

### B.  Vortex Sheet Representation

The surface may be considered as a vortex sheet of strength  k.  Its motion is determined from the Biot-Savart integral,

$$\bar{u}(p) = \frac{1}{4\pi} \int_{\partial B} \frac{K(q) \times (p-q)}{|p-q|^3}\, dq \; ; \qquad p, q \in \partial B$$

where the internal, external fluid velocity at the surface, $u_I$, $u_E$, respectively, are given in terms of the average velocity $\bar{u}$ and $K$ by the relations

$$u_I = \bar{u} - \tfrac{1}{2}(n \times K)$$

$$u_E = \bar{u} + \tfrac{1}{2}(n \times K)$$

If these relations are substituted into the momentum equation evaluated at either side of the surface and the resulting equations subtracted, then

$$\frac{d}{dt}(n \times K) + (n \times K) \cdot \nabla \bar{u} = \frac{2(\rho_E - \rho_I)}{(\rho_E + \rho_I)} \left( \frac{d\bar{u}}{dt} + \tfrac{1}{4}(n \times K) \cdot \nabla (n \times K) \right)$$

$$- \frac{2}{(\rho_E + \rho_I)} \nabla (p_E - p_I)$$

where $\frac{d}{dt}$ refers to convective derivative with velocity $\bar{u}$ and $\rho$, p are the densities and pressures (subscripts I,E refer to internal, external fluid respectively). Since $\frac{d\bar{u}}{dt}$ depends on $\frac{d}{dt}(n \times K)$ via the Biot–Savart integral, the equation for $\frac{d}{dt}(n \times K)$ is a Fredholm integral equation of the second kind. This equation may be always solved iteratively. Thus we have obtained evolution equations for the surface location and the vortex sheet strength.

We wish to emphasize the power of the vortex sheet representation. Several interfaces between fluids of constant, but different densities can each be represented by a vortex sheet. Surface forces, e.g., surface tension, can be included through the term involving $p_I - p_E$. Any potential flow can be added and, in particular, conformal mapping and the method of images can treat the presence of solid boundaries in certain cases. Finally, the inclusion of vorticity in the interior of the fluid is trivial.

In both approaches, numerical quadrature is used on the integral after the singular nature of the integrand has been removed by the addition of a suitable integrable function. For the Green's formula, a matrix equation must be solved which is $O(N^3)$ at each time step where N is the number of mesh points along the surface (here we are considering for simplicity a surface which can be parameterized by one variable). However, for the vortex sheet representation, the Fredholm integral may be solved iteratively which requires only $O(N^2)$ operations per iteration. Tests indicate only a few iterations are required to obtain a high degree of accuracy, for example for the results reported in the next section only eight iterations were needed for eight digits of accuracy. Thus the computations ran about a factor 10 faster than the Green's formula technique using 90 surface markers.

## C. Numerical Tests

Tests were conducted mainly to establish the validity and accuracy of the vortex sheet representation. The method was applied to breaking surface waves, where the Green's formula has been successfully used (Longuet-Higgins and Cokelet (1978)). Figure 1 shows the breaking wave that forms when an 80% Stokes waves is forced for a short period by a surface pressure distribution, (for details see Baker and Meiron (to be published)). The Stokes wave was first found by using the vortex sheet representation with the assumption of a steadily convecting form and solving the resultant equations by Newton iteration. Excellent agreement with the profiles calculated by Schwartz (1974) has been found.

In addition, the results for the Rayleigh-Taylor instability are shown in Figure 2 where an initial disturbance $y = 0.5 \cos(2\pi x/\lambda)$ evolves into a narrow spike falling freely into vacuum. Further details and results are available in Baker, Meiron and Orszag (1980). Direct quantitative comparison was made with the results of a conformal mapping technique applied to the Rayleigh-Taylor instability by Menikoff and Zemach (1980). The agreement was to five digits and confirms the accuracy and reliability of the vortex sheet representation.

Currently the vortex sheet representation is being applied to a variety of flow problems.

## III. THE SPECTRAL TECHNIQUE

The spectral technique was tested in the case of the Rayleigh-Taylor instability and was also implemented in the case of a closed surface in spherical geometry. We will review here the second case.

We expand the potential in the interior in a truncated series of the form

$$\phi = \sum_{j=0}^{N} A_j \phi_j \ ,$$

where $\phi_j(r,\theta)$ are solutions (regular near the origin) of the Laplace equation in spherical geometry. We assume that the potential is known at $M+1$ Lagrangian points on the surface (where $M>N$) and solve the (rectangular) system of equations for $A_j$. The surface velocities are obtained by analytical differentiation of the truncated series on the surface of interest. This procedure was extensively checked and fast convergence was obtained; particularly good results were obtained for the case $M \simeq 2N$.

All surface quantities were similarly expanded in Fourier series in terms of the polar angle $\theta$. Derivatives were obtained by analytical differentiation in $\theta$ of the series whose coefficients were obtained by least square fit at the Lagrangian points. All nonlinear terms were computed in physical space at the Lagrangian points.

## IV. DROP MODEL OF CELL DIVISION

Greenspan (1978) has revived interest in the macroscopic description of cell division as a fluid encased by surface fluid whose surface tension is modified by the presence of a chemical surfactant. The surface tension gradient must be balanced by surface viscous stresses which result from surface motion and this in turn drives a flow within the drop.

The mathematical formulation is best done in terms of tensors and the terminology in Aris (1962) has been used below.

The boundary conditions at the surface, defined by

$$x^i = x^i(x^\alpha, t) \qquad \alpha = 1,2 \ ; \ i = 1,2,3$$

( $x^\alpha$ are coordinates used to label points on a moving two-dimensional surface embedded in a three-dimensional space with coordinates $x^i$ ) are

(a) continuity of normal velocities (kinematic constraint)

$$n_i u^i_{I,E} = n_i \hat{u}^i$$

where $\hat{u}^i$ are the three dimensional space components of the surface fluid velocity

(b) balance of tangential and normal forces

(i) $\quad T^{\alpha\beta}_{\ ,\beta} = 0$

(ii) $\quad P_I = b_{\alpha\beta} T^{\alpha\beta} \qquad (f_E = P_E = 0)$

The surface stress tensor $T^{\alpha\beta}$ is composed of a surface tension term, $\sigma \, a^{\alpha\beta}$ ( $a^{\alpha\beta}$ is the conjugate metric tensor for the surface), and a viscous stress term, $\dfrac{\mu}{2} a^{\alpha\beta} \left( \dfrac{\partial a^{\alpha\beta}}{\partial t} + a_{\alpha\gamma} \hat{u}^\gamma_{\ ,\beta} + a_{\gamma\beta} \hat{u}^\gamma_{\ ,\alpha} \right)$

The comma subscript refers to surface covariant differentiation. The surface tensor $b_{\alpha\beta}$ is the one which appears in the second fundamental form of the surface. It relates to the curvature of the surface. Condition (i) determines the surface velocity while condition (ii) gives the pressure which drives the internal motion of the drop.

In addition, there exist a chemical surfactant of concentration c which convects with the surface; thus

$$\frac{\partial c}{\partial t} + (c\hat{u}^\alpha)_{,\alpha} + \frac{c}{2a}\dot{a} = 0$$

where $a = \det(a_{\alpha\beta})$. The surface tension $\sigma$ depends on c in a prescribed way. This dependency is crucial in affecting the drop's motion as can be seen from the results of a linear stability analysis.

Suppose $\sigma = \bar{\sigma}(1 + Sc^*)$ and $C = \bar{C}(1 + c^*)$ and c* is a small perturbation of the form

$$c^* = \varepsilon e^{\lambda t} P_n(\cos\theta)$$

If all other relevant quantities are perturbed with suitable forms, the following dispersion relation is found

$$\lambda^3 \beta \alpha^2 (n^2 + n - 1) + \lambda^2 \left( -\alpha |\alpha| \beta n(n+1) + |\alpha| n(n+2)(n-1) \right)$$
$$+ \lambda n(n+2)(n-1)(n^2 + n - 1 - 2\alpha) - \frac{\alpha}{|\alpha|} n^2(n+2)(n^2 - 1) = 0$$

where $\alpha = s\bar{c}/\bar{\sigma}$, $\beta = \rho R^3 \bar{\sigma}/\mu^2$ and $R$ is the undisturbed radius. The behaviour of the drop falls into two cases;

(a) $\alpha < 0$ . There is one negative real root and a pair of complex conjugate roots with negative real parts for all n.

(b) $\alpha > 0$ . There is one positive real root and a pair of complex conjugate roots with negative real parts for all n.

Physically the addition $(\varepsilon < 0)$ of chemical surfactant at the equator for $\alpha > 0$ increases the surface tension there. This pulls more surface fluid towards the equator and increases the surfactant concentration. Thus an instability results. In the other case, $\alpha < 0$ , the surface fluid flows to restore the equilibrium balance.

We obtain evolution equations for all quantities on the boundary. We find the tangential velocity from the tangential stress (global balance) equation, the normal velocity from derivatives of the potential $\phi$, the potential is updated by Bernoulli's equation, etc. Evolution equation for a visco-elastic Oldroyd type fluid were implemented in the pseudo spectral programs. It turns out that the evolution equations are stiff. The linear dispersion relation has the large n asymptotic solution.

$$\lambda \sim \frac{\alpha}{|\alpha|} , \quad -\frac{n}{\beta |\alpha|} \pm \frac{i n^{3/2}}{\sqrt{\beta} \, |\alpha|}$$

and the damped rapidly oscillating modes (related to capillary waves) cause stiffness.

Green's formula has been applied to this problem and Figure 3 shows the damped oscillation of a $P_2$ mode. An attempt was then made to follow the development of the unstable $P_2$ mode, but the code ran into difficulties soon after the total curvature at the equator changed sign. In order to explore this instability, stable conditions were chosen $(\alpha < 0)$ and surfactant concentration was added continually at the poles. In fact, this case corresponds to the conditions of the experiments done by Greenspan (1978). Once again, the code ran into difficulties when a similarly deformed profile was reached. The most deformed profile is shown in Figure 4. The difficulties are related to the emergence of a small-scale wave at the equator. Explorations to find the mechanism of the instability are planned. The case of undamped oscillation was run with the pseudo spectral method. The amplitude of the radial disturbance was .016 in the $P_2$ mode, the nonlinear interactions gave amplitudes of order $10^{-4}$ in the $P_4$ and $P_6$ modes and the frequency obtained 1.2605 is within .35% of the exact frequency obtained by Lord Rayleigh. The pseudo spectral is very flexible and short waves can be filtered easily by adjusting the cutoff frequency N. Various nonlinear cases

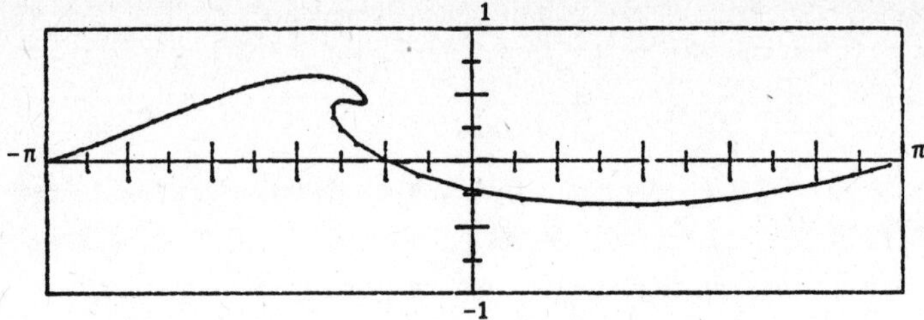

Figure 1    Profile of a breaking surface wave

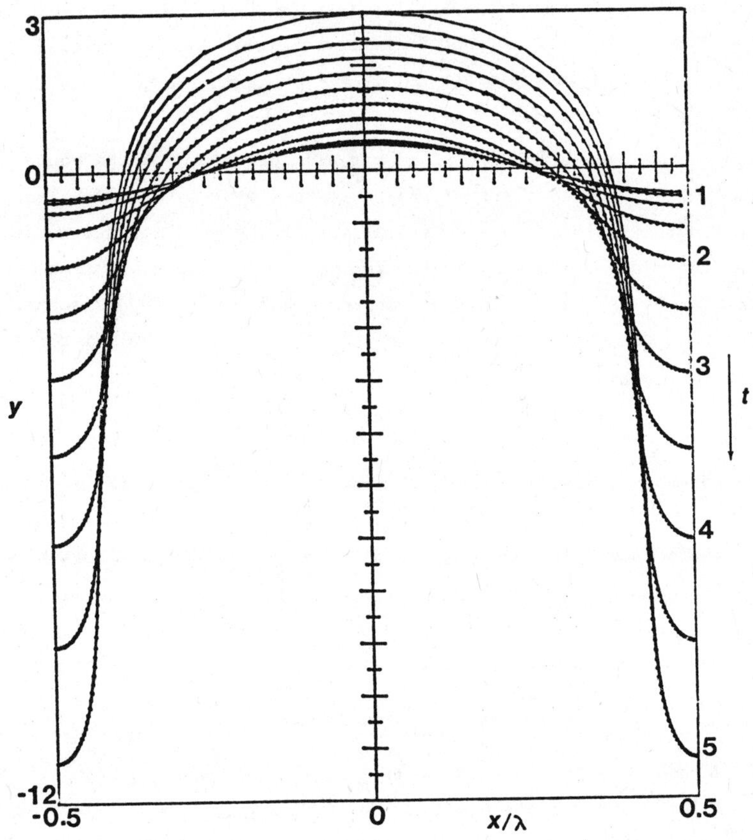

Figure 2    Time sequence of profiles for the Rayleigh-
            Taylor instability

are under investigation as well as detailed comparisons with the boundary integral methods.

ACKNOWLEDGMENTS

This work was supported in part by the Office of Naval Research under Contract No. N00014-77-C-0138, the National Science Foundation under grant ATM-7817092 and by National Science Foundation under grant MCS 78-14521.

REFERENCES

Aris, R. (1962).  Vectors, Tensors and the Basic Equations of Fluid Mechanics. Prentice Hall.

Baker, G.R., Meiron, D. and Orszag, S.A. (1980). Physics of Fluids, Vol. 23, No. 8.

Greenspan, H.P. (1978), J. Theor. Biol., 70, 125.

Longuet-Higgins, M.S. and Cokelet, F.D. (1978), Proc. Roy. Soc. London A 364, 1.

Menikoff, R. and Zemach, C. (1980), J. Comp. Physics, in press.

Schwartz, L. (1974), J.F.M. 62, 552.

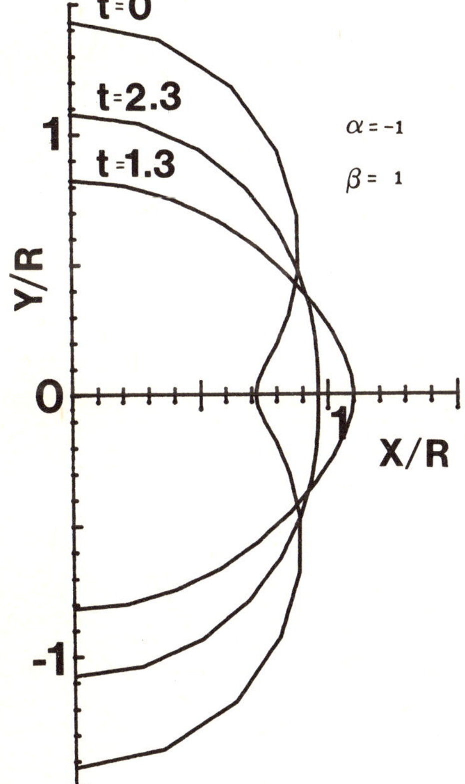


Figure 3 Radial profiles of a spherical drop at half periods while undergoing a damp capillary oscillation. The y-axis is the axis of rotation

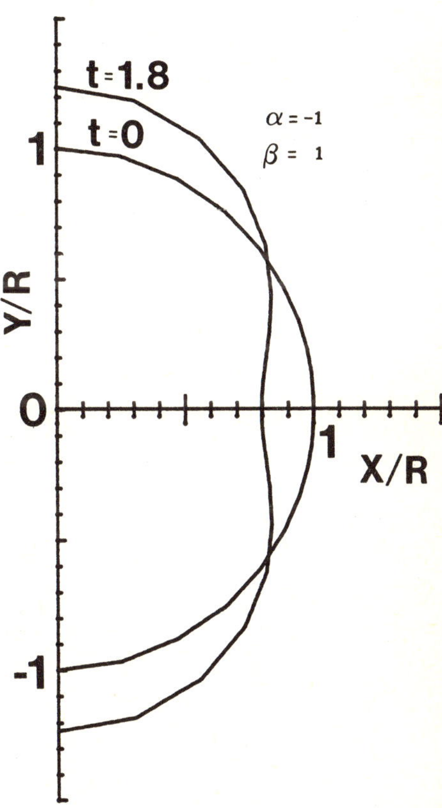


Figure 4 Radial profiles of a spherical drop deforming unstably due to continually addition of chemical surfactant at the poles

VISCOUS FLOW BETWEEN CONCENTRIC ROTATING SPHERES

Fritz Bartels*

David W. Taylor Naval Ship Research and Development Center
Bethesda, MD  USA

## Introduction

If the viscous flow in the small gap between rotating concentric spheres exceeds
a certain critical Reynolds number, Taylor vortices may appear similar to those
observed in cylindrical gaps.  However, in spherical gaps, several different
Taylor vortex configurations (modes) exist simultaneously Savatzki, Zierep (1970).
The solution depends not only on the steady state boundary conditions but also
on the choice of the initial conditions, Bonnet (1976), Bartels (1978) and the
angular acceleration leading to the steady external conditions, Bartels (1978).
The resulting Taylor vortices may be steady or unsteady. For example, at a gap
width of s = 0.15, nondimensionalized by the radius of the outer sphere, three
steady axisymmetric and two unsteady non-axisymmetric flow modes were found,
Savatzki (1970).  Each mode has its own transition point and can exist only in a
certain range of Reynolds numbers.  Transition from one mode to another is possible,
Wimmer (1974).  This problem is therefore of great interest for testing accuracy
and convergence of finite difference solutions to the Navier-Stokes equations.  In
the following, the major results of an investigation, which is described in detail
by Bartels (1978), are presented.  Therein the behavior of a finite-difference
approximation of the Navier-Stokes equations, restricted to axially symmetric
flows, was investigated for spherical gaps.  It could be shown that consistency and
stability of the locally linerized difference equation are not sufficient to ensure
always convergence to the solution belonging to the boundary conditions prescribed.

## Method of solution

For axially symmetric flows the time-dependent vorticity-stream function formulation
written in spherical coordinates can be employed.  For the computation of the circumferential velocity component the corresponding momentum equation must be included
in the solution.  The boundary conditions are the Stokes no-slip condition and the
symmetry conditions for the equatorial plane and the axis of rotation.  The boundary
condition for the vorticity is determined from the Poisson-equation for the stream
function.  Several finite difference approximations, Pearson (1967), Israeli (1971),
Bonnet (1976), Bartels (1978), as well as spectral methods, Astafeva et al. (1977)
and series expansion methods, Herbert (1978) for solving this problem have been
discussed by Roesner (1977).  Here the alternating direction method is employed for
the solution of the momentum and vorticity transport equation and successive over-
relaxation for the Poisson-equation.  Test calculations showed that a change in step
sizes can result in a completely different flow field although the profiles of the
flow variables were smooth and remained stable, indicating that the stability was
maintained throughout the integration.  It was therefore necessary to determine

---

* Postdoctoral research scientist from University Aachen, Germany, presently
  supported by the Deutsche Forschungsgemeinschaft.

the step sizes prior to the integration in order to ensure convergence to the correct solution.

## Influence of the spatial and temporal errors

It was shown that if vortices are present in the vicinity of the equator the behavior of the flow variables is almost periodic and the wave length in the colateral direction is proportional to the gap width  s . Close to the equator the flow variables can then be approximated by truncated Fourier series

$$\phi(\tau,\vartheta) = \sum_{m=0}^{N} \hat{\phi}_m(\tau) \exp\left[im\,\frac{\tilde{n}\,\tau}{s}\left(\frac{\bar{n}}{2}-\vartheta\right)\right] \tag{1}$$

For a second order difference scheme the truncation errors can be written as

$$\varepsilon_\vartheta = \frac{1}{\phi_\vartheta} \sum_{\ell=1}^{\infty} \frac{\Delta\vartheta^{2\ell}}{(2\ell+1)!}\,\phi_\vartheta^{(2\ell+1)} = -\frac{1}{3!}\left(\frac{\Delta\vartheta\,\tilde{n}\,\tau}{s}\right)^2\left[1-\frac{1}{20}\left(\frac{\Delta\vartheta\,\bar{n}\,\tau}{s}\right)^2+\cdots\right] \tag{2}$$

It follows that $\Delta\vartheta = \sqrt{6\,\varepsilon_\vartheta}\,s/\tilde{n}\tau$   where $\varepsilon_\vartheta$ is the maximum permissible error in collateral direction.  A similar estimate can be obtained for the radial directiion.  The most important wave length in the radial direction is one half of the gap width and yields $\Delta\tau = \sqrt{6\,\varepsilon_\tau}\,s/4\tilde{n}$ .   The error bounds $\varepsilon_\vartheta$ and $\varepsilon_\tau$ were determined in a test calculation by decreasing the step sizes and by comparing the flow modes calculated with those observed in experiments, Savatzki (1970). Figure 1 shows the stream line pattern calculated for s = .15 and Re = 1500 with $\varepsilon_\vartheta = 0.02, \varepsilon_\tau = 0.07$ in the upper part and $\varepsilon_\vartheta = 0.17, \varepsilon_\tau = 0.26$ in the lower part. Both are approximate solutions of the differential problem but the one obtained with the larger error bounds belongs to an acceleration of the inner sphere different from that described.  It was found that the flow field was correctly predicted if the error bounds were chosen $\varepsilon_\vartheta \leq 0.05$ and  $\varepsilon_\tau \leq 0.1$.

The magnitude of the time step affects the phase between and the damping of the Fourier components.  The finite-difference formulation is first order in time as e.g. chosen by Bonnet (1976).  For a gap width s = 0.15 two steady flow modes (two and no vortex cells) were calculated for different angular accelerations of the inner sphere.  With increasing values $\Delta$t the acceleration has to be decreased if the same steady state flow is to be maintained, as shown in Figure 2.  In order to look into this behavior and to isolate the major parameters affecting the calculation the linearized Burgers equation

$$\frac{\partial u}{\partial t} = -a\,\frac{\partial u}{\partial x} + b\,\frac{\partial^2 u}{\partial x^2} \tag{3}$$

was considered.  Here  a  is the constant wave speed and  b  the constant viscosity.  For a periodic initial condition  u(o,x) = $\sum A_m$ exp(imx) the solution of (3) is

$$u(x,t) = \sum_{m=0}^{\infty} A_m \exp(-bm^2) \exp\left[im\,(x-at)\right] \tag{4}$$

where  $-bm^2$ is the damping factor and (x−at) is the phase function.  A differential representation of a finite-difference scheme can be written as

$$V_t + \sum_{\ell=2}^{\infty} \beta_\ell V_t^{(\ell)} = -a V_x + a \sum_{\ell=2}^{\infty} \gamma_\ell V_x^{(\ell)} + b V_{xx} + \sum_{\ell=3}^{\infty} \delta_\ell V_x^{(\ell)} \tag{5}$$

Using the same initial conditions as for equation (3) one obtains

$$V(x,t) = \sum_{m=0}^{\infty} A_m \exp(\alpha_r t) \exp\left[i(mx + \alpha_i t)\right] \tag{6}$$

where $\alpha_r$ is the damping factor. A forward approximation in the time derivative and central difference approximation in space for constant step sizes results in

$$\frac{\alpha_T}{-bm^2} = \frac{(-1)^j}{2d(m\Delta x)^2} \, \ell n\left[1 + (-1)^j 4d \sin^2\left(\frac{m\Delta x}{2}\right) + c^2 \sin^2(m\Delta x)\right] \tag{7}$$

for the damping factor $\alpha_r$ and

$$\frac{\alpha_i}{-am} = \frac{1}{c(m\Delta x)} \, \tan^{-1}\left[\frac{c \sin(m\Delta x)}{1 + (-1)^j 4d \sin^2\left(\frac{m\Delta x}{2}\right)}\right] \tag{8}$$

for the dispersion coefficient $\alpha_i$ . The quantities $c = a\Delta t/\Delta x$ and $d = b\Delta t/\Delta x^2$ denote the Courant number and the diffusion number respectively. For $j = 1$ the solution represents an explicit and $j = 0$ an implicit first order difference scheme. The terms $\alpha_r/-bm^2$ and $\alpha_i/-am$ are the ratios of the numerical to analytical damping factors and the wave speeds respectively of the Fourier components. For $b = 0$ and limes $\Delta t \to 0$ equation (7) and (8) reduce to the relationship between phase error and step sizes $\Delta x$ for a hyperbolic differential equation derived by Kreiss and Oliger (1972). For an implicit scheme all Fourier components in the difference equations are faster damped than the analytical solution. This can be seen from the expansion of equation (7) and (8) for values $(m\Delta x) \ll 1$

$$\frac{\alpha_r}{-bm^2} = 1 + (-1)^j \frac{c^2}{2d} + \frac{d}{2}(m\Delta x)^2 + \cdots \tag{9}$$

and

$$\frac{\alpha_i}{-am} = 1 + (-1)^{j+1} d(m\Delta x)^2 + \cdots \tag{10}$$

Note that the numerical Fourier components have phase velocities that are faster for explicit and slower for implicit formulations than the analytical Fourier modes. If the physical damping is not to be falsified it is necessary that $c^2/2d \ll 1$ or $\Delta t \ll 2b/a^2$. In the case described in Figure 2 corresponding to the values $a = 0.1$ and $b = 1000$, which lie in the range of the critical Reynolds number, the time step must be smaller than $\Delta t \ll 0.2$ otherwise the formation of the vortices is suppressed by excessive damping. For the flow problem discussed here this result implies that the finite-difference method yield meaningful approximation of the flow only if the Fourier components with large wave numbers do not influence the solution.


## Results

The error bounds discussed in the previous analysis were adhered to the results reported here. For a gap width $s = 0.15$ two different axisymmetric vortex flows

were calculated when different accelerations from rest  of the inner sphere
and stationary outer sphere were used.  All attempts to determine the third
flow mode with one Taylor vortex as observed in experiments by Savatzki (1970)
failed until the symmetry condition at the equator was relaxed.  By changing
the application of the symmetry condition at $\vartheta$ = 90° to $\vartheta$ = 90.7° the calculation predicted the one-vortex mode.  This mode persisted above Re $\gtrsim$ 700 even
after the symmetry condition was properly restored.  The stream lines of these
three modes are shown in Figure 3 for Re = 1500.  The flow with two vortices is
obtained by an acceleration $\omega_i^*$ = 0.1 and no vortex is obtained for $\omega_i^*$ = 0.12.
These modes are stable only in a certain range of Reynolds numbers as can be
seen in Figure 4, where the torque coefficient $c_m$ is plotted versus the Reynolds number.

For Re $<$ 700 only one stable solution with no vortex exists.  If the Reynolds
number is increased the flow pattern changes continuously to the mode with two
vortices.  The flow with no vortex also exists above Re $\geq$ 1300.  If the Reynolds
number is decreased from 1300 to 1200 the flow changes into the flow with two
vortices.  The one-vortex mode generated by disturbing the symmetry is only
stable above Re = 700.  The finite-difference solution seems to differ from the
experiment, Wimmer (1976), by the restriction to axisymmetric flows, where small
temporarily three-dimensional disturbances could have been present.  The same
was also observed by Astafeva (1977) for a gap width s = 0.1.

The influence of the Reynolds numbers and the gap width was shown by Bartels (1978).
For gap widths $s \leq$ 0.09 no steady flow could be found and the solution remains un-
steady and periodic near the critical Reynolds number.  For small gap widths
periodic generation and destruction of vortex cells is reflected in the solution.
This causes a time-dependent variation of the torque coefficient as shown in
Figure 5.  The vortex formation is limited between equator $\vartheta$ = 90° and $\vartheta$ = 70°.
An evaluation of the results showed that the maximum number of vortices in a
spherical gap is about  n = 0.4 (1-$s$)/s.  At $\vartheta$ = 70° vortex cells are periodically
generated and near the equator the vortex cells are destroyed.  Both processes
differ in frequency and phase and depend on Reynolds number and gap width.  Although the generation and destruction of vortex cells in the very narrow gap
s = 0.025 is strong periodic the flow motion in general is rather irregular as
indicated in Figure 5.  This occurs because the number of vortices is limited and
their size is about the gap width.  The dimensionless time in which two vortices
are formed is much smaller ($t^+$=3.4) than the time for destruction of two vortices
near the equator ($t^+$ = 9).  Therefore several vortex cells collapse into a larger
cell somewhere between $\vartheta$ = 70° and the equator.  The number of vortex cells varied
between 13 and 16.  This was also observed in the experiments by Savatzki (1970),
although in the experiments the flow was no longer axisymmetric near $\vartheta$=70°.

These are some of the significant differences between the three-dimensional flow
of experiments and the axisymmetric flow in spherical gaps.  At the present time
the extension of the computational method to truly three-dimensional flows seems
to be unattainable because the accuracy conditions require a very large amount of
storage and computation time.

## References


1. Astafeva, N.M., Vvedenskaya, N.D., Yavorskaya, I.M. (1977) Nonlinear axisymmetric fluid flows in spherical layers Rep.No. 385, Institut of Cosmic Research of AN SSR, Mowcow

2. Bartels, F. (1978) Rotationssymmetrische Stroemungen im Spalt konzentrischer Kugeln Diss. RWTH Aachen, Germany

3. Bonnet, J.-P., Alziary de Roquefort, T. (1976) Ecoulement entre deux spheres concentriques en rotation Journal de Mecanique, Vol. 15, No. 3

4. Herbert, T.H.(1978) Berechnung der Stroemung zwischen konzentrischen rotierenden Kugelflaechen. ZAMM 58, T275

5. Israeli, M. (1971) Nonlinear motions of a confined rotating fluid, Proceedings IUTAM Symposium Unsteady Boundary Layers, Quebec

6. Kreiss, H.-O., Oliger, H. (1972) Comparison of accurate methods for the integration of hyperbolic equations, Tellux XXIV, Vol. 3, p. 199

7. Pearson, C.E. (1967) A numerical study of the time-dependent viscous flow between two rotating spheres, JFM, Vol. 28, p. 323

8. Roesner, K.G. (1977) Numerical Calculation of Hydrodynamic Stability Problems with Time-Dependent Boundary Conditions, Proceedings of the Sixth International Conference on Numerical Methods in Fluid Dynamics, Tbilisi 1977

9. Savatzki, O., Zierep, J. (1970) Das Stromfeld im Spalt zwischen zwei konzentrischer Kugeln, Acta mechanica, 9, p. 13

10. Wimmer, M. (1974) Experimentelle Untersuchungen der Stroemung im Spalt zwischen zwei konzentrischen Kugelm, die beide um einen gemeinsamen Durchmesser rotieren, Diss. TH Karlsruhe, Germany


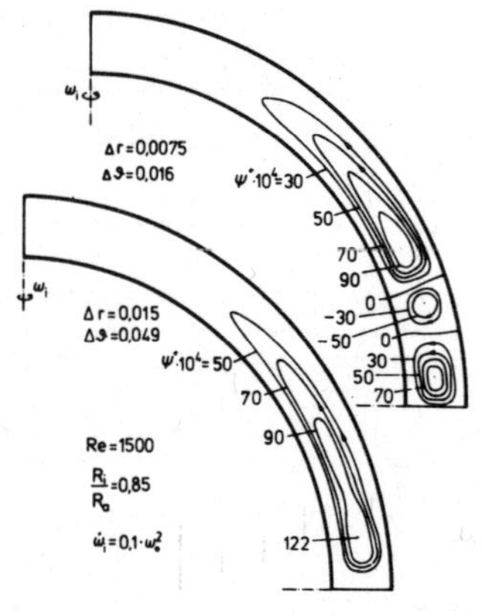


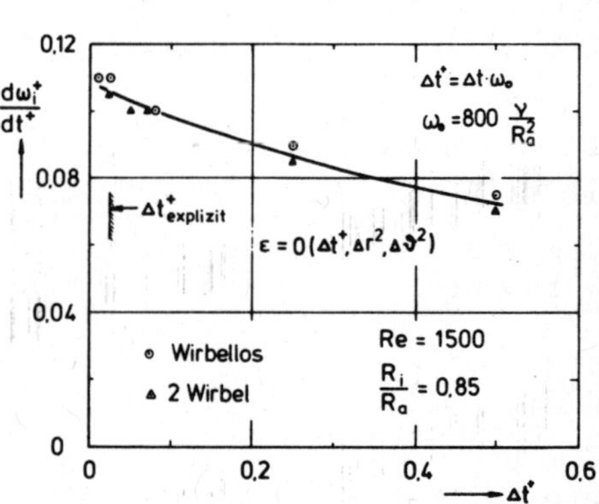


Fig. 1 Influence of the spatial resolution. Streamlines of the secondary motion.

Fig. 2 Influence of the temporal resolution

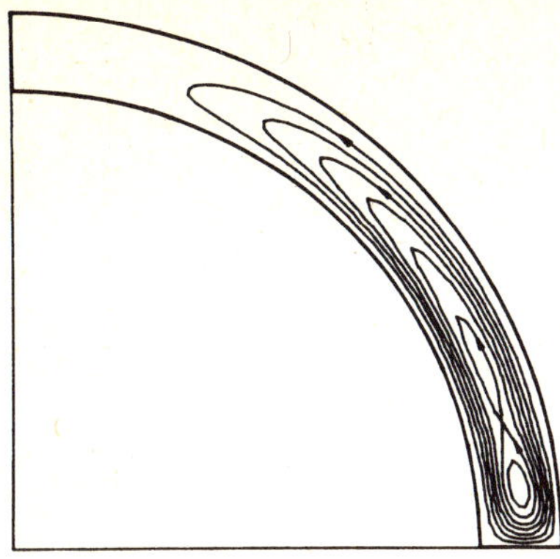

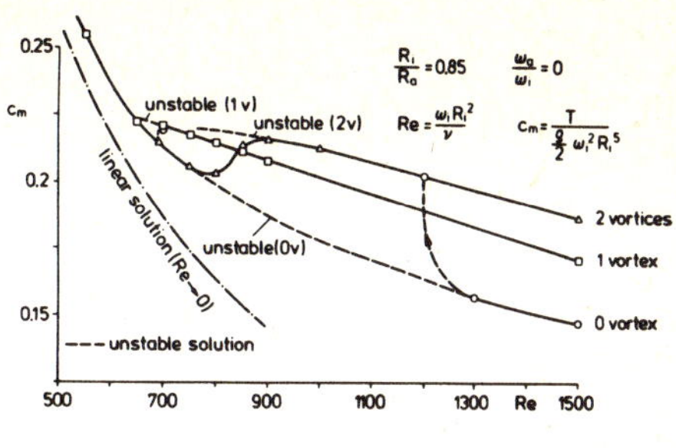

Fig. 4 Torque-coefficient and transition
between various flow modes

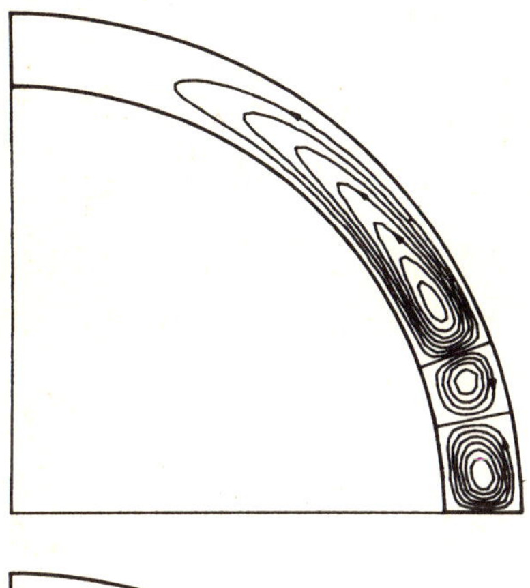

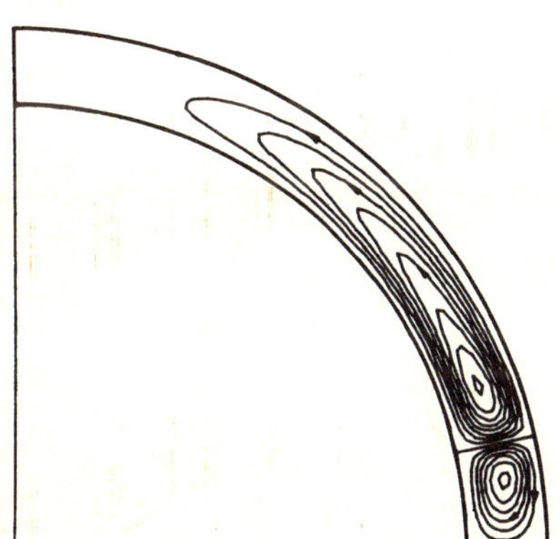

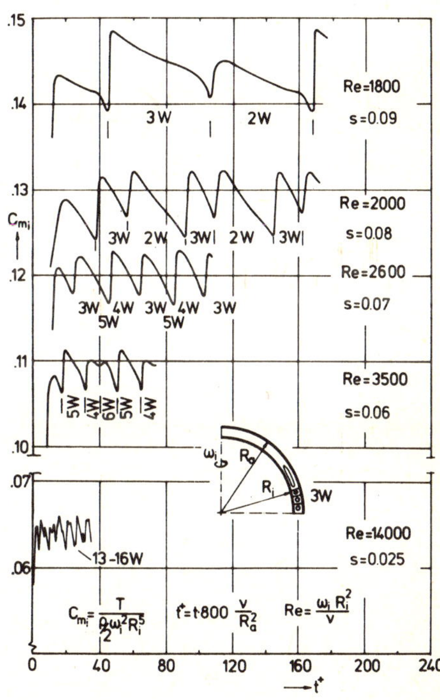

Fig. 3 Streamlines for three different
modes. Re=1500 ; s=1-$R_i/R_a$=0.15

Fig. 5 Time-dependent torque-coefficient
for several Renolds numbers and
gap widths $c_m$=f(Re,s)

NUMERICAL SOLUTIONS FOR A MOVING SHEAR LAYER
IN A SWIRLING AXISYMMETRIC FLOW

H. R. Baum, M. Ciment, R. W. Davis, E. F. Moore
National Bureau of Standards
Washington, D. C. 20234

## I. Introduction and Model Problem

This paper presents both a new model problem for unsteady, incompressible viscous flow and a new numerical method for modeling flows in cylindrical geometries. The model problem, an exact solution to the three-dimensional axisymmetric Navier-Stokes equations, represents a moving shear layer of rotating fluid whose thickness is a function of Reynolds number. An asymptotic steady-state is reached which consists of a potential vortex with a viscous core. This steady-state result was found by Rott (1958). The numerical method is a fundamental solution technique for cylindrical coordinates similar in derivation to the cartesian scheme of El-Mistikawy and Werle (1978). This new method and several other numerical schemes are tested on the model problem. Their relative performances in modeling both the transient and the steady-state are discussed.

The exact solution to the axisymmetric Navier-Stokes equations will now be described. Consider the velocity vector $\underset{\sim}{q} = (u_r, v, u_z)$, where the velocity is resolved into radial, tangential, and axial components in a cylindrical reference frame. Let $u_r = - \alpha r/2$ and $u_z = \alpha z$ (i.e., stagnation-point flow, with $\alpha = $ constant), thus satisfying the continuity equation $\nabla \cdot \underset{\sim}{q} = 0$. The equation for the tangential velocity component, $v(r,t)$, is then

$$\frac{\partial v}{\partial t} - \frac{1}{2} \frac{\partial (rv)}{\partial r} = Re^{-1}[\frac{1}{r} \frac{\partial}{\partial r}(r \frac{\partial v}{\partial r}) - \frac{v}{r^2}] \tag{1}$$

where t is time and Re is the Reynolds number $(\alpha r_o^2/\nu)$ based on characteristic length $r_o$. The equation for the swirl, $w = rv$, is

$$\frac{\partial w}{\partial t} - \frac{1}{2}[\frac{\partial (rw)}{\partial r} - w] = Re^{-1}[r \frac{\partial}{\partial r}(\frac{1}{r} \frac{\partial w}{\partial r})]. \tag{2}$$

As will be discussed, use of eq. (2) is preferable when obtaining numerical solutions, and thus the model problem will be described in terms of the swirl $w(r,t)$. The initial and boundary conditions for eq. (2) are $w(r,0) = \phi(r)$ and $w(0,t) = 0$, $w(1,t) = g(t)$.

The analytical approach is to solve a pure initial value problem with $\phi(r)$ defined for all $r \geq 0$. Due to the radial influx of swirl, the choice of $\phi(r)$ for $r > 1$ determines the boundary condition $g(t)$ at $r = 1$. The initial conditions $[\phi(r) = 0, r < 1; \phi(r) = 1, r > 1]$ give $g(t) = 1$. The physical problem then consists of a fluid spin-up inside the domain $r < 1$. This spin-up is due to a combination of convection by the stagnation-point flow and viscous diffusion, the latter effect becoming confined to a narrow layer at high Reynolds number. The solution to eq. (2) satisfying the spin-up problem boundary conditions for $t > O(1/\sqrt{Re})$ can be shown to be

$$w(r,t) = \frac{1}{2} \int_0^r \frac{Re \, \rho e^t}{(e^t - 1)} \exp[\frac{-Re(1 + \rho^2 e^t)}{4(e^t - 1)}] \, I_0[\frac{Re \, \rho e^{t/2}}{2(e^t - 1)}] d\rho \tag{3}$$

where $I_0$ is a modified Bessel function of the first kind. As $t \to \infty$, the following steady-state solution (Rott (1958)) is approached:

$$w(r,\infty) = 1 - \exp(- \frac{Re \, r^2}{4}). \tag{4}$$

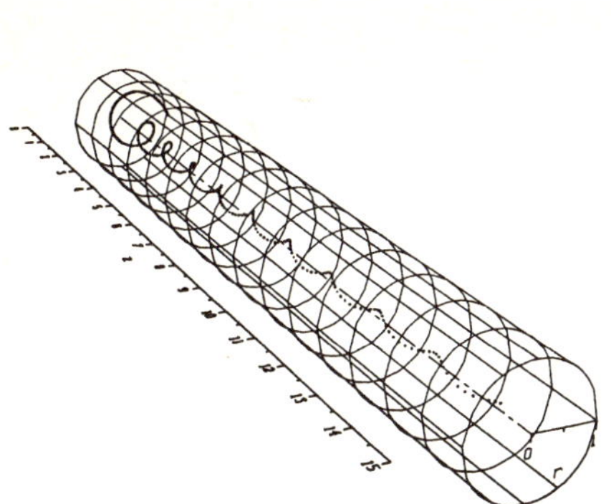

Fig. 1.   Particle Trajectory During
Spin-Up; Re = 500.

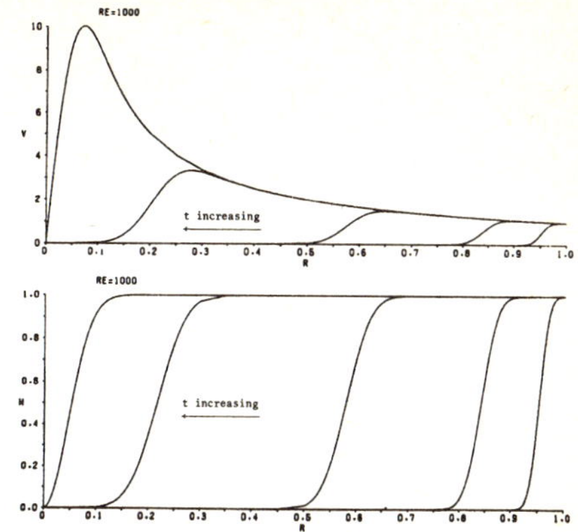

Fig. 2.   Exact Solution of Spin-Up
Problem; Re = 1000.

The Burgers vortex, eq. (4), represents vortex stretching by the stagnation-point
flow outside of a viscous core at the axis.  The transient solution, eq. (3),
represents a moving shear layer which separates rotating from nonrotating fluid.
The thickness of this shear layer decreases as Reynolds number increases.  The
primary significance of the model problem lies in exactly this fact.  In the numerical simulation of internal flows over steps or around obstacles, one must deal with
nonstationary shear layers which become very thin at high Reynolds number.  This
model problem can thus be utilized as a testbed for developing numerical schemes
which can deal with such phenomena.  The Reynolds number regimes in which methods
can cope can be distinguished from those in which they cannot.  Also, since a
steady-state is ultimately reached, the performance of a method in modeling both
steady and unsteady flows can be assessed.

A particle trajectory during spin-up is presented in Fig. 1.  The particle enters the domain at $r = 1$ and spirals in toward the axis, with the spacing between
consecutive particle locations directly proportional to particle velocity.  The
radial location of the particle corresponds to the location of the moving shear
layer.  Figure 2 presents the transient solution, eq. (3), at $Re = 10^3$ in terms of
both v and w.  The shear layer can be seen to move from right to left across these
plots, the leftmost curves being the steady-state solution, eq. (4).

Finally, the solution to eq. (2) has also been obtained for arbitrary initial
conditions $\phi(r)$ for all $r \geq 0$.  This is possible because the Green's function for
eq. (2), which is closely related to eq. (3), has been found.  Results for a rotating annulus problem will be presented later in the paper.  Also, the stagnation
-point flow can be made an arbitrary function of time, i.e., $\alpha = \alpha(t)$.  This results in an effectively time-dependent Reynolds number.


II.  <u>Cylindrical Fundamental Solution Scheme</u>

In this section we briefly outline the motivation for and derivation of a spatial difference scheme using local basis functions or fundamental solutions.  Generally, finite difference techniques for diffusion-convection problems involving
spatial operators of the form

$$Lu \equiv \nu u_{xx} - bu_x - cu$$

experience serious difficulties when the local cell Reynolds numbers, $bh/\nu$, become large. Recent investigations employing linear fundamental mode analysis have generated families of higher-order, cell-Reynolds-number-stable spatial methods (Ciment (1978), Berger, et al. (1980)).

Additional accuracy (fourth-order), generality, and stability is achieved in a computationally scalar tridiagonal context by allowing an <u>operator compact implicit</u> (OCI) tridiagonal approximation to a two-point boundary value problem for

$$Lu = f \tag{5}$$

of the form

$$\frac{\nu}{h^2} RU = Q(LU) = Qf \tag{6}$$

where Q and R are tridiagonal difference operators. Note, the reference to implicit in OCI merely refers to the fact that Q need not be the identity matrix as by classical methods.

For such spatial implicit tridiagonal schemes one can derive fourth-order accurate generalizations of stable upwinding type methods. These generalized OCI methods are, however, not convergent in the following sense. As $h \to 0$, in general, the rate of convergence depends very strongly on whether $\nu$ is fixed or is going to zero in some manner. Our concern here is not a mere mathematical exercise, but rather, convergence as both $\nu$ and $h$ tend to zero is adopted as a test standard for numerical methods for diffusion-convection problems at extremely high Reynolds numbers. When the solution converges as both $h$, $\nu \to 0$, we say the method is <u>uniformly</u>, in cell Reynolds number, convergent. Such methods do exist and have been discussed (Berger, et al. (1980)). Briefly, to achieve such uniform schemes for the cartesian problem it is essential to incorporate the local linearized fundamental solution character – here the exponential solutions (El-Mistikawy and Werle (1978)).

A uniformly second-order method can be obtained by using the local exponential solution over three neighboring points by implicitly patching in the following way. Let $V(x)$ be the $C^1$ solution (with continuous first derivatives in the interval $[x_{j-1}, x_{j+1}]$) of eq. (5), where b, c, and f are now defined by averaging over each subinterval $[x_{j-1}, x_j]$ and $[x_j, x_{j+1}]$. Then the scheme is derived in the form of eq. (6) above by defining $U_j = V(x_j)$. For cylindrical coordinate problems one cannot average at the origin because of singular coefficients.

In the spirit of the above approaches, we have derived an implicit tridiagonal scheme for the linearization of the spatial operator associated with the tangential momentum component of the Navier-Stokes equations

$$L(w) \equiv \nu r \frac{\partial}{\partial r}[\frac{1}{r} \frac{\partial w}{\partial r}] - b \frac{\partial w}{\partial r} = f(r) \tag{7}$$

where only b and f are averaged. The resulting implicit cylindrical fundamental solution (CFSS) scheme of the form eq. (6) is derived by using the following fundamental solutions $\{1, e^{ar}(1 - ar); a = b/\nu\}$ associated with eq. (7) in a manner analogous to the El-Mistikawy-Werle scheme. By numerical experiments the CFSS scheme has been determined to be, for fixed $\nu$, fourth-order accurate for $f(r) \equiv 0$ and second-order accurate for general nonzero $f(r)$. However, even this accuracy is not uniform in terms of $\nu$, $h \to 0$ for the simulation of the steady-state solution eq. (4). Perhaps this may be explained by observing that for this model problem the steady-state internal boundary layer is not a simple exponential of a linear function, but rather, a Gaussian. The underlying simple exponential solutions, although providing an accurate simulation of the Gaussian, appear to be unable to <u>uniformly</u> approximate this sharper gradient.

To solve parabolic problems of the form $u_t$ = Lu using OCI schemes one merely uses conventional temporal discretizations to $u_t$ while treating $1/h^2\, Q^{-1} RU$ as an approximation to Lu.  Thus, the forward time implicit method would be derived from

$$\frac{U_j^{n+1} - U_j^n}{\Delta t} = \frac{(Q^{-1}R)^{n+1}}{h^2}\, U_j^{n+1}.$$

Collecting terms for the unknown, with $\lambda = \Delta t/h^2$, yields the following tridiagonal system

$$[Q - \lambda R]^{n+1}\, U_j^{n+1} = Q^{n+1}\, U_j^n. \tag{8}$$

Various families of implicit-explicit schemes can be represented in this OCI tridiagonal format, including most well-known second-order methods (Berger, et al. (1980)). Moreover, these OCI methods are directly applicable to multi-dimensional time-dependent problems by a standard time-splitting method.


## III.  Numerical Results

In addition to testing the new CFSS scheme on the model problem, three other schemes have been tested.  These are first-order upwind differencing (UDS), second-order central differencing (CDS), and a quadratic upstream differencing scheme (QUICKEST) proposed by Leonard (1979).  All schemes except QUICKEST utilize fully implicit forward time differencing in the OCI tridiagonal format, eq. (8).  QUICKEST uses an explicit, Leith-type time differencing (Roache (1976)) and third-order upwinding on the convective derivatives yielding a four-point scheme for this problem. If the convective velocity $u_r$ were not unidirectional, then QUICKEST would become a five-point scheme.

Figures 3, 4 and 5 present comparisons among the four numerical schemes for Reynolds numbers of 500, $4 \times 10^3$ and $3 \times 10^4$ with the exact solution at time $t_o$ = 0.10 being the initial conditions for the various schemes.  The spatial and temporal mesh sizes, $\Delta r$ and $\Delta t$, are both 0.05.  Figures 6, 7 and 8 show results for a moving annulus of rotating fluid at the same Reynolds numbers starting at time $t_o$ = 0.10. The spatial and temporal mesh sizes here are 0.025.  The width of the annulus decreases with time as the outer shear layer overtakes the inner one.  Eventually the shear layers coalesce resulting in mutual annihilation, i.e., $w(r,\infty) = 0$.  It is seen from Figs. 3, 4 and 5 that the new CFSS scheme most accurately models the steady-state solution.  The absence of time-dependence in the fundamental solutions of this scheme may explain its poor performance during the transient.  The QUICKEST method most accurately models the transient at Re = 500 and $4 \times 10^3$.  At Re = $3 \times 10^4$, no tested scheme can reasonably track the motion of the thin shear layers.  It is also evident that upwind differencing (UDS) does poorly at all Reynolds numbers considered.

As noted previously, results for the numerical solution of eq. (2), as opposed to eq. (1), have been presented. Use of eq. (1) results in oscillatory behavior for both QUICKEST and CDS at high Reynolds numbers, with no reasonable solution obtainable for the latter at Re = $3 \times 10^4$.  The CFSS and UDS schemes are insensitive to choice of equation.

In conclusion, it is clear that the numerical modeling of transient shear layers at high Reynolds numbers remains inadequate.  The exact Navier-Stokes solution presented here can act as a very versatile and useful benchmark against which to test the performance of computational schemes in various Reynolds number regimes.

REFERENCES


Berger, A. E., Solomon, J. M., Ciment, M., Leventhal, S. H., and Weinberg, B. C., Generalized OCI Schemes for Boundary Layer Problems, to appear in Math. Comp., July 1980.

Ciment, M., Leventhal, S. H., and Weinberg, B. C., The Operator Compact Implicit Method for Parabolic Equations, J. Comp. Phy., 28, p. 135 (1978).

El-Mistikawy, T. M. and Werle, M. J., Numerical Method for Boundary Layers with Blowing - The Exponential Box Scheme, AIAA J., 16, p. 749 (1978).

Leonard, B. P., A Stable and Accurate Convective Modelling Procedure Based on Quadratic Upstream Interpolation, Comp. Meth. Appl. Mech. and Eng., 19. p. 59 (1979).

Roache, P. J., Computational Fluid Dynamics, Hermosa, Albuquerque (1976).

Rott, N., On the Viscous Core of a Line Vortex, ZAMP, 9, p. 543 (1958).


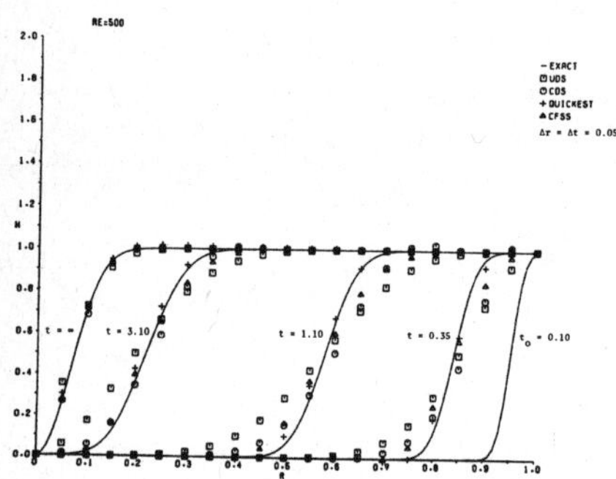

Fig. 3.   Scheme Comparison on Spin-Up Problem; Re = 500.

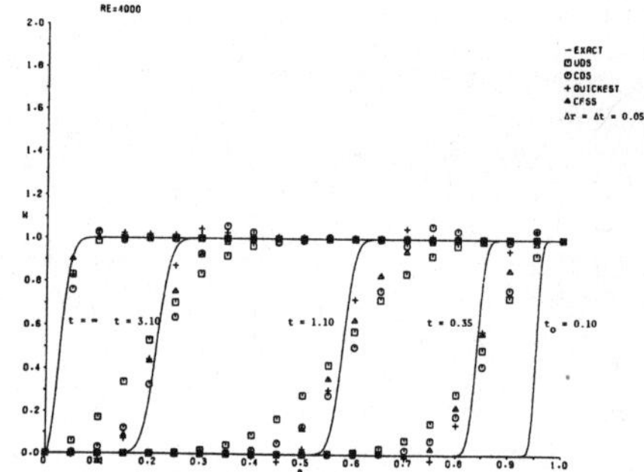

Fig. 4.   Scheme Comparison on Spin-Up Problem; Re = $4 \times 10^3$.

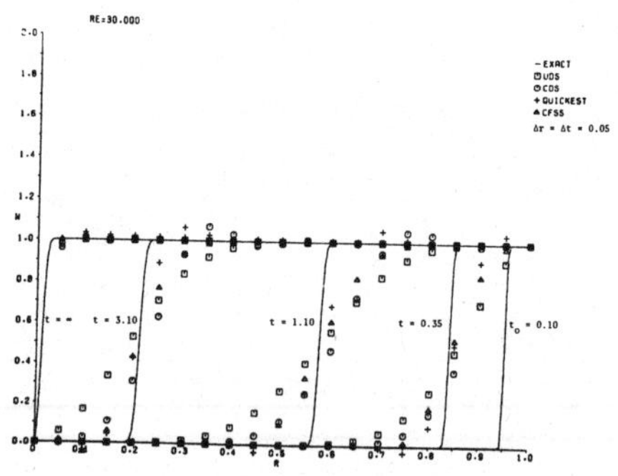

Fig. 5.   Scheme Comparison on Spin-Up Problem; Re = $3 \times 10^4$.

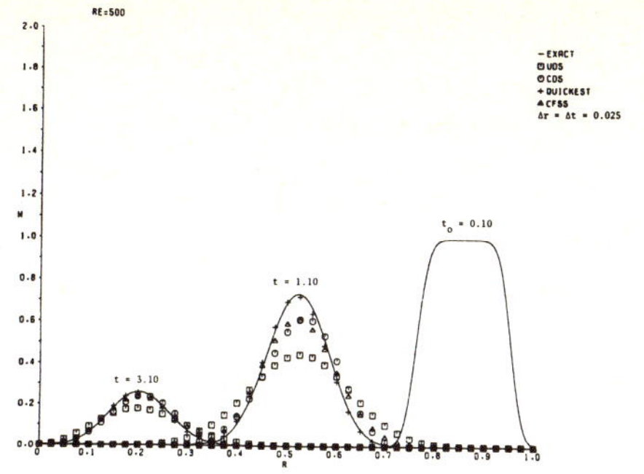

Fig. 6.   Scheme Comparison on Moving Annulus; Re = 500.

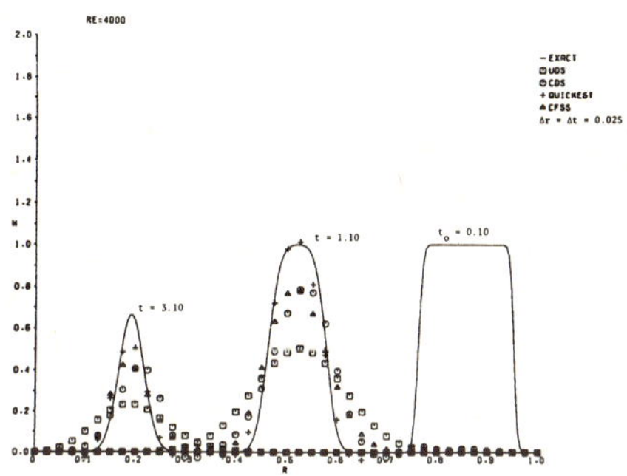

Fig. 7.   Scheme Comparison on Moving Annulus; Re = $4 \times 10^3$.

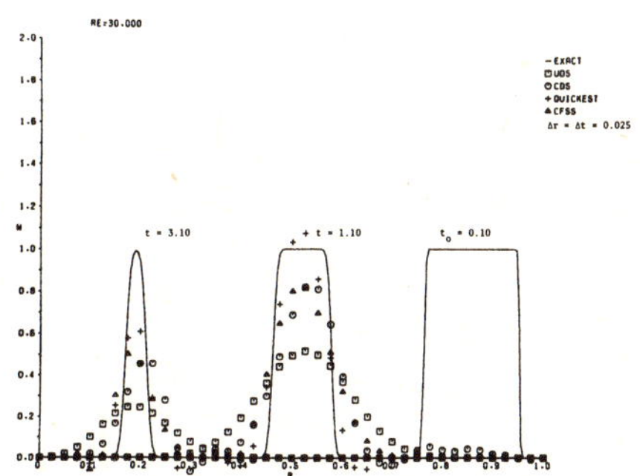

Fig. 8.   Scheme Comparison on Moving Annulus; Re = $3 \times 10^4$.

SEVENTH INTERNATIONAL CONFERENCE
ON NUMERICAL METHODS IN FLUID DYNAMICS
June 23-27, 1980
Standford University/NASA Ames, U.S.A.


FLOW INDUCED BY A JET IN A CAVITY
MEASUREMENTS AND 3D NUMERICAL SIMULATION


by J.P. BENQUE
   Y.   COEFFE
   R.   HERLEDAN


Electricité de France - Laboratoire National d'Hydraulique


## INTRODUCTION

An experimental and numerical study of the flow induced by a jet in a cavity
is described. Velocity measurements made by laser anemometry are presented.
The results of computations made with a code solving the instationnary Navier
Stokes equations in three dimensions are then compared to measurements. Two
simulations of turbulence are investigated, one by introduction of a turbulent
viscosity and the other by direct computation of the large eddies with
fluctuating boundary conditions and simulation of the small eddies by a
viscosity.


## I - FLOW DESCRIPTION

A cubic box (side a = 50 cm) is included in a loop of cold water. The flow
enters the box through the window $F_1$ (see fig. 1) and exits the box through
the window $F_2$. $F_1$ and $F_2$ are square windows (side b = 15 cm) and are
such that there is no symmetry plane for the flow. The flow consists of a jet
entering the box and impinging the opposite wall, generating on each side of
the jet a recirculating flow which goes down and progressively moves towards
the exit area near $F_2$.

The flow rate is kept as constant as possible by a free overfall control. The
entrance mean velocity is $U_0 = 24$ cm/s and the entrance Reynolds number is
Re = 36,000.

Several flow visualizations are made by taking pictures of particles
illuminated when passing through a plane light beam. These visualizations show
that except for the planes cutting the jet along its main axe, no big and
stable structures are present but only middle sized (5 to 10 cm) eddies of
relatively short life.


## II - MEASUREMENTS

Laser doppler anemometry measurements are performed to obtain simultaneously
two components of the velocity in one point of the box. The signals are then
treated to get an average value or a filtered value but also the standard
deviation and the spectrum of the fluctuations.

## III - NUMERICAL SIMULATION

The instationnary Navier-Stokes equations are discretized in finite difference on two grids, one for the velocity components and the other for the pressure (pressure is thus defined in the center of a "velocity cube"). Three main steps are introduced for the resolution :

. step I      :    convection of momentum
. step II     :    diffusion of momentum
. step III    :    continuity

This method is on the same pattern as the one proposed by CHORIN [1].


Boundary conditions :

A constant or fluctuating velocity may be imposed at each point of the entrance. For the exit the imposed velocities are derived from the exit profile after the convection step and the continuity condition.

On the walls, an adherence condition or a stress deduced from the logarithmic velocity profile near the wall, may be imposed.

Turbulence modelling :

Two turbulence models are investigated : one is an algebraic turbulent viscosity which is supposed to represent the effects of all the velocity fluctuations on the mean flow, the second consists of solving the equations of the filtered flow by calculating directly the low frequency fluctuations and modelling the effects of high frequency fluctuations on the filtered flow by a turbulent viscosity.


## IV - NUMERICAL RESULTS AND COMPARISON WITH MEASUREMENTS

Concerning the second turbulence model, the computation made did not reproduce correctly one important aspect observed in the measurements which is the increase by the jet of turbulent energy level from the entrance to the inside of the box. Several aspects are now being studied to get a better result and in particular the number of grid points is increased.

Fig. 1 gives a sample of comparison between measurements and numerical results in the case of an algebraic turbulent viscosity model. Fig. 2 gives the same comparison for a case with no such viscosity. The full comparison between measurements and numerical results shows that the main aspects of the flow are present even in the no viscosity case ; the turbulent viscosity makes the comparison better in the jet region but some discrepancies remain in the recirculating area under the jet.


## REFERENCES

[1] CHORIN - The numerical solution of the Navier Stokes equations for an incompressible fluid - Bull Amer. Math. Soc. 73, 928-931, 1967.
[2] J.P. BENQUE - Modélisation d'écoulements turbulents isothermes - rapport E.D.F. - LNH n° HE041/78.12.
[3] A.J. YULE - Large structure in the mixing layer of a round jet - J. Fluid Mech. (1978), vol. 89, part 3 pp. 413-432.
[4] W.C. REYNOLDS - Computation of turbulent flows 1975 - report n° TF4 Mechanical Engineering Department Standford University.
[5] A. LEONARD - On the enegy cascade in large eddy simulations of turbulent fluid flows - Advances in geophysics Vol. 18 A p. 237.

Horizontal plane
Vertical plane
x
y
F₁
F₂
Location of the planes of comparison
FLOW INDUCED BY A JET IN A CAVITY
Comparison between measurements and numerical results
Numerical computation caracteristics :
- Grid 21 x 21 x 27
- Algebraic turbulent viscosity
1 2 4 6 8 10 12 14 16 18 20 21
Vertical plane
Vertical plane
Calculated velocity field
Comparison between measurements (·) and calculations (+)
for several profiles
U₀ = 24 cm/s
1 2 4 6 8 10 12 14 16 18 20 21
Horizontal plane
Horizontal plane

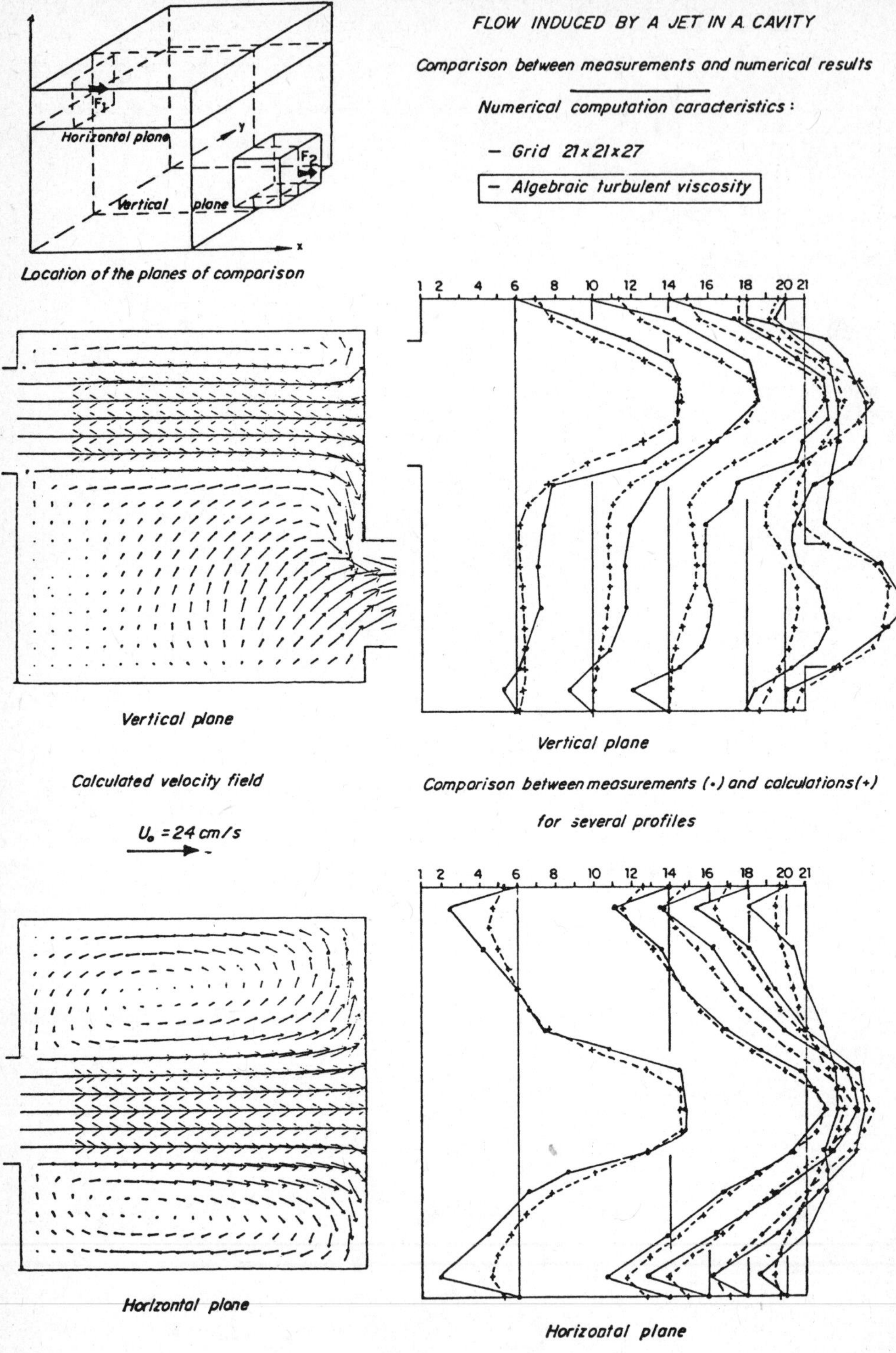

Fig. 1

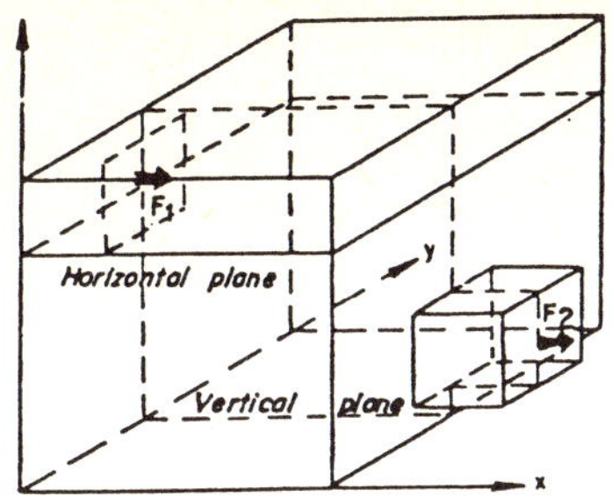


Location of the planes of comparison

FLOW INDUCED BY A JET IN A CAVITY

Comparison between measurements and numerical results

Numerical computation caracteristics:

— Grid 21x21x27

— Without turbulent viscosity

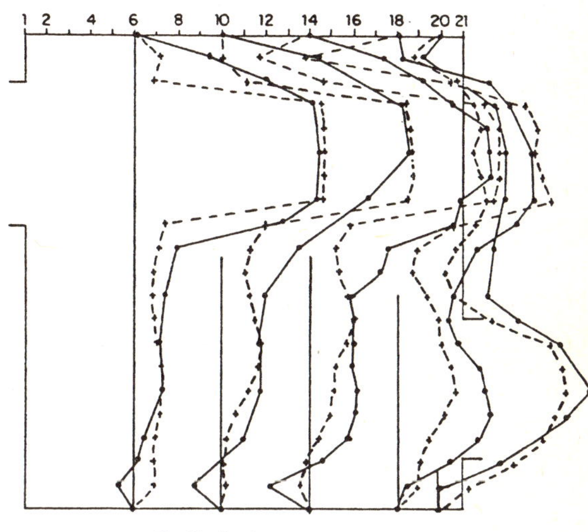


Vertical plane

Comparison between measurements(·) and calculations(+)

for several profiles

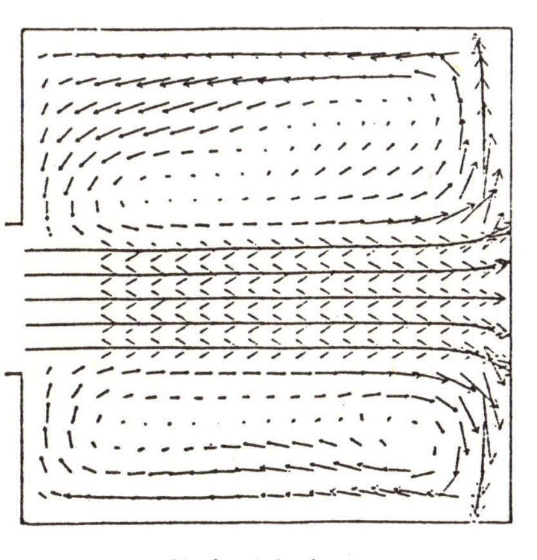

Vertical plane

Calculated velocity field

$U_o = 24\,cm/s$

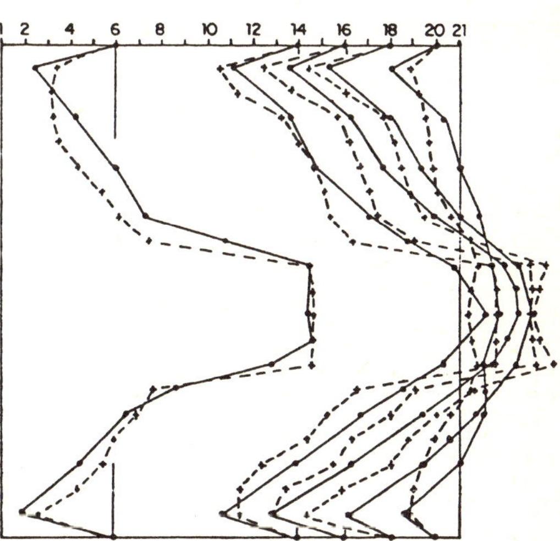


Horizontal plane

Fig.2

SIMULATION OF COMPLEX SHOCK REFLECTIONS

FROM WEDGES IN INERT AND REACTIVE GASEOUS MIXTURES[#]

D. Book, J. Boris, A. Kuhl*, E. Oran, M. Picone**, S. Zalesak
Naval Research Laboratory
Washington, D. C.  20375

The Flux-Corrected Transport (FCT) technique for solving fluid equations reduces numerical diffusion, permitting calculations with Reynolds numbers considerably in excess of the cell Reynolds number.  Recent advances in FCT, including a multidimensional flux limiter and a dynamic adaptive rezone, are illustrated in the problem of transient reflections of planar shocks from wedges in inert and reactive media.  Results are obtained with high resolution which are in quantitative agreement with experiments.

1.  Advances in FCT Techniques

In this paper we describe new adaptations of the Flux-Corrected Transport (FCT) algorithms developed by Boris and Book (1973, 1976) for solving fluid equations, and discuss their application to multidimensional shock reflections in inert and reacting gaseous mixtures.  In particular, we consider planar constant-velocity shocks reflecting from wedges.  Under certain circumstances double Mach stems are formed.  Historically these have proven to be difficult to calculate with high accuracy, although many schemes have been available to analyze compressible flow on a computationally discretized mesh: the method of characteristics, spline techniques, Glimm-type random choice schemes, and finite element, finite difference, and spectral methods.  We believe that the calculational difficulties experienced on this problem were the result of excessive numerical diffusion, especially in the region of the contact surface.

In any Eulerian calculation, numerical diffusion arises because material which has just entered a computational cell, and is still near one boundary, becomes smeared over the whole cell.  FCT minimizes this effect.  FCT algorithms can be constructed as a weighted average of a low-order and a high-order finite-difference scheme.  If the fluid equations are written in conservative form, both schemes are implemented using transportive fluxes.  Each flux describes the transfer of mass (or some other extensive quantity) from one point to a neighboring point.  The procedure for assigning weights involves limiting or "correcting" the fluxes at certain points.  The higher-order scheme is used to the greatest possible extent, consistent with avoiding the introduction of dispersive ripples (undershoots and overshoots).  The weights for the low-order scheme are chosen to be just sufficient to eliminate these ripples, thus assuring the property of "monotonicity" or "positivity."  The result is an algorithm which effectively reduces to the higher-order scheme wherever the fluid properties change gradually.  Near sharp discontinuities, however, enough diffusion is supplied to retain monotonicity.  At shock fronts this procedure automatically produces the correct local viscous heating.

The prototype second-order finite-difference formula

$$\rho_j^{n+1} = \rho_j^n - \tfrac{1}{2}\varepsilon_{j+\frac{1}{2}}(\rho_{j+1}^n + \rho_j^n) + \tfrac{1}{2}\varepsilon_{j-\frac{1}{2}}(\rho_j^n + \rho_{j-1}^n) + \nu_{j+\frac{1}{2}}(\rho_{j+1}^n - \rho_j^n) - \nu_{j-\frac{1}{2}}(\rho_j^n - \rho_{j-1}^n)$$

illustrates the procedure.  Here j labels grid position, n denotes time level and $\varepsilon_{j+\frac{1}{2}} = v_{j+\frac{1}{2}}\delta t/\delta x_{j+\frac{1}{2}}$ and $\nu_{j+\frac{1}{2}}$ are dimensionless advection and diffusion coefficients, respectively.  We write $\nu \sim \tfrac{1}{2}|\varepsilon|(|\varepsilon|+c)$, where c is a "clipping factor" measuring the extra diffusion added to achieve positivity.  When c = 0, the above scheme is second-order; in the vicinity of shocks $c \sim 1$ and it effectively reduces to first order.

A numerical diffusion Reynolds number $(Re)_{ND} = 2L/c\delta x$ can be defined, where L is the characteristic size of a structure in the flow.  Even the most accurate spectral simulations require setting c = 1 to guarantee positivity linearly.  This gives rise to the usual definition of the numerical Reynolds number, $2L/\delta x$.  Algorithms such as FCT which

*    R & D Associates, P.O. Box 9695, 4640 Admiralty Way, Marina del Rey, CA  90291
**   NRL/NRC Postdoctoral Research Associate
#    Work supported by the Defense Nuclear Agency and the Office of Naval Research

guarantee monotonicity nonlinearly can have average values $\langle c \rangle \sim 10^{-1}$—$10^{-2}$, introducing much less overall dissipation and permitting calculations with effective Reynolds numbers such that $Re \sim (Re)_{ND} \gg 2L/\delta x$.

Four advances in FCT techniques have enabled us to perform a series of shock and detonation calculations with high accuracy.  These techniques are easy to program, and they have wide applicability to general quasi-linear hyperbolic equations (i.e., equations describing continuum conservation laws).

The first of these, a generalization of FCT due to Zalesak (1979), removes the necessity of timestep splitting in multidimensional hydrodynamics.  This reduces errors associated with time splitting in regions of the flow which are nearly incompressible. The second refers to the development of FCT algorithms in which the spatial derivatives can be approximated to arbitrarily high order (fourth, sixth, eighth, etc., or pseudo-spectral).  These innovations, which relate to the transport algorithm itself, have been implemented in a two-dimensional hydrocode which utilizes the leafrog-trapezoidal (L-T) algorithm and is therefore dissipationless.  Both complex and double Mach stem structures are obtained (cf. Ben-Dor, 1978; Ben-Dor and Glass, 1978, 1979).

The third new technique, adaptive rezoning, is an extension to two dimensions of the dynamic rezoning employed in detailed one-dimensional reactive flow simulations by Oran, et al. (1979).  This concentrates needed spatial resolution in the vicinity of moving shocks, contact discontinuities and reactive surfaces.  The technique is illustrated with shock calculations using a time-split code (FAST2D).  In air for $M \approx 5$ and $\theta_w \approx 45°$, the results fall very close to the boundary between regular and Mach reflection. The calculated wall pressures are in detailed agreement with the results of Bertrand (1972).  The fourth technique is a generalization of the induction time approximation used in earlier flame, ignition and shock work (Oran et al, 1980a, b).  This provides a simple, efficient, yet reasonably accurate global chemical kinetics package to be used in connection with these comprehensive two-dimensional hydrodynamics calculations.

Section 2 describes the results of calculations in which a planar shock is reflected from a wedge in an inert gas.  In Section 3 we present the results of calculations of detonations initiated by shock reflections in stoichiometric mixtures of $H_2$ in air at low pressure.  Section 4 summarizes our conclusions.

2.  Shock Reflections in Air

The utility of these advanced FCT methods has been demonstrated by applying them to transient reflections of planar shocks from wedges for various shock strengths M and wedge angles $\theta_w$.  For nonreacting flows at Mach numbers greater than about 2.5 and wedge angles between 20 and 50 degrees, double Mach stems can develop.  Numerical schemes previously used for this problem reproduce qualitatively the wave structure and shape, but have difficulty making accurate predictions of flow details such as density contours (a conclusion drawn by Ben-Dor and Glass, 1978) even in the single Mach stem case.  To our knowledge, successful calculations of the double Mach stem case have not yet been published.  In this paper we discuss the series of calculations summarized in Table 1.

Open boundary conditions are used on the left, right, and top edges of the mesh, i.e., density, pressure and velocity are set equal to their pre-shock or post-shock values, depending on whether the incident shock front has passed that point.  Reflecting conditions are imposed on the bottom of the mesh, which corresponds to the wedge surface. Examples of the calculated density contours and wave structure for the double and complex Mach reflection cases are shown in Fig. 1.  The incident shock, I, the contact surface, CS, and the first and second Mach stems, $M_1$ and $M_2$, are indicated in Fig. 1a.  Note in particular the forward curl of the contact surface near the wall and the small region (4 by 7 mesh points) of high-density gas just to the left of the point where the contact surface impacts the wall.  The latter causes a second peak in the pressure and density distribution on the wall, as shown in Fig. 1c.  The accuracy of the calculations has been verified by comparison with experimental density distributions along the wall, as shown in Fig. 2, and with experimental pressure measurements (Bertrand, 1972).  Note that FCT provides adequate resolution of the key surfaces (contact surface and second Mach stem) in regions as small as 5 by 5 cells.

Two additional cases were calculated with larger values of $\theta_w$ (Fig. 3). As the wedge angle increases, the Mach stem develops more slowly, being separated from the wedge by a triple-point angle of only one or two degrees. [The triple-point angle $\alpha$ is the angle subtended by the Mach stem as viewed from the end of the wedge at which the shock was first incident (Fig. 4.] To reduce the size of the mesh needed, it is necessary to calculate in the frame of the Mach stem and to rezone. For these calculations we employed the time-split code FAST2D (with a 150 x 50 mesh), which incorporates an automatic continuous ("sliding zone") regridding procedure (Oran et al., 1979). For the small $\theta_w$ cases discussed above, where regridding is necessary, FAST2D yielded results very similar to those obtained with the L-T code. The cases with wedge angles of 44° and 46.5° constitute a severe test of the numerical algorithm because of the small triple-point angle. Because $\alpha$ is approximately equal to 2.8° and 1.5°, respectively, considerable spatial resolution and a large amount of running time are usually necessary to get accurate flow fields. Figure 4 illustrates our adaptive rezone technique on a grid of 60 x 40 cells with varying cell dimensions, $\delta x$, $\delta y$. This method requires one-fifth the number of cells required in a uniform grid calculation. A uniform region consisting of the smallest cells covers part of the incident shock front, the Mach stems, and the reflected shock structure. Outside this finely gridded region, we have transition zones in which the cell dimensions increase smoothly to their maximum values, $10\delta x_{min}$ and $10\delta y_{min}$.

We have investigated the accuracy of the numerical simulation by comparing the results with experimental data (Bertrand et al, 1972). Because the cases $\theta_w = 44°$ and 46.5° are so similar, we will discuss only the latter. The computed value of $\alpha$ for $\theta_w = 46.5°$ is approximately 2.5° for a real-air equation of state and approximately 3.2° for an ideal gas with $\gamma = 1.35$. Both agree with the measurements to within the experimental errors, $\sim 2\%$. In Fig. 5, we compare the calculated (using a real-air equation of state) and experimental values of the pressure at the surface of the wedge for $\theta_w = 46.5°$. The agreement in the shapes of the pressure curve is striking—the values of the lower pressure peak, corresponding to the Mach stem, are nearly identical. The calculated value of the second pressure peak is 11% lower than the experimental value and is thus within experimental uncertainty. Figure 6 shows the history of the pressure on the wedge calculated for an ideal gas with $\gamma = 1.35$. We note that the curves have much the same shape as for the real-air simulations; however, the first pressure peak is again 11% lower than the experimental value. Figure 7 shows that the double Mach reflection shock structure is well resolved in the simulation.

3. Detonations

We have also considered analogous shock reflections in reactive gases (stoichiometric mixtures of $H_2$ in air) at low pressure (0.1 atm). The induction time hypothesis (e.g., Oran et al, 1980) represents the chemistry through a composite process, in which reactants begin to combine into combustion products only after a finite time has elapsed. The rate at which the energy-releasing reactions proceed depends upon a single parameter, the induction time. This in turn is a function of the local thermodynamic variables.

Figure 8 shows the time development of a detonation initiated by a weak reflecting shock. The incident shock was chosen so that the pressure behind it is too low to cause detonation to take place within the time of the calculation. As with the calculation of Section 2, we have used open boundary conditions at the sides and top of the system. The sequence of six pressure contour plots traces the evolution of a detonation wave initiated by complex Mach reflection at the surface of the wedge. Figure 8a shows the pressure contours corresponding to an incident shock with $\theta_w = 25°$ and M = 4.0 which has just begun to reflect. The Mach stem is initially too small to be resolved. In Fig. 8b the Mach stem becomes discernible, but as yet no apparent reaction has occurred. By frame (c) the material has begun to ignite at a position well behind the current location of the Mach stem. When the Mach stem passed that position the pressure increase heated the mixture sufficiently to cause ignition after a short induction time characteristic of the $H_2$—air mixture. In frames (d) and (e), at later times, the pressure at the Mach stem continues to grow, leading to shorter characteristic induction times for material between the Mach stem and the original ignition point. Thus we see the ignited region accelerate along the wedge surface toward the Mach stem. Because more energy is being released as the burning continues, the boundary of the ignited region also accelerates in the direction of the reflected shock front, compressing and heating the material into which the burned gases expand.  In the last frame the burn front has overtaken the reflected shock and Mach stem, as we see from the decrease in the separation of the pressure contours near both locations.  A stable detonation pattern has not yet emerged, however.  This is evident from the bending out of the Mach stem and the lower density of contours between the Mach stem and the reflected shock/detonation front.

We anticipate that shock tube experiments will confirm this wave structure and that such reactive flow calculations will be extremely useful in quantitatively explaining the experimentally observed multicell structure of detonations  (Oppenheim 1970).

## 4.  Conclusions

Our calculations of complex and double Mach reflection are in close agreement with measurements for shocks reflecting in air from wedges.  Because of the accuracy and speed of FCT algorithms and the effectiveness of adaptive rezoning, the calculations are accurate and economical even when the Mach stem develops very slowly.  All of the important features (location of surfaces of discontinuity, pressure loading on the wedge surface, density contours) are correctly predicted.  The results do not depend sensitively on whether the L-T or FAST2D code is used.  Of the advances discussed in Section 1, multidimensional flux limiting and the adaptive regridding technique seem to be  the most efficacious for reflections in nonreactive media.  We conclude that FCT algorithms reduce numerical diffusion dramatically, assuring qualitative improvements in accuracy.  We believe that  to achieve comparable accuracy and efficiency, other hydrocodes must employ similar nonlinear algorithms and rezoning techniques.

Our calculations in reactive gas mixtures show that detonations tend to begin where a secondary pressure peak arises as the slip surface approaches the wedge. Because of the finite induction time in our kinetics model, the detonation begins somewhat behind this pressure peak.  The high resolution our calculations achieve enables us to follow multiple reflections and is capable of providing quantitative predictions of detonation phenomena.


## References

Bertrand, B. P., _Measurement of Pressure in Mach Reflection of Strong Shock Waves in a Shock Tube_, Ballistic Research Laboratories Report BRL-MR-2196 (1972).

Ben-Dor, G., _Regions and Transitions of Nonstationary Oblique Shockwave Diffractions in Perfect and Imperfect Gases_, UTIAS Report 232, August 1978, 61 pages + appendices.

Ben-Dor, G., and Glass, I. I., "Nonstationary Oblique Shock Reflection:  Actual Isopycnics and Numerical Examples" AIAA J. $\underline{16}$, 1146 (1978).

Ben-Dor, G., and Glass, I. I., "Domains and Boundaries of Nonstationary Oblique Shock Wave Reflections:  1. Diatomic Gas," J. Fluid Mech. $\underline{92}$, 459 (1979).

Boris, J. P. and Book, D. L., "Flux-Corrected Transport: I. SHASTA, A Fluid Transport Algorithm that Works," J. Comp. Phys. $\underline{11}$, 38 (1973).

Boris, J. P. and Book, D. L., "Solution of Continuity Equations by the Method of Flux-Corrected Transport," in _Methods in Computational Physics_, Vol. 16, Ed. by J. Killeen (Academic Press, Inc., New York, 1976).  See also Boris, J. P., _Flux-Corrected Transport Modules for Solving Generalized Continuity Equations_, NRL Memorandum Report 3237, March 1976.

Oppenheim, A. K., _Introduction to Gasdynamics of Explosions_, International Centre for Mechanical Sciences, Courses and Lectures No. 48, Springer-Verlag, pp 24-34 (1970).

Oran, E. S., Young, T. R., and Boris, J. P., "Application of Time-Dependent Numerical Methods to the Description of Reactive Shocks," _Seventeenth Symposium (International) on Combustion_, p. 43 (The Combustion Institute, Pittsburgh, 1979).

E. S. Oran, J. P. Boris, T. R. Young, T. Burks, M. Flanigan, and M. Picone, Numerical Simulations of Detonations in $H_2$-air on $CH_4$-air Mixtures, _Proceedings of the 18th Symposium (International) on Combustion_, The Combustion Institute, Pittsburgh, PA, 1980a (in press); "Simulations of Gas Phase Detonations: Introduction of the Induction Parameter Model," NRL Memorandum Report 4255, 1980b (in press).

Zalesak, S. T., "_Fully Multidimensional Flux-Corrected Transport Algorithms for Fluids_, J. Comp. Phys. 31, 335 (1979).

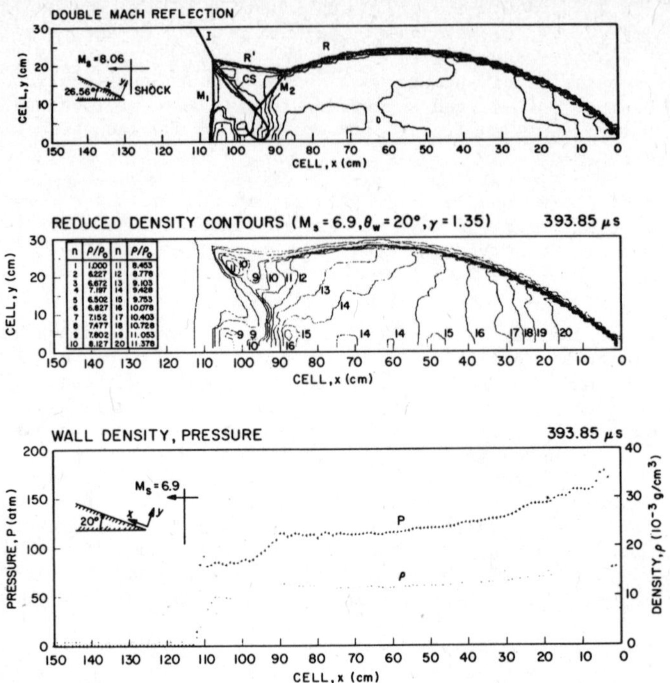

Fig. 1(a)  Wave structure and density contours for double Mach reflection from a wedge; (b)  Reduced density contours ($\rho/\rho_0$) for complex Mach reflection with levels chosen to agree with those of Ben-Dor and Glass (1978); (c)  corresponding pressure and density profiles on the wedge, plotted against cell number.

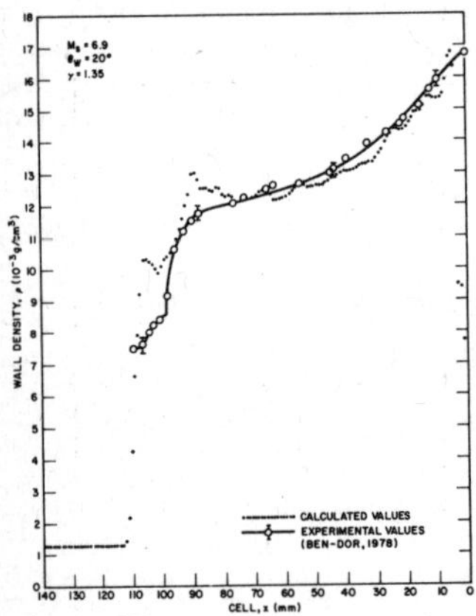

Fig. 2  Comparison of the calculated density profile on the wedge (Fig. 1b) with measured values (Ben-Dor and Glass, 1978).

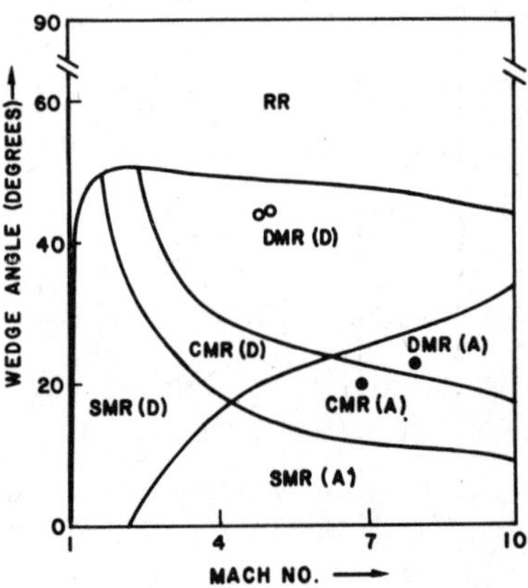

Fig. 3  Types of shock reflections; RR, CMR, SMR, and DMR denote regular reflection and single, complex, and double Mach reflection, respectively.  The D and A refer to detached and attached shocks.

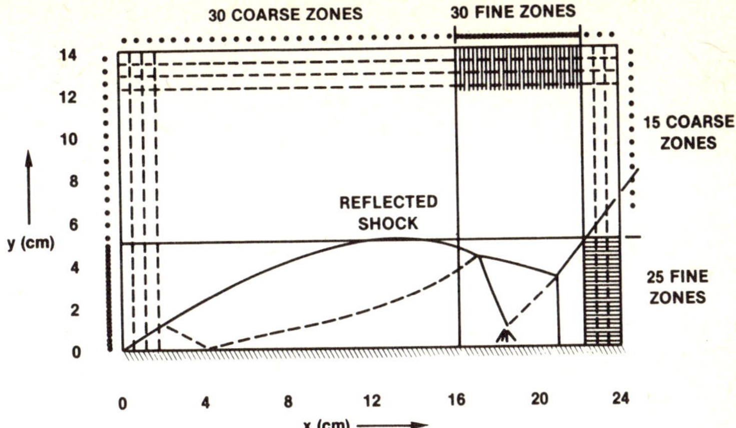



Fig. 4  Gridding for complex shock reflection problems with FAST2D code. Shown are incident, reflected and Mach shocks (solid lines) and slip surfaces (dotted lines) for incident shock coming from the left.

Table 1

| CODE | MACH NO. | WEDGE ANGLE | TYPE OF SHOCK | E.O.S. $\gamma$ = | REZONE | # STEPS | EXPERIMENT |
|---|---|---|---|---|---|---|---|
| L–T | 6.9 | 20° | CMR (A) | 1.35 | NO | 1,000 | BEN-DOR |
| L–T | 8.06 | 26.56° | DMR (A) | 1.35 | NO | 800 | BEN-DOR |
| FAST2D | 8.06 | 26.56° | DMR (A) | 1.35 | NO | 800 | BEN-DOR |
| FAST2D | 5.173 | 44° | DMR (D) | 1.35 | YES | 600 | BRL |
| FAST2D | 5.074 | 44° | DMR (D) | REAL AIR | YES | 1,000 | BRL |
| FAST2D | 5.148 | 46.5° | DMR (D) | 1.35 | YES | 1,000 | BRL |
| FAST2D | 5.046 | 46.5° | DMR (D) | REAL AIR | YES | 1,500 | BRL |

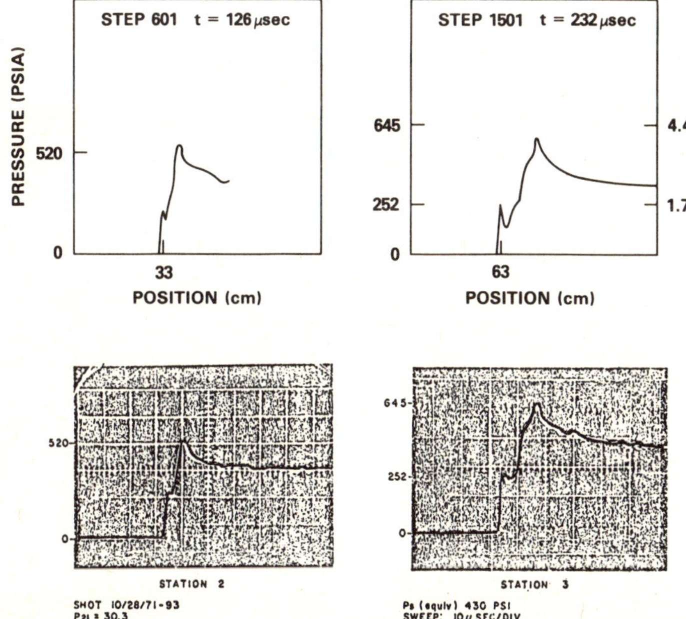



Fig. 5  Upper and lower diagrams show pressure in PSIA on the wedge as a function of position for $\theta_w$ = 46.5° (real air equation of state) at two times in the simulation and as a function of time at two stations in the experiment, respectively.

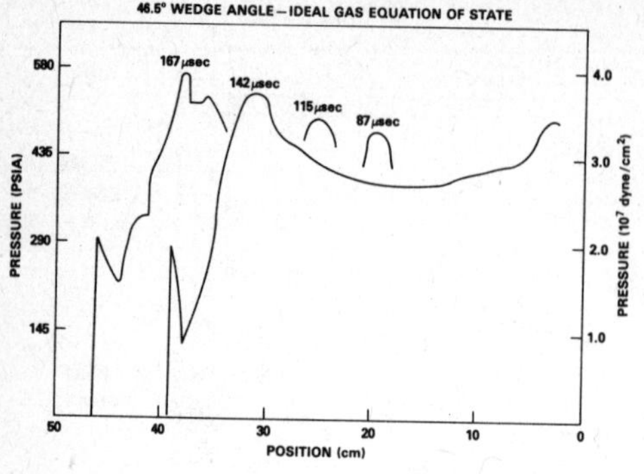

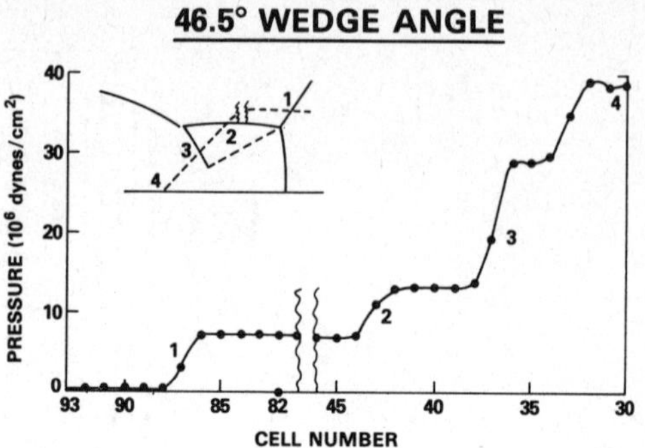

Fig. 6  Structure of the calculated shock ($\theta_w$=46.5°, M=5.15, $\gamma$=1.35) as a function of position for several times.

Fig. 7  Plot of pressure vs distance along the trajectory 1-2-3-4 shown in the inset, which intersects all surfaces of discontinuity normally.

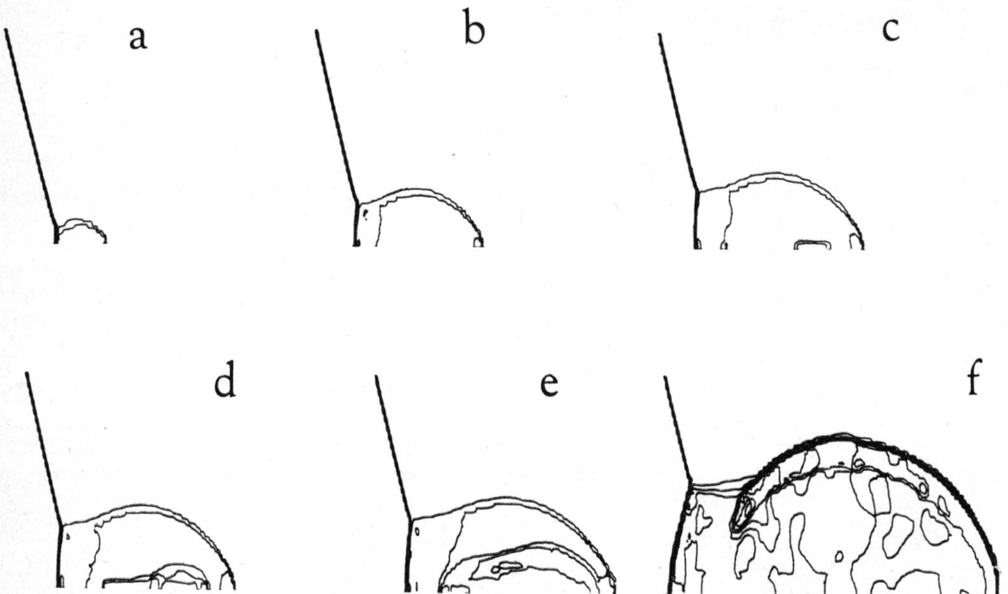

Fig. 8  Sequence of six pressure contour plots tracing the development of detonation in a stoichiometric $H_2$--air mixture at 0.1 atm due to complex Mach reflection from wedge.

COMPUTATION OF
THREE-DIMENSIONAL HORSESHOE VORTEX FLOW
USING THE NAVIER-STOKES EQUATIONS


W. R. Briley and H. McDonald
Scientific Research Associates, Inc.
P.O. Box 498, Glastonbury, CT    06033


## ABSTRACT


Numerical solutions of the compressible Navier-Stokes equations are presented
for a laminar horseshoe vortex flow created by the interaction of a boundary layer
on a flat surface and an elliptical strut leading edge mounted normal to the flat
surface.  The computational approach utilizes "zone embedding", surface-oriented
elliptic-cylindrical coordinates, interactive boundary conditions, and a consistently-
split linearized block implicit (LBI) scheme developed by the authors.  Mesh resolution tests are performed, and the horseshoe vortex flow is discussed.


## INTRODUCTION

The present study considers three-dimensional leading-edge horseshoe vortex flow
representing the interaction of a two-dimensional wall boundary layer approaching an
elliptical strut leading edge mounted normal to a flat surface.  An example of this
type of flow occurs near the junction of an airfoil or wing and its supporting surface.  Another example is present in axial compressors and turbines, where boundary
layers which develop on the annular surfaces of the axial flow passage encounter rows
of stationary and rotating blades.  The flow is of interest in connection with flow
degradation, its tendency to cause high local heat transfer rates, and its role as
the origin of corner flows.  Numerous experimental flow visualization studies (e.g.,
[1-3]) have established that the flow consists of a three-dimensional boundary layer
separation in front of the obstruction followed by a vortex flow which wraps around
the obstruction.  Little is available in the way of detailed flow measurements, however, particularly downstream of separation.  Previous analytical studies have considered the three-dimensional boundary layer flow upstream of separation (e.g., [4])
and have used rotational inviscid flow theory to estimate secondary flows [5].  The
horseshoe vortex flow is treated here by numerical solution of the compressible
Navier-Stokes equations for laminar flow at moderate Reynolds number Re and low Mach
number M.  The study has as its goals an improved understanding of the horseshoe
vortex region and formation of corner flows, and also the development of flow prediction techniques.

## ZONE EMBEDDING AND INTERACTIVE BOUNDARY CONDITIONS

The compressible Navier-Stokes equations in general orthogonal coordinates are
solved using analytical coordinate data for an elliptic-cylindrical coordinate system which fits all solid surfaces within the computational domain but is not aligned
with the direction of the free stream flow.  In selecting the computational domain,
a "zone embedding" approach is adopted whereby attention is focused on a subregion
of the overall flow field in the immediate vicinity of the leading edge horseshoe
vortex flow.  A perspective view of the geometry, coordinate system $(\xi,\eta,z)$ and a
representative computational grid is shown in Fig. 1.  The elliptic coordinates $\xi,\eta$
are related to Cartesian coordinates x,y by x=cosh $\xi$ cos $\eta$, y=sinh $\xi$ sin $\eta$, and the
(conformal) metric scale factor h is given by $h^2=\cosh^2\xi-\cos^2\eta$.  The computational
domain is defined by $0.2\leq\xi\leq1.6$, $\pi/2\leq\eta\leq3\pi/2$, $0\leq z\leq1$.

Symmetry conditions are applied at z=1, so that the flow represented is that
past an elliptical cylinder (of major semi-axis a=1.02 and minor semi-axis b=0.2)
between parallel flat plates with spacing H=2.  Velocity boundary conditions at no-
slip and symmetry surfaces are straightforward and self-explanatory.  The remaining
condition applied at these boundaries is $\partial p/\partial n=0$, where p is pressure and n denotes
the normal coordinate direction.  The condition $\partial p/\partial n=0$ at a no-slip surface is cor-

rect to order $\text{Re}^{-1}$ for viscous flow at high Reynolds number.

The treatment of inflow and outflow conditions is the principal obstacle to be overcome within the present zone embedding approach. At curved boundaries located within the free stream region, interactive boundary conditions are derived from an assumed flow structure and physical approximations which permit inflow of a boundary layer and inviscid free stream, and which permit outflow in the presence of shear layers, corner flow, streamwise vorticity and a nonuniform free stream. The initial conditions and interactive boundary conditions are devised from the incompressible potential flow velocity $\bar{U}_I(\xi,\eta)$ about the ellipse, together with two-dimensional estimates of the boundary layer thicknesses on the endwall flat plate, $\delta_1(x)$, and on the ellipse, $\delta_2(\eta)$, and finally from an

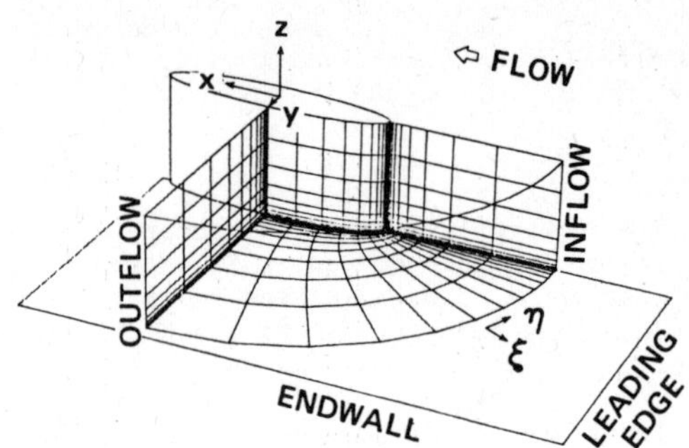


Fig. 1 - Geometry, Coordinate System, and Representative Grid

estimate of the blockage correction factor B(x) for the core flow velocity due to the endwall boundary layer growth. The complex potential W for flow past an ellipse at incidence $\alpha$ with circulation k is given by Milne-Thomson [6] as $W=U_r(a+b)\cosh$ $(\zeta-\xi_o-i\alpha)$, where $U_r$ is a reference free stream velocity, $\xi=\xi_o$ is the ellipse surface, and $\zeta$ is defined by $x+iy=c\cosh\zeta$, where $i^2=-1$ and $c^2=a^2-b^2$. This defines the incompressible potential flow velocity $\bar{U}_I$. The boundary layer thicknesses $\delta_1$ and $\delta_2$ are approximated by the Blasius value $5(s/\text{Re})^{1/2}$ where s is distance from the flat plate or ellipse leading edges, as appropriate. The blockage factor is given by $B(x)=[H/2-\delta_1^*(x)]^{-1}$. Finally, boundary layer velocity profile shapes $f_1(z/\delta_1)$ and

$f_2(\xi,\delta_2)$, $0\leq f_1,f_2\leq 1$ are defined from vonKarman-Pohlhausen polynomial profiles. The initial velocity vector $\bar{U}$ at t=o is defined by

$$\bar{U}(\xi,\eta,z) = \bar{U}_I(\xi,\eta)\ B(x)\ f_1\left[z/\delta(x)\right]\ f_2\left[(\xi-\xi_0)/\delta_2(\eta)\right]$$

$$(1)$$

At the inflow boundary $\xi=\xi_2$, a "two-layer" boundary condition is devised such that stagnation pressure $p_o$ is fixed at the free stream reference value $(p_o)_r$ in the core flow region $(z>\delta_1)$, and the Cartesian velocity u is set by $u=u_e(\eta,t)f_1(z/\delta_1)$ for $z\leq\delta_1$. Here, $u_e$ is the local free stream velocity consistent with $p_o$ and the local wall static pressure (assumed constant across the shear layer), which is determined as part of the solution and updated after each time step. The remaining inflow conditions are $v=\alpha U_r f_1(z/\delta_1)$, $\partial^2 w/\partial n^2=o$, and $\partial^2 c_p/\partial n^2=g(\eta,z)$ where g is the distribution of this quantity at t=o with $c_p$ defined as $(1-B^2\bar{U}_I\cdot\bar{U}_I)$, its value from the potential flow corrected for estimated blockage. The velocities v and w are the Cartesian velocity components normal to the plane of the ellipse chord and normal to the flat plate, respectively. The angle $\alpha$ is the approach flow incidence angle relative to the chord of the ellipse. For outflow conditions, second normal derivatives of each Cartesian velocity component are set to zero and $c_p$ is imposed and updated after each time step from an interaction model relating the computed outflow velocities and the _a priori_ potential flow solution. The imposed $c_p$ is obtained by integrating an inviscid normal momentum equation $\partial p/\partial n = -q^2/R_I$ along the outflow symmetry lines $(z=1, \eta=\pi/2, 3\pi/2)$ beginning with a fixed wall pressure. Here, q is the local computed flow speed $(q^2 \equiv \bar{U} \cdot \bar{U})$ and $R_I$ is the local potential flow streamline radius of curvature. The resulting static pressure distribution is imposed at the outflow surface assuming no variation normal to the endwall.

The foregoing interactive inflow-outflow boundary conditions are designed to permit the mass flux through the computational domain to adjust to both the imposed downstream static pressure and to viscous losses present in the flow, while maintaining a specified flow structure based on physical assumptions consistent with the flow problem under consideration. Various refinements in the interactive boundary conditions are possible, such as including the effect of local pressure gradient on boundary layer growth and profile shape.

DIFFERENCING PROCEDURES

The differencing procedures used are a straightforward adaptation of those used by the authors [7] in Cartesian coordinates for flow in a straight duct. The compressible time-dependent Navier-Stokes equations are written in general orthogonal coordinates, and for economy the stagnation enthalpy is assumed constant. The definition of stagnation enthalpy and the equation of state for a perfect gas can then be used to eliminate pressure and temperature as dependent variables, and solution of the energy equation is unnecessary. The continuity and three momentum equations are solved with density and the $\xi, \eta$ and z velocity components as dependent variables. Three-point central differences were used for spatial derivatives, and second-order artificial dissipation terms are added as in [7] to prevent spatial oscillations at high cell Reynolds number. This treatment lowers the formal accuracy to first order but does not seriously degrade accuracy in representing viscous terms in thin shear layers. Analytical coordinate transformations were used to redistribute grid points and thus improve resolution in shear layers and near the leading edge. Derivatives of geometric data were determined analytically for use in the difference equations.

SPLIT LBI ALGORITHM

The numerical algorithm used is the consistently-split "linearized block implicit" (LBI) scheme developed by the authors [7, 8] for systematic use in solving systems of nonlinear parabolic-hyperbolic partial differential equations (PDE's). To illustrate the algorithm, let

$$(\phi^{n+1} - \phi^n)/\Delta t = \beta D(\phi^{n+1}) + (1-\beta) D(\phi^n) \tag{2}$$

approximate a system of time-dependent nonlinear PDE's (centered about $t^n + \beta\Delta t$) for the vector $\phi$ of dependent variables, where D is a multidimensional vector spatial differential operator, and t is a discretized time variable such that $\Delta t = t^{n+1} - t^n$. A local time linearization (Taylor expansion about $\phi^n$) is introduced, and this serves to define a _linear_ differential operator L such that

$$D(\phi^{n+1}) = D(\phi^n) + L^n(\phi^{n+1} - \phi^n) + O(\Delta t^2) \tag{3}$$

Eq. (2) can thus be written as the linear system

$$(I - \beta\Delta t L^n)(\phi^{n+1} - \phi^n) = \Delta t D(\phi^n) \tag{4}$$

The multidimensional operator L is divided into three "one-dimensional" sub-operators $L=L_1+L_2+L_3$ (associated here with the three coordinate directions), and Eq. (4) is split as in the scalar development of Douglas & Gunn [9] and is written as

$$(I - \beta \Delta t L_1^n)(\phi^* - \phi^n) = \Delta t D(\phi^n) \tag{5a}$$

$$(I - \beta \Delta t L_2^n)(\phi^{**} - \phi^n) = \phi^* - \phi^n \tag{5b}$$

$$(I - \beta \Delta t L_3^n)(\phi^{***} - \phi^n) = \phi^{**} - \phi^n \tag{5c}$$

$$\phi^{n+1} = \phi^{***} + O(\Delta t^3) \tag{5d}$$

If spatial derivatives appearing in L are replaced by three-point difference formulas, then each step in Eqs. (5a-c) can be solved by a block-tridaigonal "inversion". Eliminating the intermediate steps in Eqs. (5a-d) results in

$$(I - \beta \Delta t L_1^n)(I - \beta \Delta t L_2^n)(I - \beta \Delta t L_3^n)(\phi^{n+1} - \phi^n) = \Delta t D(\phi^n) \tag{6}$$

which approximates Eq. (4) to order $\Delta t^3$. Complete derivations are given by the authors in [7, 8]. It is noted that Beam & Warming [10] have reformulated this algorithm as a widely-used "delta form" approximate factorization scheme whose two-level form is identical to Eq. (5).

## MESH REFINEMENT IN TWO AND THREE DIMENSIONS

Solutions were computed for three-dimensional horseshoe vortex flow past an elliptical leading edge for free stream angle of incidence $\alpha$ of zero and five degrees. These solutions have the following flow parameters: 5:1 ellipse, chordal Reynolds number of 400, Mach number 0.2, and flat plate leading edge located 1.25 chords upstream of the leading edge of the ellipse. For reasons of economy, these solutions were computed using a coarse computational grid, the finest being 14x28x14 $(\xi,\eta,z)$ for $\alpha=5°$. However, the geometry, flow conditions, and grid distributions were carefully adjusted to provide the best resolution possible within this constraint. Most length scales expected to be present in this type of flow receive at least modest resolution. Specifically, care was taken to provide resolution of boundary layers on the endwall and ellipse, of the Heimenz layer at the leading edge stagnation point, and of the flow region within one leading edge radius of both the leading edge and the endwall. The only potentially-relevant unresolved length scale is an $O(Re^{-3/8})$ streamwise distance predicted by two-dimensional "triple deck" theory near a separation point.

The truncation error associated with the present flow and computational grid was examined by mesh refinement and by comparison with an incompressible solution of Lugt and Ohring [11] for two-dimensional flow past a 10:1 ellipse at zero incidence and Re=200. The latter comparison for streamwise velocity at the present outflow boundary ($\eta=\pi/2$) is shown in Fig. 2. The agreement is quite reassuring in light of the respective differencing procedures and meshes. Good agreement was also obtained for surface pressure distributions. Additional two-dimensional solutions for the forward half of a 5:1 ellipse ($\alpha=0°$,Re=200,$\eta=0.2$) were computed assuming symmetry about the chord line $\eta=\pi$ and using both 14x14 and 28x28 grids. The streamwise velocity U along the stagnation streamline is a sensitive indicator of mesh dependence and is shown on a logarithmic scale in Fig. 3. The 14x14 mesh is identical to the 14x28 mesh used for $\xi,\eta$ in the three-dimensional solutions without symmetry about $\eta=\pi$, and the small amount of mesh dependence in Fig. 3 indicates that this mesh should be adequate, at least away from the endwall. Finally, resolution normal to the endwall (z direction) was examined for three-dimensional flow at $5°$ incidence using both 14x28x10 and 14x28x14 grids. The velocity profile at x=-1.06,y=0 near separation is the most sensitive indicator and is shown in Fig. 4. The finer mesh improves resolution considerably very near the endwall flow reversal, but makes little difference elsewhere.

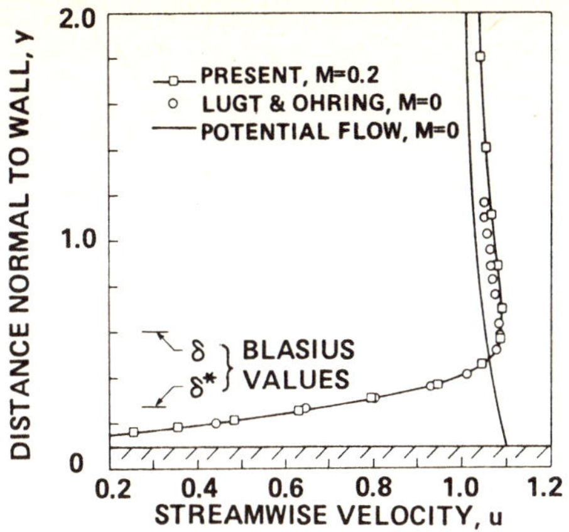

Fig. 2 - Velocity Profiles at Outflow
Boundary ($\eta=\pi/2$) for 10:1
Ellipse, $\alpha=0^\circ$, Re=200.

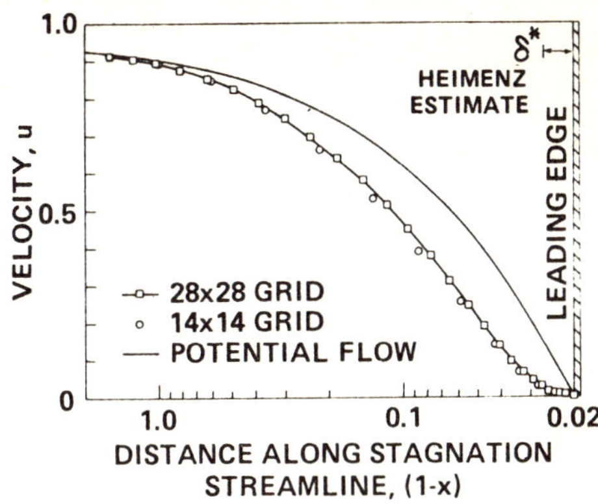

Fig. 3 - Velocity Along Stagnation Streamline
($\eta=\pi$) for 5:1 Ellipse, $\alpha=0^\circ$, Re=400.

The foregoing results provide a limited but informative assessment of truncation error associated with the present calculations. The solutions may be characterized as having both qualitative significance and reasonable quantitative accuracy. Any potential concern that failure to provide high resolution near separation would cause serious error (beyond a local smoothing of the flow) may be unwarranted in light of these resolution tests.

### HORSESHOE VORTEX FLOW

Representative results from solutions for three-dimensional horseshoe vortex flow at zero and five degree incidence are shown in Figs. 5a-f. These solutions converged in about 80 time step iterations and with a 14x28x14 grid ($\alpha=5^\circ$) required about 20 minutes of CDC 7600 run time. A 14x28x10 grid was used for $\alpha=0^\circ$. In Figs. 5a-b, vector plots of velocity in a plane one grid point away from the no-slip endwall surface are shown.

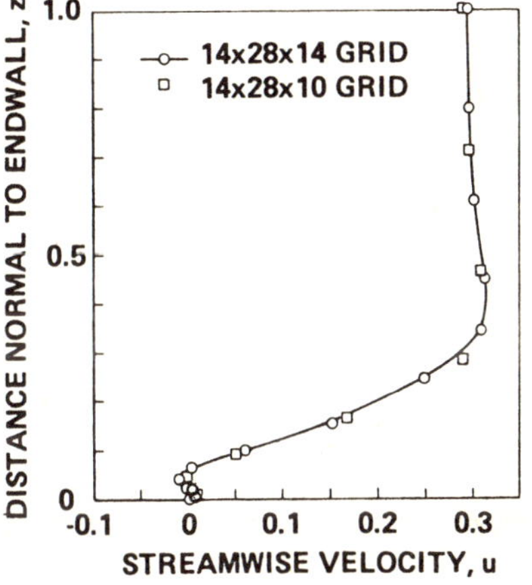

Fig. 4 - Streamwise Velocity Profiles
Near Separation for 5:1 Ellipse,
$\alpha=5^\circ$, Re=400.

Here, a saddle-point type of flow separation is evident upstream of the leading edge and, in the case of $\alpha=5^\circ$, toward the (upper) high pressure surface of the ellipse. The remaining plots in Fig. 5 show flow velocities in the plane z=0.15, located approximately in the center of both the horseshoe vortex flow region and the approaching boundary layer. Here, the velocity vector plots are not much different from the two-dimensional flow region near z=1 (not shown), and the most significant feature of the flow is seen in the contours of velocity normal to the endwall (and to the page) in Figs. 5e-f. A strong downward flow toward the endwall is present near the leading edge (behind the saddle-point separation), with maximum downward velocity of 28 percent of the freestream reference velocity for $\alpha=0^\circ$ and 32 percent for $\alpha=5^\circ$. Further results for $\alpha=5^\circ$ are shown in Figs. 6a-f.

Contours of total pressure loss coefficient $\Delta p_o/(\rho_r U_r^2/2)$ for z=0.15 are shown in

Fig. 6a.  In Fig. 6b, the velocity in the stagnation plane normal to and containing the ellipse leading edge again shows the strong downward flow toward the endwall near the leading edge.  The cross-flow velocity in outflow planes is shown in Figs. 6c-d. A moderately strong secondary flow pattern (peak velocity 20% of $u_r$) indicative of a streamwise corner vortex is clearly in evidence near the suction surface in Fig. 6d. Finally, a "limiting" surface velocity vector plot and contours of pressure coefficient are shwon in Figs. 6e-f for the surface one grid point away from the surface of the ellipse ("unwrapped" to lie in a plane).  Here there is evidence of the distorted stagnation line near the endwall, flow toward the endwall near the stagnation line, and finally the formation of a streamwise vortex visible mainly on the low pressure or suction surface near the endwall.

## CONCLUDING REMARKS

Although no other analytical results or experimental measurements of the three-dimensional horseshoe vortex flow are available for comparison, the present computed results are consistent with flow visualization studies of related leading edge vortex flows.  Computed results in three dimensions have illuminated several aspects of the flow structure.  For flows having nontrivial endwall boundary layer development upstream of the leading edge, the horseshoe vortex structure consists of an inertially-dominated rotational flow except very near the solid boundaries, where viscous effects occur within thin layers generated locally by the leading edge interaction.  The rotational inviscid portion of the vortex structure both upstream of the leading edge and in the downstream corner flow region scales with the approaching endwall boundary layer thickness.  Viscous effects are confined to thin regions near the surfaces, having a thickness which is apparently not strongly dependent on the approaching boundary layer.  The overall flow consists of a saddle-type flow separation on the endwall upstream of the leading edge, a strong spanwise flow toward the endwall near the leading edge, and streamwise vortices in the corner region downstream of the leading edge.  The strength of the streamwise corner vortex is significantly increased on the suction surface corner for flow at nonzero incidence.  Regarding the present use of "zone embedding" and interactive boundary conditions to minimize the computed flow region, the general conclusion is that while boundary conditions cannot be treated with complete rigor when located within an elliptic region of nonuniform flow, careful treatment can lead to quite reasonable results which appear completely adequate for the present goal of horseshoe vortex analysis.

## ACKNOWLEDGMENT

This work was supported by the Office of Naval Research.

## REFERENCES


1.  Belik, L., Aero. Quart., Feb. 1973, p. 47.
2.  Baker, C. J., J. Fluid Mech., 95, 1979, p. 347.
3.  Barber, T. J., J. Aircraft, 15, 1978, p. 676.
4.  Dwyer, H. A., AIAA Paper No. 68-740, 1968.
5.  Hawthorne, W. R., J. Aero. Sci., 21, 1954, p. 588.
6.  Milne-Thomson, L. M., Theoretical Hydrodynamics, Macmillan Co., New York, 1960, p. 164.
7.  Briley, W. R. and McDonald, H., J. Comp. Physics, 24, 1977, p. 372.
8.  Briley, W. R. and McDonald, H., J. Comp. Physics, 34, 1980, p. 54.
9.  Douglas, J. and Gunn, J. E., Numer. Math., 6, 1964, p. 428.
10. Beam, R. M. and Warming, R. F., AIAA Journal, 16, 1978, p. 393.
11. Lugt, H. J. and Ohring, S., Physics of Fluids, 18, 1975, p. 1.

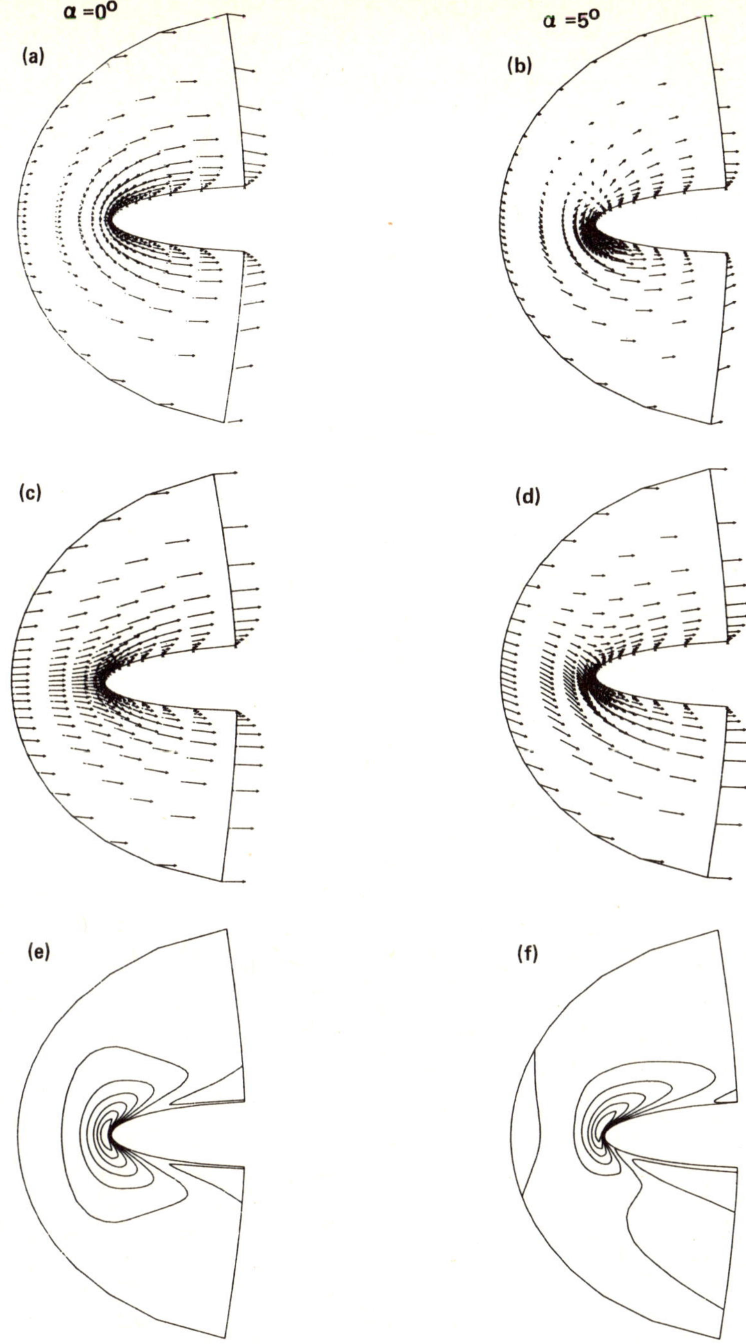

Fig. 5 – Detail of Computed Velocity for 5:1 Ellipse with α=0° and 5°, Re=400, M=0.2. (a,b): the plane of grid points adjacent to endwall surface; (c,d): the plane z=0.15 (near center of approaching boundary layer); (e,f): contours of velocity w normal to endwall in the plane z=0.15.

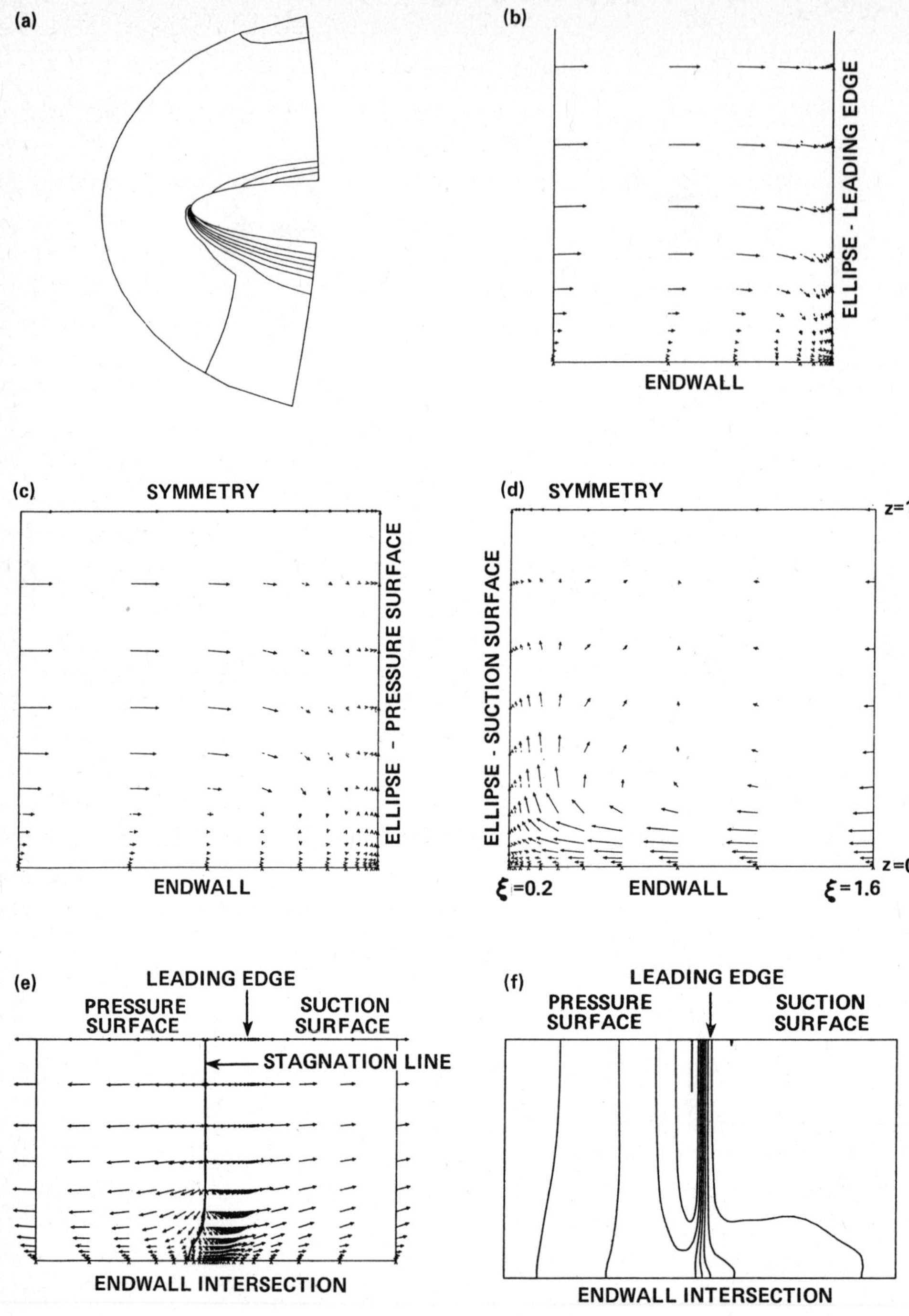



Fig. 6 – Detail of Horseshoe Vortex Flow for 5:1 Ellipse with $\alpha=5^{\circ}$, Re=400, M=0.2. (a)-Total pressure loss coefficient in z=0.15 plane; (b)-Velocity in plane containing leading edge; (c,d)-Secondary velocity in outflow surfaces; (e,f)-Velocity and contours of pressure coefficient in "unwrapped" surface of points adjacent to ellipse.

# NUMERICAL ANALYSIS OF STABILITY IN MAGNETOHYDRODYNAMICAL PROBLEMS

K.V. Brushlinsky, V.V. Savel'ev, N.M. Zueva

M.V. Keldysh Institute Appl.Math., USSR Acad.Sci., Moscow

The numerical simulation of physical processes and corresponding computations play an essential role in the studies of plasma flows and equilibrium configurations. The stability investigation also attracts numerical methods. These questions are of particular interest in connection with the plasma confinement problem. Macroscopic processes in a dense plasma are considered, so the mathematical models are based on MHD-equations. Two approaches in stability investigations may be noted. The first one is restricted to the linear approximation. In this case the investigation amounts to solving an eigen-value problem numerically for appropriate differential operators. The second approach is to investigate the nonlinear stage of processes, that is especially interesting in unstable cases. Here the problem may be solved by direct numerical integration of MHD-equations.

An example of the problem where the both approaches are utilized is the stability analysis for the plasma cylinder equilibrium in a helical magnetic field. We suppose the cylinder ends to be identified, i.e. it is a mathematical model of tore. The equilibrium has the cylinder symmetry $(\partial/\partial z \equiv \partial/\partial \varphi \equiv 0)$ . The MHD motion equation $\vec{\nabla} p = (\vec{\nabla} \times \vec{H}) \times \vec{H}$ is a single scalar one and it has in general many solutions. Usually $H_z = \text{Const}$ is fixed, and some distribution $j_z(r) = \frac{1}{r}\frac{\partial(r H_\varphi)}{\partial r}$ is given. After that the pressure $p(r)$ is determined. Note that in the case of a finite conductive plasma the equilibrium requires in addition that $\vec{\nabla} \times (\gamma \vec{j}) = 0$ and $\gamma j^2 = 0$ ( $\vec{j} = \vec{\nabla} \times \vec{H}$, $\gamma = 1/Re_m$ is the resistivity). The first requirement is satisfied by restricting the choice of $H_\varphi(r)$ . The Joule heat in the second one may be neutralized by any given energy loss, e.g. by radiation. But often one may neglect these corrections because $\gamma \ll 1$ in interesting cases.

The stability problem is three-dimensional irrespective of the original symmetry. Small perturbations are the functions of time $t$ and three space variables: $r$ , $\varphi$ , $z$ . Linearizing the MHD-equations on the equilibrium mentioned above, we get a linear equation system with coefficients dependent only on $r$ . Using the variable separation method we obtain that a typical Fourier-component of its solution has the factor $\exp(im\varphi + ikz) = \exp(im\theta)$ where $\theta = \varphi - \alpha z$,

$\alpha = -k/m = 2\pi/L$ , $L$ is a cylinder $z$ -period. Such perturbations are constant along helical lines $\theta = Const$ , therefore it is often said of the helical stability (or instability) of a plasma column. The solution dependence on the time is given by the factor $exp(\lambda t)$ . The exponent $\lambda$ is an eigen-value of a differential operator generated by the equations (with differentiation only with respect to $r$ ) and boundary conditions of the problem. It was shown in [1] that the stability is determined by the right-hand boundary of the operator spectrum irrespective of the eigen-function completeness. Two types of numerical techniques were employed for solving the eigen-value problem. The first one is the so-called shooting method: the differential equations with one-side boundary conditions are numerically solved for several values of $\lambda$ and then we choose among these solutions the one that satisfies conditions at the other boundary. This method is successful if the plasma is infinitely conductive $(\nu = 0)$ . In this case we can exclude all the variables except one (the radial velocity $v_r$ ) from the linearized MHD-equations and get the single second-order equation obtained by L.S.Solov'ev [2] (see also [4, 6]). The other method is as follows. The linear partial differential MHD-equations (before the selection of the factor $exp(\lambda t)$ ) are numerically integrated until the solution becomes dependent on $t$ exponentially [9, 10] .

Let us give some results concerning the isothermal infinitely conductive plasma. It is known that perturbations with $\alpha = \mu \equiv H_\varphi/r H_z$ (constant along magnetic field lines) are the most dangerous for the stability. Such perturbations are named resonance. The computations confirm this fact and show that the instability domain is larger than the one where the resonant values of $\mu$ exist. That is seen in a series of computations with fixed $\alpha$ and different typical values of $\mu$ . When $\mu$ increases the instability arises before the resonance appears at the center ( $r = 0$ ), and it doesn't disappear at once after the resonance reaches the cylinder boundary $r = 1$. The mode $m = 1$ is excited the earliest. At the same time the instability does not follow the resonance without fail. So, if $\alpha$ is small, perturbations with $m \geqslant 2$ do not increase even at the resonance surface.

But they become unstable if we include the plasma resistivity $(\nu > 0)$ in the computation. The destabilizing role of resistivity is observed in computations with different values of $\alpha$ , $\mu$ , $m$ : the instability increment $\lambda$ increases with $\nu$ , especially at small values of $\alpha$ , if $\nu$ is not so large. The development of

helical instability in a finite conductive plasma is often named
the "tearing-mode".

If $H_z = 0$ in equilibrium the configuration considered above is
the $Z$ -pinch. The analysis of its stability in the linear approach
has shown (in addition to the well-known fact of instability) that
the helical perturbation increment $\lambda(m, k)$ increases with the
growth of the wave frequency $k$ . In this case the helical lines
$\theta = Const$ come nearer to the magnetic lines, i.e. the instability
is connected again with a tendency to resonance.

The results presented here are partly published [4, 6] and partly obtained recently by the authors together with A.P. Shatanov.

In cases when the plasma configurations or flows are unstable,
a real picture of instability development may be seen only through
numerical solving the nonlinear MHD-equations. Economizing the computations and considering some physical reasons a symmetry is ugually
assumed. In the problem on a plasma cylinder it is the helical one:
two space variables are $r$ and $\theta = \varphi - \alpha z$ , where $\alpha = {}^{2\pi}\!/\!_L$ ,
i.e. the helical instability is considered again. The computation
results may be presented in terms of magnetic surfaces consisting of
magnetic field lines that are important in the plasma confinement
problems. Under the equilibrium these surfaces are cylinders $r = Const.$
The instability as stated above, develops near the resonance$\left(\alpha = {}^{H_\varphi}\!/\!_{r H_z}\right)$
and its type depends on the plasma conductivity. If $\nu = 0$ the magnetic
surfaces only change their form in time. Thin layers with strong
electric current are generated, the plasma tends to turn out on the
cylinder periphery, the flow is of a convective type. In the case of
finite resistivity $\nu > 0$ the change of magnetic surface topology can
be observed. They form a fibrous structure because of the magnetic
field line reconnection [5, 6, 11].

Computations give also some information on the stability of
steady-state plasma flows. The authors made a lot of computations of
two-dimensional flows in channels between two coaxial electrodes
[3, 7]. The problem is solved in the time-dependent MHD-model or its
modifications. The solution becomes steady-state in time. The stabi-
lization effect proves that this solution is stable, but only with
respect to perturbations of the same symmetry. A more detailed analysis of perturbations, their oscillations and damping may be fulfilled
in the same numerical simulation if we give a small density variation
at the duct entry and look after its dynamics [7].

Computations have shown that the two-dimensional plasma flow in
the channels is stable in the MHD-model. But if we simulate the Hall

effect, the flow becomes unstable near the anode. The computations permit to see the development of instability, which is of an explosion type $[3]$ .

A particular attention is given to the stability of the gas ionization front in the channel. Let the neutral gas enter in the duct. It is ionized by a gas discharge near the channel inlet. We simulate this phenomenon by means of a stepped conductivity dependence on the temperature: if $T < T^*$ we have the gas and its conductivity is small $(\sigma \ll 1)$ ; the ionization takes place when $T = T^*$; and if $T > T^*$ , the conductivity $\sigma = \sigma_1 T^{3/2}$ corresponds to the plasma. The computations in such a model (quasi-one-dimensional $[3]$ and two-dimensional $[8]$ yielded the following result: if the plasma conductivity is large enough $\left( \sigma_1 \sim Re_m \gg 1 \right)$ , the solution of the problem does not reach a steady-state regime but an oscillating, periodic one.

## References:


1. К.В. Брушлинский. Изв.АН СССР, серия матем. 1959, т.23, № 6, 893-912.

2. Л.С. Соловьев, В сб. "Вопросы теории плазмы" п/р М.А. Леонтовича, вып. 3, М. Атомиздат, 1963, 245-289.

3. К.В. Брушлинский, А.И. Морозов. ibid., вып. 8, 1974, 88-163.

4. Н.И. Герлах, Н.М. Зуева, Л.С. Соловьев. Магн. Гидродинамика, 1978, № 4, 49-54.

5. Н.М. Зуева, Л.С. Соловьев. Магн.Гидродинамика, 1977, № 3, 5-22.

6. Н.М. Зуева, В.В. Палейчик, Н.Н. Ченцов. В сб. "Двумерные численные модели плазмы" п/р К.В. Брушлинского, изд.ИПМ АН СССР, 1979, 67-119.

7. К.В. Брушлинский, А.И. Морозов, В.В. Савельев. ibid., 7-66.

8. В.В. Савельев. Письма в Журн.Техн.Физ. 1976, т.2,вып.13, 593-596.

9. G.Bateman, W.Schneider, W.Grossmann. Nucl.Fusion, 1974, v.14, N 5, 669-683.

10. J.A. Dibiase, J.Killeen. J.Comp.Phys. 1977, v.24, 158-185.

11. J.A. Wesson. Comp.Phys.Comm. 1976, v.12, 53-65.

Compressible Flow Simulation Using Hamilton's Equations
and Clebsch-type Vortex Parameters

By O. Buneman

Institute for Plasma Research, Stanford University, Stanford, CA 94305

It has been shown[1,2] that the Clebsch representation of the velocity field
$(\vec{v} = \lambda\nabla\mu - \nabla\phi)$ results in Hamilton's equations for the four state variables $\phi$ $\rho$ $\mu$
and $\sigma = -\rho\lambda$ . The momentum density is:

$$\vec{p} = \rho\vec{v} = -\sigma\nabla\mu - \rho\nabla\phi$$

and the energy density:

$$H = C\rho^{\gamma} + \tfrac{1}{2}p^2/\rho = C\rho^{\gamma} + (\sigma\nabla\mu + \rho\nabla\phi)^2/2\rho$$

Forming functional derivatives of the latter one obtains

$$\frac{\partial\rho}{\partial t} = -\frac{\partial H}{\partial\phi} = -\nabla\cdot\rho\vec{v} \quad \text{(conservation of mass)}$$

$$\frac{\partial\phi}{\partial t} = \frac{\partial H}{\partial\rho} = C\gamma\rho^{\gamma-1} - \tfrac{1}{2}(\frac{\sigma}{\rho})^2(\nabla\mu)^2 + \tfrac{1}{2}(\nabla\phi)^2 \quad \text{(Bernoulli's equation)}$$

$$\frac{\partial\mu}{\partial t} = \frac{\partial H}{\partial\sigma} = -\vec{v}\cdot\nabla\mu \quad \text{(vortex label } \mu \text{ follows the flow)}$$

$$\frac{\partial\sigma}{\partial t} = -\frac{\partial H}{\partial\mu} = -\nabla\cdot\sigma\vec{v} \quad \text{(vortex label } \lambda \text{ follows the flow)}$$

The benefit of this description of (ideal) compressible flow lies in the fact
that the state variables fall into two groups, potentials $(\phi,\mu)$ and densities
$(\rho,\sigma)$ , which can be updated mutually by leap-frogging them over each other. Even
better, one can stagger the records of potentials and densities in space so that
gradients and divergences are always available as central differences.

In terms of four indices, i, j, k, $\ell$, for the space-time grid: potential information $(\phi,\mu)$ is kept where and when the sum $i + j + k + \ell$ is even while density
information $(\rho,\sigma)$ is kept where and when that sum is odd. At even times ($\ell$ even),
one then has potential gradients where $i + j + k$ is odd and this is where $(\rho,\sigma)$
data are also available. Thus $\rho\vec{v}$ , $\sigma\vec{v}$ and everything needed for $\partial\phi/\partial t$ and $\partial\mu/\partial t$
can be calculated at the very places where $\phi$ and $\mu$ are on record from the previous
(odd) time. These values can therefore be overwritten, in a properly time-centered
step, by $\phi,\mu$ values at the next (odd) time. See Figures 1 and 2 for the data
arrangements at two successive time steps.

On the other hand, divergences of $\rho\vec{v}$ and $\sigma\vec{v}$ can be formed, by central differencing, at places where $i + j + k$ is even. This is just where $\rho$- and $\sigma$-data were
kept at the previous time, so that these can now be overwritten by values updated to
the next (odd) time, again in properly time-centered fashion.

One notes that the total information to be stored is just one "grid-ful" of data

for each of the four state variables.  The updating is entirely local:  for advancing
the potentials at any point, one needs the local densities and the six closest neighboring potential pairs.  For advancing the densities, one needs the six closest densities, the local potentials and the six potential pairs two spaces removed along
the co-ordinate axes.  The "computational molecule" is an octahedron.

Updating can take place in a small arithmetic unit of very limited storage
through which the state variables are pipe-lined.  Divisions are called for wherever
densities occur:  along with $\rho$ and $\sigma$ , one also needs $\sigma/\rho$ .  A short table look-
up, or a rational approximant, is needed for $\rho^{\gamma-1}$ .

In a machine with large fast-access memory, it is advantageous to store several
layers temporarily in order to minimize I/O, and to eliminate re-calculating the quo-
tients $\sigma/\rho$ which are used several times over.  More specifically, one must hold in
core two complete layers of data, as follows:

the present potentials $\phi,\mu$ of layer $n + 1$

the present potentials $\phi,\mu$ of layer $n$

the present densities $\rho,\sigma$ of layer $n$

"intermediate" densities $\rho,\sigma$ of layer $n$

The latter are densities which are partially updated between the past and the future
time.  When the potentials of layer $n + 2$ are now called in, as well as the densities of layer $n + 1$ , the velocities and fluxes of layer $n + 1$ can be found.  The
divergences of layer $n$ require these fluxes and hence the updating of the densities
there had to be left incomplete.  The intermediate density values can now be fully
updated and written out, making room for the new densities that are coming in.  The
potentials of layer $n$ are also just becoming redundant during this process and can
be overwritten by those which are being read.  The full updating of potentials and
the partial updating of densities of layer $n + 1$ can also proceed at this point.
Old and new potentials are pipelined in and out through the machine, not held temporarily.

In this procedure, no item is read in more than twice:  once to be used and once
to be updated.  It is written out only once.  The I/O traffic controls the speed.
With 2:1 packing and two channels each way we estimate .5 seconds for processing a
$512^2$ layer on the CRAY-1.  A million half-words are needed to accommodate the data
to be held from layer to layer, plus generous buffer space.  For a $512^3$ domain, it
should take 4 minutes to pipeline all the layers through the machine.  It is the simplicity of the algorithm and the resulting speed, combined with the localizability of
the processing (layer by layer) which allow one to aim at such high resolution with
this scheme.  No predictor-corrector or Crank-Nicolsen routine is needed, and no
numerical instabilities (such as the odd-even discrepancy) are anticipated.

The time-step is, of course, limited by a Courant condition.  It would take at
least 512 steps for a signal to propagate across a $512^3$ domain:  we are therefore
thinking in terms of day-long runs.

One must watch the positivity of $\rho$ in these simulations, and one must, occasionally, re-label the vortices, i.e. apply a gauge transformation to the Clebsch variables, since vortices which started as close neighbors will be pulled apart by shear. How to introduce dissipation, to simulate viscosity, is being studied.

The first application of the method is to the shear layer, partly as an exercise and test, and still on a more modest scale, but with a view to obtaining high resolution of the incipient turbulence eventually. A uniform shear layer with $v_x = \omega y$ in the range $-\tfrac{1}{2}d < y < \tfrac{1}{2}d$ is initialized by taking $\sigma = -\rho\lambda = -\rho x$ , $\mu = -\omega y$ , $\phi = -\omega xy$ within that range and $\sigma = -\rho x$ , $\mu = \mp \tfrac{1}{2}\omega d$ , $\phi = \mp \tfrac{1}{2}\omega xd$ outside, ensuring continuity at the edges. If unperturbed, it evolves like $\sigma = -\rho(x - \omega yt)$ , $\mu = -\omega y$ , $\phi = \tfrac{1}{2}\omega^2 y^2 t - \omega xy$ internally and like $\sigma = -\rho(x \mp \tfrac{1}{2}\omega dt)$ , $\mu = \mp \tfrac{1}{2}\omega d$ , $\phi = \tfrac{1}{8}\omega^2 d^2 t \mp \tfrac{1}{2}\omega xd$ externally. However, the Kelvin-Helmholtz instability (either deliberately excited, or starting spontaneously from round-off) sets in at the edges and the usual roll-up of the vortices begins. It is in the subsequent evolution of more intricate flow that the high-resolution capability of the new scheme pays off.

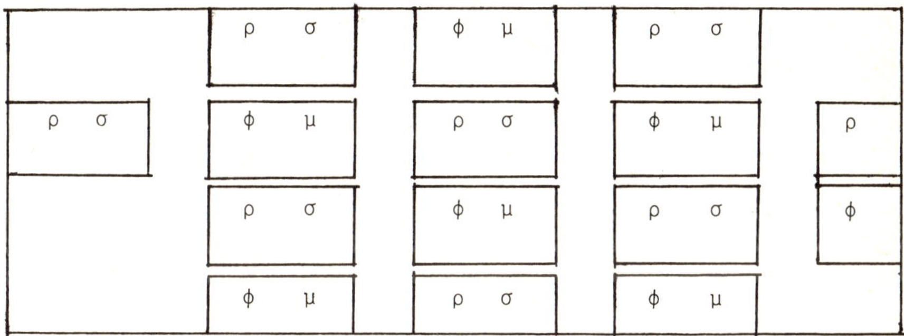

Figure 1.  Arrangement of density and potential data at timestep "n" .

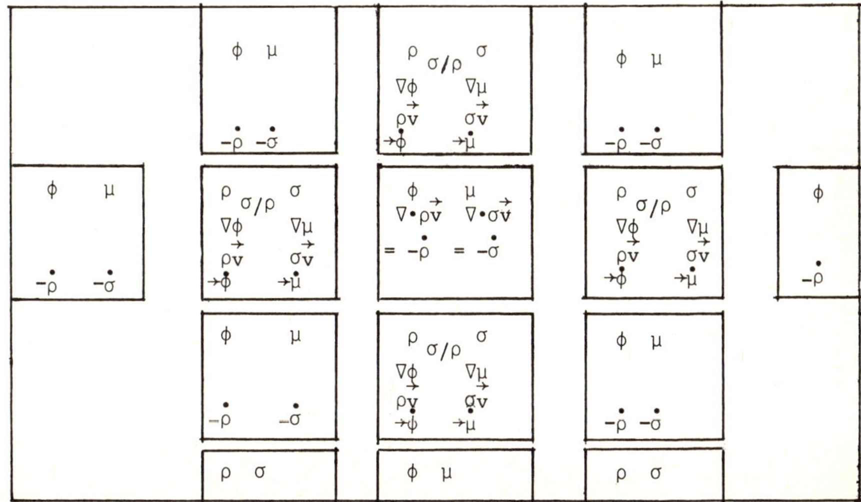

Figure 2.  Arrangement of density and potential data at timestep "n + 1" .

References.

1. O. Buneman, "Ideal Gas Dynamics in Hamiltonian Form With Benefit for Numerical Schemes", submitted to Physics of Fluids.

2. O. Buneman, "Ideal Gas Dynamics in Hamiltonian Form With Benefit for Numerical Schemes", Abstract CH4, Bull. APS, $\underline{24}$, 1139 (October 1979).

# FINITE ELEMENT CALCULATION OF STEADY TRANSONIC FLOW
## IN NOZZLES USING PRIMARY VARIABLES


J.J. Chattot[*], J. Guiu-Roux[**] and J. Laminie[**]

[*] Office National d'Etudes et de Recherches Aerospatiales (ONERA)
[**] Université de Paris-Sud, Centre d'Orsay, et Analyse Numérique (CNRS), ERA 297



## Abstract


Steady irrotational-isentropic flow of perfect fluid in a plane converging-diverging nozzle is modelled using a system of two first order partial differential equations in the primary variables. A least square formulation transforms the first-order system into an equivalent second-order system well adapted to discretization methods and allowing the use of powerful iteration algorithms such as conjugate gradient or Newton's method which yield fast convergence. The other advantage of this variational approach is the direct applicability of the finite element method which is more accurate in this case than the corresponding finite difference method.



## I. Introduction

The theoretical basis for the solution of unsteady first order systems of partial differential equations modelling transonic flows is well established. Steady solutions are obtained as asymptotic limits for large time with time independent boundary conditions of the unsteady system.  Explicit schemes are subject to a CFL condition which places a limit on the maximum permissible time step and therefore require a large amount of computing time.  To circumvent this difficulty two remedies have been proposed:

- the first consists in altering  the time derivative terms to transform the system into a pseudo-unsteady system having improved convergence properties. This approach has been extensively developped and used by Veuillot and Viviand[1] and Viviand[2]. The gains in CPU time are substantial and the simplicity of the explicit scheme makes the method robust and flexible ;

- the alternate solution consists in employing an implicit scheme. Beam and Warming [3] have proposed various implicit schemes for first order equations, but the efficiency of these schemes is partly offset  by the poor conditioning  of the associated matrices and the greater coding complexity. Recently Lerat [4] proposed a class of implicit schemes having, among other   properties, that of insuring the diagonal dominance of the associated matrix. The application of Lerat's schemes to the computation of steady transonic flow problems may allow a reevaluation of the virtues  of implicit schemes.

In contrast to the unsteady systems which are always hyperbolic in time, the relaxation methods for steady first order systems suffer from a lack of firm theoretical support.  On one hand the type of the system may vary from point to point (from elliptic to hyperbolic type) and although for purely hyperbolic situations a marching scheme proposed by MacCormack [5] is perfectly adapted, no reasonable scheme seems to apply in elliptic cases. Indeed a centered scheme would appear natural in this instance, but a centered approximation for a first derivative e.g.

$$\frac{\partial u}{\partial x} = \frac{u_{i+1} - u_{i-1}}{2\Delta x}$$ yields an ill-conditioned matrix (due to the zero coefficient for $u_i$) and thus an ill-behaved solution.

---


*Work performed with the financial support of DRET.*

This simple fact, of fundamental numerical impact, explains why most of the relaxation work dealing with steady transonic flows is associated with the potential equation, a second order partial differential equation of mixed-type for which Murman and Cole [6] for the small perturbation potential and Jameson [7] for the full potential, have constructed specific schemes and algorithms. Very few authors have dealt directly with the first order system and it is worthy of interest to mention the various approaches and their success.

The system of two first order equations in $u,v$ corresponding to the transonic small perturbation method has been solved by Steger and Lomax [8] using a mixed-scheme of Murman-Cole type for the x-derivative and a three point-alternately advanced and retarded-scheme for the y-derivative. This scheme however does not avoid some uncoupling of odd and even indexed lines corresponding to a poorly conditioned matrix of the discretized system. Blomster and Sköllermo [9] solve the system of two first order equations in $u,v$ corresponding to the full potential modelization, using a centered scheme at all points. As mentionned earlier, centered schemes do not yield a well conditioned matrix. However this is not a compulsory requirement when looking for a smooth solution such as a shock free transonic flow in a converging-diverging nozzle. Besides, the discretized system is linearized by means of Newton's method and solved by a direct method with a good initial approximation obtained from the one-dimensional theory. Thus their method seems limited to solutions without high frequency content such as shocks or corner flows.

To eliminate the conditioning difficulty Chattot [10], using the same model as Steger and Lomax, transformed the first order system into an equivalent second order system upon differentiation with respect to $y$. A centered scheme could then be used at all points for the derivative $\dfrac{\partial^2}{\partial y^2}$. Independently Johnson [11] had reached the same conclusion and started developping his surrogate equation concept described in details in his remarkably well documented Doctoral Thesis [12]. His applications concern both subsonic rotational flows using the full Euler equations and transonic flows using small perturbation assumption. An interesting discussion of the efficiency and accuracy of the surrogate equation method can be found in the Thesis. The derivation of the higher order system was made more systematic by Chattot [13] upon introduction of a variational formulation of least square type, that was applied to the quasi-one-dimensional Euler equations modelling the transonic flow in a slender nozzle. This approach was extended to a purely hyperbolic problem in two dimensions by Chattot, Guiu-Roux and Laminie [14] and solved using both finite differences and finite elements. It was noted that finite element approximation was more accurate than their finite difference counterpart on the same mesh. Fletcher [15] uses a least square formulation of the full Euler equations to solve for the subsonic flow past profiles, along with a finite element discretization.

In the work reported here, a system of two first order equations in $u,v$ corresponding to the full potential approximation is used. Although this may be inaccurate for internal flows with shock waves, for transonic shock free flow it is equivalent to the Euler equations of motion.

II. Formulation

A. First order problem (Prob. 1)

Let $\rho$, $u$, $v$ be the unknown density, and velocity components respectively inside the domain $\Omega$ representing the throat region of a converging-diverging plane nozzle.
The boundary of $\Omega$, $\partial\Omega$ is assumed locally Lipchitz continuous.
The equations to be solved are with the usual notation and non-dimensionalization :

$$(1.1) \qquad \frac{\partial \rho u}{\partial x} + \frac{\partial \rho v}{\partial y} = 0$$

$$(1.2) \qquad \frac{\partial v}{\partial x} - \frac{\partial u}{\partial y} = 0 \qquad\qquad\qquad \text{in } \Omega$$

$$(1.3) \qquad \rho = \left[(\gamma-1)\, M_\infty^2 \left(h_o - \frac{u^2+v^2}{2}\right)\right]^{\frac{1}{\gamma-1}}$$

This system corresponds to steady irrotational-isentropic flow.
The boundary conditions are :

(2.1) $\qquad p = p_{exit} = \dfrac{1}{\gamma M_\infty^2}\, \rho_{exit}^\gamma \qquad$ at the exit section if the flow is subsonic
$\qquad\qquad\qquad\qquad\qquad\qquad\qquad$ (no condition in the case of supersonic flow)

(2.2) $\qquad v = 0$ on the nozzle plane of symmetry

(2.3) $\qquad \dfrac{v}{u} = f'(x)$ on the nozzle wall of equation $y = f(x)$

(2.4) $\qquad \dfrac{v}{u} = g(y)$ at the entrance section.

We shall now define the variational formulation.

$\qquad$ B. $\quad$ <u>Associated least-square problem</u> (<u>Prob. 2</u>)

$$\text{Let} \quad I(u,v) = \int_\Omega \frac{\alpha^2}{2}\left\{\frac{\partial \rho u}{\partial x} + \frac{\partial \rho v}{\partial y}\right\}^2 dx\, dy + \int_\Omega \frac{\beta^2}{2}\left\{\frac{\partial v}{\partial x} - \frac{\partial u}{\partial y}\right\}^2 dx\, dy$$

$\qquad\qquad \alpha$, $\beta$ constant (equal to 1 in the present application).

One must find $(\tilde\rho,\tilde u,\tilde v)$ such that

(3.1) $\qquad\qquad\qquad\qquad I(\tilde u,\tilde v) = \underset{u,v}{\text{Inf}}\; I(u,v)$

(3.2) $\qquad\qquad\qquad\qquad \rho(u,v) = \rho(u^2+v^2)$ as given by the isentropic law (1.3).

Let : $H(div,\Omega) = \{(p_1,p_2) \in \mathbb{L}^2(\Omega)\,|\,div(p_1,p_2) \in \mathbb{L}^2(\Omega)\}$

$\qquad V \qquad = \{(q_1,q_2)\,|\,(\rho q_1,\rho q_2) \in H(div,\Omega)\,|\,rot(q_1,q_2) \in \mathbb{L}^2(\Omega)\}$

$V$ is a Hilbert space for the norm ($|\;|$ denotes the $L^2$ norm) :

$$\| (p_1,p_2)\|_V^2 = |p_1|^2 + |p_2|^2 + |div(p_1,p_2)|^2 + |(rot(p_1,p_2)|^2.$$

Then $I$ is a convex functional in $V$ and G-differentiable. Hence for any given $\rho > 0$ the set of solutions of (3.1) is convex. Thus :

$\qquad$ C. $\quad$ <u>Equivalence of the two problems</u>

<u>Note</u> : For any given $\rho > 0$ (1.1)-(1.2) is elliptic and has a unique solution satisfying the boundary conditions (2.1)-(2.4) which implies that the set $V$ is non-empty.

<u>Proof</u> : $\quad$ i) $\quad$ any solution to (1.1)-(1.2) is solution to (3.1) ;

$\qquad\qquad$ ii) $\quad$ the existence of the solution to (1.1)-(1.2) implies that $\text{Inf } I = 0$ .

But the only solution corresponding to $I=0$ is solution of (1.1)-(1.2) by convexity of $I$ . The equations (1.3) and (3.2) are identical. Thus Prob.1 and Prob.2 are equivalent.


III. $\quad$ <u>Finite element discretization and solution procedure</u>

$\qquad$ The domain of interest of the nozzle is approached by a polygonal domain $\Omega_h$ . $\mathscr{C}_h$ is a quadrangular finite element mesh of $\Omega_h$ . $W_h$ is the set of continuous functions which are of the form $axy + bx + cy + d$ on each quadrangle. $\mathbb{W}_h = W_h \times W_h$ is thus an internal and converging approximation of $H(div,\Omega)$ [16]. For each regular function $\omega$ , there exists a unique function $\omega_h = r_h(\omega)$ of $\mathbb{W}_h$ which coincides with $\omega$ at each node of $\mathscr{C}_h$ . The set $V$ is approached by

$$V_h = \{(u_h,v_h) \mid [(\rho u)_h,(\rho v)_h] \in W_h\} .$$

Note that $(\rho u)_h$ (resp. $(\rho v)_h$) is defined uniquely by :

$$(\rho u)_h = \sum_{i=1}^{N} \rho_i \, u_i \, \phi_i \qquad (resp. \sum_{i=1}^{N} \rho_i \, v_i \, \phi_i)$$

where
$$\begin{cases} \rho_i = \rho(u_i^2+v_i^2) \\ \{\phi_i\}_{i=1,N} \text{ is a basis of } V_h \\ N \text{ is the total number of nodes of } \mathscr{C}_h . \end{cases}$$

As in [14] the products $\rho u, \rho v$ are approximated in this manner instead of

$$(\rho u)_h = \sum_{i,j=1}^{N} \rho_i \, u_j \, \phi_i \, \phi_j$$ to insure that the numerical scheme be conservative.

The gradient of the functional has two components :

$$\frac{\partial I_h}{\partial u}(u_h,v_h)\cdot\sigma_h = \int_\Omega \alpha^2 \{\frac{\partial(\rho u)_h}{\partial x} + \frac{\partial(\rho v)_h}{\partial y}\} \frac{\partial(\rho\sigma)_h}{\partial x} \, dx \, dy - \int_\Omega \beta^2 \{\frac{\partial v_h}{\partial x} - \frac{\partial u_h}{\partial y}\} \frac{\partial\sigma_h}{\partial y} . dx \, dy$$

$$\frac{\partial I_h}{\partial v}(u_h,v_h)\cdot\tau_h = \int_\Omega \alpha^2 \{\frac{\partial(\rho u)_h}{\partial x} + \frac{\partial(\rho v)_h}{\partial y}\} \frac{\partial(\rho\tau)_h}{\partial y} \, dx \, dy + \int_\Omega \beta^2 (\frac{\partial v_h}{\partial x} - \frac{\partial u_h}{\partial y}) \frac{\partial\tau_h}{\partial x} \, dx \, dy$$

for any test functions $\sigma_h$ and $\tau_h$ of $W_h$ .

The fixed point algorithm can be described as :
for a known couple $(u_n,v_n)$ , the new iterates $(u_{n+1},v_{n+1})$ are obtained from

$$\frac{\partial I_h}{\partial u}(u_{n+1},v_n)\cdot\sigma = 0 , \qquad \rho = \rho(u_n^2+v_n^2)$$

$$\frac{\partial I_h}{\partial v}(u_{n+1},v_{n+1})\cdot\tau = 0, \qquad \rho = \rho(u_n^2+v_n^2) .$$

Each linear system is inverted by a conjugate gradient algorithm preconditioned by incomplete Choleski (ICCG). Since each equation corresponds to an elliptic problem, the matrices are very well conditioned and the convergence is obtained in 3 to 4 iterations, and the global process converges in about 20 fixed point iterations.


## IV. Results and conclusion

The nozzle and the mesh used are presented fig.1. Subsonic flows have been computed for values of the exit density $\rho_e$ of 1.5 and 1.4 and are reported on the figures 2-3. For $\rho_e = 1.355$ a supersonic shock free bubble appears on the curved wall near the throat (fig.4). By lowering the exit value of the density to $\rho_e = 1.35$ the flow chokes, a sonic line spans the flow from wall to wall and the supersonic zone is terminated by a shock wave (fig.5). In order to capture the shock wave an artificial viscosity equivalent that used by Jameson [7] is introduced by "upwinding" the density according to

$$\tilde{\rho} = \rho - \mu\delta\rho \quad , \text{ where } \mu = \sup \{0,1 - \frac{1}{M^2}\} .$$

For a value $\rho_e<1$ the flow becomes supersonic in the exit section and the downstream boundary condition disappears to leave a shock free supersonic flow (fig.6).
A least square method has been developped to solve the irrotational-isentropic

flow in a converging-diverging nozzle using primary variables. Due to the good
properties of the matrices fast convergence is obtained and the method can be extended
to treat mixed flows with shocks. Studies are in progress for alternate choices of
the artificial viscosity.


References

[1]  Veuillot J.P. and Viviand H. - *A Pseudo-Unsteady method for the computation of
     transonic Potentiel Flows*, 11th AIAA Fluid and Plasma Dynamics Conference,
     Seattle, WA, july 10-11, 1978. AIAA Paper n°78-1150.

[2]  Viviand H. - *Pseudo-unsteady methods for transonic flow computations*, 7 ICNMFD,
     Stanford, Calif. june 23-27, 1980.

[3]  Beam R.M. and Warming R.F. - *Numerical calculations of two-dimensional unsteady
     transonic flows with circulation*, NASA TN D-7605, feb. 1974.

[4]  Lerat A. - *Une classe de schémas aux différences implicites pour les ysstèmes
     hyperboliques de lois de conservation*, C.R. Acad. Sc. Paris, t.288, Série A,
     18 juin 1979.

[5]  Mac Cormack R.W. and Warming R.F. - *Survey of computational methods for three-
     dimensional supersonic inviscid flows with shocks*, in Adv. in Num. Fl. Dyn.,
     AGARD LS-64, 1973.

[6]  Murman E. and Cole J.D. - *Calculation of plane steady transonic flows*, AIAA
     Journal, jan. 1971.

[7]  Jameson A. - *Iterative solution of transonic flows over airfoils and wings,
     including flows at Mach 1*, Comm. Pure and Applied Math., vol.27, 1974, p.
     288-309.

[8]  Steger J.L. and Lomax H. - *Generalized relaxation methods applied to problems in
     transonic flow*, Lecture Notes in Physics, vol.8, Springer-Verlag, p.193-198,
     1971.

[9]  Blomster J. and Skollermo G. - *Finite difference computation of steady transonic
     nozzle flow*, Uppsala Univeristy, Department of Computer Science Report n°66,
     jan.1977.

[10] Chattot J.J. - *Une méthode de relaxation entièrement conservative pour les
     écoulements transsoniques*, La Recherche Aérospatiale N°1975-6, P.339-346.
     English version available as ONERA TP 1975-93 E.

[11] Johnson G.M. - *The computation of non-potential transonic flows by relaxation
     methods*, Von Karman Institute, Belgium, Project Report n°1975-3, june 1975.

[12] Johnson G.M. - *A numerical method for the iterative solution of inviscid flow
     problems*, Doctoral Thesis, Applied Sciences Department, Université Libre de
     Bruxelles, Belgium, dec. 1979.

[13] Chattot J.J. - *Relaxation approach to the steady Euler equations in transonic
     flow*, AIAA Paper N°77-636.

[14] Chattot J.J. Guiu-Roux J. et Laminie J. - *Résolution numérique d'une équation de
     conservation par une approche variationnelle*, Proc. 6th ICNMFD, Tbilissi,
     USSR, june 20-25, 1978, Springer-Verlag 1979.

[15] Fletcher C.A.J. - *A primitive variable finite element formulation for inviscid,
     compressible flow*, Journal of Comp. Physics, 33, 301-312, 1979.

[16] Temam R. - *Navier-Stokes equations*, Studies in mathematics and its applications,
     North-Holland, 1977.

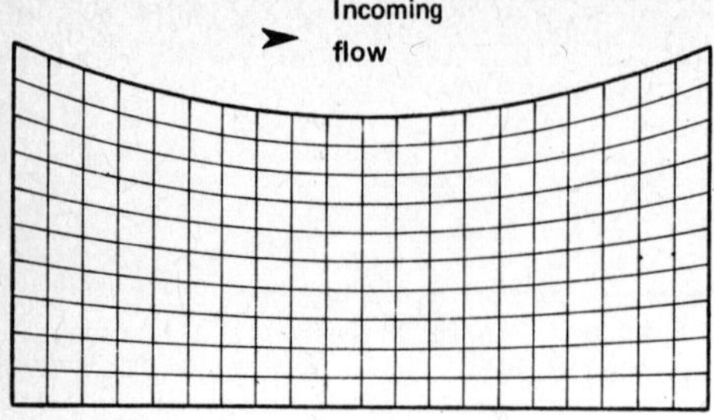


*Fig. 1 — Nozzle geometry and mesh.*

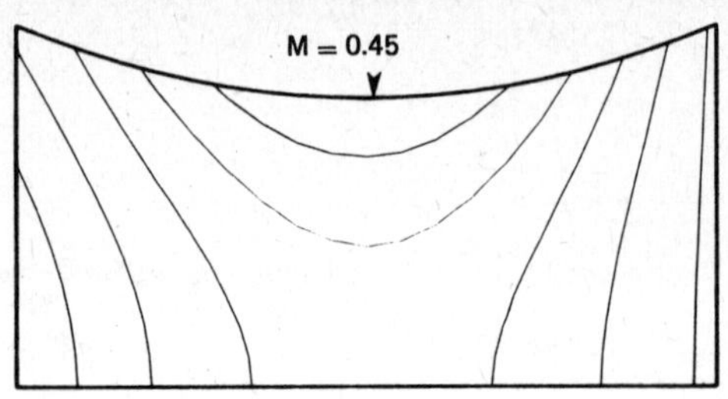


*Fig. 2 — Iso-Mach lines for an exit density = 1.5*
*$\Delta M = 0.025$*

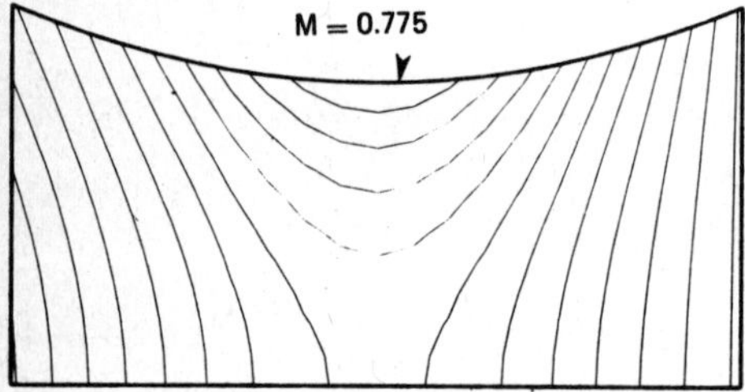


*Fig. 3 — Iso-Mach lines for an exit density = 1.4*
*$\Delta M = 0.025$*

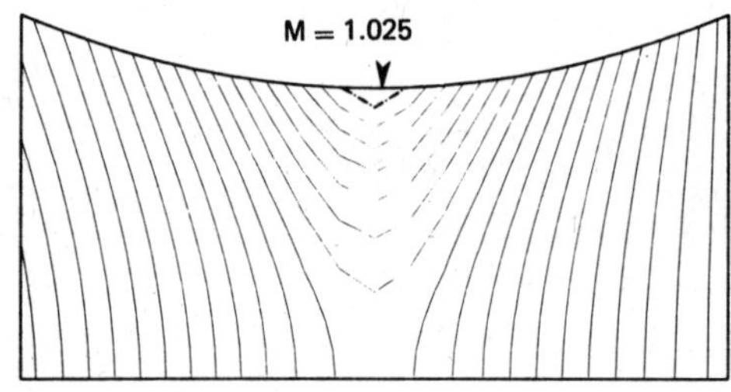


*Fig. 4 — Iso-Mach lines for an exit density  = 1.355*
*$\Delta M = 0.025$    — · — sonic line*

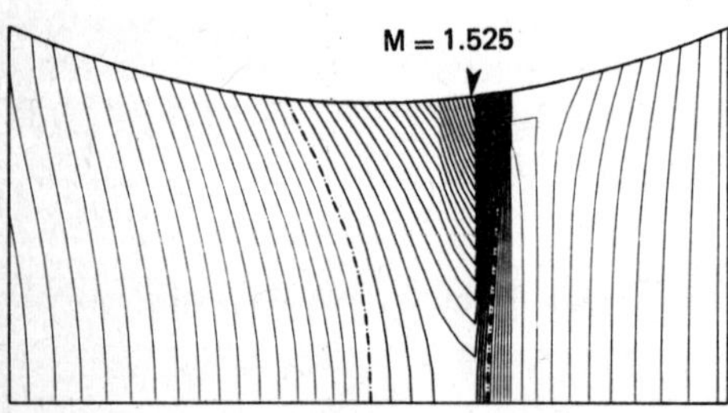


*Fig. 5 — Iso-Mach lines for an exit density = 1.35*
*$\Delta M = 0.025$    — · — sonic line*

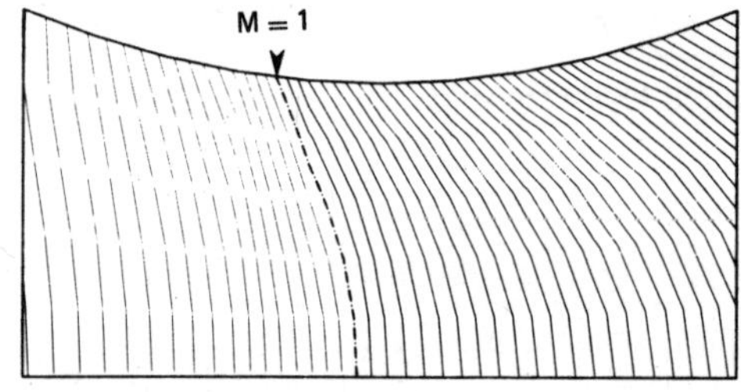


*Fig. 6 — Iso-Mach lines for the shock free choked flow*
*$\Delta M = 0.025$   — · — sonic line*

Improved Surface Velocity Method for Transonic
Finite-Volume Solutions

Hai-chow Chen
Boeing Commercial Airplane Company
Seattle, Washington, 98124 U.S.A.

## I.  SUMMARY

A new surface velocity method was developed which provides an improved prediction
of the flow field near the surface and the aerodynamic forces on the body. The original
surface velocity method produces strong flow normal to the wing and the fuselage. This
leads to pronounced error in the leading edge region, and low prediction of drag and
overprediction of the lift-to-drag ratio. Application of the method to two heavily used
transonic potential flow finite-volume programs, FLO27 and FLO28, are presented to
illustrate the improvement.

## II.  INTRODUCTION

The transonic finite-volume wing/body codes, FLO27 and FLO28, of Jameson and
Caughey (1977) and Caughey and Jameson (1979) have been widely used for
configuration development, as reviewed by Dillner and Koper (1979) and Hicks (1979).
Accurate treatment of the boundary conditions when solving the exact potential flow
equation has been studied by many investigators, e.g., Holst (1979) and a collection of
review papers by Wirz and Smolderen (1978). Similar treatment for computing the
surface velocity and reports of its significance are less noticeable.

The finite-volume method of Jameson and Caughey uses two types of interlocking
volume cells (Fig. 1). The primary cell is a trilinear isoparametric element and the
velocity potential is solved at its vertices. The faces of the secondary cells span the
midpoint of the primary cells and the flux is balanced in the secondary cell (or flux cell).
The flux cell adjacent to the boundary surface is reflected (Fig. 2) and the cells on both
sides of the boundary are assembled to produce a full flux cell. Normal to the boundary,
assuming a linear variation of the normal flux in the full cell, implies that the
impervious boundary condition is satisfied.

In the original surface velocity method of FLO27 and FLO28, in the off-body direction,
one-sided difference is used in the isoparametric mapping procedure. The resulting errors
can be severe near the wing leading edge where the flow varies greatly. The behavior of
the leading edge flow may be qualitatively described by an ideal flow around a corner
(Fig. 3) where the velocity (or flux) varies linearly in the direction normal and tangent to
the oncoming flow. The behavior of the flow is accurately represented in the finite-
volume flux formulation. This study improves the surface velocity calculation and uses a
consistent degree of approximation.

## III.  METHOD OF IMPROVEMENT

For a node point at the boundary, four neighboring cells are reflected and the resulting eight cells are assembled to produce a superelement (Fig. 4). In the special case of the element being regular, within the element the velocity is required to vary linearly and the potential quadratically in the directions normal and tangent to the boundary. An isoparametric triquadratic expansion is assumed within the element (Fig. 4).

$$f \;=\; \sum_i \sum_j \sum_k N_i M_j L_k f_{ijk} \tag{1}$$

$$N_1 \;=\; 0.5X(X\text{-}1) \quad N_2 \;=\; 1 - X^2 \quad N_3 \;=\; 0.5X(X\text{+}1)$$

$$M_1 \;=\; 0.5Y(Y\text{-}1) \quad M_2 \;=\; 1 - Y^2 \quad M_3 \;=\; 0.5Y(Y\text{+}1)$$

$$L_1 \;=\; 0.5Z(Z\text{-}1) \quad L_2 \;=\; 1 - Z^2 \quad L_3 \;=\; 0.5Z(Z\text{+}1)$$

where X, Y and Z are the local coordinates and repeated indices imply summation from one to three. Isoparametric mapping allows one to substitute f by the Cartesian coordinates x, y or z, or the velocity potential $\phi$. The X, Y and Z differencing of f can be computed by differentiating the polynomial expression in Eq. 1. The x, y and z derivatives of f can be computed by inverting Eq. 2

$$\begin{Bmatrix} f_X \\ f_Y \\ f_Z \end{Bmatrix} = J^T \begin{Bmatrix} f_x \\ f_y \\ f_z \end{Bmatrix} \qquad J = \begin{Bmatrix} x_X & x_Y & x_Z \\ y_X & y_Y & y_Z \\ z_X & z_Y & z_Z \end{Bmatrix} \tag{2}$$

Each element of the Jacobian matrix J is computed by differentiating the polynomial expressions given in Eq. 1, in which f is replaced by the corresponding variable, e.g., x. The superscript T denotes the transposition of a matrix.

In Fig. 4, corresponding to the centered node (where X=Y=Z=0), only seven nodes are involved in Eqs. 1 and 2, in which only one node is beneath the surface. To define this one node, in the physical plane, the straight line which is parametrically defined by X=Z=0 is linearly extended beneath the Y=0 surface so that the extended line is bisected by the Y=0 surface.

On the boundary surface, which is parametrically represented by Y=0 (Fig. 4), it can be shown that the impervious boundary condition of normal velocity component $q_n$ equal to zero can be expressed as

$$\begin{Bmatrix} \phi_X \\ \phi_Y \\ \phi_Z \end{Bmatrix}^T (J^T J)^{-1} \begin{Bmatrix} 0 \\ 1 \\ 0 \end{Bmatrix} = 0 \tag{3}$$

The superscript -1 denotes the inverse of a matrix and Eq. 3 can then be used to solve for $\phi_Y$. If a second boundary surface is locally represented by $Z=0$ (Fig. 5), then one has

$$\begin{Bmatrix} \phi_X \\ \phi_Y \\ \phi_Z \end{Bmatrix}^T (J^T J)^{-1} \begin{Bmatrix} 0 \\ 0 \\ 1 \end{Bmatrix} = 0 \tag{4}$$

Along a line of surface intersection, which is parametrically described by $Y=0$ and $Z=0$, Eqs. (3) and (4) are solved simultaneously to compute $\phi_Y$ and $\phi_Z$.

## IV.  RESULTS

The improved method has been tested on FLO 27 by analyzing a Boeing research wing without body (Fig. 6). Table I shows that the improved method increases the wing drag coefficient, $C_D$, due to the elimination of artificial suction near the leading edge. Figs. 6 and 7 compare the wing surface normal velocity component, $q_n$, and the body surface normal velocity component, using the original and the improved methods at the wing/body junction. These results show an inconsistency in the original surface velocity method in which the impervious boundary condition is not satisfied; this inconsistency is completely solved by the improved method. Figs. 8 and 9 are suction loop comparisons at the wing/body junction and the mid-semispan. The improvement in pressure coefficient $C_p$, shown in Fig. 8, is pronounced. The change in wing lift coefficient and pitching moment are insignificant, because the pressure coefficient errors are highly concentrated in the leading edge region and the errors on the lower and upper surfaces compensate.

## V.  CONCLUSION

Investigation of the two finite-volume programs indicates that the inconsistency in the surface velocity calculations leads to low prediction of drag. This inconsistency is completely solved by the improved surface velocity method. The improvement is pronounced for blunt-nosed configurations, especially when the mesh is coarse. Numerical experiments on typical transonic transport configurations substantiate this study. The improved method provides a more accurate prediction of drag and lift-to-drag ratio, and thus provides the configuration aerodynamicists a better tool for transonic aircraft analysis and design.

## VI.  ACKNOWLEDGEMENT

The author wishes to express his appreciation to Antony Jameson for his helpful discussion, and encouragement to the author to publish this study.

## VII. REFERENCES

Caughey, D. A. and Jameson, A., 1979; "Recent Progress in Finite-Volume Calculations for Wing-Fuselage Combinations", AIAA Paper 79-1513.

Jameson, A. and Caughey, D. A.,1977; "A Finite-Volume Method for Transonic Potential Flow Calculations", AIAA Paper 77-635.

Dillner, B. and Koper, C., 1979; "The Role of Computational Aerodynamics in Airplane Configuration Development", AGARD Flight Mechanics Panel Symposium, Munich, Germany.

Hicks, R., 1979; private communication, NASA-Ames.

Holst, T. L., 1979; "A Fast,Conservative Algorithm for Solving the Transonic Full-Potential Equation", AIAA Paper 79-1456.

Wirz, H. J. and Smolderen, J. J., eds. 1978; Numerical Methods in Fluid Dynamics, Hemisphere Publishing Corporation.

Table I

Comparison of Wing $C_D$

| Mesh | Original | Improved | Increment |
| --- | --- | --- | --- |
| Coarse   (1280 cells) | 0.0093 | 0.0166 | 0.0073 |
| Medium   (10240 cells) | 0.0126 | 0.0154 | 0.0028 |
| Fine   (81920 cells) | 0.0155 | 0.0165 | 0.0010 |

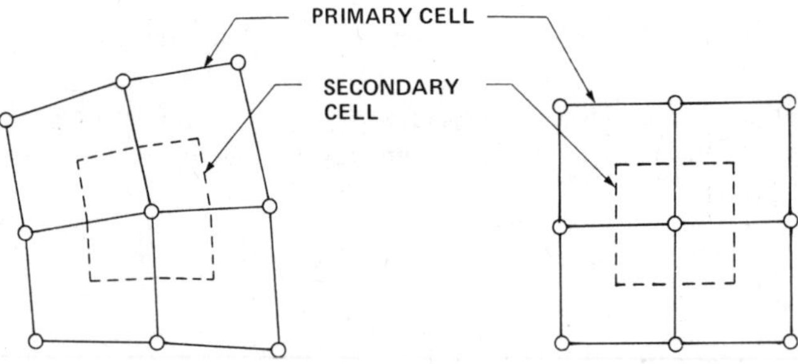


Figure 1. — Primary and Secondary Cells

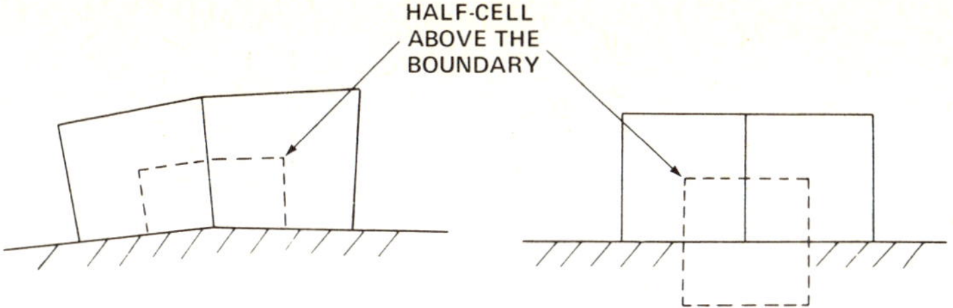



Figure 2. — Boundary Cell

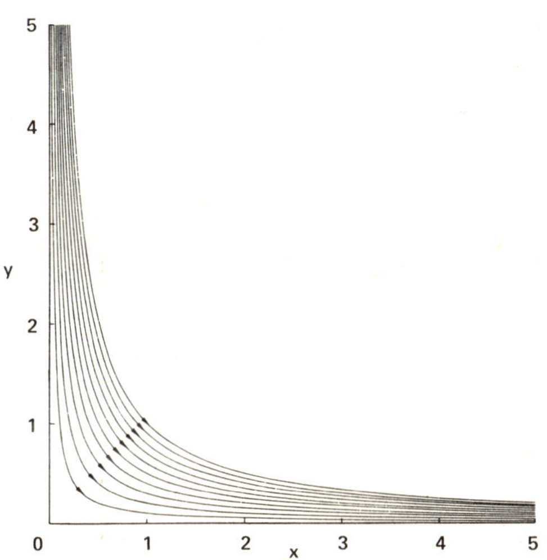



Figure 3. — Ideal Flow Around a Corner

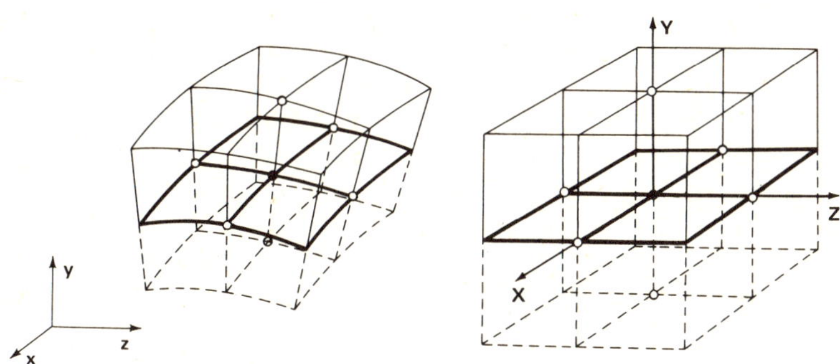



Figure 4. — Three-Dimensional Superelement

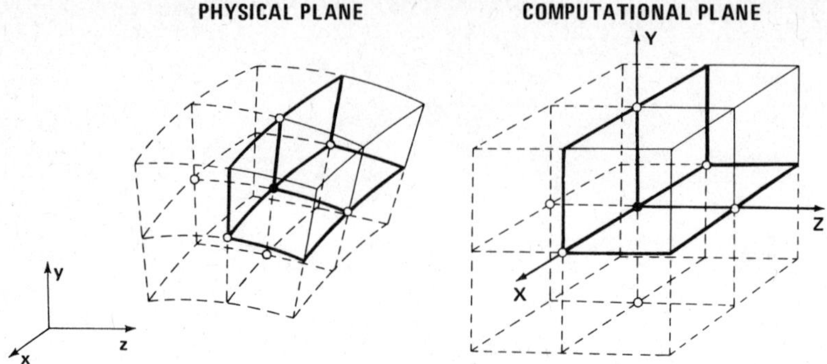



Figure 5. — Superelement at Wing/Body Junction

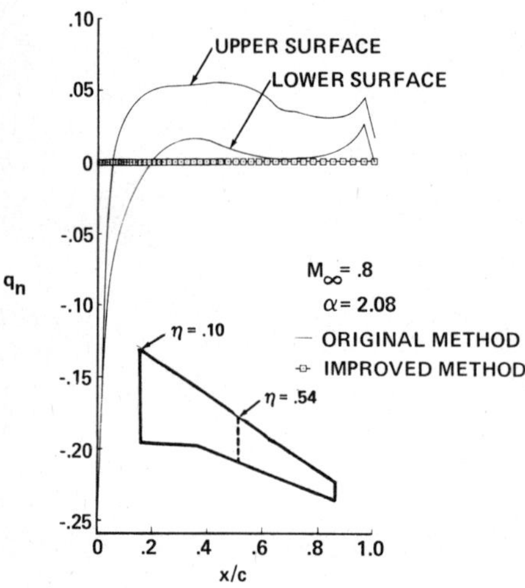



Figure 6. — Wing Surface Normal Velocity Error at $\eta = .10$

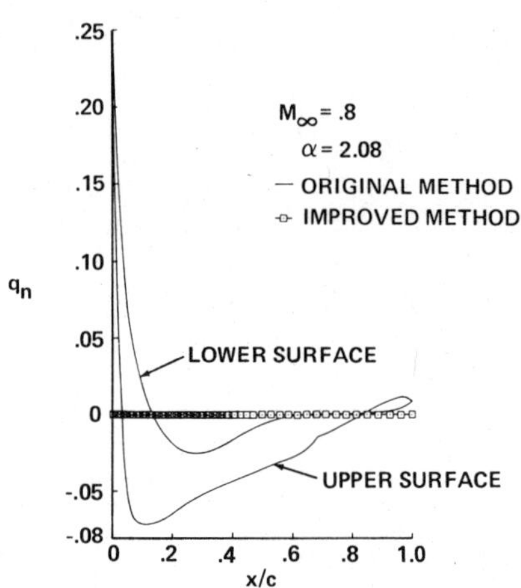



Figure 7. — Body Surface Normal Velocity Error at $\eta = .10$

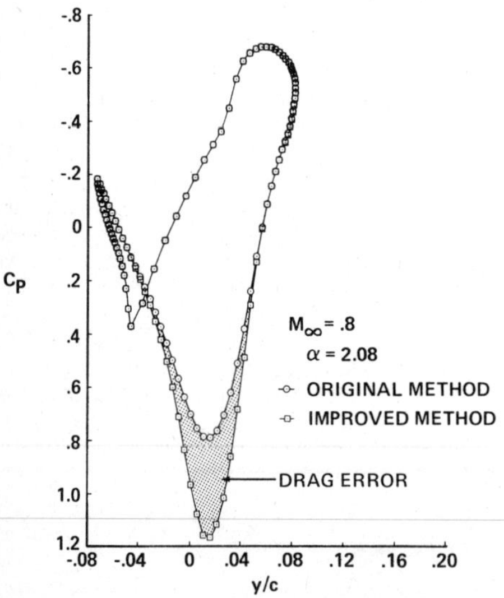



Figure 8. — Suction Loop Comparison at $\eta = .10$

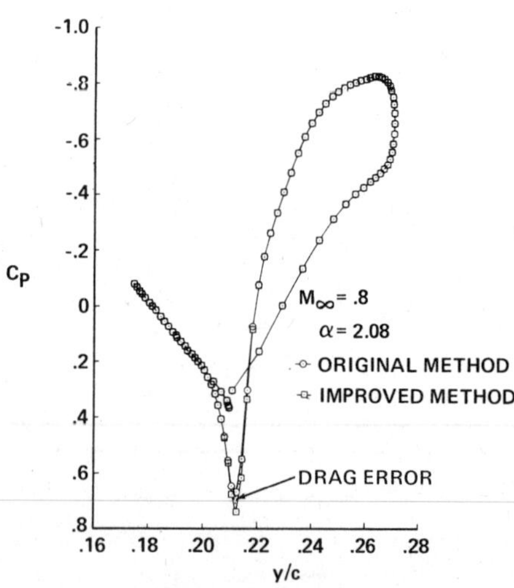



Figure 9. — Suction Loop Comparison at $\eta = .54$

NUMERICAL SIMULATION OF REACTIVE FLOW IN

INTERNAL COMBUSTION ENGINES

L. D. Cloutman, J. K. Dukowicz and J. D. Ramshaw
Theoretical Division, Group T-3
Los Alamos Scientific Laboratory
University of California
Los Alamos, NM  87545


## I.  INTRODUCTION

Multidimensional numerical simulations of the reactive fluid flow in an internal combustion engine cylinder are useful in helping engine designers obtain insight into the physical mechanisms governing efficiency and pollutant formation. For background information and references to other work in this area, the reader is referred to the proceedings of a recent symposium (Mattavi and Amann 1980).  This paper describes a comprehensive numerical model for internal combustion engine cylinder simulations that has been developed at Los Alamos.  The model is currently embodied in a two-dimensional (axisymmetric) computer code called CONCHAS-SPRAY. Work is in progress on a three-dimensional code with the same features.  A detailed discussion of all aspects of the model would be inappropriate here.  We therefore emphasize selected topics of particular interest.  More detailed discussions of some of these topics are available elsewhere (Butler et al. 1979, Dukowicz 1979, Dukowicz 1980, Ramshaw and Cloutman 1980).


## II.  FINITE-DIFFERENCE PROCEDURE

The fluid dynamical part of our model is a time-marching finite-difference procedure that is based on the combined use of the ICE (implicit continuous-fluid Eulerian) and ALE (arbitrary Lagrangian-Eulerian) methods.  The ICE method (Harlow and Amsden 1971) is a partially implicit temporal differencing procedure in which the pressure gradient in the momentum equation and the divergence term in the continuity equation are simultaneously evaluated at the advanced time level.  This procedure removes the Courant sound-speed stability restriction on the time step, thereby allowing the efficient computation of flow at low Mach number.  The ALE method (Trulio 1966; Hirt, Amsden, and Cook 1974) is a spatial differencing procedure that utilizes a mesh composed of arbitrary quadrilaterals whose corner locations may be arbitrarily specified as functions of time.  This method allows the convenient representation of curved and/or moving boundary surfaces, such as the moving piston.

The implicit part of the ICE scheme obtains the advanced time pressure and velocity fields by an iterative procedure similar to the method of successive overrelaxation.  Such methods are notoriously inefficient in effecting a change in the overall pressure level, because of the intrinsically slow relaxation of long wavelength errors.  (The relaxation rate goes to zero as the square of the wave number.) This is a serious drawback in the present context, where substantial changes in the pressure level result both from the essentially adiabatic compression caused by the piston motion, and from the heat release that accompanies the combustion of the fuel. It was therefore necessary to develop a special technique to accelerate the iteration procedure.

The basic idea of the iteration acceleration technique is to estimate the change $\Delta p$ in the overall pressure level that will occur on the time step in question, and to add that change into the entire pressure field at the beginning of the iteration procedure.  In the present scheme, the advancement of the dependent variables through a time step is computed in three stages.  The first stage is an explicit Lagrangian calculation, during which the heat release due to chemistry and/or phase change occurs.  The second stage is the pressure iteration.  The third stage is the rezone calculation, during which the mesh points are moved to their final locations and the associated transport is computed.  Since the pressure iteration is Lagrangian and isentropic, overall pressure levels are obtained by a Lagrangian isentropic  pressure equilibration within the constraint of the available total volume.  We thereby obtain the formula

$$\bar{p} = \left[ \frac{1}{V} \sum_{ij} \left( p_{ij}^0 \right)^{1/\gamma} v_{ij}^0 \right]^\gamma \tag{1}$$

for the pressure level $\bar{p}$, where $\gamma$ is the specific heat ratio, $p_{ij}^0$ and $v_{ij}^0$ are the pressure and volume of cell ij prior to equilibration, and V is the total volume in which the equilibration occurs.  This formula neglects spatial variations in $\gamma$, which is a reasonable approximation that was necessary to obtain $\bar{p}$ in closed form. The pressure change $\Delta p$ is estimated by $\bar{p}^I - \bar{p}^n$, where superscripts n and I refer respectively to the previous time level and to the first stage of the current time step.

Another problem of efficiency arises in connection with the finite-difference cells in the "squish region" between the piston face and the cylinder head.  As the piston approaches top dead center, these cells become very small and the explicit diffusional stability limit on the time step becomes very restrictive.  This problem is avoided by means of a "chopper" algorithm which automatically combines smaller cells into larger ones once a certain mimimum cell size is reached.  The mass, momentum, and energy of the original cells are apportioned among the new cells in a conservative manner.  As the piston recedes from top dead center, the inverse procedure is used to restore the original cells again.

### III.  BOUNDARY LAYER TREATMENT

Under most circumstances, the velocity and temperature boundary layers in an engine cylinder will be too thin to be explicitly resolved in a practical computing mesh.  Since these boundary layers will ordinarily be turbulent, this difficulty may be circumvented by matching to the turbulent law of the wall (e.g., Schlichting 1968), which may be written in the form

$$\frac{u}{u_*} = \frac{1}{K} \ln \left( \frac{y u_*}{\nu} \right) + B \quad . \tag{2}$$

Here u is the magnitude of the fluid velocity at a perpendicular distance y from the wall, $\nu$ is the kinematic viscosity, $K \cong 0.41$ is the Karman constant, B is a constant that depends on wall roughness (for a smooth wall, B = 5.5), and $u_*$ is the shear speed, which is related to the wall shear stress $\tau$ and the density $\rho$ by $\tau = \rho u_*^2$.  To apply Eq. (2) in a finite-difference calculation, one assumes that the first mesh point away from the wall lies in the law-of-the-wall region.  The values of u and y at this point are then substituted into Eq. (2); this determines $u_*$ and hence $\tau$. The value of $\tau$ thus determined is then imposed as a boundary condition.

Unfortunately, Eq. (2) is a transcendental equation for $u_*$, and this is not very convenient.  We therefore use an approximation to Eq. (2) obtained by replacing $u_*$ in the argument of the logarithm with its 1/7-law value.  For a smooth wall, this yields

$$\frac{u}{u_*} = 2.19 \ln \left( \frac{y u}{\nu} \right) + 0.75 \quad , \tag{3}$$

which is almost as accurate as Eq. (2) and can be solved explicitly for $u_*$.

The thermal boundary layer is handled in a similar way using the modified Reynolds analogy formula

$$q = \frac{1}{Pr_t} \left(\frac{\tau}{u}\right) C_p (T - T_w) \quad . \tag{4}$$

Here q is the heat flux to the wall, $C_p$ is the specific heat at constant pressure, $T_w$ is the wall temperature, T is the fluid temperature at the location of u, and $Pr_t \cong 0.89$ is the turbulent Prandtl number. Equation (4) may be obtained in a variety of ways; here we merely note that it results from appropriate simplification of a formula used by Launder and Spalding (1974).

## IV.  CHEMICAL REACTIONS

A ubiquitous problem in reactive fluid dynamics is the presence of chemical time scales that are very short in comparison to typical fluid dynamical characteristic times.  The temporal resolution of these short time scales would require the use of impractically small time steps.  This problem may be circumvented by assuming that the fast reactions are always in equilibrium, while the slower reactions proceed kinetically (Ramshaw 1980).  In the present scheme, the kinetic reactions are treated by a linearly implicit procedure used in earlier reactive fluid dynamics codes (Rivard, Farmer, and Butler 1975, Ramshaw and Dukowicz 1979), while the equilibrium reactions are treated by a new quadratic iterative procedure (Ramshaw and Cloutman 1980).  The latter procedure enforces the equilibrium constraints on the species concentrations by computing a sequence of progress increments in the equilibrium reactions.  Each such increment is based on the instantaneous deviation of the corresponding reaction from equilibrium, and attempts to nullify this deviation. The procedure converges rapidly because the deviations from equilibrium are of the order of the time step, except perhaps at the beginning of the calculation.

## V.  PARTICLE-FLUID MODEL FOR EVAPORATING FUEL SPRAYS

Fuel-injected internal combustion engines typically employ injectors producing finely atomized sprays with complex distribution patterns and time-varying injection rates.  Such strongly interacting, transient sprays are difficult to model using conventional techniques.  The difficulties stem from the fact that the spray contains a wide range of droplet sizes, length scales, and time scales.

Our approach is based on a Monte Carlo statistical model.  The injector distributions in droplet size, velocity, etc., are statistically sampled and these discrete droplets are followed along Lagrangian trajectories as they interact with the surrounding gas.  Each computational droplet represents a number of similar physical droplets.  The fundamental advantage of this approach is that all relevant time and spatial scales can be resolved in a computationally efficient manner, since the number of particles required for satisfactory accuracy is not excessive.

The following description will be limited to the essential features of the method.  Additional details may be found elsewhere (Dukowicz 1979, 1980).  The droplets interact with the gas by interchange of mass, momentum, and energy.  Volume displacement effects are neglected for simplicity.  We have also neglected the droplet-droplet interactions which would occur in a thick spray.  In spite of the assumption of thin sprays (void fraction $\cong 1$), the spray may contain a large amount of mass and momentum compared to the gas, because of the large ratio of liquid to gas density.  Thus the spray can strongly affect the gas flow, and vice versa. Further, because of the rapid rates of exchange of momentum and energy of small droplets, the coupling with the gas must be implicit.  The required iteration for momentum coupling can fortunately be incorporated into the ICE pressure iteration so that essentially no additional computation is required.

An important property of sprays is the turbulent diffusion of droplets.  This is modeled by a random choice method in which turbulent velocity components are added to the gas velocities, which in turn produce turbulent droplet velocities.  This requires the turbulence intensity and length scales (or time scales) to be specified.

Phase change (primarily evaporation) is based on a quasi-steady evaporation model. The evolution of each droplet's temperature and radius is followed in time. At the present time, droplet internal motions and temperature distributions are ignored. Droplets affect each other during evaporation by modifying the ambient vapor density and gas temperature. This is modeled by evaporating droplets sequentially, assuming an isentropic, constant pressure process in each cell. The sequence of droplet evaporation is shuffled randomly on each time step. The resulting mass and internal energy changes are then added to each cell prior to the isentropic pressure iteration.


## VI.  SAMPLE CALCULATION

The sample calculation is a typical simulation of a direct injection stratified charge (DISC) engine. On the intake stroke, pure air is drawn into the cylinder through a shrouded valve, causing the air to swirl about the center line of the cylinder. At some point during the compression stroke, liquid fuel (n-octane in this example) is sprayed into the cylinder, followed by ignition near top dead center. Conventional power and exhaust strokes follow. In our example, the bore is 9.844 cm, the stroke is 9.55 cm, the compression ratio is 10:1, the swirl ratio is 4:1, the engine speed is 1600 rpm, and the cup in the piston has a diameter of 4.92 cm and a depth of 3.34 cm. The calculation is started at the beginning of the compression stroke with the air at 0.097 MPa and 308 K, corresponding to a highly loaded engine. Because the model is axisymmetric, the fuel injector and spark plug are located at the same point in the center of the cylinder head.

Typical computational results are shown in Figures 1 and 2. At 15.4 ms after bottom dead center (ABDC) on the compression stroke, liquid n-octane is injected into the compressed air. The left panel of Figure 1 shows the gas velocity at 16.65 ms. The incoming droplets entrain the air and create a downward jet along the symmetry axis. Air is also being pushed out of the squish region to the right by the upward motion of the piston. The dotted lines are contours of three different values of the swirl velocity. Air entrained by the entering droplets is pulled toward the axis, creating a region of high swirl, denoted by H, near the injector orfice. Radial transport of angular momentum in the squish region plus boundary layer drag produce a local high swirl rate at the entrance to the squish region, with a lower swirl rate inside.

The right hand panel of Figure 1 shows the droplets in the spray and contours of equivalence ratio. The central contour is the location of a stoichiometric mixture, and the left and right contours correspond approximately to the rich and lean flammability limits, respectively. Most of the droplets are small and confined to the core of the spray, where their evaporation contributes to fuel rich conditions. The droplets outside the contours are a few large ones (up to 30 μm) that couple loosely to the gas.

At 16.9 ms injection ends. Ignition at the injector orifice is at 17.1 ms. Figure 2 shows the solution at 19.11 ms. The left hand panel shows isotherms (solid lines) and mass fraction of NO (dashed lines). The band of high temperature shows where the flame has passed as it traveled down the flammable region at the edge of the spray. At this time, the flame has not reached the tip of the spray. The dashed lines show mass fractions of NO, with the highest value near the H. At all times, the regions that have been hottest for the longest contain the most NO.

Some of the droplets persist, as shown in the right hand panel of Figure 2. The droplets away from the axis evaporate quickly as the flame approaches. The particles near the axis evaporate very slowly because of the low temperature and high fuel vapor concentration. The ragged character of the equivalence ratio contours is an artifact of having tiny oxygen and fuel concentrations along the upper two thirds of the mesh. These contours show the sharp separation between the oxygen and fuel. The temperature is high in this region indicating the presence of a slowly burning diffusion flame.

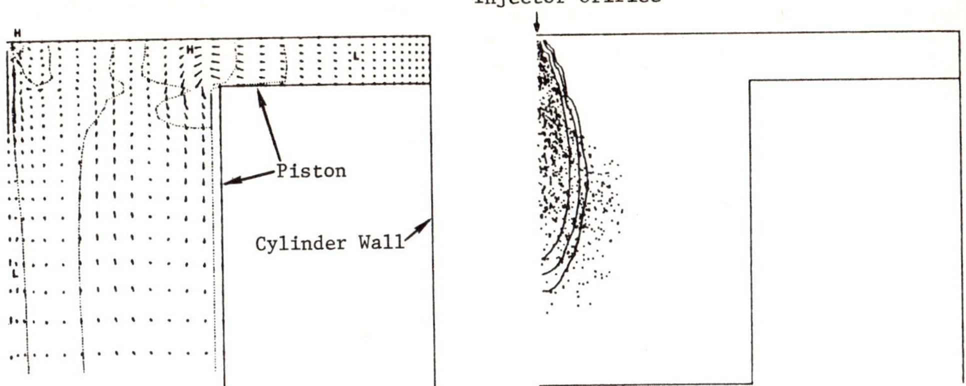



Fig. 1.  The solution at 16.65 ms ABDC.  The top of each panel is the cylinder head, and the left hand side is the symmetry axis.  Maximum speed in the velocity vector plot is 4338 cm s$^{-1}$.  The swirl velocity contours (dotted lines) have values of 1635, 3563, and 5492 cm s$^{-1}$, and the maximum value is 6777 cm s$^{-1}$.  In the spray plot, the contours of equivalence ratio have values of 2, 1, and ½ from left to right.

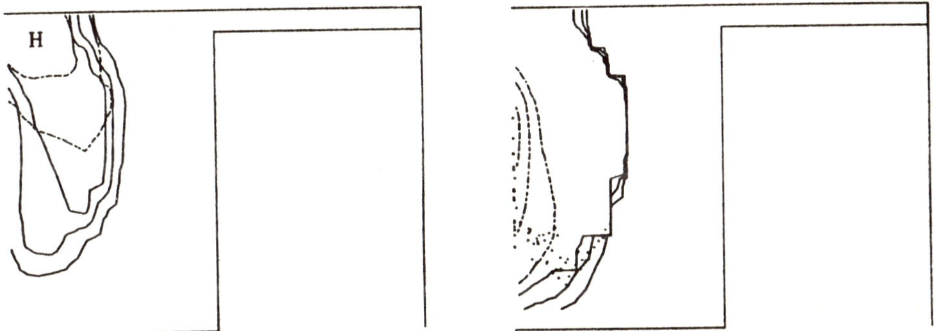



Fig. 2.  The solution at 19.11 ms ABDC.  The left panel shows isotherms (solid lines with values of 1000, 1500, and 2000 K and NO mass fraction contours (dashed lines) with values of $1 \times 10^{-5}$ and $2 \times 10^{-5}$.  The temperature reaches a maximum of 2316 K, and the NO mass fraction reaches a maximum of $2.6 \times 10^{-5}$ at the H.  In the right hand panel, a few droplets persist. The solid lines are are equivalence ratio contours with values of 2, 1, and ½ from left to right.  The dashed lines are contours of fuel mass fraction with values of 0.3, 0.2, and 0.1 from left to right.

REFERENCES

Butler, T. D., Cloutman, L. D., Dukowicz, J. K., and Ramshaw, J. D. 1979, "CONCHAS: An Arbitrary Lagrangian-Eulerian Computer Code for Multicomponent Chemically Reactive Fluid Flow at All Speeds," Los Alamos Scientific Laboratory report LA-8129-MS.

Dukowicz, J. K. 1979, "Quasi-Steady Droplet Phase Change in the Presence of Convection," Los Alamos Scientific Laboratory report LA-7997-MS.

Dukowicz, J. K. 1980, J. Comp. Phys. 35, 229.

Harlow, F. H. and Amsden, A. A. 1971, J. Comp. Phys. 8, 197.

Hirt, C. W., Amsden, A. A., and Cook, J. L. 1974, J. Comp. Phys. 14, 227.

Launder, B. E. and Spalding, D. B. 1974, Comp. Meths. Appl. Mech. Eng. 3, 269.

Mattavi, J. N. and Amann, C. A., eds. 1980, Combustion Modeling in Reciprocating Engines, Phenum Press, New York.

Ramshaw, J. D. and Dukowicz, J. K. 1979, "APACHE: A Generalized-Mesh Eulerian Computer Code for Multicomponent Chemically Reactive Fluid Flow," Los Alamos Scientific Laboratory report LA-7427.

Ramshaw, J. D. 1980, Phys. Fluids 23, 675.

Ramshaw, J. D. and Cloutman, L. D. 1980, "Numerical Method for Partial Equilibrium Flow," J. Comp. Phys., in press.

Rivard, W. C., Farmer, O. A., and Butler, T. D. 1975, "RICE: A Computer Program for Multicomponent Chemically Reactive Flows at All Speeds," Los Alamos Scientific Laboratory report LA-5812.

Schlichting, H. 1968, Boundary-Layer Theory, 6th ed., McGraw-Hill, New York.

Trulio, J. G. 1966, "Theory and Structure of the AFTON Codes," Air Force Weapons Laboratory report AFWL-TR-66-19.

Mixing Layer Simulation by an Improved
Three-Dimensional Vortex-in-Cell Algorithm

by B. Couët* and A. Leonard**
*Institute for Plasma Research, Stanford University, Stanford, CA 94305
**NASA Ames Research Center, Moffett Field, CA 94035

Introduction.

In this paper, we present an extension of the three-dimensional vortex-in-cell (VIC) method that allows for the simulation of incompressible turbulent flows such as plane mixing layers and wakes with exact boundary conditions in the normal direction. Here the vorticity is assumed to be confined between two parallel planes and the turbulence is assumed to be homogeneous in the two directions parallel to the planes.

As in previous work by Couët, Buneman and Leonard (1980), the vorticity field is represented by a set of vortex filaments which move under the influence of the velocity field that these filaments create. As a consequence, sharp gradients of vorticity can be tracked accurately and not diffused by numerical effects. To treat a large number of filaments at a reasonable computer cost, the velocity field is not calculated directly by the Biot-Savart law of interaction, like Leonard (1980), but by creating a mesh record of the vorticity field using best fit quadratic spline interpolation then integrating a Poisson's equation to generate a mesh record of the velocity field. The VIC method has been tested by Couët (1979) on fully periodic flows and no undesirable grid effects or numerical instabilities were found. Excellent comparisons were achieved with a spectral method calculation of the inviscid Taylor-Green problem and with a pure Lagrangian computation of the propagation of a periodic system of vortex rings. In the present work, Fourier transformation, imposing periodic boundary conditions, is used only in the two homogeneous directions  x and  z  leading to a system of uncoupled second order ODE's in the normal direction y  for each Fourier component of the velocity field. As discussed in the next section, discretization of each ODE produces an infinite pentadiagonal system which is then transformed into an equivalent finite pentadiagonal system. As demonstrated below, the solution is invariant (within roundoff) to translations in the  y-direction as long as the vorticity remains within the computational domain.

As an application, we study the evolution of the plane mixing layer subjected to random two- and three-dimensional initial disturbances. In a related numerical study of the physics of the forced mixing layer, Riley and Metcalfe (1980) used pseudospectral methods and free-slip boundary conditions in the normal direction.


Computational Description.

We assume an unbounded incompressible flow where the vorticity field, $\vec{\omega} = \nabla \times \vec{u}$, consists of a collection of vortex filaments with gaussian cross-section. This field may be represented as follows:

$$\vec{\omega}(\vec{r}) = \sum_i \Gamma_i \oint G(\vec{r}-\vec{r}_i(\xi)) \frac{\partial \vec{r}_i}{\partial \xi} d\xi$$

where $\xi$ is a parameter which traces each filament along its length at any instant in time, $\vec{r}_i(\xi)$ are the space curves describing the filaments with circulations $\Gamma_i$ and $G$ indicates the gaussian profile of the filaments. The summation is over individual vortex filaments. The governing dynamic equation for the vorticity in these filaments is

$$\frac{D\vec{\omega}}{Dt} = \vec{\omega}\cdot\nabla\vec{u} + \nu\nabla^2\vec{\omega}$$

where $\nu$ is the kinematic viscosity and the velocity field is determined kinematically from

$$\nabla^2\vec{u} = -\nabla\times\vec{\omega} \tag{1}$$

Note that the computed scales of motion are assumed to be essentially inviscid. Any viscous or subgrid scale dissipation effects are modeled through the filter function $G$ .

By using FFT's in the streamwise x-direction and the spanwise z-direction, we automatically impose periodic boundary conditions in these directions. Eq. (1) becomes

$$\frac{\partial^2 \vec{u}}{\partial y^2}(k_x,y,k_z) - c\vec{u}(k_x,y,k_z) = - \vec{L\omega}(k_x,y,k_z) \tag{2}$$

where $c \equiv k_x^2 + k_z^2$ and $\vec{L}$ is the linear operator defined by

$$\vec{L\omega} = \left(\frac{\partial\omega_z}{\partial y} - ik_z\,\omega_y\right)\vec{e}_x + \left(ik_z\,\omega_x - ik_x\,\omega_z\right)\vec{e}_y + \left(ik_x\,\omega_y - \frac{\partial\omega_x}{\partial y}\right)\vec{e}_z . \tag{3}$$

$\vec{e}_x, \vec{e}_y, \vec{e}_z$ are the unit vectors and $i \equiv \sqrt{-1}$ . Note also that, from now on, $\vec{u}$ and $\vec{\omega}$ stand for the bi-transformed fields in x and z . Through a variational principle derivation, it is easy to show that Eq. (2) is the necessary condition for the integral.

$$I(\vec{u}) = \int \frac{1}{2}\left[\left(\frac{\partial\vec{u}}{\partial y}\right)^2 + c\vec{u}^2 - 2\vec{u}\,\vec{L}\,\vec{\omega}\right] dy \tag{4}$$

to be a minimum. A finite element approximation of the solution $\vec{u}$ in the y-direction can be obtained by using the piecewise quadratic spline functions that interpolate the fields from an onto the mesh. The approximate vorticity and velocity fields are then of the form

$$\vec{\omega}(k_x,y,k_z) = \sum_{\ell=-\infty}^{\infty} \vec{\omega}_\ell(k_x,k_z)\,\phi_\ell(y) \quad \text{and}$$

$$\vec{u}(k_x,y,k_z) = \sum_{\ell=-\infty}^{\infty} \vec{u}_\ell(k_x,k_z)\,\phi_\ell(y) ,$$

where $\phi_\ell$ are the basis functions and $\vec{\omega}_\ell$ and $\vec{u}_\ell$ are the spline coefficients obtained at the mesh points $y_\ell$ . Considering, for example, the x-component of the fields, the stationary value to the functional $I$ is given by

$$\frac{\partial}{\partial u_j} I\left(\sum_{\ell=-\infty}^{\infty} u_\ell\,\phi_\ell(y)\right) = 0 , \quad -\infty < j < \infty ,$$

leading to

$$\sum_{\ell=-\infty}^{\infty} u_\ell \int \left[ \left(\frac{\partial\phi_\ell}{\partial y}\right)\left(\frac{\partial\phi_j}{\partial y}\right) + c\,\phi_\ell\,\phi_j \right] dy =$$

$$\sum_{\ell=-\infty}^{\infty} \left[ \omega_{z\ell} \int \phi_j \left(\frac{\partial\phi_\ell}{\partial y}\right) dy - ik_z\,\omega_{y\ell} \int \phi_\ell\,\phi_j\,dy \right] .$$

Straightforward evaluation of the integrals yields the equation

$$\alpha(u_{j+2} + u_{j-2}) + \beta(u_{j+1} + u_{j-1}) + \gamma\,u_j =$$

$$\frac{5}{12}(\omega_{zj+1} - \omega_{zj-1}) + \frac{1}{24}(\omega_{zj+2} - \omega_{zj-2})$$

$$- \frac{ik_z}{120}\left[ 66\,\omega_{yj} + 26(\omega_{yj+1} + \omega_{yj-1}) + \omega_{yj+2} + \omega_{yj-2} \right] ; \quad -\infty < j < \infty ,$$

where $\alpha = \dfrac{c}{120} - \dfrac{1}{6}$ , $\beta = \dfrac{26c}{120} - \dfrac{1}{3}$ , $\gamma = \dfrac{66c}{120} + 1$ . $\hfill (5)$

Similar equations for the two other components can be obtained using (3). The mesh spacing is 1 .

For practical purposes, one needs to solve the above pentadiagonal linear systems in a finite domain, thus requiring proper boundary treatment. In our case, with a 32×33×32 mesh and the use of FFT's in the x- and z-direction, we assume no vorticity outside our computational domain in the y-direction. Therefore the boundary conditions necessary to solve the finite pentadiagonal systems, given by Eq. (5) with $|j| \leq 16$ , are obtained by considering the homogeneous systems of equations:

$$\alpha(u_{j+2} + u_{j-2}) + \beta(u_{j+1} + u_{j-1}) + \gamma\,u_j = 0 , \quad \text{for } |j| > 16 .$$

These equations admit. Z-transform solutions of the form $\rho^j$ . The $\rho$'s are functions of $\alpha, \beta$ and $\gamma$ . Only the exponentially decaying solutions are kept to satisfy solution at infinity. Subsequent reduction of the infinite system to the finite computational system of equations results in the modification of the four upper and four lower terms of the pentadiagonal matrix with symmetry preserved. The singular case where $c = 0$ $(k_x=k_z=0)$ is treated by solving a reduced system and then adding a component of the homogeneous solution to obtain the desired value of $u$ at $y \to \pm\infty$ . Solvability of the singular case is assured by the numerical equivalent of

$$\int_R \frac{\partial\omega_x}{\partial y}\,dy = \int_R \frac{\partial\omega_z}{\partial y}\,dy = 0 , \quad \text{where } R \text{ is the computational range in } y .$$

To investigate the truncation error due to the mesh and test the non-periodic condition in the y-direction, we computed the velocity field due to an infinite row of equidistant line vortices, each of strength $\Gamma$ , whose (x,y)-coordinates are (0,0), (±a,0), (±2a,0), .... . The exact solution for this system of singular vortices

is given by Lamb (1945): 
$$u = \frac{-\Gamma}{2a} \frac{\sinh(2\pi y/a)}{\cosh(2\pi y/a) - \cos(2\pi x/a)}$$

$$v = \frac{\Gamma}{2a} \frac{\sin(2\pi x/a)}{\cosh(2\pi y/a) - \cos(2\pi x/a)} \tag{6}$$

In our calculation, we need to set a single filament at $(0,0)$: periodic conditions in $x$ and $z$ provide us with the infinite row of equidistant vortices infinite in $z$ . Thus $a = 32$; we set $\Gamma = 2$ . Figures 1a and 1b respectively display $u$ and $v$ in the fourth quadrant of the $(x-y)$-plane. The full lines indicate the exact solution given by (6). Close to the boundary of the computational domain, at $y = -15.5$ , the maximum error between theory and calculations is $0.00089\%$ for $u$ and $0.015\%$ for $v$ . Closer to the filaments, the error increases due to the fact that the mesh does not resolve the high wavenumber components of the exact solution for the singular filaments. To test the translation invariance of the numerical solution, the initial vortex was moved along the $y$-axis up to $y = 13$ . The results were exactly the same within roundoff. When we place the vortex at $y = 14$ , some discrepancies occur due to the fact that vorticity is spread outside the computational domain by the spline functions.

In our simulation of the mixing layer, the initial vorticity field consists of a uniform layer of vorticity infinite in the spanwise direction $z$ . The layer, extending from $y = -1$ to $y = 1$ , is made of 256 discrete line vortices, each vortex resolved into 64 nodes. The velocity difference across the layer is $\Delta u = 16$ and the initial momentum thickness is $\theta = 0.71$ . Fourteen seconds for each time step are necessary to process these 16384 vortex nodes on a CDC 7600 ($\Delta t = .03$) . The initial perturbation in the layer was generated by adding a random component to the $y$-component of the vortex nodes.

As a first case, we investigate a streamwise $(x)$ perturbation of the filaments identical for all $z$ . This purely two-dimensional problem has no spanwise $(z)$ dependence for all times. Only the $z$-component of vorticity is non-zero. Plots of the vorticity contours (Figures 2a,b,c) show the roll-up into two vortices as observed in many laboratory experiments. Plotting the $y$-velocity component at a later time in Figure 3 illustrates our good treatment of the $y$-boundary compared to free-slip which sets $v = 0$ there.

A full three-dimensional perturbation was set up next. Figure 4 is a non-dimensionalized plot of the mean velocity versus the momentum thickness. As similarity theory predicts, the data collapse very well. The same result was observed in the 2-D case. In Figure 5, as found by Browand and Weidman (1976), we notice that the vertical fluctuation is always larger than the longitudinal one. In the last group of figures (6a,b,c,d,e,f), two stations in time are shown. For each time, an $(x-y)$-slab of particles displays the roll-up of the large eddies while the small scales, partly due to stretching of the vortices, create very intense streamwise vorticity seen in the $(y-z)$ contour plots and in the motion of an $(x-z)$ plane of particles in the layer.

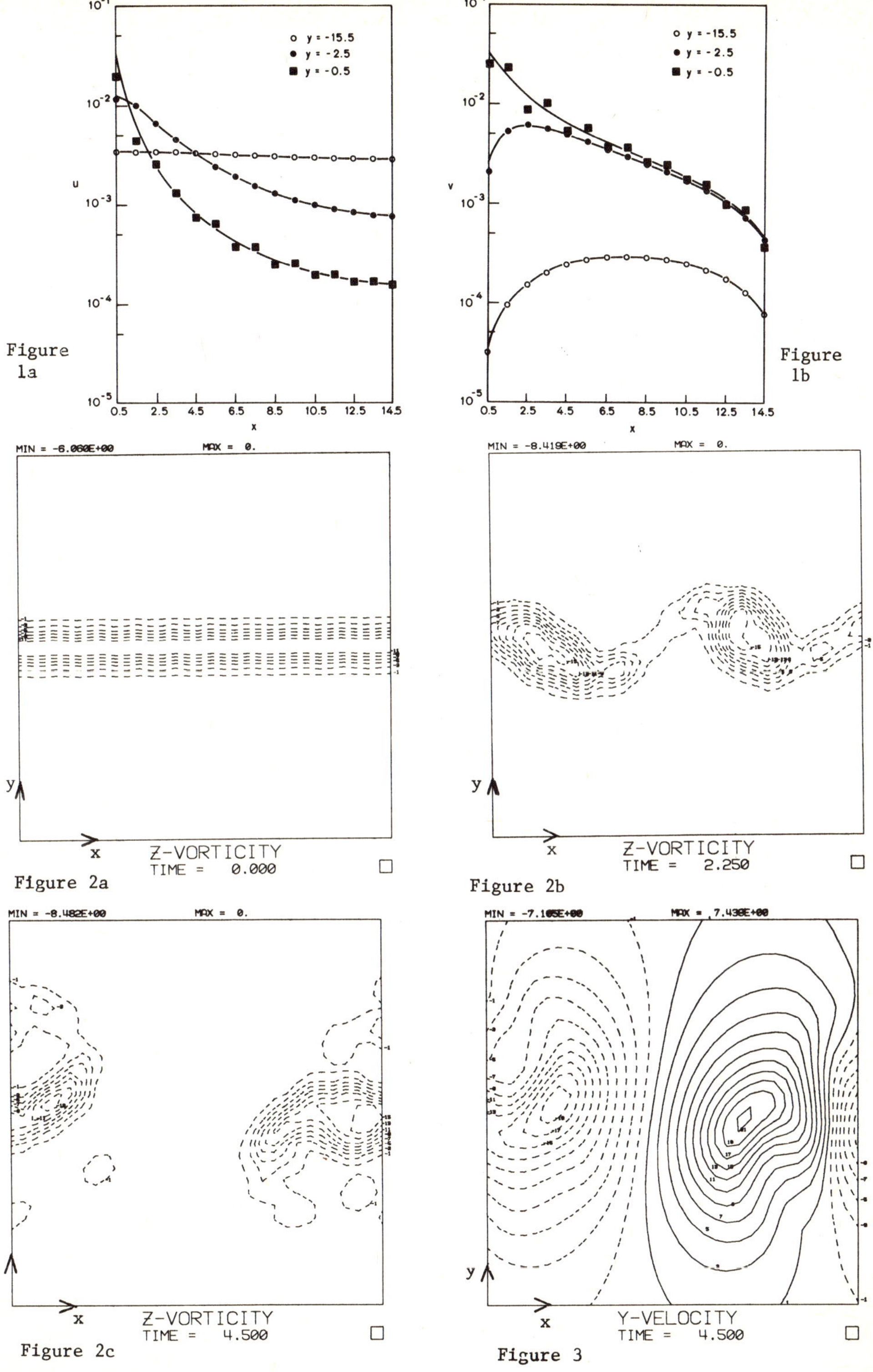

Figure 1a

Figure 1b

Figure 2a

Figure 2b

Figure 2c

Figure 3

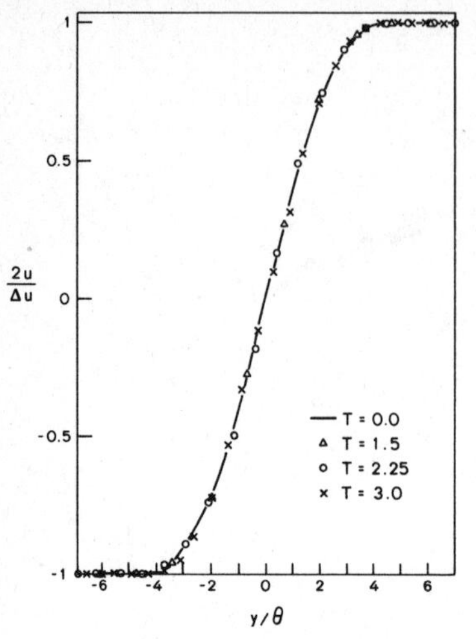

Figure 4

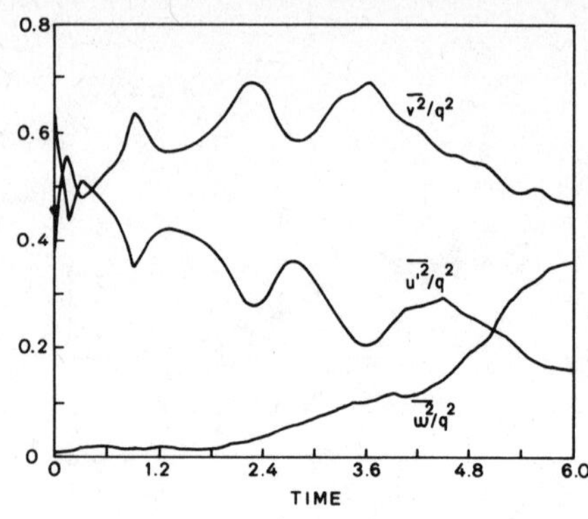

Figure 5

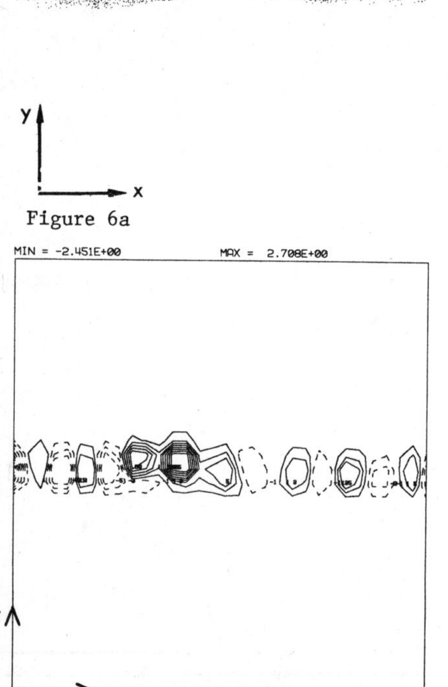

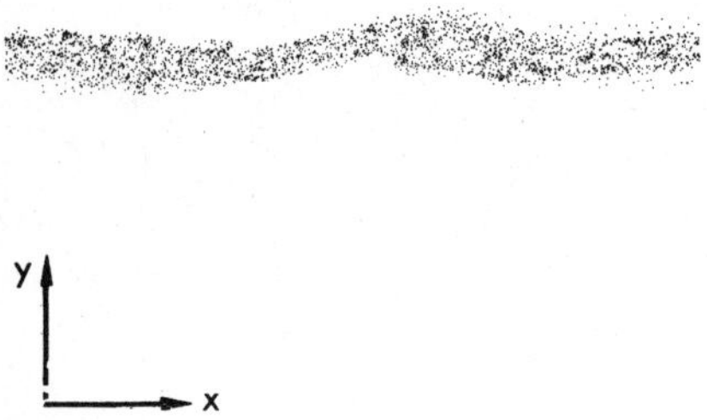

Figure 6a

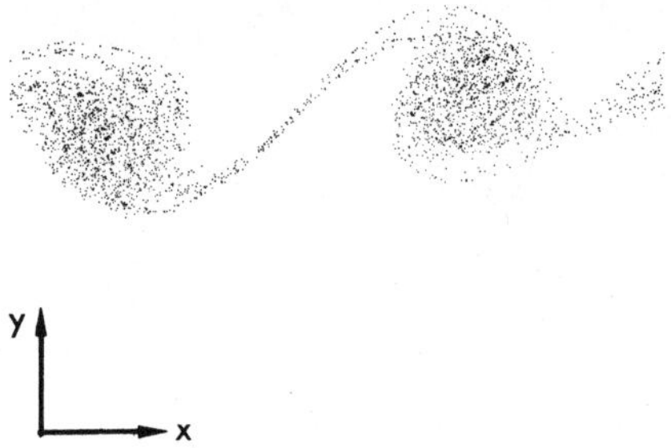

Figure 6b

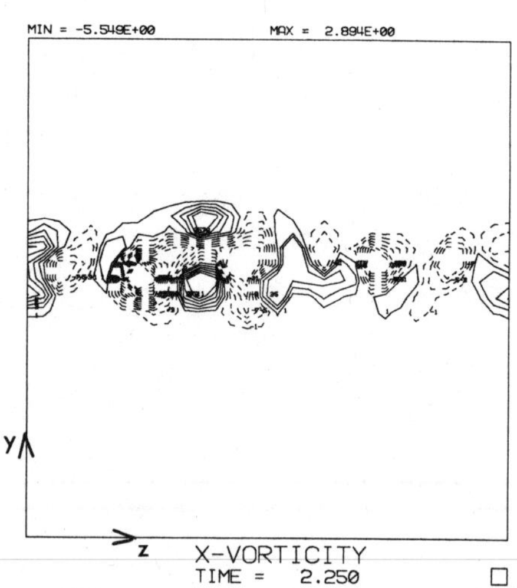



Figure 6c

Figure 6d

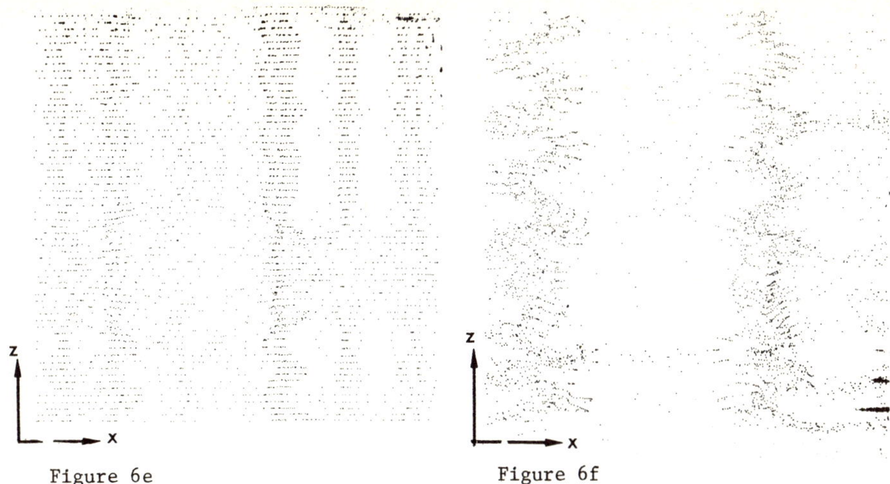


Figure 6e                              Figure 6f

References.

Browand, F.K. and Weidman, P.D. 1976. Large Scales in the Developing Mixing
    Layer. D. Fluid Mech. 76, pp. 127-144.

Couët, B. 1979. Evolution of Turbulence by Three-Dimensional Numerical Particle-
    Vortex Tracing. Stanford University Institute for Plasma Research Report 793.

Couët, B, Buneman, O. and Leonard, A. 1980. Simulation of Three-Dimensional
    Incompressible Flows with a Vortex-in-Cell Method. J. Comput. Phys. (to appear).

Lamb, H. 1945. Hydrodynamics. Dover.

Leonard, A. 1980. Vortex Methods for Flow Simulation. J. Comput Phys. (to appear).

Riley, I. J. and Metcalfe, R.W. 1980. Direct Numerical Simulation of a Perturbed,
    Turbulent Mixing Layer. AIAA-80-0274.

C. Cuvelier
University of Delft, Mathematics Department, Delft, The Netherlands.

## I.   INTRODUCTION

In this paper we will discuss two numerical methods for the calculation of a free
boundary in a fluid which is governed by the Navier-Stokes equations and where the
surface tension on the free boundary has to be taken into account. As a model we can
think of the following two dimensional situation.

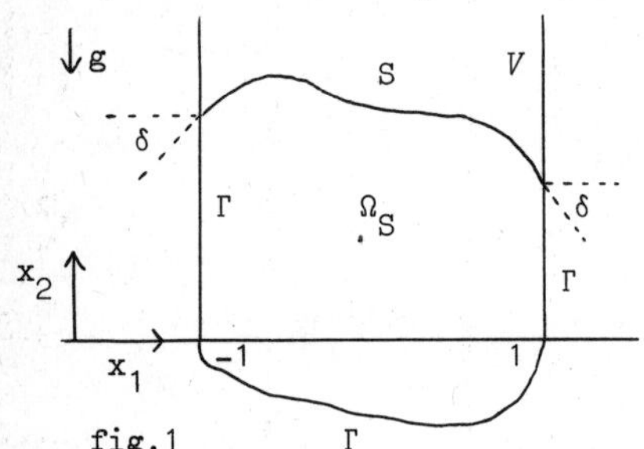



A fluid is contained in an open vessel $V$
(fig. 1) placed in the field of gravity g. On
the smooth boundary Γ of the vessel, Dirichlet
type boundary conditions are prescribed. In the
the stationary situation the surface S of the
liquid is not known a priori and is called a
free boundary. The object is to find the shape
of this meniscus

We assume that the liquid is incompressible (density ρ) with constant viscosity μ
and surface tension T. The problem can be formulated in the following non dimensional
form:

Find a velocity vector $u = \{u_1, u_2\}$, a pressure field p and an interface S
(given by $\phi : x_1 \to x_2 = \phi(x_1)$, $-1 \le x_1 \le 1$, with given contact angle δ at
$x_1 = -1, +1$), such that

(I.1)   $\quad -\Delta u + \alpha \sum_{i=1}^{2} u_i \frac{\partial u}{\partial x_i} + \text{grad } p = 0 \quad$ in $\Omega_S$ (Navier-Stokes eq.$^S$)

(I.2)   $\quad \text{div } u = 0 \quad$ in $\Omega_S$ (incompressibility condition)

with the following boundary conditions

(I.3)   $\quad u(x) = \text{Re } h(x) \quad$ on Γ

(I.4)   $\quad \sigma_\tau = 0 \quad$ on S

(I.5)   $\quad u_n = u.n = 0 \quad$ on S

(I.6)   $\quad -\left(\dfrac{\phi'(x_1)}{\sqrt{1+(\phi'(x_1))^2}}\right)' + \beta\phi(x_1) = -\gamma\,\sigma_n \qquad$ on S $\qquad (\phi' = \dfrac{d\phi}{dx_1})$

$\quad$ with $\phi'(-1) = -\phi'(1) = \text{tg } \delta$

where $\sigma_\tau$ and $\sigma_n$ are the tangential and normal component of the stress tensor $\sigma_{ij}$

$$\sigma_{ij} = -p\delta_{ij} + \left(\frac{\partial u_j}{\partial x_i} + \frac{\partial u_i}{\partial x_j}\right), \quad \sigma_\tau = \frac{\partial u_\tau}{\partial n} - \frac{\partial u_n}{\partial \tau}, \quad \sigma_n = -p + 2\frac{\partial u_n}{\partial n}$$

with $n$ the unit normal and $\tau$ the unit tangent on S and $u_\tau = u.\tau$.
The positive parameters Re, $\beta$ and $\gamma$ are defined by

$$Re = \frac{\rho H}{\mu} = \text{'Reynolds' number} \qquad (H = \text{length scale})$$

$$\beta = \frac{\rho\, g\, H^2}{T} = \text{Bond number} \qquad \gamma = \frac{\mu^2}{\rho\, T\, H} = \text{Ohnesorge number}$$

The parameter $\alpha$ is to distinguish between the full Navier-Stokes eq.[s] ($\alpha=1$) and the Stokes eq.[s] ($\alpha=0$).

Both numerical algorithms which will be studied in this paper are based on the same principle. A shape of the free boundary is assigned and the flow field within that shape is calculated after disregarding one of the three boundary conditions on S. Next a new meniscus shape is computed which satisfies as closely as possible the boundary condition that was relaxed. This procedure is  iterated until convergence is attained.

In the first method the kinematic condition (I.5) is discarded and we use techniques of optimal control theory to calculate a new approximation of the free boundary. The second method is based on an existence and uniqueness result for the free boundary. For the calculation of the flow field the normal stress balance (I.6) is relaxed and is used for the computation of a new meniscus shape.

* * *

## II.  OPTIMAL CONTROL TECHNIQUE

For simplicity we consider the case $\alpha = 0$, i.e. the Stokes eq.[s] in stead of the Navier-Stokes eq.[s].

For $S \in \mathcal{S} = $ class of admissible boundaries, problem (I.1),...,(I.4), (I.6) admits a unique solution $\{u(S),p(S)\}$ (see section III). We try to select a boundary $S_{opt} \in \mathcal{S}$ for which (I.5) is satisfied in the $L_2(S)$ sense. The problem of optimal control (design) can be formulated as follows:

$$(II.1) \quad \left| \text{Find } S_{opt} \in \mathcal{S} \text{ such that } E(S_{opt}) = \inf_{S\in\mathcal{S}} E(S), \quad E(S) = \int_S |u_n|^2\, ds \right.$$

For solving this problem a descent method in a variable domain is used. Suppose $S = \{x_2 = \tilde{\phi}(s) \mid s \in \emptyset\} \in \mathcal{S}, s = $ curve length, and let $S_\alpha = \{x_2 = \tilde{\phi}(s) + \alpha(s)n(s) \mid s \in \emptyset\} \in \mathcal{S}$ be close to S, where $n(s)$ is the outward normal on S. Using techniques of optimal control theory (see LIONS [1], PIRONNEAU [1]), we can calculate the first order variation $\delta E = E(S_\alpha) - E(S)$. Introducing the adjoint state $\{\psi,q\}$ by

$$(II.2) \quad \left| \begin{array}{l} -\Delta\psi + \text{grad } q = 0, \quad \text{div } \psi = 0 \quad \text{in } \Omega_S \\[4pt] \psi = 0 \text{ on } \Gamma, \quad \sigma_n(q,\psi) = u_n \text{ on } S, \quad \sigma_\tau(\psi) = 0 \text{ on } S \end{array} \right.$$

and using the following result for the curvature $\frac{1}{R}$ :

$$(\text{II.3}) \quad \frac{1}{R}(S_\alpha) - \frac{1}{R}(S) = \frac{\alpha}{R^2} + \ddot{\alpha} + o(\alpha,\dot{\alpha},\ddot{\alpha}) \ , \ (\dot{\alpha} = \frac{d\alpha}{ds} \text{ etc})$$

it can be proved (see CUVELIER [1]) that

$$\delta E = 2 \int_S \alpha[(\frac{\partial u_n}{\partial n} - \frac{u_n}{2R}) u_n - \tfrac{1}{2} \sum_{i,j=1}^{2} (\frac{\partial u_i}{\partial x_j} + \frac{\partial u_j}{\partial x_i})(\frac{\partial \psi_i}{\partial x_j} + \frac{\partial \psi_j}{\partial x_i}) +$$

$$+ \frac{1}{\gamma R}\frac{\partial \psi_n}{\partial n} + \frac{1}{\gamma}\frac{\partial^2 \psi_n}{\partial s^2} - \frac{\beta}{\gamma}((\frac{\partial \psi_n}{\partial n} - \frac{\psi_n}{R})\widetilde{\phi}(s)+\cos(n,x_2)\psi_n)]ds + o(\alpha,\dot{\alpha},\ddot{\alpha})$$

$$\equiv 2 \int_S \alpha \, A \, ds + o(\alpha,\dot{\alpha},\ddot{\alpha})$$

Choosing now $\alpha = -\varepsilon(A - \frac{1}{|S|} \int_S Ads)$ where $|S|$ = length (S) and $\varepsilon$ small, with the condition that $\dot{\alpha}(x_1=-1) = \dot{\alpha}(x_1=1) = 0$, we obtain

$$E(S_\alpha) < E(S) + o(\varepsilon)$$

The algorithm for solving the free boundary problem is defined as follows:

(i) $\quad S_o \in \mathcal{S} = \{S| \text{ volume } (\Omega_S) = \text{given}, \ \phi'(-1) = -\phi'(1) = \text{tg } \delta \text{ with S given}$
$\qquad\qquad$ by $x_1 \to x_2 = \phi(x_1)\}$

(ii) $\quad$ Assume $S_1,\dots,S_n = \{x_2 = \widetilde{\phi}(s)| \ s \in \emptyset\}$

(iii) $\quad$ Calculate $u(S_n)$; calculate $\psi(S_n)$

(iv) $\quad$ Calculate $\alpha$

(v) $\quad S_{n+1}$ is given by $S_{n+1} = \{x_2 = \widetilde{\phi}(s) + \alpha(s) \, n(s)| \ s \in \emptyset\}$

* * *

## III.  NORMAL STRESS ITERATION.

In this case we shall deal with the full Navier-Stokes eq.[s], i.e. in (I.1) we take $\alpha = 1$. The numerical algorithm is based on an existence and uniqueness result of the free boundary problem (cf. SOLONNIKOV [1]) in weighted Hölder spaces $C_s^\ell(\Omega)$ whose elements have bounded $|.|_{s,\Omega}^{(\ell)}$ norm defined by

$$|u|_{s,\Omega}^{(\ell)} = |u|_\Omega^{(s)} + \sum_{s<|\alpha|<\ell} \sup_{x\in\Omega} \rho^{|\alpha|-s}(x)|D^\alpha u(x)| + \sum_{|\alpha|=|\ell|} \sup_{x,y\in\Omega} R(x,y)\frac{|D^\alpha u(x)-D^\alpha u(y)|}{|x-y|^{\ell-[\ell]}}$$

with

$$|u|_\Omega^{(s)} = \sum_{|\alpha|<s} \sup_{x\in\Omega} |D^\alpha u(x)| + \sum_{|\alpha|=[s]} \sup_{x,y\in\Omega} \frac{|D^\alpha u(x)-D^\alpha u(y)|}{|x-y|^{s-[s]}} , \quad |u|_\Omega^{(s)} = 0 \text{ for } s < 0$$

$$\rho(x)=\min\{|x-x^1|,|x-x^2|\}, \ x^1=\{-1,\phi(-1)\}, \ x^2=\{1,\phi(1)\}, \ R(x,y)=\min\{\rho^{\ell-s}(x),\rho^{\ell-s}(y)\}$$

Similar definition for the space $C_s^{(\ell)}(-1,1)$, with $\rho$ replaced by $\rho_o(t) = \min\{t+1,1-t\}$.

## THEOREM III.1

$\exists\ R_o > 0$, $V_o > 0$, $\nu_o \in (0,1]$ such that for $V = \int_{-1}^{1} \phi(x_1)\ dx_1 \geq V_o$, $Re \leq R_o$, $\nu \leq \nu_o$

problem (I.1),...,(I.6) has a unique solution $\{\phi,u,p\} \in C_{1+\nu}^{3+\nu}(-1,1) \times C_\nu^{2+\nu}(\Omega) \times C_{\nu-1}^{1+\nu}(\Omega)$

for all $h \in C^{2+\nu}(\Gamma)$ with $\int_\Gamma h.n\ d\Gamma = 0$ and $h(x^i).n(x^i) = 0$, $i=1,2$ .

* * *

The proof of this theorem, for which we refer to SOLONNIKOV [1], is based on the following.

## LEMMA III.1

For $\phi$ fixed in $C_{1+\nu}^{3+\nu}(-1,1)$ and with the conditions of theorem III.1, problem (I.1),...,(I.5) has a unique solution $\{u,p_o\} \in C_\nu^{2+\nu}(\Omega) \times C_{\nu-1}^{1+\nu}(\Omega)$ with $\int_\Omega p_o\ dx = 0$

* * *

The problem is to move $\phi$ in such a way that the normal stress balance (I.6) is satisfied.

For $Re = 0$, the unique solution $\{\phi_o,\bar{u},\bar{p}\}$ of problem (I.1),...,(I.6) is given by

$$(\text{III.1}) \quad \bar{u} = o \qquad \bar{p} = \frac{1}{\gamma}\left(\frac{\beta V}{2} - \sin \delta\right)$$

$$(\text{III.2}) \quad \begin{cases} -\left(\dfrac{\phi_o'}{\sqrt{1+(\phi_o')^2}}\right)' + \beta\,\phi_o = \dfrac{\beta V}{2} - \sin \delta \qquad \text{on } (-1,1) \\[2mm] \phi_o'(-1) = -\phi_o'(1) = \operatorname{tg} \delta \end{cases}$$

We introduce $\omega = \phi - \phi_o$ and it can be proved that $\omega$ satisfies

$$(\text{III.3}) \quad \begin{cases} -\left(\dfrac{\omega'}{\sqrt{1+(\phi_o')^2}}\right)' + \beta\,\omega = G \qquad \text{on } (-1,1) \\[2mm] \omega'(-1) = \omega'(1) = 0 \end{cases}$$

with

$$G = \gamma\left(p_o - 2\frac{\partial u_n}{\partial n}\right) - \frac{1}{2}\int_{-1}^{1}\left(p_o - 2\frac{\partial u_n}{\partial n}\right)dx_1 + \frac{1}{R(\phi)} - \frac{1}{R(\phi_o)} - \left(\frac{\omega'}{\sqrt{1+(\phi_o')^2}}\right)' = G(\phi,u,p_o)$$

We define an operator $L : C_{1+\nu}^{3+\nu}(-1,1) \to C_{1+\nu}^{3+\nu}(-1,1)$ by

$$L : \hat{\omega} \to \hat{\phi} \to G(\hat{\phi},u(\hat{\phi}),p_o(\hat{\phi})) \xrightarrow{(\text{III.3})} \omega$$

and it is proved that $L$ is a contraction operator, which proves the existence and uniqueness of an element $\phi$ in theorem III.1.

$$* \quad * \quad *$$

The normal stress algorithm is the following

(i)     $\phi^o = \phi_o$  (rest state)

(ii)    Assume $\phi^1, \ldots, \phi^n$

(iii)   Solve problem (I.1), $\ldots$, (I.5) with $\phi^n$; solution $\{u^n, p_o^n\}$

(iv)    Solve (III.3) with $G = G(\phi^n, u^n, p_o^n)$; solution $\omega^{n+1}$

(v)     $\phi^{n+1} = \omega^{n+1} + \phi_o$

$$* \quad * \quad *$$

## IV   NUMERICAL ANALYSIS AND APPLICATION

The optimal control method as well as the normal stress method are discretized by a finite element method in the case of the Stokes eq.$^s$. The main point in the algorithms is the numerical solution of the Stokes eq.$^s$. This has been done using the augmented Lagrangian method. Its principle is to replace the Stokes equations

$$- \Delta u + \text{grad } p = f \qquad \text{div } u = 0$$

by the iterative procedure

$$- \Delta u^{n+1} - \sigma \text{ grad div } u^{n+1} + \text{grad } p^n = f \qquad p^{n+1} = p^n - \sigma \text{ div } u^{n+1}$$

It can be proved that $\{u^n, p^n\} \to \{u, p\}$, and for $\sigma$ big ($\approx 10^6$) one iteration is sufficient for an accurate approximation.

In the finite element method the velocity is approximated by an extended quadratic element (i.e. quadratic polynomials plus one third order term) and the pressure by a non conforming linear element. It is proved in GIRAULT - RAVIART [1] that this method is of the second order. In fig. 2 (resp fig. 3) we give the results of a numerical experiment with the optimal control method (resp. normal stress method), for other applications we refer to CUVELIER [2].

Finally let us remark that in the optimal control method two Stokes problems must be solved per iteration, but no pressure calculation is necessary (at least for 1 iteration with the augmented Lagrangian method). The normal stress method needs the solution of one Stokes problem per iteration, moreover  a convergence proof can be given.

$$* \quad * \quad *$$

## BIBLIOGRAPHY

C. CUVELIER [1]. A free boundary problem in hydrodynamic lubrication governed by the Stokes eq.$^s$. Proc. $9^{th}$ IFIP Conference, 1979, Warsaw p. 375-384.

[2]. Capillary free boundary problems governed by the Navier-Stokes eq.[s]. (to appear).

V. GIRAULT, P.A. RAVIART [1] Finite element approximation of the Navier-Stokes eq.[s]. Lecture Notes in Mathematics 749, 1979.

J.L. LIONS [1]. Sur le contrôle optimal des systèmes gouvernés par les equations aux dérivées partielles. Paris-Dunod, 1968.

O. PIRONNEAU [1]. Sur les problèmes d'optimisation de structure en mécanique des fluides. Thèse de doctorat. Paris VI. 1976.

V.A. SOLONNIKOV [1]. Solvability of the boundary value problem describing the motion of a viscous incompressible capillary fluid in an open vessel in the two dimensional case. Izvestia AN SSR, 1979. 43' p. 203-236 (in Russian).

* * *

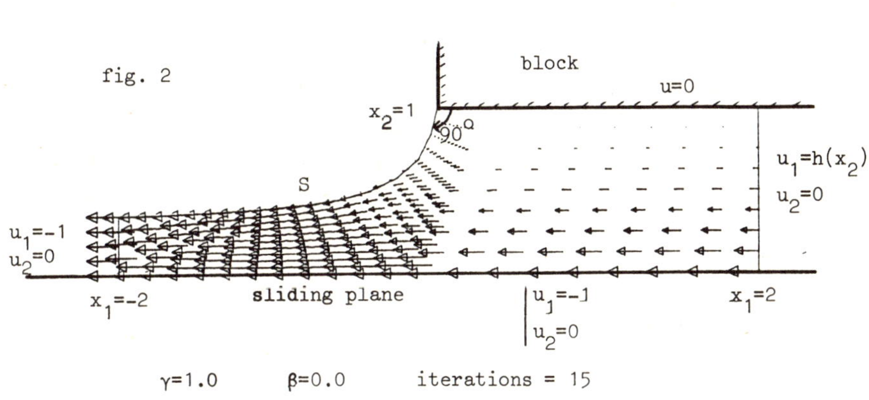


$\gamma=1.0 \qquad \beta=0.0 \qquad$ iterations = 15

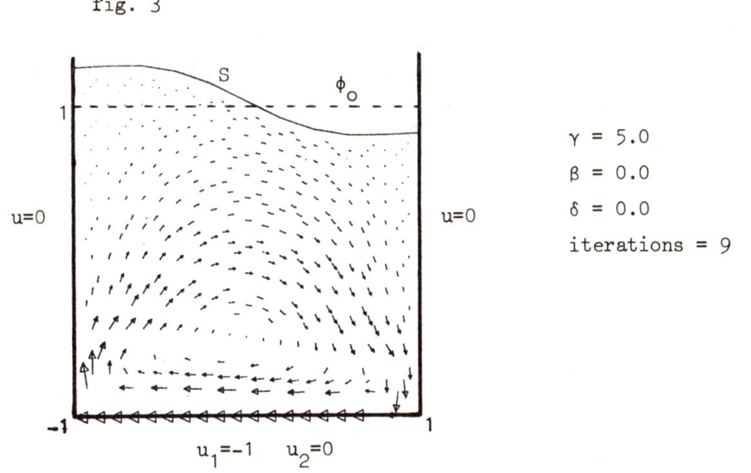


$\gamma = 5.0$

$\beta = 0.0$

$\delta = 0.0$

iterations = 9

* * *

TRANSONIC FLOW CALCULATIONS WITH HIGHER ORDER FINITE ELEMENTS


Herman DECONINCK                                    Charles HIRSCH
Research assistent IWONL                                Professor

Vrije Universiteit Brussel
Dept. of Fluid Mechanics, Brussels, Belgium


## ABSTRACT


Galerkin Finite Element (FE) techniques for the solution of the transonic potential equation in its quasi elliptic artificial density form are investigated with emphasis on the element choice and its consequences on the iterative solution with successive line relaxation (SLOR) and approximate factorization (ADI) methods.

The calculations presented show an improved accuracy for the biquadratic Lagrange element when compared to its bilinear FE or second order FD equivalent, without substantial increase in computational cost.


## INTRODUCTION

Galerkin FE methods for the solution of the transonic potential flow equation using bilinear quadrilateral elements have been presented in the recent past with SLOR (Eberle[1] , Deconinck and Hirsch[2,3]) as well as ADI[3] techniques for the iterative solution algorithm.

However the numerous experiences with subsonic FE calculations made clear that the most performant elements are biquadratic quadrilaterals giving the best compromise between calculation time and accuracy.  In subsonic calculations the element choice has no influence on the calculation procedure except for the residual calculation due to the fact that generally a direct inversion method is applied for the inversion of the linearized system of equations.  This is different in the transonic case since direct methods usually diverge and the element choice closely interacts with the iterative solution method applied, leading f.e. to a sequence of pentadiagonal systems for SLOR and ADI with quadratic elements instead of tridiagonal with bilinear FE or classical central FD shemes.  From this increased implicitness in the iteration scheme better convergence properties may be expected while still very effective pentadiogonal inversion methods can be applied.

Our numerical experiments confirm the higher accuracy of the solution as is expressed by a remarquable approximation of the velocity distribution around the stagnation point.  This is known to be very inportant in cascade computations where inaccurate  solution of the nose region causes severe problems f.e. when a leading edge shock is present.

Results are presented of applications of the bilinear and biquadratic Lagrange Elements to the transonic flow around a NACA 0012 profile and in a channel flow with a circular bump with SLOR and ADI iterative schemes.


## I. ARTIFICIAL COMPRESSIBILITY FORM OF THE EQUATION AND FE DISCRETIZATION

The potential equation in conservative form is given by

(1) $$\partial_x (\rho \, \phi_x) + \partial_y (\rho \, \phi_y) = 0$$

where x and y are Cartesian coordinates.  The density is obtained from the isentropic relation ($\rho_t$ and $T_t$ stagnation density and temperature )

(2) $$\rho = \rho_t \left[ (1 - \frac{\gamma - 1}{\gamma \, r \, T_t} (\phi_x^2 + \phi_y^2)] \right.^{1/\gamma-1}$$

In transonic calculations artificial viscosity (A.V.) is introduced in the supersonic points either implicitly in the discretization scheme (f.e. by using upwind finite differences in supersonic points, Murman and Cole) or in an explicit way by adding some explicit A.V. terms to the equation (f.e. Jameson).

A proper choice of these A.V. terms allows to include them in the density expression (Eberle[1], Holst[4], Hafez[5]) and in this way a fully elliptic discretization with classical Galerkin FE technique is possible leaving the A.V. treatment to the density updating after each iteration of the non linear iteration process. The following artificial density form proved succesfull as well in FD as in FE calculations[2][3].

$$(3) \qquad \nabla(\tilde{\rho} \, \nabla \, \phi) = 0 \qquad\qquad\qquad \tilde{\rho} = \rho - \mu \, \rho_{\overleftarrow{s}} \, \Delta s$$

where $\rho_{\overleftarrow{s}}$ is the upwind difference in the streamwise direction s and $\mu$ a switching function controlling the amount of A.V. (zero in subsonic points)

$$(4) \qquad\qquad \mu = \mathrm{Max}(0 \, , \, M^2 - 1)$$

The general iterative solution procedure is given by

$$(5) \qquad C^n \, \delta\phi = - \, \tau \, R^n \qquad\qquad \text{followed by} \qquad \phi^{n+1} = \phi^n + \delta\phi$$

where $C^n$ is the iteration operator and $R^n$ the residual at iteration number n, discretized with a standard FE weighted residual method, at meshpoint ij

$$(6) \qquad\qquad R_{ij}^n = \sum_{k,l} \phi_{kl}^n \, K_{ij}^{kl^n} - f_{ij}$$

where $K^n$ is the classical stiffness matrix (dropping the iteration index in the notation)

$$(7) \qquad\qquad K_{ij}^{kl} = \int_S \tilde{\rho}^n \, \nabla N_{kl} \, \nabla N_{ij} \, dx \, dy$$

and f the contribution of the boundaries

$$(8) \qquad\qquad f_{ij} = \int_{s_n} (\rho \, \tfrac{\partial\phi}{\partial n})^o \, N_{ij} \, ds$$

S is the flow domain, $s_n$ the part of the boundary where a Neumann condition $(\rho \, \tfrac{\partial\phi}{\partial n})^o$ is specified and $N_{ij}$ (x,y) the weight function choosen equal to the shape function of the meshpoint (Galerkin technique) while $\phi$ is discretized in the standard way

$$\phi = \sum_{k,l} \phi_{kl} \, N_{kl}$$

It is typical for FE applications that the residual automatically contains the contributions of boundary conditions, eq. (8), which are zero for material walls and no further attention is needed to this part of the problem.

Depending on the choice of $C^n$ different solution methods are recovered (direct, explicit, semi implicit or fully implicit), some of which are discussed below. In all cases a grid refinement technique is applied in order to speed up the convergence rate starting with a converged solution on a coarse grid as initial solution for the next finer grid. Work is in progress on the implementation of a full multigrid technique in which the different iterative solution procedures only serve as smoothing mechanism for the high frequency error components in the solution on a certain grid.

## II. LINE OVERRELAXATION (SLOR)

SLOR is the standard device for transonic FD calculations due to the implicit presence of upwind differences in streamwise direction in the operator $C^n$ which work stabilizing in supersonic points if the proper sweep direction is applied.

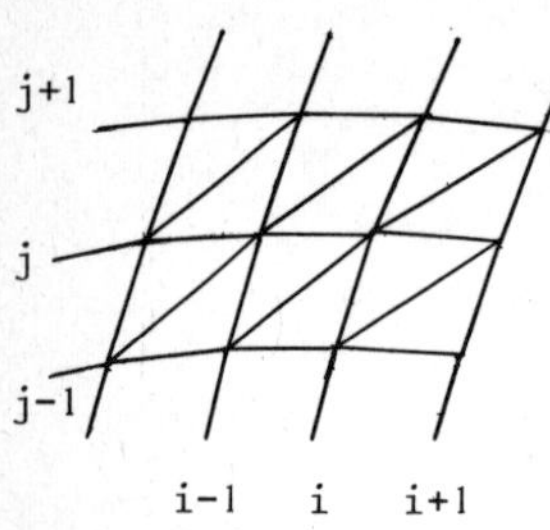

Fig. 1

For bilinear quadrilaterals or linear triangles in the configuration of fig. 1, the SLOR leads to the following scheme :

$$(9) \qquad \sum_{l=j-2}^{j+2} \delta\phi_{il} \, K_{ij}^{il} = -\omega [R_{ij}^n + \sum_{l=j-2}^{j+2} \delta\phi_{i-1,1} \, K_{ij}^{i-1,1}]$$

where the coefficients and the residual are identical for the triangles to those obtained with a classical 5-point difference star except for boundary contributions and as far as the grid is uniform. For instance, Laplace equation with Dirichlet B.C. on a uniform mesh yields for triangles or 5-point difference star the same scheme :

$$(-1 \ 4 \ -1) \begin{pmatrix} \delta\phi_{i,j+1} \\ \delta\phi_{i,j} \\ \delta\phi_{i,j-1} \end{pmatrix} = -\omega \, [\, R_{ij} + (0 \ -1 \ 0) \begin{pmatrix} \delta\phi_{i-1,j+1} \\ \delta\phi_{i-1,j} \\ \delta\phi_{i-1,j-1} \end{pmatrix}] \quad \text{where } R_{ij} = \begin{array}{c} -1 \\ -1 \ \boxed{4} \ -1 \\ -1 \end{array}$$

The bilinear elements lead to a similar scheme containing however contributions from points $(i\pm1, j\pm1)$ in the residual and in the r.h.s. of the equation

$$(-1 \ 8 \ -1) \begin{pmatrix} \delta\phi_{i,j+1} \\ \delta\phi_{i,j} \\ \delta\phi_{i,j-1} \end{pmatrix} = -\omega \, [\, R_{ij} + (-1 \ -1 \ -1) \begin{pmatrix} \delta\phi_{i-1,j+1} \\ \delta\phi_{i-1,j} \\ \delta\phi_{i-1,j-1} \end{pmatrix}] \quad \text{and } R_{ij} = \begin{array}{c} -1 \\ -1 \ \boxed{4} \ -1 \\ -1 \end{array} + \begin{array}{c} -1 \quad -1 \\ \boxed{4} \\ -1 \quad -1 \end{array}$$

In all these cases tridiagonal systems along coordinate line i have to be solved.

For 9 node biquadratic Lagrange element[6], fig. 2 shape functions extend over three nodes in one dimension giving the following pentadiagonal system for each i-line :

$$(10) \qquad \sum_{l=j-2}^{j+2} \delta\phi_{il} \, K_{ij}^{il} = -\omega \, [\, R_{ij} + \sum_{l=j-2}^{j+2} K_{ij}^{i-1,1} \, \delta\phi_{i-1,1} + \sum_{l=j-2}^{j+2} K_{ij}^{i-2,1} \, \delta\phi_{i-2,1}]$$

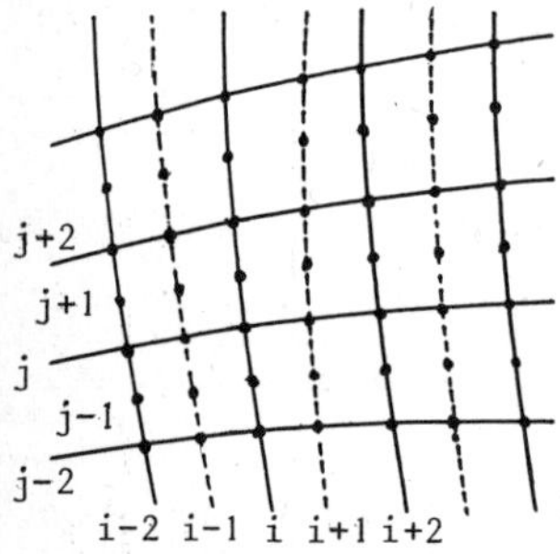

Fig. 2

The residual involves now contributions from 9 surrounding meshpoints for the central node up to 25 for a corner node.

For coordinate lines consisting of midside and central meshpoints (dotted lines) the terms in $K_{ij}^{i-2,1}$ are zero. Equations written for midside or center nodes are tridiagonal namely $K_{ij}^{i,j\pm2} = 0$, resulting in considerable reduction of the cost for inversion of the pentadiogonal system. Application of the 8 node biquadratic serendipity element[6] obtained by dropping the center node leads to alternating pentadiagonal and tridiagonal systems for the coordinate lines consisting of midside nodes.

The use of quadratic elements does not increase the calculation time in a substantial way as was known from subsonic calculation experience, the number of elements being four times lower than in the bilinear case for the same number of meshpoints.

The previous schemes can be modified in order to control the amount of implicitness by introducing Hopscotch or red and black ordering of the meshpoints, f.e. with the bilinear elements a vectorizable scheme very similar to Hafez' Zebra-SLOR[7] is obtained by coloring the meshpoints red and black respectively on odd and even columns and sweeping first over the red and afterwards over the black points using the

solution of the previous sweep in points of the other color.
In similar way the pentadiagonal systems can be reduced to tridiagonal.

## III. APPROXIMATE FACTORIZATION (ADI)

Splitting schemes are considered among the most powerfull iterative solution methods for p.d.e. and their application on transonic flow calculations was a significant step towards efficient algorithms. In the past[2,3] finite element equivalents have been presented for the well known AF1 and AF2 schemes[4,5,8], the former obtained by the choice

$$(11) \qquad C^n = (1 - \sigma \, \partial_x \, \tilde{\rho} \, \partial_x)(1 - \sigma \, \partial_y \, \tilde{\rho} \, \partial_y) = 1 - \sigma \, \nabla \, \tilde{\rho} \, \nabla + O(\sigma^2)$$

on a rectangular mesh. On a curvilinear body fitted mesh the splitting must act on the computational $\xi$-y coordinates requiring

$$(12) \qquad C^n = (1 - \sigma \, \partial_\xi \, \tilde{\rho} \, \frac{A_1}{|J|} \, \partial_\xi)(1 - \sigma \, \partial_\xi \, \tilde{\rho} \, \frac{A_3}{|J|} \, \partial_\eta)$$

$$= 1 - \sigma(|J| \nabla \, \tilde{\rho} \, \nabla - \partial_\xi \, \frac{\tilde{\rho} A_2}{|J|} \, \partial_\eta - \partial_\eta \, \frac{\tilde{\rho} A_2}{|J|} \, \partial_\xi) + O(\sigma^2)$$

where $A_1$, $A_2$, $A_3$ are the components of the metric tensor and J the Jacobian of the isoparametric transformation

$$x = \Sigma \, x_{ij} \, N_{ij}(\xi,\eta) \qquad\qquad y = \Sigma \, y_{ij} \, N_{ij}(\xi,\eta)$$

Now equation (5) is splitted as :

$$(13) \qquad (1 - \sigma \, \partial_\xi \, \frac{\tilde{\rho} A_1}{|J|} \, \partial_\xi)g = - \sigma \, \tau \, R^n \qquad\qquad (1 - \sigma \, \partial_\eta \, \frac{\tilde{\rho} A_3}{|J|} \, \partial\eta)\delta\phi = g$$

and both equations can be treated with a Galerkin FE method if the shape functions factorize

$$N_{ij}(\xi,\eta) = N_i(\xi) \, N_j(\eta) \quad :$$

$$(14) \qquad (M_\xi + \sigma \, K_\xi)g = - \sigma \, \tau \, R^n \qquad \text{and} \qquad (M_\eta + \sigma \, K_\eta)\delta\phi = g$$

where the massmatrices are given by

$$(15) \qquad (M_\xi)_{ik} = \int N_i(\xi) \, N_k(\xi) \, d\xi \qquad \text{and} \qquad (M_\eta)_{ik} = \int N_i(\eta) \, N_k(\eta) \, d\eta$$

and the one dimensional stiffness matrices by

$$(16) \qquad (K_\xi)_{ik} = \int p \, \partial_\xi \, N_i \, \partial_\xi \, N_k \, d\xi \qquad \text{and} \qquad (K_\eta)_{ik} = \int q \, \partial_\eta \, N_i \, \partial_\eta \, N_k \, d\eta$$

An analogous FE formulation of the AF2 scheme has also been developed[2].

Considering now more in detail the structure of equations (14) in the case of the Laplace equation on a rectangular grid : Triangular linear elements and central FD both result in identical schemes except for the mass matrix which is diagonal in the FD case and except for eventual B.C. contributions in the residual, f.e. for equation (14.a) the following tridiagonal system along coordinate line $\xi = \xi_i$ is obtained

$$\begin{matrix} FE : (\frac{1}{6} \ \frac{4}{6} \ \frac{1}{6}) \\ FD : (0 \ \ 1 \ \ 0) \end{matrix} \begin{pmatrix} g_{ij+1} \\ g_{ij} \\ g_{ij-1} \end{pmatrix} + \sigma(-1 \ 2 \ -1) \begin{pmatrix} g_{ij+1} \\ g_{ij} \\ g_{ij-1} \end{pmatrix} = - \sigma \, \tau \, R_{ij} = - \sigma \, \tau \, \frac{\begin{matrix} -1 \\ -1 \mid 4 \ -1 \\ -1 \end{matrix}}{}$$

Bilinear elements yield again identical equations as the triangular except for the residual which now contains contributions of the 8 surrounding meshpoints (cfr. section II).

Biquadratic Lagrange elements (the Serendipity shape functions do not factorize) lead to pentadiagonal systems composed of alternating pentadiagonal equations along odd numbered coordinate lines and tridiagonal equations along even numbered ones, f.e. for equation (15.a) along $\xi = \xi_i$ lines :

$$\begin{matrix} i = \text{odd} : (-4 \quad 8 \quad 32 \quad 8 \quad -4) \\ i = \text{even} : (0 \quad 8 \quad 64 \quad 8 \quad 0) \end{matrix} \begin{pmatrix} g_{ij-2} \\ g_{ij-1} \\ g_{ij} \\ g_{ij+1} \\ g_{ij+2} \end{pmatrix} + \sigma \begin{matrix} (10 \quad -80 \quad 140 \quad -80 \quad 10) \\ (0 \quad -80 \quad 160 \quad -80 \quad 0) \end{matrix} \begin{pmatrix} g_{ij-2} \\ g_{ij-1} \\ g_{ij} \\ g_{ij+1} \\ g_{ij+2} \end{pmatrix} = -60 \, \sigma \, \tau \, R_{ij}$$

where the residual contains again contributions of 9 to 25 surrounding nodes as discussed before.

Formal identical systems are solved along $\eta = \eta_j$ lines for the second equation (15.b). The result is a fully implicit third order accurate scheme valid on an arbitrary mesh and containing the exact boundary treatment in the residual.

Recent experience with FD (Jameson [8]) indicates that ADI schemes are very attractive as smoothing component in a multigrid algorithm and work is underway in this direction.


## IV. NUMERICAL RESULTS WITH NACA 0012 AND CHANNEL GEOMETRY

Some calculations are presented here for two different testcases in order to demonstrate the feasibility of FE calculations with quadratic elements : the flow around the symmetric NACA 0012 airfoil with a freestream Machnumber M = .80, and the flow through a channel with a circular bump shown in figure 3 for an inlet Machnumber M = .85;  both flows contain a Shock and were suggested as testcases at the GAMM workshop on "Computation of Inviscid Transonic flow with Sock Waves" held at Stockholm, 1979 .  For both geometries 1825 meshpoints are used on the finest grid, with coarse grid subdivisions of 481, 133 and 40 meshpoints, the latter only when bilinear elements are applied.

For the channel flow the ADI solutions with biquadratic and bilinear elements (and also with SLOR and bilinear elements presented in[3]) show almost identical Machdistributions along the bump, fig. 4, with somewhat lower hence more accurate values at the edges of the bump (which are potential stagnation flows) in the biquadratic case.  In these solutions a residual drop of three orders was obtained in a grid refinement sequence of 3 grids with non optimised $\sigma$ and $\tau$ parameters.

Sharp shocks are found in both cases spread over three mesh points.

For the second testcase, (fig. 5), a preliminary result with quadratic elements and ADI applied in a multigrid sequence[9] of three grids is compared with our standard bilinear ADI solution.

The multigrid solution required a calculation time equivalent to 25 iterations on the finest mesh.  The higher accuracy of the quadratic solution is visible in the leading edge solution (fig. 6) : for the bilinear solutions a stagnation Machnumber of M = .22 ($C_p$ = 1.06) is obtained, to be compared with M = .05 ($C_p$ = 1.16) for the biquadratic elements.


## CONCLUSION

Transonic flow calculations with third order accurate biquadratic FE methods allow higher accuracy or a reduced number of meshpoints without any supplementary cost compared to bilinear FE.

In combination with a multigrid sequence with ADI as smoothing component a performant method is obtained conserving the inherent advantages of FE methods concerning treatment of boundary conditions and arbitrary curved geometries which are represented by piecewise parabolic approximations.


## REFERENCES


[1] EBERLE A., MBB Bericht UEE 1352(0), 1977
[2] DECONINCK H., HIRSCH Ch., in Proc.III.GAMM Conf. on Num. Meth. in Fl.Mech.Köln 79
[3] DECONINCK H., HIRSCH Ch., in Proc.IV. Int.Symp. on Comp.Sc. in Appl.Sc.&Eng.Paris 79
[4] HOLST T.L., AIAA 78-1113, 1978
[5] HAFEZ M., MURMAN E.M., SOUTH J.C., AIAA 78-1148, 1978
[6] ZIENKIEWICZ O.C., "The Finite Element Meth.in Eng. Sc.",Mc.Graw-Hill,London 77
[7] HAFEZ M.M., SOUTH J.C., "Vectorization of Rel.Meth. for Solving  Transonic F.P.E."
Flow Research Comp. 79
[8] JAMESON A., AIAA 79-1458, 1979
[9] BRANDT A., in Math. Comp. 31, pp. 333-390, 1977


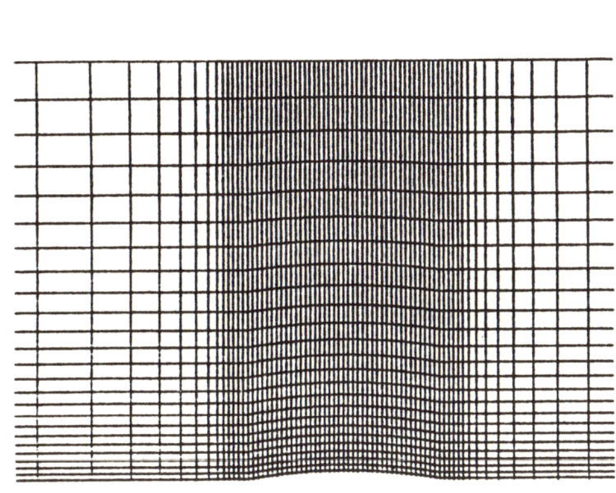

Fig. 3

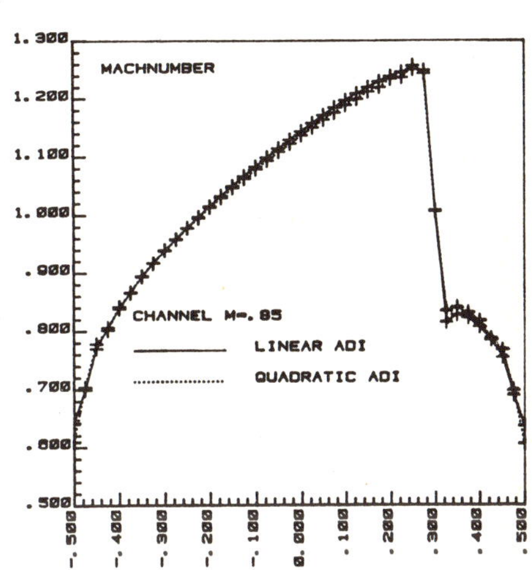



Fig. 4

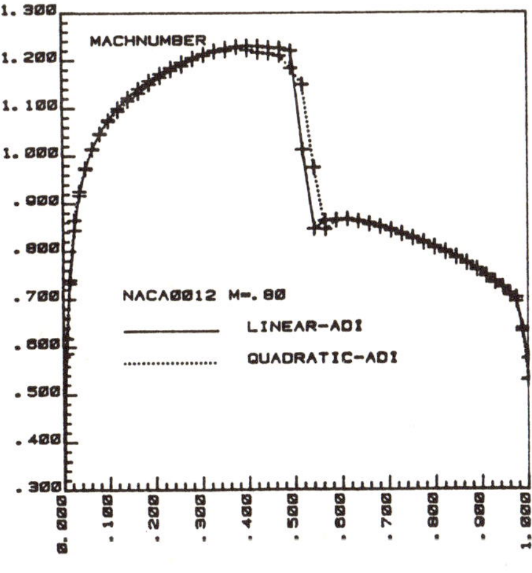



Fig. 5

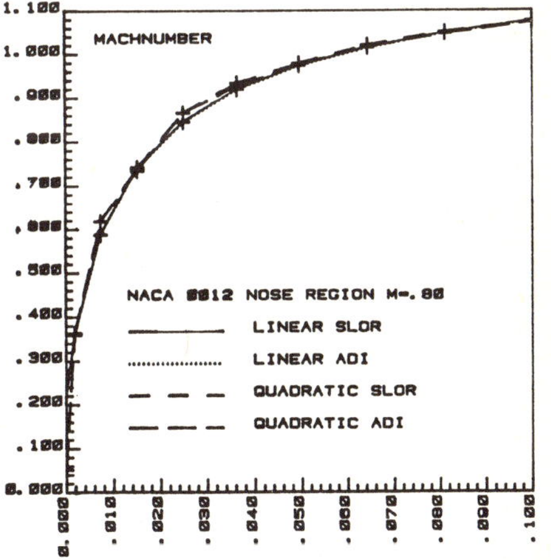



Fig. 6

A FINITE-VOLUME METHOD FOR THE PREDICTION OF
TURBULENT FLOW IN ARBITRARY GEOMETRIES

I. Demirdzic, A.D. Gosman and R.I. Issa
Mech. Eng. Dept., Imperial College,
London, SW7 2BX, England.

## I  INTRODUCTION

The growing evidence of the capabilities of existing numerical fluid dynamics
procedures, when used in conjunction with contemporary mathematical models of
turbulence, to provide acceptably-accurate predictions of complex separated flows
has accelerated interest in the application of the procedures as practical analysis
and design tools.  However most practically-interesting flows occur in or around
enclosures or bodies of complex shape, for which those methods which are formulated
for simple rectangular computing meshes are not well-suited.  Historically most
methods of the finite-difference (FD) and finite volume (FV) class have tended to
suffer from this drawback, whereas those of the more recent finite-element (FE)
variety have not, and this has given rise to the impression that the latter has a
distinct advantage over the former in this respect.

That the advantage just mentioned is more apparent than real is illustrated
by the fact that FD/FV methods have existed for some time which use curvilinear-
orthogonal grid networks possessing the necessary flexibility, such as those
described by Thom and Apelt (1961), Gosman et al (1969), Pope (1978) and Antonopoulos
et al (1978).  Such methods however possess certain drawbacks, one being the
necessity to generate an orthogonal mesh (although this is a relatively straightforward matter) and another, rather more fundamental one being the fact that the
orthogonality constraint sometimes prevents the mesh from being distributed in an
optimum manner:  for example, it may not be possible to concentrate nodes in regions
where high resolution is required without attendant increases in density elsewhere,
where it is unnecessary.

It was presumably recognition of the above drawbacks that provided some of
the incentive for the development described by Hirt (1971) of the arbitrary-mesh
'ALE' FV procedure, which possesses all of the flexibility  of the FE class and has
seen widespread application.  Subsequently, further FD/FV methods employing
arbitrary meshes have emerged,including those of Liu (1976), Cunsolo and Orlandi
(1978) and Wachspress (1979).

This paper describes the development of an arbitrary-mesh FV method by the
authors, which is eventually intended for application to complex turbulent separated
flows of the kinds encountered in reciprocating internal combustion engines,although
it is here presented in the context of steady, two-dimensional flows.


## II  Analysis
### A.  Important Considerations

There arise a number of decision points in the formulation of an arbitrary-
mesh method, some of the more important of which will now be outlined.  At the
outset it is necessary to decide whether the mesh is to be formed from co-ordinate
planes, as is conventional practice in most FD/FV formulations, or is specified
independently, as in the FE approach and certain FD methods (c.f. Chapman, 1978).
A choice must then be made about the directions in which the velocities are to be
resolved, the most popular options being Cartesian (e.g. Hirt, 1971) or covariant
or contravariant curvilinear components (Liu, 1976):  then, when the governing
differential equations are formulated, the possibility exists of 'weak' or 'strong'
conservation forms, and intermediate variants, as discussed by Vinokur (1974) and
others.

Before discretisation commences, the locations of the dependent variables on
the computational grid must be defined, some options being to locate them at common
positions (c.f. Eiseman et al, 1978) or cell-centre the pressure, as in the ALE
method (Hirt, 1971), or use fully-displaced velocity and pressure locations, as in
the work of Liu (1976) and others.  From this stage onwards, the decision points are
much the same as for an orthogonal-mesh procedure, with the important difference
that the generalised formulations usually gives rise to additional terms in the
equations, that require careful attention from both accuracy and numerical-stability
points of view.

It is not possible to give here more than a cursory evaluation of the above
alternatives, the choice from which is in any case seldom clear-cut, and usually
involves some trade-off between generality, complexity and cost.  On the matter of
computational mesh and working variables, our choice is to construct the computing
mesh from quadrilateral elements formed from co-ordinate surfaces, one advantage
being that this gives rise to discretised equations having a regular narrow-banded
matrix of coefficients.  We work with the contravariant velocities and a 'semi-
strong' conservation forumlation of the governing equations on the grounds of:
(a) relative ease of imposition of the boundary conditions, since the velocities
may easily be arranged to be normal or tangential to the boundaries;  (b) minimal
interpolation and hence minimal numerical diffusion involved in discretisation of the
equations, especially the convective transport terms;  and (c) absence of cross
derivatives in the divergence operator, although this is gained at the expense of
additional curvature 'source' terms. On the negative side, this formulation does give
rise to lengthy differential conservative equations, although the margin over apparently shorter options is not so great when the comparison is made, as it properly should,
between the discretised versions.

A fully-staggered arrangement is employed for the velocities and pressures,
as illustrated in Fig. 1, in order to avoid velocity- (or pressure-) decoupling
effects such as are observed in, for example, the ALE method.  Further details of the
present method are provided below.


## B.  Differential Equations

The equations of steady motion of a compressible Newtonian fluid, when
written in general tensor notation in terms of general co-ordinate directions $x^j$ and
associated contravariant velocity components $v^j$ become, in 'semi-strong' conservation
form

$$\frac{\partial}{\partial x^j}\left[\sqrt{g}(\rho v^i v^j - \mu g^{j\alpha} v^i,_\alpha - \mu g^{i\alpha} v^j,_\alpha)\right] + \sqrt{g}(\rho v^j v^\alpha - 2\mu g^{j\beta} v^\alpha,_\beta) \left\{ {}^{i}_{j\alpha} \right\} = -\sqrt{g}g^{ij}\frac{\partial p}{\partial x^j} - \sqrt{g}\, S_v{}^i \tag{1}$$

$$\frac{\partial}{\partial x^j}(\sqrt{g}\rho v^j) = 0 \tag{2}$$

where p, $\rho$ and $\mu$ are the pressure, density and viscosity respectively, the $g^{ij}$ are
the contravariant components of the metric tensor and g is its determinant, and the
quantities bracketed thus {} are Christoffel Symbols of the second kind.  The
conservation equation for a scalar property $\phi$ may likewise be written

$$\frac{\partial}{\partial x^j}\left[\sqrt{g}(\rho v^j \phi - g^{j\alpha}\frac{\mu_t}{\sigma_\phi}\frac{\partial \phi}{\partial x^\alpha})\right] = \sqrt{g}\, S_\phi \tag{3}$$

where $\sigma_\phi$ is the Prandtl/Schimdt number.

For turbulent flow we use the 'k-$\varepsilon$' eddy-viscosity model of Launder and
Spalding (1972) which, in the present framework, entails replacement of $\mu$ in the
above equations by $\mu_t$, calculated from

$$\mu_t = C_\mu \rho \frac{k^2}{\varepsilon} \tag{4}$$

and solution of the generalised transport equations for k and $\varepsilon$, given by

$$\frac{\partial}{\partial x^j} \left[ \sqrt{g}(\rho v^j k - g^{j\alpha} \frac{\mu_t}{\sigma_k} \frac{\partial k}{\partial x^\alpha}) \right] = \sqrt{g}(G - \rho\varepsilon) \tag{5}$$

$$\frac{\partial}{\partial x^j} \left[ \sqrt{g}(\rho v^j \varepsilon - g^{j\alpha}\frac{\mu_t}{\sigma_\varepsilon} \frac{\partial \varepsilon}{\partial x^\alpha}) \right] = \sqrt{g}(C_1 \frac{\varepsilon}{k} G - C_2 \rho \frac{\varepsilon^2}{k}) \tag{6}$$

in which $G = \mu_t(v^j,_i + g^{j\alpha}g_{i\beta}v^\beta,_\alpha)v^i,_j$ and the $\sigma$'s, $C_1$, $C_2$ and $C_\mu$ are empirical constants. The values employed for these, as well as the special 'wall functions' used to impose the no-slip boundary conditions on the above, are described in the above paper.

## C. Discretisation
## 1. General

There is a compelling attraction to exploit the full capabilities of the general equation formulation by solving in the transformed plane on a uniformly-spaced mesh, as has been done by several of the researchers mentioned earlier: for this purpose we follow Wachspress (1979) and others in fitting isoparametric parabolas to the physical co-ordinates of the grid intersections (which may therefore be specified as discrete points, rather than smooth curves) and then deducing from these the components of the metric tensor.

The discretisation is therefore performed on the regular rectangular mesh of Fig. 2, in which are shown a typical control volume or cell and associated notation. Taking the scalar transport eqn. (3) as a prototype, this is written in integral form as:

$$\sum_{i=n,s,e,w} \int_{a_i} \underbrace{(\sqrt{g}(\rho v^i \phi - g^{i\alpha}\Gamma \partial\phi/\partial x^\alpha)da_i}_{I} - \underbrace{\int_V \sqrt{g}S_\phi dV}_{II} = 0 \tag{7}$$

where the summation is over the four faces of the cell, each having 'area' $a_i$ and associated normal co-ordinate $x^i$. The integral flux expressions I each contain components which, using the Mean-Value Theorem, may be replaced by expressions of the form, e.g.

$$\left[ (\sqrt{g}\rho v^i)_w \phi_w - (g^{ii}\Gamma)_w (\partial\phi/\partial x^i)_w \right] a_i \tag{8}$$

where w denotes an average along the w face of the cell. These are then approximated by two-point hybrid differences as used by Caretto et al (1972) and others, which for the above example, would run:

$$H_w(\phi_W - \phi_P) \tag{9}$$

where $H_w$ is the (positive) hybrid-difference operator as defined in the above reference. The remaining component of the integral flux example above is a cross-derivative term arising from grid non-orthogonality, which is given and approximated by:

$$\int_{a_i} \frac{\partial}{\partial x^i}(g^{ij}\Gamma_\phi \frac{\partial\phi}{\partial x^j})da_i \approx C_w^N(\phi_N - \phi_P) - C_w^S(\phi_P - \phi_S) + C_w^{NW}(\phi_{NW} - \phi_W)$$

$$- C_w^{SW}(\phi_W - \phi_{SW}), \quad j \neq i \tag{10}$$

where the $C_i$ are central-difference operators.

Finally, if the volume 'source' integral II is approximated by the quasi-linear expression $S_1 + S_2\phi_P$, so defined that $S_2 \leqslant 0$, the full discrete counterpart of eqn. (3) is:

$$(A_P - S_2)\phi_P = \Sigma A_i \phi_i + S_1 \tag{11}$$

where the summation is over all nine nearest neighbours of P and

$$A_P = \Sigma A_i \;;$$
$$A_W = H_W - C_w^{NW} - C_w^{SW} - C_n^W - C_s^W \;;$$
$$A_{NW} = C_w^{NW} + C_n^{NW} \;;$$

with similar definitions for the remaining $A_i$'s. Although, as noted earlier, the $H_i$ are always non-zero and positive, the $C_i$ may be of either sign and vanish when the grid is orthogonal: thus eqn. (11) yields a positive coefficient matrix when the mesh is orthogonal, or nearly so, but the possibility of negative coefficients, and attendant problems, arises as the degree of non-orthgonality increases. This aspect will be discussed further below.


## 2.  Particular

The foregoing analysis applies without change to all scalar transport equations such as (3), (4) and (5) of the original set, and in slightly modified form to the momentum conservation equations (1),taking into account the fact that their control volumes are centered about the velocity locations. The latter equations also contain: (i) additional viscous flux terms arising from the stress tensor, which are approximated by central differences;  (ii) additional curvature terms resulting from the fact that momentum is conserved in a straight line – these are treated as 'sources' and linearised in the manner indicated earlier; and (iii)  pressure-gradient terms which, in the integral formulation are approximated, in,for example,the direction-$x^1$ momentum equation for the velocity $v_w^1$, as:

$$\int_V \sqrt{g}(g^{11}\frac{\partial p}{\partial x^1} + g^{12} \frac{\partial p}{\partial x^2})dV \approx B_{w_1} (p_W - p_P)$$

$$+ B_{w_2} (p_{SW} + p_S - p_{NW} - p_N) \tag{12}$$

where $B_{w2}$ is zero when the mesh is orthogonal.

The continuity relation (2) is transformed into a pressure perturbation equation in the manner of Caretto et al (1972) by integrating it over the main cells and then replacing the velocities by expressions of the form

$$v_w^1 = (v_w^1)^* + (v_w^1)' \tag{13}$$

in which the velocity perturbation $(v_w^1)'$ about the prevailing field $(v_w^1)^*$ is linked to pressure perturbations p' by:

$$(v_w^1)' \approx D_{w_1} (p_W' - p_P') + D_{w_2} (p_{SW}' + p_S' - p_{NW}' - p_N') \tag{14}$$

and the coefficients $D_{w_1}$ and $D_{w_2}$ in the latter are respectively taken as proportional to $B_{w_1}$ and $B_{w_2}$ in eqn. (12). This leads to a pressure-perturbation equation of the form:

$$A_P p_P' = \Sigma A_j p_j' - M_P \tag{15}$$

where in general the summation is over the nine neighbouring nodes, $M_P$ represents the integral divergence of the v* field, (i.e. the continuity imbalance over the cell), $A_P \equiv \Sigma A_j$ and, typically:

$$A_W = \rho_w D_{w_1} a_w$$

$$A_{NW} = -\rho_w D_{w_2} a_w$$

with similar definitions for the remaining coefficients.  Here, it should be noted that $D_{w_1}$, and hence $A_W$ are positive coefficients while $D_{w_2}$ and $A_{NW}$ may have either sign:  thus the earlier comments about the coefficient matrix associated with eqns. (11) also apply here.


## D.  Solution Algorithm

The method of solution is iterative, in which the equation set for each variable is solved in turn using a form of ADI described by Ames (1977). The sequence is to solve for v* from the momentum equations, then bring the velocities into a continuity balance by solving for p', and hence v'; and finally insert the resulting velocity field into the k  and ε equations and calculate these variables. The p' field is then used to update the pressures and the whole process is repeated until convergence is achieved.

The success of the above procedure relies in part on the coefficient matrices of the equations being diagonally dominant, or nearly so (this happens also to be a requirement for a properly-bounded solution, but this will not further discussed here).  It has already been mentioned that certain terms arising from grid non-orthogonality prevent this requirement from being unconditionally fulfilled: fortunately experience suggests that the resulting problems arise mainly in the pressure perturbation equation;  moreover they may be entirely eliminated from this by simply suppressing the terms relating to the corner nodes NW, SW, NE and SE, an alteration which, reflection will reveal, will have no effect on the final solution since for this p' is everywhere zero.  This then is the approach which has been adopted.


## E.  Application to Test Cases

The new procedure has been subjected to a programme of tests, which is still in progress at the time of writing.  In the early stages these comprised simple fully-developed inviscid and laminar flows in straight and constant-curvature ducts for which both orthogonal and non-orthogonal grids were used, thereby allowing individual facets of the discretisation to be evaluated.  In general agreement to within a few per cent of the exact analytical solutions was obtained using relatively coarse (typically 8 x 8) grids, although appreciable deterioration in accuracy occurred as the grid became highly non-orthogonal (i.e. angle between grid lines $45^{\circ}$ or less).  It is believed that this can be avoided by detailed improvements to the equations.

An example of a more recent, and slightly more complex, test case is that of plane irrotational stagnation flow, in which the streamlines are in the form of rectangular hyperbola given by $\psi = xy$, where x and y are Cartesian co-ordinates measured from the stagnation point.  This was computed on a rectangular domain with the exact velocities imposed on the boundaries and a simultaneous calculation was made of the purely convective transport of a scalar $\phi$ by solving eqn. (3) with $\Gamma_\phi$ and $S_\phi$ set to zero and a Gaussian  $\phi$ distribution imposed at the inflow boundary.

Figs. 3 and 4 provide an impression of the overall accuracy of calculation for Cartesian and non-orthogonal meshes respectively (the curved grid lines of the latter being formed from the analytical streamlines) in terms of iso-$\phi$  contours: these should of course coincide with the streamlines and the curves labelled 'exact solution' therefore represent both.  Comparison of the two figures clearly shows the advantage of a flow-oriented curvilinear computing mesh in reducing the amount

of numerical dispersion.  It must however be admitted that the main improvement
is wrought in the scalar transport calculation itself, with the curvilinear-mesh
flow-field solution being less accurate in some respects, notably the pressure
field, than that obtained on the rectangular mesh:  again the need for further
refinement is indicated.

Finally, an indication of the full capabilities of the present method is
provided in Fig. 5, showing the results of a calculation of turbulent flow through
a complex passage akin to the inlet port/valve assembly of a reciprocating internal
combustion engine.  The computing mesh used is that of Fig. 1, which is probably
not fine enough to resolve the full details of the flow in the wake of the valve
guide.  Nonetheless there is evidence of the expected separation zone there, as
well as the others provoked by the sharp curvature of the passage near the exit.
Also expected is the build-up of turbulence near the duct walls, especially in the
zone of high shear around the upper protruding corner;  and the increased spread
of the wall-generated turbulence near the base of the valve due to the mixing action
of the recirculation zone there is also reproduced.


## F.   Summary

A working method has been produced for calculating turbulent separated flows
in complex geometries using arbitrary non-orthogonal computing meshes, which may
be specified by the user in a pointwise fashion.  Although useful as it now stands,
tests on the method suggest that there is scope for improvement, particularly in
respect of accuracy, and efforts are now being made in this direction.


## G.   References

1 Antonopoulos, K., A.D. Gosman, and R. Issa. (1978)   "Flow and heat transfer in
  tube assemblies".   Proc. International Conference on Numerical Methods in Laminar
  and Turbulent Flow, University of Swansea.

2 Ames, W.F. (1977)  Numerical Methods for Partial Differential Equations  Academic
  Press, 2nd Ed.

3 Caretto, L.S., Gosman, A.D., Patankar, S.V. and Spalding, D.B. (1972).
  "Two numerical procedures for three-dimensional recirculating flows". Proc. Int'l.
  Conf. on Numerical Methods in Fluid Dynamics, Paris.

4 Chapman, M. (1979) "Two-dimensional numerical simulation of inlet manifold flow in
  a four-cylinder internal combustion engine". Society of Automotive Engineers
  SAE 790244.

5 Cunsolo, D. and Orlandi, P. (1978) "Accuracy in non-orthogonal grid reference
  frames"  First Internat. Conf. on Numerical Methods in Laminar and Turbulent Flow,
  Pentech Press, London.

6 Eiseman, P.R., Levy, R., McDonald, H. and Briley, W.R. (1978) "Development of a
  three-dimensional turbulent duct flow analysis". NASA CR-3029.

7 Gosman, A.D., Pun, W.M., Runchal, A.K., Spalding, D.B. and Wolfshtein, M. (1969).
  Heat and Mass Transfer in Recirculating Flows. Academic Press, London.

8 Hirt, C.W. (1971). "An arbitrary Lagrangian-Eulerian computing technique". Second
  Internat. Conf. on Numerical Methods in Fluid Dynamics.

9 Launder, B.E. and Spalding, D.B.(1972)  Mathematical Models of Turbulence.

10 Liu, N. (1976). "Finite-difference solution of the Navier-Stokes equations for
   incompressible three-dimensional internal flows".  Fifth Internat. Conf. in
   Numerical Methods in Fluid Dynamics.

11 Pope, S.B. (1978).  "The calculation of turbulent recirculating flows in general-
   orthogonal co-ordinates". J. Comp. Phys., 26, no. 2, 197-217.

12 Thom, A. and Apelt, C.J. (1961).  Field Computation in Engineering and Physics,
   Van Nostrand.

13 Vinokur, M. (1974). "Conservation equations of gas dynamics in curvilinear
   co-ordinate systems".  J. Comp. Phys., 14, 105-125.

14 Wachspress, E.L. (1979). "The numerical solution of turbulent flow problems in
   general geometry". General Electric Co. Rept. KAPL-4116.

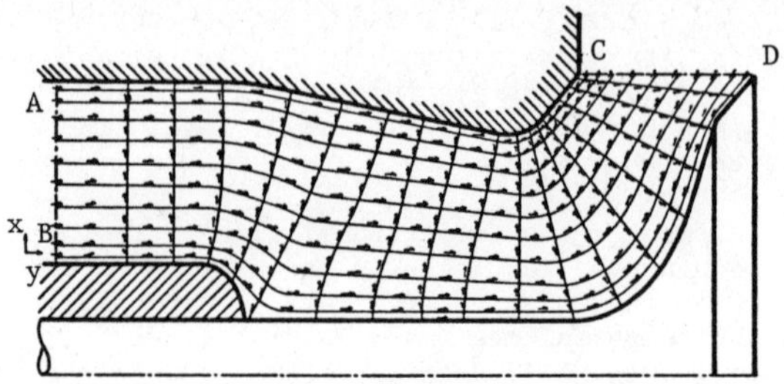

Fig. 1   Illustration of typical computing mesh (physical plane); with velocity components shown as arrows.

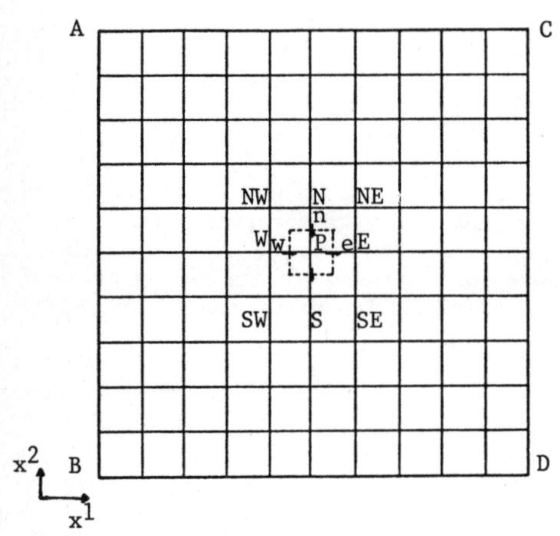

Fig. 2   Illustration of computing mesh in transformed plane, also showing grid notation.

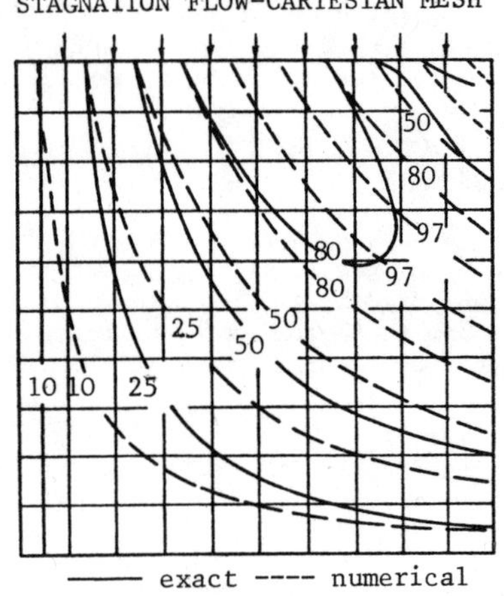



Fig. 3 Comparison between exact and computed scalar fields in irrotational stagnation flow — Cartesian mesh.

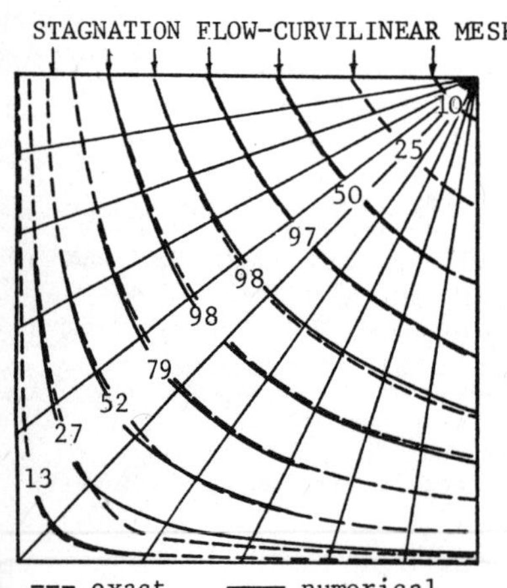



Fig. 4   Comparison between exact and computed scalar fields in irrotational stagnation flow - Curvilinear mesh.

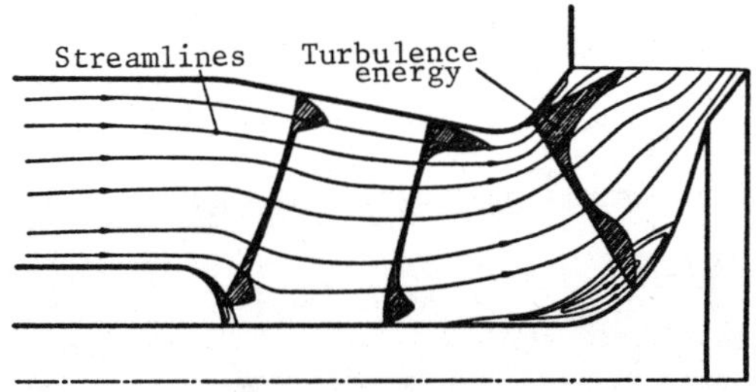



Fig. 5   Computed streamline and turbulence energy distributions in an idealised engine inlet port/valve assembly.

A FINITE DIFFERENCE METHOD FOR THE SLOW MOTION
OF A SPHERE IN A ROTATING FLUID

S.C.R. Dennis

Department of Applied Mathematics
University of Western Ontario
London, Ontario, Canada

and

D.B. Ingham

School of Mathematics
University of Leeds
Leeds, England


## ABSTRACT

A numerical method for treating certain classes of problems in the theory of rotating fluids is given. The uniform, slow motion of a sphere in a viscous fluid has been examined in the case where the undisturbed fluid rotates with constant angular velocity $\omega_0$ and the axis of rotation is taken to coincide with the line of motion. The Navier-Stokes equations can be written in the form of three coupled, nonlinear, elliptic partial differential equations. These equations are expressed in finite-difference form using a specialized technique which is everywhere second-order accurate. It is based on an expansion of a finite-difference scheme which already exists in the literature and which involves the exponential function. By expanding the exponentials in powers of their exponents an approximation is arrived at which is particularly suitable for use in obtaining numerical solutions.

Some preliminary tests of the method have been carried out. The numerical results confirm the theoretical work of Childress (1963, 1964) when both the Reynolds number and Taylor number are small. The effects of varying the mesh size, position of the outer boundary and relaxation parameters have been investigated.


## INTRODUCTION

When a sphere of radius a is towed with a speed U through a viscous fluid which is rotating as a solid mass with angular velocity $\omega_0$ about the axis of translation the phenomenon of the Taylor column is set up when the Rossby number, $R_0 = U/(a\omega_0)$, is small. In this case a column of rotating fluid fore and aft of the sphere accompanies the translatory motion, being separated from the main solid body motion of the fluid by shear layers. This phenomenon was predicted theoretically by Proudman (1916) and later studied by Taylor (1917, 1921, 1922) in a famous series of papers. However, the steady linearized inviscid equations of motion possess an infinite number of solutions all of which satisfy the boundary conditions on the sphere and all giving rise to a columnar structure of the flow.

The first attempt to remove this indeterminacy is due to Grace (1926) who solved an initial-value problem for the inviscid equations. A complete discussion of the problem was first given by Stewartson (1952). Maxworthy (1970) considered the steady, viscous problem experimentally and found that for small values of $R_0$ the drag was higher than that predicted by Stewartson by a factor of nearly two, even after an empirical endwall correction had been included. This discrepancy must cast doubt on the applicability of linearized theories and its importance has been stressed by Barnard and Pritchard (1975), who provide a valuable set of comparisons between linearized theory and experiment. Recently Hocking, Moore and Walton (1979) have also investigated this problem.

In the case of small Reynolds number, $R = Ua/\nu$, and Taylor number, $T = \omega_o a^2/\nu$, the experimental results of Maxworthy (1965) and the theoretical work of Childress (1963, 1964) are in very good agreement. It is in this parameter range that the present numerical calculations have been performed in order to provide some test of the efficiency and accuracy of the method. The results tend to confirm the theoretical work of Childress as far as comparisons can be made. It is hoped that the numerical method will be able to be used subsequently to obtain results for larger values of the Reynolds and Taylor numbers and thus shed some light on the discrepancy between the theory and experimental work in this parameter regime.

Several different finite-difference schemes can be used to approximate the Navier-Stokes equations and these vary considerably in accuracy and efficiency. If central differences are used everywhere then difficulty is often encountered in obtaining a convergent solution of the finite-difference equations and under relaxation is frequently required. This was found to be the case for the problem considered in this paper. Upwind and downwind differencing as described by Greenspan (1968), Gosman et al (1969) and Runchal, Spalding and Wolfshtein (1969) improves the convergence but at the expense of being only first-order accurate. The improvement in convergence is a consequence of the diagonal dominance of the matrix which is associated with the finite-difference equations.

Methods which have second-order accuracy and diagonal dominance have been derived by Allen and Southwell (1955), Allen (1962), Dennis (1960, 1973), Spalding (1972) and Roscoe (1975, 1976). These schemes all involve finite-difference equations with exponential coefficients. Dennis and Hudson (1978) and Dennis, Ingham and Cook (1979) have shown that by a suitable adaptation of the Allen and Southwell method, as suggested by Dennis (1960), an approximation of second-order accuracy with an associated matrix which is diagonally dominant can be obtained in Cartesian coordinates. These difference equations do not now involve the exponential function and can be looked upon as a rather more complicated version of the central-difference formulation. An extension of the method of Dennis and Hudson (1978) and Dennis, Ingham and Cook (1979) has been used in the present problem. In this case the method used is second order in accuracy but it is not now possible to show that the matrix associated with the finite-difference equations is diagonally dominant. Hence convergence is not assured. However in all the computations presented no under relaxation parameter was required. This contrasts with the use of conventional central-difference approximations in which an under relaxation parameter was required in order to obtain a convergent solution. Using these specialized finite-difference equations, some computations were performed with a relaxation parameter greater than unity and convergence was speeded up. This effect has not been analysed in detail in the present paper.

BASIC EQUATIONS

In this paper we study the steady flow of a viscous incompressible fluid past a sphere of radius a which is moving with a uniform translational velocity U. The undisturbed fluid rotates with constant angular velocity $\omega_o$ and the axis of rotation is taken to coincide with the line of motion. The flow is assumed to be symmetrical about the axis of rotation which is taken to be the direction $\theta = 0$ of spherical coordinates $(r,\theta,\phi)$ with the origin at the centre of the sphere and hence all quantities are independent of $\phi$. The velocity components $(v_r^*, v_\theta^*, w^*)$ are related to the stream function $\psi^*$ and the function $\Omega^*$ by the relations

$$v_r^* = \frac{1}{r^2 \sin\theta} \frac{\partial \psi^*}{\partial \theta}, \quad v_\theta^* = -\frac{1}{r \sin\theta} \frac{\partial \psi^*}{\partial r}, \quad w^* = \frac{\Omega^*}{r \sin\theta}. \tag{1}$$

We now introduce a dimensionless stream function $\psi$ and dimensionless function $\Omega$ given by

$$\psi^* = a^2 U\psi, \quad \Omega^* = a^2 \omega_o \Omega, \tag{2}$$

and non-dimensional velocity components $(v_r, v_\theta, w)$ which are obtained by dividing the dimensional components $v_r^*$, $v_\theta^*$ by U and $w^*$ by $a\omega_o$. If we put $\xi = \ln(r/a)$ we obtain for the velocity components

$$v_r = \frac{e^{-2\xi}}{\sin\theta} \frac{\partial\psi}{\partial\theta}, \quad v_\theta = -\frac{e^{-2\xi}}{\sin\theta} \frac{\partial\psi}{\partial\xi}, \quad w = \frac{\Omega e^{-\xi}}{\sin\theta} \tag{3}$$

and the Navier-Stokes equations for the motion are easily found to be

$$D^2\Omega = \frac{Re^{-\xi}}{\sin\theta}\left(\frac{\partial\psi}{\partial\theta}\frac{\partial\Omega}{\partial\xi} - \frac{\partial\psi}{\partial\xi}\frac{\partial\Omega}{\partial\theta}\right), \tag{4}$$

$$D^2\zeta = \frac{e^{-\xi}}{\sin\theta}\left[R\left(\frac{\partial\psi}{\partial\theta}\frac{\partial\zeta}{\partial\xi} - \frac{\partial\psi}{\partial\xi}\frac{\partial\zeta}{\partial\theta}\right) - 2R\zeta\left(\frac{\partial\psi}{\partial\theta} - \cot\theta\frac{\partial\psi}{\partial\xi}\right) + 2R^{-1}T^2\Omega\left(\frac{\partial\Omega}{\partial\theta} - \cot\theta\frac{\partial\Omega}{\partial\xi}\right)\right], \tag{5}$$

$$D^2\psi = -e^{2\xi}\zeta, \tag{6}$$

where

$$D^2 = \frac{\partial^2}{\partial\xi^2} - \frac{\partial}{\partial\xi} + \frac{\partial^2}{\partial\theta^2} - \cot\theta\frac{\partial}{\partial\theta}. \tag{7}$$

Equations (4), (5) and (6) have to be solved subject to the boundary conditions:
(i) On the surface of the sphere there is no normal or slip velocity and there is no torque. Thus

$$\psi = \partial\psi/\partial\xi = 0, \quad \Omega = \omega\sin^2\theta, \tag{8a}$$

where $\omega$ is the ratio of the angular velocity of the sphere to that of the fluid at large distances from the sphere. This parameter is not known in advance but is determined, as a function of R and T, by the requirement that the torque on the sphere be zero.
(ii) At large distances from the sphere the presence of the sphere is not felt. This gives

$$\psi \sim \tfrac{1}{2}e^{2\xi}\sin^2\theta, \quad \Omega \sim e^{2\xi}\sin^2\theta, \quad \zeta \to 0 \text{ as } \xi \to \infty. \tag{8b}$$

(iii) The flow is symmetrical about $\theta = 0$ and $\theta = \pi$, which yields

$$\psi = \zeta = \frac{\partial\Omega}{\partial\theta} = 0 \quad \text{on} \quad \theta = 0 \text{ and } \theta = \pi. \tag{8c}$$

A grid system is set up in the region $0 \le \xi \le \xi_m$ and $0 \le \theta \le \pi$, where $\xi_m$ is the value of $\xi$ at which an approximation to the boundary condition at infinity is applied. Constant angular and radial mesh sizes $k = \pi/N$ and $h = \xi_m/M$ are used, where M and N are integers. We denote all quantities at a typical set of grid points $(\xi_o, \theta_o)$, $(\xi_o + h, \theta_o)$, $(\xi_o, \theta_o + k)$, $(\xi_o - h, \theta_o)$ and $(\xi_o, \theta_o - k)$ by subscripts 0, 1, 2, 3 and 4 respectively. The method described by Dennis and Hudson (1978) and Dennis, Ingham and Cook (1979) will be extended in order to express equations (4), (5) and (6) in finite-difference form. Thus in general any of these equations can be written in the form

$$\frac{\partial^2 H}{\partial\xi^2} + (p-1)\frac{\partial H}{\partial\xi} + \frac{\partial^2 H}{\partial\theta^2} + (q - \cot\theta)\frac{\partial H}{\partial\theta} + rH = s, \tag{9}$$

where H is $\Omega$, $\psi$ or $\zeta$, p and q depend on $\psi$ only and r and s are functions which do not depend on H.

Equation (9) can be expressed in finite-difference form by using the conventional central differences but because of the presence of the terms in p and q the resulting associated matrix is not diagonally dominant and convergence of the iterative scheme is not therefore assured. We thus write equation (9) as two equations

$$\frac{\partial^2 H}{\partial\xi^2} + (p-1)\frac{\partial H}{\partial\xi} + rH = s + t(\xi,\theta), \tag{10a}$$

$$\frac{\partial^2 H}{\partial\theta^2} + (q - \cot\theta)\frac{\partial H}{\partial\theta} = - t(\xi,\theta). \tag{10b}$$

The Allen and Southwell approximation is obtained by replacing p, r and s by $p_o$, $r_o$ and $s_o$ respectively and $t(\xi,\theta)$ by $t_o$, their values at $(\xi_o,\theta_o)$, in equation (10a) and solving it as an ordinary differential equation in $\xi$ along $\theta = \theta_o$. The exact solution of this simplified equation is then used to express $t_o$ in terms of $H_o$, $H_1$ and $H_3$. A similar treatment of equation (10b) is employed along $\xi = \xi_o$ and thus $t_o$ is expressed in terms of $H_o$, $H_2$ and $H_4$. Eliminating $t_o$ from these two expressions gives a finite-difference approximation at $(\xi_o,\theta_o)$.

An alternative method was suggested by Dennis (1960) and used in Cartesian coordinates by Dennis and Hudson (1978) and Dennis, Ingham and Cook (1979). Equation (10a) is transformed locally for $\xi_o - h \le \xi \le \xi_o + h$ by the substitution

$$H = \lambda e^{-f} , \quad f = \tfrac{1}{2}\int_{\xi_o}^{\xi} p(z,\theta_o)dz \tag{11}$$

and equation (10b) is transformed locally for $\theta_o - k \le \theta \le \theta_o + k$ by the substitution

$$H = \mu e^{-g} , \quad g = \tfrac{1}{2}\int_{\theta_o}^{\theta} q(\xi_o,z)dz. \tag{12}$$

Equations (10) now become

$$\frac{\partial^2\lambda}{\partial\xi^2} - \frac{\partial\lambda}{\partial\xi} + \lambda[- \tfrac{1}{2}\frac{\partial p}{\partial\xi} + \tfrac{1}{2}p - \tfrac{1}{4}p^2 + r] = te^f + se^f, \tag{13a}$$

$$\frac{\partial^2\mu}{\partial\theta^2} - \cot\theta \frac{\partial\mu}{\partial\theta} + \mu[- \tfrac{1}{2}\frac{\partial q}{\partial\theta} + \tfrac{1}{2}q\cot\theta - \tfrac{1}{4}q^2] = - te^g. \tag{13b}$$

We now approximate the equations (13) using central differences at $(\xi_o,\theta_o)$ and eliminate $t(\xi_o,\theta_o)$. This gives

$$(1 - \tfrac{1}{2}h)e^{f_1}H_1 + (1 + \tfrac{1}{2}h)e^{f_3}H_3 + \gamma^2[(1 - \tfrac{1}{2}k\cot\theta_o)e^{g_2}H_2 + (1 + \tfrac{1}{2}k\cot\theta_o)e^{g_4}H_4]$$

$$- [2 + 2\gamma^2 + h^2\{\tfrac{1}{4}(p_o^2 + q_o^2) - p_o - q_o\cot\theta_o - r_o\}]H_o = h^2 s_o, \tag{14}$$

where the subscripts denote values at the corresponding grid points and $\gamma = h/k$. The finite-difference scheme is second order in accuracy but the associated matrix is not necessarily diagonally dominant because although all four exponential coefficients are positive for sufficiently small values of h and k it cannot be proved that the sum will not exceed the magnitude of the coefficient of $H_o$, which is the required condition. In order to avoid having exponential coefficients, Taylor series expansions for f and g in the $\xi$ and $\theta$ directions are used and then the exponential functions are expanded out. Working to second-order accuracy throughout then the finite-difference equations (14) can be simplified in a manner similar to that given by Dennis and Hudson (1978) to

$$(1 - \tfrac{1}{2}h + \tfrac{1}{2}p_o h + \tfrac{1}{8}p_o^2 h^2)H_1 + (1 + \tfrac{1}{2}h - \tfrac{1}{2}p_o h + \tfrac{1}{8}p_o^2 h^2)H_3$$

$$+ \gamma^2[(1 - \tfrac{1}{2}k\cot\theta_o + \tfrac{1}{2}q_o k + \tfrac{1}{8}q_o^2 k^2)H_2 + (1 + \tfrac{1}{2}k\cot\theta_o - \tfrac{1}{2}q_o k + \tfrac{1}{8}q_o^2 k^2)H_4]$$

$$- [2(1 + \gamma^2) + \tfrac{1}{4}h^2(p_o^2 + q_o^2) - h^2 r_o]H_o = h^2 s_o. \tag{15}$$

If equation (9) is expressed in finite-difference form using central differences throughout for the function H then equation (15) would have been obtained without any of the terms involving $p_o^2$ and $q_o^2$. These terms are second order in accuracy and therefore the finite-difference equations (15) may effectively be considered as an alternative central-difference scheme. In the case where $r_o \le 0$ it can easily be shown that the matrix associated with the finite-difference scheme (15) is diagonally dominant. This condition is satisfied by the equations for $\psi$ and $\Omega$. When $r_o > 0$ then it is not possible to prove diagonal dominance. However in all the calculations presented here there were no difficulties in obtaining convergent solutions without the use of an under relaxation parameter. When using conventional central-difference equations an under relaxation parameter must be included in order to obtain a convergent iterative scheme in some of the calculations.

The finite-difference form which was used for each of the boundary conditions (8) is now given. In each case it is assumed that the typical grid point $(\xi_0, \theta_0)$ is on the particular boundary under consideration. It then follows that

$$\psi_0 = 0, \quad \zeta_0 = -(3\psi_1/h^2 + \tfrac{1}{2}\zeta_1)/(1 + 3h/2), \quad \left.\right\} \quad \text{on } \xi = 0; \qquad (16a)$$
$$\Omega_0 = \frac{3}{4} \frac{\sin^2\theta_0}{1 + 2h + 2h^2} \int_0^\pi \Omega_1 \sin\theta \, d\theta$$

$$\psi_0 = \frac{1}{2} e^{2\xi_m} \sin^2\theta_0, \quad \Omega_0 = e^{2\xi_m} \sin^2\theta_0, \quad \zeta_0 = 0 \qquad \text{on } \xi = \xi_m; \qquad (16b)$$

$$\psi_0 = 0, \quad \zeta_0 = 0, \quad (\partial\Omega/\partial\theta)_0 = 0 \quad \text{on } \theta = 0, \pi. \qquad (16c)$$

The last condition of (16c) was satisfied by utilizing the finite-difference equations for $\Omega$ at $\theta = 0$ and $\theta = \pi$ and enforcing $\partial\Omega/\partial\theta = 0$ there in order not to introduce values outside the range of the computations. The condition for $\Omega_0$ on $\xi = 0$ has been derived by enforcing the condition of zero torque on the sphere which is

$$\text{Torque} = 2\pi\rho\nu a^3 \omega_0 \int_0^\pi (\frac{\partial\Omega}{\partial\xi} - 2\Omega)_{\xi=0} \sin\theta \, d\theta = 0.$$

All boundary conditions have been expressed in finite-difference form correct to second-order accuracy.

Initially all unknown variables $\psi$, $\zeta$ and $\Omega$ were given a linear variation between $\xi = 0$ and $\xi = \xi_m$. The order of the solution procedure was as follows. Starting from this initial solution, one complete iteration over all internal grid points was performed for $\Omega$, $\zeta$ and $\psi$ in that order. The new boundary conditions on $\Omega$ and $\zeta$ were then found. This whole sequence of operations is defined as one iteration. In each complete iterative sweep of the internal grid points the same order of the operations was preserved. This was starting at $\xi = h$, $\theta = 0$ and proceeding along all grid points of constant $\xi$. Then all grid points of constant $\xi = 2h$ were swept starting at $\theta = 0$ and so on.

The above procedure was repeated until

$$\sum |H_0^{(m)}/H_0^{(m+1)} - 1| < \varepsilon \qquad (17)$$

for $H = \Omega$, $\zeta$ and $\psi$, where the summation extends to all grid points and the superscripts refer to the number of iterations. The parameter $\varepsilon$ is an assigned tolerance which was taken to be $10^{-4}$ in most of the calculations.


RESULTS


The results given in the present section were calculated to test the method by means of a comparison with existing theoretical results. The only theoretical results available for nonzero values of T are those due to Childress (1964). The theory is valid only for small values of both R and T. A similar restriction has thus been placed on the range of the calculations. In all the calculations the iterative procedure was convergent without the need to use a relaxation parameter. Some of the calculations could have been speeded up by the inclusion of an over-relaxation parameter but all the results presented here were obtained using the simple Gauss-Seidel procedure. To test the effects of varying the mesh size and the position of application of the boundary condition at large distances, several calculations were performed when there was no rotation, i.e. T = 0. Table 1 shows these results at R = 0 and R = 0.1 for $D/D_s$, where D is the drag on the sphere and $D_s$ is the drag in Stokes flow.

The drag on the sphere is given by

$$D = -\frac{1}{3} D_s \int_0^\pi (RP\cos\theta + \zeta\sin\theta)\sin\theta \, d\theta, \qquad (18)$$

where $D_s = 6\pi\rho\nu Ua$, P is the dimensionless fluid pressure defined by dividing the dimensional pressure by $\rho U^2$ and $\rho$ is the density. The integral in (18) is to be evaluated over the surface of the sphere and the pressure is obtained by integration of the component of the Navier-Stokes equations in the $\theta$ direction. In the limit as both h and k tend to zero $D/D_s - 1$ should tend to zero when R = 0. For R = 0.1, Dennis and Walker (1971) give a value $D/D_s = 1.0337$ and Childress (1964) gives $D/D_s = 1.0375$. Table 1 shows that at these values of Reynolds number and Taylor

TABLE 1

Variation of $D/D_s$ for R = 0 and R = 0.1 with mesh size and position of the boundary condition at large distances.

| h | k | $\xi_m$ | R = 0 $D/D_s$ | R = 0.1 $D/D_s$ |
|---|---|---|---|---|
| 0.1 | $\pi/30$ | 3 | 1.0974 | 1.1018 |
| 0.05 | $\pi/60$ | 3 | 1.0983 | 1.1046 |
| 0.1 | $\pi/30$ | 4 | 1.0361 | 1.0490 |
| 0.05 | $\pi/60$ | 4 | 1.0366 | 1.0506 |
| 0.1 | $\pi/30$ | 5 | 1.0124 | 1.0369 |
| 0.05 | $\pi/60$ | 5 | 1.0125 | 1.0372 |
| 0.1 | $\pi/30$ | 6 | 1.0047 | 1.0342 |
| 0.1 | $\pi/30$ | 7 | 1.0019 | 1.0340 |

number the cruder mesh size is quite adequate but the position of the outer boundary condition is critical. None of the previous numerical investigations of the flow past a sphere in a non-rotating fluid have found it necessary to use such a large value of $\xi_m$. Most authors have used $\xi_m \leq 5$ which is clearly not adequate at low values of R. At R = 0.1 the results obtained here with $\xi_m = 7$ are in very good agreement with the numerical results of Dennis and Walker whereas the theory of Childress predicts a slightly higher value. This suggests that a value of R of order $10^{-1}$ is the limit of the applicability of the theoretical results and hence all calculations performed here have been restricted to values of R and T less than or equal to 0.5.

The work of Childress predicts theoretically that the angular velocity of the sphere relative to that of the fluid at infinity, $\omega^*$, is given by

$$\frac{\omega^*}{\omega_0} = \chi(\alpha)R + o(R) \qquad (19)$$

where $\alpha = 2T/R^2$ and $\chi(\alpha)$ is given by

$$\chi(\alpha) = \frac{3i}{8\alpha} \int_0^1 \{(t^2 + 4i\alpha t)^{\frac{1}{2}} - (t^2 - 4i\alpha t)^{\frac{1}{2}}\}(3t^3 - t)dt. \qquad (20)$$

In the case of $\alpha = 0$, i.e. T = 0, equation (20) gives $\chi(\alpha) = -0.375$. Table 2 shows the values obtained numerically for $\omega^*/\omega_0$ for various values of $\xi_m$ with h = 0.1 and k = $\pi/30$ in the case T = 0 and for the range of values R = 0, 0.05, 0.1 and 0.5.

TABLE 2

Variation of $-\omega^*/\omega_0$ for R = 0, 0.05, 0.1 and 0.5 with the position of the outer boundary condition.

| R $\xi_m$ | 0 | 0.05 | 0.1 | 0.5 |
|---|---|---|---|---|
| 4 | -0.004 | 0.012 | 0.028 | 0.099 |
| 5 | -0.002 | 0.016 | 0.034 | 0.108 |
| 6 | -0.001 | 0.018 | 0.036 | 0.112 |
| Childress | 0 | 0.01875 | 0.0375 | 0.1875 |

Again it is observed from table 2 that for values of R up to 0.1 the values obtained numerically are in excellent agreement with the values predicted theoretically by Childress.

In future work it is hoped to extend the range of the calculations in order to compare with the experimental results of Maxworthy (1965) at low values of R and T. It is then hoped to proceed to higher values of R and T in order to obtain a numerical simulation of the Taylor column and to test the discrepancy between the theoretical and experimental results in this parameter range. The work is part of a general project supported in part by a grant from NATO and in part by the Natural Sciences and Engineering Research Council of Canada. The authors would like to acknowledge discussions with Dr. S.N. Singh.

REFERENCES


Allen, D.N. De G. and Southwell, R.V. 1955 Quart. J. Mech. Appl. Math. 8, 129.

Allen, D.N. De G. 1962 Quart. J. Mech. Appl. Math. 15, 11.

Barnard, B.J.S. and Pritchard, W.G. 1975 J. Fluid Mech. 71, 43.

Childress, W.S. 1963 Jet Propulsion Laboratory, Pasadena, California, Space Programs Summary 37-18, Vol. IV, p. 46.

Childress, W.S. 1964 J. Fluid Mech. 20, 305.

Dennis, S.C.R. 1960 Quart. J. Mech. Appl. Math. 13, 487.

Dennis, S.C.R. 1973 Lecture Notes in Physics 19, 120.

Dennis, S.C.R. and Hudson, J.D. 1978 Proceedings of the International Conference on Numerical Methods in Laminar and Turbulent Flow, Swansea, United Kingdom, Pentech Press, London, p. 69.

Dennis, S.C.R., Ingham, D.B. and Cook, R.N. 1979 J. Comp. Physics 33, 325.

Dennis, S.C.R. and Walker, J.D.A. 1971 J. Fluid Mech. 48, 771.

Gosman, A.D., Pun, W.M., Runchal, A.K., Spalding, D.B. and Wolfshtein, M. 1969 Heat and Mass Transfer in Recirculating Flows. Academic Press, New York.

Grace, S.F. 1926 Proc. Roy. Soc. A 113, 46.

Greenspan, D. 1968 Lectures on the Numerical Solution of Linear, Singular and Non-linear Differential Equations. Prentice-Hall, Englewood Cliffs, New Jersey.

Hocking, L.M., Moore, D.W. and Walton, I.C. 1979 J. Fluid Mech. 90, 781.

Maxworthy, T. 1965 J. Fluid Mech. 23, 373.

Maxworthy, T. 1970 J. Fluid Mech. 40, 435.

Proudman, J. 1916 Proc. Roy. Soc. A 92, 408.

Roscoe, D.F. 1975 J. Inst. Math. Applics. 16, 291.

Roscoe, D.F. 1976 Int. J. Num. Meth. Eng. 10, 1299.

Runchal, A.K., Spalding, D.B. and Wolfshtein, M. 1969 Phys. Fluids 12, Suppl. II, 21.

Spalding, D.B. 1972 Int. J. Num. Meth. Eng. 4, 551.

Stewartson, K. 1952 Proc. Camb. Phil. Soc. 48, 168.

Taylor, G.I. 1917 Proc. Roy. Soc. A 93, 99.

Taylor, G.I. 1921 Proc. Roy. Soc. A 100, 114.

Taylor, G.I. 1922 Proc. Roy. Soc. A 102, 180.

MULTIFLUID INCOMPRESSIBLE FLOWS BY A

FINITE ELEMENT METHOD

by

A. DERVIEUX and F. THOMASSET

INRIA, B.P. 105, Rocquencourt, 78150 Le Chesnay, France

## INTRODUCTION

Many fluid flow simulations using <u>finite element</u> methods have been developped during the last decade (surveys can be found e.g. in ZIENKIEWICZ (1977), THOMASSET (1980)). The application of finite element methods to industrial problems is most appealing for the automatic mesh generation of complex geometries.

This paper deals with a new scheme for the simulation of multi-fluid, incompressible, viscous flows ; a model problem (Rayleigh-Taylor instability) is presented in §1.

The method uses an Eulerian description of the fluid : velocities are interpolated on a fixed mesh, using a first order finite element method ($P_1$ non conforming element with a divergence free basis) ; this approximation is described in §2. Although such an approximation is most suitable at low Reynolds numbers, the introduction of "upwinding" in the scheme (allowing higher Reynolds numbers) is discussed in §3. The interface-fitting technique, adapted to the velocity approximation, is described in §4. In §5, some numerical examples are given for academic problems.

## 1. A MODEL PROBLEM : RAYLEIGH-TAYLOR INSTABILITY.

Consider the two fluid configuration shown on Fig. 1, the upper fluid being heavier the problem is to predict the evolution of the interface given some initial data ; the notations are the following :

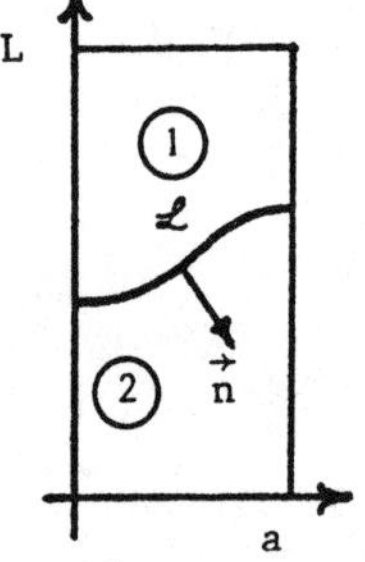


Figure 1

$\rho_\alpha$ : density of fluid $\alpha$

$\mu_\alpha$ : viscosity of fluid $\alpha$

$\vec{u}^\alpha = \begin{pmatrix} u_1^\alpha \\ u_2^\alpha \end{pmatrix}$ : velocity in fluid $\alpha$

$p^\alpha$ : pressure in fluid $\alpha$ with $\alpha = 1,2$

$g$ = gravity constant.

Then the Navier-Stokes equations are ($\delta_{ij}$ = Kronecker's symbol) :

$$\rho^\alpha \frac{\partial u_i^\alpha}{\partial t} + \rho^\alpha <\vec{u}^\alpha, \text{grad}> u_i^\alpha = \frac{\partial}{\partial x_j} \sigma_{ij}^\alpha - \rho^\alpha g\, \delta_{i2} \qquad \text{for } i=1,2 \ ; \ \alpha =1,2$$

$$\sigma_{ij}^\alpha = - p^\alpha \delta_{ij} + \mu^\alpha (\frac{\partial u_i^\alpha}{\partial x_j} + \frac{\partial u_j^\alpha}{\partial x_i}) \ ; \ \text{div } \vec{u}^\alpha = 0 \qquad 0 \le x \le a \ , \ 0 \le y \le L$$

with the boundary conditions

$$u_1^\alpha = 0, \quad \frac{\partial u_2^\alpha}{\partial n} = 0, \quad \alpha=1,2 \text{ on } x_1 = 0 \text{ and } x_1 = a$$

$$u_1^\alpha = u_2^\alpha = 0, \quad \alpha=1,2 \text{ on } x_2=0 \text{ and } x_2=L$$

and along the interface $\mathscr{L}$

$$\vec{u}^1 - \vec{u}^2 = 0 \; ; \quad (\sigma_{ij}^1 - \sigma_{ij}^2) n_j = 0 \; ;$$

the motion of the interface is defined by its normal velocity v given by

$$\langle \vec{v}, \vec{n} \rangle = \langle \vec{u}^\alpha, \vec{n} \rangle \; , \quad \alpha = 1,2 \; ;$$

the initial conditions are the following

$$\vec{u}^\alpha(x,0) = 0 \quad \alpha=1,2 \; ; \quad \mathscr{L} = \mathscr{L}_o \text{ (given)} \; .$$

## 2. VELOCITY APPROXIMATION.

For the velocity approximation, we use a nonconforming element ; given a triangulation of $\Omega$, we require for the discrete velocity $\vec{u}_h$ to lie in the fector space $V_h$ defined as follows

a) – $\vec{u}_h$ should be polynomial of degree at most one on each triangle ;

b) – $\vec{u}_h$ should be continuous at the <u>mid-side</u> points : thus the degrees of freedom of $u_h$ are its values at the mid-side points, and $u_h$ is not expected to be continuous along a side except at the mid point. CROUZEIX and RAVIART (1973) introduced this finite element and proved the following order of convergence $\| u_h - u \|_{1,\Omega} = 0(h)$ (see also TEMAM (1977)).

c) – div $\vec{u}_h = 0$ on each triangle.

The space $V_h$ is generated by a <u>divergence-free basis</u> as described in THOMASSET (1977) ; the idea of this construction is due to CROUZEIX (1976). The divergence-free basis functions fall into one of the two classes :

. a function $\vec{w}_m$ of the first class is associated to a mid-side node m :

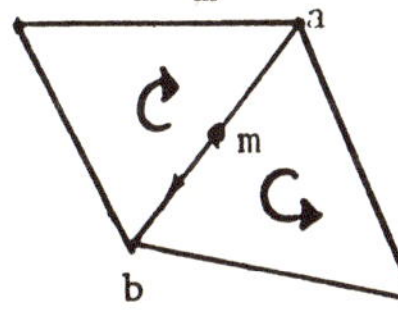

Fig. 2

$$\vec{w}_m(m) = \frac{\vec{ab}}{|ab|}$$

$$\vec{w}_m = 0 \text{ at all nodes different of m}$$

. a function $\vec{w}_v$ of the second class is associated to a <u>vertex</u> v :

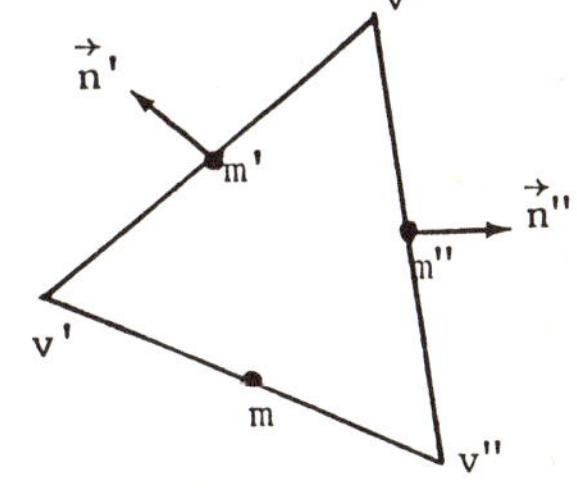

Fig. 3

$w_m = 0$ at all mid side nodes which do not belong to a common triangle with vertex v

$$w_v(m) = 0 \; ; \quad w_v(m') = \frac{-\vec{n}'}{|vv'|} \; ; \quad w_v(m'') = \frac{+\vec{n}''}{|vv''|}$$

. The case of an internal obstacle would require a special treatment for this, for the proof of completeness of the basis and for the practical implementation we refer to THOMASSET (1977).

The construction here presented is purely 2-dimensional (in fact it is equivalent to a $\Psi$-method) ; an extension to 3-dimensional space have been constructed in F. HECHT (1980-1981).

For the time discretization, we use a fully implicit scheme with linearized convective terms.

3. <u>UPWIND ADAPTATIONS</u>.

Several upwind schemes have been recently developped for first order triangular conforming elements : M. TABATA (1977) suggested a first order $L^{\infty}$ stable scheme : in order to compute the value at any vertex A of $\vec{v}\cdot\nabla\phi$ ( directional derivative of a dependent variable $\phi$), the derivative is taken on the "upstream" triangle crossed by $-\vec{v}$ (Fig. 4) :

$$\vec{\nabla}\phi(A) \simeq \vec{\nabla}\phi|_{BAC}$$

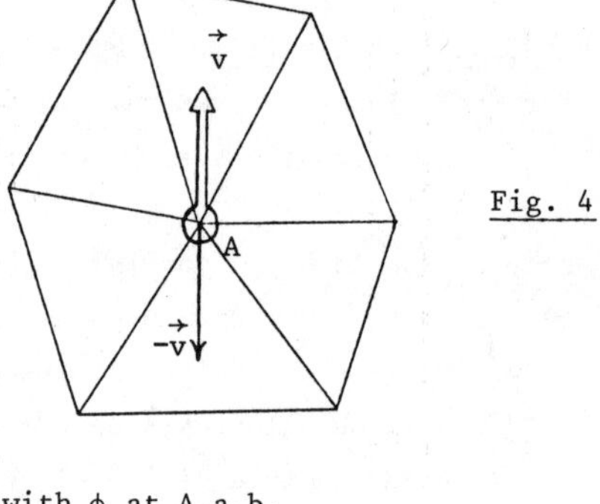

Fig. 4

A definitely less diffusive scheme (conjectured to be second order accurate) is proposed by BRISTEAU et Al. (1979) : in short, the idea is to look for two points a and b of the same half-line from node A (fig. 5) ; then the derivatives $\vec{\nabla}\phi$ are defined from the derivation at node A of a second order polynomial defined along the half line and coinciding with $\phi$ at A,a,b.

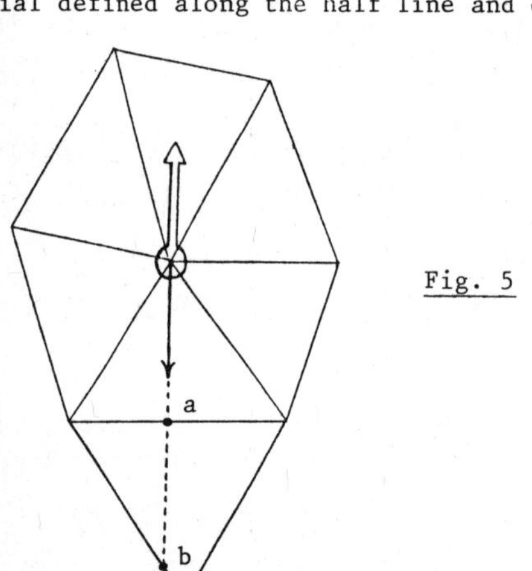

Fig. 5

An adaptation of these schemes to first degree triangular non conforming elements is the following : corresponding to each mid-side node, we construct an auxiliary triangulation such that the mid-side nodes of the initial triangulation become vertices for the auxiliary one (Fig. 6) ; then TABATA's method is applied to this new triangulation. This scheme, just as TABATA's, <u>generally</u> does not increase the bandwidth in the matrix of the resulting linear system of equations. However, for the sake of accuracy, we might choose to use extra further nodes to compute the directional derivatives (near the boundary, or to apply BRISTEAU's scheme).

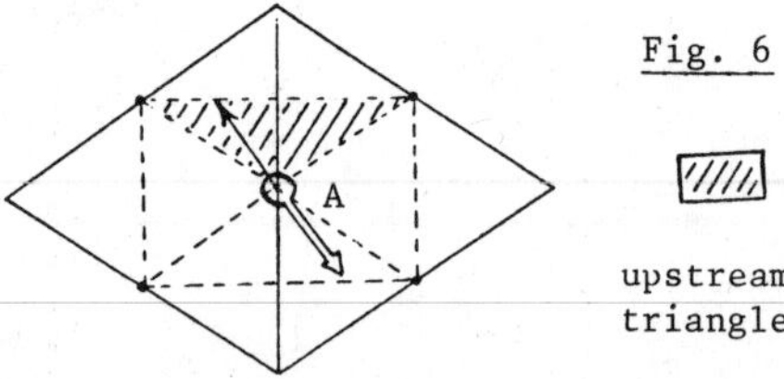

Fig. 6

## 4. <u>INTERFACE FITTING.</u>

To fit the interface, we tried to construct a method in accordance with the Finite Element Method : avoid punctual equalities (prefer integrated equalities) ; look for possibilities of extension to higher order.

Let $\chi(x,t)$ be the characteristic function of fluid 1 ( $\chi=1$ in fluid 1, 0 in fluid 2) ; then the motion of the interface is given by

$$
(1) \quad
\begin{aligned}
&\frac{\partial \chi}{\partial t} + (\vec{u} \cdot \nabla)\chi = 0 \\
&\chi(x,0) = \chi_o(x).
\end{aligned}
$$

The idea is to introduce the following variable substitution

$$
(H_o \phi)(x_1,x_2) = \begin{cases} 1 \text{ if } \phi(x_1,x_2) \geq 0 \\ 0 \text{ if } \phi(x_1,x_2) < 0, \end{cases}
$$

to solve

$$
(2) \quad
\begin{aligned}
&\frac{\partial \phi}{\partial t} + (\vec{u},\vec{\nabla})\phi = 0 \\
&\phi(x,o) = \phi_o(x)
\end{aligned}
$$

where $\phi_o$ is a smooth enough function satisfying $H_o \phi_o = \chi_o$ and to put $\chi = H_o \phi$.

This substitution is theoretically justified for convenient choices of $\phi_o$ as soon as the discretization of system (2) is a convergent one for the $L^2$ norm.

Then we may choose a $P_1$ continuous element to approximate the "pseudo-density"  ; the interface is consequently fitted as a continuous broken line with at most one segment in each triangle.

It is clear that the computation may remain accurate as long as the gradient of $\phi$ is not too large or too small ; this is related to "rezoning" problems for Lagrangian codes ; now in our computation the solution is quite simple and cheap, since we may replace function $\phi$ by the distance to the iso-line $\{\phi = 0\}$ with the sign of $\phi$. This "reinitialization" brings in a smoothing whose importance decreases rapidly with the mesh size.

In the results presented further, the triangulation for $\phi$ is the same than for the velocity : it is interesting to note the respective positions of freedom degrees for $\phi$ and u (Fig. 7)

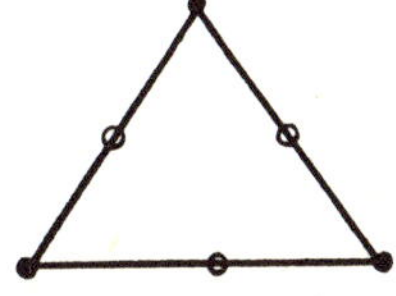

o velocity $\vec{u}$ freedom degree

• pseudo-density $\phi$ freedom degree

Fig. 7

Clearly, the above volume of each fluid is not exactly conserved, but the error observed is comparable to usual finite difference codes (0,05% per time step for a 20×20 mesh) when the boundary is not too much distorted.

It is plausible that in applications it would be better to choose for $\phi$ a sub-triangulation of the velocity one ; this would not be very expensive if we note that the $\phi$ equations needs only to be solved in a neighborhood of the interface.

We shall finish this section by noting that the pseudo density $\phi$ may be very useful to take into account surface tension effects, since the curvature K of the interface is given by

$$K = [(\frac{\partial\phi}{\partial x})^2 + (\frac{\partial\phi}{\partial y})^2]^{\frac{3}{2}} \{\frac{\partial^2\phi}{\partial x^2}(\frac{\partial\phi}{\partial y})^2 - 2\frac{\partial^2\phi}{\partial x\partial y}(\frac{\partial\phi}{\partial x})(\frac{\partial\phi}{\partial y}) + \frac{\partial^2\phi}{\partial y^2}(\frac{\partial\phi}{\partial x})^2\} .$$

## 5. NUMERICAL EXAMPLES.

We show in Fig. 8 and 9 results for the model problem of §1 with different values of the viscosity (chosen equal for both two fluids) viz resp. 0.01 (Reynolds number Re $\simeq$ 200) and 0.001 (Re $\simeq$ 2500) ; the mesh is a 20×20 (800 triangles) one and the number of degrees of freedom are respectively 441 for the pseudo-density and 1561 for the velocity.

We present in Fig. 10 results for a quite vicious problem, even at low Reynolds number (here Re $\simeq$ 1) : two fluids in a cylinder lie under gravity separated by a (horizontal) diameter ; the cylinder is set instantaneously in a constant rotation. We can see that shearings appear on the interface since the fluids have opposite motions ; the finite element discretization proves here its ability to provide a suitable approximation mesh (600 triangles, 331 degrees for the pseudo-density, 1141 degrees for the velocity).


## REFERENCES.

M.O. BRISTEAU, R. GLOWINSKI, B. MANTEL, J. PERIAUX, P.PERRIER, O. PIRONNEAU (1979) A finite Element Approximation of Navier-Stokes equations for incompressible viscous fluids. Iterative methods of solution, IUTAM Symposium, Paderborn, Approximation Methods for Navier-Stokes Problems, R. Rautmann Ed., Lecture Notes in Mathematics, Springer-Verlag, Berlin, 1980.
CROUZEIX (1976) Proceedings of : Journées "éléments finis", Université de Rennes (France)
CROUZEIX-RAVIART (1973) Conforming and non conforming finite element methods for solving the stationary Stokes equations, R.A.I.R.O.,R-3(1973), pp. 33-76.
F. HECHT (1980), Thèse de 3ème cycle, Univ. Paris VI
        (1981), Construction d'une base de fonctions $P_1$ non conforme à divergence nulle dans $\mathbb{R}^3$ (accepted by Revue Française d'Automatique Informatique et Recherche Opérationnelle/Analyse Numérique).
M. TABATA (1977), A finite element approximation corresponding to the upwind differencing, Memoirs of the Numerical Mathematics, 1 (1977), p. 47-63.
R. TEMAM (1977), Theory and Numerical Analysis of the Navier-Stokes equations", North-Holland, Amsterdam.
F. THOMASSET (1977) Numerical solution of the Navier-Stokes equations by finite elements methods ; V.K.I. Lecture Series, N° 86, Computational Fluid Dynamics, March 21-25, 1977.
F. THOMASSET (1980), Finite Element Methods for Navier-Stokes Equations, Von Karman Institute Lecture Series, Computational Fluid Dynamics, March 25-29, 1980, Rhode St Genèse, Belgium.
O.C. ZIENKIEWICZ (1977) The finite element method in engineering Science, Mac Graw Hill, New-York.

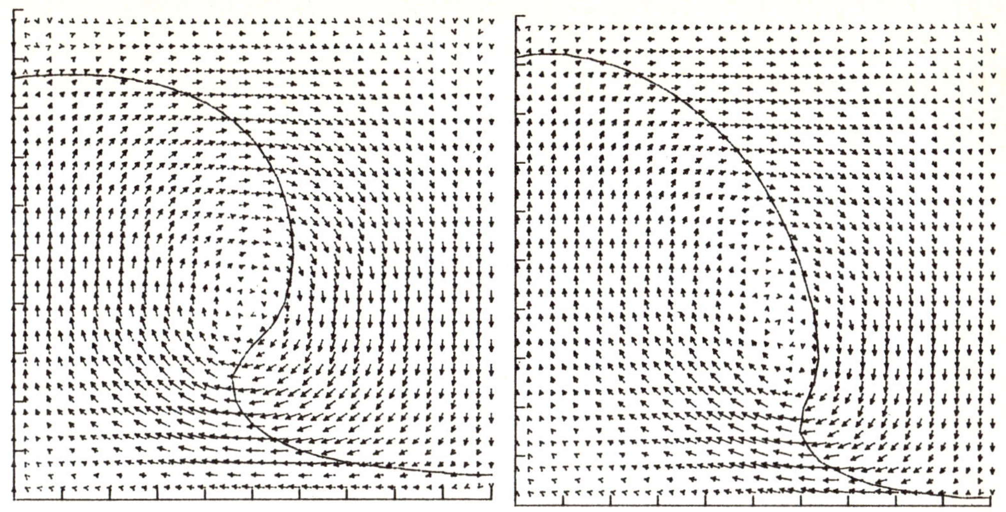

Figure 8                                    Figure 9

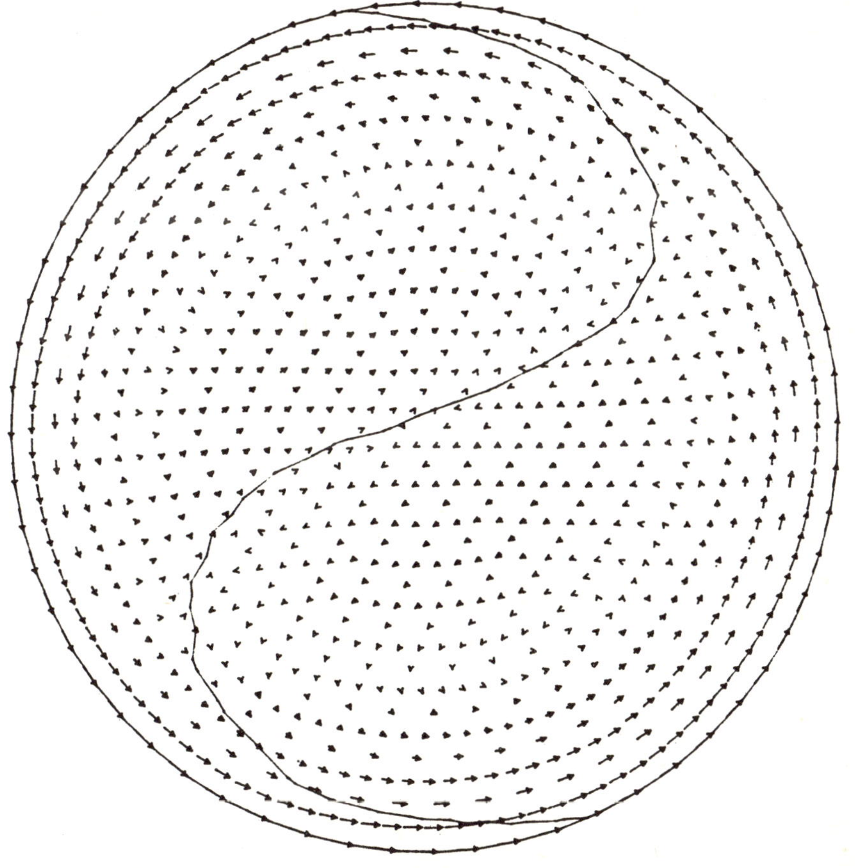

Figure 10

NUMERICAL CALCULATION OF TRANSONIC AXIAL TURBOMACHINERY FLOWS

Djordje S. Dulikravich*

National Aeronautics and Space Administration
Lewis Research Center
Cleveland, Ohio  44135

## SUMMARY

This paper presents a numerical method and the results of a computer program
for solving an exact, three-dimensional, full-potential equation that models rotating and nonrotating inviscid, absolutely irrotational, homentropic flows.  Besides
calculating the flows through an arbitrarily shaped rotor or stator blade row
mounted on an axisymmetric hub and confined in an axisymmetric duct, the computer
program is also capable of analysing flow fields about arbitrarily shaped wing-body
combinations, propellers, helicopter rotors in hover, and wind turbine rotors.  The
governing equation is solved numerically in a fully conservative form by using an
artificial time concept, a finite volume technique, rotated type-dependent differencing, successive line overrelaxation, and sequential boundary-conforming grid
refinement.  An artificial viscosity is added in fully conservative form; and an
initial guess for the potential field is applied, as determined by a two-dimensional
cascade analysis.

## INTRODUCTION

This work is based on the principles used in external transonic aerodynamics
(Jameson, 1974; Caughey and Jameson, 1977) and represents an extension of the
author's doctoral research in the field of potential, transonic turbomachinery flows
(Dulikravich, 1979).

In axial turbomachinery the pressure ratio across a stage can be substantially
increased by operating in the transonic speed regime.  Analyses of these possibly
shocked flows should account for their full nonlinearity (Rae, 1976).  The simplest
mathematical model that describes such flows exactly is the so-called full-potential
equation.  This equation can be obtained for rotating or stationary turbomachinery
geometries from the following analysis.

## ANALYSIS

The relative coordinate system $(x,y,z)$ is attached for a rotor (fig. 1) which
rotates at the constant angular speed $|\vec{\Omega}|$ about the  x  axis.  The free stream ad-
vances along the same axis at the constant speed $U_{-\infty}$.  Let

$$\vec{V}_r = u_r \hat{e}_x + v_r \hat{e}_y + w_r \hat{e}_z = q_r \hat{e}_s \tag{1}$$

be the relative velocity vector of the fluid with respect to the blade, and let the
absolute velocity vector be defined as

$$\vec{V} = \vec{V}_r + (\vec{\Omega} \times \vec{r}) \tag{2}$$

Then the sum of the inertia, centripetal, Coriolis, and pressure forces can be expressed in the form (Wu, 1952)

$$\vec{V}_r \times (\vec{\nabla} \times \vec{V}) = \vec{\nabla}I - T\vec{\nabla}S \tag{3}$$

Here  T  is the absolute static temperature, S  is the entropy, and rothalpy is defined as

$$I = h + \left(\vec{V}_r \cdot \vec{V}_r - \Omega^2 r^2\right)\!\big/2 = H - (\vec{\Omega} \times \vec{r}) \cdot \vec{V} \tag{4}$$

where  h  is the static enthalpy and  H  is the total enthalpy.

The entire flow field can be described with a single potential function $\varphi(x,y,z)$ defined as

$$\vec{V} = \vec{\nabla}\varphi \tag{5}$$

if the condition of irrotationality of the absolute velocity vector is satisfied. This requires that  $\vec{\nabla}I = 0$ and  $\vec{\nabla}S = 0$ simultaneously everywhere, or that  $\vec{\nabla}I = \vec{\nabla}S$. Consequently there should be no heat transfer between the fluid and solid surfaces, the flow should not separate, and all possible shock waves should be weak.  As already shown (Caradonna and Isom, 1972; Dulikravich, 1979), the continuity equation

$$\vec{\nabla} \cdot (\rho\vec{V}_r) = 0 \tag{6}$$

can be written in its full-potential vector operator form

$$a^2\vec{\nabla}^2\varphi - (\vec{\nabla}\varphi \cdot \vec{\nabla})(\vec{\nabla}\varphi \cdot \vec{\nabla}\varphi)/2 + 2(\vec{\nabla}\varphi \cdot \vec{\nabla})((\vec{\Omega} \times \vec{r}) \cdot \vec{\nabla}\varphi)$$

$$- ((\vec{\Omega} \times \vec{r}) \cdot \vec{\nabla})((\vec{\Omega} \times \vec{r}) \cdot \vec{\nabla}\varphi) = 0 \tag{7}$$

where  a  is the local speed of sound.  The canonical form (Dulikravich and Caughey, 1980) of equation (7) is

$$a^2\vec{\nabla}^2\varphi - q_r^2\varphi_{,ss} = 0 \tag{8}$$

where  s  is the relative streamline direction (fig. 1).  One can consider this second-order, quasi-linear, partial differential equation to be the steady-state limit of the more general (Garabedian, 1956; Jameson, 1974; Dulikravich, 1979) artificial, time-dependent equation expressed in a form suitable for the type-dependent, finite difference discretization

$$\left(a^2 - q_r^2\right)\!\left(\varphi^H_{,ss} - \varphi^E_{,ss}\right) + \left(a^2\vec{\nabla}^2\varphi^E - q_r^2\varphi^E_{,ss}\right) + (2\alpha_1\varphi_{,st} + 2\alpha_2\varphi_{,mt} + 2\alpha_3\varphi_{,nt} + \epsilon\varphi_{,t} = 0 \tag{9}$$

Here the superscript  H  designates upstream differencing (used only in the regions of locally supersonic relative flow for the purpose of numerically approximating the proper domain of dependence), and the superscript  E  designates central differencing; t  is the artificial time; and the (n,m) plane is locally orthogonal to the relative streamline direction (s  coordinate).  Equation (9) is iteratively solved by using a successive line overrelaxation technique where the iterative sweeps through the flow field are considered as successive intervals in the artificial time direction.  The mixed space-time derivatives are obtained by using a fi-

nite difference mixture of old, temporary, and new values of $\varphi$ obtained in the
iterative sweeping process (Jameson, 1976; Dulikravich, 1979).

In order of obtain a finite difference evaluation of the derivatives, the flow
field geometry, the governing equations, and the boundary conditions are transformed
from the physical space (fig. 1) into a parallelepiped-shaped computational space
(fig. 2). The periodic flow domain about a single arbitrarily shaped blade mounted
on an axisymmetric hub and confined in an axisymmetric duct is discretized by a number of intermediate axisymmetric surfaces (Dulikravich, 1980a). Every two-
dimensional periodic surface with the blade intersection contour in the middle is
then transformed (fig. 3) into a rectangular plane by using conformal mapping
(Dulikravich, 1979), elliptic polar coordinates, and coordinate stretchings and
shearings. The uniform grid in the $(X,Y)$ computational plane thus remaps back into
a periodic body-conforming, quasi-orthogonal grid in the physical space (fig. 4).
Each distorted, three-dimensional grid cell is mapped into a unit cube by using trilinear, isoparametric local mapping functions (Jameson and Caughey, 1977) of the form

$$b = \frac{1}{8} \sum_{p=1}^{8} b_p (1 + \bar{X}\bar{X}_p)(1 + \bar{Y}\bar{Y}_p)(1 + \bar{Z}\bar{Z}_p) \tag{10}$$

where the subscript $p$ refers to the value at the cube's corner; that is,

$$\bar{X}_p = \pm1 \qquad \bar{Y}_p = \pm1 \qquad \bar{Z}_p = \pm1 \tag{11}$$

and $b$ stands for any of the following: $x, y, z,$ or $\varphi$.

In order to use the finite volume technique as defined by Caughey and Jameson
(1977), equation (8) has to be expressed in terms of the computational $(X,Y,Z)$ coordinates (fig. 2). If the geometric transformation is defined as
$[J] = \partial(x,y,z)/\partial(X,Y,Z)$ and $D$ is a determinant of $[J]$, the modified contravariant
components of the relative velocity vector are

$$\lfloor U_r V_r W_r \rfloor^T = D[J]^{-1} \lfloor u_r v_r w_r \rfloor^T \tag{12}$$

If $[B] = [J]^{-1}[J^T]^{-1}$, the steady part of equation (9) can be written in the
following matrix form:

$$\left( \left( a^2 - q_r^2 \right) \Big/ \left( Dq_r \right) \right) \lfloor U_r V_r W_r \rfloor \lfloor \nabla_{XYZ}\varphi_{,s} \rfloor^T + a^2 \left( \lfloor \nabla_{XYZ} \rfloor [B] \lfloor \nabla_{XYZ}\varphi \rfloor^T \right.$$

$$\left. - (1/Dq_r) \lfloor U_r V_r W_r \rfloor \lfloor \nabla_{XYZ}\varphi_{,s} \rfloor^T \right) = 0 \tag{13}$$

where

$$\lfloor U_r V_r W_r \rfloor = D[J]^{-1} \lfloor 0 \ \Omega z \ -\Omega y \rfloor^T + D[B] \lfloor \nabla_{XYZ}\varphi \rfloor^T \tag{14}$$

and

$$\varphi_{,s} = (1/Dq_r) \lfloor U_r V_r W_r \rfloor \lfloor \nabla_{XYZ}\varphi \rfloor \tag{15}$$

Equation (6), which is actually equation (8) multiplied by $\rho/a^2$, transforms into

$$\left( \rho U_r \right)_{,X} + \left( \rho V_r \right)_{,Y} + \left( \rho W_r \right)_{,Z} = 0 \tag{16}$$

The boundary conditions in the computational space are the following: On the hub
surface $W_r = 0$, on the blade surface $V_r = 0$, and on the duct surface

$$W_r = |\vec{\Omega}|\,((z_{,X}x_{,Y} - x_{,X}z_{,Y})z - (x_{,X}y_{,Y} - y_{,X}x_{,Y})y) \tag{17}$$

The periodic boundary condition applied on the surfaces A'A"B'B" (fig. 1) is $\varphi(X + b,0,Z) = \varphi(X - b,0,Z)$, where $b \leq \pi$. If the absolute enthalpy of the fluid at upstream infinity is constant, it follows from equation (4) that the most general flow kinematics at upstream infinity can be expressed as $\vec{V}_{-\infty} = U_{-\infty}\,\hat{e}_x + (C_\theta/r)\,\hat{e}_\theta$. The blade trailing edge is enforced to be a stagnation line. The finite discontinuity in the velocity potential at the trailing edge is equal to the circulation $\Gamma(r)$ of the velocity field. This discontinuity (i.e., no jump in static pressure) is enforced at every point of an arbitrarily shaped vortex sheet (surface D'E'D"E" in fig. 2) after each iterative sweep through the flow field.

## RESULTS

A computer program has been developed on the basis of a previous analysis (Dulikravich, 1980e). This program uses a potential field generated by a two-dimensional cascade analysis as an initial guess for the three-dimensional potential field calculations (Dulikravich, 1980c). The iterative convergence rate is accelerated by using a three-level consecutive, grid refinement sequence.

The program was tested for two speed regimes. The first test case was a subsonic free rotor of a propeller-type, 100-kW wind turbine, the NASA Mod-0 type (Dulikravich, 1980b and 1980d). Numerical results showing chordwise distribution of the relative Mach number at the tip section are presented in figure 5. Because of the lack of available published results for three-dimensional, transonic rotor calculations, we decided to test an atypical ducted rotor having eight nontwisted, nontapered blades mounted on a doubly infinite cylindrical hub with a hub-tip radius ratio $r_h/r_t$ of 0.85. The blades were composed of NACA 0012 airfoil sections and had a constant twist (or setting) angle of $4^\circ$ and a relative chord length $r_t$ of 0.1. Numerical results obtained at a tip section for $|\vec{\Omega}| = 120$ rpm, $(M_x)_{-\infty} = 0.7458$, $(p_t)_{-\infty} = 16$ N/cm$^2$, and $(T_t)_{-\infty} = 325$ K are shown in figure 6. Insufficient sharpness of the isentropic shock developing on the suction surface is due to the relatively coarse grid used and the low number of iteration cycles.

## CONCLUSIONS

A general computer program has been developed for fast and accurate, fully conservative finite volume calculations of the full-potential equation for transonic, steady, three-dimensional, potential rotating and nonrotating flows. As a part of the program, three-level, refined, boundary-conforming grids are generated for arbitrarily shaped cascades of blades mounted on an axisymmetric hub and confined in an axisymmetric duct. The program uses results of the two-dimensional cascade flow calculations as an initial guess for the three-dimensional iterative procedure.

Artificial viscosity was added in conservative form, and this assures the correct strength and position of the captured isentropic shocks.

---

*National Research Council – National Aeronautics and Space Administration Research Associate.

## REFERENCES


Caradonna, F. & Isom, M. 1972 Subsonic and Transonic Potential Flow over Helicopter Rotor Blades, AIAA Journal, 10, 1606.

Caughey, D. A. & Jameson, A. 1977 Numerical Calculation of Transonic Potential Flow About Wing-Fuselage Combinations. AIAA Paper 77-677.

Dulikravich, D. S. 1979 Numerical Calculation of Inviscid Transonic Flow Through Rotors and Fans. Ph.D. Thesis, Cornell University.

Dulikravich, D. S. 1980a Fast Generation of Body Conforming Grids for 3-D Axial Turbomachinery. To be presented at the Workshop on Numerical Grid Generation Techniques for Partial Differential Equations, NASA Langley, Hampton, Va., October 6-7.

Dulikravich, D. S. 1980b Numerical Calculation of Steady Inviscid Full Potential Compressible Flow About Wind Turbine Blades. AIAA Paper 80-0607. (Also published as NASA TM-81438.)

Dulikravich, D. S. 1980c CAS2D - FORTRAN Program for Nonrotating, Blade-to-Blade, Steady, Potential Transonic Cascade Flows. NASA TP-1705, 1980.

Dulikravich, D. S. 1980d WIND - Computer Program for Three-Dimensional, Potential, Compressible Flow About the Propeller Type Wind Turbine Blades. Submitted for publication as NASA Technical Publication.

Dulikravich, D. S. 1980e CAS3D - FORTRAN Program for Three-Dimensional, Steady, Transonic, Potential, Axial Turbomachinery Flows. Submitted for publication as NASA Technical Publication.

Dulikravich, D. S. & Caughey, D. A. 1980 Finite Volume Calculation of Transonic Potential Flow Through Rotors and Fans. Cornell University, Ithaca, NY, Rep. no. FDA-80-03.

Garabedian, P. R. 1956 Estimation of the Relaxation Factor for Small Mesh Sizes. Math Tables Aids to Comput., 10, 183.

Jameson, A. 1974 Commun. Pure Appl. Math. 27, 283.

Jameson, A. 1976 Transonic Flow Calculations. Lectures Presented at Von Karman Institute.

Jameson, A. & Caughey, D. 1977 A Finite Volume Scheme for Transonic Potential Flow Calculations. AIAA Paper 77-635.

Rae, W. J. 1976 Nonlinear Small-Disturbance Equations for Three-Dimensional Transonic Flow Through a Compressor Blade Row. Calspan Corp., Buffalo, NY, Rep. no. CALSPAN-AB-5487-A-1. (AFOSR-76-1082TR, AD-A031234.)

Wu, C.-H. 1952 A General Theory of Three-Dimensional Flow in Subsonic and Supersonic Turbomachines of Axial-, Radial-, and Mixed-Flow Types. NACA TN 2604.

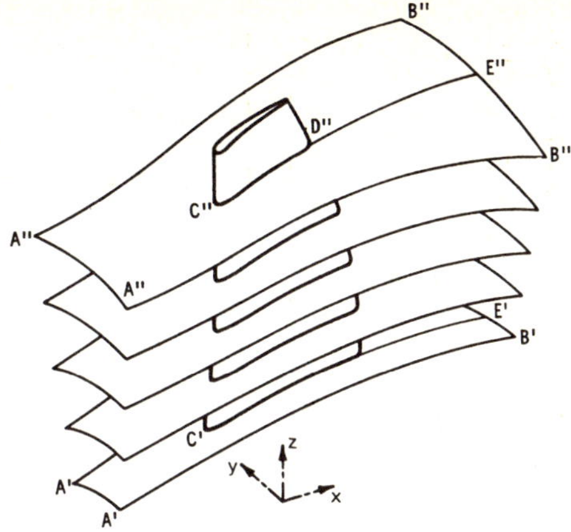

Figure 1. - Physical space.

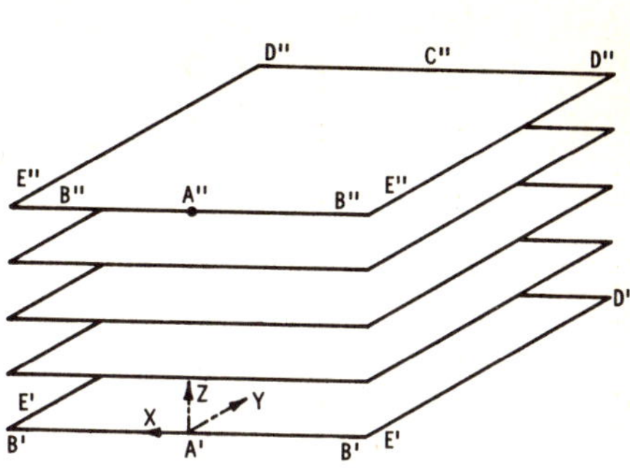

Figure 2. - Computational space.

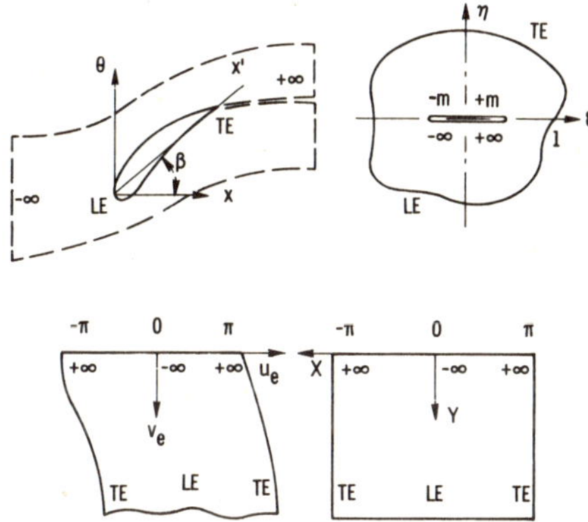

Figure 3. - Geometric transformation sequence, where TE denotes trailing edge and LE denotes leading edge.

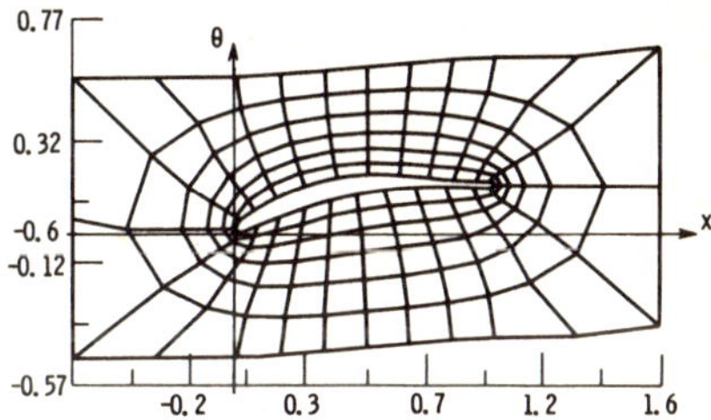

Figure 4. - Computational mesh in physical space (x, y, z).

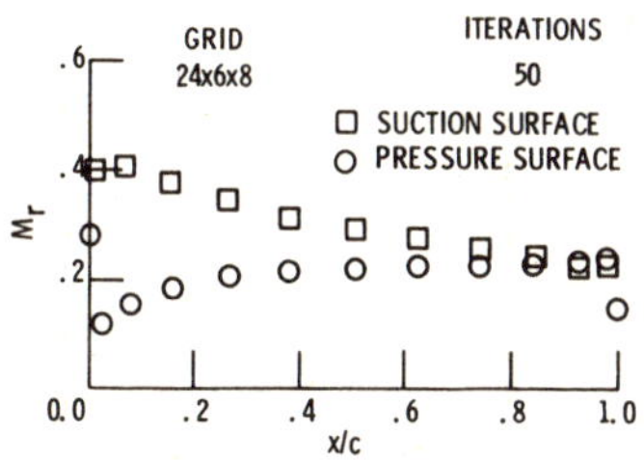

Figure 5. - Subsonic rotor.

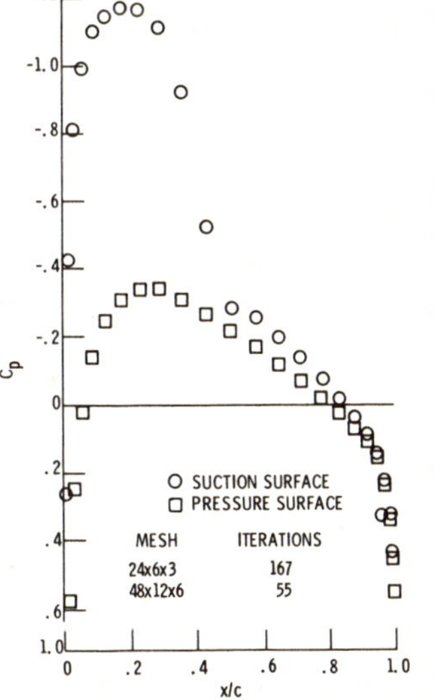

Figure 6. - Transonic rotor.

A STUDY OF REACTIVE DIFFUSION PROBLEMS WITH STIFF INTEGRATORS AND ADAPTIVE GRIDS


H. A. Dwyer - University of California, Davis

F. Raiszadeh - University of California, Davis

G. Otey - Sandia Laboratories


## Introduction

The solution of fluid dynamic flows with chemical reactions presents some very serious problems for the numerical modeler.  Two problems which are usually associated with flows and chemical reactions are stiffness and spatial scaling, and in this paper we will address these problems with adaptive gridding and implicit block solvers.  The paper will be broken up in three parts and these parts will address the following topics:  (1) Time and spatial integrators with stiff chemistry; (2) Strategies for time dependent grid adaption with moving flames; and  (3) Geometry problems with grid adaption in multi-dimensional flows. In order to make such an ambitious study tractable, a model problem has been developed (Otey, 1979) and employed.  This model problem retains the essential features of stiff chemical flows but its simplicity allows insight into the numerical difficulties involved in the solution of complex systems.

## Model Problem and Numerical Methods

In this section of the paper the model problem and the results of our investigation of numerical integration methods will be described.  The model problem itself consisted of a time-dependent, reaction-diffusion system governed by two species equations and an energy equation.  In terms of dimensionless variables, the system of equations are

$$\frac{\partial \rho_A}{\partial t} = \frac{\partial^2 \rho_A}{\partial x^2} - A_1 \rho_A e^{-\theta_1/T} \tag{1}$$

$$\frac{\partial \rho_B}{\partial t} = \frac{\partial^2 \rho_B}{\partial x^2} + A_1 \rho_A e^{-\theta_2/T} - A_2 \rho_B e^{-\theta_2/T} \tag{2}$$

$$\frac{\partial T}{\partial t} = \frac{\partial^2 T}{\partial x^2} + \frac{\beta}{1 + \beta} A_1 \rho_A e^{-\theta_1/T} + \frac{1}{1 + \beta} A_2 \rho_B e^{-\theta_2/T} \tag{3}$$

where $\rho_A$ and $\rho_B$ are the two specie density, $T$ - temperature, $\beta$ - energy distribution parameter, $\theta_1$ and $\theta_2$ - activation energy parameters, and $A_1$ and $A_2$ - pre-exponential constants.  As the temperature rises, material A will exothermically decompose by the process $A \rightarrow B \rightarrow C$ with the overall density remaining constant.  Two reactions are the minimum number necessary to produce stiffness caused by multiply time scales in the system.

The choice of numerical method is strongly influenced by the exponential nature of the reaction and the large spatial gradients of temperature and concentrations. It was found that fourth order spatial approximations to the diffusion term always gave more efficient and accurate results than second order approximations.  Therefore, fourth order approximations were employed in the comparative studies of the integration methods.  In these comparative studies five different methods were employed and they were the following:  (1) The method of lines with fourth order central five point differences MOL (Gear, 1971);  (2) The method of lines with splines PDECOL (Madsen, 1976);  (3) The method of operator splitting or fraction steps (Yanenko, 1971) with fourth order Padé differences (Kopal, 1961);  (4) An explicit/ODE method (Otey, 1979); and  (5) A linear block implicit method with fourth order Padé differences (Beam, 1977, and Brinley, 1977).

The results of this comparative study are shown in Table 1 where the normalized computer times for the five different methods are shown as a function of the reaction parameters.  The ratio of the pre-exponential constant $A_2/A_1$ has been varied over seven order of magnitude, and in all cases the linear block method was the most efficient for a prescribed level of accuracy on the flame speed $U_f$. Therefore, it was decided to use the linear block method for all other calculations with the adaptive grid.

(Note: The linear block method is said to be unconditionally stable for all step sizes $\Delta t$. However, for the problem with chemistry there are serious problems with the linearization of the exponential terms. These problems lead to a singularity in the finite difference equations for a certain time step. The details of this problem are given by Otey, 1978.)

## Adaptive Grid Methods

The second phase of the investigation involved an adaptive grid study, the purpose of which is to reduce the number of grid points necessary to carry out the flame calculations efficiently. Flames with high Damhöhler numbers introduce a very thin flame and high spatial gradients into a problem. This high gradient region is particularly difficult since the flame is moving in space and the grid must accurately resolve the diffusion inside the flame. In order to simplify the problem, only one reaction was considered in the adaptive grid part of the investigation ($A_2 = 0$), since only a high spatial gradient is necessary.

The need for an adaptive grid proceedure can easily be seen by considering the case of $A_1 = 4 \cdot 10^7$ and $\theta_1 = 4$ which requires 600 uniform grid points spread over the region $0 \leq x \leq 1$ with fourth order differences. With present-day computing machines it is not possible to extend this size of grid to multiple dimensions, and more efficient techniques must be employed. The method proposed in the present paper is to define a time and solution dependent transformation which proportions grid points on the derivative. In the computational space the grid will be uniform and the following functional dependence is desired.

$$\text{constant} = d\xi \, \alpha \left| \frac{\partial T}{\partial x} \right| dx \tag{4}$$

In order to normalize, allow for optimization, and remove singularities, Equation (4) is cast in the following integral form

$$\xi(x,t) = \frac{\displaystyle\int_0^x \left( 1 + b \left| \frac{\partial T}{\partial x} \right| \right) dx}{\displaystyle\int_0^{x_{max}} \left( 1 + b \left| \frac{\partial T}{\partial x} \right| \right) dx} \tag{5}$$

where b is an adjustable constant used for optimization, which will be discussed in detail later in the paper.

For the case $b = 0$, a uniform distribution of grid points is obtained, while for b becoming large, Equation (5) chooses an isotherm distribution.

$$\xi(x,t) = \frac{\displaystyle\int_0^x |dt|}{\displaystyle\int_0^{x_{max}} |dt|} \tag{6}$$

An entirely equivalent formulation is a generalized arclength form where the relationship between $\xi$ and $x$ is given by

$$\xi(x,t) = \frac{\displaystyle\int_0^x \left( 1 + b \left[ \frac{\partial T}{\partial x} \right]^2 \right) dx}{\displaystyle\int_0^{x_{max}} \left( 1 + b \left[ \frac{\partial T}{\partial x} \right]^2 \right) dx} \tag{7}$$

For completeness, one further form will be employed and this form attempts to place grid points in regions of high second derivatives. The $\xi - x$ relationship is

$$\xi(x,t) = \frac{\displaystyle\int_0^x \left( 1 + b \left| \frac{\partial T}{\partial x} \right| + C \left| \frac{\partial^2 T}{\partial x^2} \right| \right) dx}{\displaystyle\int_0^{x_{max}} \left( 1 + b \left| \frac{\partial T}{\partial x} \right| + C \left| \frac{\partial^2 T}{\partial x^2} \right| \right) dx} \tag{8}$$

where C is a weighting constant that controls the relative importance of second derivative influence.

In terms of the transformed variables, $\xi$ and $\tau$, the model equation becomes

$$\frac{\partial \rho_A}{\partial \tau} = \frac{-\partial \rho_A}{\partial \xi}\,\xi_t + \frac{\partial}{\partial \xi}\left(\xi_x \frac{\partial \rho_A}{\partial \xi}\right)\xi_x - A_1 \rho_A e^{-\theta/T} \tag{9}$$

$$\frac{\partial T}{\partial \tau} = -\frac{\partial T}{\partial \xi}\,\xi_t + \frac{\partial}{\partial \xi}\left(\xi_x \frac{\partial T}{\partial \xi}\right)\xi_x + A_1 \rho_A e^{-\theta/T} \tag{10}$$

Equations (9) and (10) were solved with the same linear block technique employed for Equations (1) through (3). The time dependent transformation between $\xi$ and x is solved with a simple trapezoidal rule integration of the integral equations, and linear interpolation is used to find x at the constant $\xi$ values. The formal accuracy of the solution of the transformation equation does not influence the formal accuracy of the solution of the transport equation; however, the transformation equation must locate regions of temperature gradient well for it to be effective.

The results of using the adaptive grid transformation were very successful. With fifty grid points and second order central differences the adaptive grid method gave results of equal accuracy to the 500 grid point uniform grid calculation with fourth order accuracy. Considering the order of the finite difference approximation, this represents more than one order of magnitude reduction in storage requirements and almost an order of magnitude reduction in computational time. The results of various values of b along with the arclength transformation, Eq. (7), are shown in Figure 1. Along the ordinate in Figure 1 is the ratio of the calculated flame speed to the exact value of flame speed (converged value) for three different values of pre-exponential constant $A_1$.

For $A_1 = 4 \cdot 10^7$ the uniform grid results, b = 0, gave a flame speed 60% too low. As soon as b has assumed values greater than b = .05, the adaptive grid proceedure has given values of almost 1% accuracy. As the reaction rate is decreased, the advantages of using an adaptive grid became less and less. At $A_1 = 1.98 \cdot 10^6$ there is still very substantial advantage, but at $A_1 = 2.2 \cdot 10^5$ there is essentially no advantage for employing an adaptive grid. The problem is due to the time dependent metric $\xi_t$ and not the spatial metric $\xi_x$. For $A_1 = 2.2 \cdot 10^5$ the time step for the calculation was $t = 2 \cdot 10^{-4}$, while the flame speed was $U_f = 35.5$, and for $A_1 = 4 \cdot 10^7$ the time step was $t = 1 \cdot 10^{-6}$ and the flame speed $U_f = 478.7$. These numbers imply that during one time step the flame moves $x_f = 7.1 \cdot 10^{-3}$ for the slow reaction and $\Delta x_f = 4.8 \cdot 10^{-4}$ for the fast reaction. For the slow reaction, diffusion is relatively more effective and the flame moves more than one order of magnitude in distance. Therefore, even though the gradients are quite steep for $A_1 = 2.2 \cdot 10^5$, it is more efficient to employ a uniform grid, and likewise it follows that it is not <u>always</u> advantageous to use adaptive grid techniques. However, the present problem is due solely to $\xi t$, and the conditions for adaptive gridding will be more favorable with steady problems.

Shown in Figures 2 through 5 are more results for this model problem. In Figure 2 are shown flame speed calculations for larger values of b. For $A_1 = 2.2 \cdot 10^5$, increased sensitivity to adaption only degrades the calculation. However, with $A_1 = 4 \cdot 10^7$ the calculation remains accurate for b = 50. Figures 3 and 4 show the temperature profiles in the physical and computational space, respectively, and Figure 4 shows clearly the efficient use of grid points in the computational space. The final figure in this group, Figure 5, presents $\Delta x$ or $\xi x$ in the computational space. The only problem is the sharp change in x at the ends of the flame where the second derivative is large. To remove this potential problem the transformation given by Equation (8) must be further explored.

## Multi-Dimensional Calculations

The third phase of the research was the multi-dimensional calculations with the model problem. In terms of generalized nonorthogonal coordinates and in strong conservative form (Peyret, 1975, and Vinokur, 1974) the one reaction model problem becomes

$$\frac{\partial}{\partial \tau}\left(\frac{\vec{q}}{J}\right) = -\frac{\partial}{\partial \xi}\left(\frac{\xi_t \vec{q}}{J}\right) - \frac{\partial}{\partial n}\left(\frac{n_t \vec{q}}{J}\right)$$

$$+ \frac{\partial}{\partial \xi}\left(\frac{\partial \vec{q}}{\partial \xi}\frac{\left[\xi_x^2 + \xi_y^2\right]}{J} + \frac{\partial \vec{q}}{\partial n}\frac{\left[\xi_x n_x + \xi_y n_y\right]}{J}\right) \qquad (11)$$

$$+ \frac{\partial}{\partial n}\left(\frac{\partial \vec{q}}{\partial n}\frac{\left[n_x^2 + n_y^2\right]}{J} + \frac{\partial \vec{q}}{\partial \xi}\frac{\left[\xi_x n_x + \xi_y n_y\right]}{J}\right) + A_1 R\, e^{-\theta_1/T}$$

where
$$\vec{q} = \begin{pmatrix} \rho_A \\ T \end{pmatrix} \quad \text{and} \quad \vec{R} = \begin{pmatrix} -\rho_A \\ +\rho_A \end{pmatrix}$$

and the metric quantities can be found in Peyret and Viviand (1975). The most serious problem introduced in the solution of Equation (11) is the rapid variation of the metric coefficient caused by flame shape and geometry. At the present time there is no effective way to control grid shape in an adaptive grid calculation, and this is the part of the calculation that is weak. Shown in Figures 6 and 7 is the propagation of a flame in a container with sinusoidal and straight section boundaries. The left hand part of each figure is the instantaneous coordinate system, and the right hand part the isotherm distribution for $A_1 = 2.2 \cdot 10^5$. The flame was ignited on the bottom boundary and is propagating toward the top boundary. The side walls are kept cool.

As Figures 6 and 7 show, the adaptive grid strategy of Equation (5) does provide a natural coordinate system for the problem, and the flame speed was accurate. However, a close inspection of the graphs shows that the grid cell shape changes rapidly near boundaries and with flame curvature. If a more complicated boundary is employed with the same grid, then the grid geometry problem becomes more serious.

A problem with more geometry is shown in Figures 8 and 9 where the same type of calculation is performed but with the upper boundary more complicated. As the flame propagates from bottom to top, the rapid variation of metric coefficient introduced by the geometry gradually causes a false and spurious diffusion to occur, as can be seen in the isotherm distribution. This can be overcome by adding grid points, but it is the author's belief that the most progress can be made with the development of orthogonal coordinates. At the present the research is continuing in this direction and an adaption of the method of Potter and Tuttle (1973) is being employed.

## Conclusions

In the present paper the results of a comprehensive study of the numerical problems associated with flows with chemical reactions have been presented. The problems solved are simplifications of many practical problems, but they retain the key features of complex flows. Also, the time and space scaling problems are general to most problems in fluid mechanics and the results will be applicable to many areas. The major conclusions of our study are:

1. The use of newton-linearized block solvers with higher order difference methods has given very efficient calculations for flame propagation and stiff chemistry.
2. Adaptive gridding based on gradients of the dependent variables has yielded very large efficiencies for problems with large spatial and temporal gradients. As these temporal gradients decrease there is less advantage to be gained with adaptive techniques.
3. Adaptive gridding techniques offer the promise of very large economies for multi-dimensional problems. However, the problem of grid cell geometry is not completely under control, and a need for additional research is necessary in this area.

<u>References</u>


R.M. Beam and R.F. Warming, 1977, "On the Construction and Application of Implicit Factored Schemes for Conservation Laws, SIAM-AMS Proceedings, Vol.11, April.

W.R. Birley and H. McDonald, 1977, "Solution of the Multidimensional Compressible Navier-Stokes Equations by a Generalized Implicit Method," <u>J. Computational Physics</u>, Vol.24, August, pp.373-392.

W.C. Gear, 1971, <u>Numerical Initial Value Problems in Ordinary Differential Equations</u>, Prentice-Hall, Englewood Cliffs, N.J.

Z. Kopal, 1961, <u>Numerical Analysis</u>, John Wiley and Sons, New York.

N.K. Madsen and R.F. Sincovic, 1976, "PDECOL: General Collocation Software for Partial Differential Equations," Univ. of Calif., LLL, Report UCRL-51186, May.

G.R. Otey, 1978, "Numerical Methods for Solving Reaction Diffusion Problems," Ph.D. Thesis, Dept. of Mechanical Engineering, Univ. of Calif., Davis, July.

G.R. Otey and H.A. Dwyer, 1979, "Numerical Study of the Interaction of Fast Chemistry and Diffusion," <u>AIAA J.</u>, Vol.17, No.6, June, p.606.

R. Peyret and H. Viviand, 1975, "Computation of Viscous Compressible Flows Based on the Navier-Stokes Equations," AGARD-AG212, 1975.

M. Vinokar, 1974, "Conservation Equations of Gasdynamics in Curvilinear Coordinate Systems," <u>J. Computational Physics</u> <u>14</u>(2), pp.105-125.

N.N. Yanenko, 1971, <u>The Method of Fractional Steps</u>, Springer Verlag, New York.


TABLE I    The effect of stiffness on relative efficiency ($T = 0.2$, $_1 = _2 = 4$, $= 1$)

| | | | | |
|---|---|---|---|---|
| $A_1 = 2.2 \times$ | $10^7$ | $10^5$ | $10^5$ | $10^5$ |
| $A_2 = 2.2 \times$ | $10^5$ | $10^5$ | $10^8$ | $10^{10}$ |
| $A_2/A_1$ | $10^{-2}$ | 1 | $10^3$ | $10^5$ |
| $U_f$ | 105.3 | 27.8 | 35.3 | 35.4 |
| $f$ | 0.033 | 0.084 | 0.068 | 0.068 |
| Computer time for linear block method (CDC 6600, seconds) | 89 | 4.2 | 8.2 | 8.2 |

| Method | Normalized Computer Time | | | |
|---|---|---|---|---|
| PDECOL | 2.5 | 7.1 | 13.0 | 11.0 |
| MOL | 3.1 | 4.3 | 7.2 | 6.3 |
| Explicit/ODE | 3.0 | 8.3 | 6.2 | 5.9 |
| Operator Splitting | 2.2 | 7.7 | 5.6 | 6.1 |
| Linear Block | 1.0 | 1.0 | 1.0 | 1.0 |

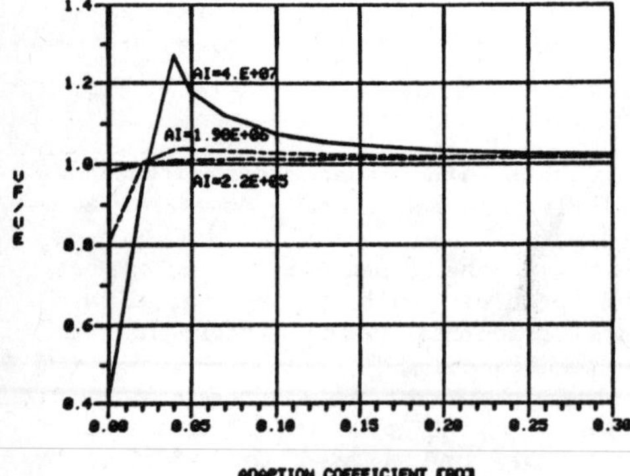


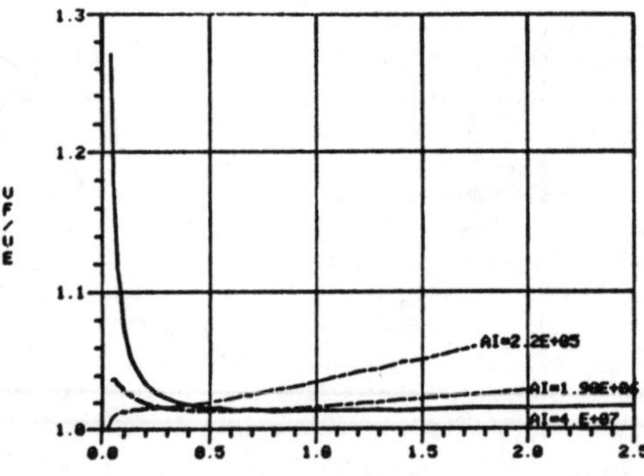


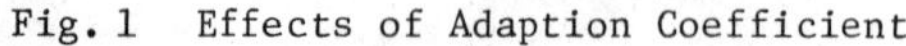

Fig. 1    Effects of Adaption Coefficient      Fig. 2    Effects of Adaption Coefficient

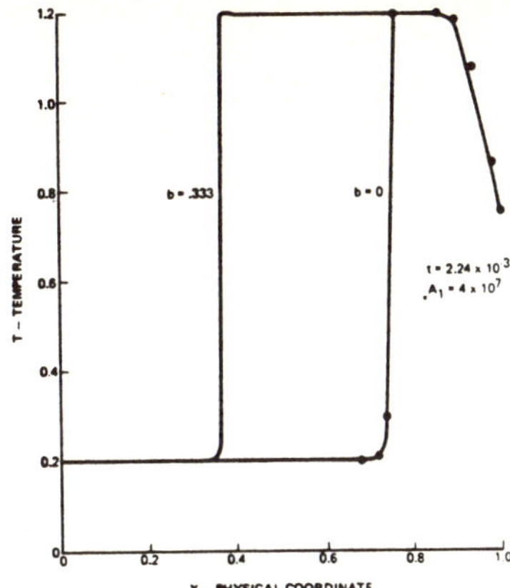

Fig. 3    Temperature Distribution
in Physical Plane

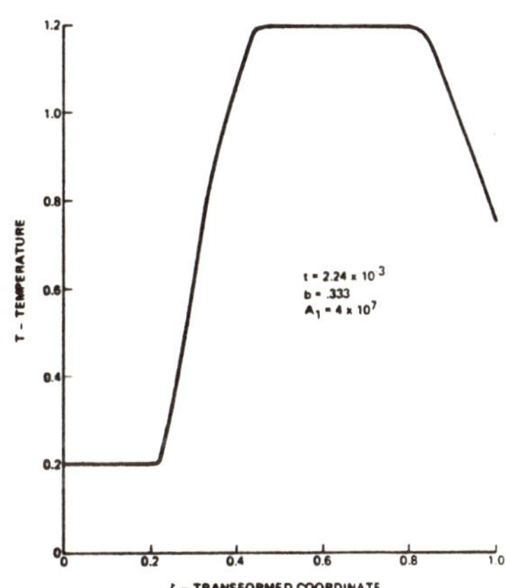

Fig. 4    Temperature Distribution
in Transformed Plane

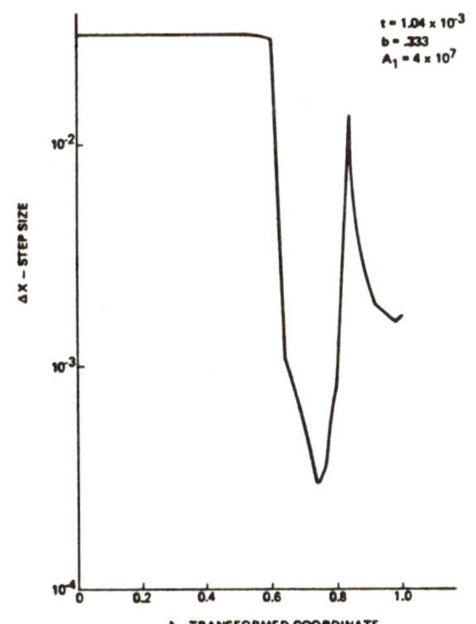

Fig. 5    Step Size Distribution
in Transformed Plane

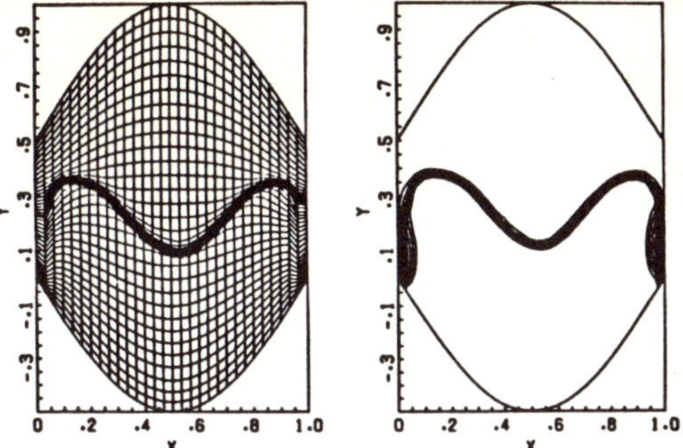

Fig. 6    Coordinate and Isotherm Distribution

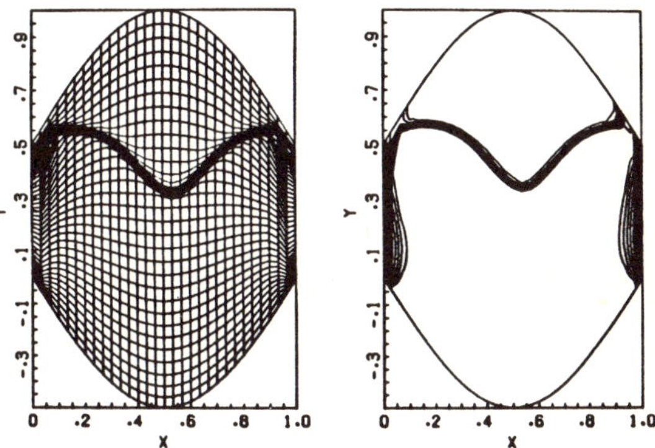

Fig. 7    Coordinate and Isotherm Distribution

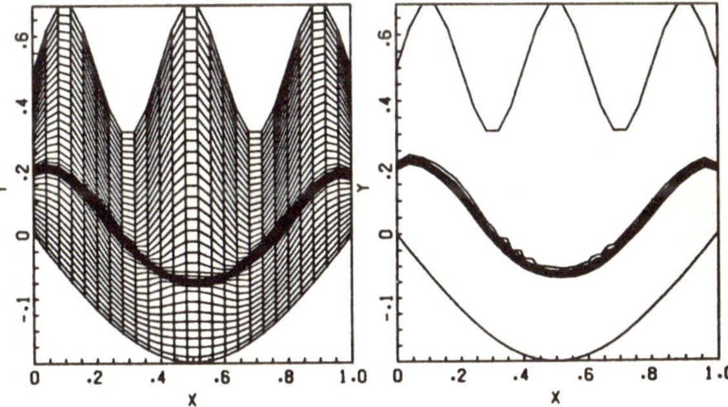

Fig. 8    Coordinate and Isotherm Distribution

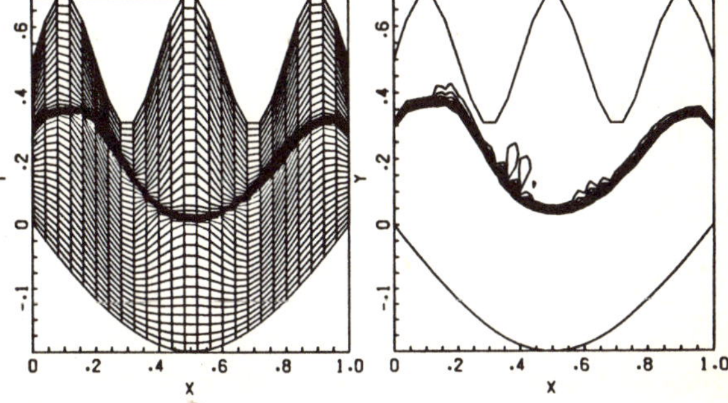

Fig. 9    Coordinate and Isotherm Distribution

COORDINATE GENERATION WITH PRECISE CONTROLS

Peter R. Eiseman

Institute for Computer Applications in Science and Engineering

A method of coordinate generation is obtained from the multi-surface transformation (Eiseman (1979)) with local interpolants. The method is computationally efficient and has more control over mesh properties than are available with previous methods. In particular, a number of important mesh properties can be simultaneously and precisely specified in advance of the mesh generation process. The properties which can be controlled can be separated into the categories of boundary specifications, uniformity specifications, and internal specifications. Boundary specifications include the basic geometry of solid objects, the junctures between distinct coordinate systems, the distribution of boundary points, the angles of coordinate curves which enter a boundary, and the rates of entry for such coordinate curves. Uniformity specifications are applied to either local or global distributions of coordinate curves or points to form a basis from which the curves or points can be redistributed by an a priori specification of a distribution function or by a solution adaptive approach, both without distortion from the underlying transformation. Internal specifications are applicable when an interior shape is needed or when an interior mesh structure such as a Cartesian or Polar system is to be smoothly embedded within a global mesh to simplify a flow field simulation in the given region.

For a sequence of surfaces $\vec{P}_1, \vec{P}_2, \ldots, \vec{P}_N$ which is ordered from bounding surface $\vec{P}_1$ to bounding surface $\vec{P}_N$, the multi-surface transformation is given by

$$\vec{P}(r,t) = \vec{P}_1(t) + \sum_{k=1}^{N-1} \frac{G_k(r)}{G_k(r_{N-1})} \, [\vec{P}_{k+1}(t) - \vec{P}_k(t)] \, , \tag{1}$$

where

$$G_k(r) = \int_{r_1}^{r} \psi_k(x)\,dx \, , \tag{2}$$

and where each $\psi_k$ is a function which vanishes at all partition points except $r_k$ for some partition $r_1 < r_2 < \cdots < r_{N-1}$ of the r-variable. In effect, $\psi_k$ is an interpolation function for the direction $\vec{P}_{k+1}(t) - \vec{P}_k(t)$ of the variable r coordinate curve at $r_k$. The intermediate surfaces $\vec{P}_2, \cdots \vec{P}_{N-1}$ are thus viewed <u>only</u> as control surfaces. When the interpolants $\psi_k$ are nonvanishing on only a local region, precise local controls over the coordinates can be obtained and will be illustrated with the local piecewise linear interpolants that are depicted in Figure 1.

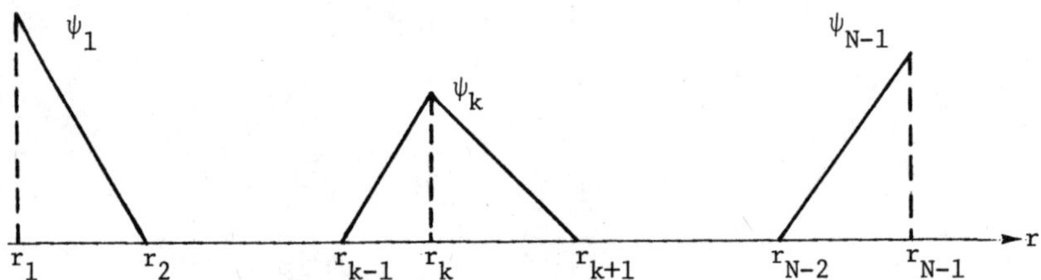


Figure 1. Piecewise linear local interpolants.

For algebraic simplicity, the analysis shall be restricted to the case with the uniform partition $r_k = k$ with the clear understanding that nonuniform partitions will

Research supported under AFOSR Contract F49620-79-C-0132 and NASA Contract NAS3-22117.

follow the same analytic pattern. Since the multi-surface transformation remains unchanged when the interpolants are scaled by any sequence of nonzero numbers, the height $\psi_k(r_k)$ of each interpolant can be chosen arbitrarily. In particular, the form of the multi-surface transformation can be simplified when the heights are adjusted so that each interpolant integrates to unity which yields $G_k(r_{N-1}) = 1$ for all $k$. The integrals are obtained from triangular areas, and by direct observation, lead to the height adjustments $\psi_1(r_1) = 2$, $\psi_k(r_k) = 1$, and $\psi_{N-1}(r_{N-1}) = 2$ in correspondence with the successive illustrations in Figure 1. Also, in correspondence, the explicit form of the normalized interpolants are given by

$$\psi_1(r) = \begin{cases} 2(2-r) & \text{for } 1 \le r < 2 \\ 0 & \text{for } 2 \le r \le N-1 \end{cases} \quad ; \tag{3}$$

$$\psi_k(r) = \begin{cases} 0 & \text{for } 1 \le r < k-1 \\ (r-k)+1 & \text{for } k-1 \le r < k \\ (k-r)+1 & \text{for } k \le r < k+1 \\ 0 & \text{for } k+1 \le r \le N-1 \end{cases} \quad ; \tag{4}$$

$$\psi_{N-1}(r) = \begin{cases} 0 & \text{for } 1 \le r < N-2 \\ 2(r-N+2) & \text{for } N-2 \le r \le N-1 \end{cases} \quad ; \tag{5}$$

and their intergrals defined in Eq. 2, by

$$G_1(r) = \begin{cases} 1 - (2-r)^2 & \text{for } 1 \le r < 2 \\ 1 & \text{for } 2 \le r \le N-1 \end{cases} \quad ; \tag{6}$$

$$G_k(r) = \begin{cases} 0 & \text{for } 1 \le r < k-1 \\ \frac{1}{2}(r-k)^2 + (r-k) + \frac{1}{2} & \text{for } k-1 \le r < k \\ -\frac{1}{2}(k-r)^2 - (k-r) + \frac{1}{2} & \text{for } k \le r < k+1 \\ 1 & \text{for } k+1 \le r \le N-1 \end{cases} \quad ; \tag{7}$$

$$G_{N-1}(r) = \begin{cases} 0 & \text{for } 1 \le r < N-2 \\ (r-N+2)^2 & \text{for } N-2 \le r \le N-1 \end{cases} \quad , \tag{8}$$

which are depicted in Figure 2.

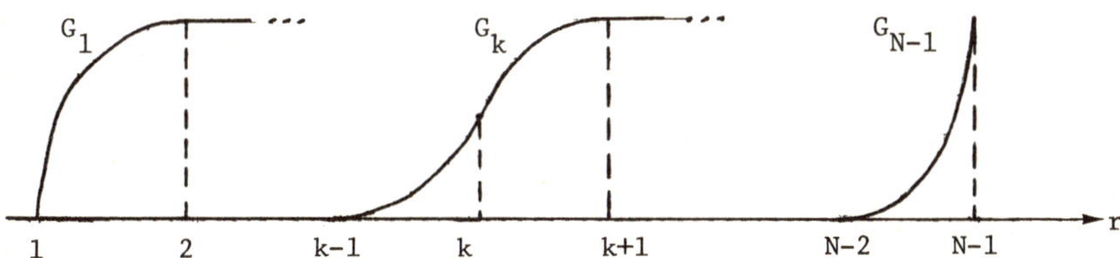


Figure 2. Integrals of normalized interpolants for the partition $r_k = k$.

On the interval $k \le r < k+1$, the integrals $G_i(r)$ which correspond to interpolants defined over <u>nonintersecting</u> intervals are either unity or vanishing depending upon whether the interval of definition precedes or follows the interval under examination. When $i = 1, 2, \ldots, k-1$ which is nonvacuous for $k > 1$, the integrals $G_i(r)$ have been evaluated over the entire domain for which the respective interpolant

$\psi_i$ is nonvanishing; hence, these preceding integrals are unity by the chosen normalization. When $i = k+2, k+3,\ldots,N-1$, the interpolants $\psi_i$ each vanish on $1 \leq r < k+1$; hence, the integrals $G_i$ also vanish there. As a consequence, $G_i$ for $i = k, k+1$ yield the only nontrivial contributions for the multi-surface transformation which reduces to

$$\vec{P}(r,t) = \vec{P}_1(t) + \sum_{i=1}^{k-1} [\vec{P}_{i+1}(t) - \vec{P}_i(t)] + G_k(r)[\vec{P}_{k+1}(t) - \vec{P}_k(t)]$$

$$+ G_{k+1}(r)[\vec{P}_{k+2}(t) - \vec{P}_{k+1}(t)] + 0 \qquad (9)$$

$$= \vec{P}_k(t) + G_k(r)[\vec{P}_{k+1}(t) - \vec{P}_k(t)] + G_{k+1}(r)[\vec{P}_{k+2}(t) - \vec{P}_{k+1}(t)] \quad ,$$

which depends upon only the three control surfaces $\vec{P}_k$, $\vec{P}_{k+1}$, $\vec{P}_{k+2}$ which can be arbitrarily selected to our advantage when they are not bounding surfaces. When $k=1$ or $k+1 = N-1$, substitutions from Eqs. 6-8 will verify that $\vec{P}(1,t) = \vec{P}_1(t)$ and $\vec{P}(N-1,t) = \vec{P}_N(t)$, an expected fit for each bounding surface. When the interval $k \leq r < k+1$ is entirely internal, the endpoint evaluations of Eq. 9 are obtained from Eq. 7 and are given by the simple averages $\vec{P}(k,t) = \frac{1}{2}[\vec{P}_k(t) + \vec{P}_{k+1}(t)]$ and $\vec{P}(k+1,t) = \frac{1}{2}[\vec{P}_{k+1}(t) + \vec{P}_{k+2}(t)]$. These averages lie on the midpoints of fixed t-value lines connecting the control surfaces. Moreover, from the general multi-surface construction, the variable $r$ coordinate curve is tangent to the connecting lines at the partition point evaluations. In graphical form, this process is depicted in Figure 3.

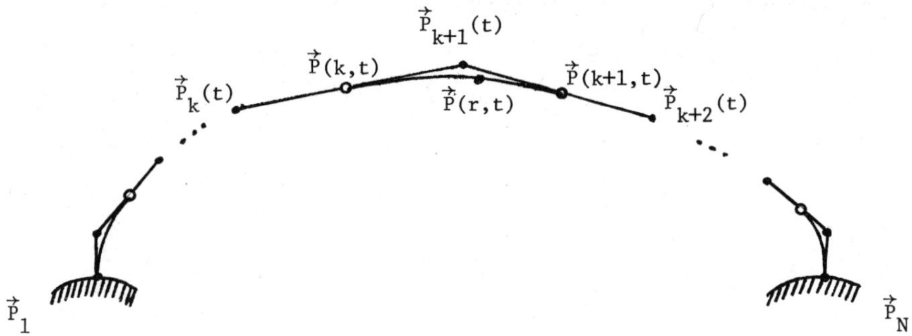


Figure 3. Coordinate curve segments for $k \leq r < k+1$

Between the control surfaces $\vec{P}_i$ and $\vec{P}_j$ for $i > j$, the distribution of constant $r$ coordinate surfaces can be controlled when uniformity can be specified along a direction of measurement $\hat{\tau}(t) = [\vec{P}_i(t) - \vec{P}_j(t)]/\| \vec{P}_i(t) - \vec{P}_j(t)\|$ for then arbitrary distribution functions can be applied relative to uniform conditions. An illustration is given in Figure 4.

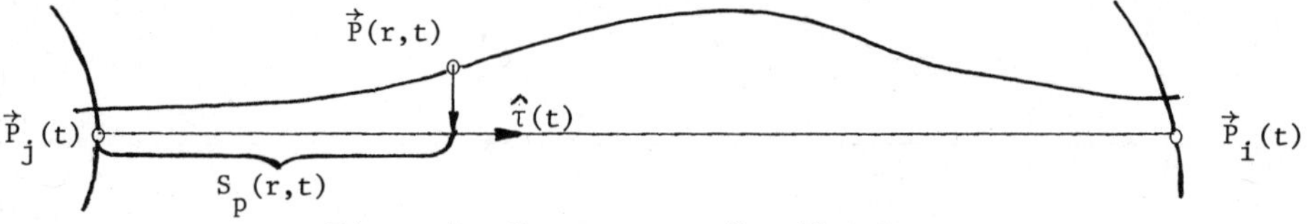


Figure 4. Measurement of uniformity

To obtain uniformity, the relative projected arc length

$$S_p(r,t) = [\vec{P}(r,t) - \vec{P}_j(t)] \cdot \hat{\tau}(t) \quad , \qquad (10)$$

must be linear in  r, or equivalently,  $\partial S_p/\partial r$  must be independent of  r.  But then from Eq. 9 and with the relative projections  $C_m(t) = [\vec{P}_{m+1}(t) - \vec{P}_m(t)] \cdot \hat{r}(t)$,

$$\frac{\partial S_p}{\partial r} = \psi_k(r)C_k(t) + \psi_{k+1}(r)C_{k+1}(t)$$

$$= \begin{cases} -2C_1(t) + C_2(t) & \text{for } k = 1 \\ -C_k(t) + C_{k+1}(t) & \text{for } 1 < k < N-1 \\ -C_{N-2}(t) + 2C_{N-1}(t) & \text{for } k = N-1 \end{cases} \quad r + \text{function of } t \quad , \tag{11}$$

where the last equality comes from Eqs. 3-5.  Hence, for  k = j, j+1,..., i-1, uniformity is obtained if  $2C_1(t) = C_2(t)$  should  k = 1,  $C_k(t) = C_{k+1}(t)$  should $1 < k < N-1$, and  $C_{N-2}(t) = 2C_{N-1}(t)$  should  k = N-1.

To illustrate the precision inherent in the method and, at the same time, to reveal the basic algorithmic steps, coordinates will be obtained for a simple transition from a purely Cartesian system into a purely Polar system.  For  $0 \leq t \leq 1$, the Cartesian coordinates will be specified below a line  $Q(t) = (2t-1,0)$; the Polar coordinates, beyond a circular arc  $\sqrt{2}\ \hat{u}(t)$  where  $\hat{u}(t) = (\cos\theta, \sin\theta)$  for $\theta = (2t+1)\pi/4$.  The line and the arc are depicted in Figure 5.

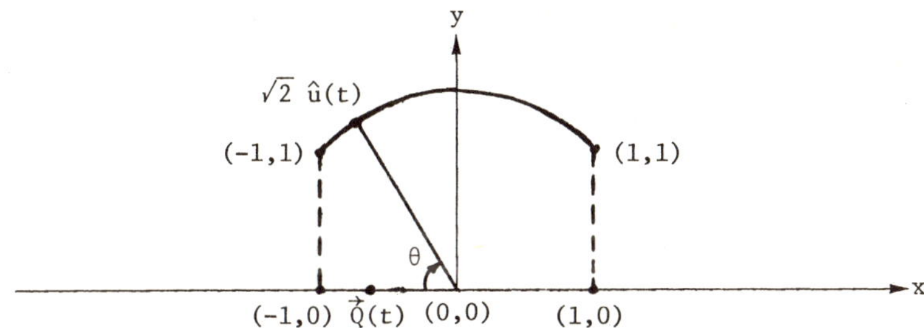


Figure 5.  Basic curves for the Cartesian to Polar transition example

To obtain uniformity near the sides  (t = 0,1) of the transitional region, the unit vertical distance will be used as a basis for displacements to establish uniformity below the line and beyond the circular arc.  For the line, let  $\vec{P}_1(t) = Q(t) - (0,2.5)$, $\vec{P}_2(t) = Q(t) - (0,2)$,  $\vec{P}_3(t) = Q(t) - (0,1)$,  and  $\vec{P}_4(t) = Q(t)$  so that for $\hat{r}(t) = (\vec{P}_4(t) - \vec{P}_1(t))/\|\vec{P}_4(t) - \vec{P}_1(t)\| = (0,1)$  we have  $C_1(t) = [\vec{P}_2(t) - \vec{P}_1(t)] \cdot \hat{r}(t)$ $= (0,\frac{1}{2}) \cdot (0,1) = \frac{1}{2}$, and similarly,  $C_2(t) = C_3(t) = 1$  which satisfy uniformity for $1 \leq r < 3$  and yields a Cartesian system  from  $\vec{P}(1,t) = \vec{P}_1(t)$  up to  $\vec{P}(3,t)$ $= \frac{1}{2}[\vec{P}_3(t) + \vec{P}_4(t)] = Q(t) - (0,\frac{1}{2})$.  For the circular arc, let  $\vec{P}_5(t) = \sqrt{2}\ \hat{u}(t)$, $\vec{P}_6(t) = (1+\sqrt{2})\hat{u}(t)$,  and  $\vec{P}_7(t) = (\frac{3}{2}+\sqrt{2})\hat{u}(t)$  be the last three surfaces so that for $\hat{r}(t) = (\vec{P}_7(t) - \vec{P}_6(t)/\|\vec{P}_7(t) - \vec{P}_6(t)\| = \hat{u}(t)$  we have  $C_5(t) = [\vec{P}_6(t) - \vec{P}_5(t)] \cdot \hat{r}(t)$ $= \hat{u}(t) \cdot \hat{u}(t) = 1$, and similarly,  $C_6(t) = \frac{1}{2}$  which satisfy uniformity for  $5 \leq r \leq 6$ which yields a Polar system between  $\vec{P}(5,t) = \frac{1}{2}[\vec{P}_5(t) + \vec{P}_6(t)] = (\frac{1}{2}+\sqrt{2})\hat{u}(t)$  and $\vec{P}(6,t) = \vec{P}_7(t)$.  The entire collection of bounding and intermediate surfaces are depicted in Figure 6.

For 21 equally spaced mesh points in  r, the evaluation of the r-dependent functions is given with two decimal places of accuracy in the table.  For a given mesh point, the interval  $k \leq r < k+1$  containing it determines the index  k  for  $G_k$ and  $G_{k+1}$  respectively in Eqs. 6-8.  Due to the uniform selection of partition points $r_k$, a repetitive pattern in the  $G_k$  evaluations can be observed and is indicative of translated versions of the same function.  When 9 uniformly  spaced mesh points are chosen for  $0 \leq t \leq 1$  and when the multi-surface transformation of Eq. 9 is evaluated for the 21 × 9 mesh, we obtain the coordinate mesh which is displayed in Figure 7. From uniformity and the table, the first 8 and the last 5 mesh points in  r  are seen to be uniformly distributed and the mesh is respectively purely Cartesian and purely Polar for those points.

| MESH INDEX | $r$ | $k$ | $G_k$ | $G_{k+1}$ | MESH INDEX | $r$ | $k$ | $G_k$ | $G_{k+1}$ |
|---|---|---|---|---|---|---|---|---|---|
| 1 | 1.00 | 1 | .00 | .00 | 11 | 3.50 | 3 | .88 | .13 |
| 2 | 1.25 | 1 | .44 | .03 | 12 | 3.75 | 3 | .97 | .28 |
| 3 | 1.50 | 1 | .75 | .13 | 13 | 4.00 | 4 | .50 | .00 |
| 4 | 1.75 | 1 | .94 | .28 | 14 | 4.25 | 4 | .72 | .03 |
| 5 | 2.00 | 2 | .50 | .00 | 15 | 4.50 | 4 | .88 | .13 |
| 6 | 2.25 | 2 | .72 | .03 | 16 | 4.75 | 4 | .97 | .28 |
| 7 | 2.50 | 2 | .88 | .13 | 17 | 5.00 | 5 | .50 | .00 |
| 8 | 2.75 | 2 | .97 | .28 | 18 | 5.25 | 5 | .72 | .06 |
| 9 | 3.00 | 3 | .50 | .00 | 19 | 5.50 | 5 | .88 | .25 |
| 10 | 3.25 | 3 | .72 | .03 | 20 | 5.75 | 5 | .97 | .56 |
|  |  |  |  |  | 21 | 6.00 | 5 | 1.00 | 1.00 |

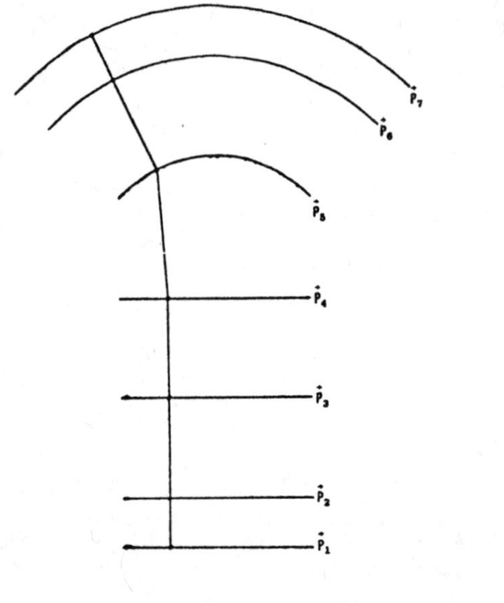


Figure 6

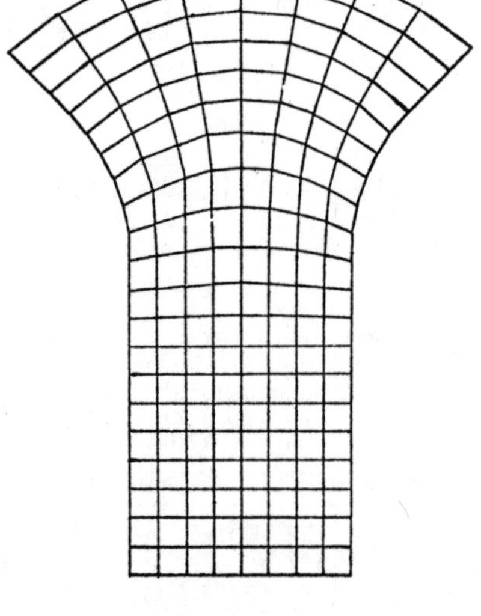

Figure 7

With the general piecewise linear interpolants depicted in Figure 1, the theoretical development is an exact parallel and was presented in Eiseman (1980). In addition, local interpolants with a higher level of smoothness (derivative continuity) can be used and are examined in Eiseman (in preparation).

## References

Eiseman, P. R.   1979  A multi-surface method of coordinate generation, *J. Computational Phys.* 33, 118-150.

Eiseman, P. R.   1980  Geometric methods in computational fluid dynamics, *von Karman Institute Lecture Series,* ICASE Report No. 80-11.

Eiseman, P. R.   (in preparation)  Coordinate generation with precise controls over mesh properties.

### AN ALTERNATING DIRECTION IMPLICIT FINITE ELEMENT
### METHOD FOR COMPRESSIBLE, VISCOUS FLOW

C.A.J. Fletcher

University of Sydney
Sydney, Australia

## I  INTRODUCTION

The majority of finite difference methods for solving the equations governing steady compressible viscous flow are based on integrating the equivalent unsteady problem until the solution no longer changes (Peyret and Viviand (1975), MacCormack and Lomax (1979)).  If an alternating direction implicit scheme is applied to the algebraic equations, e.g. Briley and McDonald (1977), an efficient algorithm results which avoids a global matrix factorisation at every time-step.

Finite element formulations for the steady Navier-Stokes equations typically seek a solution to the system of algebraic equations by repeated factorisation, e.g. Kawahara (1978), of the global stiffness matrix.  Finite element formulations for unsteady problems produce a mass matrix, multiplying the time derivative term, that must be factorised at every time-step.  Alternatively the mass matrix can be lumped onto the diagonal.  The results of Gresho et al (1978) suggests that this causes a reduced accuracy for the transient solution.

The purpose of this paper is to describe the application of a Galerkin finite element formulation to the equations governing unsteady compressible viscous flow, followed by an alternating direction implicit treatment of the algebraic equation at every time-step.  This formulation replaces the global factorisation at every time-step with the repeated solution of a tridiagonal system of equations associated with every grid line.  This is both economical in execution time and in memory requirements.  Thus the method is attractive for implementation on mini-computers dedicated to computational fluid dynamics.

## II  MODEL EQUATION

The present method (ADIFEM) can be illustrated by considering the transport equation

$$\frac{\partial T}{\partial t} + \frac{\partial}{\partial x}(uT) + \frac{\partial}{\partial y}(vT) = \alpha_x \frac{\partial^2 T}{\partial x^2} + \alpha_y \frac{\partial^2 T}{\partial y^2} . \tag{1}$$

T is the temperature and u and v are the known velocity components.

The application of a Galerkin finite element method, with linear rectangular elements, produces algebraic equations of the following form.  The term $\partial T/\partial x$ is replaced by

$$\frac{1}{36}\left|(\dot{T}_{i-1,j-1} + 4\dot{T}_{i,j-1} + \dot{T}_{i+1,j-1}) + 4(\dot{T}_{i-1,j} + 4\dot{T}_{i,j} + \dot{T}_{i+1,j}) + (\dot{T}_{i-1,j+1} + 4\dot{T}_{i,j+1} + \dot{T}_{i+1,j+1})\right|, \qquad (2)$$

where the subscript i increases in the positive x direction and j increases in the positive y direction. The term $\partial(uT)/\partial x$ is replaced by

$$\frac{1}{2\Delta x}\left|\frac{1}{6}\left\{(uT)_{i+1,j-1} - (uT)_{i-1,j-1}\right\} + \frac{2}{3}\left\{(uT)_{i+1,j} - (uT)_{i-1,j}\right\} + \frac{1}{6}\left\{(uT)_{i+1,j+1} - (uT)_{j-1,j+1}\right\}\right|. \qquad (3)$$

The terms in (3) involving the nodal groups, uT, have been obtained by assuming an analytic representation for the group of terms rather than for individual terms. This technique was used by Fletcher (1979). If the finite element formulation were applied with a non-rectangular grid then the coefficients in (2) and (3) would be different.

To implement the ADI formulation the $\dot{T}$ terms in (2) are represented by one-sided difference formulae. For the first half-time-step all terms arising from the x-derivatives and lying on the $j^{th}$ line are evaluated at the implicit time level. Terms lying on the $(j-1)^{th}$ can be obtained from the solution of the tridiagonal system of equations associated with the $(j-1)^{th}$ line. Terms on the $(j+1)^{th}$ line must be evaluated at the current time level. The evaluation of different parts of the spatial terms at different time levels causes the final algorithm to have a truncation error of $O(\Delta T)$. This is of no consequence to the steady-state solution. For the second half-time-step all terms arising from the y-derivatives and lying on the $i^{th}$ line are evaluated at the implicit time level.

A von Neumann stability analysis indicates that the basic scheme (scheme 1, Table 1) is restricted in the size of time-step that is possible. By lumping the mass matrix onto the diagonal, i.e. replacing (2) by $\dot{T}_{i,j}$ (scheme 2, Table 1), the restriction in the time-step is reduced. Scheme 2 requires only two time levels of data to be retained whereas scheme 1 requires three. The stability boundaries for schemes 1 and 2 are shown in Fig. 1. Numerical results for the entry length problem (x in Fig. 1) are consistent with the stability boundaries.

By evaluating implicitly the terms from the y-derivatives that lie on the $j^{th}$ line during the first half-time-step (and correspondingly for the x-derivatives during the second half-time-step), it is possible to obtain a computational scheme with no restriction on the time-step (scheme 3, Table 1). All the additional terms treated implicitly in scheme 3 come from the diffusive part of (1). For comparison a conventional ADI finite difference scheme is also considered (scheme 4, Table 1).

| scheme | description | $\Delta T_{max}$ | number of time steps | relative execution time | r.m.s. difference |
|---|---|---|---|---|---|
| 1 | ADIFEM | 0.0003 | 1250 | 7.73 | 0.0502 |
| 2 | ADIFEM, fully lumped | 0.002 | 247 | 1.49 | 0.0502 |
| 3 | ADIFEM, fully lumped fully implicit | – | 40 | 0.21 | 0.0502 |
| 4 | ADI finite difference | – | 332 | 1.00 | 0.0601 |

Table 1  Comparison of schemes for entry length problem

## III  ENTRY LENGTH PROBLEM

To compare the schemes shown in Table 1 the temperature distribution in the entrance of a two-dimensional duct (Fig. 2) was calculated.  To permit comparisons with the 'exact' solution a fully developed velocity distribution was assumed.  The various schemes are compared in Table 1.  All solutions were obtained with uniform step sizes, $\Delta x = \Delta y = 0.05$ on a 21×21 grid.  The "r.m.s. difference" shown in Table 1 is the root-mean-square difference between the steady-state, centre-line temperature solution and the corresponding 'exact' solution.

The results indicate that all the finite element schemes are more accurate than the finite difference scheme and that successive improvements to the finite element method (schemes 1 to 3) reduce the number of time-steps to convergence.  It can be seen that scheme 3 is more efficient than the corresponding ADI finite difference method (scheme 4).  If too large a time-step was used with scheme 4 the number of time-steps to convergence increased.  A fuller description of the application of ADIFEM to the entry length problem is given by Fletcher (1980).

## IV  VISCOUS COMPRESSIBLE FLOW

A modified form of scheme 3 has been applied to the unsteady, laminar, compressible Navier-Stokes equations as given by Peyret and Viviand (1975).  These equations have been solved for the nodal values of density, $\rho$, and the two velocity components, u and v.  The pressure has been related to the other variables by the adiabatic equation of state,

$$1 + \gamma M_\infty^2 p \;=\; \rho \left\{ 1 + \frac{\gamma - 1}{2} \cdot M_\infty^2 \left( 1 - u^2 - v^2 \right) \right\} \qquad (4)$$

where u, v, $\rho$ and p have been non-dimensionalised with respect to the appropriate freestream values.

The algebraic equations resulting from the finite element formulation have been written as a correction to the solution at the current time level.  Following Briley and McDonald (1977), the terms that need to be evaluated implicitly are linearised

about the current time level and a block tridiagonal system solved for each grid line.
This has the advantage that the right hand side is the residual of the steady-state
equations evaluated at the current time level. The r.m.s. value of the steady-state
equation residuals has been used to determine when to stop the time integration.

| $\Delta T$ \\ Re | 5 | 20 | 100 | 1000 | 1000 |
|---|---|---|---|---|---|
| 0.01 | stable | stable | stable | stable | stable |
| 0.02 | stable | stable | stable | stable | stable |
| 0.05 | stable | stable | stable | stable | stable |
| 0.20 | stable | stable | unstable | unstable | unstable |
| 1.00 | stable | stable | unstable | unstable | unstable |

Table 2   von Neumann stability for viscous, compressible flow,
$\Delta x = 0.10, \ \Delta y = 0.10$

A von Neumann stability analysis has been applied to the system of algebraic
equations and the results for $M = 0.9$ are shown in Table 2. For Reynolds numbers of
practical interest the time-step is limited by stability restrictions. The restrictions vary with spatial step-size but are relatively insensitive to nodal values or
freestream Mach number.


## V   FLOW PAST A RECTANGULAR BODY

ADIFEM has been applied to the flow about a rectangular body of aspect ratio
2.5. The computational region is shown in Fig. 3. A non-uniform rectangular mesh
was used with more points adjacent to the body surfaces ($\Delta x_{min} = 0.04$ L, $\Delta y_{min} =$
0.02 L). It was found necessary, for stability, to have a small mesh width, $\Delta x$,
adjacent to the downstream boundary. In addition it was found necessary to keep the
density fixed at the freestream value for the two lines adjacent to the upstream
boundary.

A typical flow pattern behind the rectangular body is shown in Fig. 4. A well
defined recirculating region is apparent. The corresponding pressure distribution
for the front face, top surface and rear face is shown in Fig. 5. The pressure distribution indicates a suction peak just behind the upstream corner and a smaller
suction peak near the downstream corner. Adjacent to these two peaks there is
evidence of a slight oscillation in the pressure distribution suggesting the need
for a local grid refinement.


## VI   CONCLUSION

An alternating direction implicit finite element method has been applied to the
thermal entry problem and viscous compressible flow. The method is fourth order
accurate in space and, for the thermal entry problem, the best scheme is uncondition-

ally stable.  For viscous compressible flow this scheme is subject to a time-step restriction for Reynolds numbers of practical interest.

## VII  ACKNOWLEDGEMENT

The author is grateful to the Australian Research Grants Committee for the financial assistance that made possible the purchase of the Perkin-Elmer 7/32 computer on which this research was undertaken.

## VIII  REFERENCES

BRILEY, W.R. and McDONALD, H., 1977, J. Comp. Phys., 24, 372-397
FLETCHER, C.A.J., 1979, J. Comp. Phys., 33, 301-312
FLETCHER, C.A.J., 1980, Proc. 3rd Int. Conf. on Finite Elements in Flow Problems, Calgary, Canada
GRESHO, P.M., LEE, R.L. and SANI, R.L., 1978, Finite Elements in Fluids, Vol. 3, Wiley, 335-361
KAWAHARA, M., 1978, Finite Elements in Fluids, Vol. 3, Wiley, 23-54
MacCORMACK, R.W. and LOMAX, H., 1979, Ann. Rev. Fluid Mech., 11, 289-316
PEYRET, R. and VIVIAND, H., 1975, AGARDograph No. 212

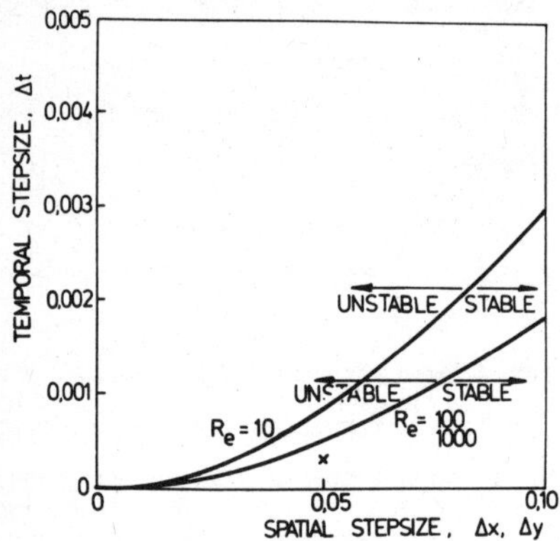


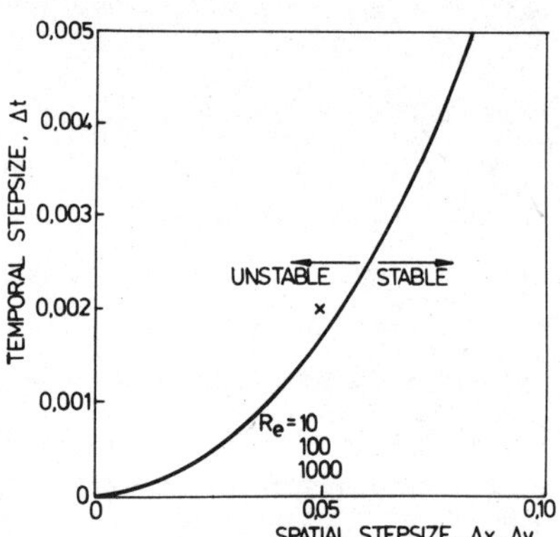


Fig. 1a    Stability boundary
for scheme 1

Fig. 1b    Stability boundary
for scheme 2

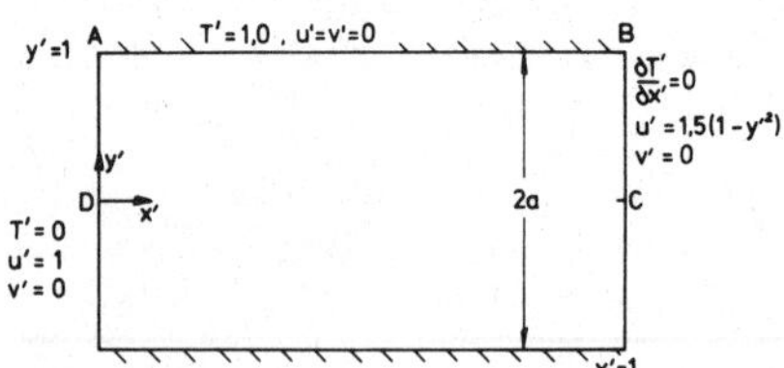


Fig. 2    Entry length problem for
flow between parallel plates

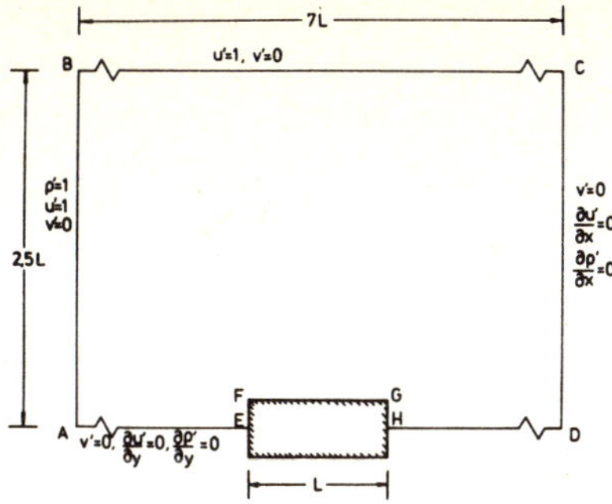

Fig. 3  Compressible viscous flow past a rectangular body

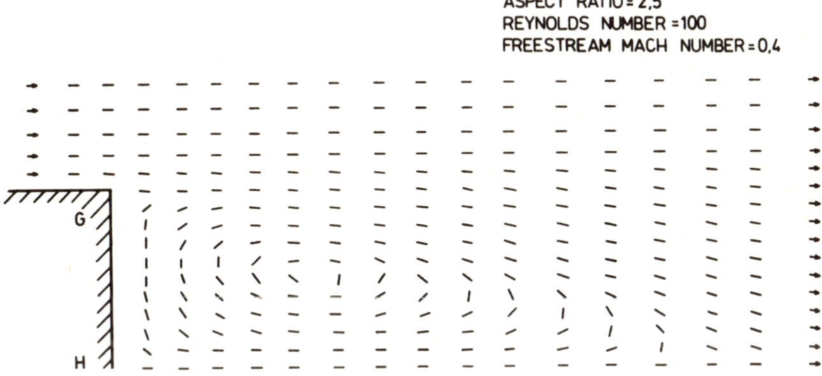



Fig. 4  Flow directions behind a rectangular body

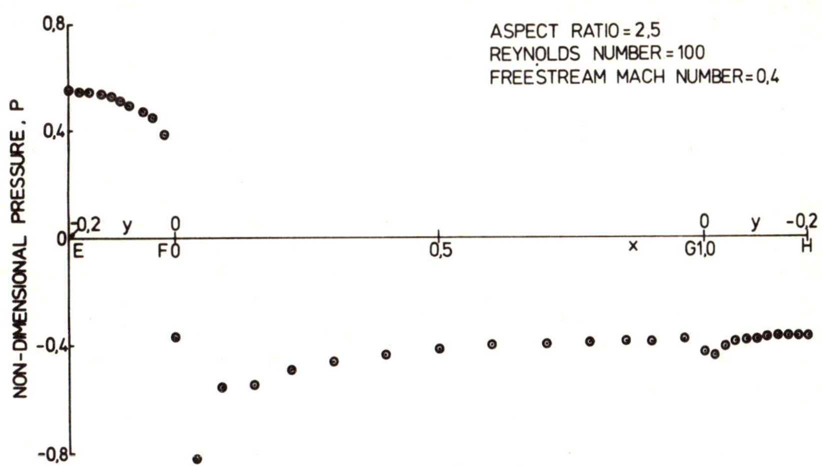



Fig. 5  Pressure distribution on a rectangular body

FINITE DIFFERENCE COMPUTATION OF THE
CAPILLARY JET, FREE SURFACE PROBLEM

J. Fromm

IBM Research Laboratory, San Jose, California

## Introduction

Scientific inquiry into surface tension driven flows began roughly a century ago, and a most appropriate reference of historical interest is the small book by C. V. Boy (1). In contrast to simple curiosity, expanded in Boy's book, today very pragmatic work on bubbles and drops is the rule (2,3). In this paper, we address the fluid dynamic problems associated with ink jets used in computer controlled printing. The extensive recent work in this regard is well covered in the review by Bogy (4), and of particular interest to numerical simuatlion is the work of Shokoohi (5).

Parameters which govern the behavior of liquid jets in the application of interest here can be reduced to four. These are the ratio of Reynolds number to Weber number, the ratio of disturbance wave length to jet diameter, the disturbance amplitude and the ratio of amplitudes of two combined disturbances. With a proper choice of the first three of these, the last provides very selective control. Figure 1, the work of Chaudhary (6) shows the nature of this control which only in case (c) is appropriate to quality ink jet printing. Chaudhary's work, which is both analytical and experimental, is the point of departure for the work described here.

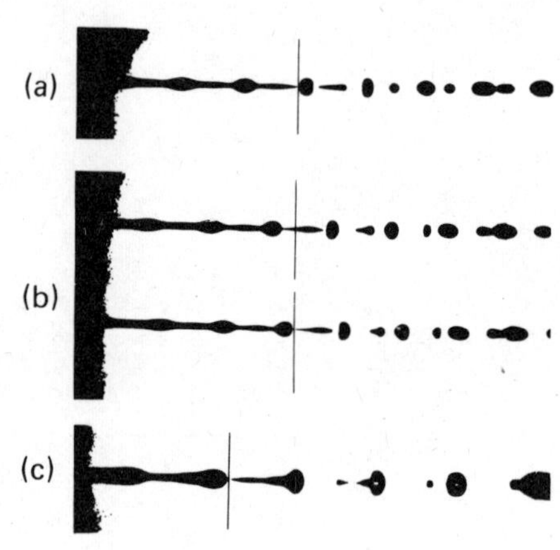


Figure 1.    Experimental Ink Jets

## Problem Description and Solution Method

The geometry considered in the present work implies what has been called the "infinite jet". That is, the jet has no beginning or end, nor is the direction of flow defined if uniform (plug) flow is assumed, and one works in a reference frame moving with this uniform flow. This point will be explored further in the interpretation of some numerical results.

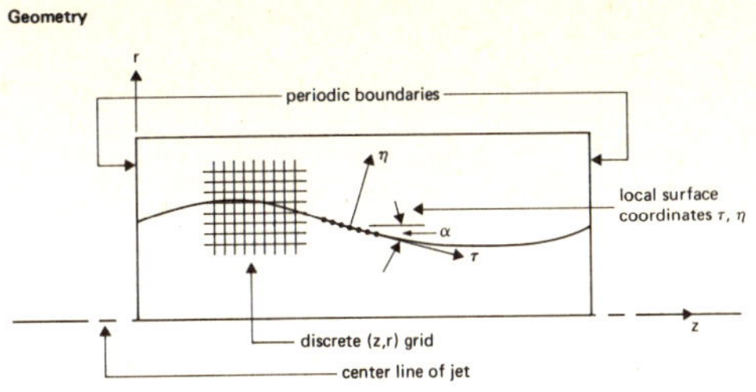



$$\text{Characteristic length}=r_0,\ \text{velocity}=\left(\frac{\sigma}{\rho r}\right)^{\frac{1}{2}},\ \text{time}=\left(\frac{\rho r^3}{\sigma}\right)^{\frac{1}{2}},\ \frac{R}{W}=\left(\frac{\sigma r}{\rho \nu^2}\right)^{\frac{1}{2}}$$

**Figure 2**

Figure 2 shows a periodic section of the axi-symmetric jet with a single wave length disturbance which we will here-after refer to as the fundamental disturbance.  The surface is actually a cylindrical shell initially and is set into motion by prescribing the stream-function as a sinusoidal function of the coordinate z.  Finite difference solution in the discrete (r,z) grid then approximates Rayleigh's invisid jet as a starting solution.  Local tangential and normal surface coordinates $(\tau,\eta)$ at angle $\alpha$ are defined at surface points (or particles) that are of a density of about 8 to 1 relative to the global (r,z) grid spacings.  The characteristic length for dimensionless computation is the initial fluid cylinder radius (set to 1).  The charactersitic velocity, defined in terms of the surface tension coefficient $\sigma$ the fluid density $\rho$, and the radius is referred to as the capillary wave velocity.  Its physical interpretation is somewhat obscure here, but it ties the work to analytical precedent and reduces the parameters by 1 so that the ratio of R/W instead of R and W separately need be considered.  The time scale of course follows from the length and velocity as defined, and thereby growth rates can be compared directly to previous work.

## The Governing Equations

1. Advance velocity fields

$$\frac{\partial u}{\partial t} + \frac{\partial u^2}{\partial z} + \frac{1}{r}\frac{\partial uvr}{\partial r} = -\frac{\partial P}{\partial z} + \nu\left[\frac{\partial^2 u}{\partial z^2} + \frac{1}{r}\frac{\partial}{\partial r}\left(r\frac{\partial u}{\partial r}\right)\right]$$

$$\frac{\partial Q}{\partial r} = ru \quad \rightarrow \quad v = -\frac{1}{r}\frac{\partial Q}{\partial z}$$

2. Obtain surface streamfunction $\quad \int_s \vec{u}\cdot d\vec{A} = 0$

3. Advance vorticity field

$$\frac{\partial \omega}{\partial t} + \frac{\partial u\omega}{\partial z} + \frac{\partial v\omega}{\partial r} = \nu\left[\frac{\partial^2 \omega}{\partial z^2} + \frac{\partial}{\partial r}\left(\frac{1}{r}\frac{\partial \omega r}{\partial r}\right)\right]$$

4. Internal streamfunction

$$\frac{\partial^2 Q}{\partial z^2} + r\frac{\partial}{\partial r}\left(\frac{1}{r}\frac{\partial Q}{\partial r}\right) = -r\omega$$

5. Reset velocity fields

6. Obtain surface velocities

$$u_\tau = \frac{1}{r}\frac{\partial Q}{\partial \eta} \, , \quad u_\eta = -\frac{1}{r}\frac{\partial Q}{\partial \tau}$$

u and v by rotational transformation

7. Update surface configuration

$$\frac{dz}{dt} = u, \quad \frac{dr}{dt} = v$$

8. Arc lengths and angles

$$\delta\tau^2 = \delta z^2 + \delta r^2$$

$$\alpha = \tan^{-1}\frac{\delta r}{\delta z}$$

9. Surface vorticity and pressure

$$\omega_s = 2\,\frac{\partial u_\eta}{\partial \tau}$$

$$P_s = \sigma\left(\frac{\cos\alpha}{r} - \frac{\partial\alpha}{\partial\tau}\right) + 2\nu\,\frac{\partial u_\eta}{\partial\eta}$$

10. Internal pressure

$$G = \frac{\partial^2 u^2}{\partial z^2} + \frac{2}{r}\frac{\partial uvr}{\partial r} + \frac{1}{r}\frac{\partial v^2 r}{\partial r}$$

$$\frac{\partial^2 P}{\partial z^2} + \frac{1}{r}\frac{\partial}{\partial r}\left(r\frac{\partial P}{\partial r}\right) = -G$$

We list the equations in differential form in order of calculation. With appropriate boundary conditions prescribed, Rayleigh's envisid solution can be expanded to all the field variables of interest, and we may proceed to step 1 in time advancement of the numerical solution.

We shall, because of space limitations, only comment on the various steps. The actual computer algorithms and logic are very extensive and are, through new experience, always subject to change. The difference equations are derived by integration over elementary volumes with special consideration for volumes on the axis. These procedures are well described in the work of Langlois (7) and will also be elaborated for the application here in a forthcoming report.

Time advanced axial velocities (u) are obtained in the first step by using the appropriate momentum equation. The radial velocities (v) may be updated by imposing continuity, i.e. be defining a streamfunction Q. In highly contorted flow, v should also be obtained through a momentum equation, but this may be done at isolated places with a proper logic control of the calculation. Radial, linewise, integration is further used in step 2 to extend streamfunction values to the surface at distributed points spaced at roughly grid mesh distances. A local surface interpolation gives Q at surface points. The local surface interpolation can be a linear one, but higher order is preferred so long as proper care can be taken to insure relatively even spacing between points where Q is given.

Mass conservation may be insured by adjusting the surface Q's so that the net surface flow is zero (in the geometry here considered). We use the criteria that any error from one time step to the next will be equally distributed among all points. In the current application, the errors are extremely small, and satisfactory results would probably occur even without the correction.

The vorticity ($\omega$) is initially everywhere zero for uniform flow, but a contribution in step 9 developes at the surface as time progresses. It is distributed throughout the fluid by the time dependent vorticity equation, step 3. If the velocity profile is initially parabolic, the tendancy is toward uniform flow rather quickly, in the range of parameters of interest here, with vorticity effectively being transferred to the surrounding medium.

With the vorticity field supplied, we can proceed to steps 4 and 5, first iterating to obtain the streamfunction everywhere and then resetting velocities to conform to the new streamfunction field. Also the velocities at the surface may be obtained. Local normal velocities follow readily from differences taken along the surface, but tangential velocities involve a procedure of finding the closest approach of the surface to specified points in the interior where Q is known.

In step 7, we update the surface position by moving surface points with their designated new velocities. Arc lengths are determined at step 8, and new local coordinate systems are prescribed at each surface point by specifying the angles $\alpha$ for the new configuration.

The derivative of the tangential velocity with respect to the normal is never calculated in the program, but the derivative of the normal velocity with respect to the
normal is calculated.  The latter uses the nearest approach method as is used in
obtaining the tangential velocity from streamfunction values.  The normal stress
condition provides the means of calculating the surface pressures ($P_s$) in step 9.
The use of the angle $\alpha$ instead of radii of curvature eliminates a number of problems
connected with the latter.

Perhaps the weakest points in the numerical method are the ways in which the continuity and the tangential stress condition enter the calculation.  We have already
discussed the former.  The latter is simply assumed to hold.  This is reflected in
determining the surface vorticity ($\omega_s$) as in step 9.  While what is done is not
entirely satisfying, the results seem to be good and seem to suggest a minor dependence on how concretely these two conditions are defined in computation.

In the last step, we compute G, the inhomogeneous part of the Poisson equation for
the pressure and then iteratively solve for the pressure.  Deformation plus voricity
are embodied in G, but it is best calculted in the conservative form as given.  It
may also be calculated at the surface but is not needed there unless some interpolation of near surface values are to be found by such means.  In this way one could
make use of the tangential stress condition in devising a formula for surface G.
Similarly, in the momentum equation (step 1), the tangential stress condition can
enter in interpolation of values near the surface, but this does not seem to be a
necessity in the present form of the programs.

<u>Results</u>

Proof testing of the numerical results
is admittedly visual in some respects.
The most apparent test is comparison of
growth rates with theory and
experiment.  Surprisingly, experimental
growth rates seem to follow Rayleigh's
envisid analysis (Figure 3), but since
the analysis is linear, and the time
during which a small disturbance can be
regarded as such is small, experimental
measurement must be extremely
difficult.  We find numerically that a
finite amplitude disturbance will
result in an intially high rate of
growth which subsides but is quickly
followed by nonlinear effects.  The nonlinearity first manifests itself in swell
growth differing from neck contraction.  Some experimentalists have used an average
of these indicators.

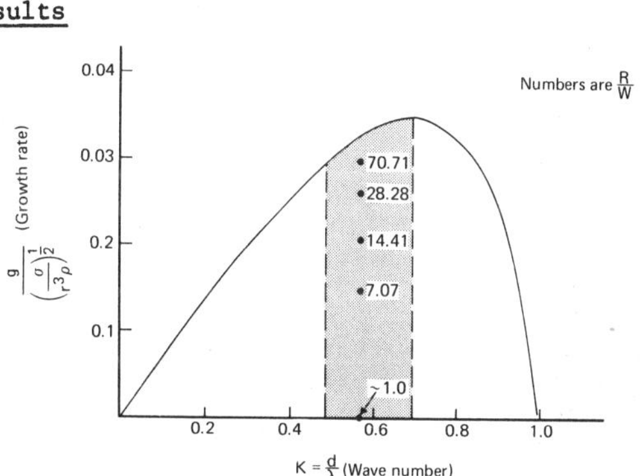


Figure 3

In Figure 3 we have indicated, somewhat qualitatively, the numerically obtained
growth rates for various R/W with reference to Rayleigh's envisid analysis for one
particular wave number.  These numbers correspond to the growth rate in a short
interval of time prior to the obvious development of higher modes but following
initial finite aplitude effects.  Further proof testing is planned.

For purposes of ink jet printing, an optimum range of parameters is sought.  The
shaded region of Figure 3 is an experimentally determined range which perhaps
provides for a rapid growth of harmonics in addition to the fundamental, thereby
leading to rapid break up times.  It is found, however, that optimum control of
break up and drop shape follows from use of appropriate harmonic disturbances

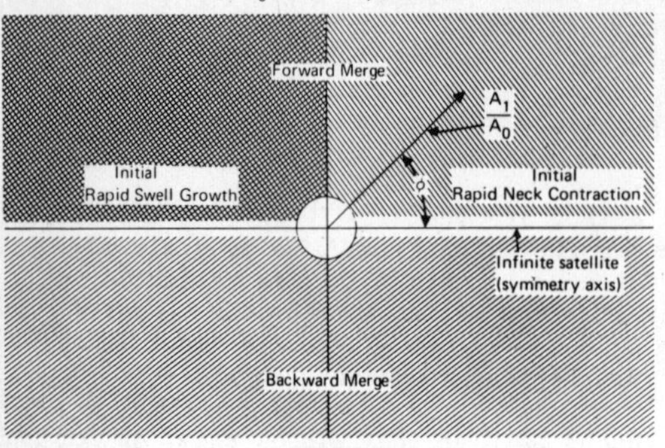



**Figure 4**

and amplitudes. In this respect, a basic set of experiments can be carried out numerically that make use of the symmetry of the "infinite jet". If we arbitrarily set one direction of the jet as the true direction of flow, then the characteristic of forward merging of satellite drops with the main drops is backward merging for the opposite choice of flow direction. With this in mind, we are led to a polar representation as in Figure 4 in which we assume that an initial disturbance of the streamfunction is equivalent to a timewise train of voltage stimulations of a transducer for a real jet. Considering only a first harmonic disturbance in addition to the fundamental, a continuously varying range of amplitude ratios $(A_1/A_0)$ and phase $(\phi)$ may be tested for drop control. With close examination of even the initial disturbances, one can anticipate the outcome roughly, but of course the nonlinearity modifies the behavior so as to leave the final flow uncertain. Nevertheless, Figure 4 gives a general picture in regard to phases of the two disturbances. Phase angles of 0° and 180° give what is called an "infinite satellite" condition for which a satellite will forever exist without merging forward or backward. The two states differ however in satellite size and sensitivity to asymmetry. $\phi$ greater than 90° but less than 180°, (initial rapid swell growth) gives the desired conditions for quality printing corresponding to (c) of Figure 1.

Finally, we include a set of late time results for selected phase angles in Figure 4. The first is the "infinite satellite" condition $\phi = 0°$. Breakup is symmetric with a relatively large satellite developing. The second result is for $\phi = 90°$ and forward merging for flow to the right (satellite not separated on the left) is evident. The last result is for $\phi = 225°$ and is backward merging for flow to the right. It is of course identical to $\phi = 135°$ for assumed flow in the opposite direction.

Streamlines are indicated in solid lines, where negative lines include dots. The axi-symmetry must be taken into account in inferring velocity from the streamline spacing. Isobars are broken lines, and we note that strong gradients in pressure exist near break up, while the main drop has almost uniform pressure.

## References


1. Boys, C. V., "Soap Bubbles", Doubleday Science Study Series, Garden City, New York (1959).

2. Sweet, R. G., "High-Frequency Oscillography with Electrostatically Deflected Ink Jets". U. S. Dept. of Commerce Clearing House, No. AD437951 (1964).

3. Collins, D. J., Plesset, M. S., Saffren, M. M., editors, Proceedings of the International Colloquim on Drops and Bubbles, Caltech and JPL, 2 volumes (1974).

4. Bogy, D. B., Ann. Rev. Fluid Mech. 11:207-28 (1979).

5. Shokoohi, F., Ph.D. Thesis, Columbia University, New York (1976).

6.  Chaudhary, K. C. and Redekopp, L. G., J. Fluid Mech. 96:257-274, (1980).

7.  Langlois, W. E., Comput. Methods Appl. Mech. Eng., in press.

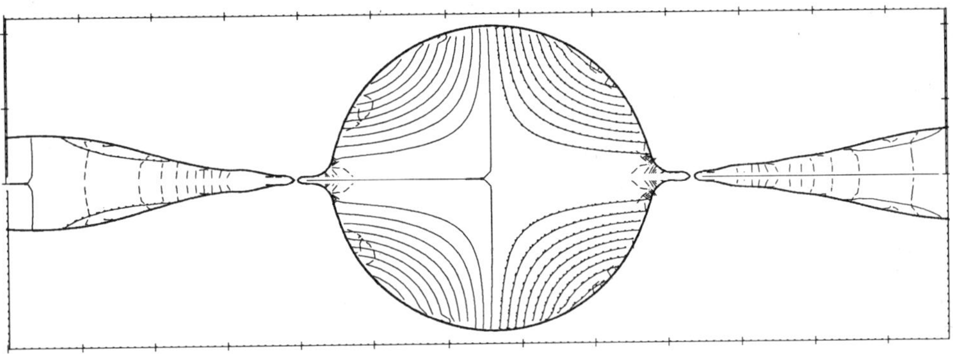

$\frac{R}{W} = 7.07, \frac{\lambda}{D} = 6.4$, Uniform Flow, $\phi = 0$, t = 26.74

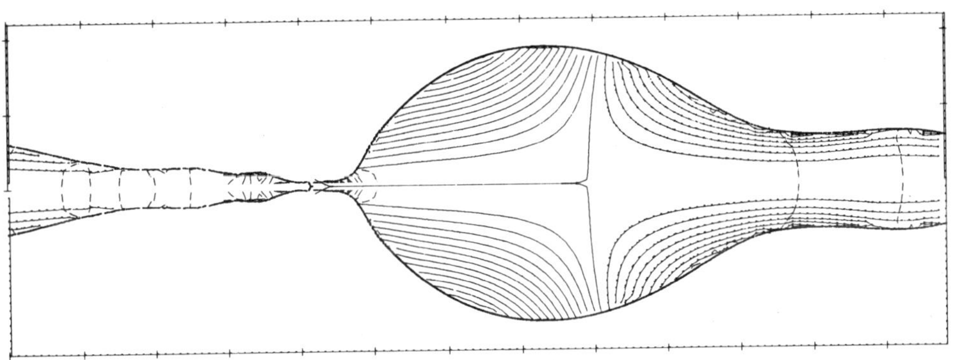

$\frac{R}{W} = 7.07, \frac{\lambda}{D} = 6.4$, Uniform Flow, $\phi = \frac{5\pi}{4}$, t = 28.45

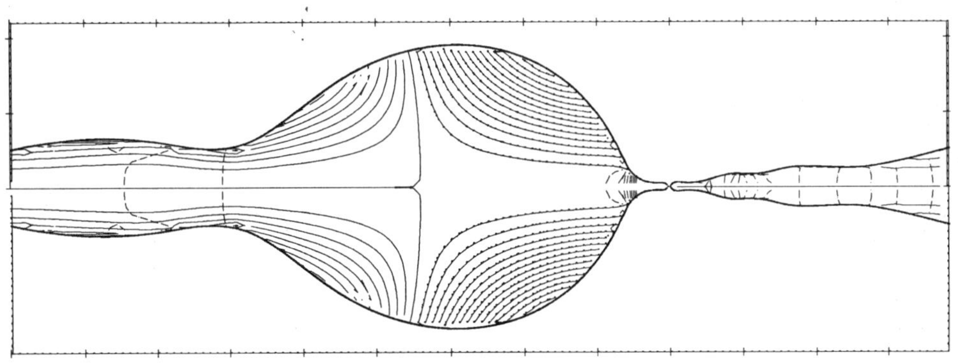

$\frac{R}{W} = 7.07, \frac{\lambda}{D} = 6.4$, Uniform Flow, $\phi = \frac{\pi}{2}$, t = 27.18

# SIMULTANEOUS SOLUTIONS OF INVISCID FLOW AND
# BOUNDARY LAYER AT TRANSONIC SPEEDS


Kozo Fujii

Institute of Space and Aeronautical Science,
University of Tokyo, Tokyo, Japan


## INTRODUCTION

Using the assumption of inviscid flow, the finite difference methods developed recently made it possible to solve the transonic flow over an airfoil accurately with the aid of high speed computers. However, the aerodynamic force cannot be predicted well without taking the viscous-inviscid interaction effects into account. An interactive scheme between an inviscid solution and a boundary layer solution seems to be simple and rapid method in which the viscous effects can be considered. In the conventional boundary layer correction method, some assumptions are needed for the solution process because of the separation singularity. In this paper is presented a newly developed method to calculate the transonic laminar viscous-inviscid interaction flow field, in which Moses' idea(1978) for the incompressible flow is modified so as to be applicable for the transonic flow. In this method, the solutions of the inviscid flow field and the boundary layer are obtained simultaneously without difficulty. There occurs no singularity at the separation point and the numerical process is simple and rapid.

## MATHEMATICAL FORMULATIONS

The equations of the steady two-dimensional compressible laminar boundary layer for perfect gas are transformed to be of the incompressible form using the Stewartson transformation. The well-known momentum and energy integral equations are derived by taking the weighted integrations across the boundary layer.

$$\{\delta_1^{(i)} \frac{dH}{da}\} \frac{da}{dx^*} + \{H\} \frac{d\delta_1^{(i)}}{dx^*} + \{\frac{(2H+1)\delta_1^{(i)}}{U_e^{(i)}}\} \frac{dU_e^{(i)}}{dx^*} = f_e \frac{P}{U_e^{(i)} \delta_1^{(i)} R_{e\infty}} \tag{1}$$

$$\{\delta_1^{(i)} \frac{dJ}{da}\} \frac{da}{dx^*} + \{J\} \frac{d\delta_1^{(i)}}{dx^*} + \{\frac{3 \delta_1^{(i)} J}{U_e^{(i)}}\} \frac{dU_e^{(i)}}{dx^*} = f_e \frac{R}{U_e^{(i)} \delta_1^{(i)} R_{e\infty}} \tag{2}$$

where $\delta_1$ is the non-dimensional displacement thickness, $U_e$ is the velocity at the boundary layer edge and H,J,P and R are the functions of the velocity profile parameter a. The superscript ,i, indicates the transformed value and the asterisk * indicates the coordinate aligned with the solid boundary. In eqs.(1) and (2), the y-dependence of the equations has been eliminated by introducing a parametric formulation for the velocity profile. The solution to the Falkner-Skan equation for the similar flow including reversed flow profile and wake-like profile was used to represent the flow quantities. Eqs.(1) and (2) contain three unknowns, $a$, $\delta_1^{(i)}$ and $U_e^{(i)}$. In the conventional boundary layer approach, $U_e^{(i)}$ is specified in advance from the inviscid flow. Then, eqs.(1) and (2) can be integrated downstream and $a$ and $\delta_1^{(i)}$ are obtained. But, when there occurs the separation, integrations cannot proceed beyond this point because of the separation singularity. In the present approach, $U_e^{(i)}$ is not specified in advance and eq.(1) and (2) with three unknowns are solved simultaneously with the inviscid equations. There occurs no singularity at the separation point in this method.

The governing equations for the inviscid flow are given in terms of the stream function, $\Psi$ and density, $\rho$ as

$$\rho(\, \Psi_{xx} + \Psi_{yy}\, ) = \Psi_x \rho_x + \Psi_y \rho_y \tag{3}$$

$$\left(\frac{\rho}{\rho_0}\right)^{\gamma-1} = 1 - \frac{\gamma-1}{2}\, \frac{\rho_0^2}{\rho^2 a_0^2}\, (\, \Psi_x^2 + \Psi_y^2\, ) \tag{4}$$

where the subscript 0 indicates the stagnation condition. Assuming that the perturbation from the uniform flow induced by the airfoil thickness is small, the following equations for the perturbation stream function, $\tilde{\psi}$ and the perturbation density, $\tilde{\rho}$ are obtained as the first order through the transonic small perturbation scheme.

$$[(1-M_\infty^2)+(\gamma+1)M_\infty^2\, \tilde{\rho}\,]\, \tilde{\psi}_{xx} + \tilde{\psi}_{yy} = 0 \tag{5}$$

$$\frac{\gamma+1}{2}M_\infty^2\, \tilde{\rho}^2 + (1-M_\infty^2)\, \tilde{\rho} + \tilde{\psi}_y = 0 \tag{6}$$

The physical variables $\tilde{\psi}$ and $\tilde{\rho}$ ( not similarity variables ) are used as the unknowns in the inviscid equations in order to constitute a simultaneous solution system with the equations for the viscous region. Similar to the perturbation potential equations for transonic flow, eq.(5) has the non-linear term in the bracket. Without this term, eq.(5) reduces to the equation for subsonic flow, containing only one unknown $\tilde{\psi}$. Eq.(5) is of the mixed type on account of the above-mentioned non-linear term and changes the type according to the value of $\tilde{\rho}$.

FINITE DIFFERENCE EXPRESSIONS AND<br>SIMULTANEOUS SOLUTION PROCESS

It is well-known that the inviscid equations of transonic flow can be easily and rapidly solved by the type-dependent method using SLOR procedures developed by Murman and Cole(1971). By using a mixed finite difference scheme, the method accounts for the mixed elliptic, hyperbolic character of the governing equations. This method is also adopted in this paper with some modification. Eq.(6) is a quadratic equation of $\tilde{\rho}$ and is solved for $\tilde{\rho}$. Each of the two solutions of $\tilde{\rho}$ corresponds to the subsonic and supersonic branches, respectively. The multiple root corresponds to the sonic point. In the numerical calculations, eq.(5) is solved for $\tilde{\psi}$ with evaluating $\tilde{\rho}$ through eq.(6). Eq.(5) is finite differenced type-dependently. Centered differences or backward differences are used for x-derivatives according to the local condition of the value of $\tilde{\rho}$. Because of the character of $\tilde{\rho}$ mentioned above, the judgement whether the point is locally subsonic or supersonic is necessary. The point is classified to be supersonic only when there is a sonic point above it on the same i=i line. Since the solution process is newly developed for the present study of the viscous-inviscid interaction, it should be noted here that the preliminary calculations for the inviscid flow by using eqs.(5) and (6) were executed in order to certify the validity of the present approach. The agreements between the results obtained and the other results were good.

In the usual solution for the inviscid potential equations by SLOR method, the unknown vector at the i-th station contains $\phi_{ij}$ (j=1,2,$\cdots$jmax). The basic idea of the simultaneous solution procedures is to modify it and to get the boundary layer parameters $U_{ei}^{(i)}$, $\delta_{1i}^{(i)}$ and $a_i$ into the unknown vector. In the present case, the inviscid equations are described with $\tilde{\psi}$ ( $\tilde{\rho}$ subsidiarily used ) so that the tridiagonal matrix system at the i-th station is written as follows:

$$
\begin{bmatrix}
\text{obtained from eqs.(1) and (2)} \\
\text{obtained from eqs.(1) and (2)} \\
\text{the matching condition} \\
\text{modified eq. from eq.(5)} \\
\\
\\
\text{eq.(5)} \\
\\
\end{bmatrix}
\begin{bmatrix}
a_i \\
\delta_{1i}^{(i)} \\
U_{ei}^{(i)} \\
\psi_{12} \\
\psi_{13} \\
\vdots \\
\vdots \\
\psi_{ijm}
\end{bmatrix}
= \mathbf{f}(i-1,i+1)
\qquad (7)
$$

where $\mathbf{f}(i-1,i+1)$ indicates the quantities at the (i-1)th and (i+1)th stations. As is noted in the system (7), two top rows of the matrix are made up from the finite difference expressions of the boundary layer integral equations, eqs.(1) and (2). The other rows except the third row are obtained from the finite difference expressions of eq.(5). For the fourth row, a slight modification is necessary because the unknowns of this row are $\tilde{\psi}_{12}$, $\tilde{\psi}_{13}$ and $U_{ei}^{(i)}$. But this equation can be easily obtained from the relation between $\tilde{\psi}_y$ and $U_{ei}^{(i)}$. The third row is the key of the simultaneous solution method. This is the matching condition between the inviscid region and the boundary layer region. This equation among $\tilde{\psi}_{12}$, $U_{ei}^{(i)}$ and $\delta_{1i}^{(i)}$ is obtained through some manipulations using the condition that the stream function at the displaced boundary is constant. This procedure is similar to that of Moses et al.(1978) and described

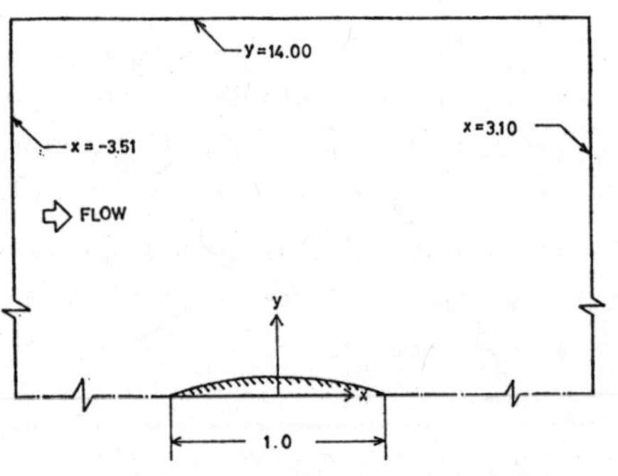



Fig.1 Computational domain

elsewhere in detail (Fujii and Karashima 1979). Thus, tridiagonal matrix at the i-th station is closed and is easily solved. The solution process from upstream to downstream is repeated until convergence is obtained.

## RESULTS AND DISCUSSIONS

The present method was applied to the flow over a 6% circular-arc airfoil. The numerical calculations were carried out using the 92×37 rectangular mesh points. In order to eliminate the instability of the initial stage, the classical boundary layer correction method is adopted from the leading edge to some forward point of the airfoil, where the system is joined to the simultaneous solution system.

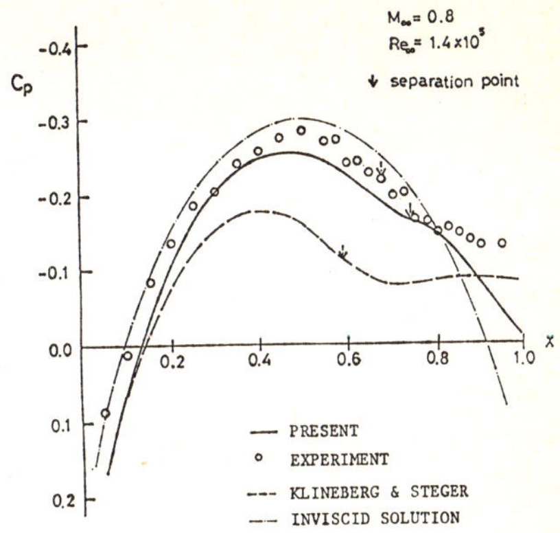



Fig.2-1  Comparison of the predicted pressure distribution with another results

Calculation results are shown in Figs.2-Fig.5. The solid line in Fig.2-1-Fig.2-3 shows the calculated pressure distribution over the airfoil. In order for comparison, the numerical results by Klineberg and Steger(1972), the inviscid results and the experimental results by Collins and Krupp(1973) are shown in some of the figures. The position indicated by an arrow is the separation point in each figure. Fig.2-1 illustrates a typical subcritical case. In the experimental result, the symmetric pressure distribution predicted by the inviscid theory is heavily distorted in the aft part of the airfoil due to the viscous effect. The present result describes this distortion fairly well except near the trailing edge. The prediction of the position of the separation point is not so good, while the plateau pressure is seen near the separation point. At higher free stream Mach number, there occurs a local supersonic region over the airfoil as is seen in Fig.2-2 and Fig.2-3. In these cases, the present method overestimates the viscous effect so that the agreement with the experimental results is not so good quantitatively. The inviscid solution exhibits the shock wave in

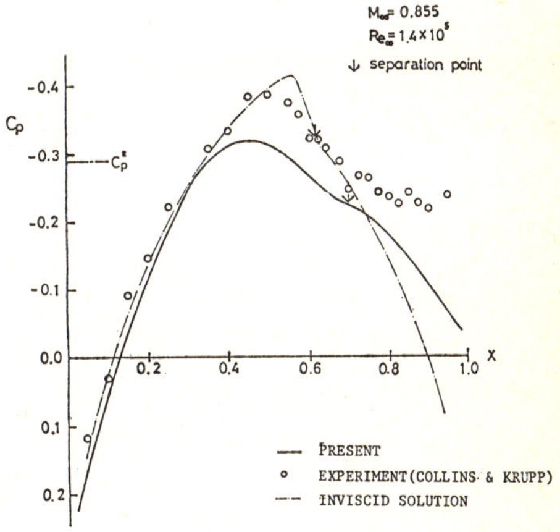



Fig.2-2  Comparison of the predicted pressure distribution with another results

Fig.2-2, while there are no clear shocks in the experiment. This trend was also predicted in the present result. The stronger shock wave is seen in the inviscid result in Fig.2-3. In the experimental result, there is also a shock wave, which is not so clear. The present method succeeded in describing the dispersion of the shock wave, but overestimates the effect of viscosity.

The boundary layer parameters obtained simultaneously with the pressure distribution are shown in Fig.3 as for the case of Fig.2-1. The trend of the variation of the parameters are almost those to be expected. The profile parameter remains almost constant near the leading edge and decreases gradually toward the trailing edge. The displacement thickness increases gradually through the separation point and has its peak value at the trailing edge.

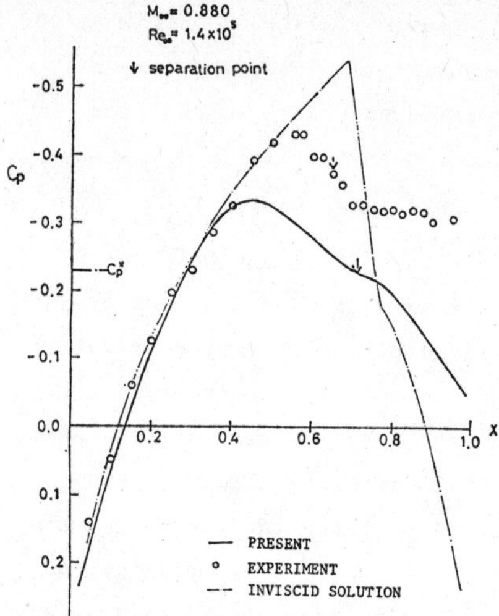


Fig.2-3  Compasison of the predicted pressure distribution with another results

The calculations are executed for several combinations of the free stream Mach numbers and the Reynolds numbers. The variation of the separation point with the free stream Reynolds numbers is illustrated in Fig.4. Quantitative agreement with the experiment is insufficient, but qualitative feature of the forward movement of the separation point with increasing Reynolds numbers is well predicted. Fig.5 shows the free stream Mach number dependence. For the lower Mach numbers, the separation point moves forward with increasing Mach numbers when the flow separates in pressure gradient mode, while rearward movement occurs for higher Mach numbers. Obviously, this rearward movement comes from the change of the separation mode. Although the existence of the shock wave is not clear in the pressure distribution ( as is seen in Fig.2 -3 ) because of the viscous effect, it is clear that flow separates in the shock induced mode. Quantitative disagreement of the position of the

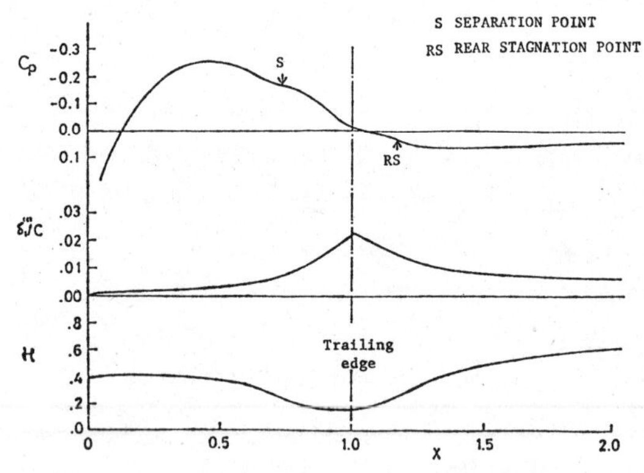


Fig.3  Distribution of the boundary layer parameters (Fig.2-1 case)

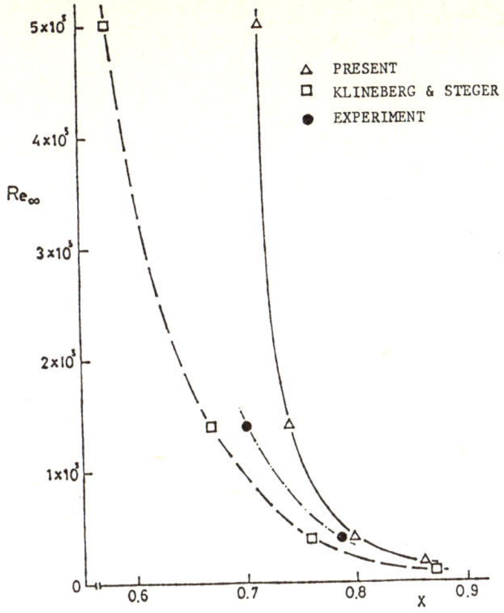
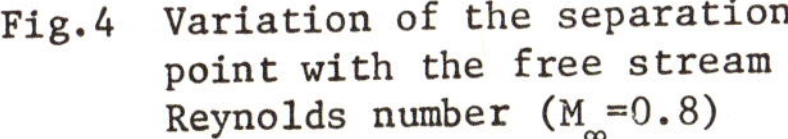

Fig.4 Variation of the separation
point with the free stream
Reynolds number ($M_\infty = 0.8$)

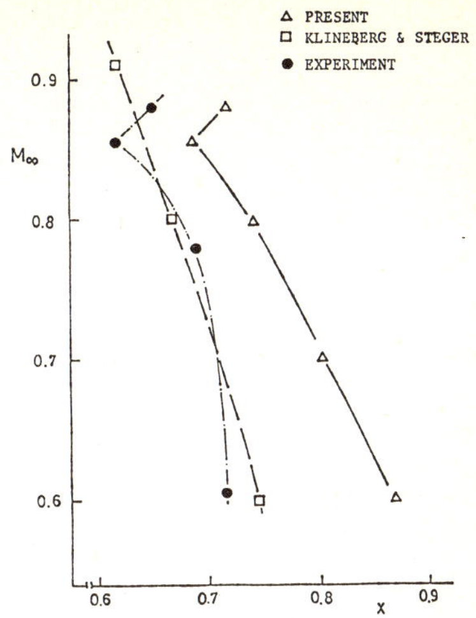

Fig.5 Variation of the separation
point with the free stream
Mach number ($Re_\infty = 1.4 \times 10^5$)

separation point with the experiment may come from the inaccurate prediction of the
pressure distribution near the trailing edge. The present theory adopted the
assumption that the stream function is constant along the displaced boundary. Errors
due to this assumption may become serious near the trailing edge. In addition to this,
the separated region becomes larger near the trailing edge, in which the boundary
layer assumptions may also become invalid. However, the present method can predict
the overall flow field quite well. The solution process is simple and the calculations
are rapid and stable, so this method can be applied also to another flow fields.

## ACKNOWLEDGEMENT

The author would like to express his gratitude to Prof. K.Karashima of ISAS,
university of Tokyo for his cordial advices and useful comments on this study.

## REFERENCES


Moses,H.L., Jones,III R.R., OBrien,Jr.W.E. & Peterson,R.S. 1978 AIAA J. 16,61-66.
Murman,E.M. & Cole,J.D. 1971 AIAA J. 9,114-121.
Fujii,K. & Karashima,K. 1979 ISAS Rept.575,Institute of Space and Aeronautical
Science, University of Tokyo.
Klineberg,J.M. & Steger,J.L. 1972 Lecture Notes in Physics, Vol.19, Springer-Verlag,
New York 162-168.
Collins,D.J. & Krupp,J.A. 1973 AIAA Paper No.73-659.

# NUMERICAL STUDIES ON NONLINEAR HYDRODYNAMIC STABILITY
## BY COMPUTER-EXTENDED PERTURBATION SERIES


Thorwald Herbert
Institut für Aerodynamik und Gasdynamik
Universität Stuttgart, FRG


## INTRODUCTION

From experience with various fluid mechanical problems, Van Dyke (1978) concluded
that the three-step scheme of extension, analysis and improvement of perturbation series is an attractive alternative to finite-difference computation, at least in simple
problems. The present work aims at verifying his statement for some rather difficult
problems in the analysis of nonlinear flow stability.

The properties of stability problems and the requirements of their analysis are different from common computational problems. Stability analysis deals with the manifold
of stable and unstable solutions of the Navier-Stokes equations at higher Reynolds
numbers where these solutions may be unsteady and exhibit three-dimensional structure.
Apart from the problem to achieve sufficient resolution, any simplification or even
the discretization of the equations risks contaminating the physical stability characteristics. Although considerable progress has been reported in the field of numerical experimentation, the basic mechanisms are as difficult to retrieve from the numerical data as from flow experiments. The multidimensional parameter space related to
these mechanisms cannot be mapped out with the few runs feasible within limited computer time. Also, the experiment fails to provide the unstable solutions of interest.

Fundamental stability studies consider flows in simple geometry and usually in an infinite domain. Periodicity of the solutions suggests the application of semi-analytical techniques, in particular of spectral methods which appear as a natural extension
of the normal-mode concept of the linear stability theory and provide at the same time
the desired high resolution. The strength of the spectral approach can be well combined with the capabilities of advanced perturbation methods in treating nonlinear
problems, resulting in a fully computational technique which retains essential parts
of the analytical structure of the solution. Computation of a single set of series coefficients yields the solution for some range of the expansion parameter rather than
a single point within reasonable computing time. Often, a properly chosen expansion
procedure directly (or easily) leads to the physical quantities of interest, avoiding
tedious data reduction. Computation using high-order perturbation series removes the
uncertainty about the validity of low-order results frequently quoted by the nonlinear
stability theory and provides a connecting link to purely numerical results.

A first obvious requirement of using this series technique is the availability of rational perturbation methods which can be carried up to arbitrary order. Such rational
methods are available for a group of stability problems strongly related to steady
equilibrium states and have been recently developed (Herbert 1980) also for problems
involving the unsteady growth of disturbances. In order to achieve our goal it remains
to be shown that series coefficients can be efficiently calculated up to sufficiently
high order and with proper accuracy for representing nonlinear solution and their
stability characteristics - directly or by means of series transformations.

## PERTURBATION METHOD

The structure of the perturbation solution is best illustrated for a nonlinear two-
dimensional solution $\underline{v}(y,z;\lambda)$ of the Navier-Stokes equations which is periodic in the
space coordinate z with wavenumber ß and tends to the basic flow $\underline{V}(y)$ through the so-
lution $\underline{v} - \underline{V} = \underline{v}'(y;\lambda_0) \exp(i\beta z) + \text{c.c.}$ of the linear stability problem as $\lambda \to \lambda_0$,
where $\lambda$ is a parameter to be suitably defined and c.c. denotes complex conjugate terms.
Fourier decomposition yields $\underline{v} = \underline{v}_0(y;\lambda) + \Sigma\{\underline{v}_n(y;\lambda) \exp(in\beta z) + \text{c.c.}\}$ where summation is for all n > o. Substituting into the Navier-Stokes equations provides a cou-
pled system of nonlinear differential equations for the coefficients $\underline{v}_n$, which after
Fourier truncation would form a first basis for numerical studies. Since a small pa-

rameter does not naturally appear in this formulation, we introduce the amplitude $A = A(\lambda)$ measuring the size of the fundamental $\underline{v}_1$ by setting $\underline{v}_1(y;\lambda) = A(\lambda)\,\underline{u}_1(y;\lambda)$ together with a proper normalization of one velocity component at $y = y_0$. Due to the quadratic nonlinearity of the Navier-Stokes equations we obtain $\underline{v}_n = A^n\,\underline{u}_n$, $n \geq o$, and the Linstedt-Poincaré method leads to the formal expansions

$$\underline{u}_n(y;\lambda) = \sum_{m=o}^{\infty} \underline{u}_{nm}(y;\lambda_0)\,\varepsilon^m, \quad n \geq o, \quad \lambda = \sum_{m=o}^{\infty} \lambda_m \varepsilon^m, \quad \varepsilon = A^2.$$

The functions $\underline{u}_{nm}$ are now determined by a sequence of linear differential equations, each of them similar to the equations of the linear stability theory, e.g. the Orr-Sommerfeld equation for parallel flows. The constants $\lambda_m$ are determined together with $\underline{u}_{1m}$ from solvability conditions or in the way discussed by Herbert (1980).

In the following we are mainly concerned with the quantity $\lambda = \lambda(\varepsilon)$ and therefore we truncate the series for $\lambda$ at $M \geq m$. Determination of $\lambda_M$ requires solving for $\underline{u}_{1M}$ which is of order $O(A^N)$, $N = 2M + 1$. Truncation at this order in the amplitude implies truncation of the Fourier series at $N \geq |n|$ since $\underline{v}_n = O(A^n)$. For calculating all terms of order $O(A^N)$ we must solve $(M + 1)(M + 2) - 1$ differential equations in succession. Thus, computer time and storage increase as $O(M^2)$. In order to keep these requirements at minimum, we apply pseudo-spectral and spectral-collocation methods, using Chebyshev polynomials in the y-direction, where no-slip boundary conditions are prescribed. Detailed studies for plane Poiseuille flow (Herbert 1977) indicated that these techniques combine high accuracy and high efficiency with the advantages of a semi-analytical form of the solution. Following the ideas of Van Dyke (1975), the tedious work of generating the numerous differential equations and collecting the nonlinear terms is carried out by the computer program simultaneously with the numerical evaluation.

Careful accuracy tests were performed with truncation at $N = 15$ because the structure of the solution process bears the danger of dramatic error propagation and the oscillatory character of higher harmonics could require very high resolution. We varied the number of Chebyshev modes and compared results of identical calculations in single precision on a Univac 1108 (27 bit mantissa) and a CD 6600 (48 bit mantissa). No accuracy problems were detected up to our usual truncation at $N = 15$ ($M = 7$) which is imposed by storage restrictions (< 64K words). The solution of 71 differential equations requires typically 36 seconds (single precision complex, Univac 1108). The programs for single-mode solutions range from 250 to 400 Fortran statements, depending on the basic flow and the type of differential equations.

The perturbation method can be straightforward extended for studies on the stability of nonlinear single-mode solutions or on nonlinear interactions of different modes (Herbert & Morkovin 1980). Using the concept of superposable N-th order interactions involving a maximum of $N + 1$ modes, also wave packets seem accessible to this method.

ANALYSIS AND IMPROVEMENT OF SERIES

With the series coefficients calculated, we cannot expect convergence of the series for all values of $\varepsilon$ in the range of physical interest. However, there are useful techniques available (Gaunt & Guttmann 1974, Van Dyke 1974) for extracting the information about the solution which may be concealed in the coefficients even in case of poor convergence. With relatively few coefficients we can neither use Domb-Sykes plots nor derive accurate results on the singularities governing the convergence domain. Nevertheless we can improve or achieve convergence by use of sequence transformations. As compared with purely numerical techniques we prefer Padé approximants in order to retain the semi-analytical form of the solution. Thus, we replace the original series $\lambda(\varepsilon) \sim [M/O] = \lambda_0 + \lambda_1\varepsilon + \ldots + \lambda_M\varepsilon^M$ by the equivalent ratio of two polynomials $\lambda(\varepsilon) \sim [I/J] = (p_0 + p_1\varepsilon + \ldots + p_I\varepsilon^I) / (1 + q_1\varepsilon + \ldots + q_J\varepsilon^J)$ with $I + J = M$ and $[M/O] = [I/J] + O\,(\varepsilon^{M+1})$.

NONLINEAR STABILITY OF CIRCULAR COUETTE FLOW

Originally, we developed the series method for parallel flows and applied it to single and interacting modes in plane Poiseuille flow. Some results of this continuing work are reported by Herbert (1980), Herbert & Morkovin (1980). Here we concentrate on the stability of circular Couette flow with the inner cylinder rotating and the outer at rest. In the first phase of these studies, we use the small-gap approximation.

The essential steps in the stability analysis of Couette flow are discussed by Stuart (1971), where also a list of references is given. The problem is mainly to find the bordering lines between the domains shown in Figure 1 for Taylor number T and axial wave number ß:

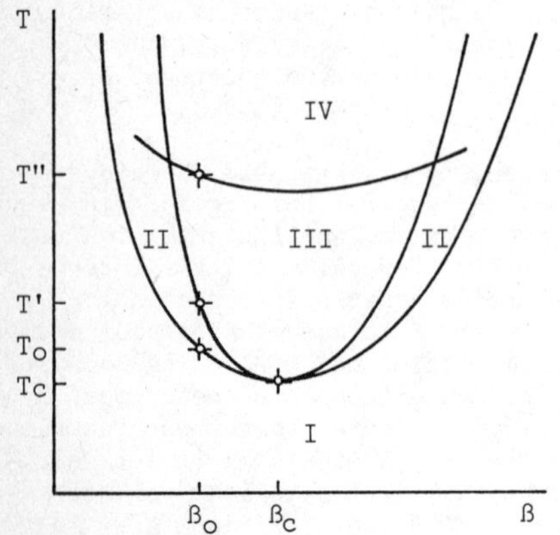



I    Couette flow is stable

II    Couette flow is unstable and Taylor vortices are unstable against vortices of another wavenumber

III    Taylor vortices are stable

IV    Taylor vortices are unstable against non-axisymmetric disturbances with k circumferential waves

Figure 1. Stability domains for circular Couette flow and the stability limits $T_0$, $T'$, $T''$ at fixed $ß = ß_0$.

More detailed analysis is required for wavenumber selection within III, for selection of $k = k(ß)$ within IV or for the properties (e.g. torque) of Taylor vortices within III or wavy vortices within IV.

Only the linear stability problems - calculation of the neutral curve between I and II, critical Taylor number $T_c$ and wavenumber $ß_c$ as well as amplification rates - are satisfactorily solved for axisymmetric and non-axisymmetric modes. The first step into nonlinearity, the study of growth and equilibration of unstable axisymmetric modes, is still a challenge for finite-difference computations with weak prospects for finding the borders between II, III and IV.

## GROWTH AND EQUILIBRATION OF TAYLOR VORTICES

By choosing $\lambda$ and $\varepsilon$ differently, we use two types of expansions for the analysis of an axisymmetric mode at fixed $ß = ß_0$. Type I is a direct approach to the steady equilibrium states similar to that suggested by Reynolds & Potter (1967). By expanding in the equilibrium amplitude $A_e$, the Taylor number T appears as expansion $T = T_0 + T_1\varepsilon_e + T_2\varepsilon^2_e + \ldots$ in $\varepsilon_e = A^2_e$ about the point $ß_0$, $T_0$ of the neutral curve. Type II is a rational extension of the approach used by Davey (1962) and Eagles (1971) for the unsteady growth of unstable modes. By expanding in the amplitude $A = A(t)$, the nonlinear growth rate $a = d(\ln A)/dt$ appears as expansion $a = a_0 + a_1\varepsilon + a_2\varepsilon^2 + \ldots$ in $\varepsilon = A^2$ about the linear growth rate $a_0$ at the point $ß_0$, $T > T_0$. Whereas type I at once provides the equilibrium states for some range of $T > T_0$, type II yields only a single state at fixed T.

Equilibrium amplitudes for the critical wavenumber $ß_c = 3.127$ are shown in the plot of $A^2_e$ versus $T/T_c$ in Figure 2. The low-order results of Davey up to $O(A^3)$ and Eagles up to $O(A^5)$ are not exactly comparable with our [1/0] and [2/0] results due to their additional approximations in determining the constants $a_m$. Our definitions introduce a factor of 2 for T, a factor of .5 for A in comparison with Davey (1962), hence $T_c = 3389.9$. The [3/3]-approximants from both types of expansions agree completely

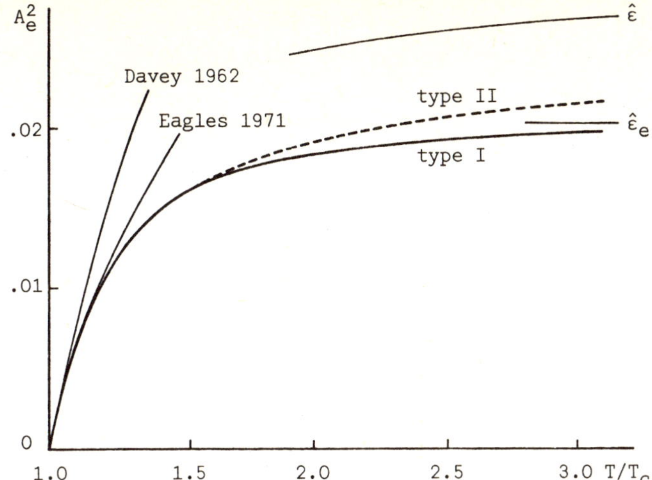


Figure 2. Square of the equilibrium amplitude $A_e$ for Taylor vortices at $\beta = \beta_c$ obtained from [3/3]-approximants

up to $T \approx 1.5\ T_c$ and differ by less than 10% up to $T = 3.0T_c$ whereas [6/0] overestimates the value of $A^2_e$ at $T = 1.5\ T_c$ by 2%. With the type I expansion, the complete curve is obtained in about 30 seconds computer time, less than needed for a single point by finite-difference computation (Booz, private communication). The sign patterns of the original series of both types indicate that convergence is restricted by a physical singularity at the positive real axis. The denominator roots $\hat{\varepsilon}$, $\hat{\varepsilon}_e$ of [3/3]-approximants are given in Fig.2. The interpretation of these singularities, which appear as a first-order pole, is different: whereas $T \to \infty$ as $\varepsilon_e \to \hat{\varepsilon}_e$, we find $a \to -\infty$ as $\varepsilon \to \hat{\varepsilon}$.

## DOMAIN OF STABLE TAYLOR VORTICES

Calculations with only a single mode of wave number $\beta_0$ taken into account provide an equilibrium state for any point in the unstable domain, $a_0 > 0$. This state may be stable or unstable with respect to an axisymmetric disturbance of different wavenumber $\beta_1 \neq \beta_0$ and small amplitude $B \ll A$. Low-order perturbation analysis (cf. Stuart 1971) indicates the existence of stable states for a subset of unstable points centred around $\beta_c$, shown by domain III of Figure 1. With the nonlinear solution for $\beta_0$ given by a series of type I or type II, the problem is to calculate the amplification rate $b = d(\ln B)/dt$ of the small disturbance in presence of a Taylor vortex of amplitude $A_c$ from the stability series $b = b_0 + b_1\varepsilon_e + b_2\varepsilon^2_e + \ldots$ . Our calculations were carried up to $O(\varepsilon^4_e)$, i.e. $O(A^9_e)$. Figure 3 shows a type I-result for $\beta_0 = 2.8$, $\beta_1 = 3.1$. The stability limit $T'$ given by $b = 0$ depends sensitively on the truncation. In agreement with Nakaya's (1974) conclusion from 5th-order analysis we find that $T'$ is underestimated by the 3rd-order calculations, in particular as $|\beta_0 - \beta_c|$ increases. Although the detailed calculations of the subdomain of stable states (which require to find the maximum of $T'$ $(\beta_1)$ for every $\beta_0$) are not yet completed, we can conclude that at given $T$ the band of realizable wavenumbers is smaller than earlier predicted but does not shrink to a single value, thus still leaving a selection problem between wavenumbers in the domain III of stable Taylor vortices.

The analytical structure of this selection problem suggests that the wavenumber of the finally surviving Taylor vortex is determined by the initial values of the unsteady problem involving two or more spectral modes. The type II-expansion for two disturbances with wavenumbers $\beta_0$ and $\beta_1$ and comparable amplitudes $A(t)$ and $B(t)$, respectively, leads to the equations

$$a = d(\ln A)/dt = a_0 + a_1 A^2 + a_2 B^2 + \ldots,$$
$$b = d(\ln B)/dt = b_0 + b_1 B^2 + b_2 A^2 + \ldots .$$

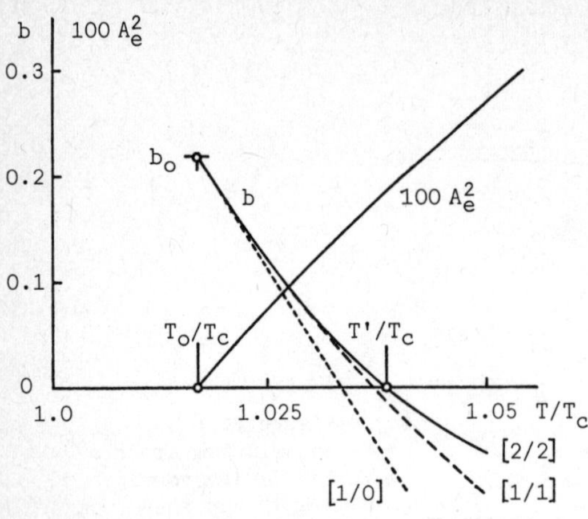

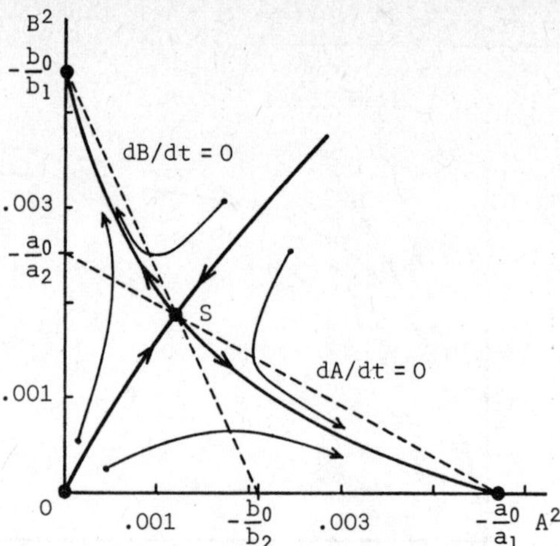

Figure 3. Amplification rate b of an axisymmetric disturbance of wavenumber $\beta_1 = 3.1$ in presence of a Taylor vortex of wavenumber $\beta_0 = 2.8$ and equilibrium amplitude $A_e$.

Figure 4. Trajectories in the phase-plane for two axisymmetric modes with wavenumbers $\beta_0 = 3.1$, $\beta_1 = 2.9$ at $T = 1.062 \, T_c$.

With the constants $a_m$, $b_m$ calculated and initial amplitudes $A(t_0)$, $B(t_0)$ given, the initial value problem can be solved by integrating a system of ordinary differential equations. The extension to more than two modes is straightforward. The decision whether 7, 8 or 9 pairs of vortices develop in a finite computational domain of 8 times the critical wavelength from given initial conditions requires the order of 10 hours finite-difference computation (Booz, private communication) but only the order of 5 minutes with the series method. Moreover, we can study the general properties of the system of differential equations, in particular the nature of stationary points (Segel 1962) and the connecting trajectories. An example for two competing modes with $\beta_0 = 3.1$, $\beta_1 = 2.9$ at $T = 1.062 \, T_c$ is shown in Figure 4, derived from 3rd-order equations. Initial conditions to the left or right of the line passing from the origin through the saddle point S lead to the stable knots $B^2_e = -b_0/b_1$, $A^2 = 0$ or $A^2_e = -a_0/a_1$, $B^2 = 0$, respectively.

## STABILITY WITH RESPECT TO NON-AXISYMMETRIC MODES

The next step of finding the border between regions III and IV of Fig. 1 at fixed $\beta_0$ again leads to a series for the nonlinear amplification rate $b = b(A^2_e)$, but now $B(t)$ is the amplitude of a non-axisymmetric mode with axial wavenumber $\beta_0$ and k circumferential waves. In accordance with Davey, DiPrima & Stuart (1968) we find the Taylor vortex stable against disturbances with the same axial phase but unstable against disturbances with the axial phase shifted by $\pi/2$. Our results of type II expansions shown in Fig. 5 basically agree with earlier work in that we find at $\beta_0 = \beta_c$ the lowest stability limit $T''(k)$ for $k = 1$ and slightly increasing limits for increasing k. The numerical values of $T''(k)$ in Table 1 differ somewhat from those obtained by Davey et al. for [1/0] and by Eagles (1971) for [2/0] since their perturbation method yields the series coefficients only to within an error of $O(a_0)$. In the critical range, the nonlinear terms are stabilizing for $k \leq 5$ but destabilizing for $k > 6$ such that instability against the mode $k = 6$ arises although this mode would decay in absence of the Taylor vortex. Coles (1965) observed with the radius ratio 0.874 the lowest stability limit for $k = 4$, slightly preferred to $k = 5$ and $k = 6$. It is likely that this observation is due to the essentially stronger amplification of waviness for larger wavenumbers k, as shown in Fig. 5.

Torque calculations for wavy vortices have as yet been carried out with 5th-order terms taken into account. In spite of some numerical differences, our results confirm the conclusion of Eagles (1974) that waviness reduces the torque of Taylor vortices. A study of the relative stability of different wavy vortices and a detailed comparison with experiments is a straightforward matter for future calculations, which should also refrain from the small-gap approximation.

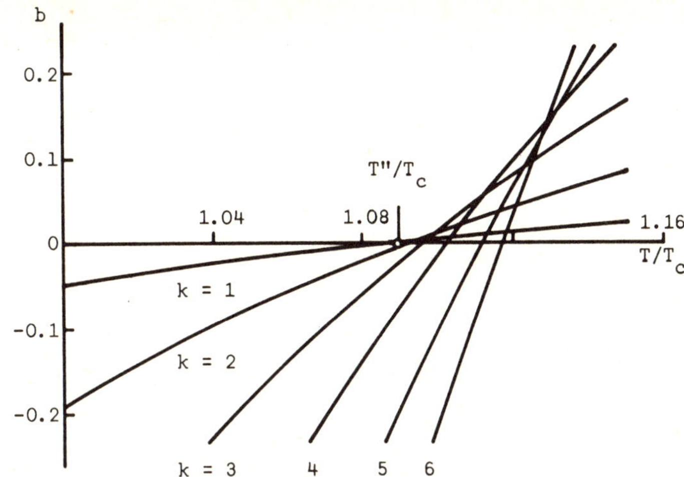


Figure 5.  Amplification rate b for disturbances with k circumferential waves in presence of a Taylor vortex of amplitude $A_e$. Wavenumber $\beta_c$, axial phase shift $\pi/2$, [1/1]-approximant.

| k | [1/0] | [2/0] | [1/1] |
|---|-------|-------|-------|
| 1 | 1.0977 | 1.0904 | 1.0890 |
| 2 | 1.0999 | 1.0929 | 1.0916 |
| 3 | 1.1036 | 1.0972 | 1.0960 |
| 4 | 1.1092 | 1.1033 | 1.1023 |
| 5 | 1.1170 | 1.1113 | 1.1115 |
| 6 | 1.1276 | 1.1213 | 1.1171 |

Table 1.  Stability limit $T''/T_c$ of Taylor vortices against disturbances with k circumferential waves

## REFERENCES


Coles, D. 1965 J.Fluid Mech. 21, 385-425
Davey, A. 1962 J.Fluid Mech. 14, 336-368
Davey, A., Di Prima, R.C. & Stuart, J.T. 1968 J.Fluid Mech. 31, 17-52
Eagles, P.M. 1971 J.Fluid Mech. 49, 529-550
Eagles, P.M. 1974 J.Fluid Mech. 62, 1-9
Gaunt, D.S. & Guttmann, A.J. 1974 In: Phase Transitions and Critical Phenomena
        (ed. C. Domb & M.S. Green), Vol. 3, 181-243, Academic Press
Herbert, Th. 1977 Habilitationsschrift, Universität Stuttgart
Herbert, Th. 1980 AIAA Journal 18, 243-248
Herbert, Th. & Morkovin, M.V. 1980 Proc. IUTAM Symposium "Laminar-Turbulent
        Transition", Stuttgart 1979, in press
Nakaya, C. 1974 J. Phys. Soc. Japan 36, 1164-1173
Reynolds, W.C. & Potter, M.C. 1967 Unpublished Stanford Univ. Paper
Segel, L.A. 1962 J.Fluid Mech. 14, 97-114
Stuart, J.T. 1971 Ann. Rev. Fluid Mech. 3, 347-370
Van Dyke, M. 1974 Quart. J. Mech. Appl. Math. 27, 423-450
Van Dyke, M. 1975 SIAM J. Appl. Math. 28, 720-734
Van Dyke, M. 1978 J.Fluid Mech. 86, 129-145

## COMPACT DIFFERENCING SCHEMES FOR ADVECTIVE PROBLEMS

Richard S. Hirsh and Ralph E. Ferguson
JAYCOR, Del Mar, California 92014

## Introduction

The use of compact differencing for the solution of partial differential equations was simultaneously introduced for parabolic/elliptic [1] and hyperbolic [2] equations. Since then the emphasis has been mostly toward improvements for the treatment of parabolic/elliptic equations of the Navier–Stokes variety [3], and the use of implicit methods [4]. There are, however, many instances in fluid dynamics where simple advection equations (possibly only one-dimensional) are sufficient to describe the phenomena under investigation, e.g., two-phase flows in nuclear reactor cooling systems. These problems require detailed knowledge of the transient behavior of the flow variables and so the ability to take large time steps, provided by unconditionally stable methods, may no longer be needed or even desired. The majority of the methods currently in use for these reactor problems utilize upwind differencing for the spatial derivatives and hence lose considerable accuracy in their solutions.

In this paper we will discuss three compact spatial differencing schemes. Two of these are new, being explicit second-order accurate in time, fourth-order accurate in space procedures. The other is an implicit, second-order in time and space, compact, upwind scheme which utilizes only two computational nodes. This scheme is not new, being identical to Keller's Box scheme [5], although its connection with compact schemes has not been mentioned previously.

## A Second-Order Compact Upwind Scheme

The derivation of this second-order upwind scheme directly follows the original derivation of compact fourth-order methods [1]. We desire an improvement in accuracy over the standard upwind formula given by

$$u_x = \frac{1}{h}\left(u_j - u_{j-1}\right) = D_- u_j \quad . \tag{1}$$

Using the example of Reference 1, the truncation error in the above equation is incorporated into the difference operator which is then recognized as an expansion, i.e.,

$$u_x = D_- u_j + \frac{h}{2} u_{xx} = D_- u_j + \frac{h}{2} D_-^2 u_j = D_-\left(1 + \frac{h}{2} D_-\right)u_j = \left[\frac{D_-}{1 - \frac{h}{2} D_-}\right] u_j$$

The final result, with $u_x = F$ gives

$$\left(1 - \frac{h}{2} D_-\right) F_j = D_- u_j \quad . \tag{2}$$

Thus, the derivative F and the function u must be determined from Equation (2) and the governing equation of the problem.

To test this method, as well as the others to be introduced, the standard simple advection problem

$$u_t + A u_x = 0 \tag{3}$$

will be employed. A von Neumann stability analysis shows that a simple explicit forward time differencing with the compact upwind scheme is unstable, so an implicit formulation is required. If the time derivative is centered, the Keller Box scheme is reproduced.

$$\frac{u_j^{n+1} - u_j^n}{\Delta t} + \frac{A}{2}\left(u_j^{n+1} + u_j^n\right) = 0 \tag{4a}$$

$$\frac{F_j^{n+1} + F_{j-1}^{n+1}}{2} + \frac{u_j^{n+1} - u_{j-1}^{n+1}}{h} = 0 \quad . \tag{4b}$$

Although this scheme is truly implicit, with amplification factor identically equal to one, for Equations (4), if A does not change sign, it can be arranged so that the computational algorithm can be accomplished explicitly, thus avoiding the need for matrix inversions. After some algebra we get

$$u_j^{n+1} = \frac{1}{1+C}\left[u_j^n + Cu_{j-1}^{n+1} - \frac{1}{2}Ch\left(F_j^n - F_{j-1}^{n+1}\right)\right] \tag{5a}$$

$$F_j^{n+1} = -F_{j-1}^{n+1} + \frac{2}{h}\left(u_j^{n+1} - u_{j-1}^{n+1}\right) \tag{5b}$$

with $A > 0$ $(C > 0)$, and boundary values for u and F given on the left side of the domain, the integration can proceed from a line of known values at time $n\Delta t$ by alternatively calculating $u_j^{n+1}$, then $F_j^{n+1}$ for increasing j.

Two Fourth-Order Explicit Compact Schemes

The solution of Equation (3) with fourth-order spatial accuracy can be achieved by direct extension of the classical Lax-Wendroff (LW) scheme. A one-step LW method and a two-step variant of the Richtmyer formulation have been developed using the compact approach. Both use three point, centered, spatial differences.

The one step procedure is a straightforward approach which simply utilizes the more accurate determination of the spatial derivatives given in Reference 1, in the time expansion of the unknown variable u which leads to the LW scheme. With A constant in Equation (3) the following expansion is obtained

$$u_j^{n+1} = u_j^n + \Delta t\, u_t^n + \frac{1}{2}\Delta t^2\, u_{tt}^n = u_j^n - A\Delta t\, u_x^n + \frac{1}{2}A^2\Delta t^2\, u_{xx} \quad .$$

With compact differences for the spatial derivatives the fourth-order analog of the Lax-Wendroff scheme is

$$u_j^{n+1} = u_j^n - ChF_j^n + \frac{1}{2}C^2h^2 S_j^n \tag{6a}$$

$$\frac{1}{6}F_{j+1}^n + \frac{2}{3}F_j^n + \frac{1}{6}F_{j-1}^n = \frac{1}{2h}\left(u_{j+1}^n - u_{j-1}^n\right) \tag{6b}$$

$$\frac{1}{12}S_{j+1}^n + \frac{5}{6}S_j^n + \frac{1}{12}S_{j-1}^n = \frac{1}{h^2}\left(u_{j+1}^n - 2u_j^n + u_{j-1}^n\right) \tag{6c}$$

Hence, the time march is done explicitly in Equation (6a) once values of $F_j^n$ and $S_j^n$ are obtained by inverting the tridiagonal systems in (6b) and (6c). As in the classical LW scheme the addition of the second derivative stabilizes an unstable forward time integration. Doing the stability analysis yields

$$c < \sqrt{\frac{5}{48}} = 0.323 \tag{7}$$

and shows the method to be conditionally stable with Courant limit less than one. However, for comparable accuracy of second-order methods, significantly larger grid sizes can be used, as will be seen below, thereby not diminishing the maximum time step allowed; although, as previously stated, 'large' time steps may not be the aim of the integration.

The two-step procedure is a novel use of the compact method in that it relates values at node points to values at mid-points. Following the two-step formulation set forth by Gottlieb and Turkel [6] we can write a two-step LW method, for Equation (3), as

$$\bar{u}_{j+1/2} = \mu \ u^n_{j+1/2} - \frac{A\Delta t}{2} \ \delta \ u^n_{j+1/2} \tag{8a}$$

$$u^{n+1}_j = u^n_j - \Delta t \ \delta \ \bar{u}_j \tag{8b}$$

where

$$\mu = \frac{1}{2} \left( E^{1/2} + E^{-1/2} \right) \quad , \qquad \delta = \frac{1}{h} \left( E^{1/2} - E^{-1/2} \right) \quad , \qquad Eu_j = u_{j+1} \tag{9}$$

Thus what is required is a compact representation of the averaging and difference operators. As before in the second-order upwind, incorporating the truncation error, and now remembering that the value at the midpoint j+1/2 is to be determined from values at node points in (8a) we can write

$$\bar{u}_{j+1/2} = U^n_{j+1/2} - \frac{A\Delta t}{2} \ F^n_{j+1/2} \tag{10a}$$

$$u^{n+1}_j = u^n_j - A\Delta t \ \bar{F}_j \tag{10b}$$

where

$$\left( 1 + \frac{h^2}{8} \ \delta^2 \right) U^n_{j+1/2} = \mu u^n_{j+1/2} \tag{11a}$$

$$\left( 1 + \frac{h^2}{24} \ \delta^2 \right) F = \delta u \tag{11b}$$

giving

$$\frac{1}{8} \ U^n_{j+3/2} + \frac{3}{4} \ U^n_{j+1/2} + \frac{1}{8} \ U^n_{j-1/2} = \frac{1}{2} \left( u^n_j + u^n_{j+1} \right) \tag{12a}$$

$$\frac{1}{24} \ F^n_{j+3/2} + \frac{11}{12} \ F^n_{j+1/2} + \frac{1}{24} \ F^n_{j-1/2} = \frac{1}{h} \left( u^n_{j+1} - u^n_j \right) \tag{12b}$$

$$\frac{1}{24} \ \bar{F}_{j+1} + \frac{11}{12} \ \bar{F}_j + \frac{1}{24} \ \bar{F}_{j-1} = \frac{1}{h} \left( \bar{u}_{j+1/2} - \bar{u}_{j-1/2} \right) \tag{12c}$$

Similar to the one-step method, this two-step explicit calculation requires the inversion of simple tridiagonal matrices, not block tridiagonal ones as is necessary in implicit methods. The algorithm requires values of derivatives and averages based on previous time step information, and the governing equation then yields the new function value by quadrature.

## Boundary Conditions

For either of the two fourth-order LW schemes, the explicitness of the method makes the determination of added boundary conditions particularly simple. In any case, the function u will be known on the boundary. It is the added boundary values of F and either S or U which must be determined in order to solve the tridiagonal matrices which arise in the compact forms. Following Reference 1, the simplest boundary conditions consistent with the already available known function values were imposed. For the one-step method this meant five point one sided differences on u to get F and S. The conditions for the two-step method actually turned out to require only four points to give consistent accuracy, and were skewed differences (not one sided) for F, and interpolations for U.

As an example, for $u_j$ known in $1 < j < N$, the two-step method requires boundary values for $U_{3/2}$ and $F_{3/2}$. These were obtained from

$$U_{3/2} = \frac{1}{16} \left( 5u_1 + 15u_2 - 5u_3 + u_4 \right) \tag{13a}$$

$$F_{3/2} = \frac{1}{24h} \left( -23u_1 + 21u_2 + 3u_3 - u_4 \right) \ . \tag{13b}$$

## Results

The results of sample calculations on Equation (3) are shown in Figures 1 through 4. In all cases $A > 0$. The initial condition was a narrow Gaussian; the boundary condition at the left was an imposed sine wave. The exact solution to which the numerical results will be compared is a traveling wave moving to the right at speed A. The four figures give results using upwind, second-order forward time central space (an LW method), and compact upwind differencing generated with a grid size of h = 1, and one of the fourth-order compact differencing schemes with h = 4.

Each figure contains four plots showing the initial conditions, the solution at two subsequent times, and the sum of the squares of the function values for all the grid points in the domain. As expected, the reference calculations showing upwind and second-order central results, Figures 1 and 2, give the standard behavior. The dissipative upwind scheme obliterates the true solution and the central scheme does a fairly good job, showing just a bit of dispersion error.

However, the compact upwind, which is a true two point upwind scheme, actually produces better overall results (see Figure 3) than the central scheme, despite the small dispersion error noticeable in the second frame of Figure 3. The fourth-order compact schemes, Figure 4 shows results from the one-step method, give extremely accurate results, even using four times larger grid spacing than the results in the first three figures. This accounts for the chopping of the waveforms in Figure 4, which does not, however, lead to any degradation of the solution.

## Conclusions

Compact differencing techniques have been shown to be applicable to simple first order advection equations. A second-order accurate compact upwind scheme has been derived which is identically equivalent to Keller's box method. Two fourth-order compact analogs of Lax-Wendroff methods have also been shown. The accuracy of any of these compact methods is superior to a standard second-order method, even for one quarter of the nodes in the case of the fourth-order schemes.

## REFERENCES


1.  Hirsh, R. S., "Higher Order Accurate Difference Solutions of Fluid Mechanics Problems by a Compact Differencing Technique," J. Comp. Phys., 19, 90-109, 1975.

2.  Ciment, M. and Leventhal, S. H., "Higher Order Compact Implicit Schemes for the Wave Equation," Math. Comp., 29, 985-994, 1975.

3.  Bontoux, P., Forestier, B. and Roux, B., "Analysis of Higher Order Methods for the Numerical Simulation of Confined Flows," Paper presented at the Sixth International Conference on Numerical Methods in Fluid Dynamics, June 20-25, 1978, Tbilisi, U.S.S.R.

4.  Leventhal, S. H., "A Two-Dimensional Operator Compact Implicit Method for Parabolic Equations," Paper presented at SIAM 1979 Fall Meeting, November 12-14, 1979, Denver, Colorado.

5.  Keller, H. B., "A New Difference Scheme for Parabolic Problems," in J. Bramble (ed.), Numerical Solutions of Partial Differential Equations, Vol. II, Academic Press, New York, 1970.

6.  Gottlieb, D. and Turkel, E., "Dissipative Two-Four Methods for Time Dependent Problems," ICASE Report No. 75-22, October 1975.


## Acknowledgement

This work was supported in part by the Electric Power Research Institute under Project RP-888-1.

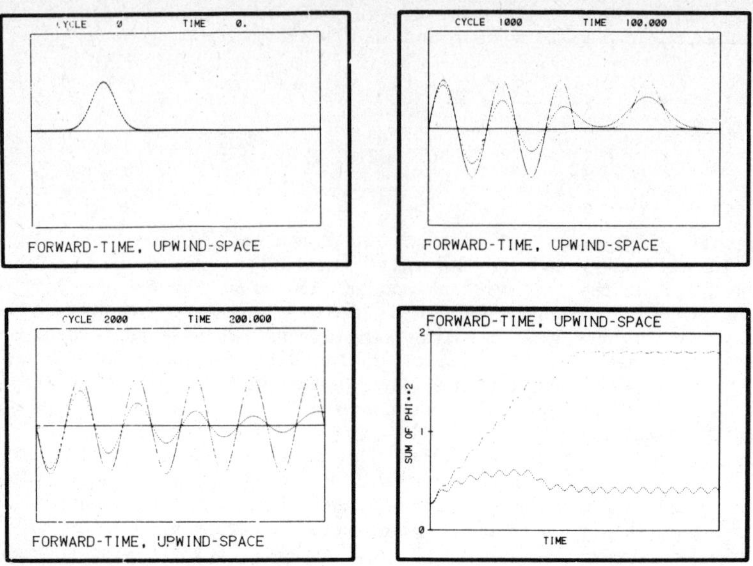

Figure 1.  Transient calculation using FTUS with h = 1.

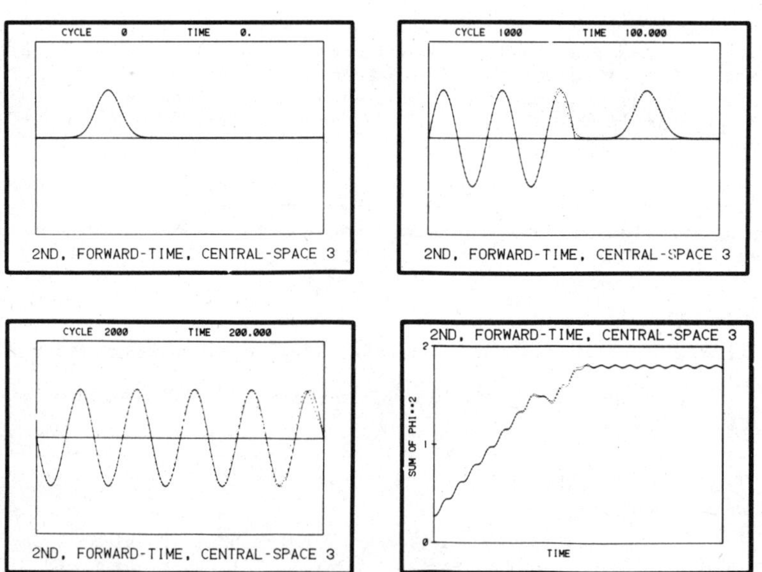

Figure 2.  Transient calculation using FTCS with h = 1.

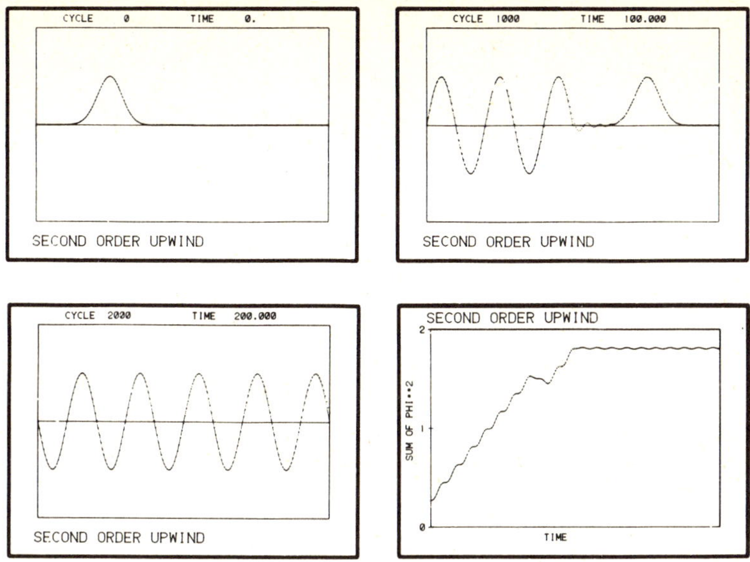



Figure 3.  Transient calculation using compact upwind with h = 1.

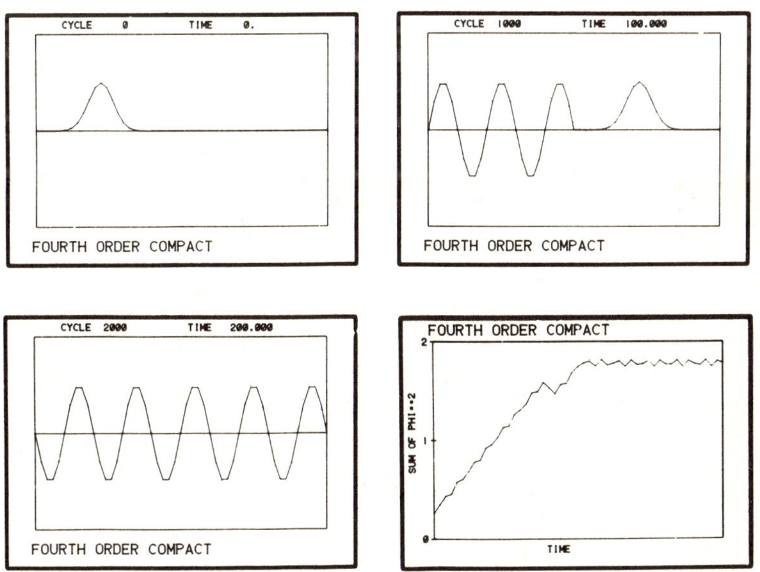



Figure 4.  Transient calculation using one step Lax-Wendroff fourth-order
compact with h = 4.

STUDY OF TWO-DIMENSIONAL FLOW PAST AN ELLIPTIC CYLINDER
BY DISCRETE-VORTEX APPROXIMATION


Kiyosi Horiuti, Kunio Kuwahara and Yuko Oshima[*]
Dept. Applied Physics, Univ. Tokyo
* Dept. Physics, Ochanomizu Univ.



Investigation of flow around an elliptic cylinder with the thickness ratio of
0.5 at an angle of attack of 45°, and 90°, having started impulsively, was carried
out using the discrete-vortex approximation.  The boundary layer along the body surface is divided into a finite number of elements and replaced by the same number of
the corresponding vortex filaments, the circulations of which are determined to be
equal to those of the elements.  Usually, the arrays of the vortex filaments thus introduced separate alternatively from the surface of the cylinder, and form Karman
vortex streets, that is, the familiar alternative separation process of the vortices
was observed.  At the angle of attack of 90°, the initial twin vortex separates
from the cylinder simultaneously, but not alternatively.  These vortex pairs eventually yield to the Karman vortex street, and this transition process successfully
simulates the experimentally-observed phenomena.  For this process it is essential
to consider the boundary layer of the rear side of the cylinder and this consideration has been successfully carried out by the use of this method.



1 Introduction

It is a familiar phenomenon that an obstacle sheds series of vortices alternatively into the wake, which form a vortex street.  In order to solve the full Navier-
Stokes equations for this phenomenon, almost prohibitively large amount of computational labour is required (Takao 1973).  Because at a high Reynolds number, the effect of viscosity is restricted within the thin boundary layer along the surface of
the body, the vorticity is assumed to be generated only in that layer.  Therefore
such a method that can duly treat the vortex motion, that is, the inviscid vortical
flow theory, is desirable.  One of such methods is the discrete-vortex approximation,
in which the flow field around the body is calculated by replacing the boundary layer
with discrete vortices (Kuwahara 1978).

In this paper, the flow patterns behind the elliptic cylinder with thickness
ratio of 0.5 and at the angle of attack of 45° and 90° were studied both numerically
by the discrete-vortex approximation and experimentally using the flow visualization
techniques, and the comparison of them was made.

## 2 Numerical simulation

In this paper, the mechanism of the generation of the vortices is treated as follows: At the instant of impulsive start, the flow field is considered to be inviscid and irrotational, and there is no vorticity in the flow field. However, in order to cancel the velocity slip on the body surface, there must be an infinitely thin boundary layer along the surface of the cylinder; it is considered as a vortex layer. In the discrete-vortex approximation used, this layer is divided into a finite number of elements, and replaced by the same number of corresponding vortex filaments, the circulations of which are determined to be equal to those of the elements. Computation was carried out using conformal mapping to transform the ellipse (on Z-plane) to a circle (on W-plane). The transformation is given by

$$Z = \frac{1}{2}(aW + 1/a(\exp(2i\gamma)W)), \quad a = \sqrt{3}$$

where $\gamma$ is the angle of attack. The boundary layer at the initial instant is approximated by 36 vortex filaments, the positions of them are given by

$$W_j = (1 + \varepsilon)\exp(2\pi i(j - \tfrac{1}{2})/2M), \quad j = 1, \cdots, 2M ,$$

and the strengths of them are determined by,

$$\Phi_1 = \Phi_2 = \cdots\cdots = \Phi_{2M}; \qquad \Phi_j = \mathrm{Re}\{f(\exp(2\pi i(j/2M)))\},$$

where 2M is the number of the vortex filaments and $f(W)$ is the velocity potential,

$$f(W) = \frac{a}{2}(W + 1/W) + i\sum_{j=1}^{n} \kappa_j \log(W - W_j) - i\sum_{j=1}^{n} \kappa_j \log(W - W_j^*), \quad W_j^* = 1/\overline{W}_j .$$

Each point vortex is convected by the velocity field due to the other vortices and their images which are taken in order to match the boundary condition and the initial potential flow. Then after a specified time step, the vortices which have compensated the circulation caused by the initial boundary layer can no longer do the new one. Therefore, the new vortices must be introduced. After the second step, the upper and lower half boundary layers of the cylinder are replaced each by a single vortex filaments, respectively. Larger number of vortex filaments seems desirable, though, considering the fact that the computation time is nearly proportional to the square of the number of the vortex filaments, so the number of vortices can not be too large. The least number to acquire a reasonable result is two. The two points of generation of the vortices were chosen at the farthest points from the center line parallel to the uniform flow. The computed results indicate, however, that the dependance of the flow pattern on the location of the generation points is not so significant, and no special consideration on the position of the nascent vortices is needed. Their strengthes are determined similarly to the first step. Repeating these steps, we can get a rotational wake in the frame work of the theory of ideal fluid.

## 3 Experiment

The elliptic cylinder of 4cm chord length with thickness ratio 0.5 was used for the experiment in a towing tank with impulsive-start capability.  Both of aluminum dust method and electrolysis method were used to visualize the stream lines and the streak lines.  Motor-driven camera was mounted on the cart which carries the model of elliptic cylinder, and also a video recorder system is adopted for continuous record.  Cart speed, U, was varied from 0.5 to 3cm/sec.  The Reynolds numbers based on the chord length and this speed were 200 to 1200.

## 4 Results and Discussions

Figure 1 is an example of the numerical results with an angle of attack of 45° at various times.  These are the stream lines and streak lines in the stationary co-ordinates.  Although this method approximates vortex region by singular point vortices, very smooth stream lines of fully developed unsteady wake were obtained within a reasonable computation time.  The computation also shows that the position of the nascent vortex needs not coincide with the separation point.  This is clearly seen on maps of the path lines [Fig.2].  The vortex moves along the cylinder surface to the separation point, and leaves there from the surface into the outer flow.  The experimental results at the corresponding time to Fig.1 are shown in Fig.3.  The drag and lift coefficients [Fig.4] obtained are generally in good agreements with the values recently reported by Izumi (1980).

In the case of 90°, due to the symmetry, somewhat peculiar process of separation of twin vortices is observed.  It is well known that in a moderately low Reynolds number, the Karman street is produced as : the attached eddies stretching farther and farther downstream, become distorted because of small disturbances and then are shed alternatively from the sides of the cylinder.  This pattern is well simulated by this approximation [Fig.5].  On the other hand, in a higher Reynolds number, the secondary vortices appear just behind the separation point, and, with the growth of these secondary vortices, and with their intrusion into the region between the elliptic cylinder and twin vortices, the initial twin vortices separate suddenly from the cylinder symmetrically if asymmetrical disturbances are carefully reduced.

In order to see this separation process it is essential to take into account of the effect of the backward flow due to the twin vortices just behind the cylinder, then at least four vortex filaments must be introduced after the second step, two for front boundary layer, two for rear boundary layer.  The flow patterns thus obtained are shown in Fig.6, in which sudden separation process of twin vortices is well simulated.  The corresponding experimental results are shown in Fig.7.

The relation of the length of the twin vortices to the non-dimensional time T = Ut/a, where a is the chord length, are plotted in Fig.8 for various Reynolds numbers. The agreements between the measured and the numerical results is excellent.

References

Izumi, K. 1980 ERC Rept. Univ. Tsukuba, vol.4.
Kuwahara, K. 1978 J. Phys. Soc. Japan, vol.45.
Takao, Y. 1973 Japan IBM Sci. Center Rept., G318-1909-0.

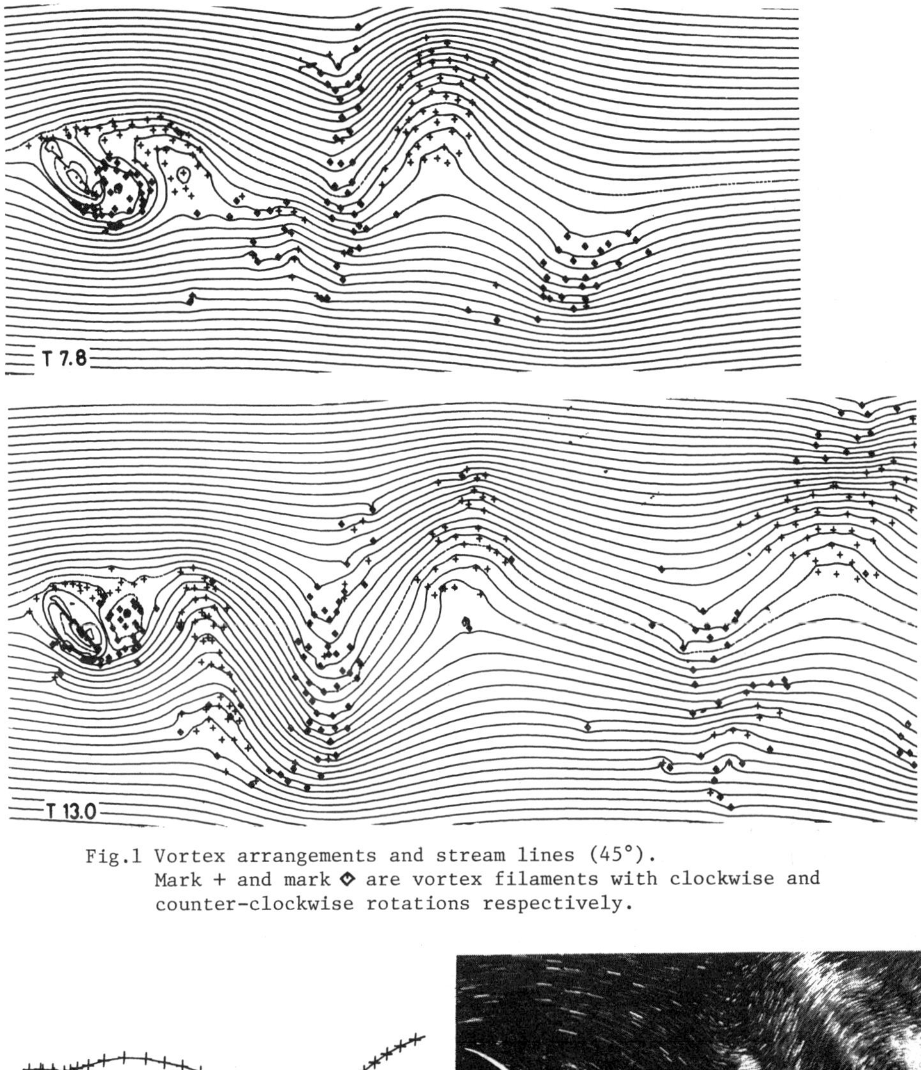

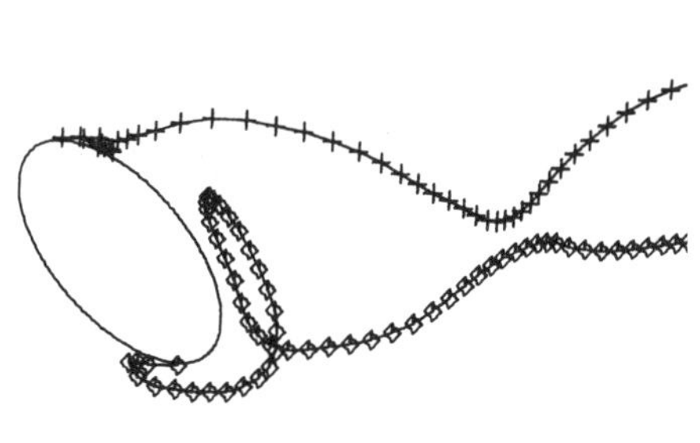

Fig.1 Vortex arrangements and stream lines (45°).
Mark + and mark ◇ are vortex filaments with clockwise and
counter-clockwise rotations respectively.

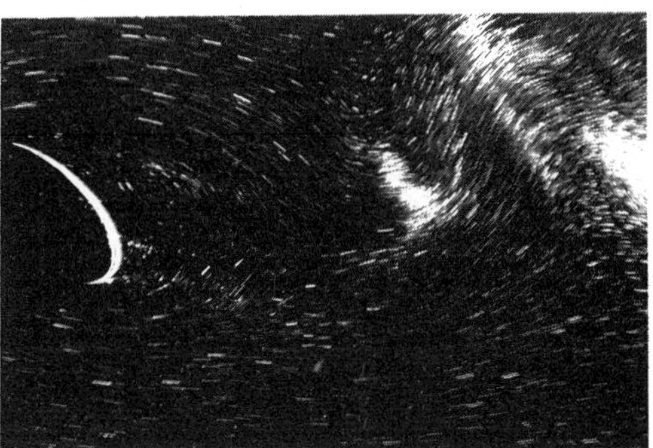

Fig.2 Particle paths (45°).

Fig.3 Stream lines (45°):
Experiment, T=7.8.

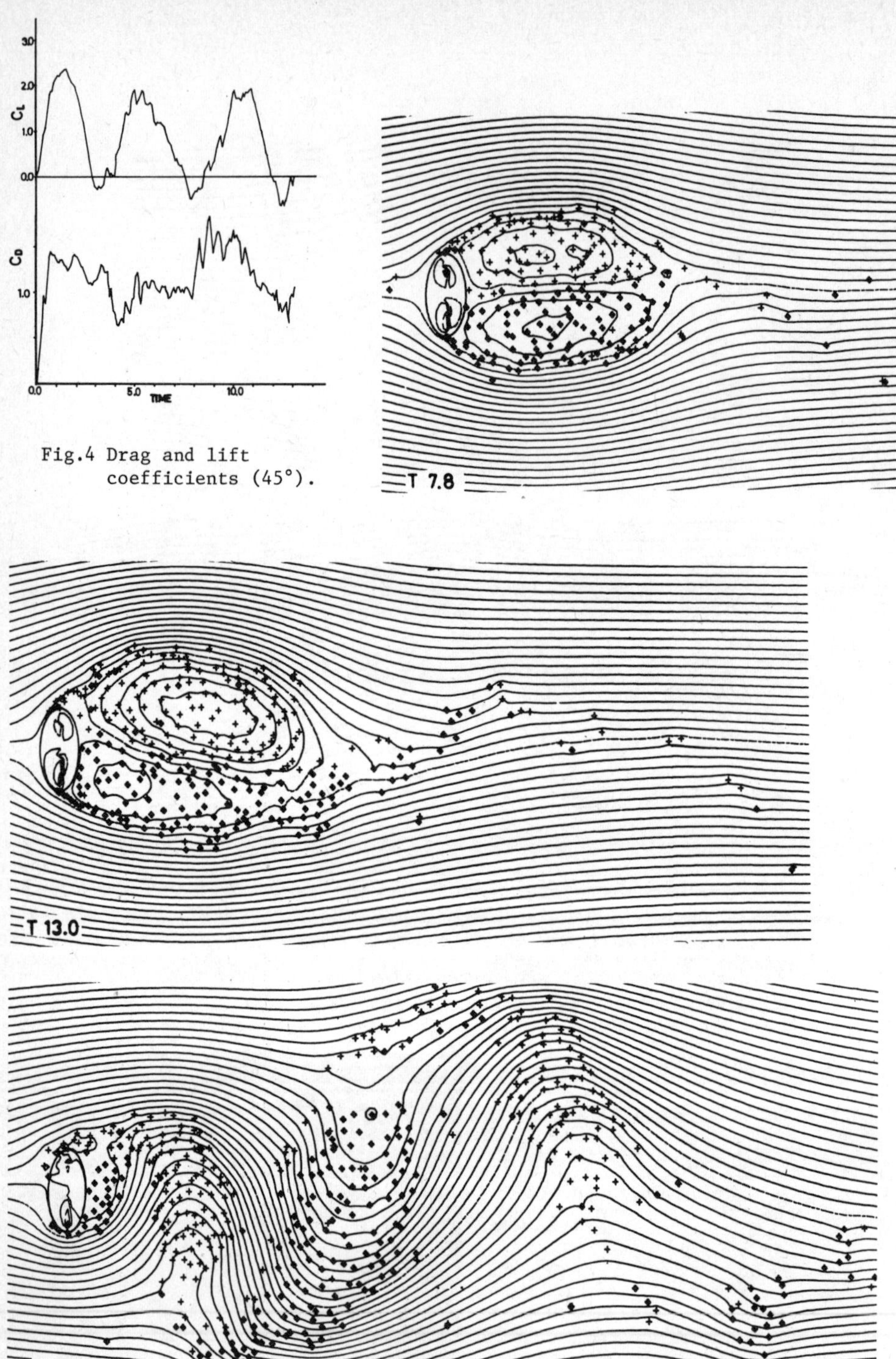


Fig.4 Drag and lift
     coefficients (45°).

Fig.5 Vortex arrangements and stream lines (90°). Simpler case.

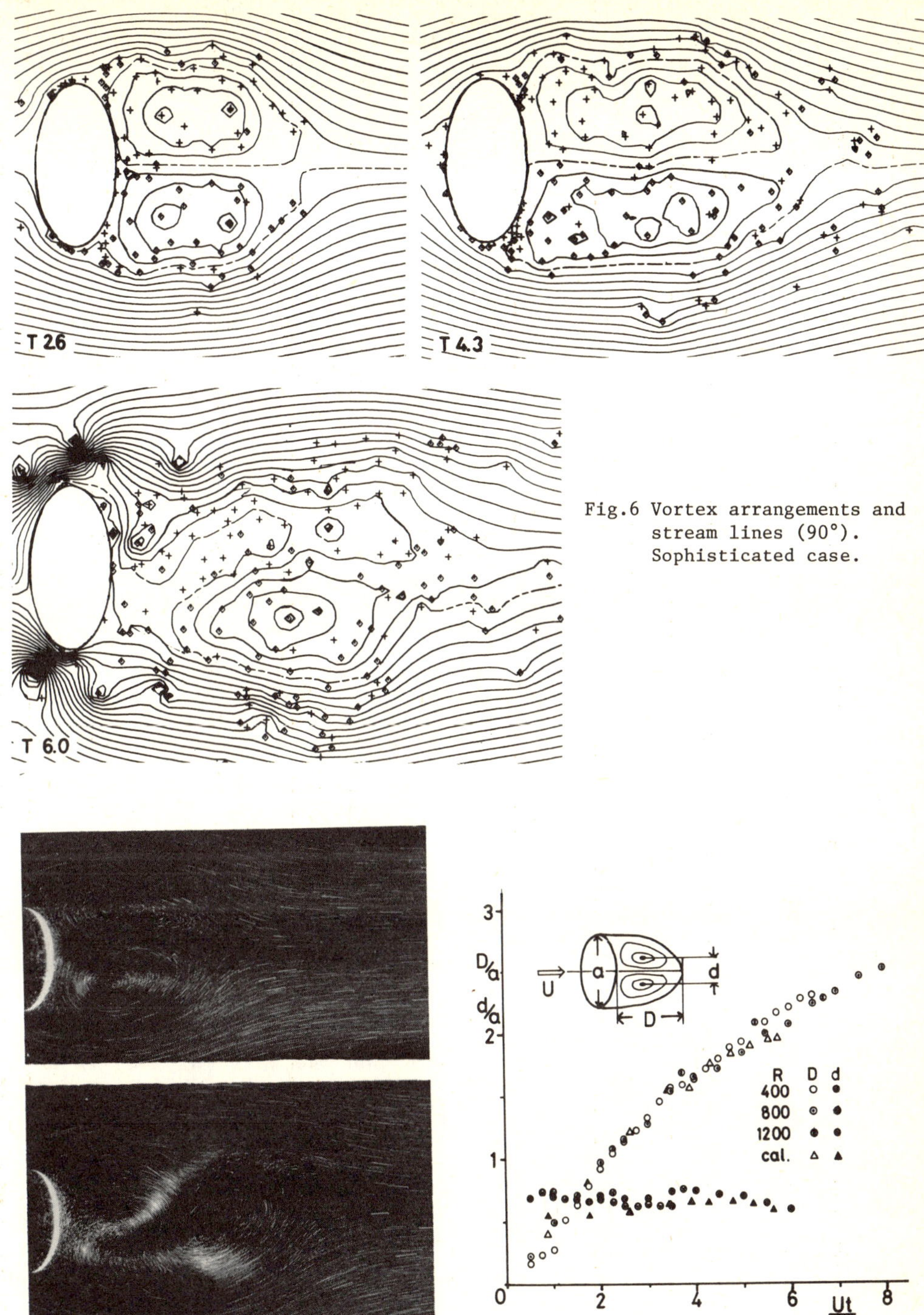


Fig.6 Vortex arrangements and
      stream lines (90°).
      Sophisticated case.



Fig.7 Stream lines:
      Experiment (90°), T=4.0, 6.0.

Fig.8 Length of twin vortices.

# TWO ANALYTICAL SOLUTIONS FOR THE REFLECTION OF

## UNSTEADY SHOCK WAVE AND RELEVANT NUMERICAL TESTS


Huang Dun      Institute of Mathematics, Beijing University
Li Yin-fan      Computing Center, Academy of Sciences Sinica
Huang Lu-ping   Department of Mathematics,Beijing University
Liu Yiu-zhi    Department of Mathematics,Beijing University
        Beijing,      People's Republic of China


## INTRODUCTION

On the ICNMFD6 V. V. Rusanov presented a rather general unsteady homoentropic test problem and emphasized the difficulty when solving gasdynamic problems with shock wave to judge the accuracy of many numerical solutions near the shock, and the need· of exact but not too simple analytical solution for testing numerical methods. In the first part of this paper two analytical solutions of the planar problem of the unsteady normal reflection of a plane strong explosion wave from a rigid wall are presented, the full unsteady nonlinear Euler's equations of gasdynamics being satisfied exactly. These solutions are matched along the reflected shock of varying intensity with the solution of L. I. Sedov (1946). The position of the shock is determined by using Rankine-Hugoniot conditions with very high accuracy, although not exactly. Our analytical solutions are neither homoentropic nor self-similar, the reflected shock is of considerable intensity.

These solutions can be used in several different aspects, e. g. for experimental calibration of equipments measuring transient pressure, for the estimate of the load on the wall.... The second part of this paper emphasizes the application of them to test numerical methods. We have been  testing several methods. Here we summarize the tests of the methods: Glimm-Chorin, Godunov, & FLIC. Several improvements of these methods are also made. The influence of the spatial steplength and other factors on the accuracy of the numerical results,  especially near the shock wave are shown by typical figures.

## FORMULATION OF THE PROBLEM AND THE ANALYTICAL SOLUTIONS

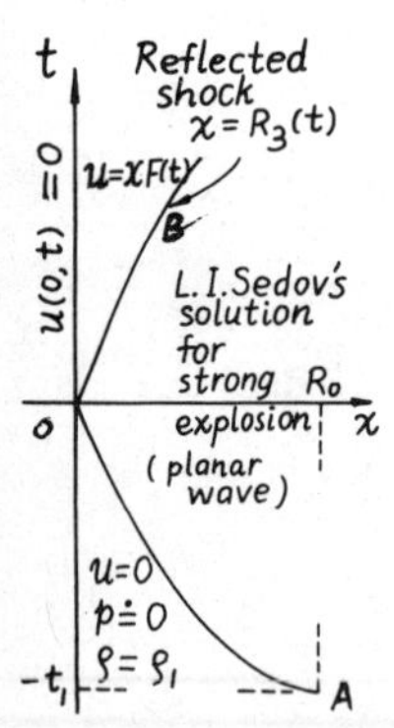


At time $-t_1 < 0$, a finite amount of energy per unit area E is suddenly  released at the plane $x=R_0$ . The unsteady explosion shock wave thus produced propagates  to   the left, and reaches the wall (at x=o) when t=o. For t > o the reflected shock of varying strength propagates to the right unsteadily. The x-t diagram is shown. We use the Euler's equations:

$$\frac{\partial u}{\partial t} + u\frac{\partial u}{\partial x} + \frac{1}{\varrho}\frac{\partial p}{\partial x} = 0 \;,\quad \frac{\partial \varrho}{\partial t} + \frac{\partial(\varrho u)}{\partial x} = 0 \;,$$

$$\frac{\partial p}{\partial t} + u\frac{\partial p}{\partial t} + \gamma p \frac{\partial u}{\partial x} = 0 \;. \tag{1}$$

where u, $\varrho$, p are respectively the gas velocity, density and pressure.

The position of the reflected shock $x=R_3(t)$ is to be determined, so that along the reflected shock wave Rankine-Hugoniot conditions should be satisfied as accurate as possible. To the right of AOB in the x-t diagram we use L. I. Sedov's solution (1946). Between OA and the t-axis, in the still air, the density is $\rho_1$, and pressure is considered as negligible. We have thus three characteristic constants: $R_0$, $t_1$ and $\rho_1$. The flow to the left of OB has no similarity, it is also not homoentropic. Particle velocity u(o,t) at the rigid wall should be zero. For this region $t>o$, $o\leq x \leq R_3$ we found two sets of solutions satisfying Euler's equations exactly. We derive the first solution by requiring $u(x,t)=xF(t)$ and $p(x,t)=p(t)$ thus obtaining:

$$u(x,t)=\frac{x}{t+e} \qquad p=p_{30}/(1+\tfrac{t}{e})^{\gamma} \qquad \rho(x,t)=\rho_{30}(1-k\frac{x}{t+e})/(1+\tfrac{t}{e}) \tag{2}$$

where $e$, $k$, $p_{30}$, $\rho_{30}$ are arbitrary constants to be determined by R-H conditions expanding all related functions asymptotically for small $t \ll t_1$, and then comparing coefficients for equal powers of t. The results for $\gamma=1.4$ and $p_{30}=800$ atm. are:

$$c_r = 2R_0/9t_1 \ , \qquad R_3(t) = c_r[\ t+\tfrac{n}{2}t^2\ ] \ll R_0 \ ,$$
$$e= 0.23\,t_1 \ , \qquad n=361/69\,t_1 \ , \qquad k=3t_1/R_0 \ , \qquad \rho_{30}=21\rho_1 \ . \tag{3}$$

We call (2), (3) the first particular solution, it is very accurate for small $t/t_1$.

We found the second particular solution for $R_3 < 0.35R_0$ , $t>o$, $o \leq x \leq R_3(t)$ retaining the restriction $u=xF(t)$, but giving up $p=p(t)$.

Introducing new variables

$$\tau(t) \equiv \int_o^{t/T_o} F(t)\,dt \qquad , \qquad \eta = t/T_o \ , \tag{4}$$

Euler's equation yields:

$$\tau\frac{\partial\rho}{\partial t} + x\frac{\partial\rho}{\partial x} + \rho = 0 \quad , \qquad \tau\frac{\partial p}{\partial t} + x\frac{\partial p}{\partial x} + \gamma p=0 \ ,$$
$$\frac{d^2F}{d\eta^2} + (\gamma+3)F\frac{dF}{d\eta} + (\gamma+1)F^3=0 \ . \tag{5}$$

Finally,

$$p(x,t)= \frac{p(o,o)}{\tau^{\gamma}} + \frac{\rho_{oo}}{\tau^{\gamma}}\frac{L_o^2}{T_o^2}\left[\frac{dF}{d\eta}+F^2\right]_{\eta=o} \cdot \int_o^{x/L_o\tau} \xi N(\xi)d\xi , \tag{6}$$

$$\rho(x,t)= \rho_{oo} N(\frac{x}{L_o\tau})/\tau , \tag{7}$$

where N is an arbitrary function, and $L_o$, $T_o$, $\rho_{oo}$, $F(o)$, $\dot{F}(o)$, $p(o,o)$ are arbitrary constants to be determined by R-H conditions along $x=R_3(t)$. We don't wish to determine $x=R_3$ and the arbitrary function and const, exactly, instead, we require global satisfaction of R-H conditions as good as possible. This is done with high accuracy by choosing

$$L_o=R_o \ , \quad T_o=e \ , \quad \rho_{oo}=21\rho_o \ , \quad \gamma=1.4 \ , \quad \rho_1=1.29 \ 9m/m^3 \ , \quad R_o/k=0.1\ m,$$
$$t_1/k = 0.2171819 \times 10^{-4}\ sec. \ , \qquad E/k=2.7749056\times10^6\ joules/sq.\,m. \ ,$$
$$1\ atm = 1.012928 \times 10^5\ newton/sq.\,m. \ , \qquad N(\xi)=1-3\frac{t_1}{e}\xi+1.49\,\xi^2 \ ,$$
$$F(\tfrac{t}{e}) = \frac{1}{1+\tfrac{t}{e}} + \frac{\gamma-1}{\gamma}1.3\left[\frac{1}{(1+\tfrac{t}{e})^{\gamma}} - \frac{1}{(1+\tfrac{t}{e})^2}\right] + \varepsilon \ ,$$

where $\varepsilon$ is a small quantity, $k$ is an arbitrary constant. We can choose $k$ as we please. $k=1$ equivalents an amount of energy equaling about 0.7 kg TNT per sq.m. With such a choice, P(0.0) behind the reflected shock is just 800 atm. If we choose $F(t/e)=1/(1+t/e)$ then the second set of solution reduces exactly to the first particular solution. The final results are as follows:

Denoting by $\quad e=0.23t_1$, $\quad T=1+t/e$, $\quad y=(0.4-T^{0.6})/T$, $\quad s=0.6+y$, $\quad w=x/0.023\hbar T$,

then for $0 \leqslant x \leqslant R_3 < 0.35R_o$ we have solution with rather high accuracy:

$$p(x,t)=\left[0.8103424\times10^{8}(1-1.3s)-6.0372(\tfrac{1}{2}-w)\left(\tfrac{x}{Te}\right)^{2}\right]/\tau^{1.4} \quad \text{newt/m}^2,$$

$$\varsigma(x,t)=27.09\left[1-3w-(1-6w)\tfrac{1.3}{1.4}s+1.49w^{2}\right]/\tau \qquad \text{kg/cu.m} \, , \tag{8}$$

$$u(x,t)=0.1\hbar w\left[1-0.4\tfrac{1.3}{1.4}\left(y+\tfrac{0.6}{T}\right)\right]/t_1 \qquad \text{m./s.}$$

The position of the reflected shock wave varies with time:

| $t\times10^{5}$ sec/$\hbar$ | $R_3/\hbar$ m. | $t\times10^{5}$ sec/$\hbar$ | $R_3$m./$\hbar$ | $t\times10^{5}$ sec/$\hbar$ | $R_3$m/$\hbar$ |
|---|---|---|---|---|---|
| 0.049676 | 0.0005385 | 0.45 | 0.00700 | 0.90 | 0.0186 |
| 0.10 | 0.00115 | 0.505 | 0.00800 | 1.00 | 0.0217 |
| 0.20 | 0.00250 | 0.60 | 0.0102 | 1.20 | 0.0288 |
| 0.30 | 0.00410 | 0.70 | 0.0126 | 1.355 | 0.0352 |
| 0.40 | 0.00600 | 0.805 | 0.0155 | | |

## THE TEST OF THE METHODS OF GODUNOV AND GLIMM-CHORIN

These two methods are all based on approximating unknown functions by piecewise constant functions of x at every time step, and then solve the enormous amount of resulting Riemann problems. The usefulness of these two methods depends on how efficiently one solves the Riemann problem. Several algorithms have been designed to solve the Riemann problem. In his book of 1976 Gudunov proposed a new procedure. For all four possible resulting wave systems we have:

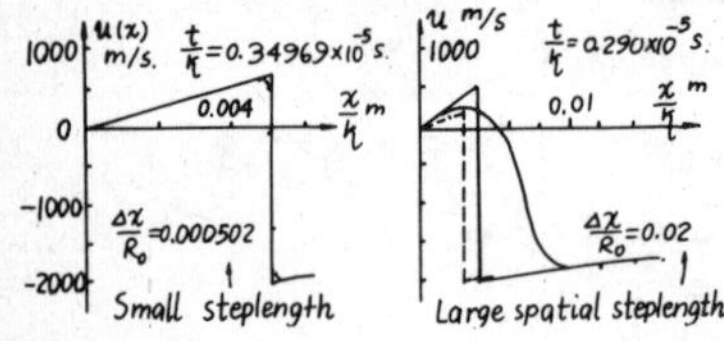


$$u_1-u_2=f(p^*,\, p_1,\, \varsigma_1\,)+f(p^*,\, p_2,\, \varsigma_2\,)=F(p^*)\,,$$

where $F(p^*)$ is increasing and convex, thus efficient Newton's iteration is applicable to compute $p^*$. This reduces the computing effort and increases the efficiency of both these two methods.

We do not linearize the solution of the Riemann problem, in order to retain the possibility to treat the formation of new shock wave in the gas, and the interaction of discontinuities.

For the method of Godunov the scheme viscosity smears the reflected wave into a transition zone. In our test runs its width increases as sq. root of the number of time layers, this coincides with the linear analysis performed by Godunov.



The left two figures show the comparison of our analytical results (full line with cusps) with the resnlt of Godunov's method (Smooth curve with smeared shock) In the right figure the result of Glimm-Chorin method is also shown (with dotted line). From these figures as well as the figures in next two pages one can see the evident influence of the spatical steplength on the numerical results for flow behind the reflected shock, especially near the shock. For small steplength the coincidence is very satisfactory, even thus for 2-3 Points near the cusp the result of method of Godunov for u is less than the more accurate analytical result by about 5%.

Chorin (1976, 1977) and Sod (1978) and Collela (1979)are developing
the Glimm-Choin method, or random choice method, which promises well.
It's outstanding feature lies in that it has neither artificial nor
scheme viscosity, and without using Rankine-Hugoniot conditions explicitly, the shock wave is sharp,i.e.not smeared. In this method a sampling procedure is used, a random variable is adopted. This method
needs further analysis and numerical experimentation. We test it by
comparing with analytical solution. The iteration in solving  Riemann
problem is modified as explained above. The number of iterations is
usually 2 or 3 and at most 5 for a few points near the shock wave. We
tested "stratified" sampling and then"van der Corput"sampling. For the
flow behind the reflected shock,$R_0$ is an important characteristic length.
Eight accompaning figures show different aspects of this method:
the profiles u(x),  $\varsigma$ (x), p(x) at different time layers, the variation
of the pressure on the wall with time p(o,t),and finally,the position
of the reflected shock $R_3$,shown by isolated black points,is compared
with the result of analytical solution drawn with full line.

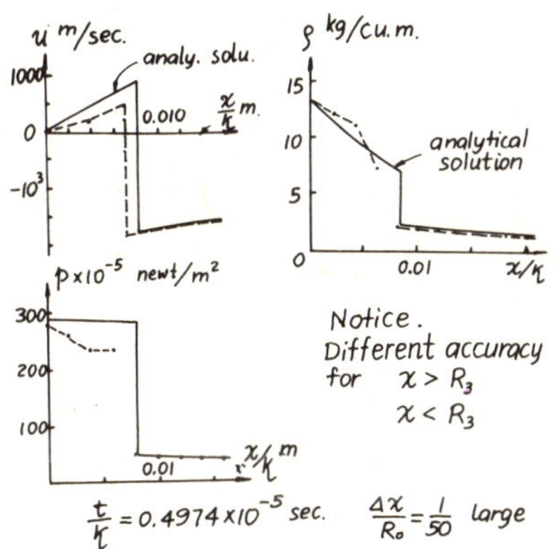


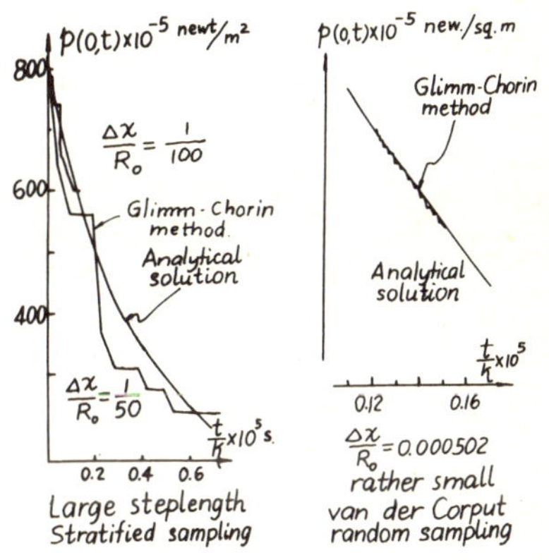


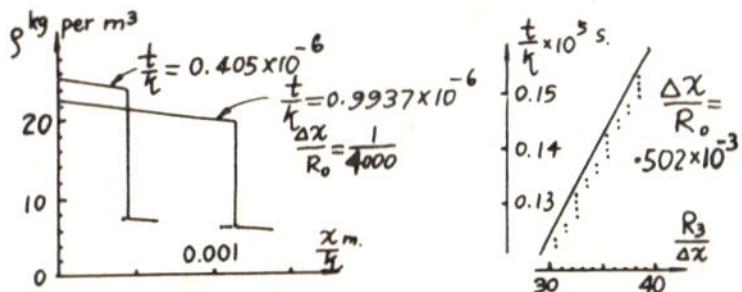


From the figures we can see,such
spatial steplength as $\Delta x/R_0$=0.02
has  to  be  considered  as not  so
small, since the roughness of the
numerical results for $x<R_3$ is
obvious. The roughness seems to be
the result of the randomness  of
the scheme. It can be reduced by
using smaller$\Delta x/R_0$as well as  by
using van der Corput sampling. We have carried many numerical experi-
mentations with different $\Delta x/R_0$. The influence of$\Delta x/R_0$ on the accuracy
of the numerical result is remarkable.

In our tests we check the relative errors of conservations of mass,
energy and momentum. The  random  number  makes  the  relative  errors
not monotonic.  It is interesting that as the relative  errors have a
relative minimum, the position of the reflected shock is  nearest to
the result of analytical solution.  It seems that we  can  select the
time layer for which the numerical results are better than the neighbouring time layers.  Using such a selection it is  possible to use a
larger spatial steplength and thus reduce the overall running time.

## ANALYSIS AND NUMERICAL TESTS OF FLIC METHOD

This method consists in two steps. As far as we know, previous linear stability analysis of Gentry and others (1966) was performed only for the first step. They wrote that the second step is of benefit to the stability. We present in this paper the linear stability analysis considering the two steps as a whole for the stagnation region ($u^2 \ll c^2$), where c is the sound speed, The stability condition for this region of small particle velocity is:

$$\Delta t/\Delta x= \begin{cases} 2(1-b)/c, & \text{if} \quad b \leq \tfrac{1}{2}; \\ 1/2cb, & \text{if} \quad b \geq \tfrac{1}{2}. \end{cases}$$

where b denotes the coefficient for artificial viscosity, which we in our tests use from 0.5 to 1. The artificial viscosity term is:

$$q = \begin{cases} -bc\varrho\Delta u, & \text{if} \quad 3u^2 < c^2; \\ 0, & \text{otherwise}. \end{cases}$$

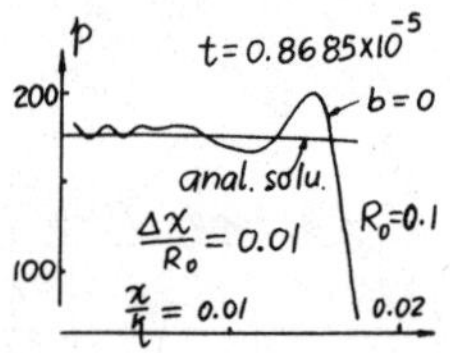


The figure on the right demonstrates the numerical result when b=o, the eigenvalue of the amplification matrix for the stagnation region being exactly one, a case of neutral stability. Analytical solution does not oscillate, but the numerical does. In our tests the fluctuation does not develope to rapid increase of amplitude, a phenomenom connected with non-linearity. On the basis of stability analysis we consciously add the term q in order to diminish the undesirable fluctuation and to increase the accuracy.

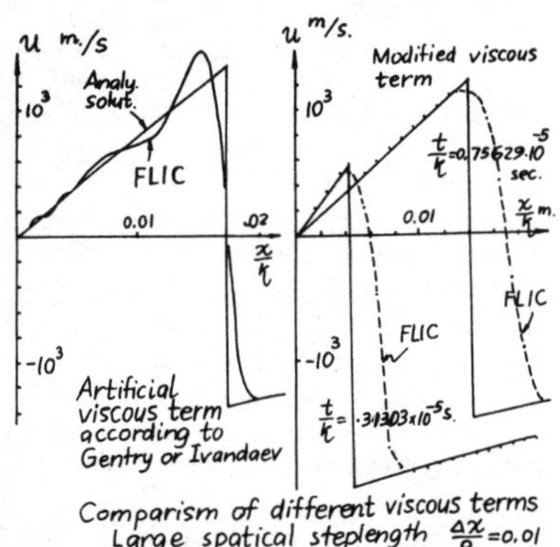


Comparism of different viscous terms
Large spatical steplength $\frac{\Delta x}{R_0}$=0.01

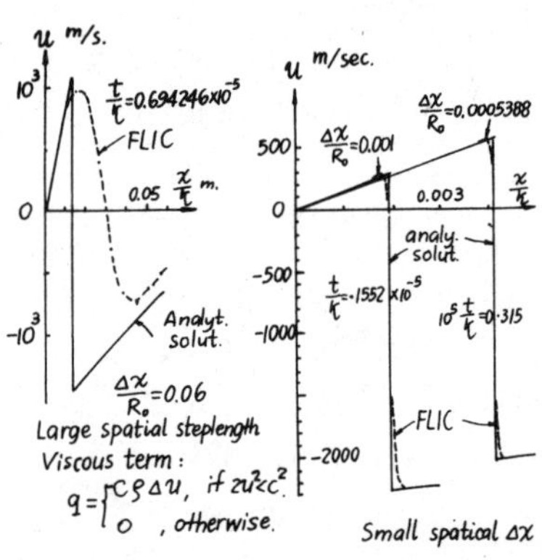


The above four figures show typical results using four different types of viscous term, namely, that of i) Gentry, ii) Belotserkovsky iii) Ivandaev, besides iv) our modification, i.e. $q=-b_1 c\varrho\Delta u$, if $Lu^2 < c^2$, or $q=-b_2 c\varrho\Delta u$ if $u_j > u_{j+1}$, $b_1$, $b_2$ being two coefficients. The last two figures show also the remarkable influence of spatial increament $\Delta x/R_0$.

## CONCLUSIONS

We have been testing methods of Godunov and Glimm-Chorin,also FLIC
method,by comparing numerical results vs. our analytical solution of
high accuracy. The analytical solution satisfies the Euler's equations
exactly. The Rankine-Hugoniot conditions along the shock is also satisfied with high accuracy. Our analytical solution is neither homoentropic, nor self-similar, besides,the intensity of shock wave varies with
time. The figures in the text show the tendency of our analytical solution. Our tests demands  more from the numerical method than those
tests, which include shock wave only of constant intensity. By suitable
arrangements,especially in choicing reasonable spatial steplength, we
can make the numerical results rather accurate in the neighbourhood of
the shock wave, and on the other hand, through these tests,we have observed the remaining small errors not only qualitatively but also quantitatively. The results of the test of other methods shall be presented
elsewhere.
 It is worth while to notice that from all the figures in this paper
one can see that the numerical results obtained  by  any of the three
methods coincide well with Sedov's solution, where it holds, even  if
large spatical steplength is used. It seems that self-similar solution
is easier to compute by numerical methods.
 For the methods of Godunov, of Glimm-Chorin, we reduce the computing
effort by an iterative procedure in solving the Riemann problem, this
is important if we generalize the methods to two dimensional unsteady
flow. For the FLIC method we modify the artifical viscosity term on the
basis of a numerical analysis considering both the two steps as a whole
for the region of small particle velocity.

## REFERENCES


1. A.J.Chorin, "Random Choice Solution..." J.Comp. Phys.,22,1976,517.
2. A.J.Chorin, "Random Choice Methods with Application to Reacting
   Gas Flow". J. Comp. Phys. 25, 1977, 253.
3. P. Collela, Ph. D.Thesis, UC Berkeley, July 1979.
4. R.A.Gentry, R.E.Martin, B.J.Daly. J.Comp.Phys. vol 1, 1966, 87-118.
5. Huang Dun, Huang Lu-ping, Liu Yu-zhi, "The calculation of unsteady
   shock wave by the method of Godunov and some remarks on the factors
   which influence the quality of solution." Report,Beijing Univ.1979.
6. Huang Dun, Huang Lu-ping, Liu Yu-zhi, "On random choice method and
   the calculation of unsteady shock wave", Jour. on Numerical Methods
   and Computer Applications,vol 1, no. 1, 1980, 34-44. Beijing, China.
7. Li Yin-fan, "A note on the stability of the FLIC method,"to appear.
8. V.V.Rusanov, "A test case for checking computational methods for gas
   flows with discontinuities," Lecture Notes in Phys. 90, 477-486.
9. G.A.Sod. Survey of finite difference methods, J.Comp. Phys,27,1978.
10. О.М.Белоцерковский,Ю.Д.Давидов ЖВМ и МФ 1971.
11. А.А.Бубайдуллин, А.И.Ивандакв ЖВМ и МФ 1977, стр 1531.
12. С.К.Годунов и др. Численное решение многомерных задач
    газовой динамики. Москва, 1976.
13. Л.И.Седов, Журнал Прикл. Мат. и Мех. т. X вып 2, 1946.

Separated Boundary Layer Flows with High Reynolds Numbers

Osamu Inoue

Institute of Space and Aeronautical Science, University of Tokyo, Tokyo, Japan

## 1.  INTRODUCTION

For high Reynolds number flows, it is not feasible, in general, to solve the
Navier-Stokes equations even with the aid of modern computers.  The boundary layer
equations cannot be solved for a prescribed pressure gradient (direct method),
because singularity occurs at separation point where the solution breaks down.  So
far several methods have been proposed to avoid this singularity of the boundary
layer equations.  They are reviewed, for an example, by Williams (1977).  Carter
(1975) solved the boundary layer equations by an inverse method, and showed that
solutions of the boundary layer equations well approximate the Navier-Stokes solutions
for a flow with separation and reattachment if the occurrence of the singularity at
separation point is avoided.  The inverse method, however, requires to primarily
specify an appropriate distribution of variables such as displacement thickness or
wall shear.  In actual applications of inverse method, experimental results  were
referred for specification of these distributions (Cebeci et al., 1979).

In the present paper, equations are employed which approximate the Navier-Stokes
system.  The approximate equations are the one quite similar to the "parabolized-
vorticity model" proposed by Ghia, Ghia and Tesch (1975).  They examined the validity
of several model equations in comparison with the Navier-Stokes equations.  Among the
models, the "parabolized-vorticity model" gave the best result. They concluded that
boundary layer type equations can predict separated flows provided that the
interaction between the boundary layer and the external inviscid flow is properly
treated.  They formulated the problem, however, in a rather special coordinate
system, and emphasized that "the success of the model depends greatly on the use of
proper variables and suitable coordinates for formulating the model".  In the present
paper, we apply the approximate equations in a Cartesian coordinate system with
ordinary variables to the retarded flow problem considered by Briley (1971).  From
comparison with the Navier-Stokes solutions, the results indicate that the solutions
closely approximate the Navier-Stokes solutions in a usual coordinate system without
requiring specialized coordinates.  Furthermore, we apply the model equations to a
flow in a divergent duct, and the effect of interaction of inner boundary layer with
outer inviscid freestream on flow separation is also analyzed.


## 2.  THE APPROXIMATE EQUATIONS

We consider two-dimensional, steady, incompressible, laminar viscous flows along
a wall.  The present analysis is based on the following approximate equations:

$$\frac{\partial u}{\partial x} + \frac{\partial v}{\partial y} = 0 \tag{1.1}$$

$$u\frac{\partial u}{\partial x} + v\frac{\partial u}{\partial y} = -\frac{\partial p}{\partial x} + \frac{1}{Re}\frac{\partial^2 u}{\partial y^2} \tag{1.2}$$

$$u\frac{\partial v}{\partial x} + v\frac{\partial v}{\partial y} = -\frac{\partial p}{\partial y} + \frac{1}{Re}\frac{\partial^2 v}{\partial y^2} \tag{1.3}$$

where Re is the Reynolds number. Equation (1) are derived by neglecting the terms $\partial^2 u/\partial x^2$ and $\partial^2 v/\partial x^2$ from the Navier-Stokes equations. This assumption is justified by the boundary layer concept; that is, $(\partial^2 u/\partial x^2)/Re$ and $(\partial^2 v/\partial x^2)/Re$, respectively, are of the order $Re^{-1}$ compared with the remaining terms in Eqs.(1.2) and (1.3). In terms of streamfunction and vorticity, Eqs.(1) are rewritten as follows:

$$u\frac{\partial \omega}{\partial x} + v\frac{\partial \omega}{\partial y} = \frac{1}{Re}\frac{\partial^2 \omega}{\partial y^2} \tag{2.1}$$

$$\omega = \frac{\partial u}{\partial y} - \frac{\partial v}{\partial x} = \frac{\partial^2 \psi}{\partial y^2} + \frac{\partial^2 \psi}{\partial x^2} \tag{2.2}$$

$$u = \partial \psi/\partial y , \quad v = -\partial \psi/\partial x. \tag{2.3}$$

These are the model equations we employ. Equation (2.1) has the same form as the boundary layer equations, so that it can be solved by any numerical method efficient for the boundary layer equations. Equation (2.2) is different from the boundary layer equations by the term $\partial v/\partial x$ (or, $\partial^2 \psi/\partial x^2$). The Poisson's equation (2.2) is known to be numerically stable for any regular function of $\omega$, so that it can easily be solved by means of global or field iteration. We note, here, that even the boundary layer equations need global iteration in cases where flows separate (Carter, 1975).

As easily seen, Eqs.(2) still possess a solution of the same form as the Navier-Stokes solution, which was evolved by Dean(1950). The only difference between the two solutions is due to neglecting the terms including $r^2$ ($r = Re^{-1}$) from the Dean's expression for the relation among the coefficients of the assumed streamfunction. Therefore, if the Navier-Stokes equations have a regular solution at separation point, Eqs.(2) are also expected to have a regular solution.

## 3. RETARDED FLOW PROBLEM

In order to examine the occurrence of singularity at separation point and the accuracy of solution, Eqs.(2) are applied to a retarded flow problem, for which the Briley's Navier-Stokes solutions are available. Boundary conditions for Eqs.(2) are as follows:

$$y = 0 \ : \ u = 0, \ v = 0 \ (3.1), \qquad y = y_e \ : \ \omega = 0, \ U_e \text{ prescribed} \ (3.2),$$

$$x = x_1 \ : \ \omega, \psi \ : \ \text{given by the Howarth boundary layer solution} \quad (3.3)$$

$$x = x_2 \ : \ \partial^2 \psi/\partial x^2 = 0 \quad (3.4)$$

As for the method of solution, we made use of the Carter's global iteration method (Carter 1975). The governing equations (2) are replaced by finite-difference expression. For forward flows ($u \geq 0$), Eqs.(2.1) and (2.3) are evaluated with Crank-Nicholson scheme at $x = x_1 + (i-1/2)\Delta x$ and $y = j\Delta y$. For reversed flows ($u < 0$), Eqs. (2.1) and (2.3) are approximated by central-difference expression, and for $\partial \omega / \partial x$ only the first order forward-difference is used. Regardless the flow is forward or reversed, Eq.(2.2) is evaluated with central-difference expression. Equations (2) thus can be wriiten as

$$A_j \omega^{k+1}_{i,j-1} + (B_j + \alpha) \omega^{k+1}_{i,j} + C_j \omega^{k+1}_{i,j+1} = D_j + \alpha \omega^k_{i,j}$$

$$E_j \psi_{i,j-1} + F_j \psi_{i,j} + G_j \psi_{i,j+1} = H_j \tag{4}$$

Since the matrix is tridiagonal, Eqs.(4) can be solved quickly with a simple algorithm. The equations (4) are iteratively solved on each line of $x = x_1 + i\Delta x$, starting from $i=1$ to $i=i_{max}$ ($x=x_2$). Then, this iteration is repeated till the solution converges. Convergence was assumed when the maximum changes in $\omega$ and $\psi$ between two successive iterations became less than $1 \times 10^{-5}$. Negative values of the parameter $\alpha$ in Eqs.(4), typically $-0.2$, were used. For positive values of $\alpha$, the solution of Eqs.(2) always diverged. The 51 x 51 grid points were used.

## 4. VISCOUS-INVISCID INTERACTION

In this section, we simultaneously solve the Laplace equation for outer inviscid freestream and the approximate equations for inner viscous boundary layer. We consider flows in a two-dimensional, symmetric divergent duct. The coordinate system $(x,y)$ is transformed into a new coordinate system $(\xi,\eta)$ such that the calculation region becomes rectangular;

$$\xi = x, \qquad \eta = (y-1)/( h(x) + \delta^*_I(x) - 1 ) \qquad \text{for inviscid flows,}$$
$$= (y-y_e)/( h(x) - y_e ) \qquad \text{for viscous flows,}$$

where $y_e$ denotes the boundary layer edge and $\delta^*_I(x)$ the effective displacement thickness ( $\delta^*_I(x) = \delta^*(x) - \delta^*_1$, where $\delta^*_1$ is the displacement thickness at the upstream boundary $x = x_1$ ). The wall shape is assumed to be described by $y_w = h(x)$. Under the assumption that $H'/H$ and $H''/H$ are of the order unity, where $H(\xi)= h(\xi) - y_e$, the approximate equations for the boundary layer flows are written as follows:

$$u \left( \frac{\partial \omega}{\partial \xi} - \frac{H'\eta}{H} \frac{\partial \omega}{\partial \eta} \right) + \frac{v}{H} \frac{\partial \omega}{\partial \eta} = \frac{1}{Re} \frac{1}{H^2} \frac{\partial^2 \omega}{\partial \eta^2} \tag{5.1}$$

$$\omega = \frac{\partial^2 \psi}{\partial \xi^2} + \frac{1+H'^2\eta^2}{H^2} \frac{\partial^2 \psi}{\partial \eta^2} - \frac{2H'\eta}{H} \frac{\partial^2 \psi}{\partial \xi \partial \eta} + \frac{2H'^2 - HH''}{H^2} \eta \frac{\partial \psi}{\partial \eta} \tag{5.2}$$

$$u = \frac{1}{H} \frac{\partial \psi}{\partial \eta} , \quad v = - \frac{\partial \psi}{\partial \xi} + \frac{H'\eta}{H} \frac{\partial \psi}{\partial \eta} \tag{5.3}$$

The boundary conditions are prescribed as follows:

$$\eta = 0 \; : \; \omega = 0, \qquad\qquad \eta = 1 \; : \; u = 0, \; v = 0$$
$$\xi = \xi_1 \; : \; \omega , \; \psi \; \text{given by the Blasius boundary layer solution,} \qquad (6)$$
$$\xi = \xi_2 \; : \; \partial^2 \psi / \partial \xi^2 = 0.$$

For the outer inviscid flows we assume that vorticity vanishes everywhere. Then, the governing equation is the Laplace equation which is the same as Eq.(5.2) with $\omega = 0$. The boundary conditions for the Laplace equation are

$$\eta = 0 \; : \; \psi = 1, \qquad\qquad \eta = 1 \; : \; \psi = 0,$$
$$\xi = \xi_1 \; : \; \frac{1}{H}\frac{\partial \psi}{\partial \eta} = 1 \; ( \; u = 1 \; ), \qquad \xi = \xi_2 \; : \; \partial^2 \psi / \partial \xi^2 = 0 \qquad (7)$$

where $H(\xi) = h(\xi) + \delta_I^*(\xi) - 1.$

Equation (5) and the Laplace equation are solved by finite-difference method, which is quite similar to that in the previous section. Solutions of Eqs.(5) and the Laplace equation are matched according to the following procedure: 1) First, we solve the Laplace equation under the assumption that displacement thickness vanishes ($\delta^*=0$). From the solution we obtain $U_e(\xi)$ at the boundary layer edge. 2) With $U_e$ obtained, we solve Eqs.(5) together with the boundary condition (6). From the solution of Eqs.(5) thus obtained, we calculate the displacement thickness $\delta^*(\xi)$. The effective wall shape is considered to be the actual wall shape expressed by $h(\xi)$ plus a correction $\delta_I^*(\xi)$. 3) With $\delta_I^*(\xi)$ thus obtained we recalculate the Laplace equation and obtain a new distribution of $U_e(\xi)$. 4) The procedure 2) to 3) is repeated till the maximum changes in all variables ( $\omega$, $\psi$, $\delta_I^*$ ) between two successive iteration become less than $1.5 \times 10^{-5}$. The grid points were 141 ( in $\xi$-direction) x 41 (in $\eta$-direction) both for the boundary layer and for the outer inviscid region.


5. RESULTS AND DISCUSSION

a) Comparison with the Briley's Navier-Stokes solutions

Parameters used by Briley are $x_1 = 0.0501$, $x_2 = 0.489$, $y_e = 0.035$ and the Reynolds number Re $= 2.08 \times 10^4$. Numerical results from the present method are compared with those by Briley in Figs.1 and 2. Both displacement thickness and skin friction show excellent agreement. The mark X in Figs.1 and 2 denotes the position where the boundary layer solution breaks down. The solution obtained from Eqs.(2) does not, at least numerically, have any singularity at separation point, as expected. The only difference of Eqs.(2) from the boundary layer equations is the addition of the term $- \partial v / \partial x$ (or $\partial^2 \psi / \partial x^2$) in Eq.(2.2). Therefore we can suggest that the occurrence of the singularity at separation point is avoided in the present approach where we retain $\partial v / \partial x$.

When we derived Eqs.(1), we neglected the terms $(\partial^2 u / \partial x^2)/Re$ and $(\partial^2 v / \partial x^2)/Re$ under the assumption that these terms are small compared with the remaining terms. The validity of this assumption can be evaluated from the resulting numerical solutions.  We examined the validity of the assumption at all the grid point.  One example is presented in Fig.3, where $I_1 = (\partial^2 u/\partial x^2)/(\partial^2 u/\partial y^2)$ and $I_2 = \{(1/Re)\partial^2 u/\partial x^2\}/Max(u\partial u/\partial x, v\partial u/\partial y)$.  They are the values along the grid line $j = 2$, which is the closest line to the wall ( $j = 1$ means the wall).  $Max(u\partial u/\partial x, v\partial u/\partial y)$ means the larger term out of $u\partial u/\partial x$ and $v\partial u/\partial y$.  The results showed that the magnitude of the terms $(\partial^2 u/\partial x^2)/Re$ and $(\partial^2 v/\partial x^2)/Re$ are respectively satisfactorily small compared with the terms in Eq.(1.2) and Eq.(1.3) at all the grid point, and thus the above assumption is well satisfied.

b)  Flows in a divergent duct.

We prescribed parameters as follows:

$$x_1 = -1.0, \quad x_2 = 6.0,$$
$$h(x) = 0.0 \; (-1.0 \leq x \leq 0.0), \qquad = Ax^2(3-2x) \quad (0.0 \leq x \leq 1.0),$$
$$= A \quad (1.0 \leq x \leq 6.0).$$

The values of constant A were taken to be 0.0 to -0.1.  The values of the Reynolds number were from 1000 to 12000.

The importance of accounting for the effect of viscous-inviscid interaction on separated flows is clearly depicted in Fig.4.  The broken lines show the results when the velocity $U_e$ is specified from the Laplace equation with $\delta_I^*(\xi) = 0$.  The results without interaction usually overestimate the magnitude of separation region very much.  Wall shear distributions are presented in Fig.5 for several Reynolds numbers with a wall shape fixed (A=-0.08).  From Fig.5 we can see that as the Reynolds number is increased, separation point shifts upstream while reattachment point shifts downstream.  Shown in Fig.6 are the velocity distributions $U_e$ along the boundary layer edge and the effective displacement thickness $\delta_I^*$, which are corresponding to Fig.5.

The author expresses his sincere appreciation to Prof. H. Oguchi, ISAS, University of Tokyo, for his kind advices and valuable suggestions.

REFERENCES

Briley, W.R., _J. Fluid Mech._, 1971, Vol.47, pp.713-736.

Carter, J.E., _NASA TR R-447_, 1975.

Cebeci, T., Khalil, E.E. and Whitelaw, J.H., _AIAA J._, 1979, Vol.17, pp.1291-1292.

Dean, W.R., _Proc. Camb. Phil. Soc._, 1950, Vol.46, pp.292-306.

Ghia, K.N., Ghia, U. and Tesch, W.A., _AGARD Conf.No.168 Flow Sep._, 1975

Williams, J.C.,III, _Ann. Rev. Fluid Mech._, 1977, pp.113-144.

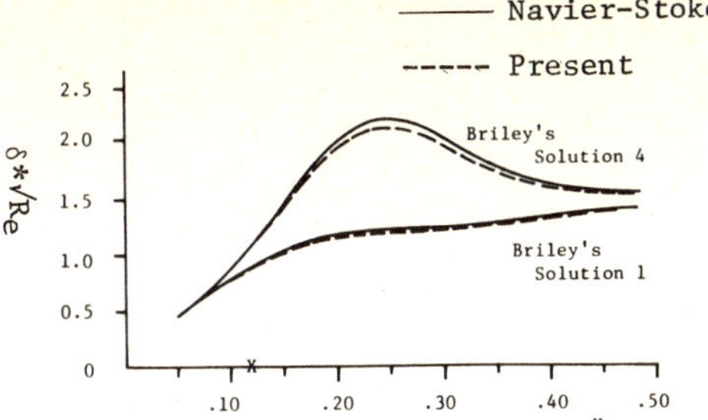


Fig.1. Displacement thickness

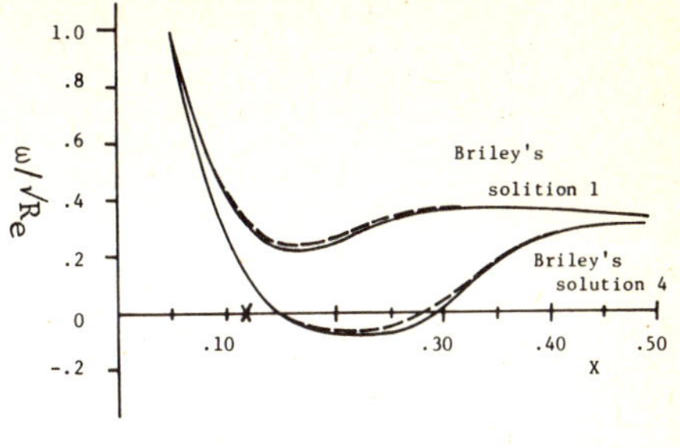


Fig.2. Wall shear distributions

—— Navier-Stokes, ---- Present

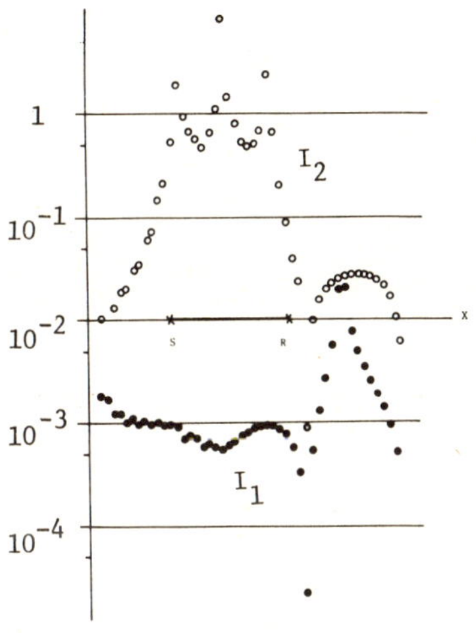


Fig.3  $I_1$ and $I_2$

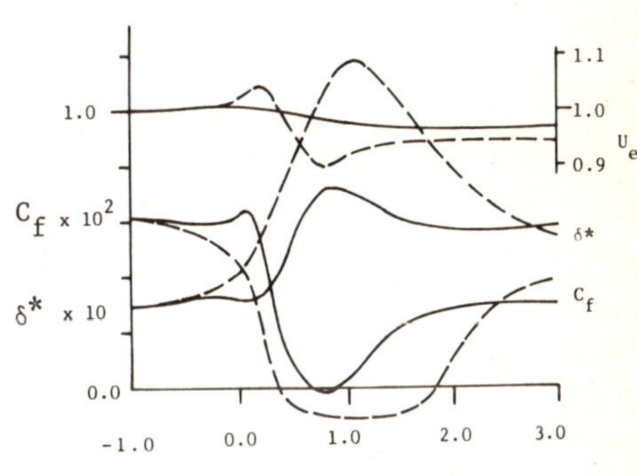


Fig. 4.  —— with interaction

---- without interaction

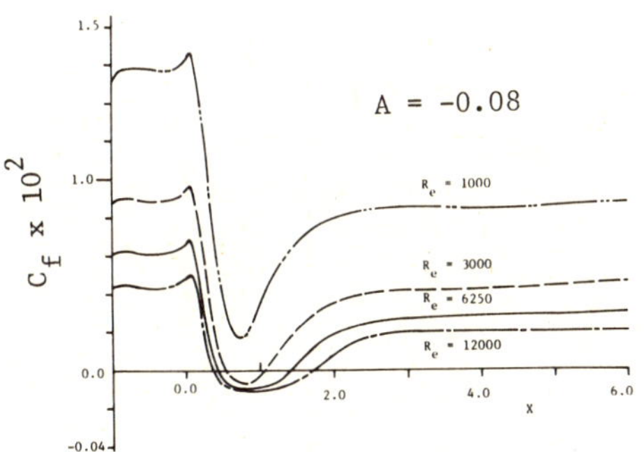


Fig.5. Wall shear distributions

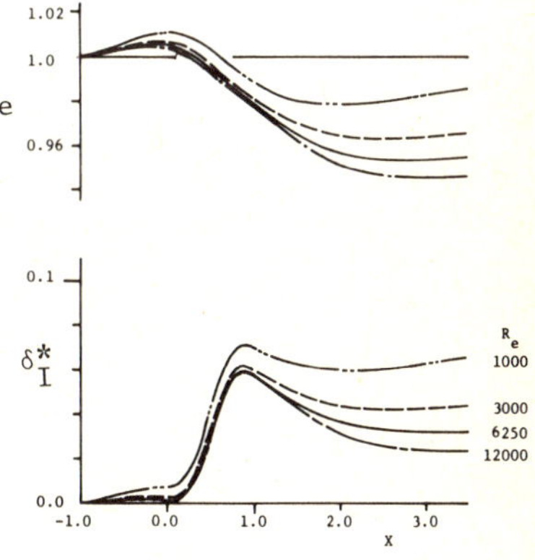


Fig.6.  $U_e$ and $\delta^*_I$

IMPROVEMENT OF NUMERICAL SCHEMES BY INCORPORATION OF
APPROXIMATE SOLUTIONS APPLIED TO ROTATING COMPRESSIBLE FLOWS

M. Israeli
Computer Science Dept.
Technion – Israel Institute of  Technology
Haifa, Israel

M. Ungarish
Interdisciplinary Committee for Applied Mathematics
Technion – Israel Institute of Technology
Haifa, Israel

## ABSTRACT

A method which reduces the truncation error in the numerical solution of differential equations by incorporating known (asymptotic) approximations is described.  The method was implemented in the solution by finite differences of rotating compressible or stratified axisymmetric flows in cylinders.  This method, called in the sequel the "booster method", is simple in principle and easy to implement in existing codes, while requiring only an insignificant amount of additional computer expenditure.

The "booster method" was applied in the case of second order ordinary differential equations and significant improvement of accuray resulted, especially in the presence of thin boundary layers.  Application of this method to the above mentioned flow problems results in a substantial reduction of the number of computational points without loss of accuracy.

## I.  THE "BOOSTER" METHOD

Consider a differential equation: $L(y) = f$ in B, $y = y_b$ on $\partial B$, where B is some region in $R^n$ with a boundary $\partial B$.

The numerical solution $y_\Delta$ is usually obtained from: $L_\Delta(y_\Delta) = f$, where $L_\Delta$ is the discretized approximation to L, (variations like replacement of  f  by  $f_\Delta$  are trivialy incorporated).

It follows that: $L_\Delta(y) - L_\Delta(y_\Delta) = L_\Delta(y) - L(y) = T(y)$, where $T(y)$ is the truncation error.  In the linear case one gets:  $L_\Delta(y-y_\Delta) = T(y)$  and it follows that

$$\| y - y_\Delta \| = \| L_\Delta^{-1} T(y) \| \leq \| L_\Delta^{-1} \| \cdot \| T(y) \| \ .$$

In the "booster method" we use an approximation $\tilde{y}$ to the solution to reduce the truncation error, the  "improved" numerical solution  $\tilde{y}_\Delta$  is obtained from: $L_\Delta(\tilde{y}_\Delta) = f + L_\Delta(\tilde{y}) - L(\tilde{y})$.  Here we get:

$$L_\Delta(y) - L_\Delta(\tilde{y}_\Delta) = [L_\Delta(y) - L(y)] + [L(y) - L_\Delta(\tilde{y}_\Delta)] = T(y) - T(\tilde{y}).$$

In the linear case we find:

$$L_\Delta(y-\tilde{y}_\Delta) = T(y) - T(\tilde{y})$$

$$\| y-\tilde{y}_\Delta \| \leq \| L_\Delta^{-1} \| \cdot \| T(y) - T(\tilde{y}) \| \ .$$

If the approximation $\tilde{y}$ is such that $T(y)-T(\tilde{y}) = k \, \varepsilon \, T$, then we get:

$$\| y-\tilde{y}_\Delta \| = k \, \varepsilon \| y-y_\Delta \| \cdot \quad \text{(The method as formulated above is not restricted to linear}$$
$$\text{problems)}$$

In particular, the methos is applicable to regular or singular perturbation problems, where approximate solutions are easy to derive, and can be used as "booster" in our method. We emphasize that when $\tilde{y}$ is known analytically the correction term in (7) can be easily computed and introduced into the computational scheme even when the equations are non linear. The solution of the new difference equations is not materially different as compared to the usual case.

Often the approximation $\tilde{y}(x)$ is itself obtained by numerical means by solving a simplified differential equation. In such cases one can solve for $\tilde{y}(x)$ on a fine grid and obtain the differential operators by differencing on that fine grid, thus the truncation error involved in this approach becomes negligible. An example of this procedure is the case of rotating flows to be discussed below. There we find the aproximate solution $\tilde{y}(r)$ by solving a second order ordinary differential equation on a fine grid, while computing $y(r,z)$ requires the solution of a non linear coupled system of partial differential equations.

The following examples were solved by means of second order central differences:

(a) $\quad \varepsilon^2 y'' - y = \sin x, \quad 0 \leqslant x \leqslant 1, \quad y(0) = y(1) = 1$

(b) $\quad y'' + y' + \varepsilon^2 e^{2x} y = 0, \quad 0 \leqslant x \leqslant 1, \quad y(0) = 0, \quad y(1) = 1.$

In both cases the "booster" $\tilde{y}(x)$ was obtained by first order (in $\varepsilon$) asymptotic approximation. We denote by $e$ the error in the usual method and by $\tilde{e}$ the error in the booster method.

In example (a) we find, for sufficiently small $h$, the following: $e$ max is found near $x = \varepsilon$ while $\tilde{e}$ max is found near $x = 0.5$. The ratio $R = \tilde{e}$ max/$e$ max is approximately $5 \times 10^3 \varepsilon^6$, therefore we obtain $R < 1$ for $\varepsilon \stackrel{\sim}{<} 0.24$. For example, for $\varepsilon = 1/10$, $R \cong 1/200$ while for $1/20$ we get $R \cong 1/13000$. Moreover, we find that while in the usual method the error in the boundary layer, at $x = \varepsilon/2$, increases like $\varepsilon^{-2}$ as $\varepsilon \to 0$, in the "booster" method the error decreases like $\varepsilon^5$ which gives an improvement which grows asymptotically like $\varepsilon^7$. For example (b) $R$, as defined above, was found to be between 0.5 and 0.02 when $\varepsilon$ changed between 0.5 and 0.1, and behaves like $\varepsilon^2$ for small $\varepsilon$.


## II.  MULTI DIMENSIONAL APPLICATION - ROTATING COMPRESSIBLE FLOWS

The multidimensional generalization of the method is straightforward and was applied to the numerical solution of the Navier-Stokes equation for the case of rotating compressible axisymetric flows in a cylinder.

This probleam was chosen because of the considerable recent interest in such flows and because of the great difficulties encountered in the numerical solution for small Ekman numbers and final rotational Mach numbers. On the other hand, good approximations can be obtained for such flows.

Consider the steady laminar flow of an ideal Newtonian gas (with a gas constant $R*$ and constant viscosity $\mu*$) in a cylinder of height $H*$ and radius $r_0^*$, (Fig. 1). The side walls and top end plate rotate with a constant angular velocity $\Omega*$ around the axis of symmetry, while the bottom plate rotates at $\Omega*(1-\varepsilon)$. The temperature of the cylinder is $T_0^*$, while $\rho_0^*$ is the density at the side wall.

The flow field is characterized by the Ekman number, $E = \mu*/\rho_0^*\Omega*r_0^2$, the Rossby number $\varepsilon$, the rotational Mach number, $M = \Omega*r_0^*/\sqrt{R*T_0^*}$, aspect ratio $H = H*/r_0^*$, and Brinkman number $B = PrM^2(\gamma-1)/\gamma$. Here $Pr = 0.72$ is the Prandtl number and $\gamma = 1.4$ is the specific heat ratio.

Rotating compressible flows at small $E$ and $\varepsilon$ were investigated recently by Sakurai and Matsuda (1974), Bark and Bark (1976) and others. They used assymptotic approximations valid for $\varepsilon M^2 \ll 1$, $\beta = \varepsilon E^{-\frac{1}{2}} \ll 1$. We remove these restrictions by expanding the logarithm of the pressure in powers of $\varepsilon$ and keeping the leading advection terms. Details of the derivation are presented elsewhere. We note, however, that the magnitude of $\beta$ has profound influence on the flow field as noted previously by Homsy and Hudson (1969) for the case of thermally forced Boussinesq fluid. We find similar behaviour for the mechanically driven compressible fluid when $B \approx 1$.

The assymptotic approximation requires the solution of a second order O.D.E. for the pressure and a Poisson equation with no boundary layer structure for the temperature and therefore is suitable for a "booster".

Comparisons between the approximate solution and a numerical solution (obtained by an Allen-cheng type scheme described in Toren and Solan (1979)) are satisfactory (Figs. 2-7). The parameters are $E = 0.0002$, $M = 2\sqrt{1.4}$, $Pr = 0.72$, $B = 1.12$, $H = 2$. The temperature on the boundary is constant and $\varepsilon = 0.2$.

The secondary flow field generated by the slow bottom plate is qualitatively similar to that in the compressible case. The main differences being the concentration of mass flux near the outer wall and the thickening of the Ekman layers near the center. Since the forcing is mechanical, the mean angular velocity defect is larger by a factor of $\approx 4/B \approx 4$ than the mean temperature deviation which is driven by the compression work ($0.1$ as compared to $0.025$). We observe that the pressure perturbation is independent of the axial coordinate and of relative magnitude $0.35$, larger by a factor of $\approx M^2/2$ than the angular velocity perturbation. The agreement between the approximation ("booster") and numerics is excellent for the pressure, good for the angular velocity and mediocre for the temperature.

We illustrate the "booster" method on the $\partial(\rho ur)/\partial r$ term of the continuity equation. The difference quotient

$$[(\rho ur)_{i+\frac{1}{2},j} - (\rho ur)_{i-\frac{1}{2},j}]/h$$

is corrected, by adding:

$$\{[\partial(\overline{\rho}\overline{u}r)/\partial r]_{i,j} - [(\overline{\rho}\overline{u}r)_{i+\frac{1}{2},j} - (\overline{\rho}\overline{u}r)_{i-\frac{1}{2},j}]/h\}$$

which is computed only once, since the "booster" fields $\overline{\rho}, \overline{u}$ are time independent. Here $h$ is a radial mesh distance.

A simple example of the use of the method is obtained by taking $B = 0$, which gives rise to isothermal flows $T = 1$. This is justified for the case of mechanical forcing when $B \approx 1$ (see above). The other parameters in this example are: $E = 0.0001$, $M = 2\sqrt{1.4}$, $H = 2$. We show four runs:

(a) Usual method, $10 \times 10$, uniform grid (x);

(b) Usual method, $10 \times 20$, grid stretched in interior ($\Delta$);

(c) Booster method, $10 \times 10$, uniform grid ($\cdot$);

(d) Booster method, $15 \times 20$, grid stretched ($\square$).

The stretching allows a ratio of largest to smallest mesh distances of $2.8$ in the radial and $5$ in the vertical direction. The initial fields are solid body rotation for the usual method and the approximate solution for the booster method.

In Figs. 8, 9, 10 we compare the perturbation pressure and azimuthal velocity (which are almost independent of the axial coordinate, z) as well as w at the midway plane for the four cases. Figure 11 illustrates the time development as a function of cpu time required.

We observe that (c) is much better than (a) by comparison with (d) which should be the nearest to exact solution. Consequently, (c) is also better than (b) which requires more storage, computer time, and stretched mesh programming efforts.

REFERENCES

Bark, F.H., Bark, T.H.: "On Vertical Boundary Layers in a Rapidly Rotating Gas,"
    J. Fluid Mech., 78, 1976.

Sakurai, T., Matsuda, T.:    "Gasdynamics of a Centrifugal Machine",  J. Fluid Mech.,
    68, 1975.

Toren, M., Solan, A.:  "Laminar Compressible Flow over a Stationary Disk Confined in
    Rotating Cylinder,"  J. Fluid Eng., 101, 1979.

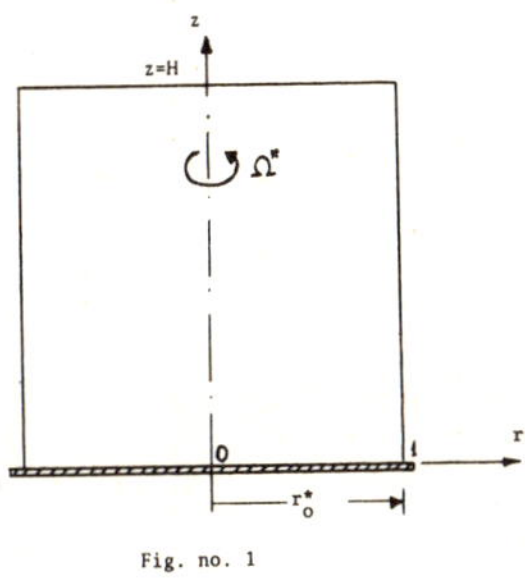


Fig. no. 1

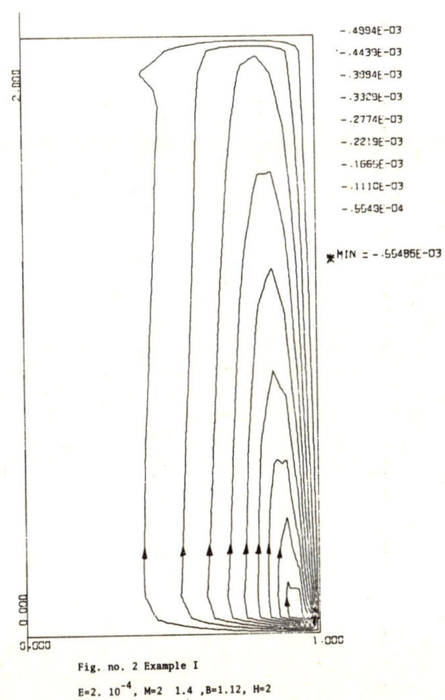


Fig. no. 2 Example I

$E = 2. \, 10^{-4}$, $M=2$  1.4, $B=1.12$, $H=2$

$\psi$,  Numerical solution

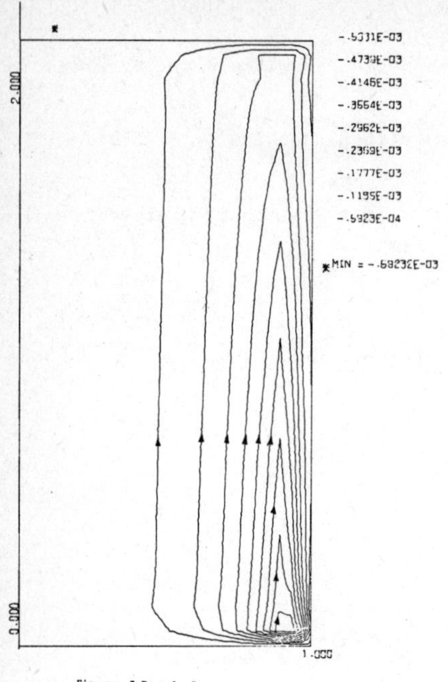

Fig. no. 3 Example I
$\psi$, Approximate solution

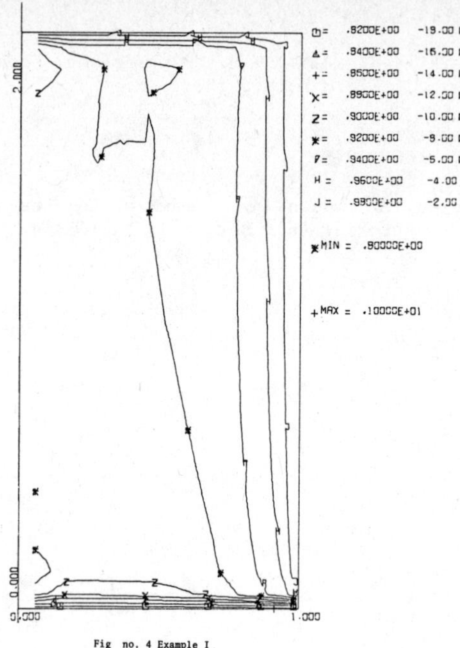

Fig no. 4 Example I
$(v/r)$, Numerical solution

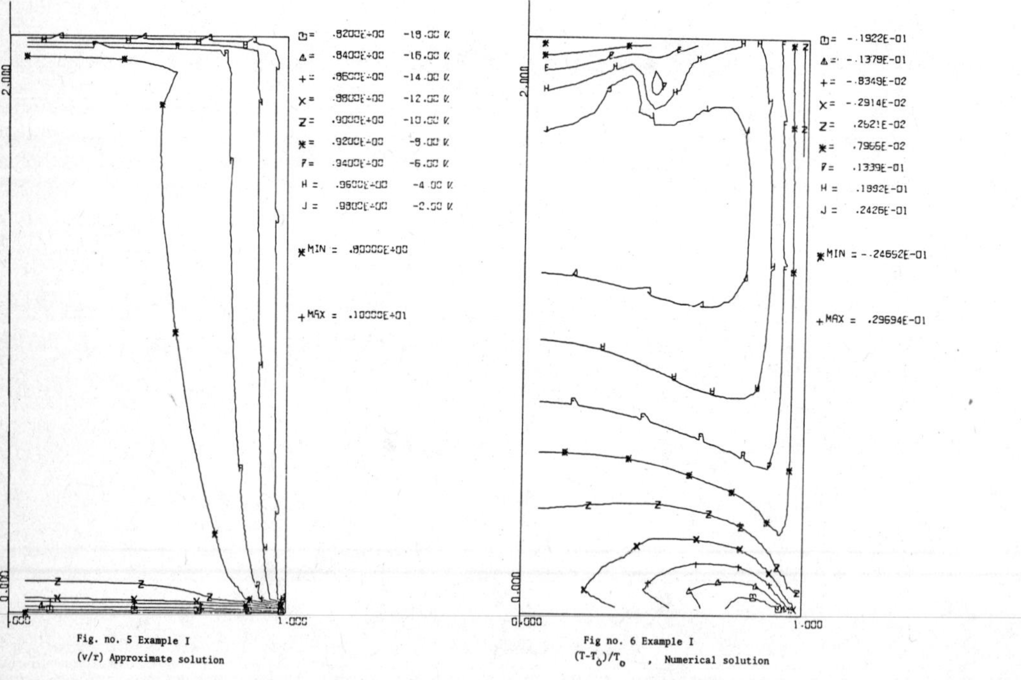

Fig. no. 5 Example I
$(v/r)$ Approximate solution

Fig no. 6 Example I
$(T-T_0)/T_0$ , Numerical solution

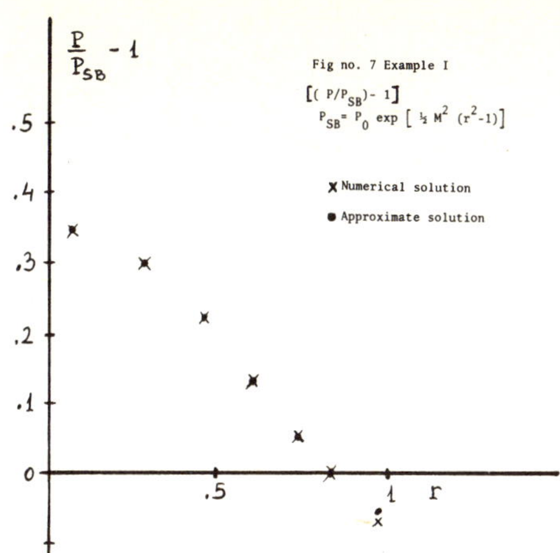


$\frac{P}{P_{SB}} - 1$
.5
.4
.3
.2
.1
0
.5
1
r
Fig no. 7 Example I
$[( P/P_{SB})- 1]$
$P_{SB} = P_0 \exp [ \frac{1}{2} M^2 (r^2-1)]$
Numerical solution
Approximate solution


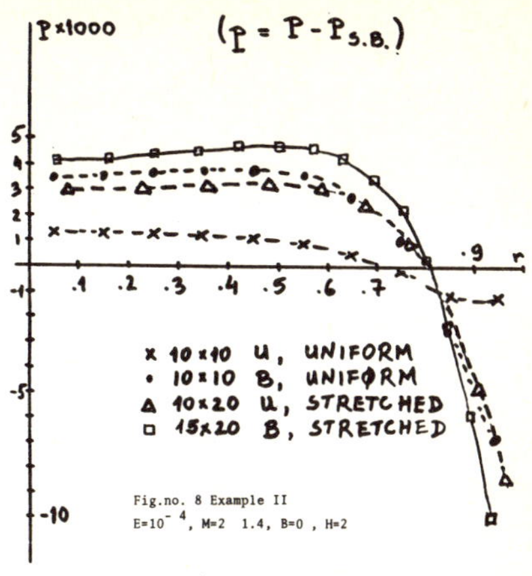


$P \times 1000$
$(\underline{P} = P - P_{S.B.})$
5
4
3
2
1
-1
-5
-10
.1 .2 .3 .4 .5 .6 .7 .9 r
× 10×10 U, UNIFORM
• 10×10 B, UNIFORM
△ 10×20 U, STRETCHED
□ 15×20 B, STRETCHED
Fig.no. 8 Example II
$E=10^{-4}$, M=2  1.4, B=0 , H=2


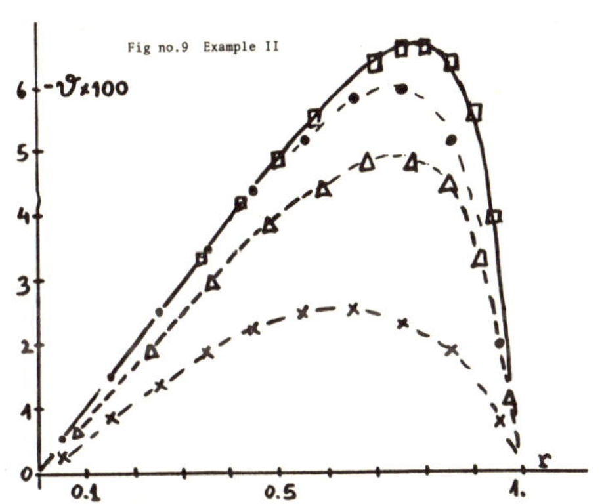


Fig no.9  Example II
$-\vartheta \times 100$
6
5
4
3
2
1
0
0.1
0.5
1.
r


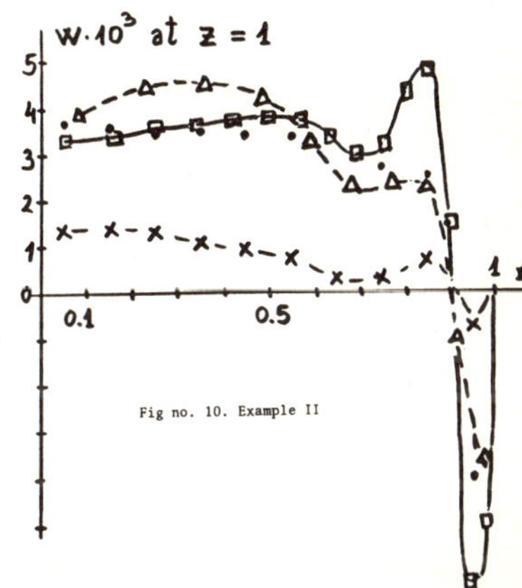


$W \cdot 10^3$ at $z = 1$
5
4
3
2
1
0
0.1
0.5
1 r
Fig no. 10. Example II


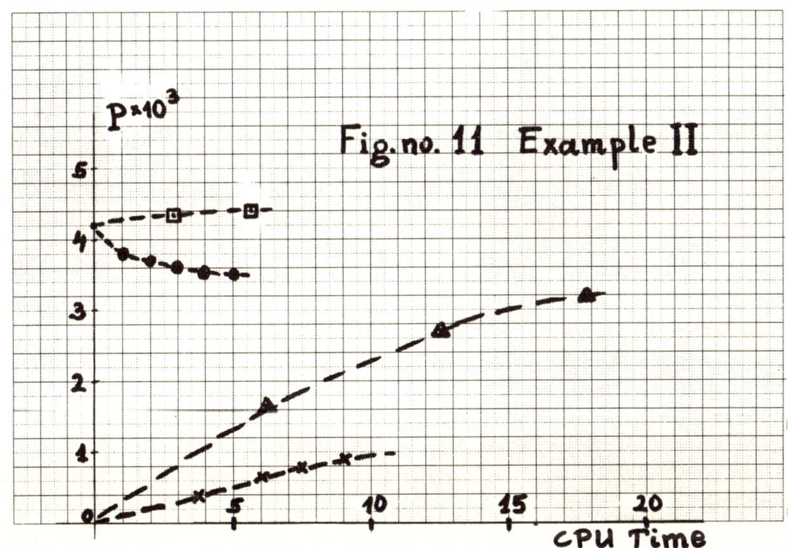


$P \times 10^3$
5
4
3
2
1
0
Fig.no. 11  Example II
5
10
15
20
CPU Time

# AN ALTERNATIVE APPROACH TO THE NUMERICAL
# SIMULATION OF STEADY INVISCID FLOW

GARY M. JOHNSON

NASA LEWIS RESEARCH CENTER

CLEVELAND, OHIO 44135, USA

## SUMMARY

A numerical procedure for the efficient simulation of steady inviscid flow is described and its utility is demonstrated. The method is uniformly valid for application in the subsonic, transonic and supersonic flow regimes. It does not rely on the introduction of additional assumptions beyond those necessary to obtain the Euler equations from the Navier-Stokes equations, nor does it make use of a time-asymptotic solution of the unsteady equations of motion.

Application of the herein-defined surrogate equation technique allows the formulation of stable, fully-conservative, type-dependent finite difference equations for use in obtaining numerical solutions to systems of first-order partial differential equations, such as the steady-state Euler equations or their various approximations.

Computational results are presented for the full Euler equations used to simulate rotational subsonic flow and for the transonic small disturbance equations. For the latter case, a computational efficiency greater than that obtained by means of the standard perturbation potential approach is indicated.



## INTRODUCTION

When written in primitive variable form, the systems of partial differential equations used to describe the steady motion of an inviscid fluid are of first order and of mixed elliptic-hyperbolic type. Common examples of such systems include the transonic small disturbance equations :

$$\beta_x + v_y = 0 \quad , \quad v_x - u_y = 0 \tag{1}$$

where u and v are the perturbation velocity components and $\beta = (1 - M_\infty^2 - \frac{\gamma+1}{2} M_\infty^2 u)u$ , and the Euler equations :

$$f_x + g_y = 0 \tag{2}$$

where $f = (\rho u, \rho u^2+p, \rho uv, (E+p)u)^T$ , $g = (\rho v, \rho uv, \rho v^2+p, (E+p)v)^T$ , $E = \frac{p}{\gamma-1} + \frac{\rho}{2}(u^2 + v^2)$ , u and v are the velocity components, $\rho$ is the density and p is the static pressure.

Because of the difficulties associated with both the formulation of robust finite difference analogs for such equations and the construction of stable iterative procedures for their numerical solution, these partial differential systems are not usually solved in the forms given above. Rather, as is well known, the transonic small disturbance equations are transformed into a scalar second-order partial differential equation by the introduction of a perturbation velocity potential function. The steady Euler equations, on the other hand, are replaced by their unsteady versions, for which a temporally-asymptotic steady solution is sought, either in real or in a pseudo-time.

Relatively few departures from these approaches are to be found in the literature. Steger and Lomax (1975) have developed an iterative procedure for solving a non-conservation form of the steady Euler equations for subcritical flow with small shear. Chattot (1976) has solved the transonic small disturbance equations by differentiating them to obtain a second-order system. This work represents a special case of the approach to be discussed here. He later adopted a variational formulation and has applied it to model problems representing the Euler equations (Chattot et al. 1979). Ozer (1977) has developed a relaxation procedure for solving the equations of motion when reformulated to yield a second-order partial differential equation in the logarithm of the pressure, together with first-order equations for the remaining variables. Blomster and Sköllermo (1977) applied Newton's method to the first-order system representing the full potential equation to solve a shockless transonic nozzle flow problem. Rizzi (1979) has extended this procedure to the steady Euler equations.

The work of these authors notwithstanding, it remains the case that contemporary numerical simulations of steady inviscid flow generally resort to either relaxation solutions of steady second-order equations in derived dependent variables or time-asymptotic solutions of unsteady first-order systems. In the former case, generality is lost, while in the latter case, the computational efficiency may be quite low.

Here we present a means by which the steady first-order, mixed-type systems of inviscid fluid flow may be readily and efficiently solved with conventional numerical techniques. The method, which we refer to as the surrogate equation technique, maintains the generality of the flow equations while allowing the use of the fully-conservative type-dependent relaxation procedures which have been developed for the efficient solution of second-order equations.

SURROGATE EQUATION TECHNIQUE

Given a first-order system, the surrogate equation technique (SET) consists of embedding this system in a second-order system (its surrogate), applying additional constraints obtained from the original system to restrict the solution set of the surrogate, and then solving the resulting partial differential problem by means of a conventional iterative procedure.

Consider a first-order system written in conservation law form, such as

$$\left[ \frac{\partial}{\partial x}(A\ ) + \frac{\partial}{\partial y}(B\ ) \right] f = 0$$ where $f$ is an n-component vector and $A$ and $B$ are $n \times n$ matrices. We embed this system in a second-order surrogate of the form

$$\left[ \frac{\partial}{\partial x}(M\ ) + \frac{\partial}{\partial y}(N\ ) \right] \left[ \frac{\partial}{\partial x}(A\ ) + \frac{\partial}{\partial y}(B\ ) \right] f = 0 \quad . \tag{3}$$

This system preserves the conservation law form of the original system. Furthermore, the jump conditions satisfied by weak solutions to the surrogate system are the same as those satisfied by the original system. The behavior of the additional character-istic directions introduced by the embedding may be controlled through the specifica-tion of the matrices  M  and  N . For example, the choice M = A , N = -B would result in the additional characteristics being the reflections of the original ones through the x-axis, while the choice  M = A$^T$ ,  N = B$^T$  would simply replicate the original characteristics, while symmetrizing the matrix coefficients of the terms of highest order in the surrogate second-order system. Having chosen  M  and  N , the problem specification is completed by requiring that, in addition to satisfying the original boundary conditions of the first-order system, the solution to the surrogate system must also satisfy the first-order equations themselves at the boundaries. This is done to insure the uniqueness of the solution.

Having thus embedded the first-order system in a second-order system, we may avail ourselves of the abundance of research results on efficient, stable iterative procedures for such equations in order to construct a suitable numerical scheme. In the interest of simplicity, we confine ourselves here to the use of fully-conservative, type-depend-ent differencing together with the well-established successive line relaxation method. We stress, however, that SET may also be used with other iterative solution methods.

RESULTS

In the course of developing SET (Johnson 1980), we have applied it to obtain solu-tions to the two-dimensional steady Euler equations for purely supersonic and purely subsonic nozzle flows and for rotational subsonic flow through bends. Application has also been made to the two-dimensional transonic small disturbance equations for both subcritical and shocked supercritical flow.

Consider the Euler equations (2), rewritten in the form

$$\left[ \frac{\partial}{\partial x} + \frac{\partial}{\partial y} (T\ ) \right]\ f = 0 \tag{4}$$

where  $T = BA^{-1}$  with  A  and  B  being the usual Jacobian matrices. If we choose the matrices  M  and  N  in the SET formu-lation (3) such that  M = I  and  N = -T we obtain the second-order system shown in Fig. 1. There the application of this system to a rotational subsonic bend flow is schematically illustrated. The bound-ary conditions used are also shown. At the left-hand boundary, the inflow is completely specified. While this con-stitutes an over-specification, such treatment is considered to be adequate

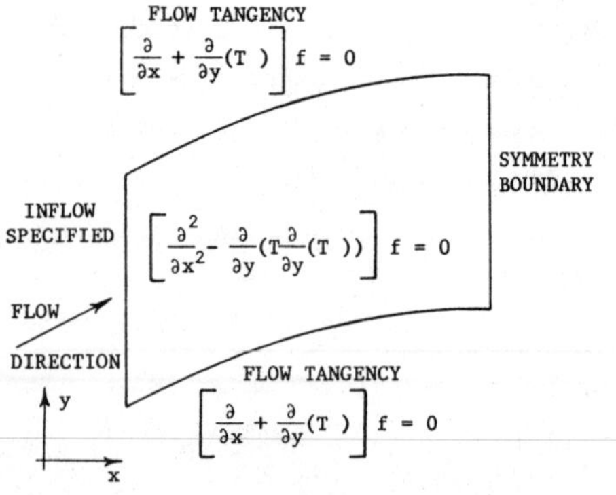


Fig. 1   SUBSONIC BEND PROBLEM

for the present, illustrative, purpose. Flow symmetry is required at the outflow boundary. On the walls, in addition to the usual flow tangency condition, satisfaction of the original first-order Euler system (4) is required. One should note that while, for simplicity, the equations presented here are written in Cartesian coordinates, the computations reported below were carried out using a slightly different form of the equations, written in sheared coordinates.

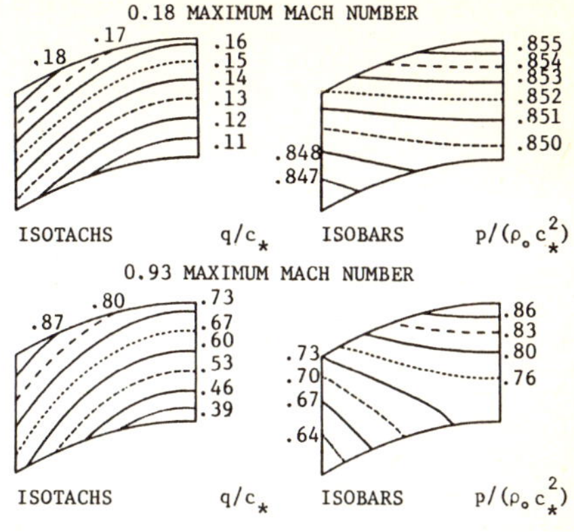



Fig. 2   SAMPLE BEND FLOW RESULTS

Shercliff (1977) presented an analytic solution to the incompressible analog of the bend flow considered here. We have solved the compressible flow problem over a range of subsonic Mach numbers. Selected results for two such Mach numbers are shown in Fig. 2. The computational domain illustrated is a section of a 90° symmetric bend, whose symmetry axis is the right-hand boundary. In its passage through the entire bend, the flow transitions between two asymptotic flows which are rectilinear shear flows. Also, as the bend cross-section reaches its maximum at the symmetry axis, the flow is decelerating in the section shown.

This case illustrates the feasability of obtaining solutions to the full Euler equations for rotational subsonic flow by means of SET. Because of the form of the matrix $T$ used in the above formulation, it is not suitable for use in computing transonic flow. This does not appear to be an insurmountable difficulty, but, for the present, we confine our transonic flow discussion to the small disturbance equations.

Given the transonic small disturbance equations (1), we choose $M$ and $N$ in the SET formulation (3) such that $M = AB^{-1}$ and $N = -I$ where

$$A = \begin{bmatrix} \partial\beta/\partial u & 0 \\ 0 & 1 \end{bmatrix} \quad \text{and} \quad B = \begin{bmatrix} 0 & 1 \\ -1 & 0 \end{bmatrix} \; .$$

This choice yields the surrogate second-order system shown in Fig. 3. This figure also illustrates the test case to which we have applied the system. We consider the flow in a two-dimensional channel with uniform inlet conditions. A circular arc airfoil surface is mounted on the lower channel wall.

The boundary conditions applied in our primary formulation of the problem (SET 1) are also shown in Fig. 3.

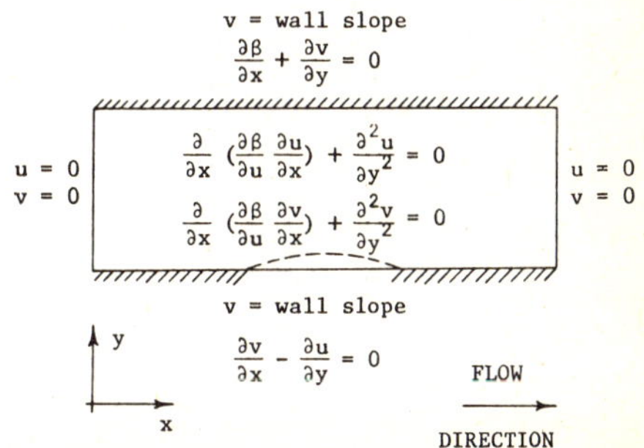



Fig. 3 TRANSONIC SMALL DISTURBANCE PROBLEM

We require both perturbation velocity
components to vanish at the channel
entrance and exit. The  v  perturbation velocity is set equal to the wall
slope on both channel walls. The  u
perturbation velocity is obtained from
the irrotationality condition on the
lower wall and from the mass conservation equation on the upper wall. In an
effort to probe the necessity of these
boundary conditions, a secondary formulation of the problem (SET 2) was created
in which the mass conservation equation
on the upper wall was replaced by the
irrotationality condition while all
other aspects were held fixed.

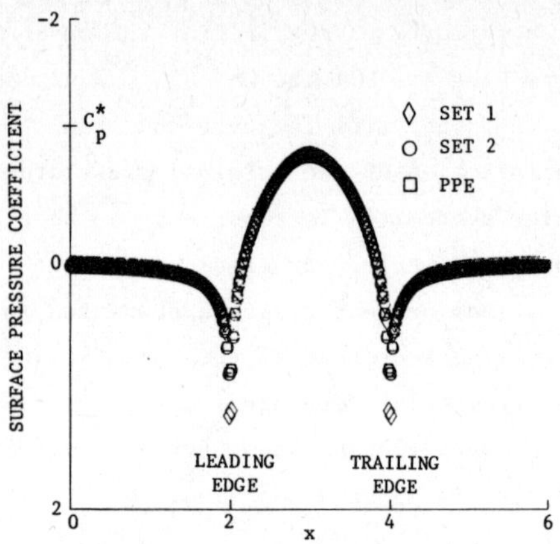

Fig. 4   SUBCRITICAL FLOW

For a subcritical flow case, the lower surface pressure coefficients resulting
from both SET formulations and from the conventional Murman and Cole (1971) perturbation potential formulation (PPE) are compared in Fig. 4. All three formulations produce the same result, with SET 1 providing slightly better resolution of the stagnation points. A similar comparison for a shocked supercritical flow case is presented
in Fig. 5. In this case, the agreement is also excellent, except in the immediate
vicinity of the shock. Here both SET 1 and SET 2 are in agreement and both produce a
very sharp shock (2 points as opposed to 4 points for PPE). However, the shock
strength is underpredicted. This anomaly is presently the object of further study.

Some interesting observations may be made concerning the relative efficiencies
of the three algorithms. PPE and SET 2
have roughly equivalent operations
counts while that of SET 1 is approximately 1.5 times as great. An examination of convergence behavior, however,
reveals a strength of the SET formulations. Fig. 6 illustrates the behavior
of the maximum residual as a function of
iteration level for all three formulations under the same conditions. From the
data we may estimate the spectral radii
for the three formulations. These, in
turn, imply relative asymptotic convergence rates of 1.0, 2.3 and 8.1 for PPE,
SET 2 and SET 1, respectively. Hence, we

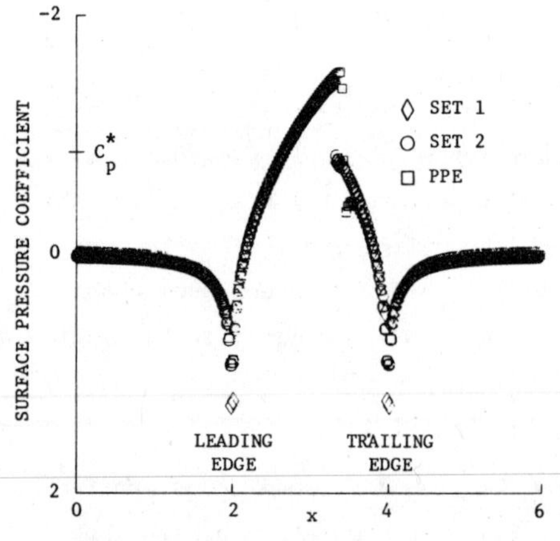

Fig. 5   SUPERCRITICAL FLOW

estimate the asymptotic computational efficiencies of PPE, SET 2 and SET 1 to be, respectively, 1.0, 2.3 and 5.4 .

## CONCLUSIONS

We have shown that it is possible to obtain a numerical solution to a system of first-order partial differential equations by solving a problem consisting of a surrogate second-order system together with the original boundary conditions and supplementary relations obtained from the first-order system.

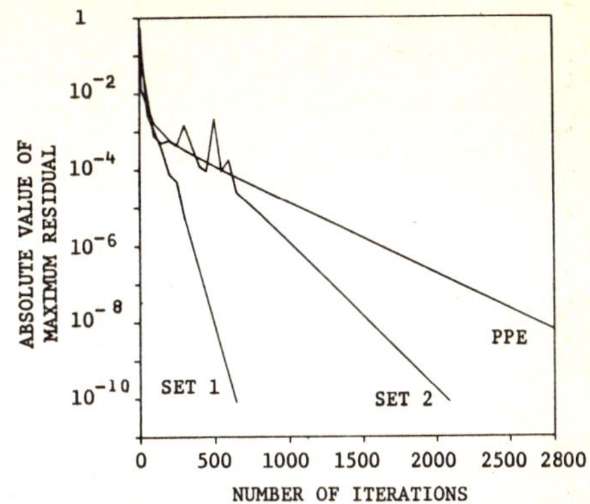


Fig. 6   CONVERGENCE BEHAVIOR

SET provides a means for formulating problems involving first-order equations describing steady inviscid flow in such a way as to allow the use of fully-conservative type-dependent differencing and iterative solution procedures. Hence, we may solve such problems without resort either to a velocity potential or stream function or to an unsteady formulation.

An application of SET to the transonic small disturbance equations results in algorithms which, on the basis of the computational experimentation reported here, appear to have computational efficiencies which are several times greater than that of the standard perturbation potential algorithm.

In view of the successful development of iterative procedures for the solution of both the full Euler equations for subsonic and supersonic flows and the small disturbance equations for transonic flow, it appears that further applications of SET are merited.

## ACKNOWLEDGEMENT

The assistance of Mr. Aaron Snyder is most gratefully acknowledged.

## REFERENCES


BLOMSTER, J. & SKÖLLERMO, G. 1977 Uppsala Univ., Comp. Sci. Dept., Report No. 66.
CHATTOT, J.J. 1976 Proc. Symp. Trans. II, Springer, pp 533-540.
CHATTOT, J.J., GUIU-ROUX, J. & LAMINE, J. 1979 Lect. Notes in Phys., Vol.90,
    Springer, pp 121-129.
JOHNSON, G.M. 1980 NASA TP to appear.
MURMAN, E.M. & COLE, J.D. 1971 AIAA J., Vol.9, pp 114-121.
OZER, J.M. 1977 Ph.D. Thesis, Liège Univ.
RIZZI, A. 1979 Lect. Notes in Phys., Vol. 90, Springer, pp 460-467.
SHERCLIFF, J.A. 1977 J. Fluid Mech., Vol.82, pp 687-703.
STEGER, J.L. & LOMAX, H. 1975 NASA SP 347, Vol.2, pp 811-838.

REFORMULATION OF THE METHOD OF
CHARACTERISTICS FOR MULTIDIMENSIONAL FLOWS


Czeslaw P. Kentzer

Purdue University, West Lafayette, IN 47907, USA


## ABSTRACT


The method of characteristics is reformulated so as to simplify it
*conceptually* and thus render it easily programable for multidimensional
computations.  A transformation to characteristic coördinates and a
change in notation allows one to coach the theory of characteristics in
the language of geometrical wave theory or of the particle mechanics.
The dependent variables $(\vec{u},p,\rho)$ are given uniquely in terms of characteristic coördinates, while the latter satisfy the Hamiltonian equations.
Computations with the method of characteristics are reduced to the determination of Hamiltonian trajectories in space-time, a task not unduely
complex even in multidimensional cases.


## I.  INTRODUCTION

The advantages of the Method of Characteristics (MOC) are that, 1° MOC is based
on the *exact* Theory of Characteristics, 2° MOC follows closely the physics of the
problem, and, 3° MOC defines and uses the natural coördinate system independently of
the particular choice of the computational grid employed.

With the advance in computational methods use is made of more complicated computational meshes.  Often the meshes depend on the solution and are governed by geometrically complicated boundaries.  It was argued by Kentzer (1971) that MOC affords
exact means of incorporating algebraic boundary conditions into a finite-difference
solution of p.d.e.'s.  He suggested to differentiate the boundary conditions in the
plane tangent to the boundary and to solve the resulting system of p.d.e.'s simultaneously.  This procedure becomes unsatisfactory whenever the transformation from
the computational to the physical coördinates becomes singular, as often is the case
with boundaries with sharp corners or with intersecting boundaries.

There appears that, if MOC is reformulated so as to use the characteristic rays
(bicharacteristics) as *coördinate lines* rather than characteristic surfaces as *coördinate surfaces*, there results a conceptual simplification of the method of particular
importance in applications to multidimensional flows with complicated boundaries.
Further, the ray formulation of MOC becomes analogous to the geometrical (ray) wave
theory or to classical particle mechanics.  Thus a hyperbolic problem in m dependent
and n independent variables becomes equivalent to, or as easy as, a problem of motion
of m particles in an n-dimensional space-time.  The objective of this work is a re-
formulation of MOC so as to maximize its computational advantages.

## II.  GENERAL THEORY OF CHARACTERISTICS

A general treatment of the theory of characteristics for the system of quasilinear equations of gasdynamics may be found in Rusanov (1963).  In particular, Rusanov investigated conditions for the existence of exactly $m$ real and independent characteristic relationships (compatibility conditions) which replace the system of $m$ conservation equations of gasdynamics.  One particular set of independent characteristic relationships for the pressure $p$, velocity $\vec{u}$, and density $\rho$ ($m=5$) may be written as

$$\frac{D}{Dt}\,(p/\rho^{\gamma}) = 0, \qquad \frac{D}{Dt} = \frac{\partial}{\partial t} + \vec{u}\cdot\nabla \tag{1}$$

$$\rho a\vec{n}_i\cdot\left[\frac{D\vec{u}}{Dt} + a\vec{n}_i\nabla\cdot\vec{u}\right] + \left[\frac{Dp}{Dt} + a\vec{n}_i\cdot\nabla p\right] = 0, \quad i = 1,\ldots,4 \tag{2}$$

where $\gamma$ = adiabatic index, $a$ = speed of sound.  The five independent relationships at any field point are obtained using Eq.(1) once and Eq.(2) four times in three space dimensions, three times in two, and twice in one space dimension.  Proper choices of the unit vectors $\vec{n}_i$ will be discussed later.

From the computational point of view, Eq.(1) is already in the conventient form of a Riemann invariant, $p/\rho^{\gamma}$= const. on $d\vec{x}/dt = \vec{u}$ or in the direction of a ray $\vec{V} = (\vec{u},1)$ in four-dimensional space-time.  As a consequence, we will concentrate on Eq.(2) representing a system of four characteristic relationships for the four unknowns, three velocity components $u,v,w$ and the speed of sound $a$.  We observe that Eq.(2) is an elliptic p.d.e. on a characteristic hyperplane with a normal $\vec{N} = (\vec{n},N_t)$, where the space component of the characteristic normal $\vec{n}$ is normalized to unity, $|\vec{n}| \equiv 1$, and the time component $N_t$ is determined by setting the quadratic factor of the characteristic determinant equal to zero,

$$(N_t + \vec{u}\cdot\vec{n}_i)^2 - a^2(\vec{n}_i\cdot\vec{n}_i) = 0, \quad \text{or} \quad N_t + \vec{u}\cdot\vec{n}_i + a|\vec{n}_i| = 0. \tag{3}$$

The characteristic ray corresponding to the relation (2) is $\vec{V} = (\vec{u}+a\vec{n}_i,1)$ and we observe that pressure $p$ is differentiated along the ray, while the components of velocity are differentiated along vectors laying in the characteristic plane but not parallel to the ray $\vec{V}$.

The apparent reason why the Method of Characteristics (MOC) has not yet been applied to time-dependent, three-dimensional flow problems (to the best of the present writer's knowledge) is the conceptual difficulty of numerical integration of Eq.(2) on intersecting characteristic hypersurfaces in 4-dimensional space-time, such hypersurfaces being determined by Eq.(3) for four choices of the unit normal $\vec{n}_i$.  A great simplification of the computational task was accomplished by Kentzer (1978) through the generalization of the concept of Riemann invariants.  In the present paper different approach will be used, viz., it will be convenient to consider the factored root of the characteristic determinant, Eq.(3), as playing the role of a Riemann invariant thus simplifying the implementation of MOC conceptually.

### III.  THEORY OF CHARACTERISTICS IN RAY FORM

One observes that Eq.(3) is homogeneous in the components of the characteristic normal.  As a consequence, a multiplication of (3) by a suitable constant with proper dimensions may be used to obtain equivalent results,

$$N_t + \vec{u}\cdot\vec{n}_i + a|\vec{n}_i| = 0, \qquad \text{(theory of characteristics)} \qquad (4a)$$

$$-\omega + \vec{u}\cdot\vec{k}_i + a|\vec{k}_i| = 0, \qquad \text{(geometrical wave theory)} \qquad (4b)$$

$$-H + \vec{u}\cdot\vec{p}_i + a|\vec{p}_i| = 0, \qquad \text{(particle mechanics)} \qquad (4c)$$

where $\omega$ = frequency, $\vec{k}$ = wavenumber vector, $\vec{p}$ = pseudo-particle momentum, $H$ = pseudo-particle Hamiltonian (total energy), $\vec{n} = \vec{k}/|\vec{k}| = \vec{p}/|\vec{p}|$ .

Making use of the mathematical analogies expressed by Eqs.(4) permits one to adopt either the wave or the particle point of view.  To the space component of the characteristic normal, $\vec{n}$, there corresponds either the wavenumber vector $\vec{k}$ or the momentum vector $\vec{p}$, and to the negative of the time component of the characteristic normal, $-N_t$, corresponds either the frequency $\omega$ or the Hamiltonian (energy) $H$.  The characteristic ray $\vec{V} = (\vec{u}+a\vec{n},1)$ defines the direction of propagation of wave energy or particle energy.  The characteristic hypersurface becomes either the hypersurface of constant phase or of constant action.

Considering $(\vec{k},-\omega)$ or $(\vec{p},-H)$ as coördinates in a space conjugate to $(\vec{x},t)$, one may solve for the dependent variables $(\vec{u},a)$ by, e.g. writing (4b) for four nonparallel vectors $\vec{k}^i$:

$$\begin{bmatrix} k_x^1, & k_y^1, & k_z^1, & k^1 \\ k_x^2, & k_y^2, & k_z^2, & k^2 \\ k_x^3, & k_y^3, & k_z^3, & k^3 \\ k_x^4, & k_y^4, & k_z^4, & k^4 \end{bmatrix} \begin{Bmatrix} u \\ v \\ w \\ a \end{Bmatrix} = \begin{Bmatrix} \omega^1 \\ \omega^2 \\ \omega^3 \\ \omega^4 \end{Bmatrix} \qquad (5)$$

where $k^i = |\vec{k}^i|$.  If the determinant of the square matrix $[k_j^i]$ is different from zero, there exists a unique solution

$$\begin{Bmatrix} u \\ v \\ w \\ a \end{Bmatrix} = [k_j^i]^{-1} \begin{Bmatrix} \omega^1 \\ \omega^2 \\ \omega^3 \\ \omega^4 \end{Bmatrix} \qquad (6)$$

Equation (6) is in the nature of a hodograph transformation giving the dependent variables $u,v,w,a$ explicitly and uniquely in terms of characteristic coördinates, either $(\vec{n},N_t)$, or $(\vec{k},-\omega)$, or $(\vec{p},-H)$, which are conjugate to the physical coördinates $(\vec{x},t)$.  In order to connect formally the physical and the conjugate coördinates one may use the Legendre contact transformation,

$$s(\vec{x},t) = \vec{x}\cdot\vec{n} + tN_t + \sigma(\vec{n},N_t)$$

$$= \vec{x}\cdot\vec{k} - t\omega + \sigma(\vec{k},-\omega)$$

$$= \vec{x}\cdot\vec{p} - tH + \sigma(\vec{p},-H).$$

The above may be interpreted, see p. 32 of Courant-Hilbert (1962), as representing a characteristic surface $s(\vec{x},t)$ by its tangent plane coördinates $\vec{n},N_t$ instead of by point coördinates $\vec{x},t$. The Legendre transformation is not convenient for our purposes. However, as discussed by Lighthill (1965), the assumption that the bilinear phase function,

$$(s - \sigma) = \vec{x}\cdot\vec{n} + tN_t = \vec{x}\cdot\vec{k} - t\omega = \vec{x}\cdot\vec{p} - tH,$$

is differentiable at least twice leads to the Hamiltonian system

$$\frac{d\vec{x}}{dt} = \frac{\partial\omega}{\partial\vec{k}} = \vec{u} + a\vec{n} \ ,$$

$$\frac{d\vec{k}}{dt} = - \frac{\partial\omega}{\partial\vec{x}} = - [(\nabla\vec{u})\cdot\vec{k} + k\nabla a] \ , \tag{7}$$

$$\frac{d\omega}{dt} = \frac{\partial\omega}{\partial t} = \vec{k}\cdot\frac{\partial\vec{u}}{\partial t} + k\,\frac{\partial a}{\partial t} \ ,$$

where $d/dt = \partial/\partial t + (\vec{u}+a\vec{n})\cdot\nabla$ is a total derivative along the ray. We have used here the wave picture and Lighthill's notation. The Hamiltonian system permits one to construct the desired transformation in the form $\vec{k} = \vec{k}(\vec{x},t)$, $\omega = \omega(\vec{x},t)$. An inversion of Eq.(5) gives then the flow variables explicitly by Eq.(6).

The problem of solving for the physical variables $\vec{u}(\vec{x},t)$ and $a(\vec{x},t)$ is equivalent to an integration of the dynamical system (7) for the motion of four pseudo-particles when the initial momenta $\vec{p}_i$ are chosen arbitrarily and the initial energies $H_i$ are determined from Eq.(4c). The ray formulation, either in the wave or in the particle picture, as developed here could be, after transformation of the Euler's equations, related to the approach taken on pp. 139-145 of Courant-Hilbert (1962). The consequence of the formulation is the replacement of $m$ p.d.e's in $n$ independent variables by an $m\times n$ system of o.d.e.'s. There is also an apparent similarity to the method of Riemann invariants where the characteristic network must be constructed first, to be followed by a solution for the flow variables from a linear algebraic system in terms of Riemann invariants determined by initial conditions. Clearly, in multidimensional case the use of rays (bicharacteristics) instead of characteristic surfaces has obvious advantages. In two independent variable case the characteristic surfaces and rays coincide and one may use the standard method of invariants.

## IV.  COMPUTATIONAL PROCEDURE

### A.  Interior Points.

First, we observe that if $\vec{k}^i$ and $\omega^i$ in Eq.(5) were pure constants then, knowing the initial conditions in the $t = t_o$ time plane, choosing four vectors $\vec{k}^i$ and calculating $\omega^i$ from (4b) with initial values of $\vec{u}$ and $a$ known, one obtains the solution at a point $(\vec{x}, t_o + \Delta t)$ directly from (6). What is needed to evaluate the initial conditions $\vec{u}_i, a_i$ in the $t = t_o$ plane is to calculate the coördinates $\vec{x}_i$ of the points of intersection of the four rays $\vec{V}_i$ which pass through the field point P, $(\vec{x}, t_o + \Delta t)$, and intersect the initial plane at $(\vec{x}_i, t_o)$. For this we use the first of Eqs.(7). Since $\vec{k}^i$ and $\omega^i$ do not remain constant along the ray, we use the remaining equations to improve accuracy. Thus a simultaneous integration of the Hamiltonian system (7) gives a predicted solution in the $t = t_o + \Delta t$ plane, and a repeated integration with the right-hand-sides of Eqs.(7), averaged between their initial and predicted values, would yield a second order accuracy with respect to time.

The question of the existence of a unique solution of system (5) at an interior point was answered by Rusanov (1963). When the magnitudes $k^i$ of the four vectors $\vec{k}^i$ in (5) are factored out, then, in terms of unit vectors $\vec{n}^i$, the determinant of the matrix equated to zero gives an equation of a plane. This implies that the four unit vectors $\vec{n}^i$ should be chosen so that, when laid off from a common origin, their end points on a unit sphere should not fall in a common plane. Of course, no two vectors $\vec{n}^i$ should be parallel.

### B.  Examples of Boundary Points.

When the boundary is space-like, the point may be treated as an interior point. When the boundary is time-like, the boundary intersects the characteristic hypercone generated by the rays $\vec{V} = [\vec{u} + a\vec{n}, 1]$ dividing it into two semicones, the semicone of dependence (generated by rays approaching the boundary) and the semicone of influence (generated by rays leaving the boundary for increasing time). The fact that the point on the boundary depends on the past and influences the future, requires that at least one ray be retained on each of the semicones. In three space dimensions Eq.(2) admits four characteristic relations. The corresponding four rays may be distributed then so that one, two or three rays are assigned to the semicone of dependence of the boundary point requiring three, two, or one boundary condition as initial conditions on the semicone of influence, respectively. For each ray of influence we omit one of the equations of system (5) and replace it by an appropriate boundary condition. If the boundary condition is linear in $u, v, w, a$, the existence of a unique solution is assured if the determinant of the resulting matrix is different from zero. Nonlinear boundary conditions must be differentiated resulting in a linear system for the derivatives, see Kentzer (1971). Below we give several examples of linear boundary conditions.

1. <u>Solid Wall</u>.  The flow tangency condition on a moving wall is $\vec{u}\cdot\vec{M} = u_n(\vec{x},t)$ where $u_n$ is given and $\vec{M}$ is a vector normal to the wall.  We replace the last row of the matrix $k_j^i$ in Eq.(5) by $\{M_x,M_y,M_z,0\}$ and obtain for the determinant

$$k^1k^2k^3(\vec{n}^1\times\vec{n}^2 + \vec{n}^2\times\vec{n}^3 + \vec{n}^3\times\vec{n}^1)\cdot\vec{M} \neq 0.$$

The sum in parentheses represents a vector perpendicular to the plane of the triangle formed by the end points of unit vectors $\vec{n}^i$;  its magnitude is twice the area of the triangle which is nonzero if the $\vec{n}^i$ are non-coplanar and nonparallel.  All $\vec{n}^i$ point in the same direction (toward the wall), thus the plane of the triangle cannot be orthogonal to the wall.  Under these conditions the solution exists and is unique.

2. <u>Subsonic Outflow Boundary</u>.  As an example of two boundary conditions we specify the outflow direction by $\vec{u}\times\vec{s} = 0$, $\vec{s}$ = arbitrary vector.  Introducing two vectors orthogonal to $\vec{s}$, such that $\vec{s} = \vec{W}^1\times\vec{W}^2$, we replace two rows of the matrix $k_j^i$ in Eq.(5) by $\{W_x^1,W_y^1,W_z^1,0\}$ and $\{W_x^2,W_y^2,W_z^2,0\}$ and obtain for the determinant

$$k^1k^2(\vec{n}^1 - \vec{n}^2)\cdot(\vec{W}^1\times\vec{W}^2) = k^1k^2(\vec{n}^1 - \vec{n}^2)\cdot\vec{s} \neq 0.$$

Thus a unique solution exists if components of unit characteristic normals along the prescribed outflow direction are not equal.

3. <u>Subsonic Inflow Boundary</u>.  In addition to specifying the direction of inflow as $\vec{s} = \vec{W}^1\times\vec{W}^2$, we add a third boundary condition (given pressure) written by means of Eq.(1) as $a = a(\vec{x},t)$.  Two rows of the matrix $k_j^i$ are replaced by $\{W_x^1,W_y^1,W_z^1,0\}$ and $\{W_x^2,W_y^2,W_z^2,0\}$, and the third row by $\{0,0,0,1\}$.  The uniqueness condition becomes

$$\vec{k}\cdot(\vec{W}^1\times\vec{W}^2) = \vec{k}\cdot\vec{s} = k\,\vec{n}\cdot\vec{s} \neq 0,$$

and the unit normal $\vec{n}$ should be chosen so that its component in the direction of inflow does not vanish.  We may add that, if one specifies instead all three components of inflow velocity, a unique solution exists unconditionally for any choice of $\vec{n}$.

The above examples illustrate the simplicity of checking the existence of unique solutions *locally* on arbitrary boundaries.  The Example 2 is at variance with Oliger & Sundström (1978) who studied *global* well-posedness of initial boundary value problems (with linear boundary conditions) defined as boundedness of perturbations in weighted $L^2$ norms.

## REFERENCES


COURANT, R. & HILBERT D.  1962 *Methods of Mathematical Physics. II.* Interscience.
KENTZER, C.P.  1971 *Lecture Notes in Physics.* 8, 108–113, Springer-Verlag.
KENTZER, C.P.  1978 *Arch. Mech (Arch. Mech. Stos.).* 31, 2, 295–299.
LIGHTHILL, M.J.  1965 *J. Inst. Maths Applics.* 1, 1, 1–28.
OLIGER, J. & SUNDSTRÖM, A.  1978 *SIAM J. Appl. Maths.* 35, 419–446.
RUSANOV, V.V.  1963 *Zh. Vych. Mat. i Mat. Fiz.* 3, 3, 508–527.

A CONJUGATE GRADIENT ITERATIVE METHOD

by

P.K. Khosla and S.G. Rubin
Department of Aerospace Engineering and Applied Mechanics
University of Cincinnati, Cincinnati, Ohio 45221

## 1.  INTRODUCTION

Finite-difference approximations to fluid flow equations lead to large, sparse sets of algebraic equations.  These can be solved either by direct   or iterative methods of solution.  Direct methods are generally based on Gaussian elimination. For large systems, these methods are computationally inefficient and suffer from round-off accumulation.  More efficient direct solvers due to Buneman,[1] Sweet and Swartztrauber,[2] Banks,[3] etc. are designed for Poisson equations and require considerable programming effort.

Iterative methods are usually simpler to program and are more versatile.  They are applicable to algebraic systems arising from arbitrary nonlinear elliptic or parabolic equations.  Point SOR, SLOR, ADI and strongly implicit, are the methods most commonly applied for this purpose and fall in the general category of factorization techniques.  The relative rate of convergence depends upon the amount of implicitness and the size of the matrix system.  Most of these relaxation methods suffer from one drawback; viz., the convergence rate drastically slows after the high frequency components of the iterative process have converged.  For large systems arising from two and three dimensional flow problems, this reduction in the convergence rate represents a considerable increase in the computational effort.

A variety of methods to improve the rate of convergence have been proposed in the literature.  When a dominant eigenvalue exists, over-relaxation or Chebyshev acceleration have been used to improve the rate of convergence of the iterative process.  Both methods require apriori knowledge of the largest and smallest eigenvalues of the matrix system.  Apart from a few simple problems this evaluation is difficult and in many cases impossible.  If the governing linear algebraic system arises from a nonlinear differential equation, the evaluation of the eigenvalues at each stage of global  iteration  can also represent a substantial increase in computational work.  Furthermore, these methods do not take into consideration the eigenvalue distribution of the coefficient matrix.

Another class of methods, which for some reason has received little consideration within the computational fluid mechanics community, comes under the category of quasi-Newton procedures.  For a detailed discussion, see reference 4.  Among the quasi-Newton methods only, the conjugate gradient technique has a finite-termination property.  In spite of this property, the conjugate gradient method does not perform favorably as a direct solver, but its application as an iterative procedure has recently been recognized by Reid.[5]  Meijerink and Vander Vorst[6] and Kershaw[7] have used this technique in conjunction with an incomplete Cholesky decomposition. Golub et al.[8] have used this method in conjunction with other factorization techniques.

The resulting procedures have performed quite favorably on model problems.

In the present paper conjugate gradient technque is used to accelerate the rate of convergence of iterative strongly implicit finite-difference methods. The resulting procedure is very fast and can easily be programmed. The present technique requires more storage than the more familiar iterative techniques, e.g., SOR, ADI, etc., but compares favorably with similar procedures based on incomplete Cholesky decomposition. Section 2 deals with the derivation of SIP/CG algorithm. In section 3, the procedure is extended to a coupled 2x2 system. Finally, section 3 deals with some applications to flow problems.

## 2. CONJUGATE-GRADIENT AND FACTORIZATION METHOD

In this section, we shall investigate a class of iterative procedures utilizing the desirable properties of the conjugate gradient method. Depending upon the nature of a given problem and the useable memory of a particular computer, any one of the familiar relaxation techniques can be applied. The rate of convergence will correspond to the amount of implicitness of the specific procedure. The solution of

$$AX = b \tag{1}$$

can be obtained by a variety of iterative techniques which are written as:

$$M e_{n+1} = AX_n - b = - r_n \ , \tag{2}$$

where n is the iteration index; $r_n$ is the residue and $e_{n+1} = x_{n+1} - x_n$ is the relative error. A is a positive definite matrix. The matrix M depends upon the choice of the relaxation procedure. The requirement of M is that its inverse be easily obtainable and can be related to the matrix A. For example, if M is a diagonal matrix with its elements those of matrix A, equation (2) corresponds to the point Jacobi method. If M consists of elements of A along a line, (2) corresponds to line relaxation. In a similar manner SOR, ADI, incomplete Choleski decomposition, etc. can be applied. M = A corresponds to a direct solver. The strongly implicit technique is similar to incomplete Choleski decomposition and for a sparsity pattern arising from five-point differencing can be viewed as an ADI procedure combined into a single step. Since $M^{-1}$ can easily be evaluated, (2) can be written as:

$$M^{-1} AX_n - M^{-1}b = - e_{n+1} \ . \tag{3}$$

In this form, (3) can be viewed as the solution of

$$(M^{-1}A) \ X_n = M^{-1}b \ , \tag{4}$$

with $e_{n+1}$ as the residue vector. The conjugate gradient can now be applied to the solution of (4). This leads to minimizing the Euclidean norm $e^T e$; in other words, the sum of the square of the relative error. If $C = M^{-1}A$, the following algorithm for updating the solution without explicitly evaluating C is (for details see reference 4) given as:

$$x_{i+1} = x_i + \alpha P_i \ , \qquad e_{i+1} = e_i - \alpha B_i \ ,$$

$$P_{i+1} = e_{i+1} + \beta P_i \ , \qquad B_{i+1} = C e_{i+1} + \beta B_i \ ,$$

where $\alpha$ and $\beta$ are given by

$$\alpha = \frac{p_i^T e_i}{p_i^T B_i} \qquad \text{and} \qquad \beta = - \frac{(e_{i+1}^T B_i)}{(p_i^T B_i)} \tag{5}$$

In this algorithm, the preconditioning of A with $M^{-1}$ has many desirable features. In particular, the eigenvalues are redistributed favorably for the application of the conjugate gradient method. It is this preconditioning with the strongly implicit method that is responsible for the significant increase in the rate of convergence.

## 3.  EXTENSION TO COUPLED 2x2 SYSTEM

In reference 9, the present authors extended the strongly implicit procedure to a coupled 2x2 system. The coupling allows the use of arbitrary time step $(\Delta t \sim 10^6)$ and a significant increase in the rate of convergence. In this section, the coupled 2x2 system is combined with CG procedure to accelerate the rate of convergence. The central difference or a higher-order collocation procedure leads to a system of equations of the following form:

$$\begin{bmatrix} A_1 & A_2 \\ A_3 & A_4 \end{bmatrix} \begin{bmatrix} \omega \\ \psi \end{bmatrix} = \begin{bmatrix} C_1 \\ G_2 \end{bmatrix}$$

where $A_1$, $A_2$, $A_3$, $A_4$ are NxN matrices and $G_1$ and $G_2$ are N-dimensional vectors. The strongly implicit procedure written in the form of relative error is

$$\begin{bmatrix} C_1 & C_2 \\ C_3 & C_4 \end{bmatrix} \begin{bmatrix} \omega^{n+1} - \omega^n \\ \psi^{n+1} - \omega^n \end{bmatrix} = \begin{bmatrix} A_1 & A_2 \\ A_3 & A_4 \end{bmatrix} \begin{bmatrix} \omega^n \\ \psi^n \end{bmatrix} - \begin{bmatrix} G_1 \\ G_2 \end{bmatrix} = - \begin{bmatrix} R_1 \\ R_2 \end{bmatrix}_n$$

where $C_1$, $C_2$, $C_3$ and $C_4$ are appropriate pre-conditioning matrices arising from the application of strongly implicit method and $R_1$ and $R_2$ are the residue vectors. As explained earlier, the superscipt n refers to the iterations. The SIP solution can be written as:

$$\begin{bmatrix} e_1 \\ e_2 \end{bmatrix}_{n+1} = - \begin{bmatrix} C_1 & C_2 \\ C_3 & C_4 \end{bmatrix}^{-1} \begin{bmatrix} R_1 \\ R_2 \end{bmatrix}_n$$

In the converged form $e_1$ and $e_2$ become zero. The application of confugate gradient amounts to minimizing the Euclidean norm $\begin{bmatrix} e_1 \\ e_2 \end{bmatrix} [e_1 \ e_2]$. A variety of updates of the solution vector can be prescribed. In the present paper, a particularly simple form of the update is investigated.

$$\begin{bmatrix} \omega \\ \psi \end{bmatrix}_{n+1} = \begin{bmatrix} \omega \\ \psi \end{bmatrix}_n + \alpha \begin{bmatrix} P_1 \\ P_2 \end{bmatrix}$$

This update is same as given by equations (4).

## RESULTS

The preconditioned strongly implicit algorithms derived in the previous sections have been applied to solve a variety of model two and three dimensional boundary value problems arising in subsonic inviscid and viscous flows.  It may be pointed out that operationally, the present type of methods are of $O(N^{1.25})$.  The questions regarding computational complexity of preconditioned conjugate gradient techniques has been discussed in detail by Axelsson and Gustafsson[10] and Wesseling and Sonnevld.[11]

(a)  <u>Laplace Equation</u>:  Solutions of Laplace equations in two and three dimensions have been obtained.  A comparison of the rate of convergence of SOR, LSOR, explicit LU decomposition and SIP are compared with the method of the present paper in Fig. 1.  The conventional procedures have been run for only 500 iterations on a 65x33 grid.  SIP is the only one of these which had converged to $10^{-5}$ with this number of iterations.  The rate of convergence of the conjugate gradient SIP procedure is extremely fast, i.e., an error of $10^{-5}$ can be achieved after only 25 iterations.

(b)  <u>Subsonic Potential Flow</u>:  Figures 2 depict the rate of convergence of subcritical flow computation in two and three dimensions.  In all the cases considered the algorithm for nonsymmetric coefficient matrix is utilized.  Figure 2a shows the $L_2$ error for subcritical potential flow over a 6% thick airfoil using transonic small disturbance theory.  Figure 2b depicts the average residue per point for the full potential flow equations.  Quasilinearization is used to treat the density implicitly; although, it does not improve the overall rate of convergence considerably.  Figure 2c shows the $L_2$ error for a three-dimensional rectangular wing with an aspect ratio of 2 and a 6% circular arc section.  For this case the density is treated explicitly.

(c)  <u>Incompressible Viscous Flow Over a Circular Cylinder (Re = 100)</u>:  The numerical solution for a 65x24 grid using coupled SIP/CG procedure has been obtained.  First 20 iterations are carried out with upwind differencing and a smaller $\Delta t = .1$, after which $\Delta t$ is increased to $10^6$ and the solution is corrected by K-R scheme to achieve second order accuracy. Solution converges in about another 20 iterations (Fig. 3).  The present plots of residue or error versus iterations might be somewhat misleading since each iteration contains certain inner iterations of the conjugate gradients correction.  A systematic study to determine the optimum number of inner iterations has not been considered, however, one to two passes are generally sufficient.  During the inner iterations, the non-linearities are also updated.  Therefore, the CG/SIP involves approximately 1.5 times the work of the SIP iteration.  In terms of computer time for the examples shown here this leads to a six-fold reduction for the CG/SIP procedure.  With approximately 1500 grid points this factor is surprisingly close to that resulting from the theoretical rate of $N^{1.5}/N^{1.25}$ corresponding to $\dfrac{(SOR)_{\text{optimum}}}{CG/SIP}$.

ACKNOWLEDGEMENT:  This research was sponsored under Grant No. AFOSR 80-0047.

REFERENCES

1.  Buneman, O., (1969), SUIPR Report No. 294, Stanford Univ., California.
2.  Schwarztrauber, P.N. and Sweet, R.A., (1973), SIAM J. Numer. Anal., Vol. 10, pp. 900-907.
3.  Bank, R.E., (1977), SIAM J. Numer. Anal., Vol. 5, pp. 950-970.
4.  Khosla, P.K. and Rubin, S.G., (1980), Computers and Fluids, to appear.
5.  Reid, J.K., (1971), Academic Press, pp. 231-254.
6.  Meijernik, J.A. and Vander Vorst, (1977), Math. Comp., Vol. 31, pp. 148-162.

7.  Kershaw, D., (1978), J. of Computational Physics, Vol. 26, pp. 43-65.
8.  Concus, P., Golub, G.H. and O'Leery, D.P., (1978), Computing, Vol. 19, pp. 321-339.
9.  Rubin, S.G. and Khosla, P.K., (1980), Computers and Fluids, to appear; also AIAA Preprint 79-0011 (1979).
10. Axelsson, O. and Gustafsson, I., (1979), J. Inst. Math. and Appl., Vol. 23, pp. 321-338.
11. Wesseling, P. and Sonnevld, P., (1980), Approximation Methods for Navier-Stokes Problems, Spring Verlag Publication 771, pp. 469-485.

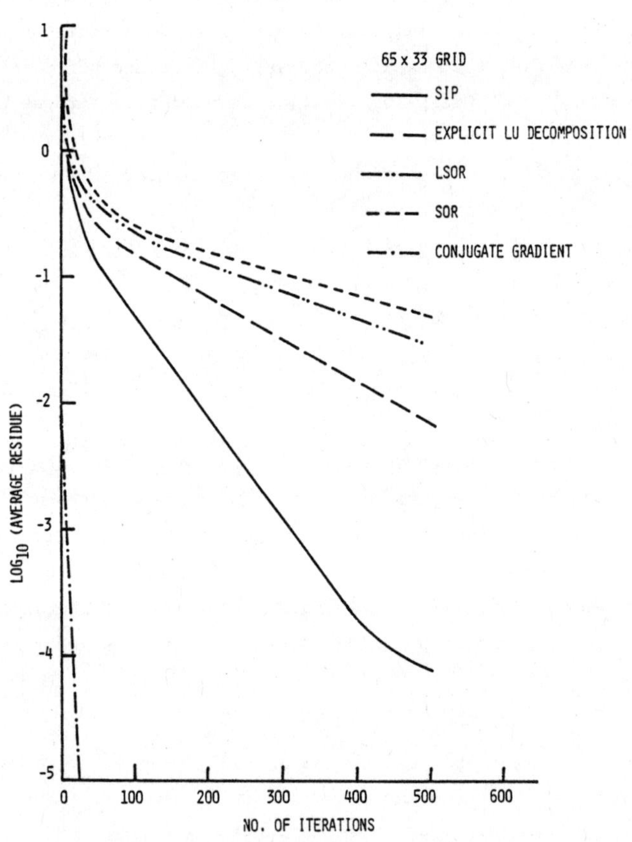


FIG 1. SOLUTION OF LAPLACE EQUATION

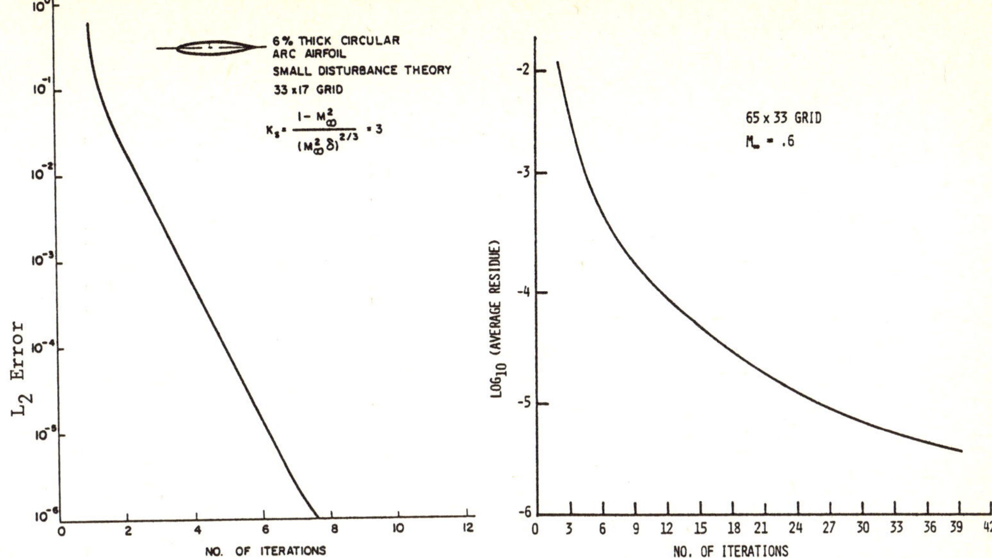



FIG. 2A. CONJUGATE GRADIENT-STRONGLY IMPLICIT SOLUTION

FIG 2B. 2-D POTENTIAL FLOW OVER A 6% CIRCULAR ARC AIRFOIL

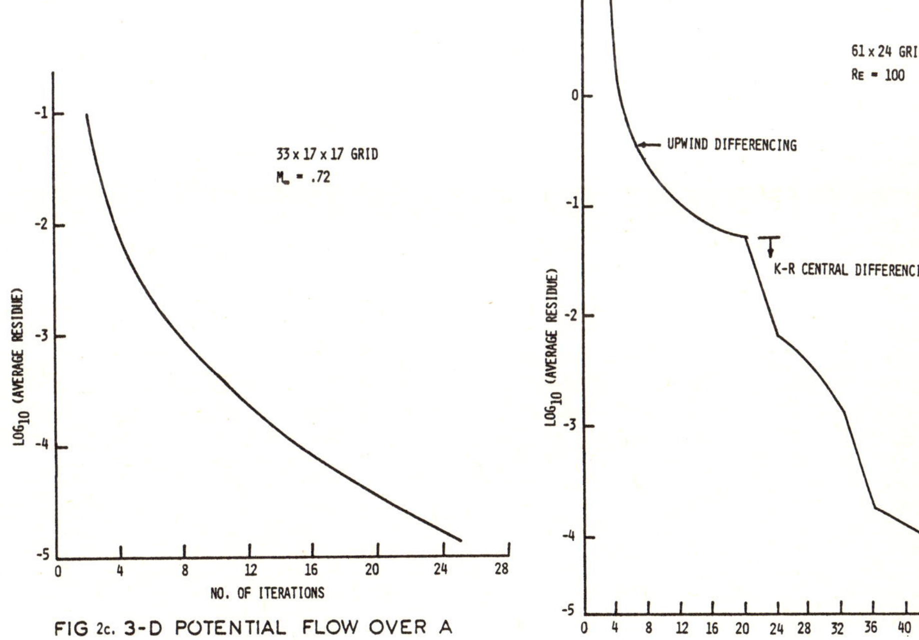



FIG 2c. 3-D POTENTIAL FLOW OVER A RECTANGULAR WING

FIG 3. INCOMPRESSIBLE FLOW OVER A CIRCULAR CYLINDER

A NUMERICAL METHOD FOR THE WAVE RESISTANCE
OF A MOVING PRESSURE DISTRIBUTION ON THE FREE SURFACE

C. Korving
Delft University of Technology
Department of Mathematics
The Netherlands

## 1. INTRODUCTION

This paper presents numerical solutions with respect to the free-surface deformation
caused by a pressure distribution which moves in a straight course over the surface
of a body of water. This deformation is, in particular, of interest in the vicinity
of the pressure distribution in view of the calculation of the wave resistance. The
moving pressure working on an elliptical area of the free surface can be considered
as representing a moving ship. If the problem is described in a reference frame
fixed to the pressure distribution, a mathematical model can be formulated in terms
of the Laplace equation, subject to non-linear free-surface conditions. The
numerical method for the two-dimensional problem has been described in Ref. [1].
The method of solution is based on an iteration procedure alternately using the
solution of the Laplace problem with fixed boundaries and the determination of an
improved position of the dynamic free surface condition.
Contrary to the approach in curvilinear coordinated in Ref. [1], the procedure to
improve the free-surface elevation presented in this paper, will be treated in
rectilinear coordinates.
Numerical solutions are presented obtained for pressure distributions of the form
shown in Figure 1. The results were found for a range of values of length-Froude
number $F = U_0/\sqrt{gl}$ and pressure coefficient $c_p = 100\ p/\rho gl$. In particular, we examine
the variation of the non-dimensional wave resistance $R\ 10^3/\rho gl^2 b$, as a function of $F$
for various values of the width/length ratio $b/l$ (Fig. 1).
The results for $b/l$ approaches to infinity are compared to the two-dimensional
results obtained by C. von Kerczek and N. Salvesen (Ref. [2]).

## 2. FORMULATION OF THE PROBLEM

In the system of coordinates fixed to the uniformly translating pressure
distribution, shown in Figure 1, the fluid far upstream, as $x \to -\infty$, is moving with
constant velocity $U_0$.

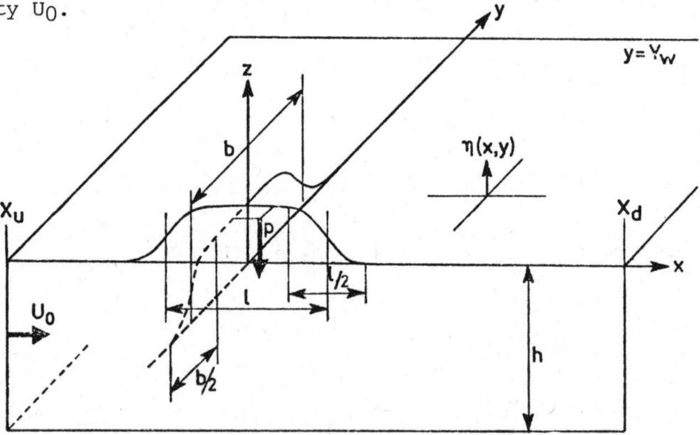


Figure 1

The origin of the coordinate system is the centre point of the pressure distribution
on the undisturbed free surface and the bottom of the stream is $z = -h$. The problem
is formulated as an incompressible potential flow in terms of the potential function
$\Phi(x,y,z)$, which satisfies the Laplace equation:

$$\Delta \Phi = 0 \tag{2.1}$$

and the boundary conditions:
On the free surface $z = \eta(x,y)$, the dynamic condition is:

$$\tfrac{1}{2}\{\Phi_x^2 + \Phi_y^2 + \Phi_z^2\} + g\eta + \frac{p(x,y)}{\rho} = \tfrac{1}{2} U_0^2 \tag{2.2}$$

and the kinematic condition is:

$$- \Phi_z + \Phi_x \eta_x + \Phi_y \eta_y = 0 \tag{2.3}$$

Far upstream the flow is uniform and far downstream a proper radiation condition is fulfilled. By applying artificial damping of waves in a small range near the downstream station $x = x_d$, the free surface remains undisturbed during the calculation at $x = x_d$, with the result that the condition of uniform flow can be supposed. Reflection of waves at fixed walls is taken at the boundary plane $y = Y_w$, which implies that $\eta_y = 0$ and $\Phi_y = 0$.
The same conditions hold for the symmetry plane $y = 0$.


3. COMPUTATION OF THE FREE SURFACE ELEVATION

In this chapter  we go into detail of the method to improve the free surface elevation after the solution of the Laplace problem:

$$\Delta \psi = 0 \tag{3.1}$$

subject to the kinematic condition

$$- \psi_z + \psi_x \eta_{0x} + \psi_y \eta_{0y} = 0 \tag{3.2}$$

at the free surface $z = \eta_0(x,y)$ and for known values of the normal derivative at the remaining boundaries of the flow domain: $- x_u < x < x_D$ , $0 < y < Y_w$ ,
$- h < z < \eta_0(x,y)$ (Fig. 1).
Since the dynamic free-surface condition cannot be fulfilled at the same time, a correction has to be performed with respect to the potential and the free surface.
So, we introduce:
$\phi = \Phi - \psi$ and
$\xi = \eta - \eta_0$.
These variables are substituted into the equations (2.1), (2.2), and (2.3), leading to:

$$\Delta \phi = 0 \tag{3.3}$$

and the linearized free-surface conditions at $z = \eta_0(x,y)$:

$$g\xi + A\phi_x + B\phi_y + C\xi_x + D\xi_y = \Omega \tag{3.4}$$

$$- \phi_z + E\phi_x + F\phi_y + G\xi_x + H\xi_y = 0 \tag{3.5}$$

where:
$$
\begin{aligned}
A &= G + E\psi_z & G &= \psi_x \\
B &= H + F\psi_z & H &= \psi_y & \quad \psi_z &= GE + HF \\
C &= G\psi_z & E &= \eta_{0x} \\
D &= H\psi_z & F &= \eta_{0y}
\end{aligned}
$$

$$\Omega = \tfrac{1}{2} U_0^2 - p/\rho - g\eta_0 - \tfrac{1}{2}(\psi_x^2 + \psi_y^2 + \psi_z^2) .$$

If the variables are seperated according to $\phi = R(x,y) * Z(z)$, the solution of the Laplace equation (3.3) in a small domain $[\Delta x, \Delta y]$ can be approximated by the expression:

$$\phi(x,y,z) = (A_0 + A_1 x + B_1 y + A_2 x^2 + B_2 y^2 + C_1 xy + \ldots)\ \cosh \alpha(z + h) \qquad (3.6)$$

Similarly, the unknown variable $\xi$ can be expressed by the polynomial:

$$\xi(x,y) = a_0 + a_1 x + b_1 y + a_2 x^2 + b_2 y^2 + c_1 xy + \ldots \qquad (3.7)$$

The expressions (3.6) and (3.7) are substituted into the equations (3.3), (3.4), and (3.5) and terms of equal power in x and y are collected. In this way, relations between the coefficients are obtained when the equations are satisfied for the various terms of equal power in x and y. Then, fifteen equations are found in twenty unknowns and an unknown value of $\alpha$.
The coefficients, which occur in the approximation for $\phi$ (3.6) are eliminated, resulting in the relations:

$$\left. \begin{aligned} a_2 + b_2 &= \frac{\alpha^2}{2}\left(\frac{\Omega}{g} - a_0\right) \\ E^* a_1 + F^* b_1 + a_2 + H^* c_1 + (H^*)^2 b_2 &= \beta g^*(1 - E^2 - F^2)\left(\frac{\Omega}{g} - a_0\right) \end{aligned} \right\} \qquad (3.8)$$

where: $E^* = -\,Eg^* + \beta(E + FH^*)$
$\qquad\quad F^* = -\,Fg^* + \beta H^*(E + FH^*)$
$\qquad\quad g^* = g/2G^2$ , $H^* = H/G$ , $\beta = \alpha \tanh \alpha h$
Trying to solve the free-surface problem, formulated in chapter 2, the first step will be the computation of the free-surface elevation from an undisturbed level. The solution of the Neumann problem simply takes the form:
$\psi = U_0 x$
and the coefficients in the equations (3.4) and (3.5), not equal to zero, are:
$A = G = U_0$ , $\Omega = -\,(p/\rho)$.
The equations (3.8) are then simplified to:

$$a_2 + b_2' = \frac{\alpha^2}{2}\left(\frac{\Omega}{g} - a_0\right) \quad , \quad a_2 = \alpha g^*\left(\frac{\Omega}{g} - a_0\right) \qquad (3.9)$$

if the case is considered that $\tanh \alpha h \simeq 1$.
Because of the wave-making action of the pressure distribution, we can imagine that a wave locally can be described by $\xi = A \cos \gamma(x \cos \theta + y \sin \theta)$, where $\theta$ represents the angle between the direction of the wave and the x-axis.
It follows that $\xi_{xx} = \xi_{yy} \times \mathrm{tg}^2\, \theta$ and the significance of the ratio $c = b_2/a_2$ of two coefficients of the polynomial (3.7) is clear, for $c = \mathrm{tg}^2\, \theta$. Substitution of the relation $b_2 = ca_2$ into the equations (3.9) leads to the local dispersion relation:

$$\alpha = 2\, g^*(1 + c) \qquad (3.10)$$

In the two-dimensional case $c = 0$ and the relation between the coefficients of the polynomial (3.7) is reduced to:

$$a_2 = \tfrac{1}{2}\, \alpha^2\left(\frac{\Omega}{g} - a_0\right)\ \text{with}\ \alpha = \frac{g}{U_0^2} \qquad (3.11)$$

In the three-dimensional version, the variation of $\xi$ in the cross direction y has to be taken into account. Figure 2 shows the approximation in one element $[\Delta x, \Delta y]$, represented by one polynomial in x and one polynomial in y.
Using a spline approximation in y direction, which implies that continuity of the function is required up to and including order 2 at the boundaries $Y = Y_j$ of the n elements, the relationship between the coefficients becomes:

$$\frac{a_0^{j+1} - 2\,a_0^j + a_0^{j-1}}{(\Delta y)^2} = \frac{b_2^{j-1} + 22\,b_2^j + b_2^{j+1}}{12} \quad , \quad j = 1, n \tag{3.12}$$

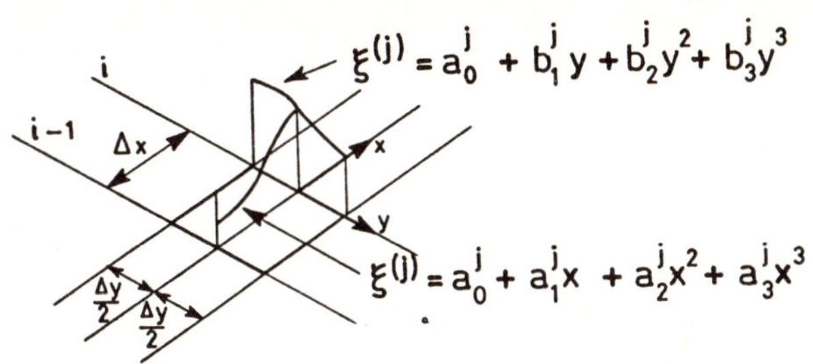



Figure 2. The approximation of $\xi$ in one element: $-\Delta x < x < 0$ and $-\dfrac{\Delta y}{2} < y < \dfrac{\Delta y}{2}$ or $X_{i-1} < X < X_i$ and $Y_j < Y < Y_{j+1}$ , $j = 1(1)n$.

The boundary conditions at $Y = Y_1 = 0$ and $Y = Y_{n+1} = Y_w$ ($\xi_y = 0$) have been arranged under this system of equations involving that:
$a_0^0 = a_0^1$ , $b_2^0 = b_2^1$ , $a_0^{n+1} = a_0^n$ and $b_2^{n+1} = b_2^n$.
If above system of equations is combined with the equations (3.9) and (3.10), written down in the following form:

$$\left. \begin{aligned} b_2^j &= c^j\, a_2^j \\ a_2^j &= \alpha^j\, g_j^* \left(\frac{\Omega}{g} - a_0^j\right) \\ \alpha^j &= 2\, g_j^* \left(1 + c^j\right) \end{aligned} \right\} \quad j = 1, n \tag{3.13}$$

an implicit relation between the coefficients $a_0^j$ and $a_2^j$ can be derived.
These equations change into the explicit relation (3.11) for the two-dimensional case ($c^j \equiv 0$). The equations (3.13) are non-linear, in this form inefficient to obtain the solution. Therefore, the equations are linearized, taking account for the significance of the quantity $c^j$ ($> 0$). If the wave pattern is globally known, f.i. from preceding calculations, $c^j$ can be estimated, writing $c^j = c_0^j + \Delta^j$.
After substitution, linearization and elimination of the variables $a_2^j$ , $\alpha^j$ and $b_2^j$ we arrive at the system of equations of the form:

$$D_j^i\, \varepsilon^j = E^i + F_j^i\, a_0^j \quad i, j = 1, n \tag{3.14}$$

where $\varepsilon^j = 2(g_j^*)^2\,(\Omega^j/g - a_0^j)\,\Delta^j$.
Together with the relation

$$a_2^j = 2(g_j^*)^2 \left\{\frac{\Omega^j}{g} - a_0^j\right\} (1 + c_0^j) + \varepsilon^j \quad j = 1, n \tag{3.15}$$

we have obtained a result comparable to the two-dimensional result of (3.11), but extended to three-dimensions.
In order to solve the equations (3.14) and (3.15), the same holds for the solution of equation (3.11), matching conditions have to be prescribed at the boundary $x = -\Delta x$ of the elements. Sufficient conditions are available if continuity of the function $\xi$ is required up to and including order two at $x = -\Delta x$ and $y = 0$ of each

element (Fig. 2).
In this way, going on in downstream direction, the entire free surface elevation can
be computed, dependent on the deviation Q of the dynamic free-surface condition.
The computation is continued with the solution of the Neumann problem and above
procedure can be started again until a desired accuracy is achieved. The speed of
convergence can be increased if the various terms in eq. (3.8), which could be
omitted in the first step, are taken into account.


## 4. RESULTS AND CONCLUSIONS

Numerical results are presented for pressure distributions of width/length ratio
$b/l = 0.5$ and $b/l = \infty$, and fixed values of the pressure coefficient $c_p = 1.2$ and
length/depth ratio $l/h = 2$.
Figure 3 shows free-surface elevations for three values of the length-Froude number
F. As far as the non-dimensional wave-resistance RD is concerned, these results are
compared with the two-dimensional results obtained by von Kerczek and Salvesen
(Figure 4). In their paper (Ref. [2]), a second and third order perturbation method
is used in order to describe the nonlinear effects more accurately. The conclusion
can be drawn that the results agree well with those obtained by the numerical method
presented in this paper.
Artificial damping of waves is applied to a small region upstream of the downstream
boundary plane, where, as a result of the undisturbed position of the free surface,
the condition of uniform flow can be imposed. If this region is chosen more than one
wave-length  away from the rear position of the pressure distribution, the influence
on the deformation of the free surface in the vicinity of the pressure distribution
is neglegible.
Figure 4 shows that the minimal wave resistance which is equal to zero in the two-
dimensional case at $F = 0.385$, is shifted to a higher value of Froude number
($\simeq 0.42$) for $b/l = 0.5$. Besides this, the wave-resistance per unit of width is
condiserably decreased for $F > 0.42$.
Figure 5 shows the effect of the position of the fixed wall being parallel to the
main direction of flow. Because of the reflection condition $\eta_y = 0$, the free surface
contours in the right-hand picture are perpendicular to the wall in contrast to the
contours in the picture on the left, where the calculations were performed for a
region twice the width of the picture. The more the wall is removed from the
symmetry plane, the lower the wave-resistance.
At the end, figure 6 shows the wave pattern when at minimal wave-resistance. In
contrast to the transverse waves, dominant in figure 5, the diverging waves are
clearly recognizable in figure 6.
The calculations were carried out on an IBM 370/158 and the computation time
required for eight iterations amounted to four minutes for a mesh of 10 points in y-
and 50 points in x-direction.
Runs of ten minutes were required for obtaining the results of fig. 5a and 6 with
18 points in y-direction.



REFERENCES

[1] C. Korving, "The Wave Resistance for Flow Problems with a Free Surface", Second
    International Conference on Numerical Ship Hydrodynamics, University of
    California, Berkeley, pp. 285-291, (1977).
[2] C. von Kerczek and Nils Salvesen, "Nonlinear Free-Surface Effects. The
    Dependence on Froude Number", Second International Conference on Numerical Ship
    Hydrodynamics, University of California, Berkeley, pp. 292-300, (1977).
[3] T. Inui and H. Kajitani, "A Study on Local Non-Linear Free Surface Effects in
    Ship Waves and Wave Resistance", 25th Anniversary Colloquium of the Institut für
    Schiffbau, Hamburg, June 1977.

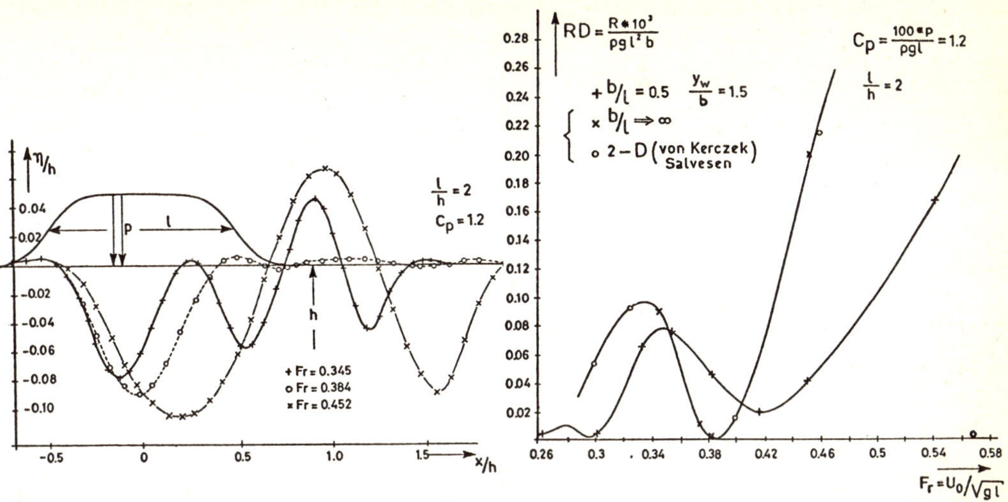


Figure 3. Free Surface Elevations for Various Froude Numbers and b/l → ∞.

Figure 4. Nondimensional Wave Resistance RD versus Froude Number F.

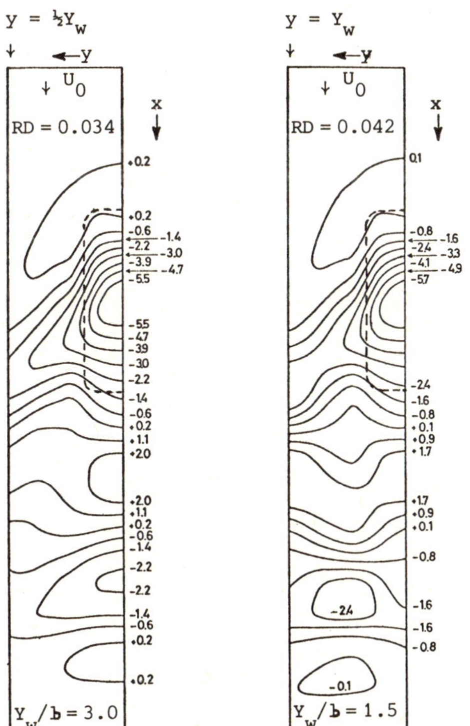


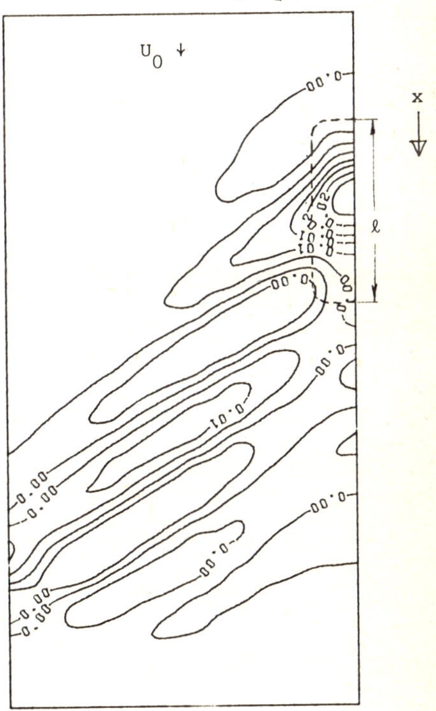


Figure 5. Free Surface Contours η/h * 100% for F = 0.452, l/h = 2, b/l = 0.5, c_p = 1.2.

Figure 6. Free Surface Contours for F = 0.417, b/l = 0.5, l/h = 1, Y_w/l = 2, c_p = 1,2.

NUMERICAL METHODS FOR SOLVING SOME FLUID
MECHANICS PROBLEMS MET IN A STRONGLY ROTATING GAS CENTRIFUGE


JP.LA HARGUE    COMPAGNIE INTERNATIONALE de SERVICE en INFORMATIQUE (C I S I)
                35 Boulevard Brune, 75683 PARIS CEDEX.14 (FRANCE)

SOUBBARAMAYER    CENTRE d'ETUDES NUCLEAIRES de SACLAY (CEN/SACLAY)
                 BP.n°2, 91190 GIF-s/-YVETTE (FRANCE)


The communication is presented in the form of a <u>computer-animation movie</u>, made from
the output data of three numerical models. These models have been built to investigate in a gas centrifuge respectively the flow, the species separation and the
optimization of the flow field. The figures heredown have been extracted from the
movie.


## I. - INTRODUCTION.

A centrifuge (Fig.1) is a cylinder of radius  a  and height  h, closed at both
ends by cover plates and rotating very rapidly about its axis of symmetry with
a constant angular velocity  $\omega$. This is a device used in the industrial process
of uranium enrichment by the ultracentrifugation method (SOUBBARAMAYER 1979).
The cylinder contains gaseous uranium hexafluoride  $(UF_6)$  which is in fact a
binary mixture of two isotopic species : $^{235}UF_6$  (molar mass  $M_1$ = 349 g)  and
$^{238}UF_6$  (molar mass  $M_2$ = 352 g). Under the action of the very strong centrifugal acceleration (practical order of magnitude : 100,000  times gravity) the
heavier species  $^{238}UF_6$  moves towards the periphery and leaves the inner core
of gas enriched in the lighter species  $^{235}UF_6$, producing a partial separation
of the two isotopes in the radial direction. The gas rotates as a rigid body
(wheel flow) with the same angular velocity as the cylinder.

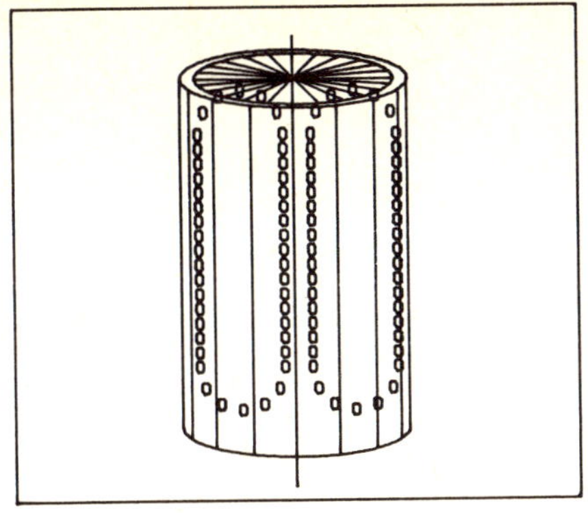

Figure 1.- A gas centrifuge

Figure 2 - Finite element mesh

In practical devices, a very small axial circulation of the gas is superimposed on the wheel flow. Such axial movement, termed in the literature by
"Countercurrent flow", changes the direction of separation from radial to
axial, increases considerably the magnitude of enrichment and permits removal
of enriched and depleted streams from the top and the bottom of the centrifuge. The countercurrent flow can be generated by imposing proper temperature
distribution (thermal drive) on the solid boundaries of the cylinder (if these boundaries are perfectly conducting) and by the presence of a stationary
object (scoop) in the stream of the gas at one of the centrifuge (friction
drive). The theoretical analysis of a countercurrent gas centrifuge includes
three parts : hydrodynamics, separative analysis and optimization of the
flow field. A numerical model has been built up to solve each of these problems and their main features are summarized in the next section.

## II. - EQUATIONS and NUMERICAL METHODS of SOLUTION.

### II.A - Flow field analysis.

The gas flow in a centrifuge is governed by the conservation laws of
continuum fluid mechanics (mass, momentum and energy) and the thermodynamic gas state law. In cylindrical coordinates with axial symmetry and
at steady state the set of six equations is written as follows :

$$\frac{\partial}{\partial z}(\varrho V_z) + \frac{1}{r}\frac{\partial}{\partial r}(\varrho r V_r) = 0,$$

$$\varrho\left\{V_z\frac{\partial V_z}{\partial z} + V_r\frac{\partial V_z}{\partial r}\right\} = -\frac{\partial p}{\partial z} + \mu\left(\nabla^2 V_z + \frac{1}{3}\frac{\partial}{\partial z}\operatorname{div}V\right),$$

$$\varrho\left(V_z\frac{\partial V_r}{\partial z} + V_r\frac{\partial V_r}{\partial r} - \frac{V_\theta^2}{r}\right) = -\frac{\partial p}{\partial r} + \mu\left[\left(\nabla^2 - \frac{1}{r^2}\right)V_r + \frac{1}{3}\frac{\partial}{\partial r}\operatorname{div}V\right],$$

$$\varrho\left(V_z\frac{\partial V_\theta}{\partial z} + V_r\frac{\partial V_\theta}{\partial r} + \frac{V_r V_\theta}{r}\right) = \mu\left(\nabla^2 - \frac{1}{r^2}\right)V_\theta,$$

$$\varrho C_v\left(V_z\frac{\partial T}{\partial z} + V_r\frac{\partial T}{\partial r}\right) + p\operatorname{div}V = \kappa\nabla^2 T + \phi_{\text{visc}},$$

$$\phi_{\text{visc}} = 2\mu\left\{\left(\frac{\partial V_z}{\partial z}\right)^2 + \left(\frac{\partial V_r}{\partial r}\right)^2 + \frac{V_r^2}{r^2}\right.$$

$$\left. + \frac{1}{2}\left[\left(\frac{\partial V_\theta}{\partial z}\right)^2 + \left(\frac{\partial V_r}{\partial z} + \frac{\partial V_z}{\partial r}\right)^2 + \left(\frac{\partial V_\theta}{\partial r} - \frac{V_\theta}{r}\right)^2\right] - \frac{1}{3}(\operatorname{div}V)^2\right\},$$

$$\operatorname{div}V = \frac{\partial V_z}{\partial z} + \frac{1}{r}\frac{\partial}{\partial r}(rV_r), \qquad \nabla^2 = \frac{\partial^2}{\partial z^2} + \frac{\partial^2}{\partial r^2} + \frac{1}{r}\frac{\partial}{\partial r}, \qquad p = \frac{R}{M}\varrho T.$$

(A)

In the above system (z, r) are the axial and radial coordinates, ($V_z$, $V_r$, $V_\theta$) the components of the velocity in the axial radial and azimuthal directions, (p, $\rho$, T) the thermodynamic variables (pressure, mass density and temperature) ; the viscosity $\mu$, the thermal conductivity $\kappa$ and the specific heat at constant volume $C_v$ are assumed to be constant. The last equation is the perfect gas law.

The system was solved numerically by JP.LA HARGUE and SOUBBARAMAYER (1978) by using <u>finite element method</u> and <u>direct resolution</u>. The meshing in quadrangular elements (Figure 2) is carried out by taking into account the boundary layers generated by rotating fluids (Stewartson layers and Ekman layers) and the very strong radial stratification of the gas density in the centrifuge. In usual cases the countercurrent flow is a small perturbation of the wheel flow : the linear approximation of (A) is sufficient for most of practical applications. The non-linear solutions are investigated, starting from the linear run and using Newton algorithm.

## II.B - <u>Diffusion - convection equation.</u>

The isotope separation is governed by the diffusion - convection equation :

$$\frac{1}{r}\frac{\partial}{\partial r}(r\Phi_r) + \frac{\partial \Phi_z}{\partial Z} = 0$$

$$\Phi_r = -\rho D\left(\frac{\partial N}{\partial r} + 2\Delta A\, r\, N\right) + \rho V_r N, \qquad \Phi_z = -\rho D\frac{\partial N}{\partial Z} + \rho V_z N \tag{B}$$

$$\Delta A = (M_2 - M_1)\,\omega/2\,RT_o$$

In (B), N  is the light isotope mole fraction and  D  the diffusion coefficient. The diffusion equation has been numerically solved by JP.LA HARGUE (1979) using the same mesh and method as for flow equations, i.e. finite elements and direct resolution. Some difficulties were to be overcome :

1)  The large aspect ratio of some cells leads to an ill- conditioned matrix. This matrix-conditioning problem has been solved by setting up a linear combination of equations on every mesh line, simulating the analytical averaging method.

2)  With a slow countercurrent, the separative power is a small second-order difference of inlet/outlet concentrations, which must be evaluated with some care.

3)  With a quick countercurrent, the centered scheme breaks down, as one knows. Fortunately the averaging method (1) repells the break point beyond usual conditions. So no upwind scheme has been used.

II.C - <u>Optimization of the flow field.</u>

The separative performance of a centrifuge is measured by a basic quantity termed separative power and expressed by :

$$\delta.U = F\theta \ V(N_p) + F(1-\theta) \ V(N_w) - F \ V(N_F)$$

$$\text{(C)}$$

$$V(N) = (2N - 1) \ln(N/1 - N)$$

In (C), $N_p$, $N_w$ and $N_F$ are respectively the light isotope mole fractions of the outlet streams $F\theta$ (enriched), $F(1 - \theta)$ (depleted) and of the inlet stream F (feed). The two quantities $N_p$ and $N_w$, derived from the solution of (B) are strongly dependent upon the flow field, which in its turn depends upon the boundary conditions generating the flow. A numerical method is developed to determine these boundary conditions in order to optimize the performance function (C). By expanding the boundary functions in a finite base the problem is reduced to minimizing a function with respect to a finite number of variables and solved by SOUBBARAMAYER and BILLET(1980) using the simplex method.

## III. - <u>CONCLUSION</u>.

The movie used for the communication is about 15' long. After an introduction to the separation mechanism in a gas centrifuge, it shows some results of numerical computations : flow fields, separation fields, optimization curves, summarizing the various problems of numerical computation in a gas centrifuge.


<u>REFERENCES</u> :

SOUBBARAMAYER, 1979

> Centrifugation, in : S.VILLANI (Ed.), Uranium Enrichment, Topics in Applied Physics Vol.35 (Springer Verlag, Heidelberg) pp.183-244

SOUBBARAMAYER, 1978

> A numerical model for the investigation of the flow and isotope concentration field in an ultracentrifuge
> Comput. Methods Appl. Mech. Eng., <u>15</u> - pp.259-273

JP.LA HARGUE, 1979

> Equation de diffusion en mailles disproportionnées. Rapport Interne CISILOG/79-031, CISI 35 Boulevard Brune, 75689 PARIS-CEDEX.14 (FRANCE)

SOUBBARAMAYER-J.BILLET, 1980

> A numerical method for optimizing the gas flow in a centrifuge. To be published in : Comput. Methods Appl. Mech. Eng.

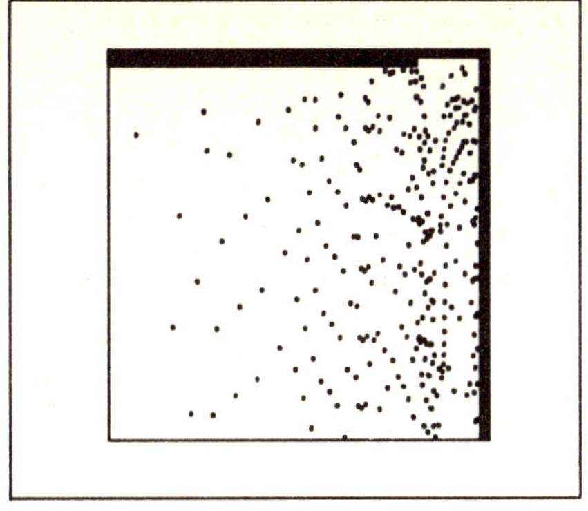

Figure 3 - Animated particles inside
the flow

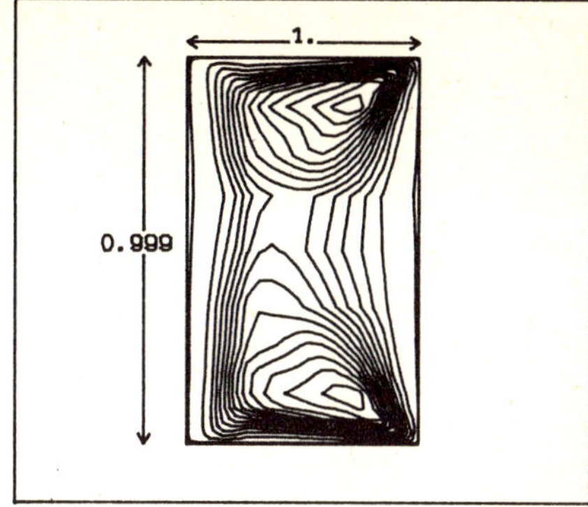



Figure 4 - An example of streamlines

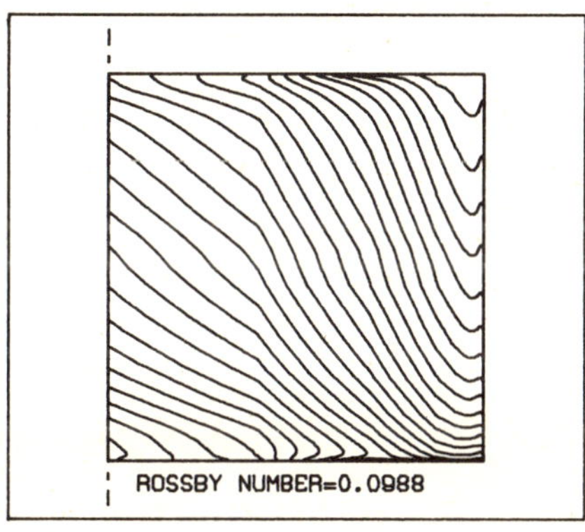



Figure 5 - Separation isolines

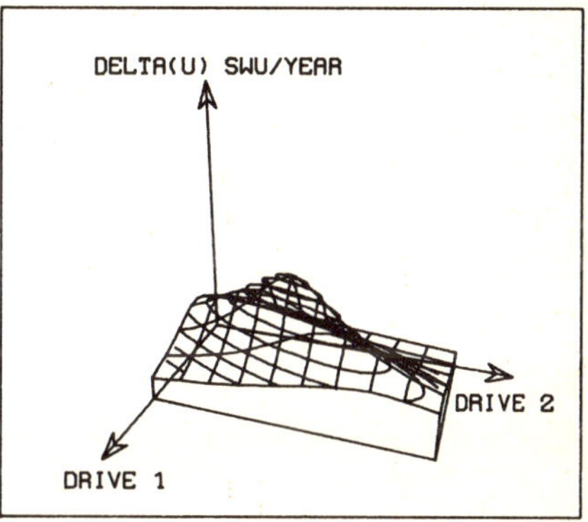



Figure 6 - Optimization

K. D. Lee and P. E. Rubbert

Aerodynamics Research Unit
Boeing Commercial Airplane Company
Seattle, Washington


## Introduction

The objective of the present work is to develop a methodology for solving the transonic full potential equation in geometrically complex domains using surface-fitted grids having acceptable density and spacing characteristics throughout the flow domain.  The present method uses a multiple block structure in which the computational (and physical) space is divided into multiple rectangular blocks, accompanied by suitable transonic algorithms that correctly solve the differential equation across block boundaries.

## Grid Generation

The approach divides the computational domain into multiple rectangular blocks, which can be defined arbitrarily to produce surface-fitted grids whose structure follows the natural lines of the configuration.  Fig. 1 shows a typical block structured grid about a wing/body/nacelle configuration.  Fig. 2 shows two types of grid structure fitted to a cross-section of a wing/body intersection; the multiple-block grid obviously provides more uniform grid densities and eliminates the "lost corner".  However, a special point termed a "fictitious corner" is introduced.  In three dimensions these generally occur along lines, as is apparent in Fig. 1.  Another type of special point is a "collapsed edge".

The present approach for generating grids within a block structured framework is related to Thompson's method (Ref. 1) in that an elliptic solver is used. However, the present approach utilizes a simpler formulation wherein the Cartesian coordinates $\vec{X} = (X,Y,Z)$ of grid points are obtained from the solution of three independent linear equations defined in the computational coordinates $(\xi,\eta,\zeta)$:

$$\vec{X}_{\xi\xi} + B\vec{X}_{\eta\eta} + C\vec{X}_{\zeta\zeta} + D\vec{X}_{\xi} + E\vec{X}_{\eta} + F\vec{X}_{\zeta} + G = 0 \qquad (1)$$

where the coefficients, B to G, are constants or specified functions used for grid control.  This approach gives good quality grids, with easy grid control, and is economical since the grid generation equations are uncoupled.

After defining the overall block structure, a one-dimensional grid generation along the block perimeters produces a perimeter discretization.  This provides boundary conditions for a subsequent two-dimensional grid generation producing grids covering the block surfaces.  These in turn serve as boundary conditions to produce three-dimensional volume grids filling each block.

## Solution Algorithms

The principle flow algorithm issues concern the treatment of the flow solution across nonanalytic grid block boundaries, and at nonstandard grid cells such as fictitious corners and collapsed edges.  A 2-D transonic full-potential code was programmed in conservation form to investigate the compatibility of flow solution algorithms with the multi-block grid structure.  The artificial compressibility method (Ref. 2) was used to provide a directional bias at transonic speeds:

$$(\tilde{\rho}DU)_\xi + (\hat{\rho}DV)_\eta = 0$$
$$\tilde{\rho} = \rho - \mu\ \delta_\xi\rho$$
$$\hat{\rho} = \rho - \mu\ \delta_\eta\rho \qquad\qquad (2)$$
$$\mu = \max\,[\,0,\ c(1-M_c^2/M^2)\,]$$

where $\rho$ = density, M = Mach number, U,V = contravariant velocity components, D = transformation Jacobian, $\mu$ = switching function, $\delta$ = difference operator, c = weighting constant, and $M_c$ = cut-off Mach number.

A cell-oriented flux formulation is used (Ref. 3, Fig. 3).  All transformation metrics are defined at cell centers while the potentials are specified at the nodes.  Potential gradients, density, and contravariant velocity components are computed at cell centers. A zero flux condition is enforced for solid boundaries. Retarded densities in a supersonic cell are calculated using the density at the upwind boundary of the cell depending on the direction of the contravariant velocity components (Fig. 4).  The density at a cell boundary is obtained from linear interpolation in the triangle which includes the upwind boundary point:

$$\rho_B = \frac{A_1\ \rho_1 + A_2\ \rho_2 + A_3\ \rho_3}{A_1 + A_2 + A_3} \qquad\qquad (3)$$

where A's are sectional areas of the triangle in physical space.

The cell-oriented formulation provides an algorithm compatible with the use of nonanalytic grids crossing block boundaries.  The flux formulation is compatible with special grid points such as the fictitious corner, real corner or boundary points.  The evaluation of artificial viscosity terms at each cell individually also provides a means for dealing with these special points in supersonic zones.

## Results

The flow algorithms were tested in two dimensions.  Flow solutions generated using a multi-block grid are compared with results from a ring-type single-block grid of Ref. 4 (Fig. 5).  Compared to the ring-type grid, the multi-block grid seems to be overly complex; however, the objective here is to demonstrate its ability to produce accurate solutions.  Its advantages in terms of adaptability to more complex geometrical shapes is obvious.

Convergent solutions were obtained with the multi-block grid structure.  Figs. 6 and 7 show cases in which the fictitious corners are located in subsonic and supersonic regions respectively.  The accuracy of the results is remarkable, the largest discrepancy being a displacement of the shock by a single grid width.  Vertical line SOR method was used for the iteration sweep from upstream to downstream.  Slow convergence and slow circulation build-up were overcome by grid refinement and extrapolation acceleration (Figs. 8 and 9).  Two additional comparisons are presented in Figs. 10 and 11.

## Conclusion

A general grid generation scheme for complex configurations is presented using a multi-block grid structure and relatively simple grid generation equations.  Transonic algorithm issues associated with the multiple block structure are resolved in two dimensions.  Extension of the present method to three dimensions will provide a capability to solve transonic flows about general configurations.

## References


1.   Thompson, J. F., Thames, F. C., and Mastin, C. M., "Automatic Numerical Generation of Body-Fitted Curvilinear Coordinate System for Field Containing any Number of Arbitrary Two-Dimensional Bodies," Journal of Computational Physics, Vol. 15, 1974, pp. 299-319.

2.   Holst, T. L., "A Fast, Conservative Algorithm for Solving the Transonic Full-Potential Equation," Proc. of AIAA 4th Computational Fluid Dynamics Conference, Williamsburg, Va., July 1979, pp. 109-121.

3.   Jameson, A., and Caughey, D. A., "A Finite Volume Method for Transonic Potential Flow Calculations," Proc. of AIAA 3rd Computational Fluid Dynamics Conference, Albuquerque, N. M., June 1977, pp. 35-54.

4.   Jameson, A., "Transonic Potential Flow Calculations using Conservation Form," Proc. of AIAA 2nd Computational Fluid Dynamic Conference, Hartford, Conn., June 1975, pp. 148-155.



The authors are pleased to acknowledge many valuable discussions and suggestions by Dr. Antony Jameson during the course of this development.

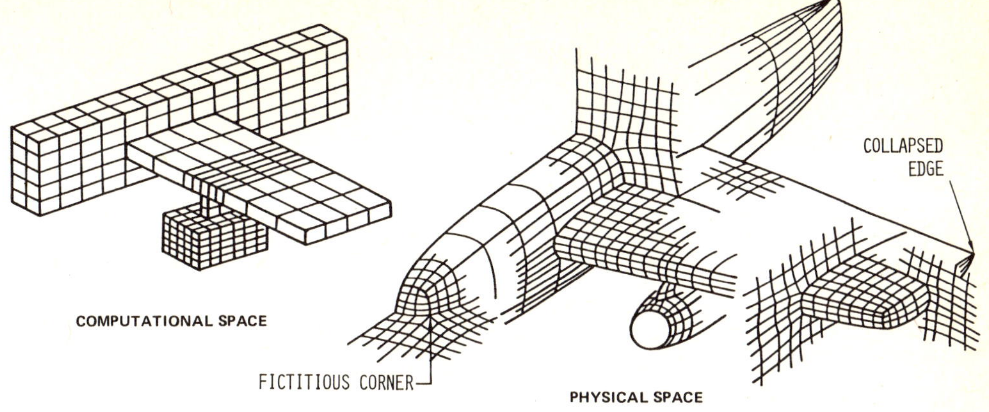


*Fig. 1.    —Block Structured Grid About a Wing/Body/Nacelle*

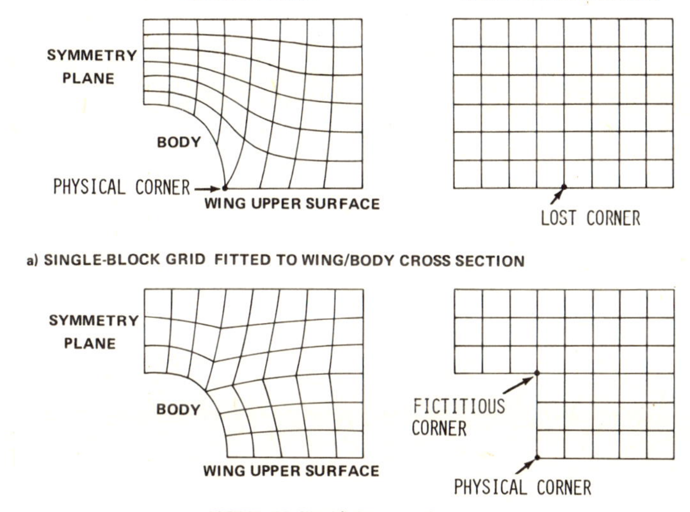


*Fig. 2.    —Comparison of Grid Structure*

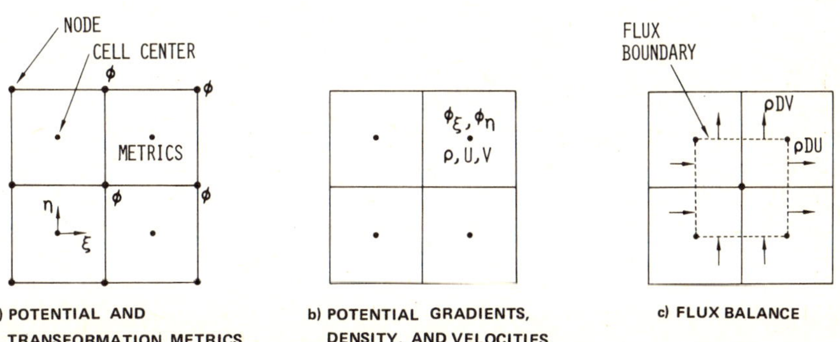


*Fig. 3.    —Formulation of Flow Solution Algorithm*

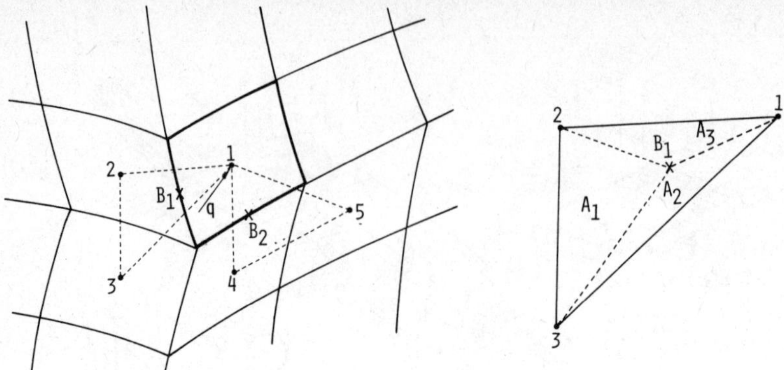

Fig. 4.    —*Computation of the Artificial Viscosity Terms at a Supersonic Cell*

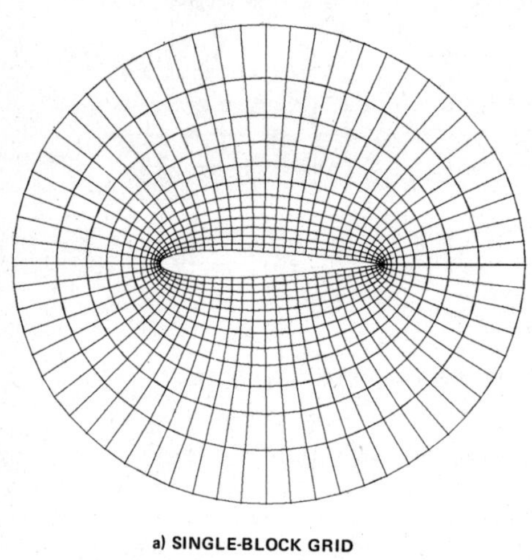

a) SINGLE-BLOCK GRID

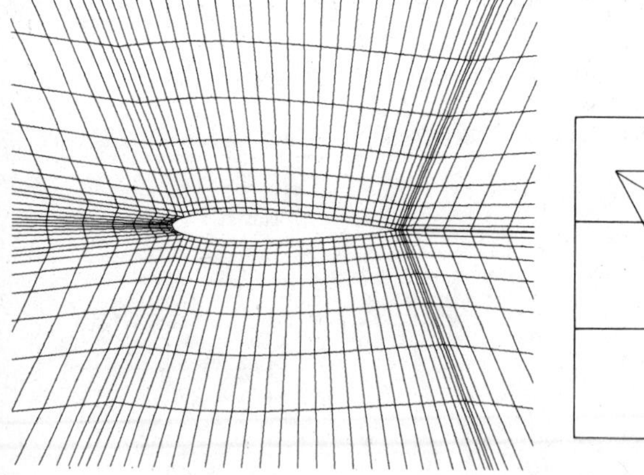

b) MULTI-BLOCK GRID

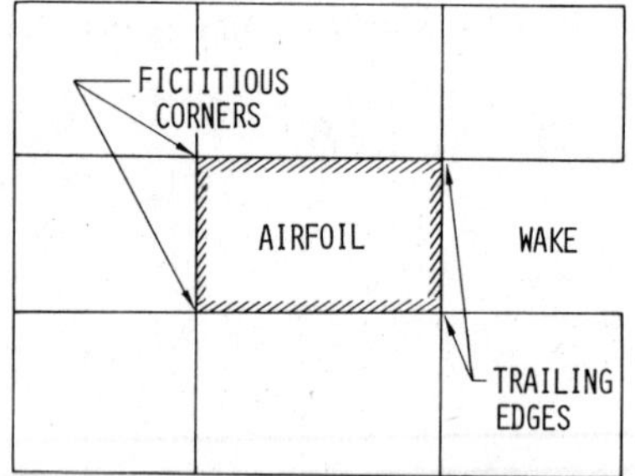


(c) MULTI-BLOCK STRUCTURE IN
COMPUTATIONAL PLANE

Fig. 5.    —*Comparison of Grids Near an Airfoil*

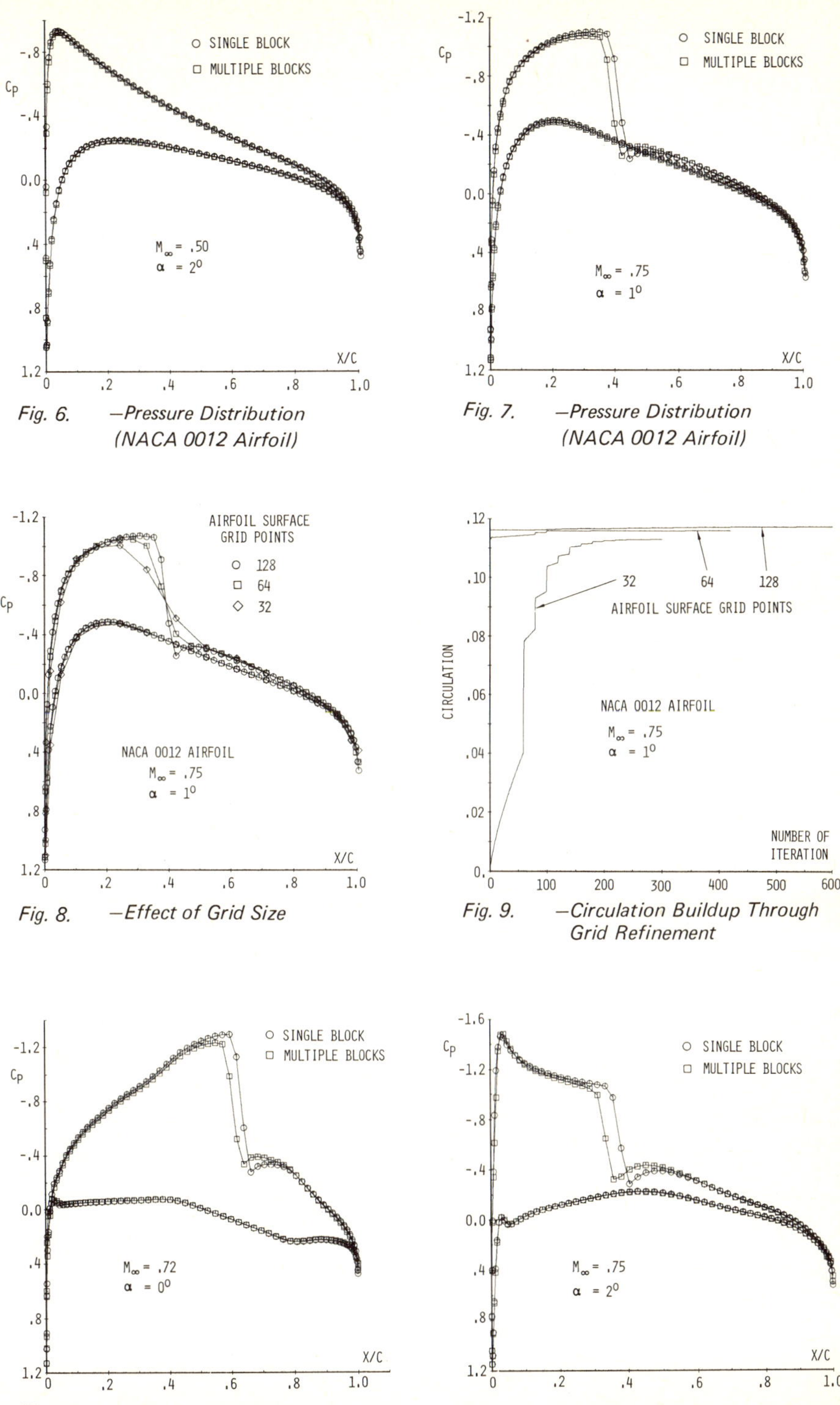



Fig. 6.  —Pressure Distribution (NACA 0012 Airfoil)

Fig. 7.  —Pressure Distribution (NACA 0012 Airfoil)

Fig. 8.  —Effect of Grid Size

Fig. 9.  —Circulation Buildup Through Grid Refinement

Fig. 10.  —Pressure Distribution (NACA 64A410 Airfoil)

Fig. 11.  —Pressure Distribution (ONERA M6 Wing Section)

$$\text{ON THE CONVERGENCE OF ITERATIVE METHODS FOR SOLVING THE}$$
$$\text{STEADY-STATE NAVIER-STOKES EQUATIONS BY FINITE DIFFERENCES}$$


Markku Lindroos

Helsinki University of Technology

SF-02150 Espoo 15, Finland


## INTRODUCTION

This paper deals with the numerical solution of the equations

$$-\frac{\partial\psi}{\partial y}\frac{\partial\omega}{\partial x} + \frac{\partial\psi}{\partial x}\frac{\partial\omega}{\partial y} + \frac{1}{Re}\left(\frac{\partial^2\omega}{\partial x^2} + \frac{\partial^2\omega}{\partial y^2}\right) = 0, \qquad (1)$$

$$\frac{\partial^2\psi}{\partial x^2} + \frac{\partial^2\psi}{\partial y^2} = -\omega, \qquad (2)$$

which govern the steady two-dimensional flow of a viscous incompressible fluid. Here, $\psi$ is the stream function, $\omega$ the vorticity, and Re the Reynolds number. The velocity components of the flow are given by

$$u = \frac{\partial\psi}{\partial y}, \qquad v = -\frac{\partial\psi}{\partial x}. \qquad (3)$$

In the search for a numerical solution of equations (1) and (2), the derivatives are replaced by consistent difference approximations, and the resulting nonlinear system of equations is solved iteratively by successive over-relaxation (SOR). The vorticity and stream-function equations have their own relaxation parameters, $r_\omega$ and $r_\psi$, and the $\omega$ and $\psi$ equation systems are iterated alternately.

Although the approach outlined is a common one, there are few theoretical studies on the convergence of the iteration. In this paper, a general theory of the iterative solution of nonlinear equations is employed for investigation of the convergence conditions. It is assumed that the domain in which solutions are sought is a rectangle, although some of the results can be extended to more general domains. The rectangle is set parallel to the coordinate axes, and is covered by a uniform grid with mesh spacings $\Delta x$ and $\Delta y$. The nodal points are identified by the indices i and j, $0 \leq i \leq M$, $0 \leq j \leq N$. Dirichlet conditions are prescribed for $\psi$ and $\omega$ along the boundaries of the domain. (In the problems encountered in practice, the boundary conditions are usually more complex.)

The Poisson equation for $\psi$ is replaced by the usual five-point scheme. Various difference approximations to the vorticity equation (1) are written in a form which offers a means for comparison of the numerical diffusion involved in them. The convergence conditions are written for these approximations.

It is shown that if the vorticity is constant in the solution of the discrete system, and if the initial estimate is sufficiently close to the solution, then the SOR iteration converges if the stream-function and vorticity iterations converge separately. The spectral radius of the vorticity iteration matrix is presented for the different approximations in the case of a constant-coefficient model problem. Finally, a discrete perturbation analysis is applied for study of the stability characteristics of the SOR iteration at a single vorticity iteration sweep through the grid.

## DIFFERENCE APPROXIMATIONS

Equation (2) is replaced by the second-order approximation

$$\psi_{i+1,j} + \psi_{i-1,j} + \gamma(\psi_{i,j+1} + \psi_{i,j-1}) - 2(1+\gamma)\psi_{ij} + (\Delta x)^2\omega_{ij} = 0, \qquad (4)$$

where $\gamma = (\Delta x/\Delta y)^2$. For the vorticity equation, consideration is given to schemes of the form

$$c_1 \omega_{i+1,j} + c_2 \omega_{i,j+1} + c_3 \omega_{i-1,j} + c_4 \omega_{i,j-1} - c_0 \omega_{ij} = 0, \qquad (5)$$

in which the coefficients $c_1$, $c_2$, $c_3$, $c_4$, and $c_0 = c_1 + c_2 + c_3 + c_4$ depend on the grid spacings, the Reynolds number, and the values of the stream function.

Some known discrete representations (5) are listed in table 1, using the abbreviations

$$p_{ij} = u_{ij} \Delta x Re \ , \qquad q_{ij} = v_{ij} \Delta y Re,$$

in which the velocity components are evaluated from

$$u_{ij} = \frac{\psi_{i,j+1} - \psi_{i,j-1}}{2\Delta y} \ , \qquad v_{ij} = - \frac{\psi_{i+1,j} - \psi_{i-1,j}}{2\Delta x} \ .$$

The subscripts i and j have been omitted from the expressions given in the table.

The scheme proposed by Dennis and Hudson (1978) is derived with A = 1/8 from the one given in the table. By means of Taylor expansions about the node (i,j) it is easy to see that the scheme presented in the table is formally accurate to the second order for any A. Moreover, the associated coefficient matrix is diagonally dominant under all circumstances if $A \geq 1/16$. For A = 0, the scheme reduces to the ordinary central-difference approximation, which yields diagonal dominance if and only if $|p_{ij}| \leq 2$ and $|q_{ij}| \leq 2$ at all nodes of the grid. The matrices associated with the upwind-difference, Allen-Southwell, and Spalding schemes are always diagonally dominant.

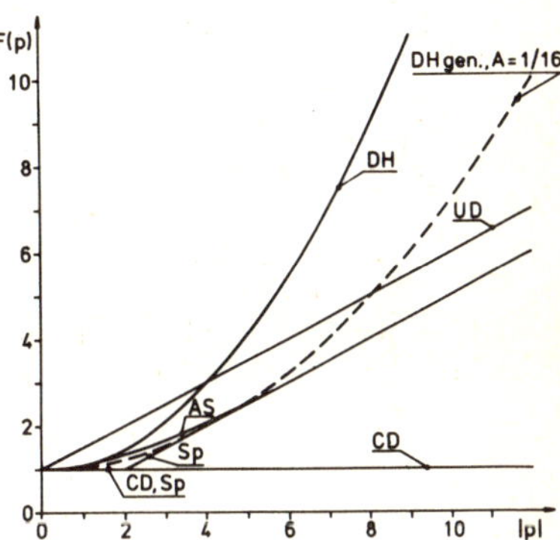



Figure 1. Function F(p).

Table 1.  Coefficients $c_\mu$ and the function F(p) for various difference analogues of the vorticity equation.

| | Central diff. | Upwind diff. | Allen and Southwell (1955) | Dennis and Hudson (1978), generalized | Spalding (1972) |
|---|---|---|---|---|---|
| | CD | UD | AS | DH, DH gen. | Sp |
| $c_1$ | $1-p/2$ | $1 + \dfrac{\lvert p \rvert - p}{2}$ | $\dfrac{p}{e^p - 1}$ | $1 - \dfrac{p}{2} + Ap^2$ | $\max\{1, \dfrac{\lvert p \rvert}{2}\} - \dfrac{p}{2}$ |
| $c_2$ | $\gamma(1-q/2)$ | $\gamma(1 + \dfrac{\lvert q \rvert - q}{2})$ | $\dfrac{\gamma q}{e^q - 1}$ | $\gamma(1 - \dfrac{q}{2} + Aq^2)$ | $\gamma(\max\{1, \dfrac{\lvert q \rvert}{2}\} - \dfrac{q}{2})$ |
| $c_3$ | $1+p/2$ | $1 + \dfrac{\lvert p \rvert + p}{2}$ | $\dfrac{pe^p}{e^p - 1}$ | $1 + \dfrac{p}{2} + Ap^2$ | $\max\{1, \dfrac{\lvert p \rvert}{2}\} + \dfrac{p}{2}$ |
| $c_4$ | $\gamma(1+q/2)$ | $\gamma(1 + \dfrac{\lvert q \rvert + q}{2})$ | $\dfrac{\gamma q e^q}{e^q - 1}$ | $\gamma(1 + \dfrac{q}{2} + Aq^2)$ | $\gamma(\max\{1, \dfrac{\lvert q \rvert}{2}\} + \dfrac{q}{2})$ |
| $F(p)$ | $1$ | $1 + \dfrac{\lvert p \rvert}{2}$ | $\dfrac{p}{2} \coth \dfrac{p}{2}$ | $1 + Ap^2$ | $\max\{1, \dfrac{\lvert p \rvert}{2}\}$ |

By means of the function F given in the last row of table 1, scheme (5) can be rearranged in the form

$$-\frac{p_{ij}}{2}(\omega_{i+1,j} - \omega_{i-1,j}) + F(p_{ij})(\omega_{i+1,j} - 2\omega_{ij} + \omega_{i-1,j})$$

$$-\gamma\frac{q_{ij}}{2}(\omega_{i,j+1} - \omega_{i,j-1}) + \gamma F(q_{ij})(\omega_{i,j+1} - 2\omega_{ij} + \omega_{i,j-1}) = 0, \qquad (6)$$

which can be interpreted as a central-difference approximation to the equation

$$-u\frac{\partial\omega}{\partial x} - v\frac{\partial\omega}{\partial y} + \frac{F(u\Delta xRe)}{Re}\frac{\partial^2\omega}{\partial x^2} + \frac{F(v\Delta yRe)}{Re}\frac{\partial^2\omega}{\partial y^2} = 0.$$

Function F is presented graphically for the different schemes in figure 1. For the central differences $F(p)$ identically equals unity. This is also true for the Spalding scheme if $|p| \leq 2$. In all other cases, $F(p)$ is greater than unity, and increases as $|p|$ increases. The overshoot of unity can be interpreted as false, or numerical diffusion of vorticity, in just the same way as that customarily applied for the upwind-difference scheme. In the Dennis-Hudson scheme, this false diffusion grows with particular rapidity with increase in the cell Reynolds numbers $|p_{ij}|$ and $|q_{ij}|$.

Representation (6) of the discrete vorticity equation suggests that many other forms of $F(p)$ might apply just as well as those listed in table 1. The resulting schemes would lead to a diagonally dominant matrix under all circumstances if and only if $F(p) \geq |p|/2$ for all p. Among the schemes which lead to a diagonally dominant matrix, the Spalding scheme is optimal in the sense that it minimizes the deviation of $F(p)$ from unity.


## CONVERGENCE ANALYSIS FOR THE NONLINEAR SYSTEM

Let us now consider the iterative solution of the nonlinear system by the SOR method, with $r_\psi$ and $r_\omega$ as the relaxation parameters. Assume that the iteration proceeds along each vertical grid line upwards and that the grid lines are taken up in order from left to right. One cycle of iteration consists of two sweeps through the grid. The values of $\psi$ are revised in the first sweep, and the values of $\omega$ in the second. In accordance with this procedure, the unknown variables are arranged in the order

$$\psi_{11}, \psi_{12}, \ldots, \psi_{1,N-1}, \psi_{21}, \ldots, \psi_{M-1,N-1},$$

$$\omega_{11}, \omega_{12}, \ldots, \omega_{1,N-1}, \omega_{21}, \ldots, \omega_{M-1,N-1},$$

and are collected as a vector $\alpha$. When the components of $\alpha$ are renamed $\alpha_i$, the nonlinear system may be written as

$$f_i(\alpha_1, \alpha_2, \ldots, \alpha_{2(M-1)(N-1)}) = 0, \qquad i = 1, \ldots, 2(M-1)(N-1).$$

If we define

$$g_i(\alpha,\beta) = f_i(\alpha_1, \ldots, \alpha_{i-1}, \beta_i + (\alpha_i - \beta_i)/r_i, \beta_{i+1}, \ldots, \beta_{2(M-1)(N-1)}),$$

$$i = 1, \ldots, 2(M-1)(N-1),$$

in which

$$r_i = \begin{cases} r_\psi, & i = 1, \ldots, (M-1)(N-1) \\ r_\omega, & i = (M-1)(N-1)+1, \ldots, 2(M-1)(N-1), \end{cases}$$

and denote

$$G(\alpha,\beta) = (g_1(\alpha,\beta), \ldots, g_{2(M-1)(N-1)}(\alpha,\beta))^T,$$

then the iteration can be written as

$$G(\alpha^{k+1}, \alpha^k) = 0, \qquad k = 0, 1, 2, \ldots .$$

Ortega and Rheinboldt (1970, p. 325) have given convergence conditions for this kind of iteration. They assume that G has continuous partial derivatives $\partial_1 G$ and $\partial_2 G$ with respect to the vector variables $\alpha$ and $\beta$, respectively, on an open neighbourhood of a point $\alpha^*$ at which $G(\alpha^*, \alpha^*) = 0$. If $\partial_1 G(\alpha^*, \alpha^*)$ is nonsingular, and if

$$\rho(-(\partial_1 G(\alpha^*, \alpha^*))^{-1}\partial_2 G(\alpha^*, \alpha^*)) < 1,$$

in which $\rho$ denotes the spectral radius of the matrix in the parentheses, then the iteration converges to $\alpha^*$ if the initial approximation is sufficiently close to $\alpha^*$.

This theorem is applicable to our problem, on the assumption that it has a solution $\alpha^*$. Let the Jacobian matrix of the functions $f_i$ with respect to $\alpha$ be partitioned into the block form

$$\left(\frac{\partial f_i}{\partial \alpha_j}\right) = \left(\begin{array}{cc} A & B \\ C & D \end{array}\right) = \left(\begin{array}{cc} L_1 + D_1 + U_1 & (\Delta x)^2 I \\ C & L_2 + D_2 + U_2 \end{array}\right) ,$$

in which A, B, C, and D are square matrices of order $(M-1)(N-1)$. Then A is the Jacobian of the functions pertaining to the discrete Poisson equation (4) with respect to the nodal variables of the stream function, and so on. Matrices A and D have further been decomposed into their strictly lower-triangular, diagonal, and strictly upper-triangular parts. The relation $B = (\Delta x)^2 I$ is a direct consequence of the form of equation (4). (I is the unit matrix of order $(M-1)(N-1)$.)

These decompositions enable us to write

$$\partial_1 G(\alpha^*, \alpha^*) = \left(\begin{array}{cc} L_1(\alpha^*)+r_\psi^{-1}D_1(\alpha^*) & 0 \\ C(\alpha^*) & L_2(\alpha^*)+r_\omega^{-1}D_2(\alpha^*) \end{array}\right) ,$$

$$\partial_2 G(\alpha^*, \alpha^*) = \left(\begin{array}{cc} (1-r_\psi^{-1})D_1(\alpha^*)+U_1(\alpha^*) & (\Delta x)^2 I \\ 0 & (1-r_\omega^{-1})D_2(\alpha^*)+U_2(\alpha^*) \end{array}\right) .$$

The eigenvalues, $\lambda$, of $-(\partial_1 G(\alpha^*, \alpha^*))^{-1}\partial_2 G(\alpha^*, \alpha^*)$ are the roots of the characteristic equation $\det(\lambda\partial_1 G(\alpha^*, \alpha^*) + \partial_2 G(\alpha^*, \alpha^*)) = 0$ or

$$\left|\begin{array}{cc} \lambda L_1(\alpha^*)+((\lambda-1)r_\psi^{-1}+1)D_1(\alpha^*)+U_1(\alpha^*) & (\Delta x)^2 I \\ \lambda C(\alpha^*) & \lambda L_2(\alpha^*)+((\lambda-1)r_\omega^{-1}+1)D_2(\alpha^*)+U_2(\alpha^*) \end{array}\right| = 0 .$$

Matrix C is the Jacobian of the functions associated with the discrete vorticity equation (5) with respect to the nodal variables of the stream function. If the values of $\omega$ are all equal in solution $\alpha^*$, then the left-hand side of equation (5) is zero, no matter what the values of $\psi$ are (in view of equation $c_0 = c_1 + c_2 + c_3 + c_4$). In this case $C(\alpha^*) = 0$, and the search for the roots of the characteristic equation is reduced to a search for the roots of equations

$$\det(\lambda L_1(\alpha^*)+((\lambda-1)r_\psi^{-1} + 1)D_1(\alpha^*)+U_1(\alpha^*)) = 0,$$

$$\det(\lambda L_2(\alpha^*)+((\lambda-1)r_\omega^{-1} + 1)D_2(\alpha^*)+U_2(\alpha^*)) = 0.$$

This means that if the initial approximation is close to $\alpha^*$, then the combined iteration converges if $r_\psi$ and $r_\omega$ are such that the $\psi$ and $\omega$ iterations converge separately.

The above result can clearly be generalized to nonrectangular regions, to a wider class of schemes than that represented by equation (5), and to nonuniform grids.


## CONVERGENCE ANALYSIS FOR A LINEAR MODEL PROBLEM

It is known that iteration of the discrete Poisson equation for $\psi$ has good convergence properties when iterated alone; for the Dirichlet problem considered here, convergence is obtained for $0 < r_\psi < 2$. The vorticity iteration also works well if the associated matrix is diagonally dominant. In this event, the presence of the first derivatives does in fact contribute to the achievement of a higher rate of convergence than that of the $\psi$ iteration (cf. figure 2 below). Difficulties may be encountered if the diagonal dominance is severely broken, as is the case for the central-difference scheme at high cell Reynolds numbers.

In general, the optimum relaxation parameters are determinable only by numerical experiments. If some idea of a good choice of $r_\omega$ is to be obtained, the vorticity iteration when the values of $\psi$ are left untouched may be considered. (The result of the previous section provides some motivation for study of this special case.)

To permit calculation of the eigenvalues of the iteration matrix explicitly, consideration is given to the model equation

$$-u \frac{\partial \omega}{\partial x} - v \frac{\partial \omega}{\partial y} + \frac{1}{Re} \left( \frac{\partial^2 \omega}{\partial x^2} + \frac{\partial^2 \omega}{\partial y^2} \right) = 0, \tag{7}$$

in which u and v are constant. The finite-difference representation is continuously assumed to be of the form (5), and may be one of those listed in table 1. The coefficients $c_\mu$ are now constant, and it is not difficult to verify that the eigenvalues of the SOR iteration matrix are obtainable from the equations

$$\lambda - r_\omega \frac{2\sqrt{c_1 c_3} \cos \frac{m\pi}{M} + 2\sqrt{c_2 c_4} \cos \frac{n\pi}{N}}{c_0} \sqrt{\lambda} + r_\omega - 1 = 0,$$

$$m = 1, \ldots, M-1, \quad n = 1, \ldots, N-1.$$

In the unaccelerated case, $r_\omega = 1$ (the Gauss-Seidel method), under all circumstances the dominant eigenvalue is

$$\mu = \left( \frac{2\sqrt{c_1 c_3} \cos \frac{\pi}{M} + 2\sqrt{c_2 c_4} \cos \frac{\pi}{N}}{c_0} \right)^2,$$

so that the spectral radius is always $|\mu|$.

If $c_1 c_3 \geq 0$ and $c_2 c_4 \geq 0$, then all of the eigenvalues of the Gauss-Seidel iteration matrix are real and non-negative. From the general theory of the SOR method (cf. Young 1971) it is known that in this case the optimum relaxation factor is

$$r_\omega^* = \frac{2}{1 + \sqrt{1-\mu}}. \tag{8}$$

The corresponding spectral radius is $\rho^* = r_\omega^* - 1$. (This case includes the upwind-difference, Allen-Southwell, and Spalding schemes for any p, q, and $\gamma$, the central-difference scheme if $|p| \leq 2$, $|q| \leq 2$, and the generalized Dennis-Hudson scheme if $1 - |p|/2 + Ap^2 \geq 0$, $1 - |q|/2 + Aq^2 \geq 0$.)

If $c_1 c_3 \leq 0$ and $c_2 c_4 \leq 0$, then $\mu \leq 0$. The optimum value of $r_\omega$ is still given by (8), but satisfies now $0 < r_\omega^* \leq 1$. The corresponding spectral radius is $\rho^* = 1 - r_\omega^*$. If $c_1 c_3$ and $c_2 c_4$ are different in sign, then $\mu$ is complex. In this case, no simple analytical expression is known for $r_\omega^*$.

The convergence properties of the SOR iteration for the schemes of table 1 are dependent upon six parameters, viz. M, N, p, q, $\gamma$, and $r_\omega$. To reduce the number of parameters, and thus to facilitate graphical representation, consideration may be given to the one-dimensional counterpart of equation (7),

$$- u \frac{\partial \omega}{\partial x} + \frac{1}{Re} \frac{\partial^2 \omega}{\partial x^2} = 0. \tag{9}$$

Numerical analogues of (9) are obtained from (6) by setting $\gamma$ at zero, and dropping the index j. The convergence properties of the SOR iteration are then dependent upon three parameters: M, p, and $r_\omega$. Figure 2 shows the spectral radii of the iteration

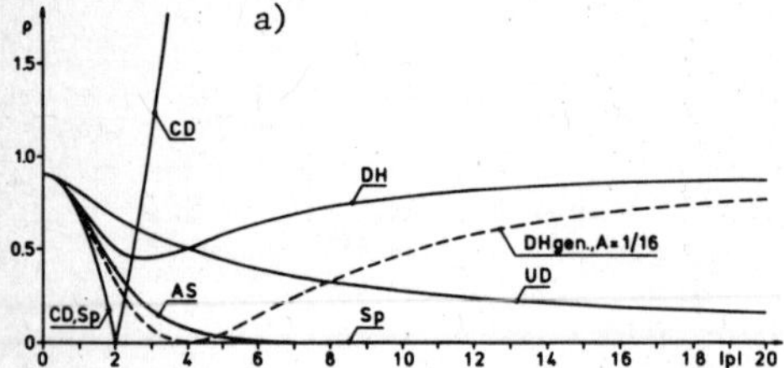
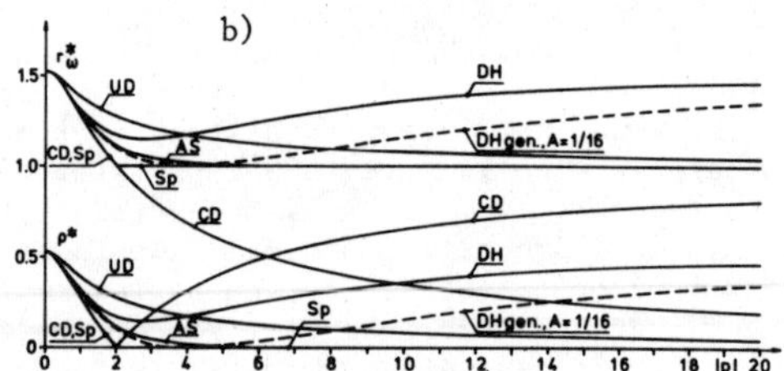

Figure 2. Spectral radius of the SOR iteration matrix for the one-dimensional model problem when M = 10. a) $r_\omega = 1$, b) $r_\omega = r_\omega^*$ ($r_\omega^*$ is also shown).

matrices for M = 10 as functions of p, when $r_\omega = 1$ and when $r_\omega = r_\omega^*$ ; $r_\omega^*$ is also presented. The figure clearly demonstrates the superior convergence properties of the uncentred schemes at high cell Reynolds numbers.

Understanding of the behaviour of the different approximations of table 1 in the SOR iteration can be complemented by application of a discrete perturbation analysis to the constant-coefficient model problem concerned. The analysis consists of the introduction of a perturbation, or error, upon a single vorticity value, and of examining its subsequent growth or decay as an individual vorticity iteration sweep proceeds forward. By means of the SOR iteration formula, it is easy to verify that a perturbation of $\varepsilon$ in $\omega_{ij}^k$ will induce a perturbation of

$$\binom{m + n}{n} \left(\frac{r_\omega c_3}{c_0}\right)^m \left(\frac{r_\omega c_4}{c_0}\right)^n \varepsilon$$

in $\omega_{i+m, j+n}^k$ , $m \geqq 0$, $n \geqq 0$ (k is the iteration index).

For the central-difference approximation, $|c_3| > c_0$ and $|c_4| > c_0$ at high cell Reynolds numbers. If the number of mesh points is large, then an explosive growth of errors will result at the very first iteration sweep, unless strong under-relaxation is employed. For the diagonally dominant schemes, it is easy to show that any perturbation diminishes for $0 < r_\omega \leqq 1$ at all cell Reynolds numbers as the distance increases from the location of the perturbation.


CONCLUDING REMARKS

Diagonal dominance is a very desirable property in iteration, and for this reason upwind differences (or other uncentred differences) have gained  considerable popularity, despite the numerical diffusion they produce. Another justification for the use of upwind differences is that at high cell Reynolds numbers the central-difference solution may display spatial oscillations, or wiggles. This has been demonstrated by Roache (1972, p. 164) for a one-dimensional model problem. Unfortunately, the one-dimensional results seem to have unduly discouraged many numerical fluid analysts from making use of central differences in two-dimensional problems. Although spatial oscillations may also occur in two dimensions at high cell Reynolds numbers, in many cases the central-difference solutions are smooth, and far more accurate than those obtained by the other schemes of table 1. Accordingly, as a rule it is worthwhile to seek the central-difference solution, although the computational effort may be high for high-Reynolds-number flows.

Of course, the central-difference solution may be determined by iterative methods other than ordinary SOR. For example, Dennis and Chang (1969) have proposed a procedure based upon the idea of deferred correction. The present author has analysed the convergence properties of this procedure and some of its variants, but by reason of the limited space available the results have to be presented elsewhere. For the results obtained by discrete perturbation analysis, and von Neumann stability analysis, refer to Lindroos (1978).


REFERENCES

Allen, D.N. de G., and Southwell, R.V. 1955 Relaxation methods applied to determine the motion, in two dimensions, of a viscous fluid past a fixed cylinder. Quart. J. Mech. Appl. Math. 8, 129-145.

Dennis, S.C.R., and Chang, G.-Z. 1969 Numerical integration of the Navier-Stokes equations for steady two-dimensional flow. Phys. Fluids 12, Suppl. II, II-88 - II-93.

Dennis, S.C.R., and Hudson, J.D. 1978 A difference method for solving the Navier-Stokes equations. In Proceedings of the First International Conference on Numerical Methods in Laminar and Turbulent Flow (edited by C. Taylor, K. Morgan, and C.A. Brebbia), pp. 69-80, Pentech Press, London.

Lindroos, M. 1978 Numerical integration of the Navier-Stokes equations for steady flow past a wavelike bulge on a flat plate. Helsinki University of Technology, Laboratory of Aerodynamics, Report No. 78-A2.

Ortega, J.M., and Rheinboldt, W.C. 1970 Iterative Solution of Nonlinear Equations in Several Variables. Academic Press, New York.

Roache, P.J. 1972 Computational Fluid Dynamics. Hermosa Publishers, Albuquerque, New Mexico.

Spalding, D.B. 1972 A novel finite difference formulation for differential expressions involving both first and second derivatives. Int. J. Num. Meth. Engng $\underline{4}$, 551-559.

Young, D.M. 1971 Iterative Solution of Large Linear Systems. Academic Press, New York.

Direct Numerical Simulations of Turbulent Shear Flows

Ralph W. Metcalfe and James J. Riley

Flow Research Company

Kent, Washington U.S.A.

## I    Introduction

The main objectives of our work have been (i) to determine how well direct
numerical simulations can predict the evolution of complex turbulent flows, and
(ii) to use these simulations to gain insight into the physics of the flow evolution.  Direct numerical simulations involve the numerical solution of the detailed
evolution of the large-scale structure of the complex turbulent velocity field.  We
have made use of very efficient and accurate pseudospectral numerical techniques
(Orszag, 1971).  No turbulence closure assumptions have been made; the calculations
are performed at Reynolds numbers such that all scales containing significant
energy are adequately resolved.  Statistical data are obtained by performing spatial,
temporal, and/or ensemble averages over the computed fields.  To the extent that
accurate simulations can be performed, this procedure is analogous to laboratory
experiments.  However it has the advantages that more precise control over the
experimental conditions is possible and more detailed statistical information can
be obtained, since the entire flow field is known at each time step.

We have chosen to study two classes of turbulent free shear flows; (1) the
wakes of axisymmetric bodies, and (2) turbulent mixing layers.  These flows, while
possessing many of the dynamic processes of all turbulent flows, are easier to
treat numerically.  In addition, these flows have been extensively studied experimentally and theoretically, so there is a good data base to which the simulations
can be compared.  We have examined the temporal evolution of these flows in the
region in which they behave in approximately a self-similar manner.  This assumes
that the flow is statistically homogeneous in the mean flow direction, which is
consistent with the experimental data and self-similarity theory.  For the wake calculations, the computed temporal evolution of the flow can be related directly to
the spatial evolution data from laboratory experiments.  The calculations are
initialized using laboratory data from a particular downstream location in the
self-preserving region, and the flow evolution is computed from that point on and
compared with the laboratory data further downstream.  This enables us to evaluate
how well the direct numerical simulation methods work as a predictive tool.


## II    Method of Approach

The nonlinear Navier-Stokes equations are solved by pseudospectral numerical
methods.  We have performed simulations on a 32 x 32 x 33 point grid on the CDC
7600 at NASA Ames which requires about 3 sec/time step (Code I).  The numerical
algorithm uses fast Fourier transforms and inverse transforms, and uses leapfrog

time differencing on the nonlinear terms, and Crank-Nicolson (implicit) time differencing on the viscous terms. (See Orszag and Pao (1974) for more details). We have also performed simulations on a 64 x 64 x 64 point grid on the Illiac IV at NASA Ames which require about 16 sec/time step (Code II). The velocity field is expanded in a triple Fourier series and fourth order Rurge-Kutta time differencing is used. (See Rogallo (1977) for details). We used no sub-grid scale models. Instead, the viscosity was chosen so that the turbulence Reynolds number $(R_\lambda)$ was typically in the range 40-60. Except for the initial energy spectrum, which was based on the laboratory data, there were no other free parameters. (See Riley and Metcalfe (1978, 1980a, 1980b) for more details). The use of two completely different numerical codes, computers, and spatial resolutions permits us to cross-correlate the results and increases our confidence in the validity of the results.

III  <u>Towed Wake</u>

These simulations were based on the laboratory data of Pao and Lin (1973). (Details of the initialization are given in Riley and Metcalfe (1978, 1980b).) Figure 1 is a plot of the temporal behavior of the maximum axial mean velocity deficit $(U_m)$ and the maximum rms turbulent axial velocity $(u_m)$ for simulations with Codes I and II. Time is non-dimensionalized using the initial values of $U_m$ and $r_m$, the mean wake radius. For these runs, $T$ and $x/D$ are related by the transformation $x/D = 36 + 9.6\ T$. Both simulations appear to be approaching the $T^{-2/3}$ decay behavior for $U_m$ and $u_m'$ as predicted by self-similarity theory, and as is consistent with the laboratory data. Although the initial values of $U_m$ and $u_m'$ for both simulations are fairly close initially, their values in the $64^3$ case drop significantly below the $32^3$ case during the first several dozen time steps. Since the initialization procedures for the two computer codes are somewhat different, it is not unexpected that the initial development period is also somewhat different. Our current initialization procedure does not introduce the proper correlations in the flow field, so that the Reynolds stresses and higher moments must build up to the proper values during the first few time steps of the calculation. This highly nonlinear, unsteady adjustment process may be very sensitive to the initial conditions.

The next four figures demonstrate the spatial behavior of the evolving flow. These data are computed by taking statistical averages over concentric shells centered on the wake axis. Self-similarity theory predicts that if the velocities are normalized by $U_m$ or $u_m$ and the lengths by $r_m$ or $r_T$, then the radial distribution of certain statistical quantities such as $U(r/r_m)/u_m$ should be the same at different times. Since the laboratory data does behave in just such a fashion, an investigation of this behavior in the calculations provides a good check on the capability of the numerical simulations to correctly model the physical processes in the flow. The normalization factors used in Figures 2-5 are obtained from the decay curves (e.g., Figure 1).

Figure 2 shows the radial distribution of the mean velocity profile at two different times in the two simulations. The scaling collapses the data quite well, although the $64^3$ simulation shows slightly more deviation from the similarity profile than the $32^3$ calculation near the origin, where statistical scatter is expected to be greatest. However, both simulations undershoot the profile for $r/r_m^* > 1.5$. This feature has also been observed in laboratory experiments, and can be attributed to intermittancy near the wake edge which decreases the turbulent diffusion there.

Figure 3 is a plot of $u'/u_m$ versus $r/r_m$ at the same two times. The behavior of the laboratory data of Pao and Lin (1973) is indicated for comparison. As this is a higher order moment, the statistical scatter in the data is expected to be larger than for $U/U_m$. Another second order moment, the normalized Reynolds stress $\overline{uw}/u_m^2$, is shown in Figure 5. This is especially interesting, since, due to the inadequacy of our initialization procedure, the Reynolds stresses are initially zero. During the adjustment process, the Reynolds stresses build up in a very consistent fashion.

An important feature of a turbulent wake is the presence of an intermittancy in the flow at the wake edge. This is manifested by the behavior of the velocity kurtosis or flatness factor, which is approximately equal to the inverse of the intermittancy factor. As with the Reynolds stresses, this high order moment was not properly initialized. During the adjustment period, this statistical quantity develops from a uniform value across the wake to the sharply peaked behavior near the wake edge as is also apparent in the laboratory data (see Figure 5). This indicates the simulations are developing a realistically contorted wake edge. Several other high order moments which we have computed, such as the velocity skewness, behave in a similar fashion, consistent with the laboratory data.

We have performed several simulations of the wake of a self-propelled body (without swirl) based on the laboratory data of Higuchi (1977). This is a more difficult problem than the towed wake since the turbulence levels are much higher relative to the mean flow, which significantly increases the statistical scatter in the mean velocity. We are currently examining the results of these calculations.

To summarize, the direct numerical simulations of the towed wake exhibit the proper temporal behavior after a short initial period of adjustment. Spatially, the data collapse quite well onto the self-similarity profiles, although there is a fair amount of statistical scatter, especially near the wake axis. The initialization procedure is not ideal, since the flow can change significantly during the adjustment period. However, the flow does develop and maintain the Reynolds stresses and higher order correlations at realistic values. In addition, the skewness and kurtosis build up near the wake edge, indicating the development of an intermittancy in the flow in this region.

IV   <u>Mixing Layer</u>

The simulations of a turbulent mixing layer were based on the laboratory data of Wygnanski and Fiedler (1970). (See Riley and Metcalfe (1980a) for details about the initialization procedure.) For the $32^3$ simulations (Code I), the computational domain was chosen to be large enough to contain the most unstable linear mode for the hyperbolic tangent mean velocity profile. In the $64^3$ simulations (Code II), the computational domain was twice as big, and could thus contain the subharmonic of the most unstable mode. Statistical data were computed by taking averages over horizontal planes. Time was non-dimensionalized with respect to $\left|d\bar{U}/dz\right|_{max}$ at $t = 0$ .

An examination of vorticity contour plots of the mixing layer simulation shows the presence of large scale features similar to the vortex cores seen in laboratory experiments (e.g., Brown and Roshko (1974)). However, the lateral (cross-track) coherence of the features appears to be rather weak. (See Riley and Metcalfe (1980) for representative plots.)

Laboratory data and similarity theory show that the growth of the mixing layer, as measured by its turbulent half width $z_{1/2}$ , should be linear. Here, $z_{1/2}$ is defined as the transverse distance from the horizontal plane of symmetry to the point where $\bar{U}$ equals one-half of its free stream value. Figure 6 shows the growth of $z_{1/2}$ with time for the $32^3$ and the $64^3$ simulations. In both cases, the proper linear growth is exhibited, although at slightly different rates. Brown and Roshko (1974) have estimated the <u>spatial</u> growth rate of $z_{1/2}$ from their laboratory experiments. If we <u>assume</u> that space and time are related by the transformation $x = \frac{1}{2}(U_1 + U_2)\,T$ , where $U_1$ and $U_2$ are free stream velocities in the laboratory experiments, then these data give $\frac{1}{U}\frac{d}{dT}z_{1/2} = 0.0275$ . Both simulations come reasonably close to this value, although this is probably somewhat fortuitous since there is no valid space-time transformation for this problem.

The spatial development of the mixing layer can be studied by looking at the development of the mean velocity, non-dimensionalized by $U$ , the mean velocity difference across the layer, as a function of transverse distance, non-dimensionalized by $z_{1/2}$ . Similarity theory and laboratory experiments show that this scaling should collapse the data. As can be seen in Figure 7, both simulations exhibit this behavior. The results of the $64^3$ simulation are quite remarkable since the thickness of the mixing layer as measured by $z_{1/2}$ grows by a factor 6 between $T = 0$ and $T = 52.3$ , yet the flow remains nearly self-similar.

There is some evidence that the $64^3$ mixing layer calculation is more realistic than the $32^3$ one. Laboratory experiments and similarity theory predict that the maximum turbulence intensity $(u'_m)$ should remain constant in time. In Figure 8, $u'_m$ is plotted as a function of time for both runs. While the $64^3$ case has approximately the correct behavior, $u'_m$ in the $32^3$ calculation slowly grows in time. It is possible that the smaller computational box in the $32^3$ run is preventing the turbulence from diffusing out from the center of the layer at its proper rate.

In summary, the direct numerical simulation of a turbulent mixing layer develops in a very realistic fashion. The mixing layer thickness and turbulent energies grow as would be expected, and the mean velocity profile develops spatially in accordance with similarity theory and laboratory experiment. These simulations are a very powerful tool in gaining insight into the physics of mixing layer evolution, as a study of a forced mixing layer has shown (Riley and Metcalfe (1980a)).

## V Conclusions

Direct numerical simulations of turbulent free shear flows have been shown to be in generally good agreement with self-similarity theory and laboratory data. The temporal development of a number of important statistical quantities has been shown to have the proper behavior, and the spatial evolution of the flow field is consistent with predictions. The main potential difficulties with the application of these methods are the inadequate initialization procedure and statistical scatter in the data. However, we do not see these problems as being insurmountable. The evolution of turbulence in a free shear flow is governed by a complex balance among the following mechanisms: generation of energy from the mean flow, turbulent diffusion by the turbulence, intercomponent transfer of energy, and dissipation. Our statistical results indicate that the direct numerical simulations are indeed faithfully representing these physical processes. We are quite optimistic about the future usefulness of direct numerical simulations as a scientific research tool.

## Acknowledgments

The authors would like to acknowledge the generous assistance of Dr. Robert Rogallo in converting his Illiac IV code to perform the $64^3$ simulations described in this paper. The authors also benefitted from useful discussions with Dr. Steven Orszag, who wrote the original version of the code used for the $32^3$ simulations. This work was supported by the NASA Ames Research Center under Contract No. NAS2-9855 and by the Office of Naval Research under Contract No. N00014-78-C-0346. Computer time was provided by the NASA Ames Research Center.

## References

Higuchi, Hiroshi (1977) PhD Thesis, California Institute of Technology.
Brown, G. L. and Roshko, A. (1974) J. Fluid Mech., 64, pp. 775-816.
Orszag, S. A. (1971) Stud. Appl. Math, 50, p. 395.
Orszag, S. A. and Pao, Y. H. (1974) Adv. Geophys., Vol. 18A, pp. 225-236.
Pao, Y.-H., and Lin, J.-T. (1973) Flow Research Report No. 12, December.
Rogallo, R. S. (1977) NASA Document No. NASA TM-73,203.
Riley, J. J. and Metcalfe, R. W. (1978) NASA document NASA CR-152282.
*Riley, J. J. and Metcalfe, R. W. (1980a) Paper No. AIAA-80-0274.
*Riley, J. J. and Metcalfe, R. W. (1980b) from Turbulent Shear Flows II, (ed. by Bradbury, et al.), Springer, Verlag.
Wygnanski, I. and Fiedler, H. E. (1970) J. Fluid Mech., 41, pp. 327-361.
*Reprints may be obtained from the authors c/o Flow Research Company, Kent, Washington 98031 U.S.A.

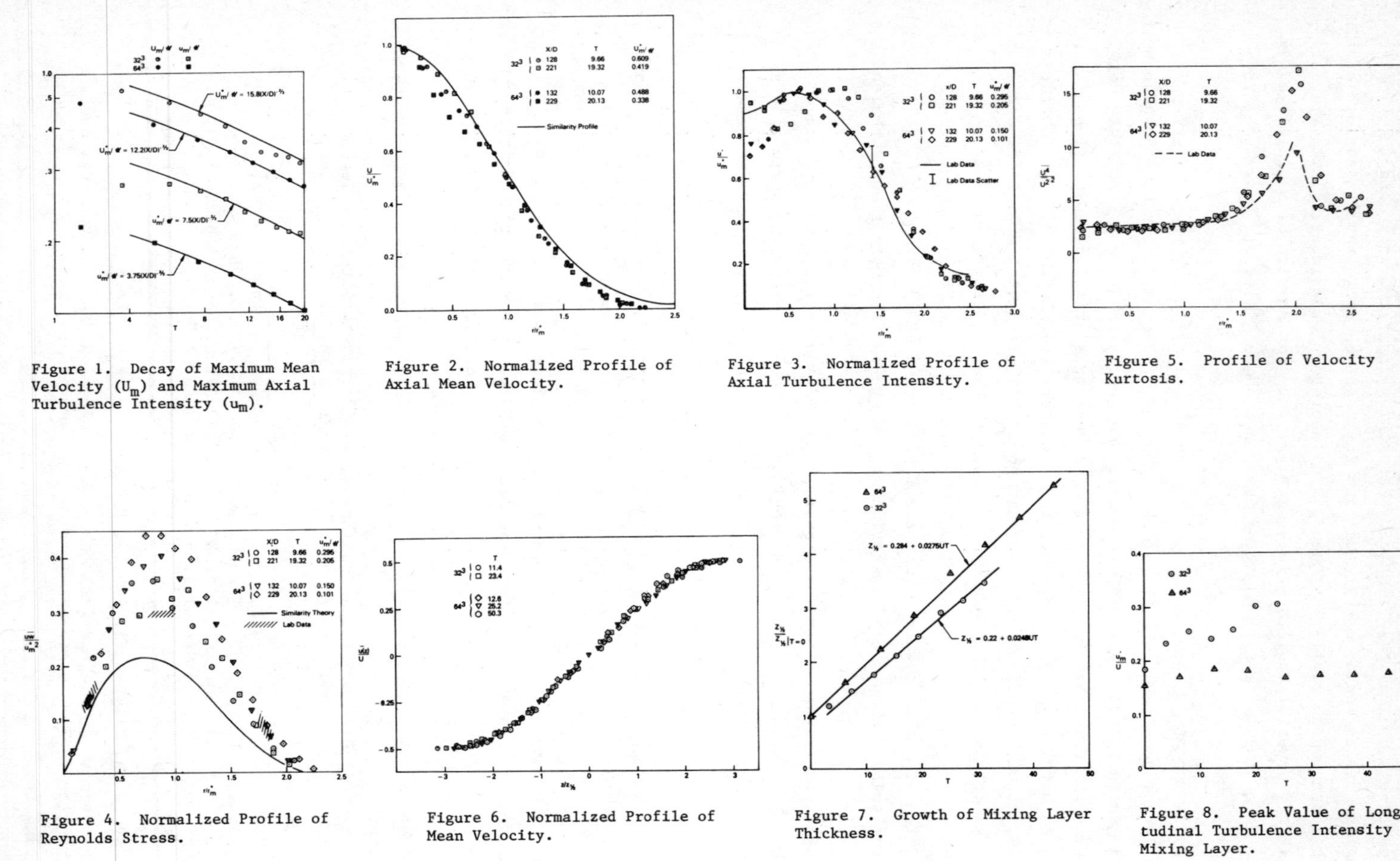

Figure 1. Decay of Maximum Mean Velocity ($U_m$) and Maximum Axial Turbulence Intensity ($u_m$).

Figure 2. Normalized Profile of Axial Mean Velocity.

Figure 3. Normalized Profile of Axial Turbulence Intensity.

Figure 5. Profile of Velocity Kurtosis.

Figure 4. Normalized Profile of Reynolds Stress.

Figure 6. Normalized Profile of Mean Velocity.

Figure 7. Growth of Mixing Layer Thickness.

Figure 8. Peak Value of Longitudinal Turbulence Intensity for Mixing Layer.

NUMERICAL SOLUTION OF THE NAVIER-STOKES EQUATIONS

BY MEANS OF A MULTIGRID METHOD AND NEWTON-ITERATION

by

W.J.A. Mol
Mathematical Centre
Amsterdam, the Netherlands

## 1. INTRODUCTION

In this paper a multigrid method for the solution of elliptic boundary value problems in a rectangle is considered. A 7-point restriction and prolongation operator is introduced, with which a Galerkin approximation can be defined as coarse grid operator. A 7-point incomplete LU-decomposition is chosen as smoothing operator. It is shown that the method is fast and robust for a large variety of problems. Especially some numerical experiments on the Navier-Stokes equations are reported: the driven cavity and the flow around a cylinder.

## 2. MULTIGRID METHODS

We consider a linear elliptic partial differential equation denoted by:

$$(2.1) \qquad Au = f$$

and valid in the unit square $\Omega = \{(x,y) \mid 0 < x < 1,\ 0 < y < 1\}$. Boundary conditions are defined on the boundary $\partial\Omega$ of $\Omega$. A computational grid $\Omega^\ell$ and a corresponding set of grid-functions $U^\ell$ are defined by:

$$\Omega^\ell = \{(x_i, y_j) \mid x_i = i.2^{-\ell},\ y_j = j.2^{-\ell},\ i = 0(1)2^\ell,\ j = 0(1)2^\ell\},$$

$$(2.2)$$

$$U^\ell = \{u^\ell \colon \Omega^\ell \to \mathbb{R}\}.$$

After discretization of (2.1) and the boundary conditions we obtain an algebraic system of equations denoted by:

$$(2.3) \qquad A^\ell u^\ell = f^\ell,$$

with $A^\ell \colon U^\ell \to U^\ell$.

The multigrid method makes use of a hierarchy of computational grids $\Omega^k$ and corresponding sets of grid-functions $U^k$, $k = \ell-1(-1)1$, defined by (2.2) with $\ell$ replaced by $k$. On the coarser grids (2.3) is approximated by:

$$(2.4) \qquad A^k u^k = f^k, \qquad k = \ell-1(-1)1,$$

with $A^k$ some suitably chosen coarse grid operator. A restriction operator $R^k$ and a prolongation operator $P^k$ are introduced:

$$(2.5) \qquad R^k \colon U^k \to U^{k-1}, \qquad P^k \colon U^{k-1} \to U^k.$$

Finally, we define a smoothing operator on each reduction level k:

$$(2.6) \qquad u^k := S(k, A^k, u^k, f^k).$$

A class of multigrid methods can be described in quasi-Algol as follows:

```
procedure multigrid (k,u,f); integer k; array u,f;
begin    integer q;
         if k=1 then u^k:=(A^k)^{-1}f^k
         else
         begin for q:=1(1)qa[k] do u^k:=S(k,A^k,u^k,f^k);
               f^{k-1}:= R^k (f^k-A^k u^k); u^{k-1}:=0;

               for q:=1(1)qc[k] do multigrid (k-1,u,f);
               u^k:=u^k+P^k u^{k-1};

               for q:=1(1)qb[k] do u^k:=S(k,A^k,u^k,f^k);
         end;
```

One execution of multigrid ($\ell$,u,f) will be defined as one multigrid iteration.

Most multigrid strategies described in the literature can be obtained as cases of the foregoing algorithm for special choices of the parameters qa[k], qb[k], qc[k] and the operators $P^k$, $R^k$, $A^k$ and $S(k,A^k,u^k,f^k)$. Our multigrid strategy will be described for the case that (2.3) is a 7-point finite difference approximation to a general second order elliptic partial differential equation (2.1) containing mixed derivatives. The difference molecule is given in the accompanying figure.

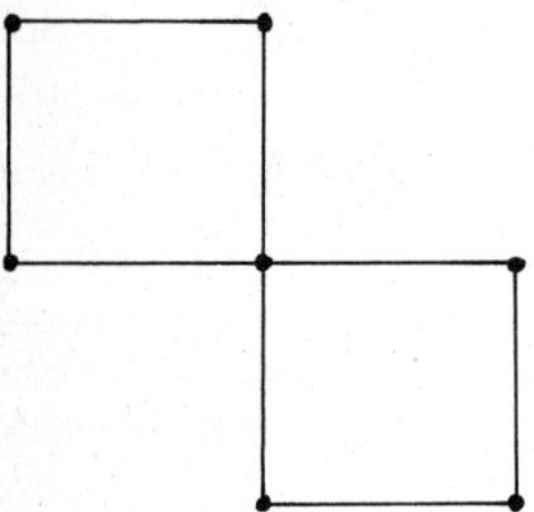

Finite difference molecule of (2.3)

Furthermore, the following choices are made ($k = \ell-1(-1)2$):

Parameters:

$$(2.7) \qquad qa[k] = 0, \quad qb[k] = 1, \quad qc[k] = 1$$

Restriction:

$$(2.8) \qquad (R^k u^k)_{i,j} = \frac{1}{4} u^k_{2i,2j} + \frac{1}{8} (u^k_{2i+1,2j} + u^k_{2i-1,2j} + u^k_{2i,2j+1} + u^k_{2i,2j-1} + u^k_{2i+1,2j-1} + u^k_{2i-1,2j+1})$$

Prolongation:

$$(2.9) \quad \begin{aligned} (P^k u^{k-1})_{2i,2j} &= u^{k-1}_{i,j}; \quad (P^k u^{k-1})_{2i+1,2j} = \frac{1}{2}(u^{k-1}_{i,j} + u^{k-1}_{i+1,j}) \\ (P^k u^{k-1})_{2i,2j+1} &= \frac{1}{2}(u^{k-1}_{i,j} + u^{k-1}_{i,j+1}); \quad (P^k u^{k-1})_{2i+1,2j+1} = \frac{1}{2}(u^{k-1}_{i+1,j} + u^{k-1}_{i,j+1}) \end{aligned}$$

Coarse grid operator:

$$(2.10) \quad A^{k-1} = R^k A^k P^k$$

Smoothing operator:

$$(2.11) \quad S(k, A^k, u^k, f^k) = u^k + B^k(f^k - A^k u^k)$$

with $B^k$ the 7-point incomplete LU-decomposition (ILU-7) of $A^k$

$$(2.12) \quad B^k = (\tilde{L}^k \tilde{U}^k)^{-1}.$$

The matrices $\tilde{L}^k$ and $\tilde{U}^k$ are constructed as described by Wesseling and Sonneveld (1980) with whom the use of ILU-decomposition for smoothing in the multigrid method originates.

Another novel feature in the present method is the use of 7-point restriction (2.8) and prolongation (2.9) operators. The use of Galerkin approximations for the coarse grid operators according to (2.10) has been considered by Frederickson (1975). Hackbush (1978), Wesseling and Sonneveld (1980) and Wesseling (1980). Brandt (1977, 1979) takes for $A^k$, $k = \ell-1(-1)1$ finite difference approximations: (2.3) with $\ell$ replaced by k.

# 3. SMOOTHING ANALYSIS AND SOME NUMERICAL EXPERIMENTS

For smoothing analyses based on Fourier mode analysis, I refer to Brandt (1977). He found for point and line Gauss Seidel applied to the usual 5-point descretization of the Poisson equation smoothing factors $\bar{\mu} = 0.50$ and $\bar{\mu} = 1/\sqrt{5} \cong 0.447$ respectively. In the same way, we can find smoothing factors for 5-point and 7-point incomplete LU-decomposition. Using his notation we obtain as convergence factor $\mu(\theta)$ for the ILU-5:

$$(3.1) \quad \mu(\theta) = \frac{a.\cos(\theta_1-\theta_2)}{2-\cos\theta_1-\cos\theta_2+a.\cos(\theta_1-\theta_2)}$$

with $a = 1 - \frac{1}{2}\sqrt{2}$ and for the ILU-7:

$$(3.2) \quad \mu(\theta) = \frac{b.\cos(2\theta_1-\theta_2)}{2-\cos\theta_1-\cos\theta_2+b.\cos(2\theta_1-\theta_2)}$$

with $b = 0.11181$. The corresponding smoothing factors for ILU-5 and ILU-7 are $\bar{\mu} = 0.204$ and $\bar{\mu} = 0.126$ respectively. In the following table we assume that these smoothing factors are representative for the general cases: the 5-point and the 7-point discretization of a general elliptic equation with variable coefficients of the same order of magnitude. The table gives smoothing factors factors $\bar{\mu}$, numbers of operations per grid point per iteration step ($n_i$, $i = 1,2,3$) and numbers of operations per grid point for $10^{-1}$ reduction of the error $(n_i/|\log \bar{\mu}|, \; i = 1,2,3)$.

| Method | Poisson | | | General 5-point | | General 7-point | |
|---|---|---|---|---|---|---|---|
| | $\bar{\mu}$ | $n_1$ | $n_1/|\log \bar{\mu}|$ | $n_2$ | $n_2/|\log \bar{\mu}|$ | $n_3$ | $n_3/|\log \bar{\mu}|$ |
| Point Gauss–Seidel | 0.50 | 5 | 16.6 | 9 | 29.9 | 13 | 43.2 |
| Line Gauss–Seidel | 0.447 | 8 | 22.9 | 14 | 40.1 | 18 | 51.5 |
| ILU–5 | 0.204 | 11 | 15.9 | 14 | 20.3 | 14 | 20.3 |
| ILU–7 | 0.126 | 15 | 16.7 | 18 | 20.0 | 18 | 20.0 |

Table 3.1. Smoothing factors and estimate of the number of operations per grid point for $10^{-1}$ reduction of the error.

On the basis of this table ILU–5 and ILU–7 are better than the two Gauss–Seidel methods for the general cases. In the case of singularly perturbed problems smoothing analysis demonstrates that incomplete LU–decomposition is less sensitive to ordering of grid points and other directional effects than Gauss–Seidel.
The number of operations in one multigrid iteration with the adopted strategy in Chapter 2 is

$$\text{Poisson: } 27\tfrac{2}{3} \text{ operations/gridpoint}$$

$$\text{General 5 or 7–point case: } 32\tfrac{2}{3} \text{ operations/gridpoint}$$

Results of numerical experiments with this multigrid method will be given. The multigrid iterations are terminated when the maximum of the difference between two iterands is smaller than $10^{-6}$

$$(3.3) \qquad |z_1^{(\sigma)}| = |(u^\ell)^{(\sigma+1)} - (u^\ell)^{(\sigma)}| < 10^{-6}$$

Furthermore, we define the average reduction factor:

$$(3.4) \qquad r_{av} = \left(\frac{|z_1^{(\sigma)}|}{|z_1^{(0)}|}\right)^{\frac{1}{\sigma}} \qquad \sigma \neq 0$$

Table 3.2 gives the average reduction factors for some elliptic problems. The functions f and the boundary conditions are chosen so that the exact solution in column 2 is approximated. The problems are valid in the unit square. The mesh width of the finest grid is $h = 1/32$. All problems are discretized by means of central differences, except $\partial\omega/\partial x$ and $\partial\omega/\partial y$ in problem 4. They are discretized with upwind differences.

| | Equation | Exact solution | $r_{av}$ |
|---|---|---|---|
| 1. | $\Delta\omega = f$ | $\omega = \sin(x) \cdot e^y$ | 0.018 |
| 2. | $\Delta\omega = 0$<br>$\Delta\psi = \omega$ | $\omega = 4\cos(x) \cdot \sinh(y)$<br>$\psi = 2x\sin(x) \cdot \sinh(y)$ | 0.017 |
| 3. | $\partial/\partial x\{(1+\sin x)\,\omega_x\} + \partial/\partial y\{(1+xy)\omega_y\} - \omega = f$ | $\omega = y(x+\cos(x))$ | 0.015 |
| 4. | $\partial\omega/\partial x - \partial\omega/\partial y = \Delta\omega/Re; \; Re = 10^4$ | $\omega = (1-e^{Rex}) \cdot (1-e^{-Rey})/(1-e^{Re})^2$ | 0.073 |

Table 3.2. $r_{av}$ for some problems

Note that the $r_{av}$ for the last, singular perturbed problem, is not much greater than for the other problems.

We can make an estimate of the number of operations for $10^{-1}$ reduction of the error for the Poisson equation:

$$27\tfrac{2}{3} \cdot 1/|\log 0.018| \cong 15.9 \text{ operation/gridpoint}$$

Compare with Brandt (1977): $\approx 28$ operations/gridpoint and Nicolaides (1979): 30-35 operations/gridpoint:

## 4. APPLICATION: THE NAVIER-STOKES EQUATIONS

Consider the driven square cavity flow with the Navier-Stokes equations in $(\omega,\psi)$-formulation:

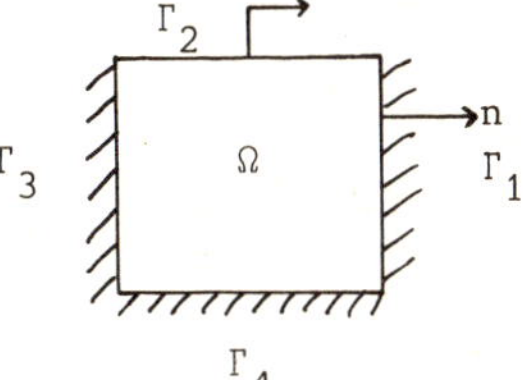

$$(4.1) \qquad \left.\begin{array}{c} \dfrac{\partial(\psi,\omega)}{\partial(y,x)} = \dfrac{1}{Re}\,\Delta\omega \\[2mm] \Delta\psi = \omega \end{array}\right\} \quad (x,y) \in \Omega,$$

with boundary conditions

$$(4.2) \qquad \psi = 0, \quad \frac{\partial\psi}{\partial n} = g$$

Fig. 4.1.

with $g = 0$ on $\Gamma_1$, $\Gamma_4$, $\Gamma_3$ and $g = 1$ on $\Gamma_2$.

An equidistant computational grid $\Omega^\ell$ is chosen:

$$(4.3) \qquad \Omega^\ell = \{(x_i,y_j)\,|\,x_i = (i+1)(2^\ell+2)^{-1}, \quad y_j = (j+1)(2^\ell+2)^{-1},$$
$$i = 0(1)2^\ell, \ j = 0(1)2^\ell\}$$

There is a slight difference with (2.2) because the boundary conditions are substituted in the difference scheme. The equations (4.1) are discretized centrally except the first derivatives of $\omega$, for instance $\partial\omega/\partial x$:

$$(4.4) \qquad \left.\frac{\partial\omega}{\partial x}\right|_{i,j} = \frac{(1+\alpha_{ij})(\omega_{i+1,j}-\omega_{ij})+(1-\alpha_{ij})(\omega_{ij}-\omega_{i-1,j})}{2h}$$

with $h = (2^\ell+2)^{-1}$ and $\alpha_{ij}$ the Il'in coefficient

$$(4.5) \qquad \alpha_{ij} = -\coth\left(\frac{Re\,\frac{\partial\psi}{\partial y}\big|_{ij}\,h}{2}\right) + \frac{2}{Re\,\frac{\partial\psi}{\partial y}\big|_{ij}\,h}\,.$$

The boundary conditions for $\omega$ are found by combining $\Delta\psi = \omega$ and $\partial\psi/\partial n = g$:

$$(4.6) \qquad \omega_w = \frac{3}{h^2}\,\psi_{w+1} - \frac{1}{2}\,\omega_{w+1} + \frac{3}{h}\,g_w.$$

w is a point of $\partial\Omega$, w+1 indicates its nearest neighbour in $\Omega^\ell$ in the direction of the normal.

The difference equations are Newton-linearized and the (linear) system in each Newton iteration is solved by the multigrid method. The termination criterium for the multigrid iterations is (3.3) and for the Newton iterations:

$$(4.7) \qquad |z_2^{(\rho)}| < |(u^\ell)^{(\rho+1)} - (u^\ell)^{(\rho)}| < 10^{-4}.$$

Experiments have been made for Reynolds numbers Re = 10, 50, 150. At Re = 10 we start with the zero solution, at the other Re-numbers with the solution of the preceding lower Re-numbers. The table 4.1 gives $n^{(\rho)}$, the number of multigrid iterations and $r_{av}^{(\rho)}$, the average reduction factor in the $\rho^{th}$ Newton iteration.

Note that $r_{av}$ does not increase as $h \downarrow 0$ and is insensitive to changes in the coefficients induced by Newton iteration. Furthermore, $r_{av}$ is comparable to $r_{av}$ for the Poisson equation.

Another example is the flow around a cylinder with radius R and a uniform flow with velocity $u_\infty$ at infinity. The non-dimensional Navier-Stokes equations are given in (4.8).

| Re | h | $n^{(1)}$ | $n^{(2)}$ | $n^{(3)}$ | $r_{av}^{(1)}$ | $r_{av}^{(2)}$ | $r_{av}^{(3)}$ |
|----|----|----|----|----|----|----|----|
| 10 | 1/6 | 4 | 2 | – | 0.027 | 0.026 | – |
|  | 1/10 | 5 | 2 | – | 0.059 | 0.033 | – |
|  | 1/18 | 5 | 2 | – | 0.056 | 0.034 | – |
|  | 1/34 | 5 | 2 | – | 0.056 | 0.034 | – |
| 50 | 1/6 | 5 | 4 | 2 | 0.031 | 0.042 | 0.049 |
|  | 1/10 | 5 | 4 | 1 | 0.052 | 0.061 | 0.056 |
|  | 1/18 | 6 | 4 | 2 | 0.082 | 0.056 | 0.051 |
|  | 1/34 | 6 | 4 | 2 | 0.083 | 0.062 | 0.052 |
| 150 | 1/6 | 7 | 6 | 3 | 0.099 | 0.092 | 0.079 |
|  | 1/10 | 6 | 5 | 3 | 0.083 | 0.084 | 0.078 |
|  | 1/18 | 6 | 5 | 2 | 0.080 | 0.083 | 0.064 |
|  | 1/34 | 6 | 5 | 2 | 0.081 | 0.082 | 0.063 |

Table 4.1. Results for square cavity flow.

$$(4.8) \qquad \left. \begin{array}{l} \dfrac{\partial(\psi,\omega)}{\partial(\eta,\xi)} = \dfrac{1}{Re}\,\Delta_{\xi\eta}\omega \\[2mm] \Delta_{\xi\eta}\psi = e^{2\xi}\omega \end{array} \right\}$$

with polar coordinates $x = e^{\xi}\cos\eta$, $y = e^{\xi}\sin\eta$. The Reynolds number is defined by:

$$(4.9) \qquad Re = \frac{u_\infty 2R}{\nu}\,.$$

The boundary conditions are:

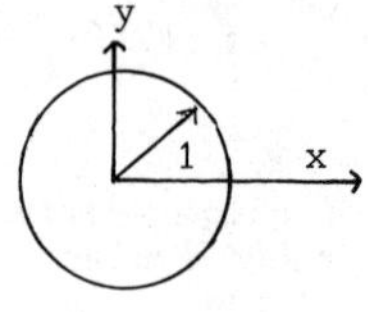

$$(4.10) \qquad \left. \begin{array}{l} \xi = 0: \ \psi = \dfrac{\partial\psi}{\partial\xi} = 0 \\[2mm] \eta = 0,\pi \text{ and } \xi \geq 0: \ \omega = \psi = 0 \\[2mm] \xi = \pi: \ \omega = 0, \ \psi = e^{\xi}\sin\eta \end{array} \right\}$$

Fig. 4.2

The computational region is $\Omega = \{(\xi,\eta)\,|\,0 \leq \xi \leq \pi,\ 0 \leq \eta \leq \pi\}$. The calculation is analogous with the calculation for the driven square cavity flow, so with Il'in upwind discretization. The results are presented in the following table:

| Re | h | $n^{(1)}$ | $n^{(2)}$ | $n^{(3)}$ | $r_{av}^{(1)}$ | $r_{av}^{(2)}$ | $r_{av}^{(3)}$ |
|----|----|----|----|----|----|----|----|
| 10 | $\pi/6$ | 7 | 5 | 1 | 0.140 | 0.110 | 0.034 |
|  | $\pi/10$ | 7 | 5 | 1 | 0.139 | 0.120 | 0.053 |
|  | $\pi/18$ | 7 | 5 | 1 | 0.141 | 0.119 | 0.064 |
|  | $\pi/34$ | 8 | 5 | 1 | 0.170 | 0.100 | 0.029 |
| 50 | $\pi/6$ | 8 | 5 | 1 | 0.178 | 0.106 | 0.042 |
|  | $\pi/10$ | 8 | 6 | 2 | 0.180 | 0.150 | 0.063 |
|  | $\pi/18$ | 8 | 6 | 3 | 0.175 | 0.153 | 0.068 |
|  | $\pi/34$ | 8 | 6 | 3 | 0.185 | 0.154 | 0.092 |
| 150 | $\pi/6$ | 8 | 7 | 4 | 0.180 | 0.201 | 0.105 |
|  | $\pi/10$ | 8 | 6 | 4 | 0.185 | 0.160 | 0.105 |
|  | $\pi/18$ | 8 | 7 | 4 | 0.186 | 0.200 | 0.108 |
|  | $\pi/34$ | 8 | 7 | 4 | 0.187 | 0.195 | 0.110 |

Table 4.2. Results for the flow around a cylinder.

Although still fast, the average reduction factors are greater than in the previous cases, but they are still insensitive to h and to changes in the coefficients induced by Newton iteration.

## 5. CONCLUSION

A multigrid method has been presented that is fast and robust in the sense that it works for a large variety of elliptic problems without needing tuning or special modifications. The use of incomplete LU-decomposition makes it possible to treat uniformly elliptic and singularly perturbed problems by one and the same method. The combination of incomplete LU smoothing and Galerkin coarse grid approximation looks very promising.

## 6. REFERENCES

BRANDT, A - Multi-level adaptive solutions to boundary value problems. Math. Comp. 31, 333-390, 1977.

BRANDT, A. - Multi-level adaptive techniques (MLAT) for singular perturbation problems. In: P.W. Hemker and J.J.H. Miller (eds.): Numerical Analysis of singular perturbation problems. London, etc.: Academic Press, 1979.

FREDERICKSON, P.O. - Fast approximate inversion of large sparse linear systems. Math. Report 7-75, Lakehead University, 1975.

HACKBUSH, W. - On the multi-grid method applied to difference equations. Computing 20, 291-306, 1978.

NICOLAIDES, R.A. - On some theoretical and practical aspects of multigrid methods. Math. Comp. 30, 933-952, 1979.

WESSELING, P. & P. SONNEVELD - Numerical experiments with a multiple grid and a preconditioned Lanczos method. In: R. Rautmann (ed.): Approximation problems for Navier-Stokes equations. Lecture Notes in Math. 771, Berlin etc.: Springer-Verlag, 1980.

WESSELING, P. - The rate of convergence of a multiple grid method. In: G.A. Watson (ed.): Numerical Analysis, Lecture Notes in Math. 773, Berlin etc.: Springer-Verlag, 1980.

ANALYTIC STRUCTURE OF HIGH REYNOLDS NUMBER FLOWS

Rudolf H. Morf, RCA Laboratories, Zurich, Switzerland

Steven A. Orszag and Daniel I. Meiron, Dept. Math., M.I.T., Cambridge, MA ,USA

Uriel Frisch, C.N.R.S. Observatoire de Nice, Nice, France

and

Maurice Meneguzzi, C.N.R.S., Section d'Astrophysique,
C.E.N., Saclay, France

## INTRODUCTION

The structure of small-scales in high-Reynolds number flows is central to the
understanding of the fundamental dynamics of turbulence. Upon taking the limit of
zero dissipation, these small-scale features are directly related to the singularities
of the inviscid flows. We have recently made progress in understanding the
analytic structure of flows using techniques borrowed from the analysis of phase
transitions in statistical mechanics. In this paper we outline some of these
applications of series analysis techniques.

We shall confine attention here to the analytic structure in time of the
solutions to various partial differential equations. High-order power series in
t are developed (using multiple precision computer codes) and then analyzed
using Pade approximants and ratio methods (Baker 1975, Bender & Orszag, 1978,
Gaunt & Guttmann,1974).

## BURGERS' EQUATION

The solution to the inviscid Burgers' equation

$$\frac{\partial u(x,t)}{\partial t} + u(x,t) \frac{\partial u(x,t)}{\partial x} = 0 \tag{1}$$

with initial condition $u(x,0) = \cos x$ satisfies the functional equation
$u = \cos(x-ut)$. To study its singularity structure we form the enstrophy

$$\Omega(t) = \int_0^{2\pi} u_x^2 \, dx = \sum_{n=0}^{\infty} a_n t^{2n}. \tag{2}$$

The radius of convergence of this series is the magnitude of the smallest singularity of $\Omega(t)$. The Domb-Sykes plot in Fig. 1 of the ratios $a_{n+1}/a_n$ of
coefficients of the power series (2) suggests that the radius of convergence of
(2) is 1 and that, near $t = 1$, $\Omega(t)$ behaves like $(1-t^2)^{-1/2}$. In fact,
it is not hard to show that the exact expression for $\Omega(t)$ for this solution
of (1) is

$$\Omega(t) = t^{-2}(1-t^2)^{-1/2} - t^{-2}, \tag{3}$$

in agreement with the result suggested by the Domb-Sykes plot.

The singularity of $\Omega(t)$ at $t = 1$ has a simple physical interpretation.
At $t = 1$, a shock forms in this solution of Burgers' equation, representing
the abrupt formation of small-scale structure.

## THREE-DIMENSIONAL EULER EQUATIONS

The classical theorems of Kelvin and Helmholtz imply that, in an inviscid incompressible fluid of constant density, vortex lines move with the fluid and vorticity is amplified proportional to the stretching of a vortex line element.  For boundary-free flow, these theorems imply that an initially smooth, inviscid flow remains smooth so long as vortex lines are stretched only a finite amount.  The spontaneous appearance of a singularity in a flow after a finite time requires that vortex lines develop infinite length in a finite time.  One way this can happen is by the vortex lines becoming intricately wound up and twisted, even though their end points have not separated more than a finite distance.  Recently, Morf, Orszag & Frisch (1980) have analyzed the Taylor-Green three-dimensional vortex flow and have found substantial evidence that a singularity does indeed occur after a finite time.  This singularity is the first manifestation of turbulence with excitation at arbitrarily small spatial scales.

The Taylor-Green vortex is that flow that develops from the initial conditions

$$u(x,y,z,t=0) = v(-y,x,z,t=0) = \cos x \, \sin y \, \cos z \tag{4}$$

$$w(x,y,z,t=0) = 0. \tag{5}$$

We define the generalized enstrophy and its power series in $t$ by

$$\Omega_p(t) = \sum_{\vec{k}} k^{2p} |\vec{u}(\vec{k},t)|^2 = \sum_{n=0}^{\infty} A_n^{(p)} t^{2n} \tag{6}$$

where $\vec{u}(\vec{k},t)$ are the Fourier expansion coefficients of the velocity field $\vec{v}(\vec{x},t)$.  Here the series expansion of $\Omega_p(t)$ involves only even powers of $t$ because of time-reversal invariance.

We have calculated $\Omega_p$ to order $t^{44}$ by recursive evaluation of the time derivatives $\partial^m \vec{u}(\vec{k})/\partial t^m |_{t=0}$ using the Navier-Stokes equation.  Several observations are crucial in writing an efficient code to do this computation.  First, with the initial conditions (4) – (5), the first nonzero time derivative of $\vec{u}(\vec{k},0)$ is of order $m \geq \max_\alpha || k_\alpha |-1|$.  Also, the quadratic nature of $\Omega_p$ in $\vec{u}(\vec{k})$ implies that computation of $A_n^{(p)}$ requires that time derivatives of $\vec{u}(\vec{k},0)$ be known only up to order $2n - \max_\alpha | |k_\alpha|-1|$.  Since the number of required computations to calculate to order $t^N$ scales as $N^8$, it turns out that the latter two constraints lead to a speedup by a factor of about 30 in addition to the very substantial computational efficiencies afforded by the symmetries of the Taylor-Green vortex (Orszag 1974).  Double-precision calculation of the series coefficients of $\Omega_p(t)$ through order $t^{44}$ required about 7h on a CDC 7600.  The same computation would require roughly double the time on a CRAY-1 computer, because vectorized double precision operations on the CRAY-1 run more than two orders of magnitude slower than single-precision calculations.

In Table I, we list $A_n^{(p)}$ for $n = 0,\dots,22$ and $p = 1,1/3$.  An analysis of the effect of roundoff error indicates that each coefficient has at least 10 significant digits.

Padé approximants of $\Omega_1$ as a function of $t^2$ suggest the existence of a real singularity at $t^2 \simeq 27$.  In order to reveal this singularity using the ratio method, Morf, Orszag & Frisch observed that the radius of convergence of the series for $\Omega_p$ is determined by an unphysical singularity at $t^2 \simeq -5$.  This unphysical singularity is found by the ratio test applied to the series coefficients listed in Table I.  Morf et al performed the Euler transformation $w = 6t^2/(t^2+5)$ to move the unphysical singularity to $w \simeq \infty$.  The possible physical singularity at $t^2 \simeq 27$ now determines the radius of convergence of the transformed series $\bar{\Omega}_p(w)$.  The Domb-Sykes plot of the ratios of the series coefficients of $\bar{\Omega}_p(w)$ in Fig. 2 for $p = 1-4$ shows that these series possess a common singularity near $\frac{1}{w} \approx 0.197$ or $t^2 \approx 27$.

In addition, analysis of the Domb-Sykes plot in Fig. 2 shows that $\Omega_p$ behaves as a power law $(t_*-t)^{-\gamma(p)}$ near the singularity time $t_*$. A plot of the critical exponents $\gamma(p)$ versus $p$ is given in Fig. 3. If the energy spectrum of the flow at $t_*$ behaved like $k^{-s}$ as $k \to \infty$, then the largest zero of $\gamma(p)$ would lie at $p = \frac{1}{2}(s-1)$. If the energy spectrum at $t_*$ satisfies the Kolmogorov $k^{-5/3}$ inertial range law then the largest zero of $\gamma(p)$ should lie at $p = 1/3$. Our current analysis of $\gamma(p)$ for small positive $p$ is consistent with $\gamma(p) > 0$ for all $p > 0$, but this result is uncertain without knowledge of the series coefficients beyond $t^{44}$. However the energy spectrum for $t_*-0$ may differ from that for times larger than $t_*$ because energy transfer may not begin until $t_*+0$. The series coefficients of $\Omega_{1/3}$ are given in Table I in the hope that they can be used later to elucidate the nature of the inertial range.

By comparison of various Padé approximants to $\Omega_p$, it is possible to piece together a map of the singularities of the Taylor-Green vortex in the complex-$t$ plane (see Fig. 4). The resolution of the Padé approximants is limited to a circle in this plane with a radius of about 5. Finally, in Fig. 5, we compare the results of direct numerical simulation, Padé approximation, and series truncation applied to $\Omega_1$. Observe that for $t \gtrsim \sqrt{5}$, the power series diverges. Also, as $t \to t_* \approx 5.2$, all the approximations diverge.

## TWO-DIMENSIONAL EULER EQUATIONS

The restriction of a flow to two space dimensions precludes vortex-line stretching so there can be no real physical singularities of the generalized enstrophies $\Omega_p(t)$. On the other hand, it is possible for complex-$t$ singularities to occur. To investigate this phenomenon and to test the ability of our series analysis methods to resolve the existence of singularities, we have calculated the series expansion coefficients for that two-dimensional flow that develops from the initial conditions

$$u(x,y,0) = - \sin y - \sin 2y, \quad v(x,y,0) = \sin x + \sin 2x \qquad (7)$$

As expected, no real singularities are found. A singularity map of the complex-$t$ plane is given in Fig. 6. Evidently, the Sobolev norms (6) are not analytic for all $t$.

## TWO-DIMENSIONAL INVISCID MAGNETOHYDRODYNAMIC FLOW

Pouquet (1978) suggested that real-time singularities may appear in inviscid two-dimensional MHD flow. Orszag & Tang (1979) found some evidence for this singular behavior by direct numerical simulation at moderately high Reynolds numbers (see Fig. 1 of Orszag & Tang). Using the initial conditions suggested by Orszag & Tang,

$$u(x,y,0) = - \sin y, \quad v(x,y,0) = \sin x, \qquad (8)$$

$$A(x,y,0) = \cos y + \frac{1}{2} \cos 2x, \qquad (9)$$

where $A$ is the magnetic potential, we have calculated the series coefficients of the total enstrophy $\Omega_p$ up to $t^{44}$.

The results obtained by double precision calculations for $\Omega_1$, $\Omega_2$, and $\Omega_3$ are listed in Table II. An analysis of these series using Pade approximants and ratio methods suggest that there is no real-time singularity. The behavior observed by Orszag & Tang seems to be due to the presence of complex singularities located near, but not on, the real axis. A singularity map of the complex-$t$ plane for this problem obtained by Padé analysis is given in Fig. 7.

Some additional evidence for the absence of a real singularity in inviscid two-dimensional MHD flow is obtained by analysis of the energy spectrum of the inviscid flow. The energy spectrum at time $t$, computed by a high-resolution spectral code, seems to behave like $e^{-\alpha(t)k}$ for large $k$. A singularity for the Sobolev norms (6) corresponds to a time at which $\alpha(t) = 0$. In Fig. 8, we plot $\alpha$ versus $t$ in this simulation. It seems that $\alpha$ behaves exponentially in time and does not achieve the value $0$ at finite time. Evidently, even though the presence of a magnetic field destroys the two-dimensional vorticity constraint and allows vortex stretching, it is not possible to stretch vorticity an infinite amount in a finite time in this flow. More details on the analysis of MHD flow will be given in a later paper by us.

## KELVIN-HELMHOLTZ INSTABILITY OF A VORTEX SHEET

The time evolution of an inviscid vortex sheet is governed by Birkhoff's equation

$$\frac{\partial z^*(e,t)}{\partial t} = \frac{1}{2\pi i} \, P \int \frac{\gamma(e')de'}{z(e,t)-z(e',t)} \tag{10}$$

where $z$ is the location of the sheet in the complex plane, $\gamma$ is the vortex sheet strength, and $e$ is a Lagrangian marker coordinate. Moore (1979) has used small-amplitude asymptotic methods to argue that a singularity appears after a finite time during vortex sheet rollup. Here we present a model problem amenable to exact finite-amplitude series analysis.

Meiron, Baker & Orszag (1980) show that if

$$z(e,0) = e, \quad \gamma(e) = 1 + a \cos e, \tag{11}$$

then the series expansion of $z(e,t)$ in powers of $t$ has the special form

$$z(e,t) = e + \sum_{n=1}^{\infty} t^n \sum_{j=1}^{n} a_{nj} \sin je. \tag{12}$$

We have computed the expansion coefficients of

$$\Omega_p(t) = \int_0^{2\pi} \left| \frac{\partial^p z}{\partial e^p} \right|^2 de = \sum_{n=0}^{\infty} C_n^{(p)} t^{2n} \tag{13}$$

up to order $t^{20}$ for various values of $a$. These coefficients have been calculated in double precision using a spectral transform method based on the representation (13) that requires order $N^6$ operations to obtain all terms through order $t^N$ (see Meiron et al 1980 for more details).

In Fig. 9, we give a Domb-Sykes plot for the series coefficients with $a = 1/4$. Evidently, the radius of convergence of $\Omega_p(t)$ is determined by a singularity on the real-time axis at $t \simeq 2.2$. As shown in Fig. 10, the variation of the singularity time $t_*$ as a function of the disturbance amplitude $\underline{a}$ is in qualitative agreement with Moore's leading order asymptotic result

$$1 + \frac{1}{2}t_* + \ln t_* = \ln 4/a \tag{14}$$

## HIGH REYNOLDS NUMBER FLOWS

When viscosity is introduced into the equations studied above, the singularity structure of the flows changes drastically. The analysis becomes much more difficult and only partial results are now available. In order to illustrate the kinds of new behavior possible, we consider the simplest example, namely Burgers' equation. At finite viscosity, the generalized enstrophies $\Omega_p(t)$ for typical solutions of Burgers' equation have zero radius of convergence in $t$. As is well known, the backwards heat equation is ill-posed for initial-value problems. It may be shown that this is the origin of the singularity of $\Omega_p(t)$ at $t = 0$ and that their series expansion coefficients $A_n^{(p)}$ grow like $n!$ as $n$ increases. Under these circumstances, Borel summation methods maybe used to investigate the structure of singulariites of $\Omega_p(t)$.

Similarly, we expect that the generalized enstrophies $\Omega_p(t)$ of solutions to the three-dimensional Navier-Stokes equations have zero radius of convergence in $t$ at finite Reynolds numbers. Analysis of the Taylor-Green problem at Reynolds numbers of 100 and 1000 is currently under way.

We would like to acknowledge helpful discussions with Drs. G. R. Baker, B. G. Nickel, A. Pouquet, and P.-L. Sulem. The computations reported in this paper were done at the Computing Facility of the National Center for Atmospheric Research, Boulder, Colorado, which is sponsored by the National Science Foundation. This work was supported by the Office of Naval Research under Contracts Nos. N00014-77-C-0138 and N00014-79-C-0478 and the National Science Foundation under Grants Nos. ATM-7817092 and DMR-7710120.

## REFERENCES

Baker Jr., G. A. 1975 Essentials of Padé Approximants, Academic Press, New York.

Bender, C. & Orszag, S.A. 1978 Advanced Mathematical Methods for Scientists and Engineers, McGraw Hill, New York.

Gaunt, D. S. & Guttmann, A. J. 1974 in Phase Transitions and Critical Phenomena (ed. C. Domb and M. S. Green), Vol. 3, p. 191, Academic Press, London.

Meiron, D.I., Baker G. R. & Orszag, S.A. 1980 to appear.

Morf, R.H., Orszag, S.A. & Frisch, U. 1980 Phys. Rev. Lett. 44, 572.

Moore, D. W. 1979 Proc. Roy. Soc. London, Ser. A 365, 105.

Orszag, S. A. & Tang, C.M. 1979 J. Fluid Mech. 90, p. 129.

Orszag, S. A. 1974, in Computing Methods in Applied Sciences (ed. R. Glowinski and J. L. Lions), part 2, p. 150, Springer, Berlin

Pouquet A. 1978 J. Fluid Mech. 88,1.

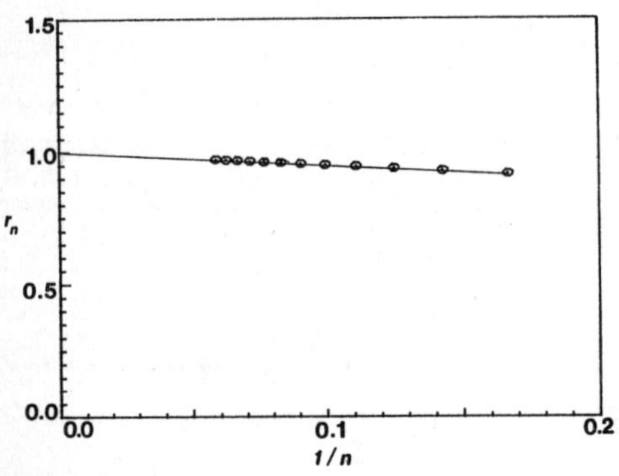


FIG. 1. THE DOMB-SYKES PLOT FOR THE ENSTROPHY SERIES OF THE INVISCID BURGERS' EQUATION WITH INITIAL CONDITION $U(X,0) = \cos X$. HERE $R_N = A_{N+1}/A_N$ WHERE $A_N$ IS DEFINED IN (2). THE EXTRAPOLATION TO $N = \infty$ INDICATES A SINGULARITY AT $T = 1$ WITH CRITICAL EXPONENT 1/2.

| N | $A_N^{(1)}$ | $A_N^{(1/3)}$ |
|---|---|---|
| 0 | $7.500000000000000000 \times 10^{-1}$ | $3.605623925768520956 \times 10^{-1}$ |
| 1 | $7.812500000000000000 \times 10^{-2}$ | $8.714850463946744026 \times 10^{-3}$ |
| 2 | $5.918560606060606061 \times 10^{-3}$ | $2.376048663073870882 \times 10^{-4}$ |
| 3 | $-2.784360642534797738 \times 10^{-4}$ | $-4.947620025036309772 \times 10^{-5}$ |
| 4 | $6.104821360841404098 \times 10^{-5}$ | $6.503543683920781261 \times 10^{-6}$ |
| 5 | $-9.636134340637909700 \times 10^{-6}$ | $-8.280036365354740656 \times 10^{-7}$ |
| 6 | $1.237645143293940934 \times 10^{-6}$ | $1.306359600675467308 \times 10^{-7}$ |
| 7 | $-1.300167339948584205 \times 10^{-7}$ | $-2.077384192774363309 \times 10^{-8}$ |
| 8 | $1.150766494021460551 \times 10^{-8}$ | $3.048913568845232405 \times 10^{-9}$ |
| 9 | $-4.068992793050291878 \times 10^{-10}$ | $-4.794463136496165291 \times 10^{-10}$ |
| 10 | $-1.572264704871744723 \times 10^{-10}$ | $8.753689208413707266 \times 10^{-11}$ |
| 11 | $5.645708146214230001 \times 10^{-11}$ | $-1.647248801819486637 \times 10^{-11}$ |
| 12 | $-1.369692878285606328 \times 10^{-11}$ | $3.066666808926901586 \times 10^{-12}$ |
| 13 | $2.992019895422864045 \times 10^{-12}$ | $-5.879032730116968158 \times 10^{-13}$ |
| 14 | $-6.199196318336325236 \times 10^{-13}$ | $1.149676993531427004 \times 10^{-13}$ |
| 15 | $1.245094572691313975 \times 10^{-13}$ | $-2.232187167120197005 \times 10^{-14}$ |
| 16 | $-2.477447982899959415 \times 10^{-14}$ | $4.338664256991685632 \times 10^{-15}$ |
| 17 | $4.925284134033088544 \times 10^{-15}$ | $-8.517733476863733154 \times 10^{-16}$ |
| 18 | $-9.771904182930658956 \times 10^{-16}$ | $1.674914676948627021 \times 10^{-16}$ |
| 19 | $1.937544476710612003 \times 10^{-16}$ | $-3.292720331968618086 \times 10^{-17}$ |
| 20 | $-3.850599084534800708 \times 10^{-17}$ | $6.508744975750743035 \times 10^{-18}$ |
| 21 | $7.670128722686259747 \times 10^{-18}$ | $-1.292344661251738540 \times 10^{-18}$ |
| 22 | $-1.530106963188030747 \times 10^{-18}$ | $2.569321323546438128 \times 10^{-19}$ |

TABLE 1. COEFFICIENTS $A_N^{(p)}$ OF THE SERIES EXPANSION OF $\Omega_p(T)$ FOR THE TAYLOR-GREEN VORTEX IN POWERS OF $T^2$.

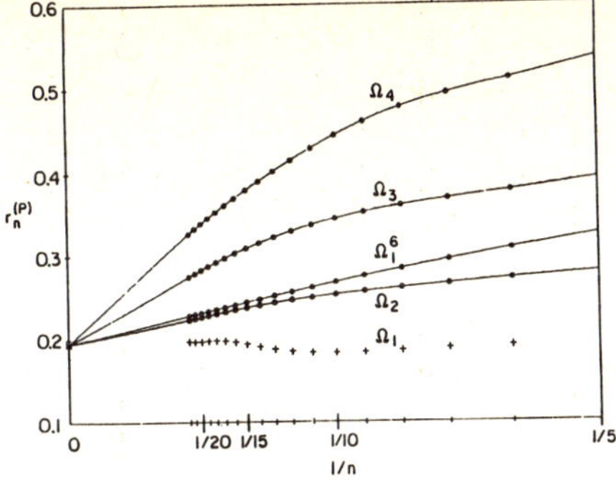

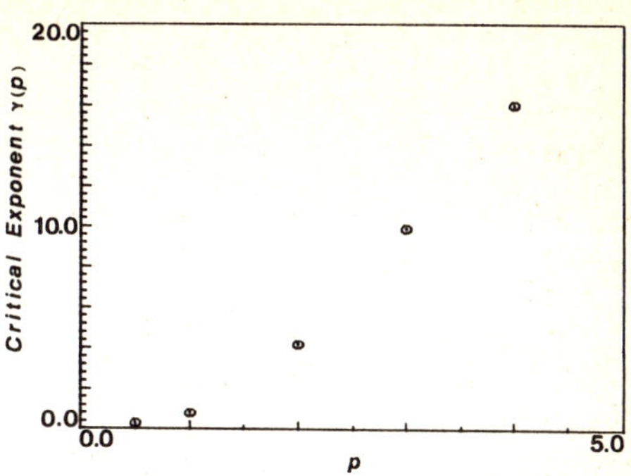

FIG. 2.   THE DOMB-SYKES PLOT FOR THE GENERALIZED ENSTROPHIES OF THE
TAYLOR-GREEN VORTEX. HERE $R_N^{(P)}$ ARE THE RATIOS OF THE
SERIES EXPANSION COEFFICIENTS OF THE EULER TRANSFORM ENSTROPHIES
$\bar{\Omega}_P(W)$. THE PIECEWISE QUADRATIC EXTRAPOLATIONS IN THIS PLOT
INDICATE A REAL-TIME SINGULARITY AT   $1/W \cong 0.197$  OR  $T_* \cong 5.2$.

FIG. 3.   A PLOT OF THE CRITICAL EXPONENT  $\gamma(P)$  OF THE GENERALIZED
ENSTROPHY FOR THE TAYLOR-GREEN VORTEX AS A FUNCTION OF  P .

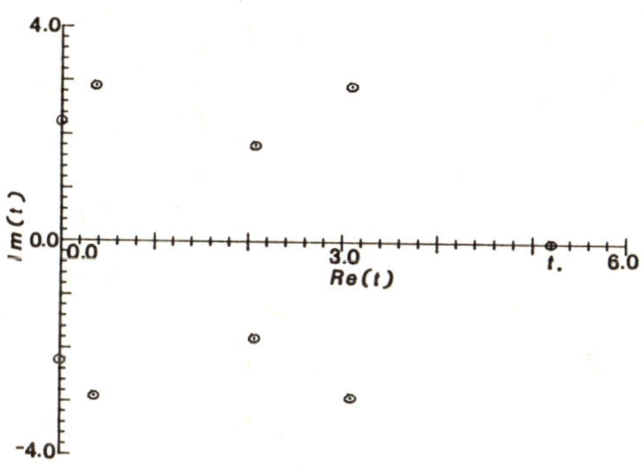

FIG. 4.   A MAP OF THE COMPLEX-T PLANE SHOWING THE SINGULARITY STRUCTURE
OF THE ENSTROPHY $\Omega_1$ FOR THE TAYLOR-GREEN VORTEX.

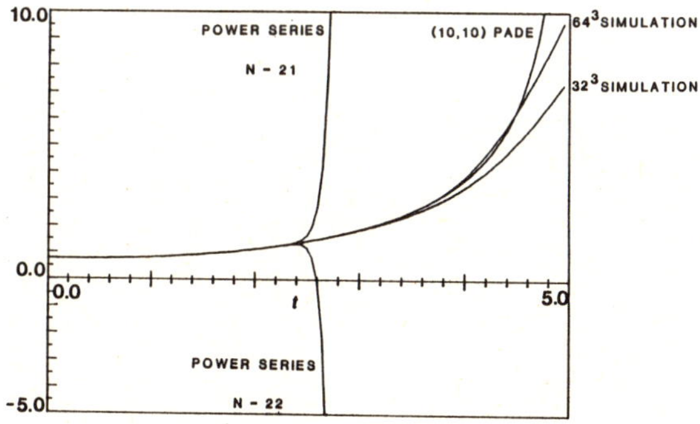


FIG. 5.   A COMPARISON OF TRUNCATED POWER SERIES, PADÉ APPROXIMATION,
AND DIRECT NUMERICAL SIMULATION FOR THE ENSTROPHY $\Omega_1$ OF
THE INVISCID TAYLOR-GREEN VORTEX.

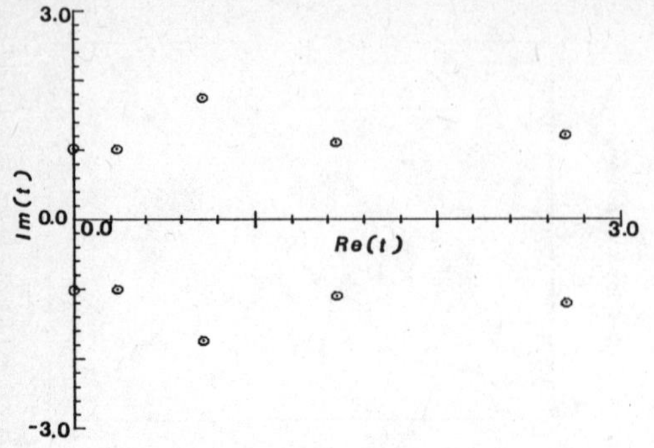

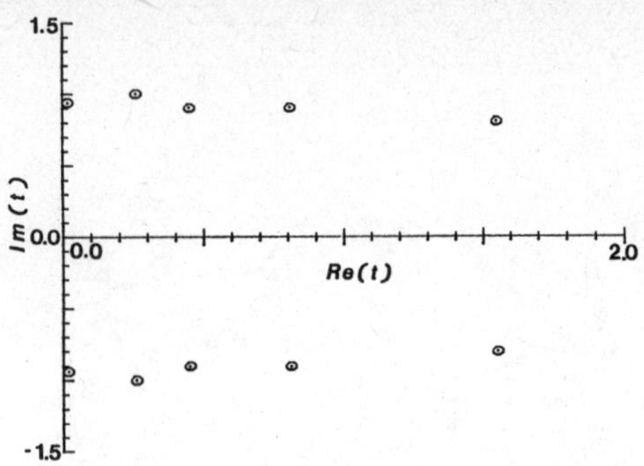

FIG. 6.   A MAP OF THE COMPLEX-T PLANE SHOWING THE SINGULARITY STRUCTURE OF THE GENERALIZED ENSTROPHY $\Omega_2(T)$ FOR THE INVISCID TWO-DIMENSIONAL VORTEX FLOW WITH INITIAL CONDITION (7).

FIG. 7.   A MAP OF THE COMPLEX-T PLANE SHOWING THE SINGULARITY STRUCTURE OF THE TOTAL ENSTROPHY $\Omega_1$ OF THE INVISCID TWO-DIMENSIONAL MHD FLOW WITH INITIAL CONDITIONS (8) AND (9).

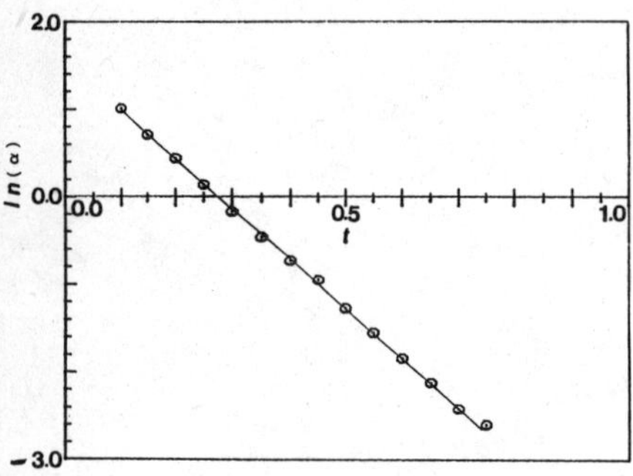

FIG. 8.   A SEMILOG PLOT OF $\alpha(T)$ VS T WHERE $\alpha(T)$ IS THE BEST FIT AT TIME T TO THE LOGARITHMIC DECREMENT OF THE KINETIC SPECTRUM OF THE INVISCID TWO-DIMENSIONAL MHD FLOW DEVELOPING FROM THE INITIAL CONDITIONS (8) AND (9).

| N | $B_N^{(1)}$ | $B_N^{(2)}$ | $B_N^{(3)}$ |
|---|---|---|---|
| 0 | $3.5000000000 \times 10^0$ | $9.5000000000 \times 10^0$ | $3.3500000000 \times 10^1$ |
| 1 | $2.5000000000 \times 10^0$ | $1.8300000000 \times 10^1$ | $1.1350000000 \times 10^2$ |
| 2 | $2.7374019608 \times 10^0$ | $6.4899176471 \times 10^1$ | $9.9534122549 \times 10^2$ |
| 3 | $-1.1551848896 \times 10^{-1}$ | $6.0275128588 \times 10^1$ | $2.2897038721 \times 10^3$ |
| 4 | $8.5127410874 \times 10^{-2}$ | $6.4635760176 \times 10^1$ | $5.3561289117 \times 10^3$ |
| 5 | $3.4295318745 \times 10^{-1}$ | $5.4056320722 \times 10^1$ | $8.1711039316 \times 10^3$ |
| 6 | $1.1344190552 \times 10^{-1}$ | $5.3410279955 \times 10^1$ | $1.2885847553 \times 10^4$ |
| 7 | $-5.5810703308 \times 10^{-1}$ | $-1.0578155527 \times 10^0$ | $1.2129563914 \times 10^4$ |
| 8 | $8.2552271494 \times 10^{-1}$ | $3.3360227325 \times 10^1$ | $1.3633718218 \times 10^4$ |
| 9 | $-1.0151418109 \times 10^0$ | $-3.3236462320 \times 10^1$ | $7.4938186271 \times 10^3$ |
| 10 | $1.1763194020 \times 10^0$ | $3.8779992057 \times 10^1$ | $6.5168953517 \times 10^3$ |
| 11 | $-1.2533115756 \times 10^0$ | $-4.0411567000 \times 10^1$ | $8.3775605637 \times 10^2$ |
| 12 | $1.2178784116 \times 10^0$ | $4.5041185973 \times 10^1$ | $2.2643409633 \times 10^3$ |
| 13 | $-1.0975472971 \times 10^0$ | $-5.0315099200 \times 10^1$ | $-1.5518870968 \times 10^3$ |
| 14 | $9.2205834075 \times 10^{-1}$ | $5.5903829422 \times 10^1$ | $2.6771651410 \times 10^3$ |
| 15 | $-6.9513617139 \times 10^{-1}$ | $-6.0773430214 \times 10^1$ | $-3.8817369457 \times 10^3$ |
| 16 | $4.1043516371 \times 10^{-1}$ | $6.3881457856 \times 10^1$ | $6.2071440004 \times 10^3$ |
| 17 | $-6.4480888002 \times 10^{-2}$ | $-6.3043207792 \times 10^1$ | $-8.6264751817 \times 10^3$ |
| 18 | $-3.5052207589 \times 10^{-1}$ | $5.4458005053 \times 10^1$ | $1.0698740173 \times 10^4$ |
| 19 | $8.5635064354 \times 10^{-1}$ | $-3.3705173906 \times 10^1$ | $-1.1812868019 \times 10^4$ |
| 20 | $-1.4876859909 \times 10^0$ | $-2.8863466127 \times 10^0$ | $1.1342220361 \times 10^4$ |
| 21 | $2.2840261308 \times 10^0$ | $5.8212396859 \times 10^1$ | $-8.6928857052 \times 10^3$ |
| 22 | $-3.2828485462 \times 10^0$ | $-1.3500906429 \times 10^2$ | $3.3912793987 \times 10^3$ |

TABLE II.   COEFFICIENTS $B_N^{(P)}$ OF THE SERIES EXPANSION OF THE TOTAL (KINETIC PLUS MAGNETIC) ENSTROPHIES $\Omega_P(T)$ FOR THE INVISCID TWO-DIMENSIONAL MHD VORTEX FLOW IN POWERS OF $T^2$.

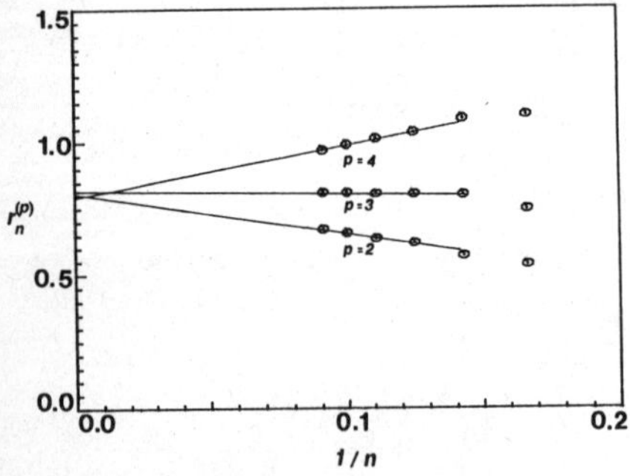

FIG. 9.   THE DOMB-SYKES PLOT FOR $\Omega_P(T)$ DEFINED IN (14) FOR THE KELVIN-HELMHOLTZ VORTEX SHEET PROBLEM EVOLVING FROM THE INITIAL CONDITION (18) WITH A = 1/4. EXTRAPOLATION TO N = ∞ INDICATES A REAL-TIME SINGULARITY AT $T_* = 2.24$.

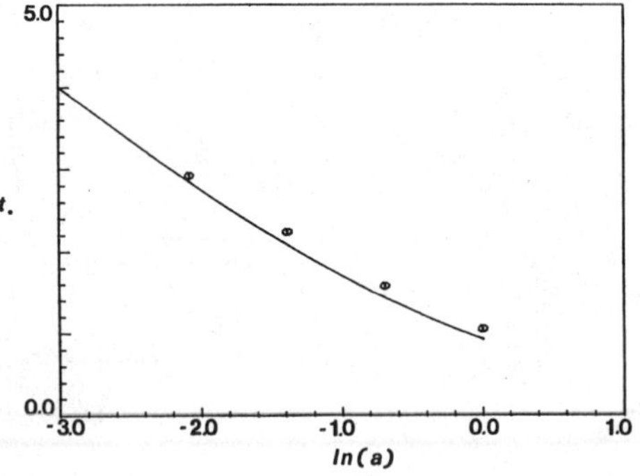

FIG. 10.   A PLOT OF $T_*$ VS LN(A) FOR THE KELVIN-HELMHOLTZ FLOW WITH INITIAL CONDITION (18). THE SOLID LINE IS MOORE'S LEADING ORDER ASYMPTOTIC APPROXIMATION TO $T_*(A)$.

# INCOMPRESSIBLE MIXED FINITE ELEMENTS FOR THE STOKES'EQUATION IN $\mathbb{R}^3$


J.C. NEDELEC

Centre de Mathématiques Appliquées (ERA/CNRS 747) - Ecole Polytechnique
91128 PALAISEAU CEDEX - FRANCE



ABSTRACT

In this paper, we study a new family of mixed finite elements for the Stokes' equation. These elements are exactly incompressible, but they are not conforming for the velocity. When studying  the Navier-Stokes' equation, this nonconformity permits us to introduce an upwind scheme.


## I – MIXED VARIATIONAL FORMULATION FOR THE STOKES' EQUATION

Let $\Omega$ be a bounded polyhedral domain in $\mathbb{R}^3$, simply connected, which boundary is connex. Let $\vec{n}$ be the normal to the boundary $\Gamma$ . The Stokes' equations are :

$$(1) \quad \begin{cases} \text{Find} \ \vec{u} = (u_1,u_2,u_3) \ \text{ and } \ p \ \text{ such that} \\[4pt] -\nu \ \Delta\vec{u} + \text{grad } p = f \quad \text{in } \Omega \ ; \\[4pt] \text{div } \vec{u} = 0 \qquad\qquad \text{in } \Omega \ ; \\[4pt] \vec{u}\big|_\Gamma = 0 \quad . \end{cases}$$

Let us consider the vorticity :

$$(2) \qquad \vec{w} = \text{rot } \vec{u} \ .$$

We have

$$(3) \quad \begin{cases} \text{rot } \vec{w} + \text{grad } p = f \quad \text{in } \Omega \ ; \\[4pt] \vec{w} - \text{rot } \vec{u} = 0 \qquad \text{in } \Omega \ ; \\[4pt] \text{div } \vec{u} = 0 \qquad\qquad \text{in } \Omega \ ; \\[4pt] \vec{u}\big|_\Gamma = 0 \ . \end{cases}$$

Let us define the following Hibert's spaces :

$$H(\text{div}) = \left\{ \vec{v} \in (L^2(\Omega))^3 \ ; \ \text{div } \vec{v} \in L^2(\Omega) \right\} \ ;$$

$$H(\text{div}^0) = \left\{ \vec{v} \in (L^2(\Omega))^3 \ ; \ \text{div } \vec{v} = 0 \ ; \ \vec{v}.\vec{n}\big|_\Gamma = 0 \right\} \ ;$$

$$H(rot) = \left\{ \vec{v} \in (L^2(\Omega))^3 \; ; \; rot \, \vec{v} \in (L^2(\Omega))^3 \right\} .$$

A variational formulation for equation (3) then is

$$(4) \quad \begin{cases} Find \; \vec{u} \in H(div^0) \; ; \; \vec{w} \in H(rot) \;\; such \; that \\[2ex] \nu \int_\Omega rot \, \vec{w} \, . \, \vec{v} \, dx = \int_\Omega \vec{f} \, . \, \vec{v} \, dx \; ; \; \forall \, \vec{v} \in H(div^0) \; ; \\[2ex] \int_\Omega \vec{w} \, . \, \vec{\pi} \, dx - \int_\Omega \vec{u} \, . \, rot \, \vec{\pi} \, dx = 0 \; ; \; \forall \, \vec{\pi} \in H(rot) . \end{cases}$$

We obtain an equivalent formulation by using a vector stream potential such that

$$(5) \quad \begin{cases} rot \, \vec{\phi} = \vec{u} \;\; in \;\; \Omega \; ; \\[1.5ex] div \, \vec{\phi} = 0 \;\; in \;\; \Omega \; ; \\[1.5ex] div \, \vec{\phi}|_\Gamma = 0 \; ; \\[1.5ex] \vec{\phi} \wedge \vec{n}|_\Gamma = 0 . \end{cases}$$

We can prove that this potential $\vec{\phi}$ is unique in the following Hilbert's space

$$\mathcal{K} = \left\{ \vec{\psi} \in H(rot) \; ; \; div \, \vec{\psi} = 0 \; ; \; \vec{\psi} \wedge \vec{n}|_\Gamma = 0 \right\} .$$

An equivalent variational formulation of (4) now is

$$(6) \quad \begin{cases} Find \; \vec{\phi} \in \mathcal{K} \; ; \; \vec{w} \in H(rot) \;\; such \; that \\[2ex] \nu \int_\Omega rot \, \vec{w} \, . \, rot \, \vec{\psi} \, dx = \int_\Omega \vec{f} \, . \, rot \, \vec{\psi} \, dx \; ; \; \forall \, \vec{\psi} \in \mathcal{K} ; \\[2ex] \int_\Omega \vec{w} \, . \, \vec{\pi} \, dx - \int_\Omega rot \, \vec{\phi} \, . \, rot \, \vec{\pi} \, dx = 0 \; ; \quad \forall \, \vec{\pi} \in H(rot) . \end{cases}$$

## II – <u>MIXED FINITE ELEMENTS</u>

We use the finite elements introduced in J.C. NEDELEC [6] for the approximation of equations (4) and (6).

When using tetrahedra, the simplest finite element conforming in the space H(rot) is the following one :

. the degrees of freedom are $\vec{u}(M_i) . \vec{t}_i$ ; $\vec{t}_i$ directed along the edge $a_i$ of the middle point $M_i$ ;

. the space of shape fucntions is

$$\mathcal{R} = \left\{ \begin{bmatrix} \alpha_1 \\ \beta_1 \\ \gamma_1 \end{bmatrix} \oplus \begin{bmatrix} \alpha_2 \\ \beta_2 \\ \gamma_2 \end{bmatrix} \wedge \begin{bmatrix} x_1 \\ x_2 \\ x_3 \end{bmatrix} \right\}$$

This element is unisolvent.

Let $\mathcal{T}_h$ be a triangulation of the domain $\Omega$ by tetrahedra. A finite dimensional subspace of $H(rot)$ then is

$$W_h = \left\{ \vec{w}_h \in H(rot) \; ; \; \vec{w}_h|_K \in \mathcal{R} \; ; \; \forall K \in \mathcal{T}_h \right\} .$$

It is completely determined by the degrees of freedom $\vec{u} \cdot \vec{t}$ on each edge of the triangulation.

The simplest finite element conforming in the space $H(div)$ is the following one

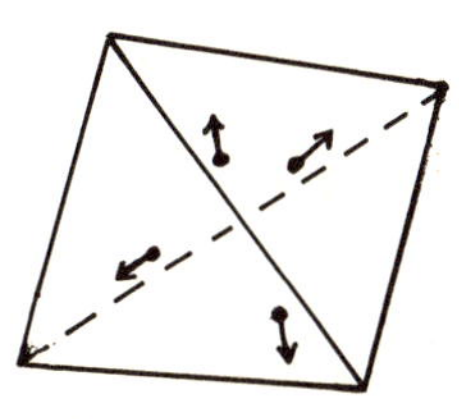

. the degrees of freedom are $\vec{u}(G_i) \cdot \vec{n}_i$ ; $\vec{n}_i$ is the unit normal to the face $f_i$ which center of gravity is $G_i$ ;

. the space of shape functions is

$$\mathcal{D} = \left\{ \begin{bmatrix} \alpha_1 \\ \alpha_2 \\ \alpha_3 \end{bmatrix} \oplus \beta \begin{bmatrix} x_1 \\ x_2 \\ x_3 \end{bmatrix} \right\}$$

This element is unisolvent.

A finite dimensional subspace of $H(div)$ then is

$$V_h = \left\{ \vec{v}_h \in H(div) \; ; \; \vec{v}_h|_K \in \mathcal{D} \; ; \; \forall K \in \mathcal{T}_h \right\} .$$

It is completely determined by the degrees of freedom $\vec{u} \cdot \vec{n}$ on each face of the triangulation.

Let us set

$$U_h = V_h \cap H(div^0) .$$

We have an approximate problem, associated to the variational formulation (4) :

$$(7) \quad \begin{cases} \text{Find } \vec{u}_h \in U_h \; ; \; \vec{w}_h \in W_h \quad \text{such that} \\[2mm] \nu \int_\Omega rot \, \vec{w}_h \cdot \vec{v}_h \, dx = \int_\Omega \vec{f} \cdot \vec{v}_h \, dx \; ; \; \forall \vec{v}_h \in U_h \; ; \\[2mm] \int_\Omega \vec{w}_h \cdot \vec{\pi}_h \, dx - \int_\Omega \vec{u}_h \cdot rot \, \vec{\pi}_h \, dx = 0 \; ; \; \forall \vec{\pi}_h \in W_h \; . \end{cases}$$

As in the case of equations (4) and (6), we can introduce an equivalent formulation. Let us define

$$\Theta_h^0 = \left\{ \theta_h \in H_0^1(\Omega) \; ; \; \theta_h|_K \in \mathbb{P}_1 \; ; \; \forall K \in \mathcal{T}_h \right\} .$$

This usual space of finite elements is completely determined by the values of $\theta_h$ at the interior vertex of the triangulation.

Let us set

$$W_h^0 = \left\{ \vec{w}_h \in W_h \; ; \; \vec{w}_h \wedge \vec{n}|_\Gamma = 0 \right\} .$$

We have

*LEMMA* : *Let* $\vec{u}_h \in U_h$ . *There exists a unique* $\vec{\phi}_h$ *in the space* $W_h^0$ *such that*

$$(8) \qquad \text{rot } \vec{\phi}_h = \vec{u}_h \; ;$$

$$(9) \qquad \int_\Omega \vec{\phi}_h \cdot \text{grad } \theta_h \, dx = 0 \; ; \; \forall \, \theta_h \in \Theta_h^0 \; . \qquad \blacksquare$$

Using this lemma, we can easily prove that equation (7) is equivalent to

$$(10) \qquad \begin{cases} \text{Find } \vec{w}_h \in W_h \; ; \; \vec{\phi}_h \in W_h^0 \quad \text{such that} \\[2mm] \nu \int_\Omega \text{rot } \vec{w}_h \cdot \text{rot } \vec{\psi}_h \, dx = \int_\Omega \vec{f} \cdot \text{rot } \vec{\psi}_h \, dx \; ; \; \forall \, \vec{\psi}_h \in W_h^0 \; ; \\[2mm] \int_\Omega \vec{w}_h \cdot \vec{\pi}_h \, dx - \int_\Omega \text{rot } \vec{\phi}_h \cdot \text{rot } \vec{\pi}_h \, dx = 0 \; ; \qquad \forall \, \vec{\pi}_h \in W_h \; ; \\[2mm] \int_\Omega \vec{\phi}_h \cdot \text{grad } \theta_h \, dx = 0 \; ; \qquad\qquad\qquad \forall \, \theta_h \in \Theta_h^0 \; . \end{cases}$$

## III – APPLICATION TO THE NAVIER-STOKES' EQUATION

A variational formulation of the Navier-Stokes' equation corresponding to (4) is

$$(11) \qquad \begin{cases} \nu \int_\Omega \text{rot } \vec{w} \cdot \vec{v} \, dx + \sum_{i,j=1}^3 \int_\Omega u_j \frac{\partial u_i}{\partial x_j} v_i \, dx = \int_\Omega \vec{f} \cdot \vec{v} \, dx \; ; \; \forall \, \vec{v} \in H(\text{div}^0) \; ; \\[2mm] \int_\Omega \vec{w} \cdot \vec{\pi} \, dx - \int_\Omega \vec{u} \cdot \text{rot } \vec{\pi} \, dx = 0 \; ; \qquad\qquad \forall \, \vec{\pi} \in H(\text{rot}) . \end{cases}$$

The nonlinear term can be written as

$$(12) \qquad \sum_{i,j=1}^3 \int_\Omega u_j \frac{\partial u_i}{\partial x_j} v_i \, dx = \int_\Omega [w_1(u_2 v_3 - v_2 u_3) + w_2(u_3 v_1 - v_3 u_2) + w_3(u_1 v_2 - u_2 v_1)] dx$$

and this expression leads to a centred scheme of discretization.

It can be more interesting to use an upwinding. This is a way of doing so :
for any tetrahedron K with boundary $\partial K$ , we introduce

$$\begin{cases} \partial K_- = \left\{ x \in \partial K \; ; \; \vec{u} \cdot \vec{n}_K < 0 \right\} \; ; \\[2mm] \partial K_+ = \left\{ x \in \partial K \; ; \; \vec{u} \cdot \vec{n}_K > 0 \right\} . \end{cases}$$

The nonlinear term is then discretized as

$$\sum_{m=1}^3 \sum_{K \in \mathcal{T}_h} \left[ \int_{\partial K_-} \vec{u}_h \cdot \vec{n}_K (u_{h,m}^i - u_{h,m}^e) \, v_{h,m}^i \right] d\gamma \; ,$$

where $u^i$ and $u^e$ are respectively the interior and exterior value of u on K .

COMMENTS AND CONCLUSION

Our finite element method is the exact generalization to the three-dimensional case of the $(\phi,w)$ method introduced in the two-dimensional case by R. GLOWINSKI [4] and P.G. CIARLET-P.A. RAVIART [1].

The idea of upwinding is due to M. FORTIN [2].

We have only presented here a particular case of a general family of finite elements using polynomials of degree $k$. In the general case, we can prove error estimates in $h^{k-1}$.

There exist also two families of finite elements conforming in $H(rot)$ and $H(div)$ respectively, which are associated to cubes.


REFERENCES

[1]     CIARLET, P.G., RAVIART, P.A., *A mixed finite element method for the biharmonic equation,* in "Mathematical Aspects in Finite Element Equation", C. de Boor (ed.), Academic Press, New York, 1974, 125-145.

[2]     FORTIN, M., *Résolution numérique des équations de Navier-Stokes par des éléments finis de type mixte,* 2nd International Symposium on Finite Element Methods in Flow Problems, S. Margherita Ligure, Italy, 1976.

[3]     GIRAULT, V., RAVIART, P.A., *Finite element approximation of the Navier-Stokes equations,* Lecture Notes on Mathematics n° 749, Springer Verlag, Berlin-Heidelberg-New York, 1979.

[4]     GLOWINSKI, R., *Approximations externes par éléments finis d'ordre un et deux du problème de Dirichlet pour* $\Delta^2$, "Topics in Numerical Analysis I", J.J.H. Miller (ed.), Academic Press, London, 1973, 123-171.

[5]     LIONS, J.L., *Quelques méthodes de résolution des problèmes aux limites non linéaires,* Dunod, Paris, 1969.

[6]     NEDELEC, J.C., *Mixed finite elements in* $\mathbb{R}^3$, Rapport Interne n° 49, Centre de Mathématiques Appliquées, Ecole Polytechnique, Palaiseau, 1979.

[7]     RAVIART, P.A., *Méthodes d'éléments finis pour les équations de Navier-Stokes,* Cours de l'Ecole d'été EDF-CEA-IRIA, 1979.

[8]     TEMAM, R., *Navier-Stokes equations,* North Holland, Amsterdam, 1977.

[9]     SCHOLZ, R., *A mixed method for 4th order problems using linear finite elements,* RAIRO, Analyse Numérique, 12 (1978), 85-90.

# A FRACTIONAL VOLUME OF FLUID METHOD FOR FREE BOUNDARY DYNAMICS[*]


B. D. Nichols, C. W. Hirt, and R. S. Hotchkiss
Theoretical Division, Group T-3
University of California
Los Alamos Scientific Laboratory
Los Alamos, NM  87545


## I.   INTRODUCTION

In this paper we describe an exceptionally versatile Eulerian method to track free boundaries that undergo large deformations.  The method follows regions of fluid, defined by a volume of fluid function, in contrast to a direct representation of free boundaries, such as marker particle chains.  This new scheme is superior to previous free boundary methods because it requires minimum computer storage, avoids logic problems associated with the creation or destruction of disjoint fluid regions, is applicable to arbitrarily contorting flows, and is directly extendible to three-dimensional problems.

## II.   THE VOF METHOD

The volume of fluid (VOF) method is based on a function F whose value is unity at any point occupied by fluid and zero otherwise.  The average value of F in a computational cell represents the fractional volume of the cell occupied by fluid.  In particular, a unit value of F corresponds to a cell full of fluid, while a zero value indicates that the cell contains no fluid.  Cells with F values between zero and one must then contain a free boundary.

In addition to defining which cells contain a boundary, the VOF method defines where fluid is located in a boundary cell.  The normal direction to the boundary lies in the direction in which the value of F changes most rapidly.  Because F is a step function, however, its derivatives are computed in a special way.  Finally, knowing both the normal direction and the value of F in a boundary cell, a line cutting the cell can be constructed that approximates the interface there.  This boundary location can then be used in setting boundary conditions.  In addition, surface curvatures can be computed from the F distribution when surface tension forces must be considered.

The time dependence of F is governed by a kinematic equation stating that the F values flow with the fluid.  If standard finite-difference approximations were used to compute the advection of F, excessive numerical smearing of the F function would occur and interfaces would lose their definition.  Fortunately, the fact that F is a step function with values of zero or one permits the use of a special flux approxi-

---


[*]This work was supported by the Electric Power Research Institute under contract RP-965-3, and in part by the Office of Naval Research, contract NA-onr-6-75, NR 062-455, with the cooperation of the U. S. Department of Energy.

mation that preserves its discontinuous nature.  In particular, a type of donor-acceptor flux approximation is used that uses information about F downstream as well as upstream of a flux boundary.  In this way a crude interface shape is established that can be used to compute the flux of F.

III.  THE SOLA-VOF CODE

The VOF method is applicable to Eulerian or the more general Arbitrary-Lagrangian-Eulerian (ALE) numerical formulations.  We have verified the accuracy and versatility of the method by incorporating it into a two-dimensional, finite-difference, Eulerian scheme that uses a solution algorithm (SOLA) by Hirt et al. (1975) based on the well-known Marker-and-Cell (MAC) method (Harlow and Welch, 1965).  The combined code, SOLA-VOF, (described by Hirt and Nichols, 1980 and Nichols et al., 1980), uses an Eulerian mesh of rectangular cells having variable sizes $\delta x_i$ for the ith column and $\delta y_j$ for the jth row.  The fluid equations solved are the Navier-Stokes equations in either Cartesian or cylindrical coordinates.  These equations are supplemented with either the continuity condition for incompressible fluids or the continuity condition for fluids with limited compressibility effects (e.g., acoustic waves).

The SOLA-VOF code also has a variety of additional options.  Any combination of cells in the mesh can be defined as obstacle cells into which fluid cannot flow.  Instead of one fluid with a free surface, two fluids can be defined with different density ratios separated by a free interface.  Surface tension forces are optional at the fluid interface for both the one and two fluid cases.

The basic procedure for advancing a solution through one increment in time consists of three steps:

(1) Explicit approximations of the momentum equations are used to compute the first guess for new time-level velocities using the initial conditions or previous time-level values for all advective, pressure, and viscous accelerations.

(2) To satisfy the incompressible fluid continuity equation, pressures are iteratively adjusted in each cell and velocity changes induced by each pressure change are added to the velocities computed in step 1.  An iteration is needed because the change in pressure needed in one cell to satisfy the continuity equation will upset the balance in the four adjacent cells.  On the other hand, when the limited compressibility option is used the SOLA-VOF program automatically, and continuously, switches from an implicit to an explicit solution for pressures as the time step is reduced below the Courant stability limit.  This feature permits more accurate results to be obtained with less computational work!

(3) Finally, the F function defining fluid regions must be updated to give the new fluid configuration.

Repetition of these steps will advance a solution through any desired time interval.  At each step, of course, suitable boundary conditions must be imposed at all mesh and free-surface boundaries.

## IV.  SAMPLE PROBLEMS

We have chosen several calculational examples to illustrate the capabilities of the SOLA-VOF method.

### A.  Broken Dam Problem

A simply executed problem, for which experimental data is available, is the "broken dam" problem.  An initially rectangular block of fluid, in hydrostatic equilibrium, collapses under the force of gravity and flows across a dry horizontal floor, as shown in Fig. 1.  A comparison of the experimental data reported by Martin and Moyce (1952) with the calculated leading edge of the fluid as a function of time, Fig. 2, shows the greatest deviation is everywhere less than a calculational cell width.

### B.  Collapse of a Cylindrical Fluid Column

The collapse of a cylindrical column of fluid is similar to the "broken dam" problem, but uses the cylindrical coordinate system.  Several additional features of the SOLA-VOF code are also illustrated by this calculation.  Flow visualization is typically realized by plotting the velocity field with velocity vectors drawn from the center of each mesh cell containing fluid and by depicting the free surfaces with the volume fraction ($F=1/2$) contour, as seen in Fig. 1.  However, marker particles can be used to follow the fluid flow as in Fig. 3.  In addition, this calculation demonstrates the use of obstacle cells.  The capability of the code to handle highly contorted fluid configurations is exceptionally well illustrated here.

### C.  A Reactor Safety Application

Many boiling water reactors use a large pool of water to condense steam should a major steam leak occur.  In some designs, steam would be forced into the pool through vertical pipes extending several pipe diameters below the surface of the pool.  Before steam enters the pool, however, air initially in the pipes must be pushed out.  The ejection of this noncondensable air forms large bubbles in the pool and displaces the pool surface upward.  Several small scale experimental programs have been conducted to understand the hydrodynamic forces generated during the process.  We have numerically calculated these hydrodynamic forces and compared with laboratory test data (Nichols and Hirt, 1980).

To model these experiments it was necessary to supplement the SOLA-VOF code with calculations for the gas pressure in the pipe and for the pressure in the closed space above the pool surface.  These pressures are then used as free surface boundary pressures.  A sequence of calculated results illustrating the fluid dynamics associated with the air-clearing process are contained in Fig. 4.

### D.  Instability of a Liquid Column

For some applications, such as the breakup of a thin liquid jet, surface tension forces must be considered.  A classic problem of this type concerns the instability of a cylindrical column of fluid.  When its free surface is perturbed by radial displacements, an exponential growth in perturbation amplitude may result that eventually causes the cylinder to break up into a series of discrete drops. Figure 5 illustrates a SOLA-VOF calculation of this type of surface tension driven instability. The cylinder is initially perturbed with an axisymmetric displacement of its free surface that is sinusoidal in the axial direction. Two axial wave lengths, $\lambda$, are followed, with $\lambda = 4.598$ D, where D is the diameter of the undisturbed liquid column. The initial perturbation amplitude is 0.001 D. A comparison of the computed amplitude growth with linear theory shows the interesting result that the linear theory is valid for large amplitude displacements. Nevertheless, nonlinear effects are important, for they are the cause of the small satellite drops that develop between the large drops.

Many additional calculations have been performed that validate various capabilities of the SOLA-VOF code not mentioned in the above examples. Included in these are a study of bubble growth and collapse, which made use of the limited compressibility feature, and the passage of an immiscible liquid drop through a constriction in a tube, in which the two-fluid and surface tension options were utilized together.

## SUMMARY

We present the volume of fluid (VOF) technique as a simple and efficient means for numerically treating free boundaries embedded in a calculational mesh of Eulerian or Arbitrary-Lagrangian-Eulerian cells. It is particularly useful because it uses a minimum of stored information, treats intersecting free boundaries automatically, and can be readily extended to three-dimensional calculations.

## REFERENCES

Harlow, F. H. and Welch, J. E., 1965, "Numerical Calculation of Time-Dependent Viscous Incompressible Flow," Phys. Fluids $\underline{8}$, 2182; Welch, J. E., Harlow, F. H., Shannon, J. P., and Daly, B. J., 1966, "The MAC Method: A Computing Technique for Solving Viscous, Incompressible, Transient Fluid Flow Problems Involving Free Surfaces," Los Alamos Scientific Laboratory report LA-3425.

Hirt, C. W., Nichols, B. D., and Romero, N. C., 1975, "SOLA - A Numerical Solution Algorithm for Transient Fluid Flows," Los Alamos Scientific Laboratory report LA-5852.

Hirt, C. W. and Nichols, B. D., 1980, "Volume of Fluid (VOF) Method for the Dynamics of Free Boundaries," J. Comp. Phys., accepted for publication.

Martin, J. C. and Moyce, W. J., 1952, Phil. Trans. Roy. Soc., London, A244, 312.

Nichols, B. D. and Hirt, C. W., 1980, "Numerical Simulation of BWR Vent-Clearing Hydrodynamics," Nuc. Sci. Engr. $\underline{73}$, 196.

Nichols, B. D., Hirt, C. W., and Hotchkiss, R. S., 1980, "SOLA-VOF: A Solution Algorithm for Transient Fluid Flow with Multiple Free Boundaries," Los Alamos Scientific Laboratory report LA-8355.

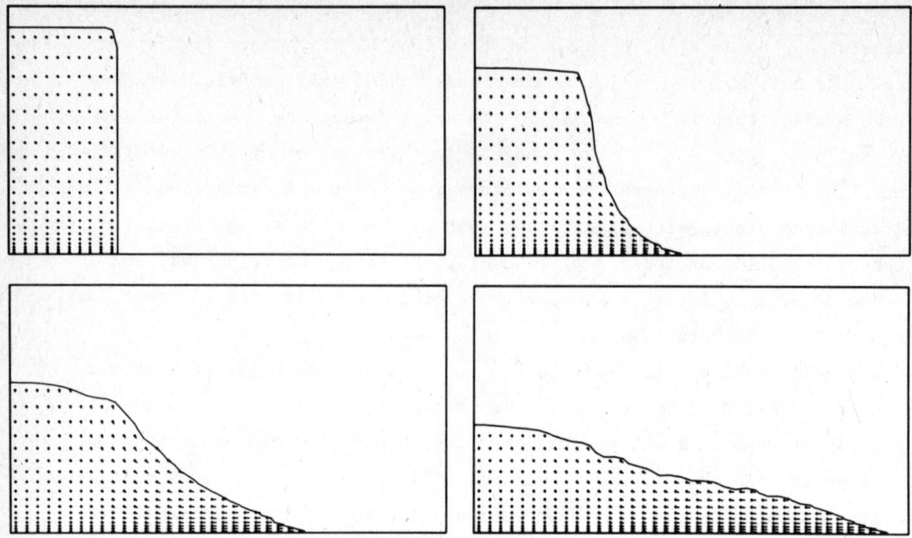

Fig. 1. Velocity vectors and free surface configurations computed for the broken dam problem at times 0.0, 0.9, 1.4, and 2.0.

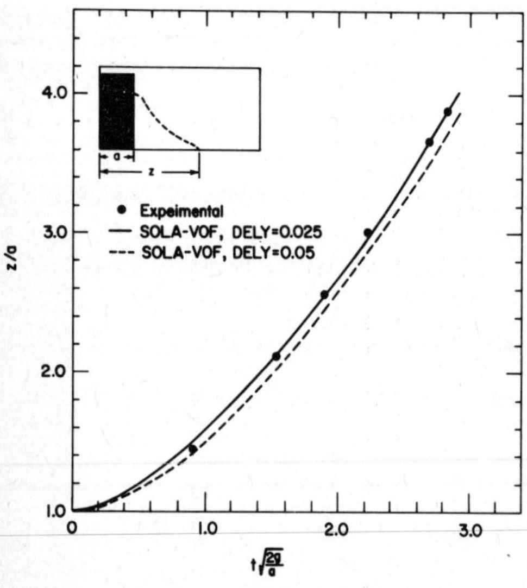


Fig. 2. Comparison of calculated results with experimental data for the broken dam problem.

Fig. 3.  Calculational results showing the collapse of a cylindrical column of fluid surrounded by a low retaining wall.  Times shown are 0.0, 1.6, 2.5, 3.6, 4.2, and 4.6

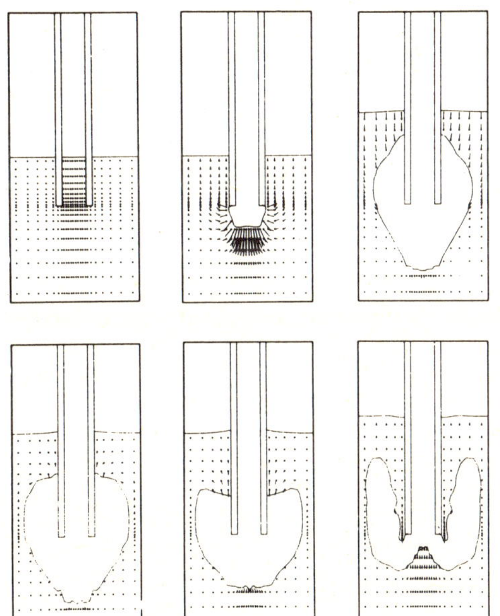

Fig. 4.  Velocity vectors and free surface configurations computed when air is forced through submerged vent pipe.

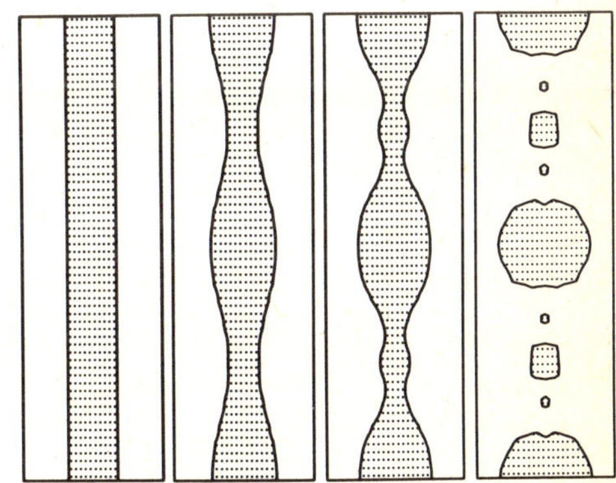

Fig. 5.  Fluid configurations calculated for a surface tension driven instability of a cylindrical column of fluid.  Times are approximately 0.0, 6.03, 6.49, and 7.0 in units of $\sqrt{D^3/\sigma}$.

NUMERICAL ANALYSIS OF DYNAMIC STALL PHENOMENA OF AN OSCILLATING
AIRFOIL BY THE DISCRETE-VORTEX APPROXIMATION.


Kiyoaki Ono*, Kunio Kuwahara**,
Koichi Oshima***
* Coll. Sci. & Tech., Nihon Univ.
** Dep. Appl. Physics, Univ. Tokyo
*** ISAS, Univ. Tokyo


The flow field around an oscillating airfoil under pitching and/or heaving
motion was analized using the discrete-vortex approximation.  The maximum angle of
attack during oscillation was taken to be larger than the static stall one.  The
process of the formation of the leading edge separation vortex, its succeeding
convection along the wing surface and its shedding into the wake was simulated.
The dynamic lift stall curve was obtained when the reduced frequency defined with
the half chord length was smaller than 0.25.  When it was larger than 0.5, the
dynamic lift stall did not occur.

1 Introduction

An oscillating airfoil in pitch and/or in heave with its maximum angle of
attack larger than the static stall one, experiences peculiar hysteresis in drag,
lift and moment coefficients.  It is believed that these phenomena are caused by
the formation of the leading edge separation vortex, its convection along the wing
surface and its shedding into the wake (Carr et al., 1977).

The linearized unsteady airfoil theory, developed by Theodorsen(1935) and
Karman & Sears (1938) and much improved since then by including the airfoil thickness and the wake deformation etc., has been widely used for the analysis of the
smaller amplitude oscillation than the dynamic stall angle.  However in order to
treat the dynamic stall, the unsteady separated flow theory is needed.

The flow field around an oscillating NACA0012 airfoil in pitch was analized
by solving numerically the two dimensional Navier-Stokes equations for fairly large
Reynolds numbers (Mehta 1977).  In this calculation the process of the dynamic stall
was clearly presented.  But this calculation was quite computer-time consuming.  The
leading edge separation was taken into account by Ham(1968) in the potential theory,
in which the strength of the shedded vortex was determined using the Kutta condition
at the leading edge and the trailing edge.  But some phenomenological argument based
on experiments was needed, and the satisfactory results were not obtained.

In the present analysis the boundary layer formed on the surface of the wing

was replaced by a finite number of discrete vortices at each time step and their
interactions were calculated to determine their convected positions.  Thus the
process of formation and convection of separated flow mass could be simulated.
In order to determine the strength of newly generated vortices, the "Non-Slip"
condition was introduced instead of the usual Kutta condition (Kuwahara 1978).
So this theory could be applied to the analysis of the dynamic stall without further
assumptions.

2 Mathematical Formulation

Let us consider a two dimensional NACA airfoil (its chord length 4a) oscillating in pitch and/or in heave.  The pitching angle $\alpha$ (positive in the clockwise
rotation) and the vertical displacement of the pitching axis $Y_h$ (assumed on the
Y-axis) vary following the next formulae:

$$\alpha = \alpha_0 - \bar{\alpha} \cos\omega t,$$
$$Y_h = H \sin(\omega t + \beta). \tag{1}$$

where $\beta$ is arbitrary (Fig. 1).  The flow field around the oscillating airfoil
(z-plane) is conformally mapped outside the static unit circle ($\zeta$-plane) by the
following transformations: (Fig. 2)

$$z = [a'(\zeta + \gamma + \frac{c^2}{\zeta+\gamma}) - b]e^{-i\alpha} + iY_h = G(\zeta,t), \tag{2}$$

$$\gamma = \xi_0 + i\eta_0,$$
$$c = (\xi_0 + (1- \eta_0^2))^{\frac{1}{2}}(1-\delta). \tag{3}$$

Parameters $a'$, $c$, $b$, $\xi_0$, $\eta_0$, $\delta$ are determined according to Mehta (1977).

Laplace equation must be solved under the boundary condition that the normal
component of the difference between the velocity of the body surface $(\dot{X},\dot{Y})$ and
that of the fluid particle attached to it $(u,v)$ is zero:  (Fig. 3)

$$(u-\dot{X})\cdot\cos\theta + (v-\dot{Y})\cdot\sin\theta = 0. \tag{4}$$

On the unit circle in the $\zeta$-plane it is represented as

$$\text{Re}[\zeta\cdot F_\zeta] = \text{Re}[\zeta\cdot G_\zeta\cdot\overline{G}_t] \quad, \qquad |\zeta| = 1. \tag{5}$$

So the complex potential at time step N is as follows:

$$F = a'u(\zeta e^{-i\alpha} + e^{i\alpha}/\zeta) + i\dot{\alpha}a'^2(A/\zeta + B/(\zeta+\gamma))$$
$$- i\dot{Y}_h a'(e^{i\alpha}/\zeta + c^2 e^{-i\alpha}/(\zeta+\gamma))$$
$$+ i\sum_{j=1}^{N}\sum_{k=1}^{M} \kappa^{jk}(\ln(\zeta - \zeta^{jk}) - \ln(\zeta - 1/\overline{\zeta}^{jk})), \tag{6}$$

$$A = \gamma - b/a' + c^2/\gamma \ , \quad B = c^2(\gamma - 1/\gamma) - bc^2/a + c^4\gamma/(\gamma^2-1),$$

where M is the number of discrete vortices generated at each time step and $\zeta^{jk}$ is the location of the k-th vortex generated at time step j.

The "Non-Slip" condition is introduced to determine the strength of vortices. The non-slip condition for the viscous fluid is represented as

$$(u-\dot{X})\cdot\sin\theta - (v-\dot{Y})\cdot\cos\theta = 0. \tag{7}$$

It is also expressed on the unit circle in the $\zeta$-plane as follows:

$$\mathrm{Im}(\zeta\cdot F_\zeta) = \mathrm{Im}(\zeta\cdot G_\zeta\cdot\overline{G}_t) \quad, \qquad |\zeta| = 1. \tag{8}$$

Within the framework of discrete-vortex approximation, it is impossible to satisfy eq.(8) on the entire circle. So by integrating eq.(8) along the unit circle, the following equation is obtained for the velocity potential $\Phi$:

$$\Phi_{k+1} - \Phi_k = -\int_{\theta_k}^{\theta_{k+1}} \mathrm{Im}(\zeta\cdot G_\zeta\cdot\overline{G}_t)\,d\theta, \tag{9}$$

$$\theta_k = \arg(\zeta_k). \tag{10}$$

Eq.(9) means that no net flow exists on the average sense along the wing surface between two points $\zeta_k$ and $\zeta_{k+1}$. This is why eq.(9) is called the "Non-Slip" condition. In order to fulfil eq.(9), a discrete vortex is generated just outside the unit circle between $\zeta_k$ and $\zeta_{k+1}$ at each time step (Fig. 4).

The induced velocity of the k-th vortex generated at time step j is as follows:

$$\frac{dz^{jk}}{dt} = \left( \frac{\overline{dF^{jk}}}{dz} \right)_{z = z^{jk}} , \tag{11}$$

$$F^{jk} = F - i\kappa^{jk}\ln(\zeta-\zeta^{jk}). \tag{12}$$

So in the $\zeta$-plane it is represented as

$$\frac{d\zeta^{jk}}{dt} = \left[ \left( \left( \frac{\overline{dF^{jk}}}{d\zeta} \cdot \frac{d\zeta}{dz} \right) - \frac{dz}{dt} \right) \cdot \frac{d\zeta}{dz} \right]_{\zeta = \zeta^{jk}} . \tag{13}$$

The new convected position is determined at time $t_{N+1}=t_N+\Delta t$ with the following equations:

$$\zeta_{N+1}^{jk} = \zeta_N^{jk} + \left( \frac{d\zeta^{jk}}{dt} \right)_N\cdot\Delta t . \tag{14}$$

The drag D, the lift L, and the moment M are calculated using the following formulae:

$$D - iL = -i\oint p\,d\overline{z} ,$$

$$M = -\frac{1}{2}\mathrm{Re}[\oint p\overline{z}\,dz] , \tag{15}$$

$$p = p_0(t) -\frac{\partial\Phi}{\partial t} - \frac{1}{2}\frac{dF}{dz}\left(\overline{\frac{dF}{dz}}\right) .$$

## 3 Results

Numerical calculations were carried out chiefly for the pitching oscillation of $\alpha_0=10°-15°$, $\bar{\alpha}=10°-15°$. The reduced frequency $K=\omega(2a)/U$ was 0.125-2.45. The airfoil surface was divided into two parts, one was the forward half and the other was the rear half of the wing. Two discrete vortices were generated near the leading edge and the trailing edge. The computer used was HITAC-8700 at Nihon Univ.

In Fig. 5 the vortex distribution and the instantaneous streamlines are shown on the case K=2.45. TIME means t/T, in which t is the elapsed time from the beginning of the oscillation and T is the period of the oscillation. The process of the formation of the leading edge separation vortex, its convection along the wing surface and its shedding into the wake is clearly demonstrated. The reattaching process near the leading edge is also shown.

In Fig. 6 drag, lift and moment coefficients versus the angle of attack $\alpha$ are shown for the case K=0.5 and 0.25. The results of the linearized theory are also shown by dotted lines. When the angle of attack $\alpha$ is small, the difference between the calculated results and those of the linearized theory is small. But when $\alpha$ is large, their fluctuations become remarkable and their discrepancy is significant. Judging from the instantaneous streamlines, the leading edge separation bubbles were formulated in both cases, but in the case K=0.5, the lift stall does not occur, though in the case K=0.25, the sharp dynamic stall of lift takes place. This is probably because when the reduced frequency is large, the growth of the leading edge separation vortex is not sufficient on account of the quick decrease of the angle of attack.

## 4 Conclusion

The unsteady separated theory by the discrete-vortex approximation was developed and was applied to the flow field of an oscillating airfoil in pitch and/or in heave. The process of the formation of the leading edge separation vortex, its convection along the wing surface and its shedding into the wake was presented. When the reduced frequency defined with the half chord length was larger than 0.5, the lift stall did not occur. On the contrary when K was less than 0.25, the sharp lift stall curve was obtained.


### References

Carr, L.W., McAlister, K.W. & McCroskey, W.J. 1977 NASA TND-8382.

Ham, N.D. 1968 AIAA Journal, vol.6, No.10.

von Karman, T. & Sears, W.R. 1938 J.Aero. Sci., vol.5, No.10.

Kuwahara, K. 1978 J. Phys. Soc. Japan, vol.45.

Mehta, U.B. 1977 AGARD Paper 23.

Theodorsen, T. 1935 NACA Report 496.

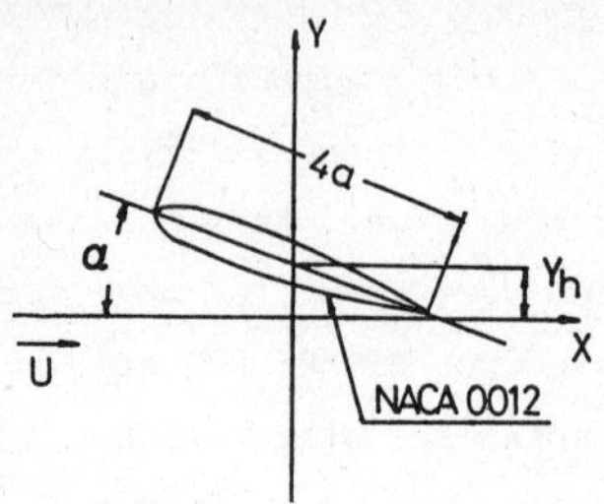

Fig. 1 A reference of frame in the physical plane. (z-plane)

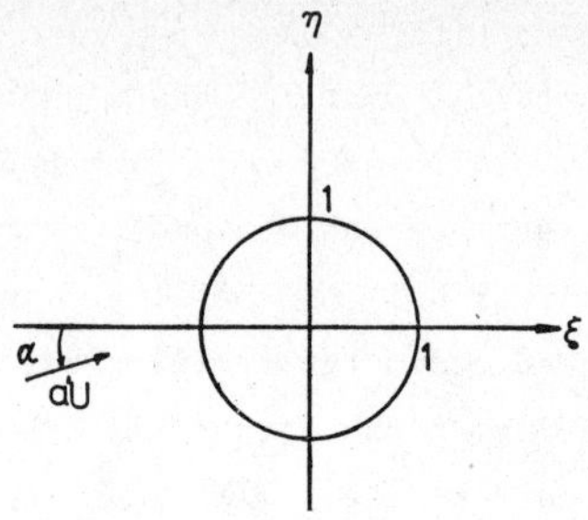

Fig. 2 A reference of frame in the mapped plane. (ζ-plane)

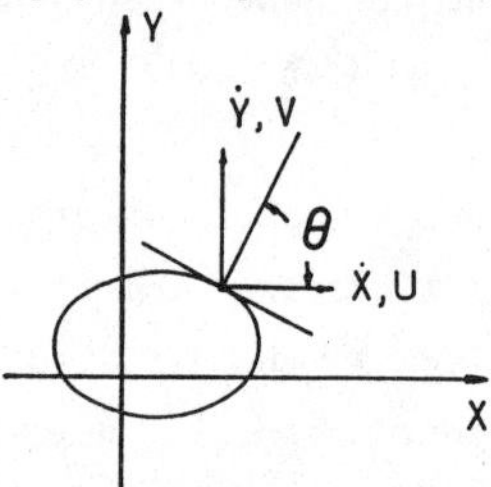

Fig. 3 The definition of $u,v,\dot{X},\dot{Y}$ and $\theta$.

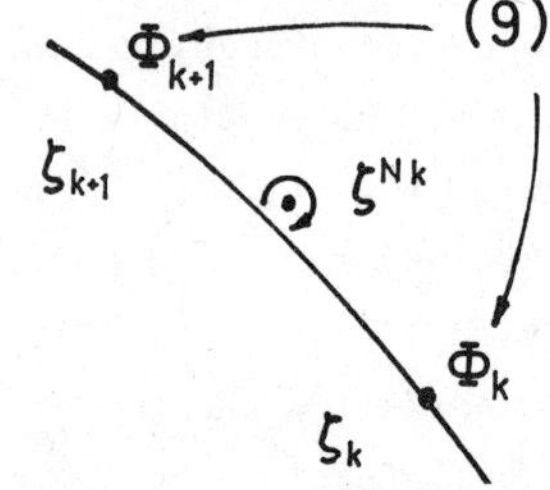

Fig. 4 The "Non-Slip" condition.

(a)  TIME = 1.0

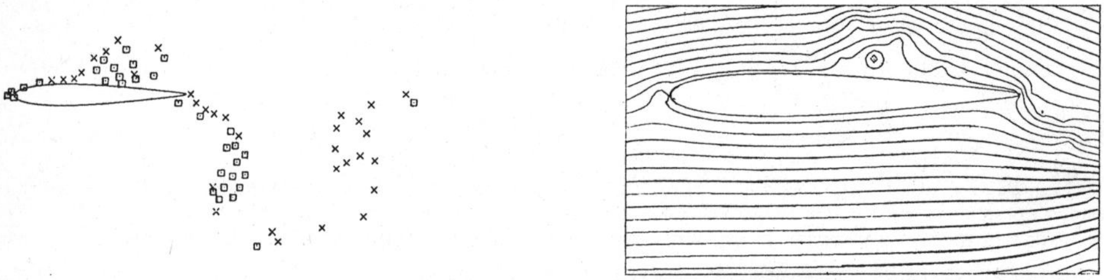

(b)  TIME = 1.25

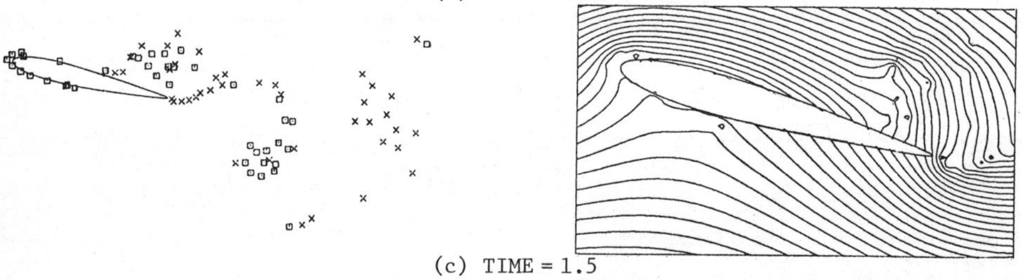

(c)  TIME = 1.5

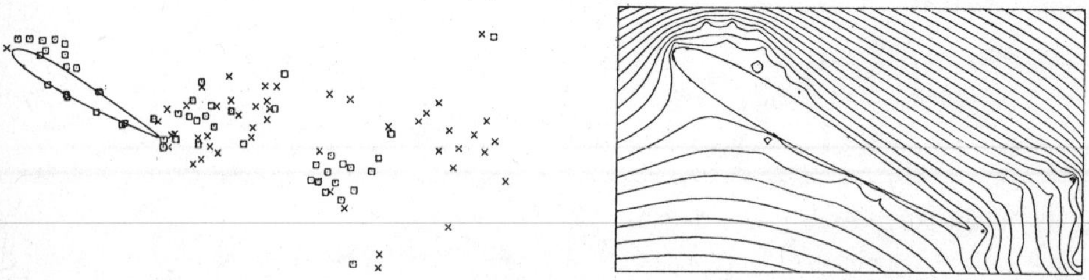

(to be continued)

(d) TIME = 1.75

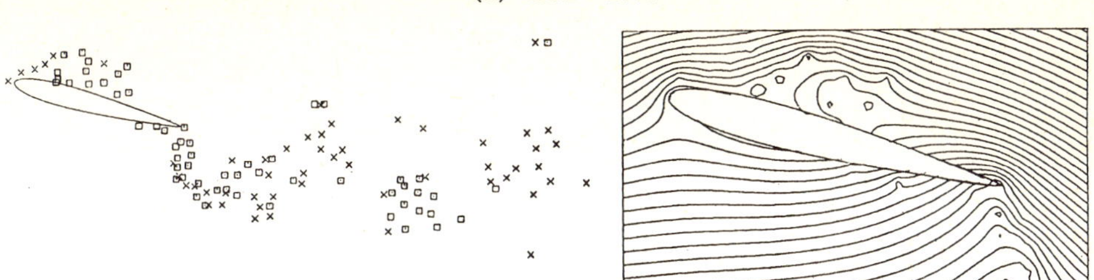



Fig. 5 The vortex distribution and the instantaneous streamlines
(U=2cm/s, T=2.56s, 4a=4cm, α=15°-15°cosωt, K=2.45, TIME=t/T)

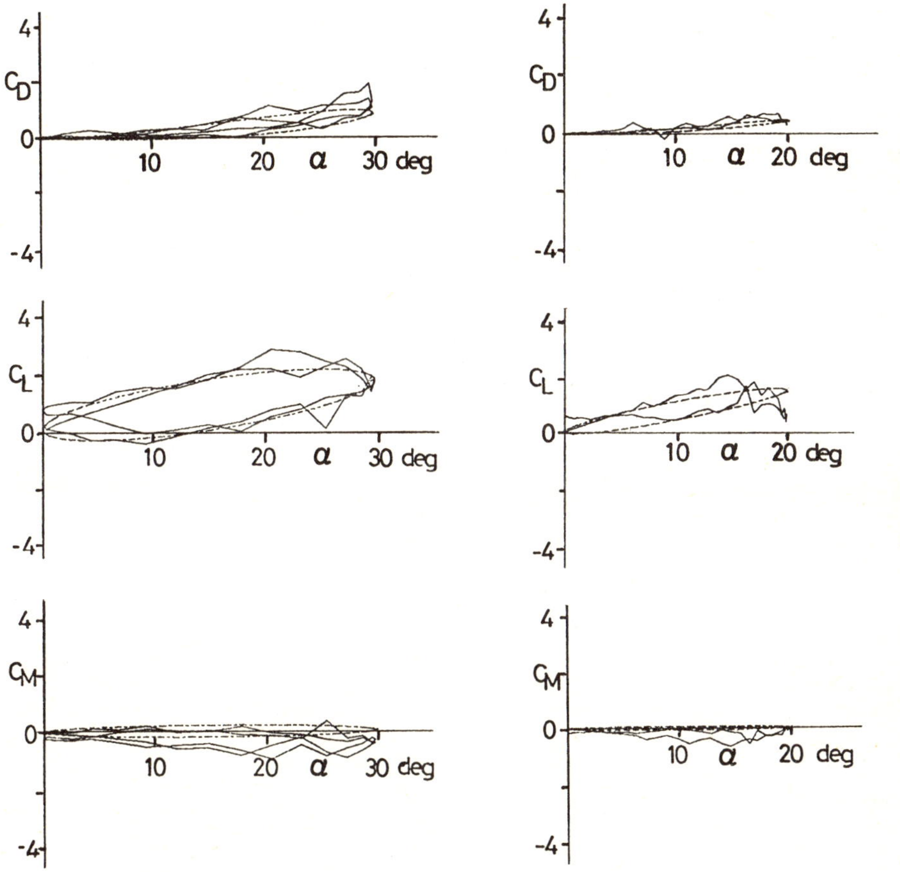

(U=25cm/s, T=1s, 4a=4cm, α=15°-15°
cosωt, K=0.5)

(U=25cm/s, T=2s, 4a=4cm, α=10°-10°
cosωt, K=0.25)

Fig.6 The drag, lift and moment coefficients versus the angle of attack

IMPLICIT NON-ITERATIVE SCHEME FOR TURBULENT UNSTEADY BOUNDARY LAYERS

P. Orlandi

Istituto di Aerodinamica dell'Università di Roma

## INTRODUCTION

To solve steady and unsteady turbulent boundary layers it is worthwhile to use
implicit schemes both for stability reasons and for having a fast coupling of the
mean velocities and the turbulent quantities.  Iterative procedures have been usually employed to solve the non-linear system of equations; these procedures require as
more iterations as more complex boundary layers are to be considered.  To overcome
the iterative time consuming scheme, an implicit non iterative scheme can be used;
it consists on linearizing the system of equations with the same accuracy employed
to discretize the derivatives respect to which the equations are parabolic.  This
algorithm has been developed by Beam and Warming (1978) to solve the compressible
Navier-Stokes equations, and the linearization procedure was applied to the time derivative.  The boundary layers equations are parabolic both in time and in the downstream direction, thus the linearization procedure can be applied to both the directions; in a previous paper, Orlandi & Reynolds (1979) showed that it is worthwhile
to apply the linearization to the downstream direction.  But in such a case the general formulation can not be generalized to retain the full variety of schemes of the
temporal difference approximation.

To test the validity of the turbulence model used, my opinion is to compare with
the available experiments not only the global quantities, such as the skin friction,
the shape factor and the displacement thickness, but also each term of the turbulent
energy balance.  It is well known that the region where the turbulent energy balance
is more critical is the viscous layer: there all the terms are of the same order of
magnitude except the advection one.  Thus models which take into account the viscous
layer should be employed.  In a previous work Orlandi & Reynolds modified and applied
to solve steady and unsteady zero pressure gradient boundary layers the Norris-Reynolds
one-equation turbulence model.  Excellent agreement was found in the steady case with
the experimental results; in the unsteady case, due to the lack of validated experimental results, a definite answer about the goodness of the model can not be drawn.
In the conclusion of the previous paper it was claimed that the one-equation turbulence
model should not give a whole picture of the boundary layer, being not able to predict
the behavior of the mixing length measured by Cousteix et al. (1978).  Thus a two-
equation turbulence model should give a better representation of the unsteady boundary
layer.  In this paper the two-equation model has been used and the results have been
compared with the previous one obtained with the one-equation model.

The greater difficulty encountered on using the two-equation model consists on
assigning the initial profile of the energy dissipation.  When dealing with the one-
equation, the turbulent energy initial profile was calculated by solving the turbulent energy equation with the Newton-Rapson scheme, neglecting the advection term
and assigning the velocity profile.  The same procedure has been applied to the stiff
system of the two-equation model, but a good convergency was not found; thus the
initial condition have been calculated starting with the initial profiles yielded
by the one-equation model and solving the full system of the steady equations for a
short downstream distance until  the velocity and the turbulent quantities are in a
good agreement with the experimental ones.

## GOVERNING EQUATIONS AND NUMERICAL SCHEME

The two-equation turbulence model here used has the same form of the $k$-$\varepsilon$ model developed at the Imperial College by Jones and Launder (1973), but it differs in modelling the viscous layer.  The momentum and turbulent quantities transport equations can be written in the form

$$(1) \qquad \frac{D\vec{Z}}{Dt} = \vec{P} + \frac{1}{Re} \frac{\partial}{\partial y} \left(1 + \nu_T \, C_{4i} \, \frac{\partial}{\partial y}\right) \vec{Z} - \vec{E}$$

where

$$(2) \qquad \vec{Z} = \begin{vmatrix} U \\ Q \\ D \end{vmatrix} , \quad \vec{P} = \begin{vmatrix} U_e \dfrac{\partial U_e}{\partial x} + \dfrac{\partial U_e}{\partial t} \\[2mm] \dfrac{2\nu_T}{Re} \cdot \left(\dfrac{\partial U}{\partial y}\right)^2 \\[2mm] \dfrac{C_1}{Re} \dfrac{D}{Q} \nu_T \left(\dfrac{\partial U}{\partial y}\right)^2 \end{vmatrix} , \quad \vec{E} = \begin{vmatrix} 0 \\ 2\,D \\ C_7 \dfrac{D^2}{Q} \left(1 - e^{-C_6 Re\, y\, Q^{\frac{1}{2}}}\right) \end{vmatrix}$$

The relationship between the eddy viscosity and the mean turbulent quantities is given by

$$(3) \qquad \nu_T = C_2 \, C_3 \, Re \, \frac{Q^2}{D} \left(1 - e^{-C_8 Re\, y\, Q^{\frac{1}{2}}}\right)$$

The effect of the low turbulence Reynolds number has been modelled in a different manner with respect to the Jones and Launder one.  A damping function has been introduced in the eddy viscosity expression, to simulate its behavior proportional to $y^n$ , with $n = 4$ , in the viscous layer.  Moreover the same damping function has been introduced in the dissipation term of the D transport equation, in order to take into account the finite constant value of D at the wall.  Thus there is no need to introduce the extra-terms considered by Jones and Launder in the $k$ and $\varepsilon$ equations.

The values of the constants in the turbulent energy equation and in the equation (3) maintain the same values previously obtained by Orlandi and Reynolds; the values of $C_1$ and $C_7$ are the same as the Jones and Launder model, the value of $C_6$ has been determined in a heuristic way in order to have the best agreement with the experiments of Klebanoff (1955).

In the one-equation turbulence model a production term has been introduced, related to the isotropic dissipation in the region less than $y^+ \simeq 20$ ; it looks like the pressure velocity correlation term calculated by the large eddy simulation of a pipe flow by Moin et al. (1978), and deduced from the experiments of Laufer (1951). This term brings to values of $Q^+$ in a better agreement with the experimental ones of Klebanoff (1955).  I did a first tentative to model the pressure-velocity correlation terms in the two-equation model, but their introduction brings to negative turbulent energy at a distance of the order of 1 in wall coordinate dimension. I think this is due to the difficulty to have a good initial distribution of D , thus more work should be done to introduce these very important terms.

Associated with the system of equation (1) there is the continuity equation; it has been differentiated with respect to y in such a way to have the matrix associated to the vertical velocity V , with a diagonal dominance.

If the near wall region (viscous layer) has to be computed without introducing any assumption, the boundary conditions associated to the system of equation (1) and to the differentiated continuity equation are

$$y = 0 \qquad U = V = Q = \frac{\partial D}{\partial y} = 0$$

(4)

$$y \to \infty \qquad U = U_e \, , \qquad \frac{\partial V}{\partial y} = - \frac{\partial U_e}{\partial x} \, , \qquad \frac{\partial Q}{\partial y} = \frac{\partial D}{\partial y} = 0$$

The assumption of $\partial D/\partial y = 0$ at the wall is satisfactory because it is known, from the experiments, that in the viscous layer the $U'$ and $W'$ fluctuations behave as $U' = a_1 y + a_2 y^2 + a_3 y^3 + O(y^4)$ and $W' = b_1 y + b_2 y^2 + b_3 y^3 + O(y^4)$ ; the coefficients multiplying the higher order exponents are much smaller than $a_1$ and $b_1$ . Thus it can be derived that

$$\left(\frac{\partial D}{\partial y}\right)_{y=0} = \frac{6}{Re} \, (\overline{a_1 a_2} + \overline{b_1 b_2}) \simeq 0$$

To solve the system of equations (1) it is useful to normalize the normal coordinate with respect to the boundary layer thickness. It has been chosen to introduce a coordinate transformation function of the boundary layer thickness at the old time step. Thus the independent variables have been transformed as

(5)
$$y = \delta(x, t - \Delta t) \, \eta(x_2) \, , \qquad x = \xi(x_1)$$

where $x_1$ and $x_2$ are the new variables with a constant spacing and $\xi(x_1)$ and $\eta(x_2)$ are stretching functions introduced to have more computational points in the regions where higher gradients of the physical quantities occur; for example the function $\eta(x_2)$ is chosen in such a way to have at least seven points between $y^+ = 0$ and $y^+ \simeq 10$ .

To solve the non linear system of equations (1), transformed by the relationships (5), the linearization procedure, developed by Beam and Warming for the temporal derivatives, has been applyed to the downstream derivatives. Proceeding as in the previous paper of Orlandi and Reynolds (1979), the system of equations (1) becomes

(6)
$$\left[\Delta(\vec{M}_{x_1}) + (\vec{M}_{x_1})^i\right] \frac{\Delta x_1}{1+\zeta} = U^i \, \Delta \vec{Z} - \frac{\zeta}{1+\zeta} \, U^i \, \nabla \vec{Z} + \left(1 - \zeta - \frac{1}{2}\right) O(\Delta x_1) + O(\Delta U \, \Delta \vec{Z})$$

where the $\Delta$ operator means $\vec{Z}^{i+1} - \vec{Z}^i$ , and $M_{x_1}$ represents the system of equations (1) solved with respect to $U(\partial U/\partial x_1)$ ; $\zeta$ gives the accuracy of the method, $\zeta = 0$ represents the Euler implicit scheme and $\zeta = 1/2$ the three points backward. The operator $\Delta$ applyed to one of the non linear terms of the equation (1) gives

(7)
$$\Delta \vec{P} = \left(\frac{\partial \vec{P}}{\partial U}\right)^i \Delta U + \left(\frac{\partial \vec{P}}{\partial U_{x_2}}\right)^i \frac{\partial \Delta U}{\partial x_2} + \left(\frac{\partial \vec{P}}{\partial Q}\right)^i \Delta Q + \left(\frac{\partial \vec{P}}{\partial D}\right)^i \Delta D + O(\Delta x_1^2)$$

Thus if the operator $\Delta$ is applyed to each term of $M_{x_1}$ and if the differentiated continuity equation is discretized in the downstream direction in such a way to assume a form like the equation (6), and the derivatives with respect to $x_2$ are expressed by a centered finite difference scheme, a block tridiagonal linear system is yielded

(8)
$$A P_j \, \Delta \vec{Z}_{j+1} + A C_j \, \Delta \vec{Z}_j + A M_j \, \Delta \vec{Z}_{j-1} = D C_j$$

which is solved by an efficient elimination procedure developed by Moin et al. (1978). The coefficients of the equation (8) at $j = 1$ and $j = jM$ have been derived from boundary conditions (4) expressed in function of the quantities $\Delta \vec{Z}$ .

The initial conditions at $t \neq 0$ and $x = 0$ have been related to the conditions at $t = 0$ and $x = 0$ as suggested by Singleton (1974).

## RESULTS AND CONCLUSIONS

The turbulence model has been tested for the steady zero pressure gradient turbulent boundary layer, starting from the Klebanoff's condition until a $Re\,x = 10^7$. The constants in the damping factors contribute in an appreciable manner on the results, for example with $C_6 = .05$ an excellent agreement was found, and giving to it a value equal to $.06$, $C_{f/2}$ diminished of $10$ percent.

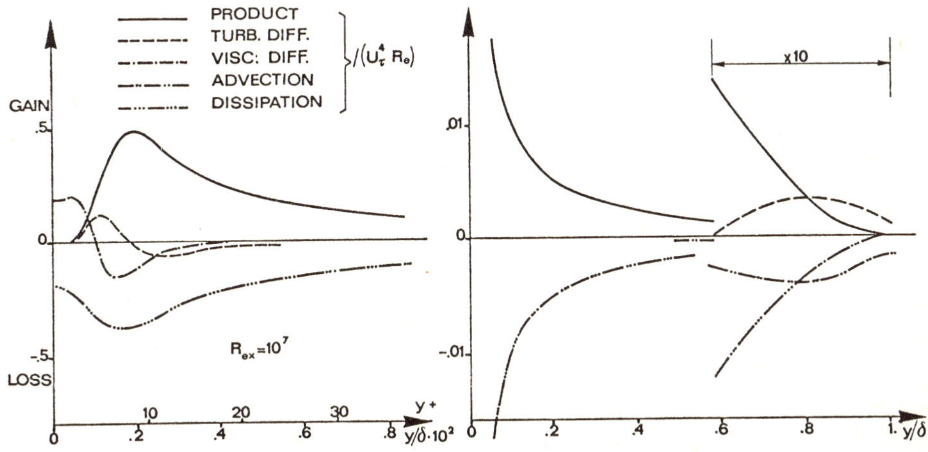



Fig. 1 - Turbulence energy balance.

In Fig. 1 the turbulence energy balance in dimensionless form obtained by dividing for $U_\tau^4\,Re$, is plotted at $Re\,x = 10^7$ for the steady B-L; as before mentioned the pressure work term has not been explicitly modelled, thus in the near wall region the dissipation does not exceed the production. This behavior is not in agreement with the theoretical assumption that the near wall region is a dissipating region where the energy is brought into from the outer region by the pressure work term. As a consequence, smaller values of the viscous diffusion are obtained at the wall. The production and the turbulent diffusion show the same behavior and maintain the same values yielded with the one-equation model, with the pressure work term explicitly modelled. Due to this change of the turbulence energy balance the

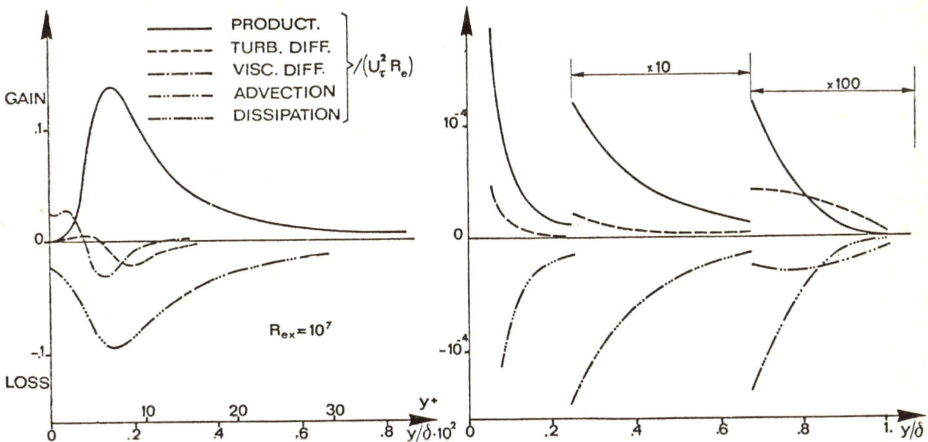



Fig. 2 - Isotropic dissipation balance.

maximum $Q^+$ value calculated with the Q-D model is slightly smaller than the value obtained with the one-equation model. The turbulence energy balance in the region $y/\delta > .1$ shows the same behavior experimentally observed by Klebanoff. Two regions can be recognized: the equilibrium region, where there is equilibrium between production and isotropic dissipation, and the region near the outer edge, where the balance is between the turbulent diffusion and the advection.

In Fig. 2 the energy dissipation balance at $Re\,x = 10^7$ is shown: in the wall region it looks like the turbulent energy balance, the term which differs more is the turbulent diffusion; in the energy balance the gains and the losses of it are almost of the same order, whereas in the dissipation balance the losses are greater; thus in the outer part the turbulent diffusion of the dissipation accounts more with respect to the production and dissipation; the dissipation itself is diffused toward the edge of the boundary layer. As well as the energy balance, at the edge, the advection and the turbulent diffusion are predominant with respect to the other terms.

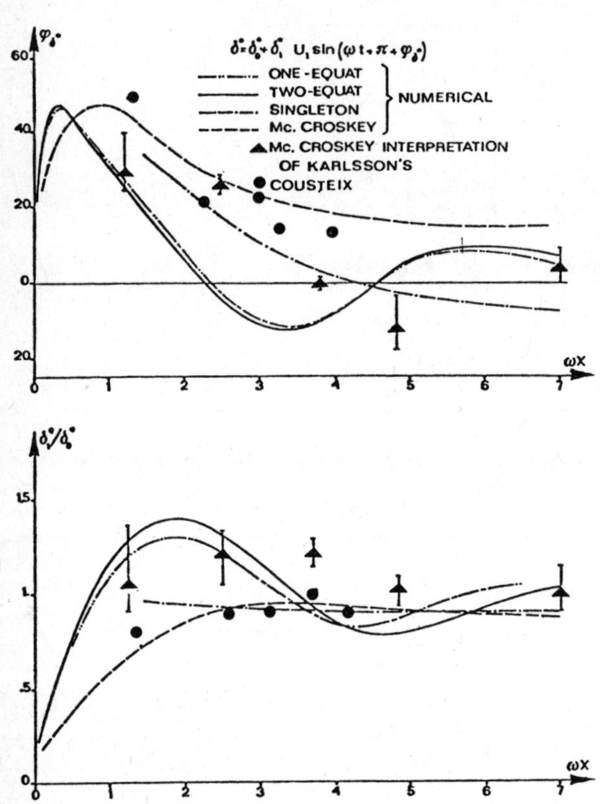


Fig. 3 - Phase and amplitude of the
displacement thickness.

The two-equation model, giving good predictions for the steady case, has been carried over to the unsteady case, since the assumption that there is no interaction between the deterministic and random fluctuations has been done. This assumption is valid if the time scale of the unsteadiness is much larger than the time scale of the random fluctuations. In Fig. 3 the phase angle between the displacement thickness and the edge velocity ($U_e = 1 + U_1 \sin \omega t$) is plotted in function of the reduced frequency, compared with the behavior obtained with the one-equation model. Both predict the qualitative trend of the Karlsson's experiments interpreted by McCroskey (1975), that is a change of phases at intermediate reduced frequencies. The two-equation model is able to predict the behavior of the mixing length; in Fig. 4 the profiles of the mixing length at different points of the cycle are shown and compared with the measured ones of

Cousteix (1979), the agreement is not so good, may be due to the fact that they are evaluated at different conditions.

Fig. 5 shows the profile of the inphase and out-of-phase velocities obtained both with the two-equation and one-equation model, the first give a better representation but does not predict the experimentally observed overshoot of the inphase velocity.  I think it depends on having not explicitly modelled the pressure velocity correlation terms; thus much more work should be done in this direction.

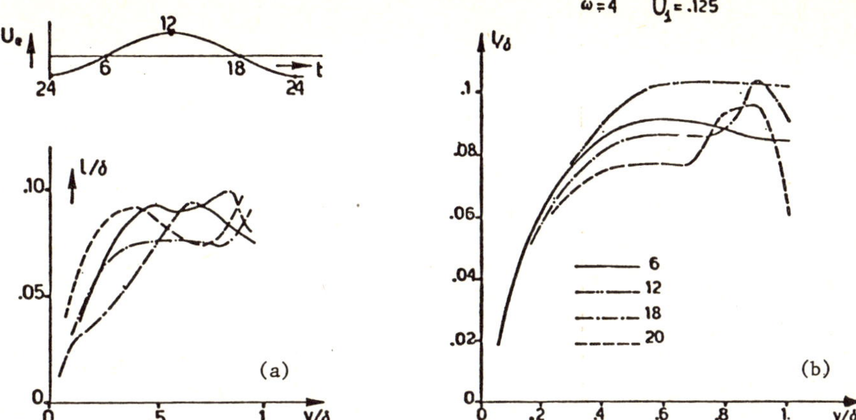

Fig. 4 - Instantaneous mixing length profiles: a) Cousteix experiments
b) present calculation.

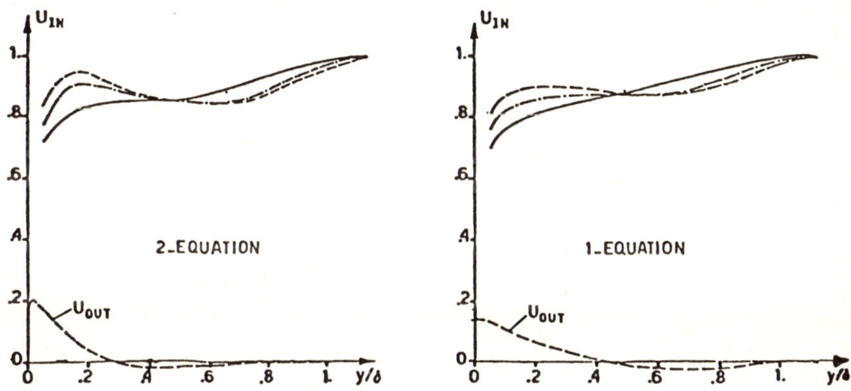

Fig. 5 - Inphase and out-of-phase velocity profiles:
———— $\omega x = .7$; –·–·– $\omega x = 1.5$; – – – – $\omega x = 3.0$ .

## REFERENCES


Beam R.M. and Warming R.F., 1978, AIAA Journal,vol. 16, pp. 392-402.
Cousteix J., Houdeville R. and Desopper A., 1978, AGARD CP-227.
Jones W.P. and Launder B.E., 1973, Int. Journal Heat and Mass Transfer, vol. 16, pp. 1119-1130.
Klebanoff P.S., 1955, NACA Report 1247.
Laufer J., 1954, NACA Report 1174.
Mc Croskey W.J. and Philippe J.J., 1975, AIAA Journal, vol. 13, pp. 71-79.
Moin P., Reynolds W.C. and Ferziger J.H., 1978, Mech. Eng. Dept. Stanford University, Report n° TF-12.
Orlandi P. and Reynolds W.C., 1979, Report of the Mech. Eng. Dept. Stanford University, in printing.
Singleton R.E. and Nash J.F., 1974, AIAA Journal, vol. 12, pp. 590-595.

A PHYSICAL APPROACH TO SOLVE NUMERICALLY COMPLICATED HYPERBOLIC FLOW PROBLEMS

by

M. PANDOLFI  (*)

L. ZANNETTI  (**)


## INTRODUCTION

We report in this paper on the last stages of the investigation we are carrying on
to find out an efficient methodology for the numerical solution of the partial differential equations, which describe multidimensional hyperbolic phenomena and we are
here concerned with unsteady compressible inviscid flows. As *efficient* we mean a methodology which allows for reliable computations and, at the same time, for the minimum amount of computational points in order to save both time and storage in the
computer.
The first steps of this investigations have been reported in $\underline{/\,1\,\underline{/}}$, where some rough
ideas were developed with reference to a particular 2D unsteady flow.
Then the principles have been settled down and extended to steady supersonic flows.
This second stage and several numerical examples have been reported in $\underline{/\,2\,\underline{/}}$.
More recently further improvements have been indicated in $\underline{/\,3\,\underline{/}}$, where a slightly new
formulation of the problem and the extension to 3D unsteady flows have been reported.
Finally several examples have been presented in $\underline{/\,4\,\underline{/}}$ with reference to the calculation of transonic steady flows about airfoils or in ducts achieved through a time
dependent technique.
We point out that we are mainly interested in investigating *real unsteady flows* and
therefore we are not dealing here with those methods called as pseudo-time-dependent
techniques or quasi-natural time-dependent methods, or similar, where the main interest is the fast computation of steady configurations obtained by numerical devices
to speed up the convergence to the final solutions. In these approaches the transients
before the achievement of the steady solution are pure numerical transients which have
nothing to do with the actual physical unsteadiness.
In fact, in the numerical examples, we report hereafter, the steady solutions have
been achieved through a *real time dependent process* and the numerical transients describe conceivable physical transients. Investigations of unsteady flows *per se* will
be reported in a short time in $\underline{/\,5\,\underline{/}}$, where the unsteady flow in vibrating cascades
will be examined.
We will point out, in the following, the two distinct, but strictly connected, aspects
of the methodology: first, a particular formulation of the governing Euler equations,
where new variables will be introduced and which appears as the generalization of the
Riemann invariants, well known in the classical 1D unsteady flow theory. Then we propose the numerical algorithm, which is based on a finite difference approximation. It
is strictly related to the physical interpretation of each term appearing in the new
formulation of the unsteady flow equations.
We refer shortly on the computation at boundaries such as inlet, outlet and solid
walls. However we focus our attention on the explicit numerical treatment of the vortical line leaving the trailing edge of a lifting profile during a transient. The correct
treatment of such a line appears as an imperative requirement for computing unsteady
flows about airfoils or blades in a turbomachine.
Finally we present some examples of application of this methodology. They refer to
transonic flow past airfoils, in ducts and through cascades.

---


(*)   Visiting professor at the Institut de Thérmique Appliquée, Ecole Federale Poly
      technique de Lausanne, Switzerland. On leave of absence from the Politecnico
      di Torino, Italy.

(**)  Associate professor, Istituto di Macchine e Motori per Aeromobili, Politecnico
      di Torino, Italy.

## THE FORMULATION OF THE EQUATIONS

We confine the following analysis to the 2D unsteady compressible inviscid flow. However there are no limitations in extending the methodology to the 3D flows, as shown in $\underline{/\,3\,\underline{/}}$. Furthermore we will consider here the flow as homoentropic; this assumption is not restrictive at all and makes only easier the formal writing of the equations. In fact one of the following numerical examples has been computed by taking into account also the entropy gradients generated behind a shock wave. Finally we develop our analysis for cartesian coordinates (x,y,t) and the corresponding velocity components (u,v). Neverthless in the practical applications, a transformation more or less simple will be required and even the velocity components may be chosen depending on that.

The methodology we propose wants to be *consistent* with the physical nature of the problem we are dealing with: the description of the wave propagation phenomenon, whose the mathematical model is given by a set of hyperbolic partial differential equations. From a general point of view, the solution of these equations is achieved by integrating in time, at a given point P, the local derivatives in time, provided by the unsteady terms in the governing equations, once the space derivatives are approximated through finite differences.

From the physical point of view, these time derivatives result as the interaction of the *signals* carried on bicharacteristics which converge at that given point P and at the given time. It is interesting to note that such signals are represented by particular combinations of the primitive dependent variables (velocity components and speed of sound) which in the 1D unsteady flow result as the Riemann invariants. The variables defined by these combinations appear to be really the *most significant parameters* of the unsteady flows.

The full unsteady inviscid compressible flow equations are:

$$P_t + \bar{q} \cdot \nabla P + \gamma \nabla \cdot \bar{q} = 0 \qquad \text{(continuity)}$$

$$\bar{q}_t + (\bar{q} \cdot \nabla)\bar{q} + T\nabla P = 0 \qquad \text{(momentum)}$$

where "P" represents the logarithm of the pressure, "T" the temperature, and "$\bar{q}$" the velocity.

Let us consider, in the (x,y,t) frame, the characteristic conoid which leaves the point P (Fig. 1). The line "m" is the intersection of the characteristic surface, touching the cone on the bicharacteristic "1", with a constant time, space like, surface. Compatibility equations on characteristic surfaces can be obtained by the primitive equations:

$$(1) \qquad \underline{/\,\bar{q}_t + (\bar{q}\cdot\nabla)\bar{q} + T\nabla P\,\underline{/}} \cdot \bar{\xi} + \frac{a}{\gamma} \underline{/\,P_t + \bar{q}\nabla P + \gamma\nabla\cdot\bar{q}\,\underline{/}} = 0$$

where $\bar{\xi}$ is the unit vector, normal to $\bar{m}$, and running on the constant time plane, and "a" represents the speed of sound, according to the normalization reported in $\underline{/\,1\,\underline{/}}$. Eq. (1) may be now written as the compatibility equation over a general characteristic surface defined by the bicharacteristic "1" and the line "m" which, in turn, are identified by the unit vector $\bar{\xi}$ or its angle $\delta$:

$$(2) \qquad \underline{/\,\frac{d}{dt}(\bar{q}\cdot\bar{\xi})\,\underline{/}}_1 + \frac{a}{\gamma} \underline{/\,\frac{d}{dt}(P)\,\underline{/}}_1 + a\frac{\partial}{\partial m}(\bar{q}\cdot\bar{m}) = 0$$

where:

$$\underline{/\,\frac{d}{dt}(\ )\,\underline{/}}_1 = \frac{\partial(\ )}{\partial t} + (u + a\cos\delta)\frac{\partial(\ )}{\partial x} + (v + a\sin\delta)\frac{\partial(\ )}{\partial y}$$

$$\frac{\partial(\ )}{\partial m} = -\sin\delta\frac{\partial(\ )}{\partial x} + \cos\delta\frac{\partial(\ )}{\partial y}$$

By defining:

$$Q = \frac{2}{\gamma-1} a + \bar{q}\cdot\bar{\xi}$$

we may write Eq. (2) shortly as:

(3)
$$\left[\frac{d}{dt} Q\right]_1 = - a \frac{\partial}{\partial m} (\overline{q} \cdot \overline{m})$$

By following now the suggestions given in $\lfloor 6 \rfloor$, it is convenient to select the four bicharacteristics $l_1$, $l_2$, $l_3$, $l_4$ defined by the corresponding angles $\delta_1 = 0$, $\delta_2 = \pi$, $\delta_3 = \pi/2$, $\delta_4 = 3/2\pi$ .
The compatibility equations over these four bicharacteristics are then:

$$\left[\frac{dC}{dt}\right]_{l_1} = - av_y \qquad (\delta_1 = 0)$$

$$\left[\frac{dD}{dt}\right]_{l_2} = - av_y \qquad (\delta_2 = \pi)$$

(4)

$$\left[\frac{dE}{dt}\right]_{l_3} = - au_x \qquad (\delta_3 = \pi/2)$$

$$\left[\frac{dF}{dt}\right]_{l_4} = - au_x \qquad (\delta_4 = 3/2\pi)$$

where the particular values of Q over these lines are:

(5) $\qquad C = \frac{2}{\gamma-1} a + u \; ; \qquad D = \frac{2}{\gamma-1} a - u \; ; \qquad E = \frac{2}{\gamma-1} a + v \; ; \qquad F = \frac{2}{\gamma-1} a - v$

The sketch of Fig. 2 represents the top view of the physical plane (x,y) and the projections of the lines $l_1$, $l_2$, $l_3$, $l_4$, which are originated at four definite points (black circles) at the time t, and converge on a given point (center of the cross) at the new time t + Dt.
We may now combine linearly Eqs. (4) and, by using the continuity equation, we get the time derivatives of the primitive variables:

$$u_t = - \frac{1}{2}\lfloor (u+a)C_x + vC_y - (u-a)D_x - vD_y \rfloor$$

(6) $\qquad v_t = - \frac{1}{2}\lfloor uE_x + (v+a)E_y - uF_x - (v-a)F_y \rfloor$

$$P_t = - \frac{\gamma}{2a}\lfloor (u+a)C_x + (u-a)D_x + (v+a)E_y + (v-a)F_y \rfloor$$

It may be noticed that all the space derivatives in Eqs. (6) refer to the *signals* C,D,E,F. Each of these derivatives represents the space component of the derivative along the corresponding bicharacteristic, as it may be seen in Fig. 2. All the derivatives along the space line "m" have been dropped out.


THE NUMERICAL ALGORITHM


Once the derivatives in time of the primitive variables (components of the velocity and pressure) are provided by the Eqs. (6), we may proceed to the integration in time to get the description of the unsteady flow.
We use a two levels in time, predictor-corrector, integration scheme, quite similar to the well known suggested by MacCormack many years ago. However in this last scheme the space derivatives are approximated as finite differences in the x and y direction, alternatively forward and backward, without any regard to the sign of the velocity components and their magnitude with respect to the speed of sound. All the four neighboring computational points (see Fig. 2) are then used to evaluate the flow evolution at a given point. However in the case of transonic flow, the correct domain of influence may be violated.
In fact the formulation given by Eqs. (6) shows that each of the space derivatives should be evaluated, in the finite difference approximation, as one-sided derivative, according to the sign of the component of the propagation velocity along the corresponding bicharacteristic:

$$u + a \quad \text{and} \quad v \quad \text{for } C_x \text{ and } C_y \text{ along } l_1$$
$$u - a \quad \text{and} \quad v \quad \text{for } D_x \text{ and } D_y \text{ along } l_2$$

$$u \quad \text{and} \quad v + a \quad \text{for} \quad E_x \quad \text{and} \quad E_y \quad \text{along} \quad l_3$$
$$u \quad \text{and} \quad v - a \quad \text{for} \quad F_x \quad \text{and} \quad F_y \quad \text{along} \quad l_4$$

By following these indications, the domain of dependence for each space derivative is taken into account correctly.
The finite difference approximation for the x and y derivatives is based on a two or three points discretization, as indicated in $/\overline{\phantom{x}}1,2\_/$ in order to ensure che second order of accuracy.


## THE BOUNDARIES

Depending on the physical problem, we may find different kinds of boundaries.
In the case of an isolated airfoil, at the computational points representing the physical infinity, the flow properties (pressure and velocity components) are considered as unperturbed.
In the case of internal flow, we have permeable boundaries, as inlets and outlets. In this case we follow the modeling of these boundaries reported in $/\phantom{x}7\_/$, which has been well experimented in a large variety of problems.
In the case of solid walls, the physical boundary condition requires the component of the velocity normal to the wall as equal to the corresponding wall velocity. This value may be prescribed zero or different if the wall is moving. The methodology here is slightly different with respect to the interior points. First the original equations are written in a local frame of reference with the axis and velocity components $(\breve{u},\tilde{v})$ tangential and normal to the wall (in many problems the general transformation of coordinates satisfies already to this requirement). Then the following new variables are defined:

$$\tilde{C} = \frac{2}{\gamma-1}\, a + \breve{u} \; ; \qquad \tilde{D} = \frac{2}{\gamma-1}\, a - \breve{u} \; ; \qquad \tilde{E} = \frac{2}{\gamma-1}\, a + \tilde{v} \; ; \qquad \tilde{F} = \frac{2}{\gamma-1}\, a - \tilde{v}$$

At this point the tangential velocity $\breve{u}$ is computed by integrating the derivative $\breve{u}_t$ which comes from an equation similar to the first of Eqs. (6).
The normal component $\tilde{v}$ is determined by the wall boundary condition. Finally the pressure derivative in time is evaluated by combining the compatibility equations relative to the tangential velocity $(\tilde{C},\tilde{D})$, that compatibility equation related to the normal component of the velocity whose the bicharacteristic runs against the wall, taken twice, and the continuity equation. In such a way one gets the term $P_t$ as depending only on space derivatives of the new variables $(\tilde{C},\tilde{D},\tilde{E}$ or $\tilde{F})$ just as at the interior point, without any arbitrary violation of the domain of dependence. The reader may find in $/\phantom{x}3\_/$ more details on this analysis.
All the above mentioned boundary computations have been worked out by adding in the computer code, to the Fortran statements needed for the interior points, some further statements related to each boundary, by following the suggestions on the *post-correction technique* proposed in $/\overline{\phantom{x}}8\_/$.


## THE VORTICAL LINE BEHIND A PROFILE

The unsteady flow past an airfoil generates a lift variable in time and related to the vorticity left in the wake, behind the trailing edge.
This wake may be assumed as a discontinuity line floating in the flow field downstream of the airfoil. The pressure and the velocity component normal to this line is continuous across it. However the tangential component can show a jump which represents the circulation per unit lenght of the wake $/\phantom{x}\gamma(1,t)$ in Fig. $3\_/$. Any computational procedure for unsteady flows about airfoils or blades in a turbomachine must take into account such a discontinuity.
This modeling of the wake and its explicit numerical treatment have been worked out indipendently for transonic flow in truncated nozzles or for the solution of inverse problems in $/\phantom{x}9,10\_/$ and with specific reference to blades in turbomachinery in $/\phantom{x}11\_/$. The present methodology may incorporate this modeling into the algorithm. A very interesting check has been done to verify the soundness of the numerical results in the case of the unsteady flow generated in a vibrating cascade (Fig. 3). Let us consider

the flow picture at the time t. We can compute the circulation around the blade as:

$$(7) \qquad \Gamma_b = \oint_b q(s,t)ds$$

or we can roughly evaluate its derivative in time by considering the flow fields at t and t + Dt:

$$(8) \qquad \frac{d\Gamma_b}{dt} = \frac{\Gamma_b(t+\Delta t) - \Gamma(t)}{\Delta t}$$

On the other hand the variation in time of the vorticity spread in the wake downstream of the blade is given by:

$$(9) \qquad \frac{d\Gamma_w}{dt} = \frac{1}{2} \underline{/}^- q_2^2(t) - q_1^2(t) \underline{\_/}$$

and the circulation around the wake is given by:

$$(10) \qquad \Gamma_w(t) = \int_o^t \frac{d\Gamma_w}{dt} \, dt$$

Simple considerations show that the time rates of change of the two circulations as well as the two instantaneous circulations have to be equal.
The plot of Fig. 4 shows the comparison between Eq. (8) and Eq. (9) as they have been computed. The agreement between the two curves is remarkable and very important to check the reliability of these unsteady flow computations. More details on this matter will be presented in $\underline{/}^-5\underline{\_/}$.


## THE RESULTS

We present here some results on different problems to test the capabilities and the accuracy of the methodology and to allow for a judgement on its *efficiency* in terms of economy on the computer time and storage and reliability of the numerical outputs.
It would have been interesting to show comparisons with exact solutions or experimental results on unsteady flows. Unfortunately we do not know exact solutions for 2D unsteady flows and we have to confine our checks to considerations as those presented in Fig. 4, which however turn out to be quite interesting. We will try in the future to find out experimental results for comparison with numerical predictions.
Therefore we show examples of steady flows achieved through a real time-dependent procedure.
The first two examples refer to the flow past the NACA 0012 airfoil. The computational grid is obtained through a classical conformal mapping and presents 60 intervals all over around the airfoil and 15 between the airfoil and the infinity. The governing equations have been written with respect to the orthogonal curvilinear coordinates given by the mapping. The integration in time is then carried out on a set of equations similar to Eqs. (6) with minor modifications which account for the modulus of the derivative of the map function. The results for a subcritical flow are shown in Fig. 5 (black and white circles), and compared with the solution given in $\underline{/}^-12\underline{\_/}$ (solid lines). A supercritical case is reported in Fig. 6, where the solid line is given in $\underline{/}^-13\underline{\_/}$ and comes from the fully conservative relaxation for the potential flow equation.
It may be noted the capability of the methodology to *capture* the shock in only one mesh interval, even if the shock is not tracked correctly during the transient. The reader may refer to $\underline{/}^-2,3\underline{\_/}$ for comments on this remarkable feature.
The next example represents an internal transonic flow where a very simple and rough, but efficient, shock fitting has been provided. The example concerns the transonic flow in a channel with a circular arc bump. The computational grid is very large: 6 intervals in y and 28 in x. Fig. 7 reports the pattern of the isobars and the shock location. It is interesting to mention that by using a rougher mesh 6 x 14, the flow picture do not change remarkably, despite the very low number of computational points. The reader may find in $\underline{/}^-4\underline{\_/}$ more details on these examples and comparisons with numerical results achieved about the same problems by different methods and researchers.
The transonic shockless flow proposed by Ringleb has been used as benchmark for several computations $\underline{/}^-1\underline{\_/}$. Fig. 8 reports the pressure as computed (black circles) on the wall of a channel described with two streamlines given by the theoretical solution, whereas the solid line refers to the exact values.

Finally the results obtained in the transonic shock free cascade proposed by Hobson
$/\overline{14}/$ are reported in Fig. 9. The computational grid corresponds to lines along the
peripheral direction and lines along the blade-to-blade passage. Therefore the grid
is not orthogonal and far from those beautiful grids obtained through conformal map-
pings. However the need to solve unsteady flows in turbomachines with closely packed
bladings does not allow for the use of such interesting transformations.
The governing equations have been here written with the velocity components $\tilde{u}$ along
the streamwise coordinate and $\tilde{v}$ normal to it. In this manner the corresponding *signals*
$(\tilde{C},\tilde{D},\tilde{E},\tilde{F})$ are carried on bicharacteristics which come from four points almost symmetrical with respect to the local streamline direction and some of the inconvenients
due to the distorted grid are overcome. The agreement of the computed results (black
circles) with the exact solution (solid line) is good, except in the narrow high Mach
number region on the suction side. However let us point out the size of the grid, with
10 intervals along the blade and 6 from blade-to-blade.

AKNOWLEDGMENT   -   Part of this research has been supported by the Consiglio Nazionale
                    delle Ricerche, Italy, (Contract N. 115.6799 78 02427.07) and part
                    has been performed at the Ecole Polytechnique Federale de Lausanne, CH.

REFERENCES

$/\overline{1}/$   M. PANDOLFI and L. ZANNETTI, "Some tests on finite difference algorithms for
         computing boundaries in hyperbolic flows", Notes on Numerical Fluid Dynamics,
         Vol. 1, Vieweg, 1978.

$/\overline{2}/$   G. MORETTI, "The $\lambda$ scheme", Computers and Fluids, Vol. 7, N. 3, September, 1979.

$/\overline{3}/$   L. ZANNETTI and G. COLASURDO, "A finite difference method based on bicharacte-
         ristics for solving multidimensional hyperbolic flows", Istituto di Macchine,
         Politecnico di Torino, Report PP217, 1979.

$/\overline{4}/$   L. ZANNETTI, G. COLASURDO, L. FORNASIER and M. PANDOLFI, "A physically consi-
         stent time-dependent method for the solution of the Euler equations in transo-
         nic flow", GAMM Workshop on Numerical Methods for the Computation of Unviscid
         Transonic Flow with Shock Waves, Stockholm, September, 1979.

$/\overline{5}/$   M. PANDOLFI, "Numerical experiments on unsteady flows through cascades", paper
         to be presented at the Second International Symposium on Aeroelasticity in
         Turbomachines, Lausanne, Switzerland, September, 1980.

$/\overline{6}/$   D.S. BUTLER, "The numerical solution of hyperbolic system of partial differen-
         tial equations in three independent variables", Proc. of the Royal Society,
         Series A, April, 1960.

$/\overline{7}/$   M. PANDOLFI and L. ZANNETTI, "Some permeable boundaries in multidimensional
         unsteady flows", Lecture Notes in Physics, Vol. 90, Springer Verlag, 1979.

$/\overline{8}/$   G. MORETTI and T. DE NEEF, "Shock fitting for everybody", Third GAMM Confe-
         rence on Numerical Methods in Fluid Dynamics, Köln, 1979.

$/\overline{9}/$   L. ZANNETTI, "Transonic flow field in two-dimensional or axisymmetric con-
         vergent nozzle", Second GAMM Conference on Numerical Methods in Fluid Dyna-
         mics, Köln, 1977.

$/\overline{10}/$   L. ZANNETTI, "A time dependent method to solve the inverse problem for in-
         ternal flow", AIAA Paper 79-0013, 1979.

$/\overline{11}/$   J.I. ERDOS, E. ALZNER and W. McNALLY, "Numerical solution of periodic tran-
         sonic flow through a fan stage", AIAA, Vol. 15, N. 11, November, 1977.

$/\overline{12}/$   R.C. LOCK, "Test cases for numerical methods in two-dimensional transonic
         flows", AGARD Report N. 575, 1970.

$/\overline{13}/$   E.M. MURMAN, "Analysis of embedded shock waves calculated by relaxation me-
         thos", AIAA Computational Fluid Dynamics Conference, July, 1973.

$/\overline{14}/$   D.G. HOBSON, "Shock-free transonic flow in turbomachinery cascades", Univer-
         sity of Cambridge, Dept. of Eng., Report CUED/A Turbo/TR 65, 1974.

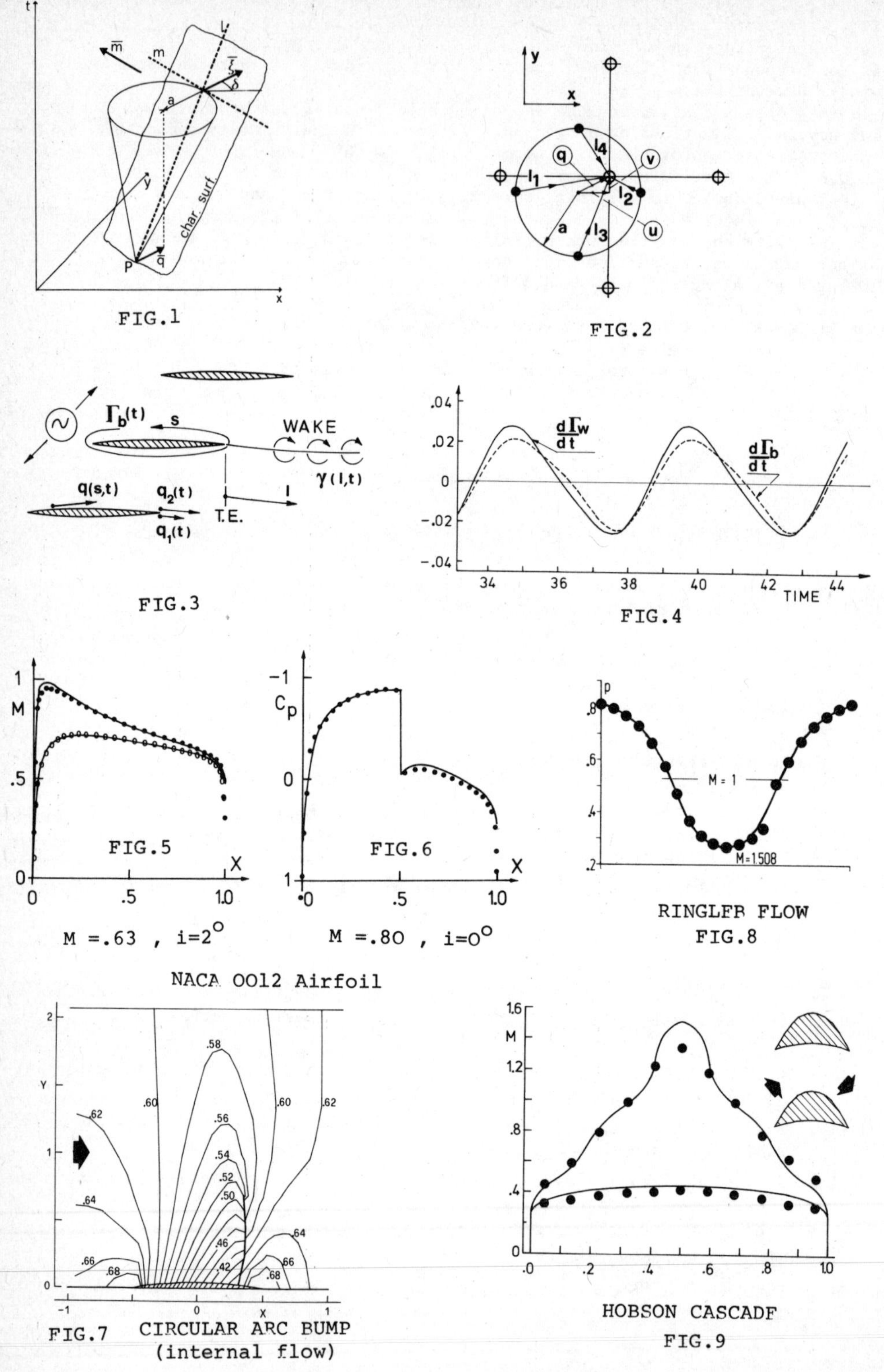


FIG.1
FIG.2
FIG.3
FIG.4
FIG.5
M =.63 , i=2°
NACA 0012 Airfoil
FIG.6
M =.80 , i=0°
RINGLER FLOW
FIG.8
FIG.7  CIRCULAR ARC BUMP
(internal flow)
HOBSON CASCADE
FIG.9

TRANSITION AND TURBULENCE IN PLANAR CHANNEL FLOWS

Anthony T. Patera and Steven A. Orszag
Department of Mathematics
Massachusetts Institute of Technology
Cambridge, Massachusetts  02139


## INTRODUCTION

In this paper we describe some recent work on the numerical simulation of transition and turbulence in planar channel flows by direct solution of the three-dimensional incompressible Navier-Stokes equations.  Transition to turbulence has been studied in plane Poiseuille and plane Couette flow with good agreement with experimental observation.  Orszag & Kells (1980) showed that transition Reynolds numbers of order  1000  are obtained when finite-amplitude three-dimensional perturbations are considered.  Here, we present a mechanism that leads to a simple explanation of the transition phenomenon.  We have discovered a class of two-dimensional quasi-equilibrium states (in which production and dissipation nearly balance) that are strongly unstable to infinitesimal three-dimensional perturbations. This strong three-dimensional instability persists down to Reynolds numbers of order 1000.

We shall also present results on the direct numerical simulation of turbulent channel flow.  These simulations are performed with no sub-grid scale or turbulence modelling up to  Reynolds numbers of order  5,000.  The principal result is that a nearly statistically steady state is achieved.  An asymptotic wall layer is obtained with evidence of a logarithmic layer with von Karman's constant  $\kappa \sim 0.3$. No adjustable constants or parameters are available in this computation.


## NUMERICAL METHODS

The Navier-Stokes equations

$$\frac{\partial \vec{v}}{\partial t} = \vec{v} \times (\vec{\nabla} \times \vec{v}) - \vec{\nabla}\Pi + \nu\nabla^{2}\vec{v} \tag{1}$$

$$\vec{\nabla} \cdot \vec{v} = 0 \tag{2}$$

are solved using spectral methods (Gottlieb & Orszag 1977, Orszag & Kells 1978) in space and finite difference methods in time.  In the  x  and  y  directions, periodic boundary conditions are applied so Fourier series are used.  In the  z direction, rigid no-slip boundary conditions are applied so Chebyshev polynomial expansions are employed.  Equations for the spectral expansion coefficients are obtained by pseudospectral (collocation) methods.

Two quite different techniques have been used to impose the incompressibility constraint and viscous damping.  First, uniform second-order accuracy in time has been achieved by a combination of Adams-Bashforth time differencing of the nonlinear terms and Crank-Nicolson time differencing of the pressure and viscous terms. The resulting implicit scheme requires the solution of a fourth-order equation in space.  This equation can be factored into two nearly tridiagonal equations for the spectral expansion coefficients (Gottlieb & Orszag 1977).

An alternative, and simpler, method is to apply a fractional-step method like that used by Orszag & Kells.  Here the incompressibility constraint and viscous damping terms are accounted for in separate fractional steps.  Each fractional step requires the solution of a second-order equation, again nearly tridiagonal for the spectral expansion coefficients.  Because the pressure projection operator and viscous damping terms are noncommutative operators when rigid no-slip boundary conditions are applied, the splitting method gives errors of order  $0(\Delta t^{2}) + 0(\nu\Delta t)$, even though the nonlinear terms are included by the second-order Adams-Bashforth scheme.

The first method (employing the fourth-order equation) has been implemented in the computer code  CHEER  on the CRAY-1 computer with up to  $64 \times 64 \times 65$  spatial resolution.  The fractional step method is implemented in the  SPLITP  code that provides up to  $32 \times 32 \times 33$  spatial resolution.  The SPLITP  vectorized code runs in fast memory on the CRAY-1 in about  0.4 s/time step with  $32 \times 32 \times 33$  spatial resolution.  In other words, the computer code runs at about  $5\mu s$  per degree of freedom per time step.

Our spectral codes have a number of advantages over more conventional methods. First, they are infinite-order accurate when applied to problems with smooth solutions.  In particular, they are essentially free of phase errors (except for time-differencing effects) which is essential for accurate simulation of transition problems.  Second, the use of a polynomial-spectral representation of the flow in the  z  direction gives very high resolution near the solid boundaries. This property is quite useful in resolving both the transition flows and the wall layer of the turbulent flows.  Third, our methods for implementing the pressure and viscous terms are stable and accurate.


## TRANSITION IN PLANE POISEUILLE FLOW

While experiments show that incompressible plane Poiseuille flow may undergo transition to turbulence at Reynolds numbers (based on center-line velocity and half-channel width) of order  1000,  linear stability analysis gives a critical Reynolds number  Re = 5772 and nonlinear stability analysis, using the Landau-Stuart-Watson technique, gives  subcritical finite-amplitude two-dimensional instabilities for  $Re \gtrsim 2900$  (Reynolds & Potter 1967).  The discrepancy between theory and experiment suggests that the mechanism of transition is not properly represented by parallel flow linear-stability analysis.

The experiments of Klebanoff et al (1962) and the theory of Benney & Lin (1960) suggest that three-dimensional effects are central to the phenomenon of transition in boundary-layer flows.  Three dimensionality is necessary to produce  streamwise vorticity that produces highly inflectional instantaneous velocity profiles.  The instability of these profiles provides a mechanism for the transition to turbulence. With this motivation, Orszag & Kells (1980)  studied the susceptibility of plane Poiseuille and plane Couette flow to three-dimensional disturbances.  They found that transition could indeed occur at Reynolds numbers down to about  1000, provided that sufficiently strong finite-amplitude disturbances were imposed on the flow.  Here we provide a relatively simple mechanism to explain these results.

Let us begin by reviewing an alternative to the Stuart-Watson method for analysis of finite amplitude stability.  Zahn et al (1974)  and Herbert  (1977) investigated the structure of finite-amplitude two-dimensional equilibrium states of plane Poiseuille flow.  They sought two-dimensional travelling-wave solutions of the Navier-Stokes equations of the form

$$\vec{v}(\vec{x},t) = \vec{F}(\vec{x}-\vec{c}t) \tag{3}$$

with real wavespeed  $\vec{c}$  that are periodic in  x  with period  $2\pi/\alpha$  and independent of  y.  Numerical solution of the equations for  $\vec{F}$  show that such solutions exist for  $Re > 2900$  in plane Poiseuille flow.

For each choice of periodicity wavenumber $\alpha$,  there are either  1  or  3 types of solutions (3).  The solution  $\vec{v} = 0$  is always linearly stable at sub-critical Reynolds numbers.  There are also two possible types of finite-amplitude solutions, called lower branch (LB) and upper branch (UB)  solutions.  The LB solutions are unstable to small perturbations while the  UB  solutions are stable.

The evolution of orbits in phase space is illustrated in Fig. 1.  Here we fix the downstream periodicity wavenumber  $\alpha$  and follow the detailed evolution of various perturbed solutions, plotting the results as dots separated by equal inter-vals of time on a two-dimensional projection of phase space.  Here $E_1$  is the energy of that part of the solution that depends on  x  as  $e^{i\alpha x}$  while  $E_2$  is the energy of that part of the solution that depends on  x  as  $e^{2i\alpha x}$.  The point marked 'steady solution' is the appropriate point on the upper branch of the Zahn et al-Herbert neutral surface.  For initial disturbances that are sufficiently small, the solution dies out and approaches the linearly stable  E = 0  surface. This is illustrated by the orbit in the lower left-hand corner of the plot.  On the other hand, orbits of solutions with initially large energies do not follow simple curves.  Solutions with energies less than that of the  UB  solution can overshoot the  UB  energy by factors of  4  or more and then very slowly approach the  UB  solution in time.  This overshooting indicates that the single-amplitude (autonomous) Landau equations of the Stuart-Watson theory can not be valid for large-amplitude motions.  Furthermore, the typical orbit seems to quickly approach a band of quasi-equilibrium states in which production nearly balances dissipation.

In Fig. 2, we plot the time evolution of $E_1$ and $E_2$ from an initial state that lies near the  UB  solution.  Evidently, the primary mode amplitude overshoots the steady solution and then slowly settles down.  There is a quick initial adjustment on a time scale on the order of  10  to a quasi-equilibrium state and, then, on a time scale of  1000-10000,  the  UB  solution is approached.  In the quasi-equilibrium states, production balances dissipation to about one part in  1000.

The quasi-equilibrium states of plane Poiseuille flow are strongly unstable to infinitesimal three-dimensional disturbances.  In Fig. 3, we plot the evolution of an (initially small) three-dimensional disturbances superposed on finite-amplitude two-dimensional motions.  Evidently, the three-dimensional disturbance quickly (on a time scale of order  10)  achieves a form that grows approximately exponentially in time for  $Re \gtrsim 1000$.  Here  $\alpha = 1.3231$  gives nearly the most dangerous disturbances.  The growth rate of the three-dimensional disturbances is rapid with their amplitude increasing by a factor  10  in a time of about  10.  This short time scale for three dimensional growth should be contrasted with the long time scale of order  1000  for evolution of the quasi-equilibria.  There is strong evidence that this instability is a physically relevant one in that it is fairly insensitive to intial conditions and has small threshold two-dimensional energies.

In Fig. 4, we show the effect of varying the downstream periodicity wavenumber $\alpha$  on the growth of three-dimensonal disturbances.  We have also studied the effect of energy of the two-dimensional quasi-equilibrium states on the growth rate of the three-dimensional disturbances.  An analytical theory has been developed that can be used to predict the results of these three-dimensional stability analyses.  More details on the results and theoretical analysis will be published elsewhere.


## TRANSITION IN PLANE COUETTE FLOW

Linear stability analysis of plane Couette flow shows that this flow is stable to infinitesimal disturbances at all Reynolds numbers.  On the other hand, non-linear stability analysis suggest the existence of finite-amplitude two-dimensional instabilities even at Reynolds numbers well below  1000  (Davey & Nguyen 1971).  We have performed a variety of two-dimensional calculations searching for finite-amplitude two-dimensional instabilities at these low Reynolds numbers.  No such instabilities have been found.  For example, in Fig. 5, we plot the evolution of a two-dimensional disturbance at  Re = 500,  close to a disturbance predicted to be unstable by the nonlinear stability theory.  Even though the two-dimensional states decay in time, they are still subject to strong three-dimensional instability. In Fig. 6, we plot the evolution of the energy of initially very small three-dimensional disturbances superposed on finite-amplitude two-dimensional perturbations of plane Couette flow.  Evidently, there are strong three-dimensional instabilities down to about  Re = 1000.  The growth rate of these disturbances is again about a factor  10  in a time of about  10.  In contrast to the marked difference between the linear and nonlinear stability theory of plane Couette flow and plane Poiseuille flow, there seems to be no qualitative difference in the three-dimensional finite-amplitude evolution of these two plane shear flows.


## TURBULENT PLANE POISEUILLE FLOW

In this paper, we report the results of two direct numerical simulations of turbulent channel flow using the spectral code CHEER with $32 \times 32 \times 33$ spatial resolution. The first run involves a simulation of turbulence evolving from "natural" transition, while the second run involves initial turbulence intensities much larger than eventual equilibrium values. The second run evolves much quicker to an interesting turbulent state. The natural transition run is performed by continuing one of the transition runs discussed above, using the evolved three-dimensional flow as an initial condition.

In Fig. 7, we plot the normalized Reynolds stress $\tau$ for the natural transition run. Here $\tau = -\overline{uw}/u_*^2$, where the friction velocity $u_*$ is defined by the mean wall shear stress and the overbar indicates an average over $x$ and $y$ (but not $z$). Several features are noteworthy. First, there is good agreement with the asymptotic theory of high Reynolds number turbulent channel flow, especially in the outer region where viscous effects are negligible. Second, the spatially averaged Reynolds stress is quite stable in time with only moderate statistical fluctuations, demonstrating that good statistics are achieved in our simulation. Third, the quick equilibration of the Reynolds stress to values of order $u_*^2$ (which is itself of order $\nu$ since the mean pressure gradient is fixed at its laminar value) results in very slow development of the mean velocity profile (on a time scale of order $1/\nu$). The resulting mean profile is plotted in Fig. 8. Note the quick development of the wall region compared to the rather slow deformation of the core.

The slow development of the core in the "natural" turbulence run suggests that more rapid evolution can be achieved by means of strongly forced initial conditions which provide better mixing between the wall and core regions. In Fig. 9, the Reynolds number in wall units, $Re_*$, is plotted versus time $t$, for this second run. Notice that $Re_*(t)$ overshoots its long time value $Re_*(0)$ with a quasi-stationary state being achieved after the initial transient. Therefore, one would expect results at intermediate times to be in better agreement with asymptotic high Reynolds number theories than at the end of the run. The rapid initial growth of $Re_*$ seems to reflect the rapid development of an asymptotic wall layer. A plot of the mean velocity profile in channel coordinates at several times in the run is given in Fig. 10. Notice how a typically blunt "turbulent" profile develops from an initially laminar parabolic profile. In Fig. 11, instantaneous (unaveraged) velocity profiles are plotted together with the mean profile, showing the large turbulent fluctuations that are removed by averaging.

In Fig. 12, we plot the mean velocity profile in wall coordinates when $Re_* = 125$ in the second run. At this intermediate time, the viscous sublayer and buffer region are in good agreement with experimental data obtained at very much larger Reynolds numbers (Laufer 1954, Comte-Bellot 1965). As can be seen in Fig. 12, there is a modest region in which the mean velocity profile fits the logarithmic law of the wall with a von Karman constant $\kappa \sim 0.3$. As the wall Reynolds number decreases later in the run, the logarithmic layer tends to merge with the buffer region. It is important to note that, even as the logarithmic layer shrinks later in this run, an asymptotic wall layer is achieved in good agreement with observations. Finally, in Fig. 13, we plot the Reynolds stress in channel coordinates when $Re_* = 125$. Here there is very good agreement with both the asymptotic outer solution and the prediction of Prandtl's mixing length model. Also notice the relative stability of the spatial averages over time intervals long compared to typical eddy turnover times.

These turbulent shear flow simulations provide just a glimpse into the nature of such flows. We expect that significant new insights into turbulent shear flows should be possible by extension on these calculations over the next several years. We emphasize that the calculations reported above have been successful using no sub-grid scale or turbulence modelling. Higher Reynolds number simulations would require some kind of modelling in the interior of the flow, in order that excited small scales not affect the accuracy of the simulation. It is not clear, at least to the present authors, that available sub-grid scale and turbulence models are sufficiently reliable for this purpose. Fundamental new insights maybe required to calculate accurately at Reynolds numbers much larger than those reported here.

The computations reported in this paper were performed at the Computing Facility of the National Center for Atmospheric Research, Boulder, Colorado, which is supported by the National Science Foundation. This work was supported by the

Office of Naval Research under Contracts N00014-77-C-0138 and N00014-79-C-0478.

## REFERENCES

Benney, D. J. & Lin, C.C. 1960 Phys. Fluids 3, 656.

Comte-Bellot, G. 1965 Écoulement turbulent entre deux parois parallèles. Publications Scientifiques et Techniques du Ministère de l'Air, No. 419.

Davey, A. & Nguyen, H.P.F. 1971 J. Fluid Mech 45, 701.

Gottlieb, D. & Orszag, S.A. 1977 Numerical Analysis of Spectral Methods: Theory and Applications. NSF-CBMS Monograph No. 26, Soc. Ind. App. Math., Philadelphia.

Herbert, T. 1977 In Laminar-Turbulent Transition, AGARD Conf. Proc. No. 224, p. 3-1

Klebanoff, P.S., Tidstrom, K.D., & Sargent, L.M. 1962 J. Fluid Mech. 12,1.

Laufer, J. 1951 Investigation of turbulence in a two dimensional channel, N.A.C.A. Tech. Rep. No. 1053.

Orszag, S. A. & Kells, L. C. 1980 J. Fluid Mech. 96, 159.

Reynolds, W. C. & Potter, M. C. 1967 J. Fluid Mech. 27, 465.

Zahn, J. P., Toomre, J., Spiegel, E.A., & Gough, D.O. 1974 J. Fluid Mech. 64, 319.

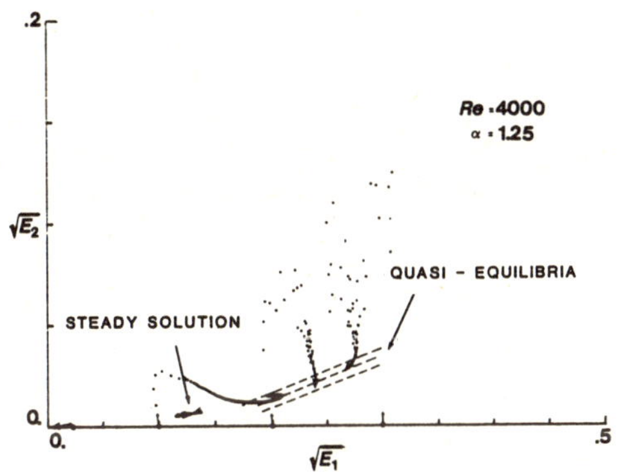


FIG. 1. TRAJECTORIES OF FINITE-AMPLITUDE DISTURBANCES OF PLANE POISEUILLE FLOW ARE PLOTTED IN A TWO-DIMENSIONAL PHASE SPACE FOR A SUB-CRITICAL REYNOLDS NUMBER. ALTHOUGH SOLUTIONS TEND TO A UNIQUE EQUILIBRIUM STATE, INDICATED BY ARROWS, THERE IS A BAND OF QUASI-EQUILIBRIA IN WHICH PRODUCTION NEARLY BALANCES DISSIPATION.

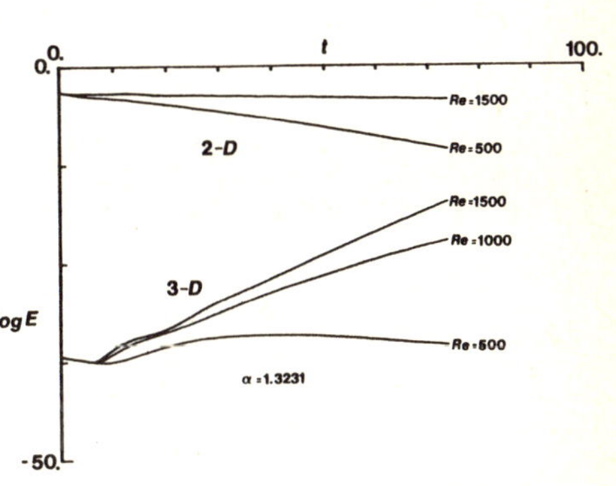


FIG. 3. A PLOT OF THE GROWTH OF SMALL THREE-DIMENSIONAL PERTURBATIONS ON TWO-DIMENSIONAL FINITE-AMPLITUDE STATES IN PLANE POISEUILLE FLOW. OBSERVE THE STRONG INSTABILITY OF THE THREE-DIMENSIONAL PERTURBATIONS FOR REYNOLDS NUMBERS LARGER THAN ABOUT 1000.

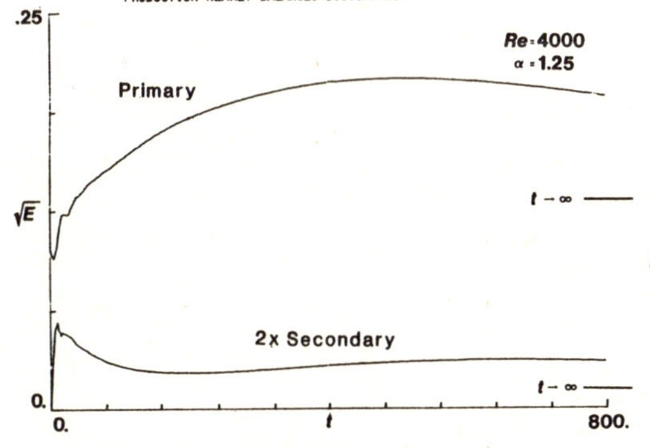


FIG. 2. A PLOT OF THE PRIMARY AND SECONDARY DISTURBANCE AMPLITUDES VS T FOR A SUB-CRITICAL DISTURBANCE IN PLANE POISEUILLE FLOW. OBSERVE THE SEPARATION OF TIME SCALES.

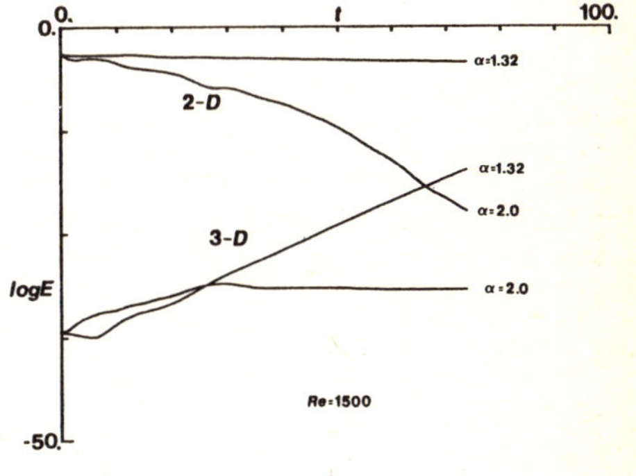


FIG. 4. A PLOT OF THE GROWTH OF THREE-DIMENSIONAL PERTURBATIONS WITH STREAMWISE WAVE NUMBERS α = 1.32 AND α = 2.

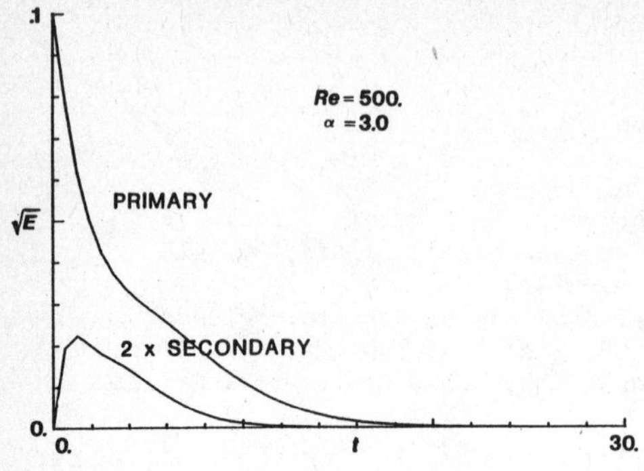



FIG. 5.  A PLOT OF THE EVOLUTION OF THE PRIMARY AND SECONDARY TWO-DIMENSIONAL DISTURBANCES IN PLANE COUETTE FLOW.

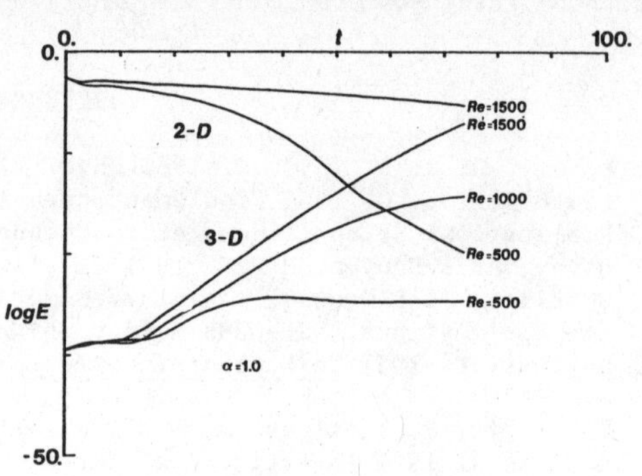



FIG. 6.  A PLOT OF THE GROWTH OF SMALL THREE-DIMENSIONAL PERTURBATIONS ON FINITE-AMPLITUDE TWO-DIMENSIONAL STATES IN PLANE COUETTE FLOW.  OBSERVE THE STRONG THREE-DIMENSIONAL GROWTH RATES FOR REYNOLDS NUMBERS LARGER THAN ABOUT 1000.

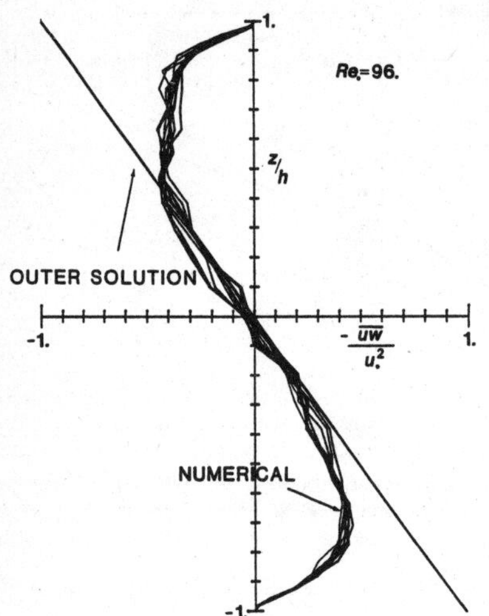



FIG. 7.  REYNOLDS STRESS DISTRIBUTION AT A TIME OF 90 AFTER IMPOSING VERY SMALL THREE-DIMENSIONAL PERTURBATIONS ON A FINITE-AMPLITUDE TWO-DIMENSIONAL STATE AT AN INITIAL REYNOLDS NUMBER OF 2500.  NOTE THE STABILITY OF THE SPATIAL AVERAGES AS WELL AS THE GOOD AGREEMENT WITH THE HIGH REYNOLDS NUMBER OUTER SOLUTION.

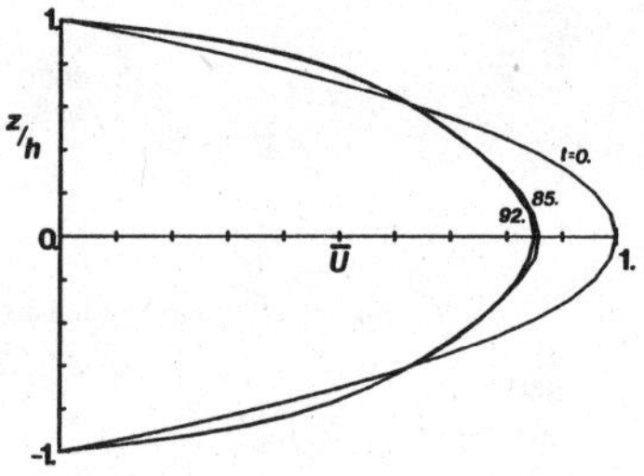



FIG. 8.  A PLOT OF THE MEAN VELOCITY PROFILE IN THE "NATURAL" TURBULENCE RUN.  OBSERVE THE VERY SLOW EVOLUTION OF THE MEAN VELOCITY PROFILE AFTER AN INITIAL TRANSIENT DURING TRANSITION.

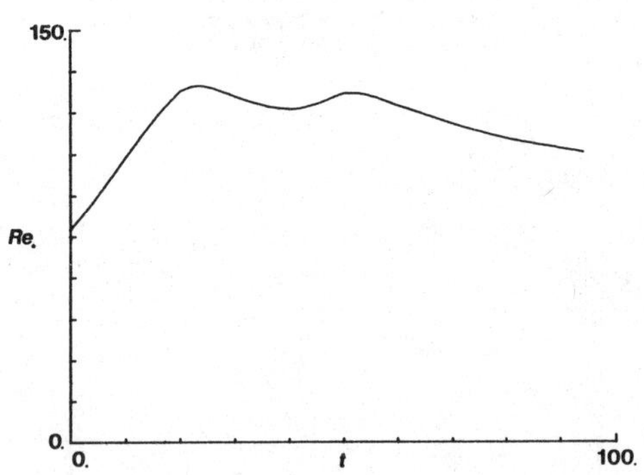



FIG. 9.  A PLOT OF THE REYNOLDS NUMBER IN WALL UNITS FOR THE FORCED TURBULENCE RUN.  FOLLOWING AN INITIAL TRANSIENT, THE FLOW ACHIEVES A QUASI-STATIONARY STATE.  FOR LARGE TIME, THIS REYNOLDS NUMBER MUST APPROACH ITS INITIAL VALUE.

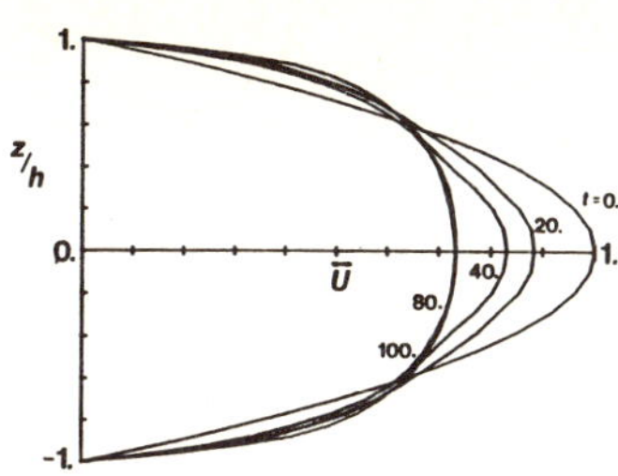

FIG.10. A PLOT OF THE MEAN VELOCITY PROFILE AS A FUNCTION OF CHANNEL COORDINATE IN THE STRONGLY FORCED TURBULENCE RUN.

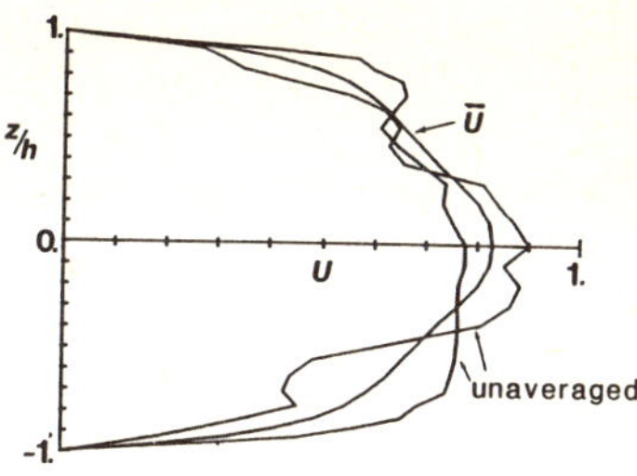

FIG. 11. A PLOT OF THE INSTANTANEOUS (UNAVERAGED) X-VELOCITY PROFILES AT $T = 40$ IN THE STRONGLY FORCED TURBULENCE RUN. THE MEAN VELOCITY PROFILE AT THIS TIME IS PLOTTED FOR COMPARISON. OBSERVE THE STRONG TURBULENT FLUCTUATIONS.

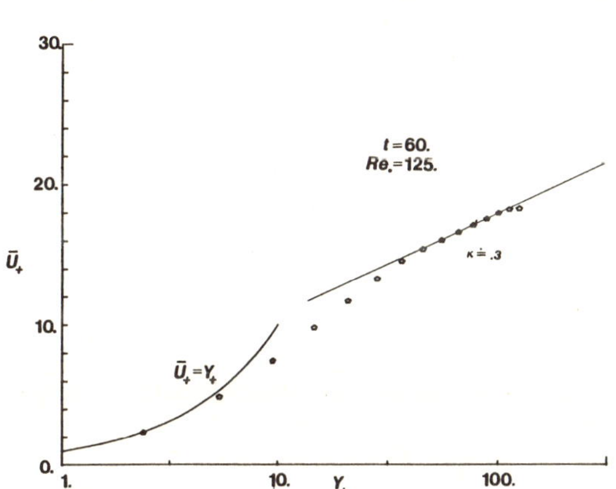

FIG. 12. A PLOT OF THE MEAN VELOCITY PROFILE IN WALL COORDINATES FOR THE STRONGLY FORCED TURBULENCE RUN WHEN $Re_* = 125$. NOTE THE WELL DEVELOPED VISCOUS SUB-LAYER, BUFFER REGION, AND MODEST LOGARITHMIC LAYER. THE BEST FIT OF THE "LOGARITHMIC" LAYER GIVES A VON KARMAN CONSTANT $\kappa \doteq 0.3$.

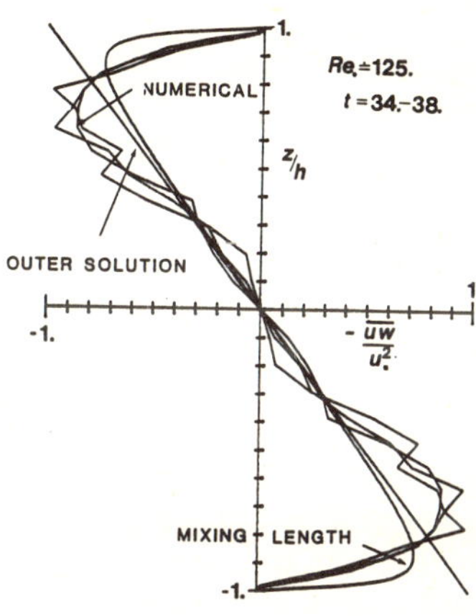

FIG. 13. A PLOT OF THE REYNOLDS STRESS FOR THE STRONGLY FORCED TURBULENCE RUN WHEN $Re_* = 125$. NOTICE THE GOOD AGREEMENT WITH BOTH THE ASYMPTOTIC OUTER SOLUTION AND PRANDTL'S MIXING LENGTH MODEL.

SOME MATHEMATICAL PROBLEMS OF AIR BASIN  PRESERVATION
U.G.Pirumov, M.B.Prokhorov, Y.A.Ryzhov
Moscow Aviation Engineering Institute


The history of technics was developing in the direction of ener-
getic efficiency being not only the main  engine  characteristics,
but also the main characteristics of the devices which transform
energy from one form into another, of technological cycles and so on.
The influence of these objects on the environment had not been
taken into account as the waste of any singular object is rather
small if compared to the natural emission even nowadays. But  the
concentration of such objects which takes place in the cities, or
the long-term influence of one object can lead and does lead to the
substantial changes in the dwelling region (e.g. the waste lands
which appear around the thermoelectric power stations after several
decades work, lead-poisoned lands along the highways and so on).
This results in introducing one more index, that is the speed of the
biosphere processing the waste, or the upper limit of the concentration of the toxic components in the combustion products, or the combination of both of them. The methods for toxic components control
can differ,  but the most popular are those based either on the neutralization of toxic components with the help of various catalysts,
or on organising the technological process so as to escape the
toxic components emerging.
But it must be mentioned that the efficiency of the unit is not
decreased notably.
The main sources of the toxic components (such as carbon mono-
xide, nitrogene oxides, sulphur oxide, a number of carcinogenic elements and aerosols) are the high temperature processes taking place
in power engineering, metallurgy, internal combustion engines and
rocketry.

To make qualitative and quantitative prognosis of these processes, to create methods for the toxic components and aerosols control,
to foretell the way of their  transformation in the atmosphere after
the exhaust equipment, it is necessary to create the physical model
taking into account the main peculiarities of the phenomenon,  after
that to make the conformable  mathematical model and to work out the
efficient methods for the solution of the corresponding equations.


I   The Physical Model


Most of toxic matters (named above) are generated at high temperatures when burning the fuel as the result of physical-chemical
transformations being non-equilibrium. This process can be presented
as the flow of tough heat-conducting gas along the complex curvilinear channel in the general case, and alongside in the presence  of
non-equilibrium processes and multiphaseness with heat extraction
into the wall. It is necessary to take the multiphaseness into
account because in the process of burning there can appear some

aerosols and solid particles (e.g. soot). After having been initiated into the atmosphere the toxic elements are transformed further. They are drifted by the wind, they diffuse and are transmuted chemically. For example, under solar radiation the nitrogene oxides produce smog. When interacting with water vapour in the air they produce nitric acid. It is obvious that the process of spreading the impurities in the atmosphere can also be presented as the flow of tough heat-conducting gas in the periterrestrial boundary layer (taking into account the physical-chemical transformations).

II  The Mathematical Model

In the general case the non-stationary flow of tough heat-conducting reacting gas can be described by the Navier-Stokes equation system. Having added to them the balance equation for energy and the equation of continuity we shall obtain the system which describes the flows. We can allow for the phase interactions by initiating the additional members and equations into the system. But this system is rather awkward and difficult to solve that is why it is necessary to simplify it in every concrete case using the physic qualities of the process or the object. Thus when calculating the flows of reacting mixture (in steam generators, in chambers of internal combustion engines, in MHD-generators and so on) it is possible to obtain substantial results if using the one-dimensional approximation without taking into account the gas being tough and heat-conducting. The uniformity of the mixture across the flow is usually provided by the very way of initiating the fuel. Its aim is to mix the fuel and the oxidizer as well as possible. On the other hand very often the one-dimensional approximation produces the results near the smoothed values across the nonplanar stream; and the main aim of this research is to define the toxic components flow through the exhaust. Besides, the one-dimensional approximation is the basis for the preliminary analysis to find the qualitative characteristics and the following specification of the model. Thus the system of the flow equations for the non-viscous non-heat-conducting reacting gas in the one-dimensional stationary case looks like

$$w\,dw + \frac{1}{\varrho}\,dp = 0 \qquad\qquad p = \frac{\varrho R T}{\mu}$$

$$\frac{w^2}{2} + H(T) = H_o \qquad\qquad H(T) = \sum_{i=1}^{N} h_i(T)\gamma_i$$

$$\frac{d\gamma_i}{dt} = \varphi_i - \gamma_i \phi_i \qquad\qquad \frac{1}{\mu} = \sum_{i=1}^{N} \gamma_i$$

$$\varrho w F = Const$$

($C_{ij}$ - stoichiometric mixture ratio matrix; $\gamma_i$ - $i^{th}$ component concentration; W - velocity of the mixture flow; p - pressure; T - temperature; $H_o$ - total enthalpy of the mixture; $h_i(T)$ - individual enthalpy of the $i^{th}$ component; N - the number of chemically different components; $N_R$ - the number of the considered chemical reactions;

$$\varphi_i = \sum_{r=1}^{N_R} \left(C_{ri}^- - C_{ri}^+\right)\left(\frac{P}{RT}\right)^{\sum_{k=1}^{N} C_{rk}^+ + a_r^+ - 1}\left(\frac{1}{\mu}\right)^{\sum_{k=1}^{N} C_{rk}^+ - 1} K_r^+(T) \prod_{k=1}^{N} \gamma_k^{C_{rk}^+}$$

- the velocity of the $i^{th}$ component initiation;

$$\gamma_i \phi_i = \sum_{r=1}^{N_R} \left(C_{ri}^- - C_{ri}^+\right)\left(\frac{P}{RT}\right)^{\sum_{k=1}^{N} C_{rk}^- + a_r^- - 1}\left(\frac{1}{\mu}\right)^{\sum_{k=1}^{N} C_{rk}^- + a_r^- - 1} K_r^-(T) \prod_{k=1}^{N} \gamma_k^{C_{rk}^-}$$

- the velocity of the $i^{th}$ component decomposition.)

The main difficulty of this equation system is the "stiffness" of the kinetic equations which is provoked by the fact that the values of the rate constant of the different chemical reactions differ by several orders. The high order of the equation system solved here leads to the same phenomenon. This makes explicit difference schemes practically useless. But the implicit scheme proposed in [1] as well as the following development of the ideas of that work allowed to create some effective algorithmes for solving the systems of the mentioned type which are based on Newton method. The results of the calculations by these algorithms and the more thorough investigation of the physical model allowed to split this system and decrease the time of calculation substantionally by using some variants of the simple iterations method. At present these algorithms are realized as FORTRAN programs. The system of one-dimensional flow equations displayed above is not suitable to describe all the processes which take place in industrial plants and which are interesting from the point of view of the formation and diffusion of deleterious inpurities. It applies to processes as the processes of dispersion of fuel in the burners of thermoelectric power stations; the processes of burning down the exhaust of the aviation and rocket engines, and the diffusion of deleterious impurities in the atmosphere. All these processes are substantionally two- and three-dimensional. The mutual diffusion of different gas components, viscous dissipation, turbulence of the flow are of great importance there. The jet engine efflux is often off-design and results in taking the pressure gradients into account. The flows named above are all of the jet character. All these peculiarities define the mathematical model as the model which describes the turbulent non-isobaric jet in the presence of non-equilibrium chemical processes. As the flow influence up the stream is usually small, the equations of motion were chosen in the form of the reduced Navier-Stokes equations, which differ from the total equations by the absense of the second-longitudinal and mixed derivatives. The equation of energy is also written in this approximation. At moderate temperatures taking place in the considered cases (no more than 3000 K) gas can be adopted ideal. The equation system in the cylindrical coordinates lookes like

$$\frac{\partial}{\partial x}(\rho u z) + \frac{\partial}{\partial z}(\rho v z) = 0$$

$$\rho u \frac{\partial u}{\partial x} + \rho v \frac{\partial u}{\partial z} = -\frac{\partial P}{\partial x} + \frac{1}{z}\frac{\partial}{\partial z}\left(\mu z \frac{\partial u}{\partial z}\right)$$

$$\rho u \frac{\partial v}{\partial x} + \rho v \frac{\partial v}{\partial z} = -\frac{\partial P}{\partial z} + \frac{4}{3}\frac{\partial}{\partial z}\left(\mu \frac{\partial v}{\partial z}\right) + 2\mu\frac{\partial}{\partial z}\left(\frac{v}{z}\right) - \frac{2}{3}\frac{\partial}{\partial z}\left(\mu\frac{v}{z}\right)$$

$$\rho u \frac{\partial H}{\partial x} + \rho v \frac{\partial H}{\partial r} = \frac{1}{r}\frac{\partial}{\partial r}\left[ r \frac{\mu}{P_r}\frac{\partial H}{\partial r} + \mu r\left(1-\frac{1}{P_r}\right)u\frac{\partial u}{\partial r} - \left(\frac{1}{Le}-1\right)r\frac{\mu}{P_r}\sum_{i=1}^{N} h_i \frac{\partial \gamma_i}{\partial y}\right]$$

$$\rho u \frac{\partial \gamma_i}{\partial x} + \rho v \frac{\partial \gamma_i}{\partial r} = \frac{1}{r}\frac{\partial}{\partial r}\left(r\frac{\mu}{P_r}\frac{\partial \gamma_i}{\partial y}\right) + \rho W_i, \quad i=1,\ldots,N$$

when X = 0   $u=u_0(r), V=V_0(r), T=T_0(r), P=P_0(r), \gamma_i=\gamma_i(r)$

when r = 0   $\frac{\partial u}{\partial r}=V=\frac{\partial H}{\partial r}=\frac{\partial \gamma_i}{\partial r}=0$

when r = ∞   $u=u_\infty, V=0, H=H_\infty, \gamma_i=\gamma_{i\infty}$

This equation system can be solved numerically by approximating its algebraical with the implicit 6-point scheme of the second accuracy order. The produced algebraical system with three-diagonal coefficient matrix is solved by driving-through method. At present the FORTRAN program is created to solve this type problems. Alongside with certain advantages the model has some shortcomings.For example, the burning in the burner of the thermal station is organized in such a way that in the burning zone there is a considerable amount of oil fuel though it is ground small. Thus the exact calculation of such a flow suggests considering the multiphaseness and multiphase burning. The mathematical model of such a process ought to be elaborated. The diffusion of impurities, which outflow from the industrial chimneys, proceeds in the presence of the earth surface and cross wind. As a result the flow becomes substantially three-dimensional.On the other hand the variety of the relief (highlands, waterbasins, cities) suggests setting special boundary conditions of the rough surface. That is also a grave mathematical problem which has not been worked out enough recently. It is necessary to note that there exist enough models which allow to foretell various deleterious situations,such as: smog, dangerous concentration of toxic matters in some city regions[2] and so on. Here are some approaches to the description of impurities distribution in these models. They can be classified as: a)empirical approach; it suggests compiling tables for all weather conditions on the basis of the long-term observation; b) semiempirical approach; it uses the model of impurity distribution in the form of the finite formula of Gaussian law; c) the approach mentioned above. The first two approaches have the typical shortcomings of the statistic methods, i.e. they are true only for average values and demand great amount of initial information, which is lacked as a rule. The third approach is not for the time being adapted to the practical requirements because of the calculative difficulties, appearing with its realisation. In this connection the usage of the experience preserved when solving physical gasdynamic problems seems to be of great importance.


III Some Results Of Numerical Modelling


Among the most significant parameters which define the contents of the exhaust the following ones can be singled out: a) the contents of the fuel mixture; b) oxidizer-to-fuel ratio; c) temperature in the burning area; d) rate of mixture temperature changes. It is obvious, that they are connected with each other one way or another, but they can be analysed independently to make the analysis easy.For example, Fig.1 shows the change of the CO, NO concentration and temperature in the torch depending on the oxidizer-to-fuel ratio. To make it obvious the concentrations are measured in parts of maximum concentration limit (MCL). It is necessary to note that the temperature maximum if mixed perfectly, and also the summary concentration minimum (in MCL parts) of CO and NO is obtained at oxidizer ratio $\alpha = 0.95$.This can be explained by the fact that dissociation is already important at temperatures of about $2000^{\circ}K$ and higher. Fig.2 shows the change of temperature and CO and NO concentration when injecting water into the torch subject to water flow in respect to fuel flow. It is already seen that water injection leads to notable decrease of NO and CO concentration. This mathematical model being effective was checked up experimentally at neutralization of high-temperature ($2500^{\circ}K$) jet containing large amount of CO outflowing into the atmosphere. The calculations resulted in the scheme of burning down of CO being proposed and experimentally proved. These results are shown in Fig.3, which shows the dependance of CO flow in mass unit/per second subject to the flow of $H_2O$ and $O_2$ which are initiated into the jet.

The approach described above can be slightly transformed and used when analysing the processes in internal combustion engine. At present in zero-dimensional set the problem of defining the thermodynamic parameters and mixture composition in burning chamber is solved on the basis of the solution of the equation system of chemical kinetics and the equation system of chemical kinetics and the equation of energy in variable volume. The phases of inlet and exhaust are checked by initiating variable mass. This model makes possible the analysis (in the first approximation) of the toxic exhaust being dependent on such engine parameters as pressure ratio, revolutions per minute, synchronized ignition, oxidizer-to-fuel ratio. The received data is used afterwards to calculate the process of burning in the chamber considering the hydrodynamics of mixture flow.

All that was said above results in the fact that the mathematical methods developed in physical gasdynamics can be successfully used to solve the problem of air basin preservation. The physical model described here allows to consider the problems of toxic components formation in different industrial plants and their propagation in peri-terrestrial atmosphere layer. B esides, this model allows to propose the methods of deleterious impurities neutralization because of detailed elaboration of burning process.


LITERATURE



1. Kamzolov V.N., Pirumov U.G. The calculation of non-equilibrium flows in nozzles. AC of the USSR Press, MZhG, 1966, No 6.
2. Encyclopedia of Environmental Science and Engineering. Ed. T.R.Pfaffhin and E.N.Ziegher, V.2, N .Y. 1976.

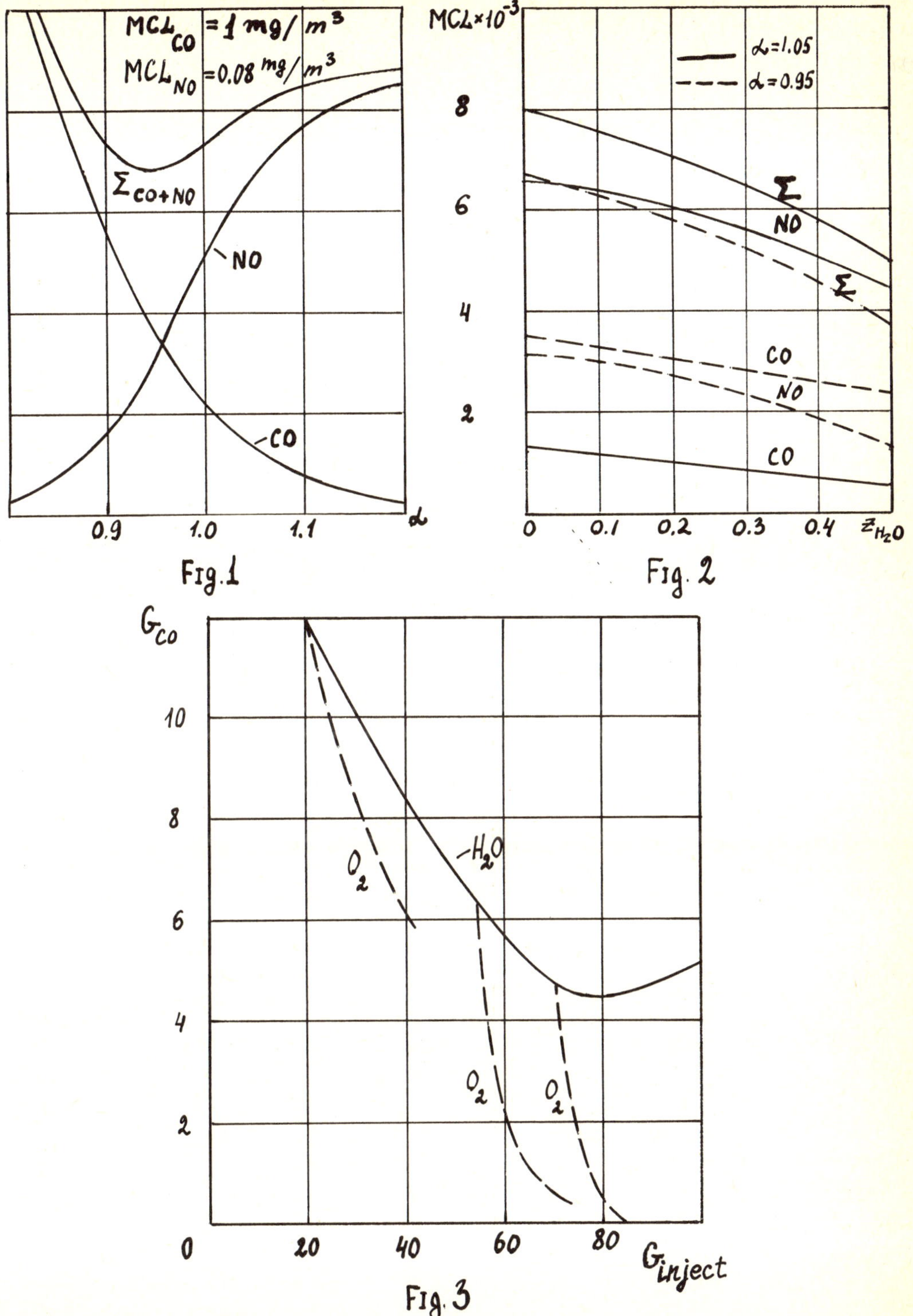
MCL_CO = 1 mg/m³
MCL_NO = 0.08 mg/m³
Σ_CO+NO
NO
CO
0.9
1.0
1.1
α
Fig.1
MCL×10⁻³
8
6
4
2
α=1.05
α=0.95
Σ
NO
Σ
CO
NO
CO
0
0.1
0.2
0.3
0.4
z_H2O
Fig. 2
G_CO
10
8
6
4
2
O₂
H₂O
O₂
O₂
0
20
40
60
80
G_inject
Fig. 3

A PROJECTION METHOD BASED ON GAUSSIAN QUADRATURES
WITH APPLICATION TO COMPRESSIBLE NAVIER-STOKES EQUATIONS*

K. C. Reddy**
Sverdrup/ARO, Inc., AEDC Division
Arnold Engineering Development Center
Arnold Air Force Station, TN  37389

*SUMMARY*

*A projection method based on Gauss-Legendre quadratures for solving hyperbolic and parabolic partial differential equations is proposed. It is comparable to spectral methods in accuracy and convergence, and is more convenient to implement for non-linear problems. Results obtained with a wave equation and Burgers equation modeling a flow with sharp gradients show the high resolution achieved for a given number of nodes by using this method. Time-split implicit techniques are used for approximating the compressible Navier-Stokes equations.*

Spectral methods have been shown to have high accuracy (they are formally infinite order accurate) and resolution by Orszag and his colleagues (Refs. 1-3) for certain classes of hyperbolic and parabolic partial differential equations. They have been efficiently implemented for solving flow problems such as turbulent shear flows (Ref. 3) and stability of incompressible Navier-Stokes equations (Ref. 4). Typically in a spectral method, unknowns are expanded in terms of a complete set of functions, such as Chebyshev polynomials, and sets of ordinary differential equations are integrated in the spectral space. In this paper a projection method is proposed based on Gaussian Quadratures, which is comparable to the spectral methods, in particular to the Legendre-Tau method; but, the ordinary differential equations are in terms of the actual unknowns at Gaussian nodes instead of the spectral coefficients. The method has some advantages in solving the non-linear fluid flow equations. Consider an initial boundary value problem

$$u_t + Lu = f(x,t), \quad -1 \leq x \leq 1, \quad t > 0 \tag{1}$$

where L is a differential operator in x such as

$$\text{(i)} \quad Lu = \frac{\partial u}{\partial x} \text{ for a wave equation (hyperbolic)} \tag{2}$$

$$\text{(ii)} \quad Lu = u \frac{\partial u}{\partial x} - \nu u_{xx} \text{ for Burgers equation (parabolic)} \tag{3}$$

$$\text{with the initial condition:} \quad u(x,o) = g(x) \tag{4}$$

$$\text{and boundary conditions:} \quad B_i u(x,t) = 0, \quad i = 1,\ldots k \tag{5}$$

$$\text{where } k = 1 \text{ for operator (i) and } u(-1,t) = b(t) \tag{6}$$

and $k = 2$ for operator (ii) and in that case

$$u(-1,t) = b_\ell(t), \quad u(1,t) = b_r(t) \tag{7}$$

Let $\{V_i(x), i = 1,2,--\}$ be a complete set of functions over $(-1,1)$, such as the normalized Legendre polynomials. Project the partial differential equation (1) over a finite dimensional subspace of the Legendre polynomial space.

$$\frac{d}{dt}(u, V_i) + (Lu, V_i) = (f, V_i) \tag{8}$$

$$i = 1,2,---, n - k$$

where $V_i(x) = P_{i-1}(x)/||P_{i-1}(x)||$ is the normalized Legendre polynomial of (i-1)th degree. It is convenient to integrate $(Lu, V_i)$ by parts, as for example, in the case of Burgers equation

$$(Lu, \ V_i) = \left(\frac{u^2}{2} - \nu u_x\right) V_i(x) \ \Bigg|_{-1}^{1} - \int_{-1}^{1} \left(\frac{u^2}{2} - \nu u_x\right) V_i'(x)\,dx \tag{9}$$

The integrals in equations (8) and (9) are now approximated by n point Gauss-Legendre quadratures.

$$(u, \ V_i) = \int_{-1}^{1} u \ V_i \ dx \ \underset{\sim}{\simeq} \ \sum_{j=1}^{n} w_j V_i(x_j) u(x_j, \ t) \tag{10}$$

where $(x_j, \ j=1,2,\text{---}n)$ are the Gauss-Legendre nodes, which are the roots of the nth degree Legendre polynomial, $V_{n+1}(x)$, and $\{w_j, \ j=1,2,\text{---}n\}$ are the corresponding weights. This approximation implies the approximation of u by an (n-1)th degree polynomial $P_{n-1}(x, \ t)$,

$$u(x, \ t) \ \underset{\sim}{\simeq} \ P_{n-1} \ (x,t) = \sum_{j=1}^{n} \phi_{n,j}(x) \ u(x_j, \ t) \tag{11}$$

where

$$\phi_{n,j}(x) = \frac{V_{n+1}(x)}{(x - x_j)V_{n+1}'(x_j)} \tag{12}$$

Thus, $u_x$ in equation (9) can be approximated as

$$u_x(x, \ t) \ \underset{\sim}{\simeq} \ \sum_{j=1}^{n} \phi_{n,j}'(x) \ u(x_j, \ t) \tag{13}$$

Using equation (11), the k boundary conditions, equation (5), are approximated to supplement the (n-k) ordinary differential equations, equation (8), for the n unknowns $u(x_j, \ t)$ at the Gaussian nodes. Initial conditions are known from equation (4). This system of non-linear ordinary differential equations are solved by the implicit trapezoidal rule (Crank-Nicholson) method coupled with the Newton iteration to take care of the non-linearity.

Much of the accuracy, convergence and stability theory developed by Gottlieb and Orszag (Ref. 2) is applicable for this method which is comparable to the Legendre-Tau method. It can be shown that the method is algebraically stable for problems with semi-bounded operators L in equation (1), in the same manner as it is done for the Legendre-Tau spectral method.

In order to resolve regions of sharp gradients, like shocks or boundary layers, a segmented version of the projection method has been developed. In this method, the spatial domain is segmented into several intervals, $(a_m, \ a_{m+1})$, m = 1,2,---M, according to the requirements of the resolution. The Legendre polynomials are suitably transformed to the $m_{th}$ segment $(a_m, \ a_{m+1})$ to define $V_i^m(x)$ and the Gauss-Legendre quadratures corresponding to the equation (10) are appropriately modified to approximate the projection integrals over the $m_{th}$ segment. The interpolation formulae corresponding to equations (11) to (13) over the $m_{th}$ segment remain structurally unaltered. The projection method is applied in each interval separately and the projection equations over all intervals are coupled with the requirements of continuity of the unknown variable and its derivatives, if necessary, at the end points of the segments. The implicit trapezoidal rule method coupled with Newton iteration is applied to solve the resulting system of ordinary differential equations with appropriate boundary conditions. In this method the number of Gaussian nodes in each segment need not be the same. The matrix of the algebraic equations to be solved at each time step, has a tri-diagonal block structure, where the order of the diagonal matrices correspond to the number of Gaussian nodes in different segments. The off-diagonal matrices have all zero elements except for a specific number of rows depending upon the continuity requirements imposed. For example, while solving Burgers equation the unknown variable and its first derivative are required to be continuous between two segments. In this case, the off-diagonal matrices have only one row of non-zero elements. The L-U decomposition method can be efficiently implemented for solving the system of algebraic equations by taking advantage of the zero element structure of the off-diagonal matrices.

The extension of the method for solving a system of partial differential equations of the type given by equation (1) is straight forward. For solving a system of multi-dimensional equations like the time-dependent Navier-Stokes equations, the projection method is combined with the time splitting or approximate factorization techniques. The objective is to find the asymptotic steady state solution of the Navier-Stokes equations accurately by solving a sequence of one-dimensional problems.

The two-dimensional compressible Navier-Stokes equations can be written in the conservation law form as

$$\frac{\partial U}{\partial t} + \frac{\partial}{\partial x}\left[F(U) - V_1(U, U_x)\right] + \frac{\partial}{\partial y}\left[G(U) - W_2(U, U_y)\right]$$

$$= \frac{\partial}{\partial x}V_2(U, U_y) + \frac{\partial}{\partial y}W_1(U, U_x) \tag{14}$$

where U is the vector of conserved variables and F, G, V and W are the flux vectors. Warming and Beam (Refs. 5,6) have developed an implicit factored scheme where a two or three level time differencing is used and second order finite differences are used for the spatial differences. We use their time-splitting technique and apply the projection method for spatial discretization. Following Warming and Beam, a spatially factored scheme based on the trapezoidal rule time differencing is written as follows:

$$\left\{I + \frac{\Delta t}{2}\left[\frac{\partial}{\partial x}(A - P + R_x)^n - \frac{\partial^2}{\partial x^2}(R)^n\right]\right\}\Delta U^* = -\Delta t\left(\frac{\partial F}{\partial x} + \frac{\partial G}{\partial y} - \frac{\partial U}{\partial x} - \frac{\partial W}{\partial y}\right)^n$$

$$+ \Delta t\left[\frac{\partial}{\partial x}(\Delta V_2)^{n-1} + \frac{\partial}{\partial y}(\Delta W_1)^{n-1}\right] = \text{RHS} \tag{15}$$

$$\left\{I + \frac{\Delta t}{2}\left[\frac{\partial}{\partial y}(B - Q + S_y)^n - \frac{\partial^2}{\partial y^2}(S)^n\right]\right\}\Delta U^n = \Delta U^* \tag{16}$$

$$U^{n+1} = U^n + \Delta U^n$$

The notation is that of Ref. (5). A, B, P, Q, R and S are the Jacobian matrices defined by

$$A = \frac{\partial F}{\partial U}, \quad P = \frac{\partial V_1}{\partial U}, \quad R = \frac{\partial V_1}{\partial U_x}$$

$$B = \frac{\partial G}{\partial U}, \quad Q = \frac{\partial W_2}{\partial U}, \quad S = \frac{\partial W_2}{\partial U_y}$$

For each of the steps in equations (15) and (16), the projection method outlined earlier is applicable for spatial discretization. The spatial resolution achieved by the projection method should be higher than that of second order or fourth order finite difference methods with the same number of nodes.

Typically the projection equations for a segment $(a_m, a_{m+1})$ in the x-direction, for equation (15), are of the form

$$\frac{\Delta t}{2} \sum_{\beta=1}^{4} \left\{ \left( A_{\alpha\beta} - P_{\alpha\beta} \right) \Delta U_\beta^* - R_{\alpha\beta} \frac{\partial \Delta U_\beta^*}{\partial x} \right\} V_i^m (x) \Bigg|_{x=a_m}^{a_{m+1}}$$

$$+ \int_{a_m}^{a_{m+1}} \left[ \Delta U_\alpha^* V_i^m (x) - \frac{\Delta t}{2} \sum_{\beta=1}^{4} \left\{ \left( A_{\alpha\beta} - P_{\alpha\beta} \right) \Delta U_\beta^* - R_{\alpha\beta} \frac{\partial \Delta U_\beta^*}{\partial x} \right\} V_i^{m'} (x) \right] dx$$

$$= \int_{a_m}^{a_{m+1}} RHS_\alpha\, V_i^m (x)\, dx \tag{17}$$

$$\alpha = 1,4; \quad i = 1, J_m - 2; \quad m = 1, M$$

where $J_m$ is the number of Gaussian nodes in the $m_{th}$ segment and M is the number of segments. Equation (17) is written at a Gaussian node of a segment in y-direction. The integrals are replaced by the Gaussian quadratures. The terms at $a_m$ and $a_{m+1}$ and the derivatives in equation (17) and the derivatives in the matrices P and Q are replaced by suitable interpolations of the form given by equations (11) – (13), appropriate for the particular x and y segments. Continuity of U and $U_x$ is imposed between two x-segments, which couple the projection equations of a segment with those of the adjacent ones. Together with the boundary conditions (non-coupled) these result in a system of algebraic equations for $\Delta U^*$ at all the x-nodes at a particular y location. The matrix of the equations has a tri-diagonal block structure with the diagonal matrices of order 4 $J_m$ and the off-diagonal matrices with only 4 rows of non-zero elements. These can be solved efficiently by the L-U decomposition method. After solving equation (15) for all the y-nodes, equation (16) is solved in the other direction for all the x-nodes. This completes the calculations for one time step and U values are updated.

## Numerical Results

Numerical results for model problems with wave equation and Burgers equation confirm the high accuracy and resolution of the present projection method. A wave equation problem:

$$u_t + u_x = 0; \quad u(-1, t) = -\sin \pi M t, \quad u(x, 0) = 0$$

which has the exact solution

$$u(x, t) = \begin{cases} -\sin M\pi\, (x + 1 - t), & x \leq t - 1 \\ 0 & , \quad x > t - 1 \end{cases}$$

has been solved with $(N + 1)$ Gaussian nodes. The $L_2$ error, $\|u_{exact} - u_{num}\|$ and the end point error $|u_{exact} - u_{num}|$ at $\dot{x} = 1$, averaged over $4 \leq t \leq 4.5$ for M = 5 are compared with the results of the Legendre-Tau method of Gottlieb and Orszag (Ref. 2) in Fig. 1. In the Legendre-Tau method $(N + 1)$ Legendre polynomials are used to approximate u.  Results are comparable and the errors decrease exponentially for $N \geq 5\pi$.  The $L_2$ error comparison of the spectral method (Chebyshev-Tau) with 2nd and 4th order finite difference methods for this problem, made by Gottlieb and Orszag (Ref. 2) are presented in Table 1 with results from the present projection method (Gauss-Legendre).  Results indicate the comparable accuracy of the spectral and projection methods, which are clearly more accurate for a given number of nodes or spectral coefficients than the 2nd or 4th order methods.

Burgers equation, $u_t + uu_x = \nu u_{xx}$ has been solved for two cases with small $\nu$

(a)  Shock propagation:  $u(x, t) = 1 - \tanh \{(x - t)/2\nu\}$
     Initial and boundary conditions are set according to the exact solution in the numerical algorithm.  Typical results are shown in Figs. 2 and 3 for $\nu = 0.1$ and 0.01.  As $\nu$ becomes smaller, there are not enough points in the shock region which cause the numerical errors.  Shock speed, however, is preserved accurately.

(b)  Shock formation:  $\nu \ll 1$
$$u(x, o) = - \sin \frac{\pi}{2} x; \quad u(-1, t) = 1, \quad u(1, t) = -1$$
     The asymptotic exact solution is $u(x, \infty) = -\tanh (x/2\nu)$.  Figure 4 shows the numerical results with $\nu = 0.01$ and 48 Gaussian nodes.  The errors in the asymptotic solution are again due to insufficient resolution of the shock region.

In order to resolve the regions of sharp gradients, like shocks or boundary layers, a segmented version of the projection method has been applied.  In this method the spatial domain is segmented into several intervals according to the requirements of the resolution.  The projection method is applied in each interval separately and the projection equations over all intervals are coupled due to the requirements of continuity up to the first derivatives at the end points of the segments.  Segmented version results presented in Figs. 5 and 6, for $\nu = 0.01$ and $\nu = 0.0001$, respectively, show the success of this method in resolving shock regions accurately.  Note the resolution of the near discontinuity in the asymptotic solution of Fig. 6 is without any visible errors.  Numerical results for the Navier-Stokes equations will be reported elsewhere.

---

*The research reported herein was performed by the Arnold Engineering Development Center, Air Force Systems Command.  Work and analysis for this research was done by personnel of ARO, Inc., a Sverdrup Corporation Company, operating contractor of AEDC.  Further reproduction is authorized to satisfy needs of the U. S. Government.

**Consultant.  Also, Professor, The University of Tennessee Space Institute, Tullahoma, Tennessee 37388.

## References

1.  Orszag, S. A.  "Numerical Simulation of Incompressible Flows within Simple Boundaries:  Accuracy."  _J. Fluid Mechanics_, 49, 1971, p. 75.

2.  Gottlieb, D. and Orszag, S. A.  _Numerical Analysis of Spectral Methods:  Theory and Applications_.  SIAM, Philadelphia, PA, 1977.

3.  Orszag, S. A.  "Numerical Simulation of Turbulent Flows."  Handbook of Turbulence, Plenum Press, New York, 1977, p. 281.

4.  Murdock, J. W.  "A Numerical Study of Nonlinear Effects on Boundary Layer Stability."  _AIAA Journal_, August 1977, p. 1167.

5.  Warming, R. F. and Beam, R. M.  "On the Construction and Application of Implicit Factored Schemes for Conservation Laws."  SIAM-AMS Proceedings, Vol. 11, Symposium on Computational Fluid Dynamics, New York, April 1977.

6.  Beam, R. M. and Warming, R. F.  "An Implicit Factored Scheme for the Compressible Navier-Stokes Equations,"  AIAA Journal, April 1978, p. 393.

Table 1

$L_2$ (rms) errors for the solution of $u_t + u_x = 0$; $u(x, 0) = 0$, $u(-1, t) = -\sin M\pi\, t$.  The errors listed are measured at $t = 5$ when the solution is smooth.  Time differencing errors are negligible and N is the number of grid points or Chebyshev polynomials or Gaussian nodes.  Observe that to achieve a 1% error, the second-order method requires $N/M \geq 40$, the fourth-order method requires $N/M \geq 15$, while the Chebyshev-tau method requires $N/M \geq \pi$.

| Second Order | | | Fourth Order | | | Chebyshev-Tau | | | Gauss-Legendre (Present Method) | | |
|---|---|---|---|---|---|---|---|---|---|---|---|
| N | M | $\varepsilon_2$ | N | M | $\varepsilon_4$ | N | M | $\varepsilon_\infty$ | N | M | $\varepsilon_\infty$ |
| 40 | 2 | 0.1 | 20 | 2 | 0.04 | 16 | 4 | 0.08 | 19 | 4 | 0.001 |
| 80 | 2 | 0.03 | 30 | 2 | 0.008 | 20 | 4 | 0.001 | 23 | 4 | 0.0002 |
| 160 | 2 | 0.008 | 40 | 2 | 0.002 | 28 | 8 | 0.2 | | | |
| 40 | 4 | 1.0 | 40 | 4 | 0.07 | 32 | 8 | 0.008 | 31 | 8 | 0.008 |
| 80 | 4 | 0.2 | 80 | 4 | 0.005 | 42 | 12 | 0.2 | | | |
| 160 | 4 | 0.06 | 160 | 4 | 0.0003 | 46 | 12 | 0.02 | | | |

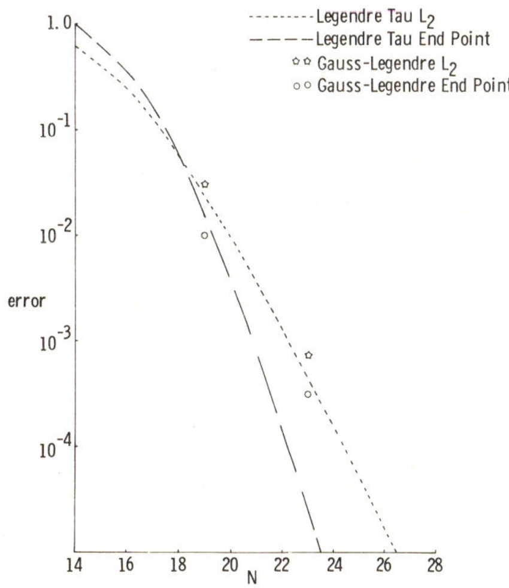


Fig. 1  A comparison of the Legendre-Tau $L_2$ and end point ($x = +1$) errors (Ref. Gottlieb and Orszag) with the $L_2$ and end point errors by the present method (Gauss-Legendre).  $u(-1,t) = -\sin 5\pi t$  Errors are averaged over the interval $4 \leq t \leq 4.5$.

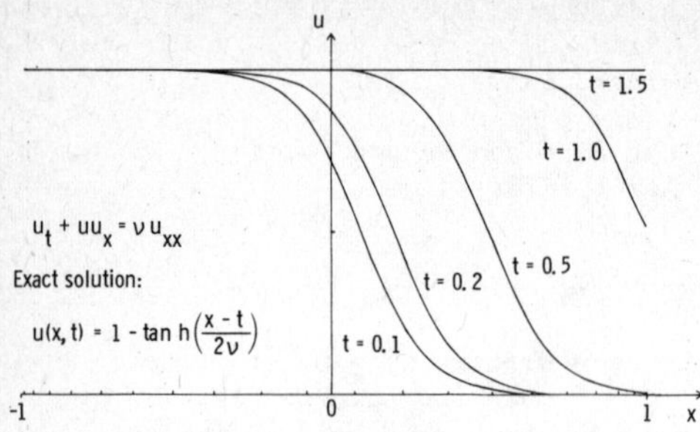


Fig. 2   Shock Propagation – Solution of
Burgers Equation, $\nu = 0.1$,
$\Delta t = 0.001$, JMAX = 20

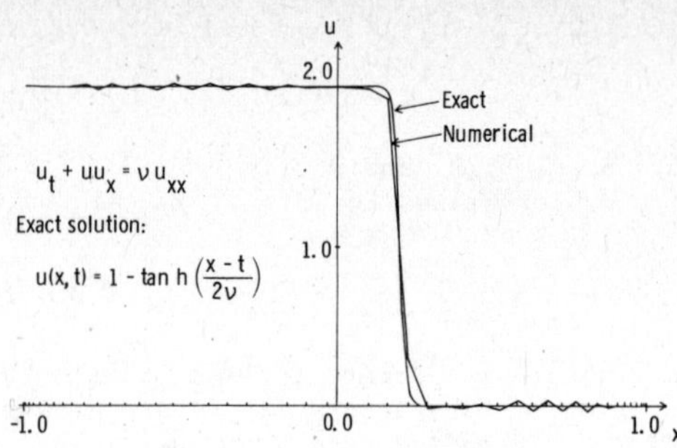


Fig. 3   Shock Propagation – Solution of
Burgers Equation, $\nu = 0.01$,
$\Delta t = 0.001$, t = 0.2, JMAX = 48

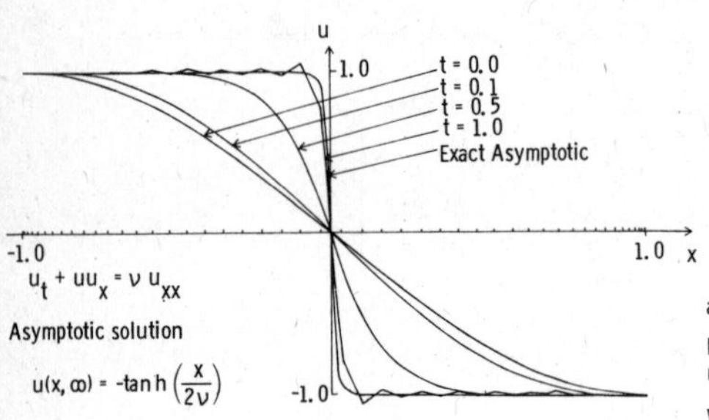


Fig. 4   Shock Formation – Solution of
Burgers Equation, Single Segment
Version, No. of Points = 48,
$\nu = 0.01$, $\Delta t = 0.01$

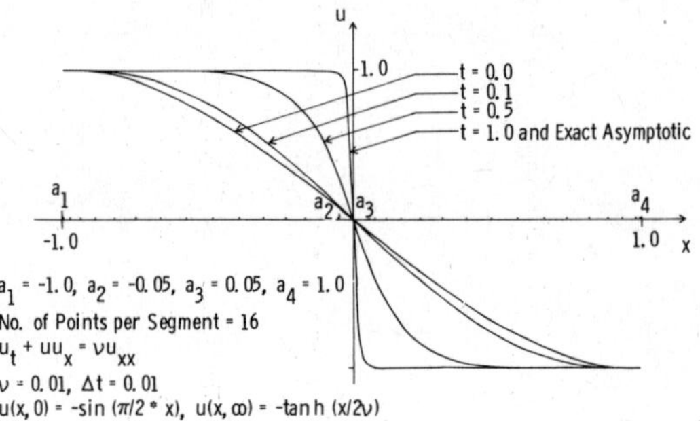


Fig. 5   Shock Formation – Solution of
Burgers Equation, 3–Segment
Version

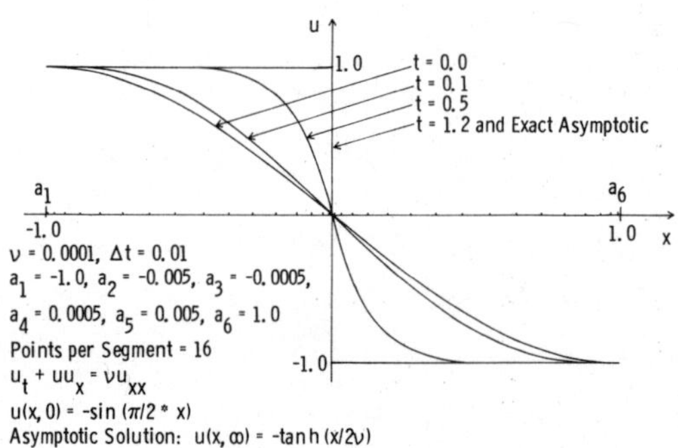


Fig. 6   Shock Formation – Solution of
Burgers Equation, 5–Segment
Version

# SEMIDIRECT SOLUTION TO STEADY TRANSONIC FLOW BY NEWTON'S METHOD

Arthur Rizzi
FFA The Aeronautical Research     &     Gunilla Sköllermo
Institute of Sweden                     Stockholm's Computing Center
P.O. Box 11021                          S-104 50  STOCKHOLM, Sweden
S-161 11  BROMMA, Sweden


## Introduction

The most commonly used methods solve the transonic flow equations in either a time-
dependent formulation or a time-like relaxation procedure.  In general their convergence to a steady-state solution is restricted by the rate at which signals generated at one grid point can propagate throughout the entire field.  Other methods,
however, have been devised to solve the steady-flow problem directly as a large implicit system so that signals are felt at all points simultaneously, but since the
problem is nonlinear these methods must incorporate some iterative means in order
to reach a solution (hence the term semi-direct).  The semidirect procedure that is
usually applied to transonic flow uses a fast elliptic solver at each step of a
Picard iteration process (Martin & Lomax, 1975).  For subsonic flow governed by nonlinear elliptic equations this is a very effective method, but it is unstable if the
equations become hyperbolic, although some progress is being made in treating this
instability.

In contrast to this elliptic-solver Picard-iteration method, Blomster & Sköllermo
(1977) developed a semidirect method which uses a Newton iteration scheme to solve
the nonlinear equations of transonic flow written as a first-order system.  On each
iteration the system of equations resulting from the linear Newton operator is
solved by Gaussian elimination.  With a computed example they demonstrated that this
iteration procedure is stable even when the equations become hyperbolic.  The
example they chose was one of shock-free subsonic to supersonic flow.  In this paper
we continue the development of the semidirect Newton method and show that with the
proper choice of boundary conditions and with some provision to satisfy the entropy
condition it can obtain successfully the weak solution to transonic flow problems
which contain shock waves.


## Physical Problem

Consider the problem of transonic flow
over a circular arc situated on the upper
wall of a parallel channel, whose reservoir conditions are such that a supersonic
region develops in the subsonic field,
separated by a sonic line and shock wave
(Fig. 1).  We model this flow by the full
potential equation written as a first-
order system

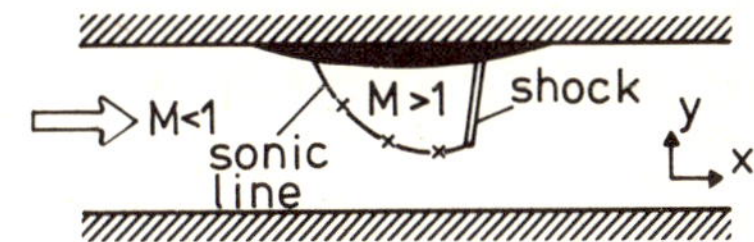


Fig. 1   Features of transonic
flow over a circular
arc on a channel wall.

$$A(q)\, q_x + B(q)\, q_y = 0 \qquad\qquad (1)$$

$$q = \begin{pmatrix} u \\ v \end{pmatrix} \qquad A = \begin{pmatrix} c^2 - u^2 & 0 \\ 0 & 1 \end{pmatrix} \qquad B = \begin{pmatrix} -2\,uv & c^2 - v^2 \\ -1 & 0 \end{pmatrix}$$

together with the physical boundary conditions  $\vec{V} \cdot \hat{n}_w = 0$  on the walls and specification of the reservoir conditions of the flow.  The motivation here is to study a
problem that can serve as a model for more complex first-order systems of mixed
elliptic-hyperbolic type.

## Semi-direct Newton Method

A shearing tranformation maps the channel to a rectangle, and fills in the zeros of matrix B in Eq (1). The solution to the transformed Eq s. (1) is approached in the following manner based on Newton's method. Let $q^n$ be some approximate solution. We seek a correction term $\Delta q$ which when added to $q^n$ will define a new improved approximation $q^{n+1} = q^n + \Delta q$. This is done by taking $q^{n+1}$ to be a solution to Eq s. (1) and linearizing the system around $q^n$. The result is a linear partial differential equation

$$L^n \, \Delta q = -D^n \, q^n \tag{2}$$

where $\quad D^n = A^n \dfrac{\partial}{\partial x} + B^n \dfrac{\partial}{\partial y} \qquad C^n = \begin{pmatrix} f & g \\ 0 & 0 \end{pmatrix} \qquad L^n = C^n + D^n$

which, when solved, determines the unknown $\Delta q$. The new estimate $q^{n+1}$ can then be calculated, and the whole process is repeated until the left side of Eq s. (2) approaches zero. The right side then represents a solution to the original problem.

Our procedure here is a somewhat novel application of Newton's method in that we apply the linearization directly to the differential equations before discretization. We are left, subsequently, to solve a <u>linear</u> differential equation using e.g. finite differences. The more usual approach is to difference the nonlinear differential problem and solve the resulting nonlinear difference equations by Newton's method. The end result probably is very similar, if not identical, but we believe that our way offers some advantage at least in the formalism of the procedure.

## Finite Difference Solution to Equations (2)

The correction $\Delta q$ is determined from a finite difference solution to Eq s (2). The flowfield first is discretized by a grid and then all derivatives at interior grid points are approximated by centered differences. The situation at boundary points, however, is not so straightforward because at the walls the physical boundary condition $\vec{V} \cdot \hat{n}_w = 0$ must be supplemented by a numerical one in order to specify both components of $q^{n+1}$. Our choice is to difference the first equation of system (2) according to the scheme indicated by the shaded pattern in Fig. 2.

The inflow and outflow boundaries are artificial terminations of the flowfield, and the flow is not known there precisely. The usual guide for appropriate conditions in such cases is to look at the characteristic directions for the corresponding time-dependent problem, which suggest that at each boundary one component of $q$ should be set and the other should be calculated from interior values. These as well as various alternatives based on the same reasoning were tried, but none led to a stable calculation. In an effort to understand this outcome, we turned for guidance to a steady-state analysis which uses the Caucly-Riemann equations as a model for our system. At inflow we set $v = 0$, at outflow $u = $ constant, and differenced the equations to find all the other unknowns according to the scheme in Fig. 2. We found that the resulting system of linear algebraic

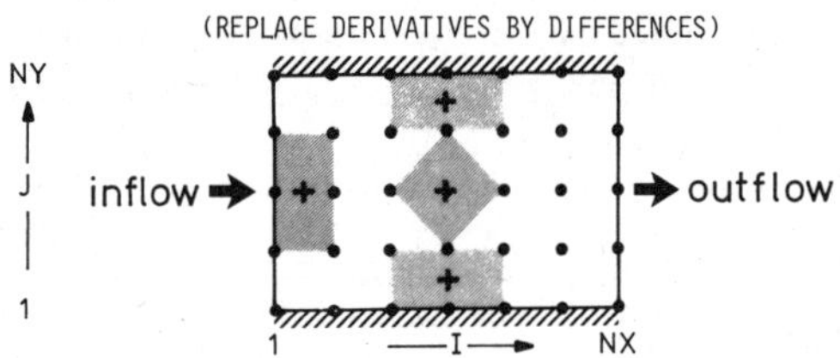


BOUNDARY CONDITIONS

— WALLS : (PHYSICAL) $\vec{V} \cdot \hat{n}_w = 0$ + (NUMERICAL) DIFF. ONE PDE.

— INFLOW/ OUTFLOW : (ARTIFICIAL TERMINATION)

- STEADY-STATE CONDITIONS FOUND STABLE
  INFLOW: $v = 0$, DIFF. ONE PDE.
  OUTFLOW: $M = M_\infty$, $v_x = 0$

Fig. 2 The discretization and boundary conditions used in the method.

equations is under-determined, but that this model is well-posed if we set both
$u$ = constant and $v$ = 0 (or alternatively $V_x$ = 0) at outflow. This type of outflow condition was tried on our full system (2) and the computation proved to be
stable. At this point we do not know whether the explanation for these results is
a fundamental feature of Eqs. (1) or whether it depends only on the particular
choice of solution procedure.

The discrete approximation
$L_k^n \, \Delta q_k^n = -D_k^n \, q_k^n$ including the
boundary conditions can ce expressed
as a superblock matrix equation as
written in Fig. 3. We solve it by
Gaussian elimination using partial
provoting and external storage.

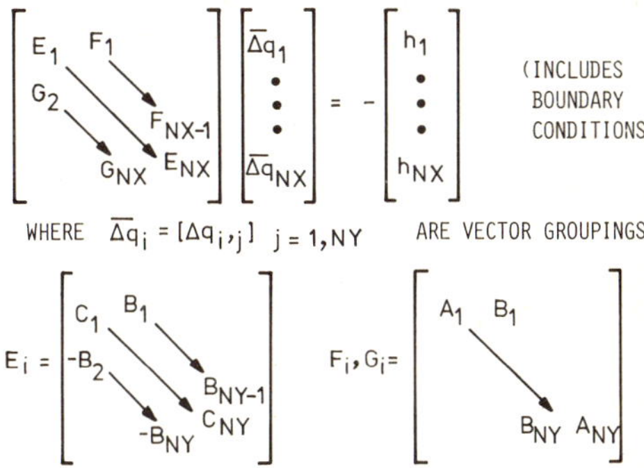

Fig. 3   The linear system $L_k^n \Delta q_k^n = -D_k^n q_k^n$
written as a superblock matrix
equation.

## Convergence of the Method

The theoretical basis for convergence of the method rests upon the
Kantorovich Theorem (see e.g.
Henrici, 1962). The requirements
are that the initial approximation
$q^0$ be "close enough" to the actual
solution and that the matrix $L^n$
be invertible. We devised two
examples to test empirically the
convergence of the method when a
shock is present in the solution.
The two flow conditions $M_\infty$ = .80 and $M_\infty$ = .85 were selected in order to produce
a weak and a strong shock respectively in the channel. In both cases we discovered,
rather surprisingly, that convergence is _not_ sensitive to the particular initial
approximation. We construct it from a piece-wise parabolic distribution for Mach
number on both the upper and lower walls, and linearly interpolate for values in
the interior. Convergence, as indicated in Figs. 4, 5 and 6, is not dependent
upon the coefficients of the parabolas (cf. initial approximations, iteration $n$ = 0).

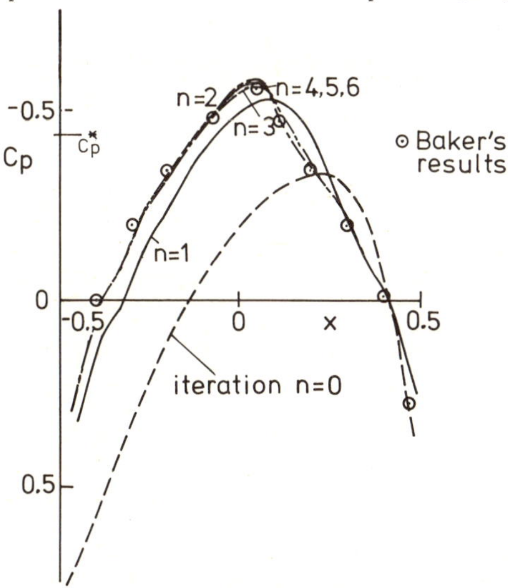


Fig. 4   Pressure distribution computed on each iteration for
$M_\infty$ = .80   (weak shock).

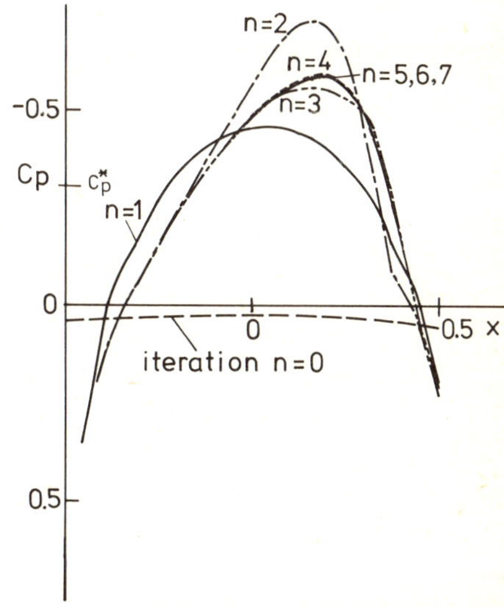


Fig. 5   Pressure distribution computed on each iteration for
$M_\infty$ = .85   (strong shock).

When applied to the flow with the weak shock wave, the method converged, as shown by the pressure distributions in Fig. 4, in the manner predicted by theory. Notice that although the initial approximation is entirely subsonic (i.e. below the value $C_p{}^*$) the first iteration develops some supersonic flow. The changes on successive iterations become progressively smaller and by iteration $n = 6$ the results agree reasonably well with those of Baker's (1980) obtained by an approximate factorization procedure.

In the strong-shock example, however, our first attempts quickly diverged. One explanation for this may be that the solution to the difference equations across the shock wave is not uniquely determined because of the ambiquity of whether the flow should expand or compress through the shock. In this situation some additional condition,

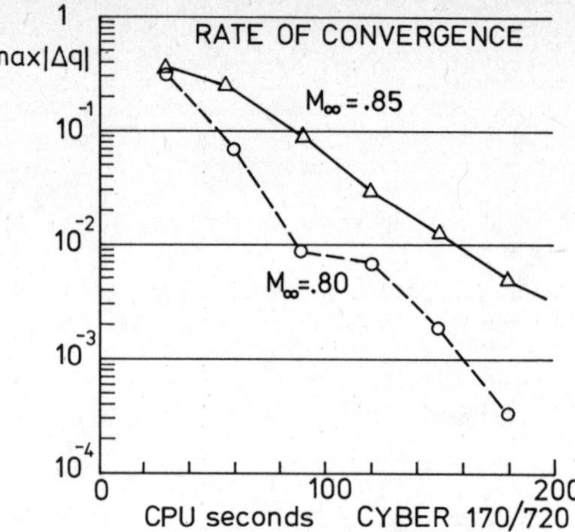


Fig. 6  Maximum change in $\Delta q$ versus computer time for both examples using a 49 x 13 grid. (Each symbol denotes one iteration.)

the so-called entropy condition (Lax, 1973), is required to determine the physically relevant solution. As a means to meet this condition, we added a dissipative term $\beta u_{xx}$ to the first equation of system (2)

where $\quad \beta = \alpha * DIM [1 - (.9/M)^2, 0. ]^2 * \Delta x^2$

is a variable switch and $\alpha$ is a constant. With $\alpha = 5/\Delta$ , we obtained the convergent sequence of pressure distributions shown in Fig. 5. With this value for $\alpha$ the dissipative term reduces the accuracy of the difference scheme in $x$ to first-order, and as well, rounds off the sharp peak in $C_p$ just ahead of the shock. At this time we are not certain whether the reduction to first-order accuracy in $x$ or the overall smoothness given to the solution by dissipation is the crucial factor that determines convergence in this case. Further study and additional tests with other means to specify an entropy condition will have to be made before we can give a conclusive explanation.


## Concluding Remarks

Based on our studies of two transonic flows, one with a weak shock and the other with a strong shock, we believe that the linear operator obtained from Newton's method is suited for the semidirect solution of transonic flow problems. It appears not to be sensitive to the initial approximation and it converges rapidly, usually in less than ten iterations. Convergence is impeded when a strong shock wave stands in the flowfield, but we have shown that adding dissipation to the system restores stability. The same outcome probably would be obtained by using a type-dependent differencing scheme in a way similar to that of Engquist and Osher (1980). And no doubt a conservation-law form would be preferable also.

Furthermore the Newton formulation Eq s. (2) offers a simple way to obtain a higher-order solution. When the procedure is nearly converged, one can use a higher-order scheme in the right-hand side, but keep the second order differencing of $\Delta q$ and proceed with the iteration. This is similar to the concept of deferred corrections, which is a convenient means of getting an improved solution to a differential equation by solving it once again using the truncation error resulting from the original scheme as a forcing function on the right-hand side. There are also many ways to reduce the actual computation time, for example by keeping the Jacobian matrix $L$ fixed for a number of steps, using a preconditioned conjugate gradient procedure, etc.

We believe, however, that the most important items to be resolved, and which are central to the understanding of our steady state approach to transonic flow problems, are:

1)  A means to analyze boundary conditions for problems of mixed elliptic - hyperbolic type

2)  the specification of a condition on entropy e.g. dissipation, artificial density, etc.

## References

Baker, T.  1980.  Personal communication.

Blomster, J., and Sköllermo, G.  1977.  Finite difference computation of steady transonic nozzle flow.  Computer Science Dept. Rep. No. 66, Uppsala Univ.

Engquist, B., and Osher, S.  1980.  Stable and entropy satisfying approximations for transonic flow calculations.  Math. Comp. 34:45–75.

Henrici, P.  1962.  Discrete Variable Methods in Ordinary Differential Equations. New York:  John Wiley. pp. 364 ff.

Lax, P.  1973.  Hyperbolic Systems of Conservation Laws and the Mathematical Theory of Shock Waves.  Philadelphia: SIAM.

Martin, E.D., and Lomax, H.  1975.  Rapid finite-difference computation of subsonic and slightly supercritical aerodynamic flows.  AIAA J.  13: 579–586.

$$\text{THE USE OF THE RIEMANN PROBLEM IN FINITE DIFFERENCE SCHEMES}$$


P L Roe

ROYAL AIRCRAFT ESTABLISHMENT, BEDFORD, UK


## I  INTRODUCTION

Given an arbitrary set of hyperbolic conservation laws for an unknown vector $\underline{u}$ $(x,t)$

$$\underline{u}_t + \underline{F}(\underline{u})_x = 0 \tag{1}$$

the Riemann problem is to solve (1) for the special case of initial data

$$\underline{u}\,(x,0) = \underline{u}_L\,(x < 0); \quad \underline{u}\,(x,0) = \underline{u}_R\,(x > 0) \tag{2}$$

A general solution to the Riemann problem serves to encapsulate a great deal of
information about whatever physical system is modelled by (1).  For that reason,
many authors have felt it to be an attractive "building block" for the construction
of numerical methods, especially those intended to provide for the automatic handling
of discontinuities.  There have been two main lines of development.  Godunov's (1959)
method has recently been extended to second order by van Leer (1979), and Glimm (1965)
devised a method with an element of random choice which makes shockwaves precise in
their structure, but uncertain in their location.  In comparison with more straightforward methods, both strategies prove expensive.  The present work leads to methods
which are cheaper, more flexible, accurate, and conservative.  The penalty paid for
these advantages is some weakening of the theoretical foundations.  The individual
Riemann problems which arise are solved by a direct, non-iterative, method which is
not in general exact, but which does have certain special properties built into it.

## II  THE LINEAR SCALAR CASE

We begin by studying the simplest possible example of (1), namely the linear
advection equation, $u_t + au_x = 0$, where u is a scalar unknown, and a is a constant.
This has the analytic solution $u = f(x - at)$.  We suppose that the solution is sought
on a regularly spaced grid $x = i\Delta x$, $t = n\Delta t$.  Any linear two-level difference scheme
for this problem can be put into the form

$$u_i^{n+1} = \sum_k c_k\, u_{i+k}^n \tag{3}$$

Godunov showed such a scheme to be $p^{th}$ order accurate if the coefficients $c_k$ satisfy

$$\sum_k k^j c_k = (-\nu)^j \quad j = 0,1, \ldots p \tag{4}$$

where $\nu$ is the Courant number, $\nu = a\Delta t/\Delta x$.  It is very easy to put existing
algorithms into the form (3), or to generate new ones by finding solutions of (4).
The starting point for the present work is the observation that any scheme of the
form (3) can be recast in an "increment form"

$$u_i^{n+1} - u_i^n = -\nu \sum_k \gamma_k\, (u_{i+k+1}^n - u_{i+k}^n) \tag{5}$$

Here the $\gamma_k$ are new coefficients, related to the $c_k$, which must likewise satisfy a set of linear constraints so as to produce specified accuracy. Specifically,

$$\Sigma\gamma_k = 1 \; ; \; \Sigma k\gamma_k = (1+\nu)/2 \; ; \; \Sigma k^2\gamma_k = (1+\nu)(1+2\nu)/6 \text{ etc.} \tag{6}$$

When put into the form (5), most well-known methods have a simple representation, for example

TABLE 1

|  | $\gamma_0$ | $\gamma_1$ | $\gamma_2$ | $\gamma_3$ |
|---|---|---|---|---|
| First order upwind | 0 | 1 | 0 | 0 |
| Lax-Wendroff (1960) | $(1-\nu)/2$ | $(1+\nu)/2$ | 0 | 0 |
| Fromm (1968) | $(1-\nu)/4$ | 1 | $(\nu-1)/4$ | 0 |
| Warming-Beam (1976) | 0 | $(3-\nu)/2$ | $(\nu-1)/2$ | 0 |
| Moretti (1979) | 0 | $(3-2\nu)/2$ | $(3\nu-1)/2$ | $-\nu/2$ |

What is more significant about (5), however, is that the method of implementing it gives rise to concepts which can be generalised fruitfully. For each interval $(i, i+1)$ in turn we compute the quantity $-\nu(u_{i+1}^n - u_i^n)$, and then divide it into fractions proportional to $\gamma_k$ (where $\Sigma\gamma_k = 1$). The $k^{th}$ fraction is added as an increment to the mesh point $(i+k)$. The process is illustrated in Fig 1. In moving from (3) to (5) there is a change in emphasis. Equation (3) answers the question "How does one find the solution at a given mesh point?". Equation (5) answers the question "What should one do with the data in a given mesh interval?". This change in emphasis foreshadows the subsequent emergence of the Riemann problem as crucial to the method.


III   THE NON-LINEAR SCALAR CASE


The same strategy can be readily applied to solve the non-linear scalar equation $u_t + F_x = 0$ where F is an arbitrary non-linear function of u. One begins by defining, for each interval, a local Courant number $\nu_{i+\frac{1}{2}} = (\Delta F\Delta t)/(\Delta u\Delta x)$. One selects an algorithm, perhaps from Table 1, and computes local weights, $\gamma_{k,i+\frac{1}{2}} = \gamma_k(\nu_{i+\frac{1}{2}})$. Then the quantity $-\nu_{i+\frac{1}{2}}\Delta u$ is split into fractions proportional to $\gamma_{k,i+\frac{1}{2}}$, and the $k^{th}$ fraction is added as an increment to the mesh point $(i+k)$. It can easily be checked that the total of all increments added at all mesh points from all sources is exactly that required to maintain conservation. It is less easy to verify that the accuracy of the non-linear algorithm is the same as that of the linear version, but in all cases checked so far this has proved to be the case.


IV   SETS OF LINEAR EQUATIONS


We now extend the method to deal with sets of linear equations $\underline{u}_t + A \underline{u}_x = 0$ where $\underline{u}$ is now a vector of m unknowns, and A is a constant m x m matrix with a complete set of real eigenvectors, $\underline{e}_1$, $\underline{e}_2$, ... $\underline{e}_m$. The general analytic solution of the initial-value problem is well known, and consists of projecting $\underline{u}$ (x,0) onto the eigenvectors thus

$$\underline{u}(x,0) = \sum_{j=1}^{m} f_j(x) \, \underline{e}_j \qquad (7)$$

then

$$\underline{u}(x,t) = \sum_{j=1}^{m} f_j(x-\lambda_j t)\underline{e}_j \qquad (8)$$

where $\lambda_j$ is the eigenvalue of A corresponding to $\underline{e}_j$. Thus, once the decomposition of the data onto eigenvectors has been accomplished, the solution, whether analytical or numerical, can proceed exactly as in the scalar case. A way of implementing the method is illustrated in Fig 2. In each interval, the measured vector difference $\Delta\underline{u}$ is split up into components associated with each wave system, ie $\Delta\underline{u} = \Sigma a_i \underline{e}_i = \Sigma(\Delta\underline{u})_i$. (In making this decomposition, we are of course solving the linear Riemann problem.) Weights $\gamma_{k,j}$ are computed from $\gamma_{k,j} = \gamma_k (\nu_j)$ and increments $\Delta\underline{u}_{k,j} = \gamma_{k,j} (\Delta\underline{u})_j$ are added at the mesh point (i+k). In this linear problem, all the wave systems are independent, and all properties of accuracy, stability, etc, are therefore carried over unchanged from the scalar case. It is worth mentioning, however, that there is no need to use the same algorithm for each wave system.

## V    SETS OF NON-LINEAR EQUATIONS

We  now combine the two preceding generalisations and study the <u>non-linear system</u> $\underline{u}_t + \underline{F}_x = 0$ where $\underline{u}$ is again a vector of m unknowns, but $\underline{F}$ is now an arbitrary vector function of $\underline{u}$. It is also convenient to write

$$\underline{u}_t + A\underline{u}_x = 0 \qquad (9)$$

where the m x m matrix A is the Jacobian $\partial F/\partial u$, and its elements are no longer constants, but functions of $\underline{u}$. We still assume, of course, that A always has a complete set of real eigenvectors.

The general strategy will be to combine the previous lines of attack. Thus, in each interval we find components of $\Delta\underline{u}$, $\Delta\underline{F}$ due to each wave system

$$\Delta\underline{u} = \Sigma a_i \underline{e}_i \;\; ; \;\; \Delta\underline{F} = \Sigma\lambda_i a_i \underline{e}_i \qquad (10a; \; b)$$

where, although the notation is as in Section IV, the eigenvalues $\lambda$ and eigenvectors $\underline{e}$ are no longer constants but locally defined in each interval. Multiplication of each $\lambda_j$ by $\Delta t/\Delta x$ produces a local Courant number $\nu_j$ appropriate to the $j^{th}$ wave. Reference to Table 1 produces weights $\gamma_{k,j}$ which can be used to distribute the increment $\Delta t \, \Delta\underline{F}_j/\Delta x$ over the mesh.

What we have so far left unspecified is the precise means by which the eigenvalues and eigenvectors are to be estimated. To provide such estimates we need to solve, at least approximately, the non-linear Riemann problem. We could of course use the exact solution but that would impose extra computing costs which do not seem to be justified by the results.  Instead, we seek an approximation having the following properties

    i    As $\underline{u}_{i+1} \to \underline{u}_i$, then (for all j) $\lambda_j$ and $\underline{e}_j$ tend to their linearised values. This is an obvious necessity for accuracy in smooth parts of the flow.

    ii    We require that $\lambda_j$, $\underline{e}_j$, $a_j$ are such that both (10a) and (10b) are satisfied <u>exactly</u> for arbitrary $\Delta u$.  This is rather a loose requirement - it gives 2 m equations in $m^2 + m$ unknowns, hence $m(m - 1)$ degrees of freedom. However, two rather nice properties flow from it.

The first is that since (10b) is satisfied exactly then the total sum of all increments will be such as to ensure conservation, as in Section III.  The second property arises whenever the states $\underline{u}_i$, $\underline{u}_{i+1}$ can be connected by a single shock or contact discontinuity, ie whenever $\Delta\underline{F} = S\Delta\underline{u}$ for some scalar S.  In that case, provided the $\underline{e}_j$ are linearly independent, we must have, for some s, $S = \lambda_s$, and $\Delta\underline{u} = \underline{e}_s$.  Thus, in this case, the method will return the exact solution.

In combination, I call properties (i) and (ii) Property U.  It ensures accuracy in the two situations most commonly encountered; that is to say the flow is either smooth, or close to a single discontinuity.  Where several discontinuities intersect, the present approximation has no special virtue (other than conservation) but in practice the error seems highly localised.  Working out an approximation having Property U for a given set of equations is not difficult in principle, but may involve tedious algebra.  A technique which partly overcomes this has been devised, and will be published elsewhere.  Here we merely note the results for the unsteady Euler equations in one space dimension.  The linearised eigenvectors and eigenvalues for that problem are

$$\underline{e}_1 = \begin{pmatrix} 1 \\ u-a \\ H-ua \end{pmatrix}, \quad \underline{e}_2 = \begin{pmatrix} 1 \\ u \\ \tfrac{1}{2}u^2 \end{pmatrix}, \quad \underline{e}_3 = \begin{pmatrix} 1 \\ u+a \\ H+ua \end{pmatrix} \tag{11}$$

$$\lambda_1 = u-a, \quad \lambda_2 = u, \quad \lambda_3 = u+a \tag{12}$$

where $H = \dfrac{\gamma p}{(\gamma-1)\rho} + \dfrac{1}{2}u^2$ and $a^2 = (\gamma-1)\left[H-\tfrac{1}{2}u^2\right]$     (13a; b)

Evidently the first part of Property U is ensured by taking u, a, H to be any valid averages over the interval.  The second part can be achieved by defining those averages more particularly.  The simplest so far discovered are

$$u_{i+\frac{1}{2}} = \frac{(\rho^{\frac{1}{2}}u)_i + (\rho^{\frac{1}{2}}u)_{i+1}}{\rho_i^{\frac{1}{2}} + \rho_{i+1}^{\frac{1}{2}}} \; ; \quad H_{i+\frac{1}{2}} = \frac{(\rho^{\frac{1}{2}}H)_i + (\rho^{\frac{1}{2}}H)_{i+1}}{\rho_i^{\frac{1}{2}} + \rho_{i+1}^{\frac{1}{2}}} \tag{14a; b}$$

and $a_{i+\frac{1}{2}}$ is obtained by substituting (14a) and (14b) into (13b).

## VI  THE PROBLEM OF SPURIOUS OSCILLATIONS

It was first proved by Godunov that any numerical method of the form (3), having second-order accuracy or better, would produce an oscillating response to step data. Since (5) is merely (3) rewritten, nothing so far said contributes directly to the removal of these "wiggles".  However, the techniques which we have described allow us to identify wiggles as originating with a particular wave system.  This seems to permit control mechanisms rather more delicate than those in currect use. Mechanisms such as "artificial viscosity" or "flux limitation" act to suppress wiggles without acknowledging the information which they carry.  Although space does not allow it to be described here, a new mechanism for the control of wiggles has been developed, based on non-linear combinations of the algorithms in Table 1. Details will be published elsewhere, but the results produced on a standard test problem are shown in the next section.

Alternatively, one accepts wiggles but seeks to reduce them.  Almost invariably this approach relies on directionally biased formulae which maintain the physically correct flow of information, eg Fromm (1968), Steger and Warming (1978), Moretti (1979), Leonard (1979).  The present work is very compatible with these ideas.

## VII  NUMERICAL RESULTS

Results have been obtained for the shock tube problem used as a test case by Sod (1978).  When the upwind algorithm from Table 1 was used in conjunction with the approximate Riemann solution, the results agreed to 4 or 5 significant figures with those from the standard Godunov method employing the exact Riemann solution. This confirms the opinion that the exact solution is not needed.  The results in Fig 3 were obtained by choosing weights corresponding to Fromm's algorithm.  They compare well with most of the results surveyed by Sod, especially when it is realised that they are obtained without any recourse to artificial viscosity. Fig 4 shows results obtained from the control mechanism, mentioned in Section VI. These results show no overshoots, and are third-order accurate in smooth regions.

## VIII FURTHER DEVELOPMENTS

The material above has concentrated on explicit two-level algorithms for one-dimensional problems.  However, the approach of analysing the data as a set of approximate Riemann problems appears to be very widely applicable.  Both multilevel schemes, and implicit schemes, have been developed in the same way.  In these cases also, it is possible to begin with the linear advection equation and go through a similar process of generalisation to treat non-linear systems.  The extension to multidimensional problems is possible via operator splitting, and has been applied to practical transonic airfoil calculations by Sells (1980).

REFERENCES

Fromm, J.E. (1968)　　　　　　　　　　　　J. Comp. Phys, $\underline{3}$, p 176
Glimm, J. (1965)　　　　　　　　　　　　　Comm. Pure Appl. Math, $\underline{18}$, p 697
Godunov, S.K. (1959)　　　　　　　　　　Mat. Sbornik, $\underline{47}$, p 271
Lax, P.D., Wendroff, B. (1960)　　　　Comm. Pure Appl. Math, $\underline{13}$, p 217
Leonard, B.P. (1979)　　　　　　　　　　Comp. Meth. App. Mech. Eng., $\underline{19}$, p 59
Moretti, G. (1979)　　　　　　　　　　　Computers and Fluids, $\underline{7}$, p 191
Sells, C.C.L. (1980)　　　　　　　　　　RAE TR 80065
Sod, G. (1978)　　　　　　　　　　　　　J. Comp. Phys, $\underline{27}$, p 1
Steger, J.L., Warming, R.F. (1978)　　NASA TM 78605
van Leer, B. (1978)　　　　　　　　　　　J. Comp., Phys, $\underline{32}$. p 101
Warming, R.F., Beam, R.M. (1976)　　　AIAA Jul., $\underline{14}$, p 1241

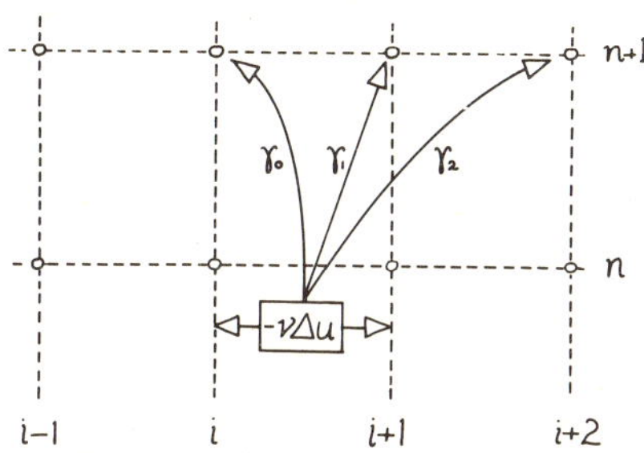


Fig 1　　Dispersal of scalar increments

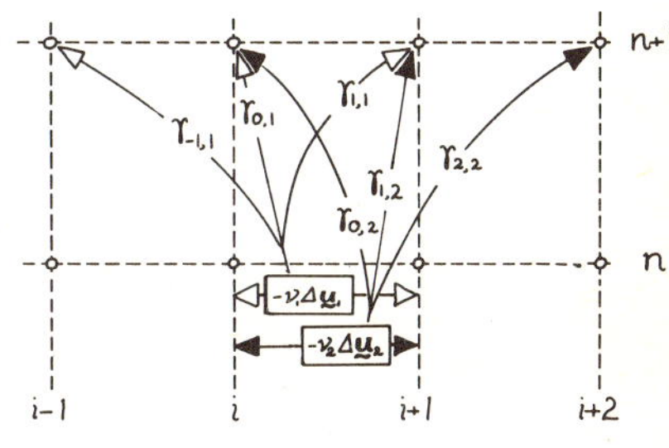


Fig 2　　Dispersal of vector increments

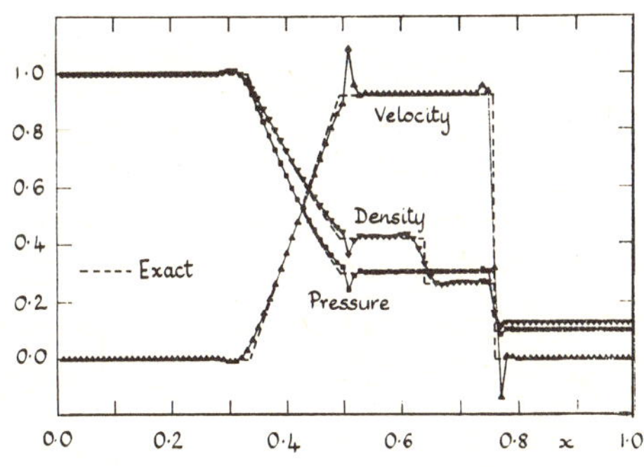


Fig 3　　Results from Fromm's algorithm

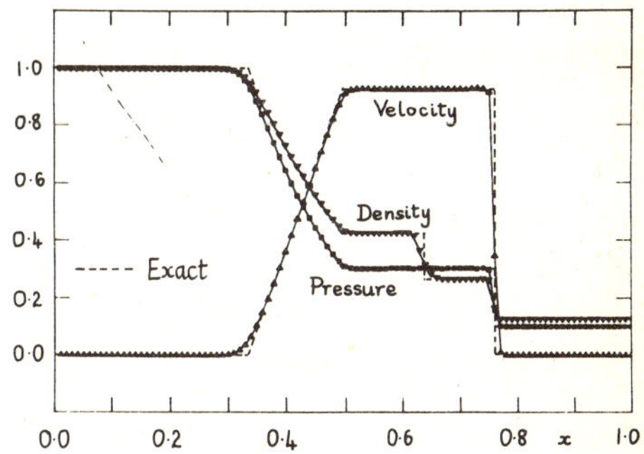


Fig 4　　Results from optimised algorithm

# STABILITY AND SEPARATION OF FREELY INTERACTING BOUNDARY LAYERS


Oleg S. Ryzhov and Vladimir I. Zhuk

Computing Center, Academy of Sciences
Moscow, USSR


## I. ABSTRACT


The triple-deck theory to describe boundary-layer free interaction and separation is carefully investigated from the viewpoint of how it predicts stability properties of a viscous flow. The linearized version of this theory gives the same results as the linear stability theory based on the Orr-Sommerfeld equation if the flow is incompressible and the wave number of small disturbances goes to zero. Hence, a rather general criterion results to fix limits for a subsonic boundary layer to be stable. On the contrary, the linear approximation to the triple-deck theory leads to discouraging conclusions when the velocity of the oncoming stream exceeds the speed of sound, since it fails to reveal the boundary-layer instability. The Prandtl equations with a self-induced pressure gradient included are used to formulate a nonlinear approach for elucidating stability properties of freely interacting boundary layers both for subsonic and supersonic cases. The shock-wave-boundary-layer interaction and separation on a moving wall is numerically studied, the formation of two recirculation bubbles being the most striking feature. With the shock strength increasing, both bubbles tend to divide into smaller vortex cells whence the nonsteady process of velocity field "breathing" stems.


## II. INTRODUCTION

In the past decade an asymptotic triple-deck theory to predict the major features of a laminar boundary layer separation has been simultaneously developed by Neiland (1969), and Stewartson and Williams (1969). Essentially the same approach was used by Messiter (1970) in his study of the flow field in the vicinity of the trailing edge of a flat plate. The further impressive achievements of this theory in elucidating many physical phenomena encourage one to examine deeply its capabilities and limitations. After time dependence has been included in the corresponding series expansions, the question arises as for the stability of freely interacting boundary layers under various conditions. Since stability is one of the most fundamental properties of any viscous fluid motion, an attempt should be undertaken to clarify the nature of the triple-deck theory from this point of view. As is well known, the conventional boundary-layer theory of Prandtl fails to treat properly the formulated problem.

## III. BASIC EQUATIONS

Let $t$ designate the time, $x$ and $y$ the axes of Cartesian coordinates, $u$ and $v$ the velocity vector components, and $p$ the pressure. Then, in accord with the triple-deck theory, the equations of fluid motion in a thin viscous sublayer adjacent to a flat plate can be written in the form

$$\frac{\partial u}{\partial x} + \frac{\partial v}{\partial y} = 0 \ , \tag{1a}$$

$$\frac{\partial u}{\partial t} + u\frac{\partial u}{\partial x} + v\frac{\partial u}{\partial y} = -\frac{\partial p}{\partial x} + \frac{\partial^2 u}{\partial y^2} \ , \tag{1b}$$

$$\frac{\partial p}{\partial y} = 0 \ . \tag{1c}$$

The limiting condition as $y \to \infty$ reads

$$u - y \to A\,(t,x) \ , \tag{2}$$

while the self-induced pressure $p$ is related to the streamline displacement $A$ by the expression

$$p = \begin{cases} \dfrac{1}{\pi} \displaystyle\int_{-\infty}^{\infty} \dfrac{\partial A/\partial X}{x - X} \ dX \ , & \text{if } M_\infty < 1 \ , \tag{3a} \\[2em] -\dfrac{\partial A}{\partial x} & , \quad \text{if } M_\infty > 1 \ . \tag{3b} \end{cases}$$

A special system of units is used here to avoid the dependence of all the coefficients in the equations and limiting condition on any gas parameters. As a result, both the Reynolds number $R$ and the Mach number $M_\infty$ drop out of the formulation of the problem.

At the flat plate surface $y = 0$, the desired functions satisfy the non-slip condition $u = v = 0$. Far upstream of an interaction region the flow is a uniform shear, in which case $u \to y$, $p \to 0$ as $x \to -\infty$. In stability problems the latter requirements obviously must be replaced by a periodicity condition in the $x$-coordinate for the pressure and velocity field.


## IV. LINEAR APPROACH

The functions $u = y$, $v = 0$, and $p = 0$ yield the solution in a viscous sublayer for undisturbed Blasius flow. Let us superimpose small perturbations on these functions and look for the following solution

$$u = y - a\,e^{\omega t + kx}\,\frac{df}{dy} \ , \tag{4a}$$

$$v = a\,k\,e^{\omega t + kx}\,f \ , \tag{4b}$$

$$p = a\, e^{\omega t + kx} \,. \tag{4c}$$

Upon the substitution of the formulae (4a) - (4c) into the system of basic equations (1a) - (1c) and their linearization with respect to the perturbation amplitude a, the third order differential equation

$$\frac{d^3 f}{dy^3} - (\omega + ky)\frac{df}{dy} + kf + k = 0 \tag{5}$$

appears for the function f, subject to the boundary conditions

$$\frac{df}{dy} \to \frac{1}{k} \quad \text{as} \quad y \to \infty , \tag{6}$$

and

$$f = \frac{df}{dy} = 0 \quad \text{at} \quad y = 0 \,. \tag{7}$$

The boundary-value problem (5) - (7) is a problem in eigenvalues. Eigenfunctions obtained by solving the problem under consideration represent internal waves that propagate through a freely interacting boundary layer. The internal waves are generated by combined action of the self-induced pressure and shear stresses; there is no external agency to provoke them.

## V. SPECTRUM

The frequency $\omega$ and the wave number k are connected through a dispersion relation

$$\frac{d\,Ai(\zeta)}{dz}\left[\int_{-\infty}^{\infty} Ai(z)\,dz\right]^{-1} = \begin{cases} \mp i k^{4/3} , & \text{if } M_\infty < 1 , & \text{(8a)} \\[2mm] - k^{4/3} , & \text{if } M_\infty > 1 , & \text{(8b)} \end{cases}$$

where $Ai(z)$ is the Airy function and $\zeta = \omega k^{-2/3}$. Using the asymptotic expansions for the Airy function, it has been shown by Zhuk and Ryzhov (1979) that an infinite (discrete) spectrum of eigenvalues Re $\zeta_j$, Im $\zeta_j$ corresponds to each given wave number k. For large j the roots $\zeta_j$ of the dispersion equations (8a), (8b) lie in the vicinity of the negative real semi-axis in the complex plane $\zeta$ with

$$|\zeta_j| = \left[\frac{3\pi}{2}\left(j + \frac{1}{4}\right)\right]^{2/3}$$

whether the boundary layer is subsonic or supersonic. Since the frequency $\omega_j = \zeta_j k^{2/3}$, internal waves corresponding to these roots are stable and decay exponentially with time.

## VI. INSTABILITY OF SUBSONIC FLOWS

In contrast to the asymptotic analysis, numerical calculations reveal the sign of the real part of the first eigenfrequency $\omega_1$ to be dependent on the magnitude of $|k|$ if a freely interacting boundary layer is subsonic. The change in the sign occurs when $|k| = k_* = 1.005$. At this point it is of advantage to come back from a special dimensionless system of units where the wave number is k to the original (also dimensionless) system with the wave number $\alpha$. Then the simple criterion

$$C^{1/2} \lambda^{-5/3} R^{-1/2} |M_\infty^2 - 1|^{-1/2} \left(\frac{T_w}{T_\infty}\right)^2 |\alpha|^{4/3} \le k_*^{4/3} \quad , \tag{9a}$$

$$\tau_w = C^{-1/2} \lambda \left(\frac{T_w}{T_\infty}\right)^{-1} \tag{9b}$$

holds for a flow being stable. Here C is the constant entering the Chapman viscosity law, T is the temperature, and $\tau$ is the shear stress; the subscript w refers to wall conditions. The opposite sense of the inequality (9a) signifies instability of a boundary layer.

It is necessary to emphasize that for an incompressible fluid the triple-deck theory under consideration leads to the same results as the linear stability theory based on the Orr-Sommerfeld equation provided a critical layer of neutral oscillations is adjacent to a wall. The internal waves in a freely interacting incompressible boundary layer are asymptotic forms of the Tollmien-Schlichting waves as $R \to \infty$, $\alpha \to 0$. Tollmien was virtually the first to introduce the notion of the critical layer as a part of the interior structure of a disturbed boundary layer in terms of singular perturbations (see Lin, 1955). A comprehensive analysis of neutrally oscillating flows within the framework of the linear stability theory has been developed by Smith (1979).

The above stated conclusions provide a reliable background to formulate an asymptotic nonlinear stability theory for subsonic boundary layers by the use of the Prandtl equations with the self-induced pressure gradient included. In contrast to the original system of Navier-Stokes equations, they do not contain any small parameter. In the nonlinear approach, as well as in the linear one, it is necessary to require that the desired functions be periodic in the x-coordinate.


## VII. SEPARATION OF SUPERSONIC BOUNDARY LAYER

The numerical results show that all the eigenfrequencies $\omega_j$ have negative real parts regardless of the magnitude of $|k|$ if a boundary layer is supersonic. This leads to a discouraging statement that the triple-deck theory fails in this case to predict stability properties of the flow, while the linear theory based on the Orr-Sommerfeld equation does treat the same properties correctly. However, the conclusion concerning stability of supersonic boundary layers may change with non-linear considerations. To verify the latter conjecture let us examine the specific problem of shock-wave-boundary-layer interaction on a moving wall, in which case the boundary condition (2) needs to be slightly modified:

$$u - (y + c_w) \to A(t,x) \quad \text{as} \quad y \to \infty \quad , \tag{10}$$

whereas the self-induced pressure obeys the expression

$$p = \theta H(x) - \frac{\partial A}{\partial x} \quad , \quad H(x) = \begin{cases} 0 \, , & \text{if } x < 0 \, , & \text{(11a)} \\ 1 \, , & \text{if } x > 0 \, , & \text{(11b)} \end{cases}$$

instead of (3b). The wall velocity is equal to $c_w$, and the scaled amplitude of an incident shock is proportional to $\theta$. Accordingly, the non-slip condition reads $u = c_w$, $v = 0$ at $y = 0$.

The nonlinear problem (1a)-(1c), (10), (11a), (11b) was solved numerically for various values of the parameters $c_w$ and $\theta$. The results of the calculations basically disagree with the so-called Moore-Rott-Sears model (see Williams, III, 1977). The most interesting feature of the phenomenon relates, however, to the formation of two closed vortex bubbles where gas particles are rotating in opposite directions. The streamline patterns are drawn in Figure 1 for $c_w = 0.2$, $\theta = 4$. As the shock wave amplitude $\theta$ increases above, say 6 to 8, both separation bubbles begin to divide into smaller vortex cells and an unsteady velocity field stems from this process. Insofar as its initial stage is concerned, a third recirculation zone is produced by stream-lines located between the two original zones; the central streamline segments rise and break through the upper bubble. The formation of other closed recirculation regions follows in a similar way.

Strong theoretical support for the numerical experiments is provided by the fact that velocity profiles associated with the separation bubbles inside the viscous sub-layer are unstable even within the framework of the linear theory. In the course of computations by the use of time-dependent finite-difference schemes, all velocity profiles developing in different cross-sections of the flow undergo continuous exam-ination with respect to their stability. After separation bubbles have attained a certain extent, they become unstable and, accordingly, commence to "breathe." The time dependence of "breathing" happens to be different under different conditions.

## ACKNOWLEDGMENTS

This work has been initiated in Moscow and completed during the stay of the first author at The University of Michigan, Ann Arbor. It is a pleasure for him to appreciate fruitful discussions of some aspects of the triple-deck and stability theories with his American colleagues. Special thanks are to Professors T.C. Adamson, Jr., A.F. Messiter, and M. Van Dyke for many improvements of the original text.



## REFERENCES

Lin, C.C. 1955. The theory of hydrodynamic stability. Cambridge Univ. Press.

Messiter, A.F. 1970. Boundary-layer flow near the trailing edge of a flat plate. SIAM J. Appl. Math., Vol. 8, N1: 241-257.

Neiland, V. Ya. 1969. Contribution to the theory of laminar boundary-layer separation in supersonic flow. Izv. AN SSSR, Mekh. Zhidk. i Gaza, N4: 53-57 (in Russian).

Smith, F.T. 1979. On the non-parallel flow stability of the Blasius boundary layer. Proc. R. Soc. London, Ser. A, Vol. 366, N1724: 91-109.

Stewartson, K. and Williams, P.G. 1969. Self-induced separation. Proc. R. Soc. London, Ser. A, Vol. 312, N1509: 181-206.

Williams, J.C., III. 1977. Incompressible boundary-layer separation. Ann. Rev. Fluid Mech., Vol. 9: 113-144.

Zhuk, V.I. and Ryzhov, O.S. 1979. On solutions of a dispersion relation from the theory of free interaction of a boundary layer. Dokl. AN SSSR, Vol. 247, N5: 1085-1088 (in Russian).

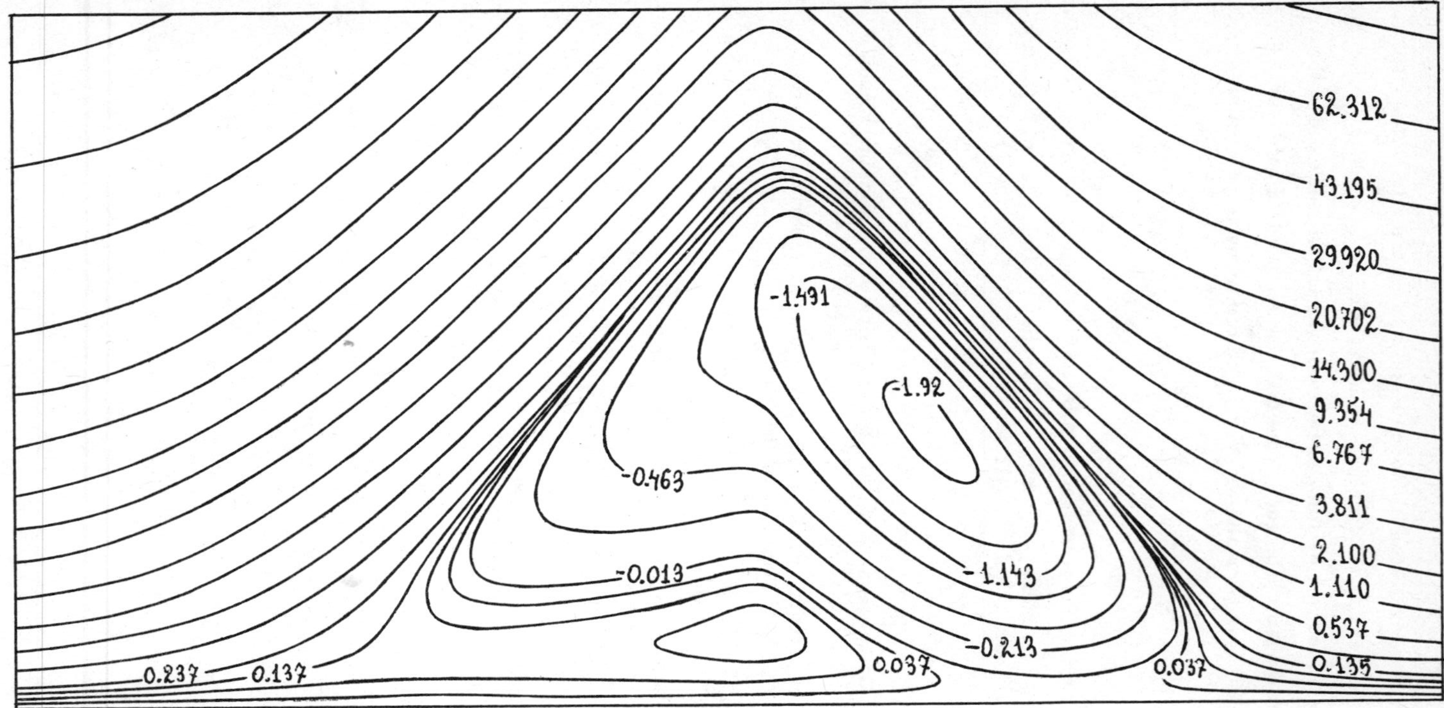



Figure 1. Typical flow pattern, for a certain range of parameters $c_w$ and $\theta$, in the thin viscous sublayer adjacent to a downstream moving wall. Two separation bubbles are clearly distinguished. There are streamlines, coming from upstream infinity, that encircle the lower bubble from the right and pass between the recirculation zones; upon encircling the upper bubble from the left, they leave the interaction region for downstream infinity.

# An Algorithm for Unsteady Transonic Potential Flow Past Airfoils


N. L. Sankar and Y. Tassa

Lockheed-Georgia Company, Marietta, Ga.  30063, USA


## INTRODUCTION

The importance and usefulness of efficient methods for unsteady transonic flow past airfoils can not be overemphasized.  In the past research workers have used a variety of computational tools to study this problem.  For example, Steger (1977) developed efficient Euler and Navier-Stokes solvers; Traci, Albano and Farr (1975) developed a perturbation procedure expanding the velocity potential in time as a harmonic series, about a mean steady flow; Ballhaus and Steger (1975), and Ballhaus and Goorjian (1977) developed and perfected the low-frequency small disturbance formulation; Goorjian (1979), Isogai (1977), and Chipman and Jameson (1979) used the full potential to study unsteady transonic flows.  Among the variety of approaches described, the unsteady full potential formulation offers a unique combination of high accuracy and low computational costs compared to others.

Because of this reason, in the present work we study the unsteady full potential formulation; to ensure accurate treatment of unsteady boundary conditions, the governing equations are cast in a non-orthogonal, time-deforming body-fitted coordinate system.  A major departure in our approach from the above works is the use of a Strongly Implicit Procedure (SIP) to solve the governing equation, in place of the well-known Alternating Direction Implicit (ADI) procedure.  The SIP was first introduced by Stone (1969), as an efficient procedure for the iterative solution of well-behaved elliptic problems, with spatially varying coefficients.  When the coefficients vary widely, the SIP was found to be superior in convergence to many other relaxation procedures including ADI.  Some research workers, for example Sampath (1977), have studied solution of incompressible flows using SIP as an iterative time-marching procedure.  The present work differs from these earlier works in the following two respects.  First, subsonic as well as embedded supersonic regions with shocks may be treated by the present procedure.  Secondly, the present procedure is a one-step non-iterative time-marching procedure.  An efficient relaxation procedure is also obtained as a useful by-product if one is interested only in steady transonic flows.  Results are presented to establish the reliability and accuracy of the method including steady and unsteady, lifting and non-lifting potential flows.

## MATHEMATICAL FORMULATION

In order to facilitate treatment of boundary conditions, the region around the airfoil, which may be moving or deforming with time, is mapped numerically onto the computational domain, so that the body surface becomes a coordinate line in the transformed plane.  The transformation may be written as,

$$\xi = \xi(x,y,t) \qquad \eta = \eta(x,y,t) \qquad \tau = t \tag{1}$$

The continuity equation in the transformed plane may be written as

$$\sigma \frac{\partial}{\partial \tau}\left(\frac{\rho}{J}\right) + \frac{\partial}{\partial \xi}\left(\frac{\rho U}{J}\right) + \frac{\partial}{\partial \eta}\left(\frac{\rho V}{J}\right) = 0 \tag{2}$$

where

$$U = \xi_t + \xi_x u + \xi_y v = \xi_t + (\xi_x^2 + \xi_y^2)\phi_\xi + (\xi_x \eta_x + \xi_y \eta_y)\phi_\eta \tag{3}$$

$$V = \eta_t + \eta_x u + \eta_y v = \eta_t + (\eta_x^2 + \eta_y^2)\phi_\eta + (\xi_x \eta_x + \xi_y \eta_y)\phi_\xi \tag{4}$$

$$\xi_t = -x_\tau \xi_x - y_\tau \xi_y \tag{5}$$

$$\eta_t = -x_\tau \eta_x - y_\tau \eta_y \tag{6}$$

$$J = \xi_x \eta_y - \xi_y \eta_x \tag{7}$$

Here $x_\tau$ and $y_\tau$ are the Cartesian components of velocity of the deforming grid in the physical plane; u and v are the Cartesian components of velocity of the fluid; J is the Jacobean of transformation. We note that the definition of velocity potential $\phi$ has already been introduced in equations (2) and (3). It is assumed for convenience that the density $\rho$ is normalized with respect to freestream value $\rho_\infty$, the velocity with respect to free stream velocity $V_\infty$ and all distances are normalized with respect to airfoil chord C. The parameter $\sigma$ separates the time accurate solution procedure ($\sigma = 1$) from the relaxation solution procedure ($\sigma = 0$).

For isentropic and irrotational flows, the density is related to the velocity potential through the relation,

$$\rho = [1 + \frac{\gamma-1}{2} M_\infty^2 (1 - 2(\phi_\tau + \phi_\xi \xi_t + \phi_\eta \eta_t)\sigma - \phi_x^2 - \phi_y^2)]^{1/\gamma-1} \tag{8}$$

Performing a direct differentiation of $\rho$ with respect to t and plugging it into equation (2), one obtains a single hyperbolic equation for $\phi$ given by,

$$-\sigma M_\infty^2 \frac{\rho^{2-\gamma}}{J} [\phi_{\tau\tau} + U\phi_{\xi\tau} + V\phi_{\eta\tau}] + \frac{\partial}{\partial\xi} \left\{ \rho \frac{\xi_x^2 + \xi_y^2}{J} \phi_\xi + \rho \frac{\xi_x \eta_x + \xi_y \eta_y}{J} \phi_\eta \right\}$$

$$+ \frac{\partial}{\partial\eta} \left\{ \rho \frac{\eta_x^2 + \eta_y^2}{J} \phi_\eta + \rho \frac{\xi_x \eta_x + \xi_y \eta_y}{J} \phi_\xi \right\} = \left( -\rho (\frac{1}{J})_\tau + \frac{M_\infty^2 \rho^{2-\gamma}}{J} [\phi_\xi (\xi_t)_\tau \right.$$

$$\left. + \phi_\eta (\eta_t)_\tau + \phi_x (\phi_\xi (\xi_x)_\tau + \phi_\eta (\eta_x)_\tau) + \phi_y (\phi_\xi (\xi_y)_\tau + \phi_\eta (\eta_y)_\tau)] \right) \sigma = R \tag{9}$$

Equation (9) is highly non-linear. Also, it is non-conservative with respect to time. Thus it would not be expected to predict the strength or speed of moving shock waves very accurately. For subsonic flows, the exact conservation is relatively unimportant, and the non-conservative form may be adequate. Before discussing the question of arriving at a conservation form, we linearize equation (9) in the following manner. The entire right hand side R is evaluated at known time "n". The coefficients $\rho^{2-\gamma}$, U, V, J, $\rho$ etc. on the left hand side are also evaluated at time "n". All the cross derivatives in space are evaluated at time "n". The terms involving exclusively the spatial derivatives are written in central difference form. The terms $\phi_{\xi\tau}$ and $\phi_{\eta\tau}$ are discretized at the new time level (n+1), using windward differences, depending on the sign of quantities U and V. The term $\phi_{\tau\tau}$ is written, for constant time steps as,

$$\phi_{\tau\tau}^{n+1} = \frac{1}{\Delta t^2} (\phi^{n+1} - 2\phi^n + \phi^{n-1}) \tag{10}$$

When the flow changes from subsonic to supersonic, some type of upwind bias must be added to equation (9) to ensure numerical stability. In the present work the density $\rho$ in the term $(\frac{\rho U}{J})_\xi$ is written with an upwind bias as follows:

$$\tilde{\rho} = \rho + \nu \frac{\partial \overleftarrow{\rho}}{\partial \xi} \Delta \xi , \quad \nu = \text{Max}(0, (M^2 - 1)C) \tag{11}$$

The arrow indicates that the term $\frac{\partial \rho}{\partial \xi}$ is written using windward differences. This technique has been used with considerable success by Goorjian (1979). Here C is a user-specified constant.

If a strong conservation form is desired, it may be done by adding certain terms to the right hand side of equation (9) to arrive at a term $R_1$ where

$$R_1 = R^n - M_\infty^2 \left( \frac{\rho^{2-\gamma}}{J} \right)^{n-1} \left[ \phi_{\tau\tau}^n + U^{n-1}\phi_{\xi\tau}^n + V^{n-1}\phi_{\eta\tau}^n + \rho^{n-1}(\frac{1}{J})_\tau^{n-1} \right] \sigma$$

$$- \sigma M_\infty^2 \left( \frac{\rho^{2-\gamma}}{J} \right)^{n-1} \left\{ \phi_\xi (\xi_t)_\tau + \phi_\eta (\eta_t)_\tau + \phi_x (\phi_\xi (\xi_x)_\tau + \phi_\eta (\eta_x)_\tau) \right.$$

$$\left. + \phi_y (\phi_\xi (\xi_y)_\tau + \phi_\eta (\eta_y)_\tau) \right\}^{n-1} - \frac{\sigma}{\Delta t} \left( \frac{\rho}{J} \right)^n + \frac{\sigma}{\Delta t} \left( \frac{\rho}{J} \right)^{n-1} \tag{12}$$

Note that we have simply added the term $\left( \frac{1}{J} \rho_\tau^n + \rho \left( \frac{1}{J} \right)_\tau^n - \frac{1}{\Delta t} \left( \frac{\rho}{J} \right)^n + \frac{1}{\Delta t} \left( \frac{\rho}{J} \right)^{n-1} \right) \sigma$ which is formally of order $O(\Delta t)$, and ensured complete cancellation of terms such as $U\phi_{\xi\tau}$ as one marches in time.

Upon introduction of the incremental quantity $\Delta\phi_{ij}^{n+1}$, equal to $(\phi_{i,j}^{n+1} - \phi_{i,j}^n)$, and performing discretizations discussed above, one obtains a system of linear equations of the form:

$$B_{ij}\Delta\phi_{i,j-1}^{n+1} + D_{ij}\Delta\phi_{i-1,j}^{n+1} + E_{ij}\Delta\phi_{i,j}^{n+1} + F_{ij}\Delta\phi_{i+1,j}^{n+1} + H_{ij}\Delta\phi_{i,j+1}^{n+1} = Q_{ij} \tag{13}$$

The boundary conditions are yet to be applied. The simplest way to apply the boundary conditions will be to apply them explicitly, allowing $\Delta\phi_{ij}^{n+1}$ at all boundaries to be zero. In practice, we have allowed boundary conditions everywhere to lag behind by one time step. except at the solid where the condition $V = 0$ has been implicitly built into (13). The explicit treatment of boundary conditions will be discussed in a separate section.

In matrix form, equation (13) may be written as,

$$[M] \{\Delta\phi\}^{n+1} = \{Q\} \tag{14}$$

A direct LU decomposition of the matrix $[M]$ is certainly possible, but time consuming. The L and U matrices that result will also be full. To avoid this difficulty, Stone introduced the SIP, which essentially involves addition of certain error terms to equation (13). While such arbitrary introduction of error terms may certainly be tolerated in relaxation schemes, care is necessary to ensure that these error terms are formally of the same order as the truncation error, when one is interested in a one-step non-iterative time marching solution of equation (13). The SIP modifies the original 5-point star into a 7-point star that includes the two additional nodes (i-1, j+1) and (i+1, j-1). Equation (13) thus becomes,

$$B_{ij}\Delta\phi_{i,j-1}^{n+1} + C_{ij}[\Delta\phi_{i+1,j-1}^{n+1} - \alpha(\Delta\phi_{i+1,j}^{n+1} + \Delta\phi_{i,j-1}^{n+1} - \Delta\phi_{i,j}^{n+1})] + D_{ij}\Delta\phi_{i-1,j}^{n+1} + E_{ij}\Delta\phi_{i,j}^{n+1}$$

$$+ F_{ij}\Delta\phi_{i+1,j}^{n+1} + G_{ij}[\Delta\phi_{i-1,j+1}^{n+1} - \alpha(\Delta\phi_{i-1,j}^{n+1} + \Delta\phi_{i,j+1}^{n+1} - \Delta\phi_{i,j}^{n+1})] + H_{ij}\Delta\phi_{i,j+1}^{n+1} = Q_{ij} \tag{15}$$

The error terms may be formally written as

$$C_{ij} \; (1-\alpha)\Delta\tau \left( \frac{\partial\phi}{\partial\tau} \right)_{i+1,j-1} + \alpha\Delta\xi\Delta\eta\Delta\tau \left( \frac{\partial^3\phi}{\partial\xi\partial\eta\partial\tau} \right)_{i+\frac{1}{2},\, j-\frac{1}{2}}$$

$$+ G_{ij} \; (1-\alpha)\Delta\tau \left( \frac{\partial\phi}{\partial\tau} \right)_{i-1,j+1} + \alpha\Delta\xi\Delta\eta\Delta\tau \left( \frac{\partial^3\phi}{\partial\xi\partial\eta\partial\tau} \right)_{i-\frac{1}{2},\, j+\frac{1}{2}} \tag{16}$$

The coefficients $C_{ij}$ and $G_{ij}$ are to be determined during the factorization and $\alpha$ is a relaxation parameter. While treating the unsteady flow problem $\alpha$ is set to 1 so that the error introduced is of order $O(\Delta\xi\Delta\eta\Delta\tau)$. Thus the formal accuracy of the scheme is not violated, and one expects the LU factorization to produce a true solution.

While using the SIP procedure purely as a relaxation procedure ($\sigma = 0$), the repeated use of $\alpha$ equal to 1 leads to instabilities. This is because the diagonal term $E_{ij}$ is not augmented by a term such as $\phi_{\tau\tau}$.  Using a Von Neumann analysis, Stone (1969) points out that large values of $\alpha$ tend to suppress the low frequency error components while small values of $\alpha$ tend to suppress the high frequency components of error.  A cyclic variation of $\alpha$ will therefore suppress uniformly all the components of error.  An alternative to varying $\alpha$ (between 0 and 1) cyclically is to use a multi-grid relaxation. This attractive alternative has not yet been fully explored by the present authors.

The L and U matrices are sparse matrices with 3 diagonals each.  SIP may be programmed so that only the U matrix need be stored, and this may be done efficiently due to its sparseness.  The details of the LU factorization, and a related procedure called the alternating SIP which has been used by the authors when the present procedure operates in the relaxation mode, are not discussed here, and the reader is referred to the excellent work of Stone (1969).

<h3 align="center">APPLICATION OF BOUNDARY CONDITIONS</h3>

In our work, the region around the airfoil is mapped onto a rectangular computational domain using the sheared parabolic coordinate transformation.  The airfoil coincides with a $\xi$-line so that $V = 0$ on the solid.  Then, to first order accuracy in time, one may write,

$$\phi_\eta^{n+1} = - \frac{n_t^{n+1} + (\xi_x \eta_x + \xi_y \eta_y)^{n+1} \phi_\xi^n}{(\eta_x^2 + \eta_y^2)^{n+1}} \tag{17}$$

The term $\phi_\xi$ is replaced by central differences, the term $\phi_\eta$ is replaced by one-sided differences so that $\phi_{ij}^{n+1}$ on the wall is linked to its immediate neighbor point inside the computational domain.

The vortex sheet is assumed to lie along the x axis.  The condition that the normal velocity and the pressure be continuous across the wake leads, upon linearization and discretization, to two simultaneous equations for the values of $\phi^{n+1}$ on either side of the vortex sheet, at the same physical location.

At the far field, everywhere except the downstream boundary, the disturbances are assumed to vanish.  (For relaxation solution, the values of $\phi$ at the far field is updated with the compressible vortex solution.)  At downstream, the condition

$$\phi_x = \phi_\xi \xi_x + \phi_\eta \eta_x = 1 \tag{18}$$

is applied, usually to first order accuracy, by applying the above condition one point ahead in the $\xi$ direction, using central differences for the spatial derivatives.

<h3 align="center">RESULTS AND DISCUSSION</h3>

The present algorithm has been tested with a number of well-documented cases, both in the relaxation mode ($\sigma = 0$) and the time-dependent mode ($\sigma = 1$).  The relaxation solution is carried out in circle plane similar to that used by Bauer et al. (1975), even though the sheared parabolic coordinate transformation should work just as well.  In Figure 1, the surface pressure distribution on a lifting NACA0012 airfoil at $M_\infty = 0.63$ and angle of attack $2°$ is shown and compared with an accurate solution by Steger (1977).  Very good agreement was observed in about 54 iterations on a 80 x 30 grid.  We have also tested other lifting transonic flows with shocks, past NACA0012 airfoil, and the supercritical airfoil 75-06-12 described by Bauer et al. (1975).  These results are not documented here for want of space.

The unsteady mode ($\sigma = 1$) has thus far been programmed only to treat the non-conservative form.  This mode was tested by computing the thickening and thinning circular arc airfoil problem at $M_\infty = 0.85$.  A simple sheared coordinate system, similar to that used by Goorjian (1979) was used, and for ease of comparison the very smoothly varying grid spacing used by Goorjian was used in the present work. The airfoil motion with time may be found in the above reference. In Figures 2 and 3, the pressure distribution in the streamwise direction is plotted and compared with the low-frequency small disturbance potential solution by Ballhaus and Steger (1975) and the full potential solution by Goorjian (1979). During the period $0 \leq t \leq 18.25$, when a shock wave forms and advances in the flow direction, the three solutions are in good agreement, even in predicting the shock motion. Indeed, at $t = 8.5$, the present solution is virtually indistinguishable from the solution by Goorjian (1979) except over a small portion on the airfoil. As the shock wave reverses its direction of motion the three schemes begin to deviate from each other in predicting the shock motion, but are in reasonable agreement everywhere else. A value of $\Delta t = .125$, which is an order of magnitude larger than that needed by explicit stability criteria, was used with no instability. It is hoped that a conservative form, which is obtained by programming equation (12) into the present code will improve the agreement between the present work, and existing solutions.

An additional lifting case has been calculated using the time-dependent mode of the present algorithm. The lift and moment about mid-chord are plotted as a function of time, for a NACA64A006 airfoil subjected to a step change in angle of attack equal to $1/10°$ at Mach Number 0.80 in Figures 4 and 5. The starting solution was the steady solution at zero angle of attack. The present procedure, and the linear theory computed by Lomax et al. (1952), both show rapid variations in $C_L$ following the start. The low frequency formulation by Ballhaus and Goorjian (1977) does not simulate these variations. The three sets of results converge to somewhat different values. At asymptotic steady state, the present procedure predicts the lift and moment about mid chord to be .0233 and .0043 respectively which may be compared with the values of .0215 and .0050 predicted by the transonic airfoil program documented by Bauer et al. (1975). A sheared parabolic coordinate system with 155 nodes in the $\xi$ direction (including 100 nodes on the airfoil) and 22 nodes in the $\eta$ direction was used. The computer time per time step was .3 sec for the above grid on the CDC 7600.

## REFERENCES

Ballhaus, W. F., and Goorjian, P. M. (1977), "Implicit Finite Difference Computations of Unsteady Transonic Flows About Airfoils," AIAA Journal, Vol. 15, No. 12.

Ballhaus, W. F., and Steger, J. L. (1975), "Implicit Approximate Factorization Schemes for the Low Frequency Transonic Equation," NASA TMX-73,082.

Bauer, F., Garabedian, P., Korn, D., and Jameson, A. (1975), "Supercritical Wing Sections II," Lecture Notes in Economics and Mathematical Systems, 108, Springer-Verlag.

Chipman, R., and Jameson, A. (1979), "Fully Conservative Numerical Solutions for Unsteady Transonic Flow about Airfoils," AIAA 12th Fluid and Plasma Dynamics Conference, Williamsburg, Virginia.

Goorjian, P. M. (1979), "Computations of Unsteady Transonic Flow Governed by The Conservative Full Potential Equation Using an Alternating Direction Implicit Algorithm," NASA CR 152274.

Isogai, K. (1977), "Calculation of Unsteady Transonic Flow Over Oscillating Airfoils Using the Full Potential Equation," AIAA Dynamics Specialists Conference, San Diego, California.

Lomax, H., Heaslet, M. A., Fuller, F. B., and Sluder, L. (1952), "Two- and Three-Dimensional Unsteady Problems in High Speed Flight," NACA Report 1077.

Sampath, S. (1977), "A Numerical Study of Incompressible Viscous Flow Around Airfoils," Ph.D. Dissertation, Georgia Institute of Technology, Atlanta, Georgia.

Steger, J. L. (1977), "Implicit Finite Difference Simulation of Flow About Arbitrary Geometries with Application to Airfoils," AIAA 10th Fluid and Plasma-dynamics Conference, Alburquerque, New Mexico.

Stone, H. L. (1969), "Iterative Solution of Implicit Approximations of Multi-Dimensional Partial Differential Equations," SIAM Journal of Numerical Analysis, Vol. 5, No. 3.

Traci, R. M., Albano, E. D., and Farr, Jr., J. L. (1975), "Small Disturbance Transonic Flows about Oscillating Airfoils and Planar Wings," AFFDL-TR-75-100.

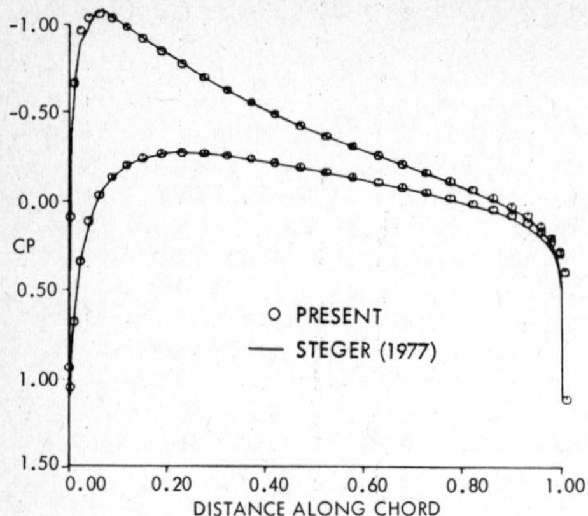

Figure 1. Surface Pressure Distribution Over a Lifting Airfoil (NACA 0012, $M_\infty = 0.63$, Angle of Attack 2°)

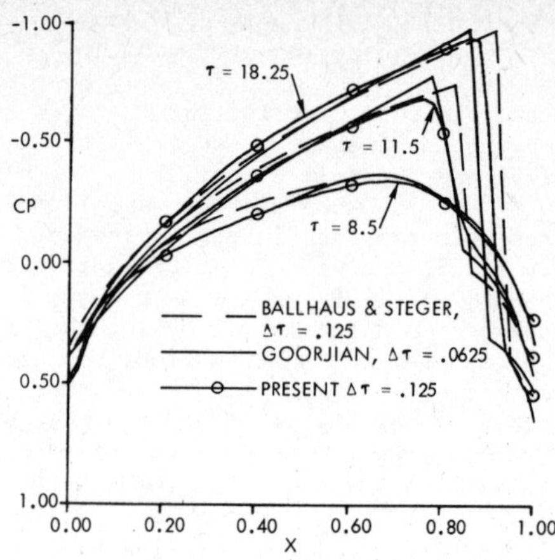

Figure 2. Surface Pressure Distribution Over Thickening Circular Arc Airfoil

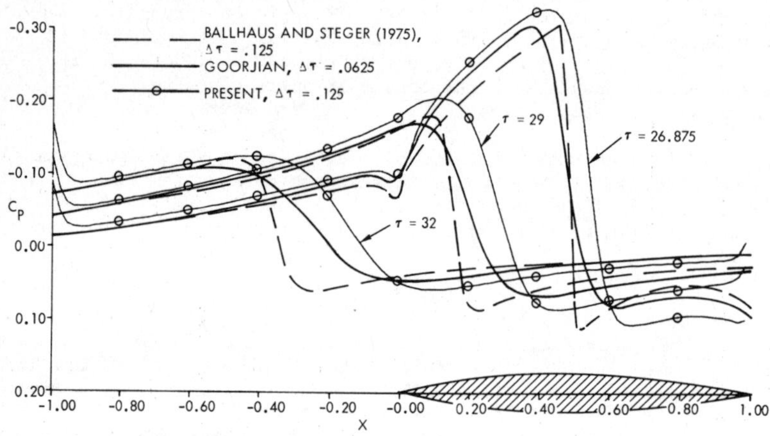

Figure 3. Chordwise Pressure Distribution During Thinning Motion

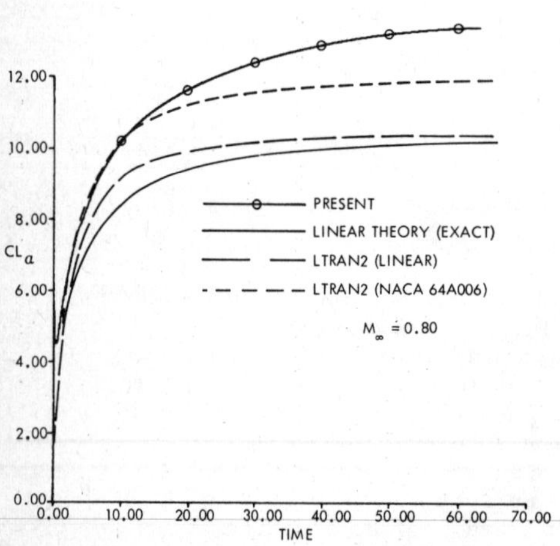

Figure 4. Lift History Following Step Change in Angle of Attack ($\Delta\alpha = .1°$)

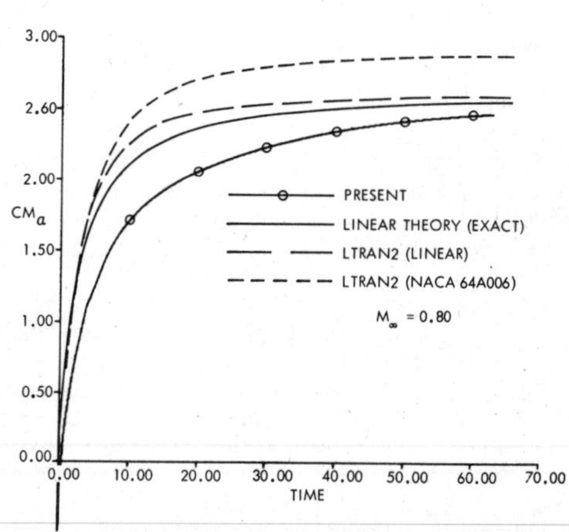

Figure 5. Moment History Following Step Change in Angle of Attack ($\Delta\alpha = .1°$)

NUMERICAL ANALYSIS OF THE ASYMPTOTIC
FLOW BEHAVIOR ABOUT THE EDGE OF A
ROTATING DISK

V. P. Shidlovsky

Computing Center, Academy of Sciences
of the USSR
40 Vavilova Str., Moscow B-333, USSR

## 1. Mathematical Formulation

The flow of incompressible viscous fluid induced by a steadily rotating disk was first studied by Karman (1921). Karman's solution was done for full Navier-Stokes system, but concerned only with idealized formulation for a disk of infinite radius. Detailed analysis of the viscous flow structure near the edge of the disk of finite radius was carried out by Shidlovsky (1977). It was found that this particular kind of flow has much in common with "triple-deck" pattern in vicinity of trailing edge of a flat plate, of which the study was initiated by Stewartson (1969) and Messiter (1970). However, the flow about the edge of a disk reveals some specific features, the most interesting of which are connected with three-dimensional character of that flow.

As it was shown by present author in the paper cited, upon the introduction of a small parameter

$$\varepsilon = \mathrm{Re}^{-1/2} = \sqrt{\nu/(\omega R^2)} \tag{1.1}$$

where $\nu$ is viscosity, $R$ - disk's radius, and $\omega$ - angular velocity of its rotation, one can single out a certain region around the disk's edge with characteristic dimension in radial direction being of the order of $\varepsilon^{6/7}$, and that in axial direction - of the order of $\varepsilon^{9/7}$. The governing equations for this region are of boundary-layer type, namely

$$\frac{\partial \psi}{\partial z} \frac{\partial^2 \psi}{\partial r \partial z} - \frac{\partial \psi}{\partial r} \frac{\partial^2 \psi}{\partial z^2} = -\Phi(0) D'''(r) + \frac{\partial^3 \psi}{\partial z^3},$$

$$\frac{\partial \psi}{\partial z} \frac{\partial h}{\partial r} - \frac{\partial \psi}{\partial r} \frac{\partial h}{\partial z} = \frac{\partial^2 h}{\partial z^2}, \quad \frac{\partial p}{\partial z} = 0, \quad p = \Phi(0) D''(r) \tag{1.2}$$

where all variables are specifically scaled, and where $\psi$ is flow function defining axial and radial components of fluid velocity, h - circular component of the same velocity, p - pressure, D(r) - special function contained also in boundary conditions, and $\Phi(0)$ - constant defined by Karman's solution ($\Phi(0) = -.05395$). Note that the origin of the coordinate system used here is shifted to the edge of a disk.

Boundary conditions for equation (1.2) are those of non-slip at the surface, symmetry and zero shear at Z = 0 away from the disk, and asymptotic matching with "outer" solutions both in radial and axial directions. We have

$$\psi = \partial\psi/\partial z = h = 0 \quad \text{at} \quad z = 0, \; r < 0,$$

$$\psi = \partial^2\psi/\partial z^2 = \partial h/\partial z = 0 \quad \text{at} \quad z = 0, \; r > 0,$$

$$\psi \to a z^2/2, \quad h \to b z \quad \text{at} \quad r \to -\infty,$$

$$\psi \to (a/2)[z + D(r)]^2, \quad h \to b[z + D(r)] \quad \text{at} \; z \to \infty, \quad (1.3)$$

$$\psi \to r^{2/3} f_0(\eta), \quad h \to r^{1/3} g_0(\eta) \quad \text{at} \quad r \to \infty$$

$$(\eta = z/r^{1/3})$$

where a and b  are constants obtained from Karman's solution ( a = 0.510, b = -0.616), while $f_o(\eta)$ u $g_o(\eta)$ are functions obtained from a solution for peripheral wake [Leslie (1972)].

Obviously, equations (1.2) are of parabolic type and they might describe some planar boundary layer with additional transversal flow of velocity h(r,Z). The main  difficulty in numerical solution of (1.2) lies in the fact that the function D(r) (and, consequently, pressure gradient) is not known in advance. The situation is thus similar to that of a flat plate's triple-deck, but for the fact that here the relationship between pressure and D(r) is of a differential, rather than of an integral nature.

## 2. Computational Procedure

To obtain a numerical solution for equations (1.2) with boundary
conditions (1.3) it is necessary to take into account a) parabolic
character of the mathematical problem, and b) inevitability of iterational
cycle because of unknown term with $D'''(r)$ in equation (1.3). Fortunately,
we were able to rely on the experience of Jobe and Burggraf (1974), who
gave sufficiently detailed description of a numerical procedure designed
to solve very similar problem of a flat plate's trailing edge.

Just as in the case of Jobe and Burggraf (1974), very important
role is played by an initial guess for $D(r)$ function and its derivatives.
To make this guess adequate, we made use of asymptotic expressions derived
in Shidlovsky (1977). It is worth to note that inappropriate choice of
initial $D(r)$ might lead to a divergence and consequent breakdown of
computational procedure.

For the sake of brevity we shall not write down here rather complicated
analytical expression for initial $D(r)$. It is sufficient to mention that
at large positive $r$ this should tend to

$$D_{r \to \infty} = 1.1165 \, r^{1/3} \tag{2.1}$$

while at large negative $r$ it was to match another asympotic formula

$$D_{r \to -\infty} = C_1 r^{-2/3} \tag{2.2}$$

with undetermined constant $C_1$ which, in its turn, was found by way
of gaining a smothest possible link between two asymptotic expressions.

To reduce difficulties in solving two-side boundary value problem,
at each position in radial direction we have used a linearized version
of the first equation from (1.2), representing this in a form

$$U^* \frac{\partial u}{\partial r} + V^* \frac{\partial u}{\partial z} = A(r) + \frac{\partial^2 u}{\partial z^2} \tag{2.3}$$

with functions $A(r)$, $U^*(r,Z)$ and $V^*(r.Z)$ considered as known. For an

initial iteration $U^*$ and $V^*$ were taken from a preceeding position in r,

while each of the next iterations consisted of putting the result of

previous one instead of old $U^*$ and $V^*$. Here

$$V^* = -\int_0^z \frac{\partial u}{\partial r}\, dz \tag{2.4}$$

The linearity of equation (2.3) permits us to use double sweep

procedure to satisfy boundary conditions in axial direction, which in

this notation are

$$u(r,0)=0 \text{ at } r<0, \quad \partial u(r,0)/\partial z=0 \text{ at } r>0, \quad \partial u(r,Z_k)/\partial z=a \tag{2.5}$$

where $Z_k$ is sufficiently large value of $Z$ (our final results were

obtained with $Z_k = 20$). Having obtained the value $U_k = U(r, Z_k)$ it is

possible to find a new value of $D(r)$:

$$D(r) = U_k/a - Z_k \tag{2.6}$$

which is to replace a previous value at the same position r. After

passing through all positions in radial direction we are able to  determine,

by means of numerical differentiation, new $A(r)$ and $P(r)$, and to carry

out another iteration. The criterium of convergence chosen here, was

$$\left| D_{new}(r) - D_{old}(r) \right| \leq 10^{-4} \text{ for all } r \tag{2.7}$$

After the iteration on $D(r)$ converge, it is possible to obtain a numerical

solution for the second equation from (1.2), which results in $h(r,z)$.

It is to be noted that in constructing a numerical procedure special

care was taken to avoid an appearance of discontinuities in $D(r)$ at $r = 0$,

with the initial guess for this function being here, once again, of primary

importance.

### 3.  Numerical Results

We have used here an implicit numerical scheme with step sizes equal to 0.02 both in radial and axial directions.  Larger step size in r was possible, but the main limitation was connected with the subsequent necessity of numerical stability of a numerical scheme.  Internal iteration loop (on U* and V*) was adjusted on fixed number of iterations, for, as it was confirmed by numerous checks, 3 iterations are sufficient for all cases of practical interest.  External iteration loop was controlled by a condition (2.7). The position of outer limiting values of arguments was determined after several trials, and final results were obtained with

$$z_k = 20, \; r_{-k} = -6, \; r_k = 8$$

The plot of function D(r) is given at Fig. 1.  It might be seen that the slope of the curve is discontinuous at r = 0.  Disturbance velocity profiles above the surface are given at Figs. 2 and 3.  The Figs. 4 and 5 show the behavior of two velocity components away from the disk, in its plane. Fig. 6 shows the plot of surface shear stress $\tau_{z\theta}$ against radial coordinate near disk's edge.  The variation of pressure might be seen from Fig. 7. The discontinuity in pressure, revealed here, can not be avoided by means of simple adjustment of constants; to resolve this singularity it is necessary to study yet another asymptotic region with the dimensions of the order of $\varepsilon^{3/2}$ in both directions (reasoning similar to that of Messiter (1970)).

The general gorm of the expression for the coefficient of the moment of viscous friction forces is ($r^o$ - dimensional coordinate)

$$C_m = -\frac{2}{\rho\pi\omega^2 R^5} \int_0^R 2\pi r^{o2} \tau_{z\theta}\, dr^o =$$

$$= -B\varepsilon - 4\varepsilon^{13/7} \int_{-\infty}^{0} \left[ \left(\frac{\partial h}{\partial z}\right)_w - B \right] dr$$

Using the results of present calculation one obtains

$$C_m = 0.616\, Re^{-1/2} + 1.428\, Re^{-13/14} \qquad (3.1)$$

As is seen from (3.1), second term of this expression corresponds to the correction for $C_m$ connected with the radius of disk being finite.  It is interesting to evaluate the relative importance of this correction:

$$\Delta C_m / (C_m)_{R\to\infty} = 2.318\, Re^{-3/7} \qquad (3.2)$$

## Acknowledgment

The important part of the calculations reported here was made at the University of Illinois at Urbana-Champaign, the program being run on the computer CDC Cyber - 175.  The author uses this opportunity to thank Professor H. H. Hilton for a possibility to carry out the research at this particular institution, Professor S. M. Yen for helpful discussions, and Mr. D. R. Hall for his help in writing and running the computer program.

REFERENCES


Jobe, C. E., Burggraf,O. R., 1974.  The Numerical Solution of the Asymptotic
Equations of Trailing Edge Flow.  Proc. of Roy. Soc. of London, Ser. A, Vol.
340, No. 1620.

Karman, T. (1921).  Ueber laminare und Turbulente Reibung.  ZAMM, Bd. 1, H. 4.

Leslie, L. M. (1972).  The Wake of a Finite Rotating Disc. Journ. Austral.
Math. Soc., Vol. 13, pt. 3.

Messiter, A. F. (1970).  Boundary Layer Flow Near the Trailing Edge of a
Flat Plate.  SIAM Journ. Appl. Math., Vol. 18, No. 1.

Shidlovsky, V. P. (1977).  The Structure of Viscous Fluid Flow Near the Edge of
a Rotating Disk.  PMM:  Apl. Math. & Mech., Vol. 41, No. 3 (in Russian).

Stewartson, K. (1969).  On the Flow Near the Trailing Edge of a Flat Plate II.
Mathematika, Vol. 16, pt. 1.

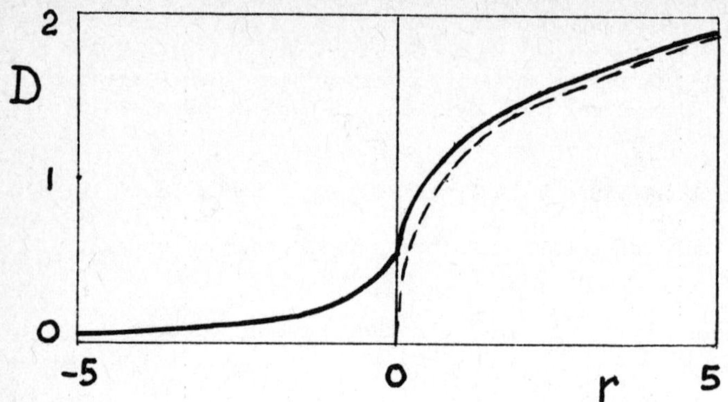

Fig.1. Plot of D(r) function. --- Eq. (2.1)

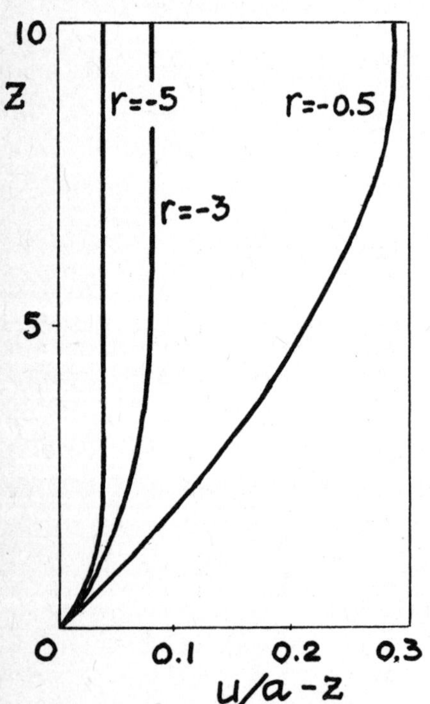



Fig.2. Disturbance of radial
velocity.

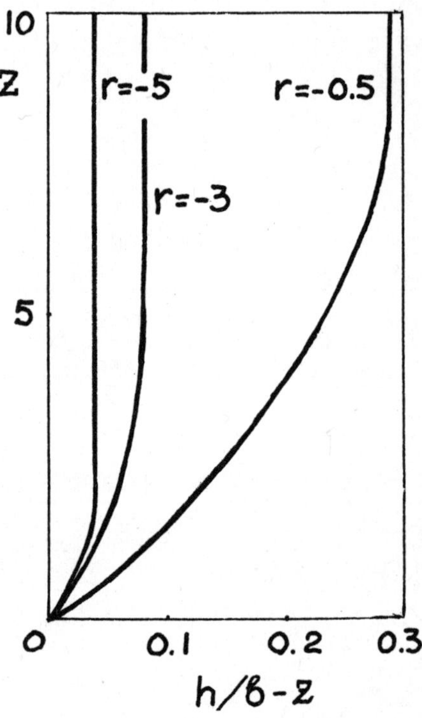



Fig.3. Disturbance of circular
velocity.

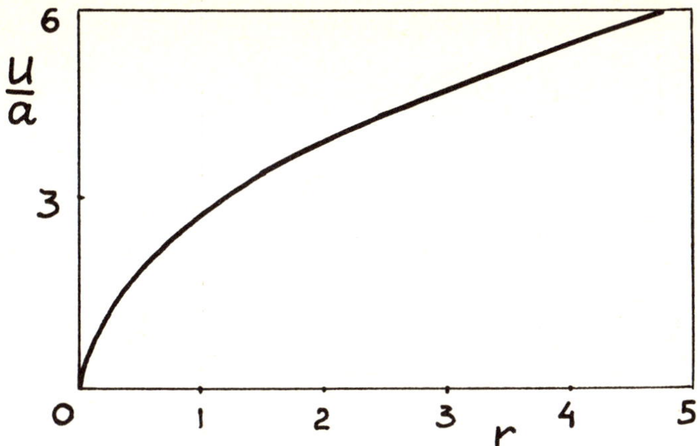

Fig.4. Radial velocity in the peripheral
region, at z=0.

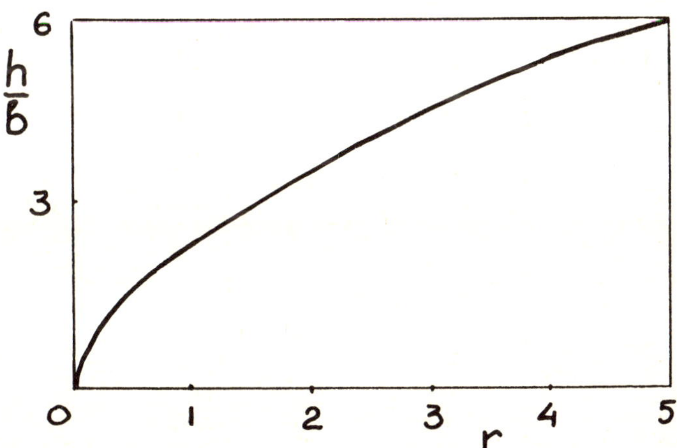

Fig.5. Circular velocity in the peripheral
region, at z=0.

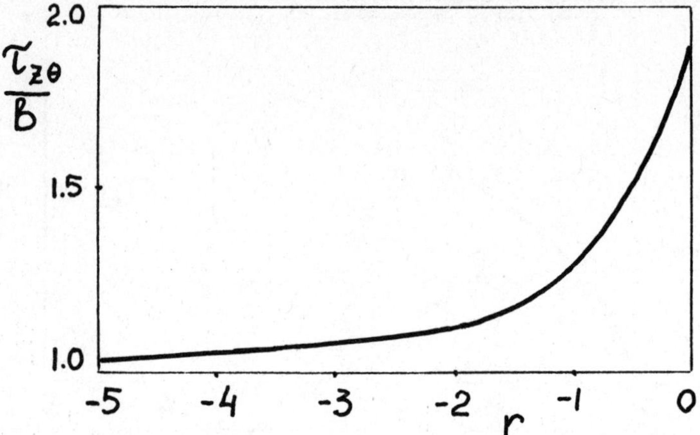

Fig.6. Plot of surface shear stress.

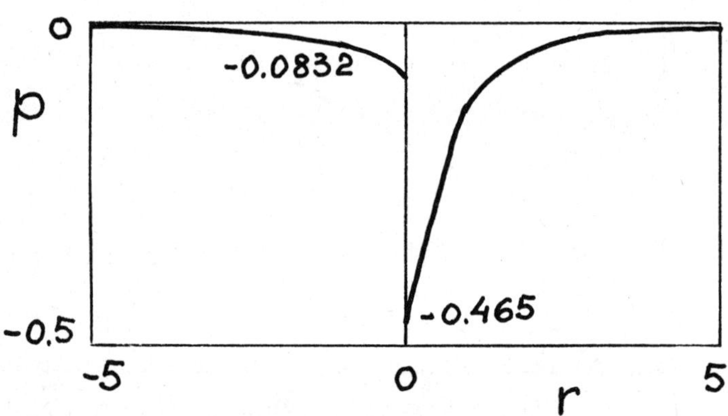

Fig.7. Pressure distribution.

# ANALYSIS OF CONSERVATIVE PROPERTIES OF THE DIFFERENCE SCHEMES BY THE METHOD OF DIFFERENTIAL APPROXIMATION


Yurii Shokin

Insitute of Theoretical and Applied
Mechanics, Novosibirsk, U.S.S.R.


1.Computational experience shows that the important properties which a difference scheme for the equations of gas dynamics must possess are those of being conservative and completely conservative ( Samarskii and Popov , 1969 , 1975). Here we will establish a connection between these properties and those of first differential approximations of difference schemes (Shokin,1979).

Consider the following family of difference schemes:

$$\Delta_0 u^n(x) + \ae \Delta_1 p^{\sigma_1}(x-h/2) = 0; \qquad (1)$$

$$\Delta_0 v^n(x+h/2) - \ae \Delta_1 u^{\sigma_2}(x) = 0; \qquad (2)$$

$$\Delta_0 \varepsilon^n(x+h/2) + \ae p^{\sigma_3}(x+h/2)\Delta_1 u^{\sigma_4}(x) = 0, \qquad (3)$$

which approximate the system of equations of gas dynamics in Lagrangian coordinates:

$$u_t + p_x = 0; \qquad (4)$$

$$v_t - u_x = 0; \qquad (5)$$

$$\varepsilon_t + p u_x = 0. \qquad (6)$$

Here $u$ is the velosity, $v$ the specific volume, $\varepsilon$ the specific internal energy, $p = p(\varepsilon, v)$ the pressure; $f^{\sigma} = \sigma f^{n+1} + (1-\sigma) f^n$, $0 \leq \sigma \leq 1$, $\ae = \tau/h$, $\Delta_0 = T_0 - E$, $\Delta_1 = T_1 - E$, where $T_0$, $T_1$ are the shift operators with respect to $t$ and $x$, respectively, and $E$ is the identity operator.

The first differential approximation of the difference

schemes (1)-(3) has the form

$$u_t + P_x = -\tau(0.5-\sigma_1)(a^2 u_x)_x = \bar{Q}_1 ; \qquad (7)$$

$$v_t - u_x = \tau(0.5-\sigma_2)P_{xx} = \bar{Q}_2 ; \qquad (8)$$

$$\mathcal{E}_t + pu_x = -\tau(0.5-d_3)a^2 u_x^2 - \tau(0.5-\sigma_4)pp_{xx} = \bar{Q}_3 ; \qquad (9)$$

$$a^2 = pp_\mathcal{E} - p_\gamma .$$

On multiplying the system (7)-(9) by the matrix

$$T = \begin{pmatrix} 1 & 0 & 1 \\ 0 & 1 & 0 \\ u & 0 & 1 \end{pmatrix},$$

we obtain the following system of equations:

$$\mathcal{W}_t + F_x = Q , \qquad (10)$$

where

$$\mathcal{W} = \begin{pmatrix} u \\ v \\ g \end{pmatrix} , \quad F = \begin{pmatrix} P \\ -u \\ up \end{pmatrix} , \quad g = \mathcal{E} + u^2/2 ,$$

$$Q = \begin{pmatrix} -\tau(0.5-d_1)(a^2 u_x)_x \\ \tau(0.5-\sigma_2)P_{xx} \\ -\tau(0.5-\sigma_1)u(a^2 u_x)_x - \tau(0.5-d_3)a^2 u_x^2 - \tau(0.5-\sigma_4)pp_{xx} \end{pmatrix}$$

The system (10) can be obtained by another method. Multiplying (1) by $u^{0.5}(x)$ and adding the result to (3), we get

$$\Delta_0[\mathcal{E}^n(x+h/2) + 0.5(u^n(x))^2] = -\mathfrak{X}\Delta_1 p^{\sigma_1}(x-h/2)u^{0.5}(x) +$$

$$+ \mathfrak{X}[(\sigma_4 - 0.5)p^{\sigma_1}(x+h/2)\Delta_0\Delta_1 u^n(x) -$$

$$- (d_1 - \sigma_3)\Delta_0 p^n(x+h/2)\Delta_1 u^{\sigma_4}(x)] . \qquad (11)$$

Equation (11), a consequence of the difference scheme (1)-
(3), approximates the equation for the total energy. Writing
out the first differential approximation of the difference
scheme (1), (2), (11), which approximates the system of equations

$$\mathcal{W}_t + F_x = 0 , \qquad (12)$$

we obtain (10).

THEOREM 1. In order for the difference scheme (1)-(3) to
be completely conservative, it is necessary and sufficient
that it possess property $K$ (Shokin, 1979) and the system (10)

of equations of the first differential approximation be a divergence system.

C o r o l l a r y. When the conditions of Theorem1 are satisfied, a difference scheme of the form (1)-(3) belongs to class $M$ (Shokin, 1979).

For the system of equations (12) consider the family of conservative difference schemes

$$\Delta_0 u^n(x) + \varkappa \Delta_1 p^{\sigma_1}(x-h/2) = 0,$$

$$\Delta_0 v^n(x+h/2) - \varkappa \Delta_1 u^{\sigma_2}(x) = 0,$$

$$\Delta_0 [\varepsilon^n(x+h/2) + (u^n(x))^2/4 + (u^n(x+h))^2/4] +$$

$$+ \varkappa \Delta_1 [(p^{\sigma_3}(x+h/2) + p^{\sigma_3}(x-h/2)) u^{\sigma_4}(x)/2] = 0 .$$

$$(13)$$

THEOREM 2. Scheme (13) is completely conservative if and only if it belongs to the class $K \cap M$ .

The families (1)-(3) and (13) of difference schemes are equivalent under the condition of being completely conservative (Samarskii and Popov, 1975).

THEOREM 3. Provided that the condition of being completely conservative is satisfied, the families (1)-(3) and (13) belong to class $I$ (i.e., they are invariant under the group of transformations admitted by the equations of gas dynamics (Shokin, 1979)).

If we introduce an artificial viscosity into a difference scheme of the form (13) or (1)-(3) in such a way that it appears as an additive component to the pressure $p$ and does not depend on $x$ , $t$ , $u$ , then when $\sigma_1 = \sigma_3$ , $\sigma_2 = \sigma_4 = 0.5$ we obtain a scheme with the property of being completely conservative and belonging to the class $I \cap K \cap M$.

The order of approximation of the schemes considered above is $O(\tau + h^2)$. When $\sigma_i = 0.5$ , $i = 1, 2, 3, 4$, we obtain a scheme having second order approximation.

2. Let us consider the class of difference schemes approximating the system of equations (4)-(6) with order $O(\tau + h)$

$$\Delta_0 u^n(x) + \varkappa \Delta_1 p^{\sigma_1}(x-h) = 0,$$

$$\Delta_0 v^n(x) - \varkappa \Delta_1 u^{\sigma_2}(x) = 0,$$

$$\Delta_0 \varepsilon^n(x) + \varkappa p^{\sigma_3}(x) \Delta_1 u^{\sigma_4}(x) = 0.$$

$$(14)$$

The first differential approximation of scheme (14) has the form

$$\overline{w}_t + \overline{A}\,\overline{w}_x = \overline{Q} + \overline{R}\,, \qquad (15)$$

where $\overline{Q}$ is defined by formulas (7)–(9) and

$$\overline{R} = \begin{pmatrix} 0.5\,h\,p_{xx} \\ 0.5\,h\,u_{xx} \\ -0.5\,h\,p\,u_{xx} \end{pmatrix}.$$

Multiplying system (15) by the matrix $T$, we obtain the system of equations

$$w_t + F_x = Q + R\,, \qquad (16)$$

where

$$R = \begin{pmatrix} 0.5\,h\,p_{xx} \\ 0.5\,h\,u_{xx} \\ 0.5\,h\,[\,(u p_x)_x - (p u_x)_x\,] \end{pmatrix},$$

THEOREM 4. For the difference scheme (14) to be completely conservative, it is necessary and sufficient that it possess property $K$ and that the first differential approximation system (16) be of divergence form.


REFERENCES

Popov Ju.P.,and Samarskii A.A.(1969). Z.Vycisl. Mat.i Mat.Fiz. 9,953.
Samarskii A.A.,and Popov Ju.P.(1975).Difference schemes of gas dynamics,"Nauka",Moscow.
Shokin Yu.I.(1979).The method of differential approximation, "Nauka",Novosibirsk.

A GENERALIZED HYBRID RANDOM CHOICE METHOD WITH
APPLICATION TO INTERNAL COMBUSTION ENGINES


G. ANDREW SOD

Princeton University

Department of Mechanical and Aerospace Engineering

Princeton, New Jersey 08540

and

Lawrence Berkeley Laboratory

University of California

Berkeley, California 94720


## INTRODUCTION

Examples were presented [Sod (1978)] in the Sixth International Conference on
Numerical Methods in Fluid Dynamics obtained by the first version of this method.
The method consisted of solving the two-dimensional equations of gas dynamics in an
axially symmetric geometry in the interior of the computational domain.  In this region (away from boundaries) the flow is considered inviscid.

Viscous effects were introduced at the boundary (where the flow is assumed incompressible) with a grid-free method, when the computational elements are pieces of
vortex sheets [Chorin (1978)] and vortex sleeves [Sod (1980 a-d)].  This method had
serious shortcomings;  the boundary layer equations break down in the region of
separation and despite the number of interactions between the vortex sheets and vortex sleeves being small, the number of sheets and sleeves continually grows.  As a
result, in the course of a long computation, the storage and computing time requirements become great.

A method is introduced that brings the effect of viscosity from the boundary
into the interior in a natural manner in such a way that most of the vortex sheets
and vortex sleeves are destroyed.  The computational elements are circular vortex
filaments [Sod (1980 b), (1980 d)].

## OUTLINE OF THE METHOD FOR THE INTERIOR

The two-dimensional equations for an inviscid, non-heat conducting flow in cylindrical coordinates with axial symmetry, and zero azimuthal velocity can be written
in the form

$$\partial_t \underline{U} + \partial_r \underline{F}(\underline{U}) + \partial_z \underline{G}(\underline{U}) = -\frac{1}{r} \underline{W}(\underline{U}), \tag{1}$$

where

$$\underline{U} = (\rho, \rho u_r, \rho u_z, e)^T, \quad \underline{F}(\underline{U}) = (\rho u_r, \rho u_r^2 + p, \rho u_r u_z, (e+p)u_r)^T,$$

$$\underline{G}(\underline{U}) = (\rho u_z, \rho u_r u_z, \rho u_z^2 + p, (e+p)u_z)^T, \text{ and } \underline{W}(\underline{U}) = (\rho u_r, \rho u_r^2, \rho u_r u_z, (e+p)U_r)^T,$$

where $\rho$ is the density, $u_r$ is the velocity in the radial direction, $u_z$ is the velocity in the axial direction, p is the pressure, e is the energy per unit volume, t is time, and r is the radial distance from the axis of symmetry.

We shall assume that the flow is isothermal, that is, c = constant, say $\sqrt{c_0}$. This leads to the equation of state

$$p = c_0 . \tag{2}$$

The first step is to use operator splitting to remove the inhomogeneous term $-\frac{1}{r} \underline{W}(\underline{U})$ from the system (1). The resulting set of equations takes the form

$$\partial_t(\underline{U}) + \partial_r f(\underline{U}) + \partial_z G(\underline{U}) = 0, \tag{3a}$$

$$d_t\underline{U} = -\frac{1}{r} \underline{W}(\underline{U}). \tag{3b}$$

The system of equations (3a) represents the two-dimensional equations of gas dynamics in Cartesian coordinates written in conservation form.

The method used to solve system (3a) is the random choice method introduced by Glimm (1965) and developed for hydrodynamics by Chorin (1976). Operator splitting is used again to reduce the system of two-dimensional equations to two sets of one-dimensional equations. Details of the method are given in the paper by Chorin (1976) and Sod (1980 b-c) and (1981).

Once system (3a) is solved, the system of ordinary differential equations (3b) is solved, where the solution of system (3a) is used to determine the inhomogeneous term $-\frac{1}{r} \underline{W}$ in (3b). We approximate equation (3b) by a modified version of the Cauchy-Euler scheme. Randomness, in keeping with the spirit of the Glimm scheme, is introduced.

## BOUNDARY LAYER TREATMENT

The boundary layer equations for axially symmetric cylinder (retaining the correct derivatives in r) may be written in the form

$$\partial_t\xi + (\underline{u}\cdot\underline{\nabla})\xi = \nu\partial_r^2\xi + \frac{\nu}{r} \partial_r\xi , \tag{4a}$$

$$\partial_z(ru_z) + \partial_r(ru_r) = 0 , \tag{4b}$$

$$\xi = -\partial_r u_z , \tag{4c}$$

where $\underline{u} = (u_z,u_r)$ with $u_z$ tangent to the boundary and $u_r$ normal to the boundary, $\xi$ is the vorticity, and $\nu$ is the viscosity. The boundary conditions are

$$\underline{u}(r_0,z) = 0 ,$$

and

$$\underline{u}(\infty,z) = U_\infty(z) .$$

Using equations (4b) and (4c) the velocity field can be computed once the vorticity field is known. The method used is one introduced by Chorin (1978) for the boundary layer equations for a flat plate in two-dimensions and extended to the boundary layer equations for a cylinder by the author (1980 b–d).  Consider a collection of N segments $S_i$ of vortex sheets (or vortex sleeves when viewed in a three-dimensional cylindrical geometry) with intensities $\xi_i$, i = 1, ... ,N.  The $S_i$ are straight line segments parallel to the z-axis having length h and center $(r_i, z_i)$.  The difference in $u_z$ across the segment $S_i$ is $\xi_i$.

The boundary layer equations are solved by moving the segments of vortex sheets.  The advection component of the system is solved deterministically where the sheets $S_i$ are advanced in time according to the velocities $v_{r_i}$ and $v_{z_i}$ of these sheets.  The diffusion of the sheets in the r-direction is treated by a random walk.  Finally, the radial term is treated by moving the sheets along the characteristic of a hyperbolic equation.

A detailed discussion of this method of solution as well as a discussion of the boundary conditions and vorticity creation may be found in Chorin (1973) and (1978) and in Sod (1980 b–d) and (1981).

COUPLING OF THE INTERIOR AND BOUNDARY LAYER METHODS

Finally we must describe how the two methods are pieced together.  The method used to couple the boundary layer calculation to the interior calculation is more complicated than originally considered by Chorin (1978).

It is essential that the boundary layer act on the interior (unless the boundary layer does not separate).  In this method the edge of the calculation is not the boundary of the domain, but rather the edge of the boundary layer.

In the problem considered below the boundary is substantially more complicated than a simple cylinder.  We shall allow for boundaries which consist of pieces of cylinders and flat plates (normal to cylinder).  The boundary layer equations for a flat plate are used for these regions normal to the cylinders.

A boundary layer calculation for the entire boundary is made every two quarter steps each with time step $\frac{\Delta t}{2}$ .  The tangential component of the velocity at the edge of the interior calculation (edge of the boundary layer) is used as the velocity at infinity for the boundary layer calculation.  For the boundary layer calculation the velocity at infinity is considered to be piecewise constant over the same intervals as in the interior calculation.  This is important since the values of r and z in the interior calculation may not be the same as the spacing h of the boundary points for the boundary layer calculation.

A method is introduced that brings the effect of the viscosity from the boundary into the interior in a more natural manner in such a way that most of the vortex sheets are not retained.  For complete details see Sod (1980 b–d) and (1981).

Consider the Navier-Stokes equations for a viscous, incompressible fluid in an axially symmetric geometry in vorticity transport form

$$\partial_t \xi + (\underline{u} \cdot \underline{\nabla})\xi = \nu\Delta\xi + \frac{\nu}{r}\partial_r\xi - \frac{\nu}{r^2}\xi,$$

$$\partial_r(ru_r) + \partial_z(ru_z) = 0,$$

$$\xi = \partial_z u_r - \partial_r u_z$$

We can define the stream function $\psi$ such that

$$u_r = \frac{1}{r}\partial_z\psi \text{ and } u_z = \frac{1}{r}\partial_r\psi.$$

Consider a collection of N circular vortex filaments in $\mathbb{R}^3$ of intensity $\xi_j$ which lie in a plane normal to the z-axis. This represents a point vortex in an axially-symmetric geometry in $\mathbb{R}^2$.

Under these conditions, we have for the j-th vortex filament with center $\underline{r}_j = (r_j, z_j)$

$$\psi_0(\underline{r} - \underline{r}_j) = -\frac{\xi_j(r - r_j)r_j}{8\pi}\int_0^\infty e^{-k|z-z_j|} J_1(k(r - r_j))J_0(kr_j)\,dk$$

where $J_0$ and $J_1$ are Bessel functions of the first kind of order zero and one respectively.

The vorticity field is given by

$$\xi(\underline{r}) = \sum_{j=1}^{N} \xi_j\xi_0(\underline{r} - \underline{r}_j)$$

where
$$\xi_0(\underline{r}) = \begin{cases} \dfrac{1}{4\pi\sigma||\underline{r}||} , & ||\underline{r}|| < \sigma, \\[2ex] 0 , & ||\underline{r}|| \geq \sigma, \end{cases}$$

and $\sigma$ is a cut-off which is taken to be $h/2\pi$. It is seen that $\xi_0$ is a smooth approximation to be Dirac delta function.

The vortex blob $\xi_0(\underline{r})$ generates its own stream function $\psi(\underline{r})$ which may be obtained by solving

$$\Delta\psi_0 - \frac{1}{r}\partial_r\psi = -r\xi_0.$$

Since $\sigma$ is small we see that for $||\underline{r} - \underline{r}_j|| < \sigma$, r is near $r_j$ (the r-coordinate of the j-th vortex blob which is the radius of the j-th vortex filament in $\mathbb{R}^3$) as a result we replace $\frac{1}{r}$ in (6) with $\frac{1}{r_j}$ . This gives a new relationship between the stream function and the vorticity

$$\Delta\psi_0 = r_j \, \omega_0 \quad ,$$

which has solution

$$\psi_0(\underline{r} - \underline{r}_j) = \frac{r_j}{4\pi\sigma} \, ||\underline{r} - \underline{r}_j|| \quad .$$

Combining this with (7) gives the stream function $\psi_0(\underline{r} - \underline{r}_j)$ generated by (8). Furthermore, it is possible to generate a velocity field $\underline{v}(\underline{r} - \underline{r}_j)$ induced by this vorticity field,

$$\underline{v}(\underline{r}) = \sum_{j=1}^{N} \xi_j \underline{v}_0(\underline{r} - \underline{r}_j) \quad ,$$

where $v_0(\underline{r} - \underline{r}_j)$ is the velocity field at $\underline{r}$ induced by a single vortex blob located at $\underline{r}_j$. By differentiating the stream function $\psi_0(\underline{r} - \underline{r}_j)$ we obtain

$$\underline{v}_0(\underline{r} - \underline{r}_j) = \begin{cases} (-\frac{1}{r_j} \partial_z \psi_0(\underline{r} - \underline{r}_j), \frac{1}{r_j} \partial_r \psi_0(\underline{r} - \underline{r}_j)), & ||\underline{r} - \underline{r}_j|| < \sigma, \\[2ex] (-\frac{1}{r} \partial_z \psi_0(\underline{r} - \underline{r}_j), \frac{1}{r} \partial_r \psi_0(\underline{r} - \underline{r}_j)), & ||\underline{r} - \underline{r}_j|| \geq \sigma. \end{cases}$$

We are now in a position to describe the coupling of the boundary layer calculation to the interior calculation using the vortex filaments and the way in which the vortex sheets are removed.

The vortex sheets and vortex filaments are very similar in structure, that is, they are determined by theirposition and intensity. The circulation around a vortex sheet of intensity $\xi$ is $\xi h$ where h is the length of the sheet. The intensity of the vortex blob must be $\xi h$, which gives a transition from vortex sheets to the vortex filaments. When a vortex sheet is converted to a vortex filament, its intensity $\xi$ is multiplied by h, the length of the sheet.

Even in separated flow there is a thin viscous sublayer near the boundary where the boundary layer equations remain valid. This is an essential fact. For this allows us to use the vortex sheets near the boundary. The vortex sheets will satisfy the boundary conditions much more accurately than the vortex filaments. See Chorin, Hughes, McCracken, and Marsden (1978).

Let L be a length chosen in such a way that $P[|n|>L] < \epsilon$ , where $\epsilon$ is some small number. Using Tscebysheff's theorem, we see that L will be a multiple of the standard deviation $\sqrt{2\nu\Delta t}$ . Any vortex sheet which is greater than a distance 2L from the boundary will be converted to a vortex filament by multiplying the intensity of the sheet by its lengths.

The velocity field induced by these vortex filaments on the interior grid is then computed using (8) where the integrals are computed using Gauss-Laguerre quadrature. This velocity field is added to the velocity field obtained by the interior calculation. Once this is done the vortex filaments are destroyed. This leaves only a small number of vortex sheets at any given time.

This approach is consistent with what happens physically for a slightly viscous flow.  Far away from the boundaries the flow may be considered inviscid.  The vorticity brought into the interior using the vortex filaments is advected with the flow through the interior method.

## APPLICATION TO CUP-IN-PISTON CHAMBER

A brief description of the remaining features of the numerical method as they relate to engine modeling will, for reasons of brevity, be void of detail.  Most realistic engines geometries are such that some of their boundaries (such as the valve seat and valve face) are oblique  to the (r,z)-coordinate system.  The method has been modified to treat geometries in which boundaries are oblique to the mesh.  Details are given in Sod (1980 e).

The computational grid used in the interior is a random moving grid (in the spirit of Glimm's method).  This provides excellent resolution as the piston approaches top dead center on the compression or exhaust stroke.  Details are given in Sod (1980 b-c).

Some current engine designs use the cup-in-piston type of chamber arrangement to help promote mixing through a process called "squish".  This is often combined with air induction-induced swirl.

The method described here has been used to study the flow in a motored cup-in-piston engine chamber during two complete four-stroke cycles.  The chamber is assumed to axially symmetric, with a single intake-exhaust valve located on the chamber axis. The crankshaft-driven piston and camshaft-actuated valve were simulated in the model. Complete details as well as results may be found in Sod (1980 e)

## REFERENCES


Chorin, A. J., 1973, Numerical Study of Slightly Viscous Flow, J. Fluid Mech. 57, 785.
Chorin, A. J., 1976, Random Choice Solution of Hyperbolic Systems, J. Comp. Phys. 22, 517.
Chorin, A. J., 1978, Vortex Sheet Approximation of Boundary Layers, J. Comp. Phys., 27, 428.
Chorin, A. J., Hughes, T. J. R., McCracken, M. T., and Marsden, J. R. (1978), Product Formulas and Numerical Algorithms, Comm. Pure Appl. Math., 31, 205.
Glimm, J., 1965, Solution in the Large for Nonlinear Hyperbolic Systems of Equations, Comm. Pure Appl. Math. 18, 697.
Sod, G. A., 1977, A Numerical Study of a Converging Cylindrical Shock, J. Fluid Mech. 83, 785.
Sod, G. A., 1978, Proceedings Sixth Internal Conference on Numerical Methods in Fluid Dynamics, Lecture Notes in Physics, Vol. 90.
Sod, G. A., 1980a, Automotive Engine Modeling with a Hybrid Random Choice Method, II, Society of Automotive Engineers Paper 800288.
Sod, G. A., 1980b, A Numerical Method for Slightly Viscous Axisymmetric Flows with Application to Internal Combustion Engines, J. Comp. Phys. (to appear).
Sod, G. A., 1980c, Computational Fluid Dynamics with Stochastic Techniques, Proceedings von Karman Institute for Fluid Dynamics Lecture Series on Computational Fluid Dynamics.
Sod, G. A., 1980d, A Random Approximation to a Boundary Layer Exhibiting Separation and Reattachment (to appear).
Sod, G. A., 1980e, A Numerical Model of a Motored Cup-In-Piston Type Engine Chamber (to appear).
Sod, G. A., 1981, Numerical Methods in Fluid Dynamics, Academic Press.

COMPUTATIONAL MODELS FOR CONVECTIVE MOTIONS
INDUCED AT FLUID INTERFACES

M. Strani, R. Piva

University of Rome, Italy

1. Introduction.

   The interface between two immiscible fluids can influence the motion of both
fluid bulks, through its geometrical configuration and the variation of surface
tension along it.  Surface tension gradients, which arise at the interface between
two fluid phases, in presence of heat and mass diffusion, act as a shear stress at
the surface itself, generating particular convective flows, besides the ones direct
ly established by temperature or concentration gradients.

   Surface tension driven flows become important in comparison with buoyancy driven
flows, for small values of the Bond number, defined as the ratio of gravitational to
surface tension forces.  Therefore these flows are significant or may even become
predominant for very small characteristic dimensions of the flow field, or for reduced gravity conditions.  Technologies based on liquid films and space processes are
among the more interesting engineering applications for which the surface tension can
not be neglected in flow fields investigations.

   Due to the intrinsic difficulties of experimental investigation of these flows,
the computational models are, at the moment, an irreplaceble tool for a deeper com
prehension and a quantitative evaluation of the described phenomena.  Considering
that the investigated motion is generated at the interface and then it propagates
in the fluid bulk, the application of appropriate boundary conditions at the interface plays a central role in the mathematical description.  Some numerical models,
appeared in the literature in the last few years, do not seem to simulate with the
required accuracy the boundary conditions at the fluid interface.  Simplified inter
face balance equations have been proposed by several authors, for instance Levich,
Krilov and Ostrach, and used as boundary conditions for analytical or numerical
solutions of the bulk equations.  The assumption of a fixed rectilinear configuration
[Chang, Wilcox (1976)] or the discrete F.D. approximation in Cartesian coordinates
of the interface [Daly (1969)], altough in the frame of sophisticated computational
procedures, seem to be not adequate for an accurate simulation of the physical pheno
mena.  In particular, to satisfy the required accuracy, it is necessary to let the
curvilinear interfacial boundary to be coincident with a portion of a coordinate sur
face.  A general formulation of such boundary conditions in vectorial form and in
orthogonal curvilinear coordinates has been recently presented by Bedeaux, Albano,
Mazur (1976) and Napolitano (1978).

   For the study of flow fields contained in enclosures, as we find in the applications,
non-orthogonal curvilinear coordinates may better fit all the boundaries, in particu
lar in the region near the contact angle between the fluid interface and the solid
walls, which has a primary influence on the overall flow.  Hence, non-orthogonal cur
vilinear systems are more suitable for accurate numerical solutions, but they require
a complete re-formulation of the balance equations at the interface in general tensor
form.

   In particular in section 2 we define the coordinate systems and related metrics
utilized in the mathematical formulation and we briefly recall the bulk conservation
equations written in tensorial form.  The expression of the interface balance equations for mass, momentum and energy — which are used as boundary conditions for the
fluid bulks — are specified in section 3 together with the constitutive relations for
the surface quantities.  In section 4 from the analysis of the complete set of equations, in its non dimensional form, we make few considerations about the order of
magnitude of the various terms, deducing some acceptable approximations for the nume-

rical solutions. We also describe the computational procedure, in particular for the treatment of the boundary conditions, and some numerical applications to show the capabilities of the proposed model.

## 2. Coordinate systems and bulk equations.

The steady state motion of two immiscible fluids separated by a surface $S$ of arbitrary shape is considered and referred to a three dimensional space $S_3$. Let $<x^k>$ be a system of orthogonal Cartesian coordinates defined in $S_3$. As usually assumed, latin indexes range from 1 to 3, while greek indexes range from 1 to 2. Assuming the surface $S$ sufficiently smooth, we may consider in $S_3$ a system of generalized coordinates $<\xi^k>$ in which the surface $S$ is a coordinate surface $\xi^3 = C$. The coordinates $\xi^k$ are related to the Cartesian coordinates $x^k$ by the known functions $x^k = \hat{x}^k(\xi^k)$, $\xi^k = \hat{\xi}^k(x^k)$. The contravariant metric tensor components $g^{ik}$ (or covariant $g_{ik}$) and the covariant base vectors $\vec{g}_i$ (or contravariant $\vec{g}^i$) define the metric in the physical space $S_3$.

On the surface $S$ we consider a generalized coordinate system $<\xi^\alpha>$, whose metric is defined, in the two dimensional subspace $S_2$, by the base vectors $\vec{\gamma}_\alpha$ (or $\vec{\gamma}^\alpha$) and by the components $\gamma^{\alpha\beta}$ (or $\gamma_{\alpha\beta}$) of the metric tensor. The steady state bulk conservation equations of mass, momentum and energy for fluids (+) and (−), respectively in the regions $\xi^3 > C$ and $\xi^3 < C$, are given in non dimensional form, by

$$(1) \qquad \left(\rho\, u^k\right)^\pm_{/k} = 0$$

$$(2) \qquad \left(\phi^{hk}\right)^\pm_{/k} = \left(\rho\, u^h u^k - \sigma^{hk} - \frac{R_\sigma}{R^2}\, B\, \psi\right)_{/k} = 0$$

$$(3) \qquad \left(\phi^k\right)^\pm_{/k} = \left(\rho\, u^k\, (U + \frac{E}{2}\, u^h u_h) + q^k - E u_h \sigma^{hk}\right)^\pm_{/k} = 0$$

where $\rho$ is the density, $\psi$ the potential of the gravity force, $U$ the internal energy, $u^k$ ($u_k$) the contravariant (covariant) velocity components, while the thermal flux $q^k$ and the stress tensor $\sigma^{hk}$ are, respectively, given by the constitutive relations

$$(4) \qquad q^k = -\frac{1}{RP}\, g^{hk}\, T_{,k}$$

$$(5) \qquad \sigma^{hk} = \left(-p + \frac{1}{R}\left(\frac{\mu_v}{\mu} - \frac{2}{3}\right) u^n_{/n}\right) g^{hk} + \frac{1}{R}\left(u_{n/m} + u_{m/n}\right) g^{hn}\, g^{km}$$

$R$, $P$, $E$, $B$ are respectively the Reynolds, Prandtl, Eckert and Bond non dimensional groups. $R_\sigma$ is the surface tension Reynolds number defined as the ratio of surface tension to viscous forces. For each bulk fluid the thermodynamic variables $U$, $p$, $T$ and $\rho$ are connected by two state equations, while $\mu$, $\mu_v$, $\lambda$ have been assumed constant.

## 3. Balance equations at the interface.

From a macroscopic point of view, the interface separating the two fluids (+) and (−) may be considered as a surface of discontinuity of mechanical and thermal variables, characterized by the surface tensors $\tilde{\phi}^\alpha$ and $\tilde{\phi}^{\alpha\beta}$, which are the equivalents, in the two-dimensional subspace $S_2$, of the tensors $\phi^k$ and $\phi^{hk}$ in equations (2) and (3). In the frame of a discontinuous model of the interface and under the hypotheses of not permeable interface and zero mass surface density $\tilde{\rho}$, the following interface balance equations in non dimensional form are obtained [Strani, Piva 1979]. The surface mass balance gives

$$(6) \qquad (u^3)^+ = (u^3)^- = v^3 = 0$$

where $v^k$ are the surface velocity components.  The surface momentum balance is
expressed by the equations

$$(7) \qquad \left[\frac{\sigma^{33}}{g^{33}}\right]_-^+ + \frac{R_\sigma}{R^2}\, \tilde{\sigma}^{\beta\alpha}\, b_{\beta\alpha} = 0$$

$$(8) \qquad \left[\frac{\sigma_\alpha^3}{\sqrt{g^{33}}}\right]_-^+ + \frac{R_\sigma}{R^2}\, \tilde{\sigma}^\beta_{\alpha//\beta} = 0$$

where the symbol $\left[\ \ \right]_-^+$ stays for the discontinuity jump, across the surface $\xi^3 = C$,
of the quantities inside the square parenthesis, $(\ )_{//\alpha}$ means covariant differentia
tion with respect to $\xi^\alpha$ over the surface $\xi^3 = C$, $b_{\beta\alpha}$ are the curvature tensor com
ponents and $\tilde{\sigma}^\beta_\alpha$ is the surface stress tensor.

The surface energy balance gives the equation

$$(9) \qquad \frac{1}{\sqrt{g^{33}}}\left[q^3 - R\,P\,E\,\sigma_\alpha^3\left(u^\alpha - v^\alpha\right)\right]_-^+ + \left(\tilde{q}^\alpha + R\,P\,\tilde{U}\,v^\alpha\right)_{//\alpha} - \frac{P\,R_\sigma\,E}{R}\,\tilde{\sigma}^{\delta\alpha}\,v_{\delta//\alpha} = 0$$

with $\tilde{q}^\alpha$ thermal surface flux and $\tilde{U}$ internal energy surface density.

In the hypothesis of linear relations between thermodynamic forces and fluxes,
the following set of constitutive relations in non dimensional form may be derived,
from the expression of surface entropy production [Strani, Piva (1979)], by assuming zero the Onsager Casimir cross coefficients

$$(10) \qquad \tilde{\sigma}^\alpha_\beta = \left(\sigma + \frac{R}{R_\sigma}\left(\mu_{sv} - \mu_s\right)v^\delta_{//\delta}\right)\delta^\alpha_\beta + \frac{R}{R_\sigma}\left(v^\alpha_{//\beta} + v_{\beta//}^{\ \ \alpha}\right)$$

$$(11) \qquad \tilde{q}_\alpha = -\lambda_s\,\theta_{,\alpha}$$

$$(12) \qquad \frac{1}{2\sqrt{g^{33}}}\left(\left(\sigma_\alpha^3\right)^+ + \left(\sigma_\alpha^3\right)^-\right) = \beta\left(\left(u_\alpha\right)^+ - \left(u_\alpha\right)^-\right)$$

$$(13) \qquad \frac{1}{\sqrt{g^{33}}}\left(\left(\sigma_\alpha^3\right)^- - \left(\sigma_\alpha^3\right)^+\right) = \alpha_{33}\left(v_\alpha - \frac{1}{2}\left(\left(u_\alpha\right)^+ + \left(u_\alpha\right)^-\right)\right)$$

$$(14) \qquad \frac{1}{2\sqrt{g^{33}}}\left(\left(q^3\right)^+ + \left(q^3\right)^-\right) = -L_{22}\left(\frac{1}{T^- + T_c} - \frac{1}{T^+ + T_c}\right)$$

$$(15) \qquad \frac{1}{\sqrt{g^{33}}}\left(\left(q^3\right)^+ - \left(q^3\right)^-\right) = -L_{33}\left(\frac{1}{\theta + T_c} - \frac{1}{2}\left(\frac{1}{T^+ + T_c} + \frac{1}{T^- + T_c}\right)\right)$$

where $\mu_s$, $\mu_{sv}$ and $\lambda_s$ are the non dimensional surface diffusivity coefficients, $\theta$
is the surface temperature and $T_c$ the non dimensional reference temperature.

## 4. Computational procedure and numerical applications.

From the analysis of the complete set of equations described in sections 2 and
3, we may easily verify that the number of boundary conditions is sufficient to
solve the system of differential equations.  In fact after substitution of the constitutive relations (10, 11), the balance equations (6, 8) plus a combination of the
constitutive relations (12, 13) are the boundary conditions at $\xi^3 = C$ for the six

velocity components $(u^i)^+$ and $(u^i)^-$, while the balance equation (9) plus a combination of the constitutive relations (14 , 15) are the boundary conditions at $\xi^3 = C$ for the temperatures $(T)^+$ and $(T)^-$ of the two fluids. The remaining surface balance equation (7) determines the unknown curvature tensor $b_{\beta\alpha}$ of the surface S , and with the appropriate boundary conditions at the contact angle, its geometrical shape $x^k = \hat{x}^k_s(\xi^1,\xi^2)$ in the Cartesian coordinate system $\langle x^k \rangle$ . As a consequence, all the metric coefficients of space and surface coordinate systems $\langle \xi^k \rangle$ and $\langle \xi^\alpha \rangle$ , appearing in the field and boundary equations, may be determined from the known functions $\hat{x}^k_s$ .

The complexity of the resulting set of equations requires the adoption of numerical integration procedures, if we want to maintain the main features of the physical model without undergoing to drastic approximations. The illustrated mathematical formulation in generalized curvilinear coordinates is particularly appropriate for finite difference methods, being all the field boundaries coincident with coordinate surfaces as required for accurate solutions.

The computational model has been applied here, without any loss of generality, to a simple physical test case, sketched in fig. 1, based on the following hypotheses:
- two dimensional plane field ($\xi_1 = $ const, latin indexes assume values 2 and 3, greek indexes the value 2);
- negligeable gravity forces, that is very small Bond numbers;
- negligeable values of $\sigma^3_2$ and $q^3$ for fluid (+) at the surface $\xi^3 = C$ , that is only fluid (-) is considered in the solution.

In particular the finite difference computational procedure, adopted for the numerical applications, is based on the following main features:
- the coordinate transformation is generated analytically by assuming

$$(16) \qquad x^2 = f(\xi^2) \qquad\qquad x^3 = H\big(f(\xi^2)\big) \cdot t(\xi^3)$$

where $f(\xi^2)$ and $t(\xi^3)$ are stretching functions of the mapping and $H(x^2)$ is the shape function of the interface;
- the contravariant velocity components $u^2$ , $u^3$, the pressure p and the temperature T are assumed as variables in the solution of the field equations (1 , 2 , 3); in the discretization scheme, these variables are localized, for the accuracy of the solution, on independent meshes, properly shifted with respect to a reference cell [Di Carlo, Piva, Guj (1980)];
- the velocity and temperature boundary conditions at the interface are given respectively for the velocity by the solution of eq.(8) and equation

$$(17) \qquad \frac{\sigma^3_2}{\sqrt{g^{33}}} = \frac{\alpha}{\gamma^{22}} \left(v^2 - u^2\right)$$

obtained by a combination of eqs. (12) and (13) in the assumed hypotheses; and for the temperature by the solution of eq.(9) and equation

$$(18) \qquad \frac{q^3}{\sqrt{g^{33}}} = L\left(\frac{1}{\theta + T_c} - \frac{1}{T + T_c}\right)$$

obtained by a combination of eqs. (14) and (15); the discretization of equations (8 , 17) and (9 , 18) brings to coupled tridiagonal systems respectively in the unknowns $u^2$ , $v^2$ and T , $\theta$ which are solved by a modified tridiagonal algorithm;
- the interface configuration $H(x^2)$ is determined together with the reference pressure $p_o$ , by the solution of equation (7) and related boundary conditions - on contact angles and overall volume conservation - which in the considered case may be obtained through a semi-analytical procedure;
- an iterative false-transient procedure is adopted to obtain the steady stata velocity and temperature fields; at each iteration step the geometrical configuration and the boundary conditions at the interface are obtained consistently with the approximate bulk field.

For the numerical simulation of the physical case under consideration we have assumed $R_\sigma$ coincident with R (in order to have velocity gradients normal to the interface of the same order of temperature gradients along the interface) ranging from 1 to 100 , negligeable values of E and B , constant value of $P = 1$ and values of the aspect ratio $a = h^*/l^*$ equal or lower than one.

Let consider first the influence of the interface displacement on the flow field in the bulk. At this purpose we compare the flow fields for $R = 20$ and $a = .5$ obtained: (a) with the present model (stream line plots in Fig.2) and (b) by assuming a flat undeformable and not diffusive interface. The bulk velocity and temperature vertical profiles, which are compared in Fig. 3 at the enclosure center line, show that the convective motions are largely over-extimated by the adoption of a simplified physical and computational model as the one adopted by Chang, Wilcox (1976). This results could be of relevant importance for the above mentioned tecno logical applications where, in general, such convective motions are not desired.

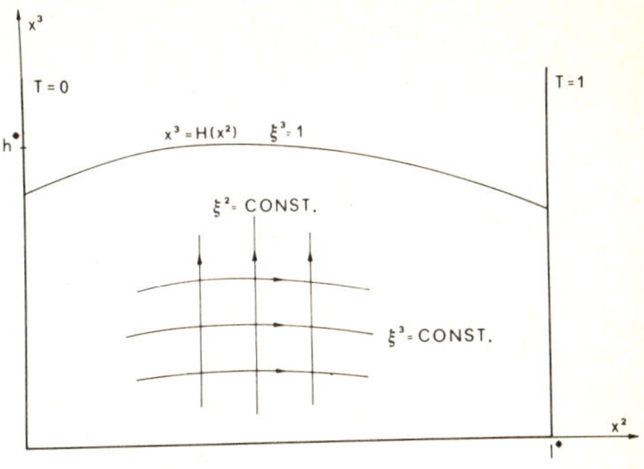



Fig. 1 - Sketch of the 2D test case.

The influence of diffusive surface phenomena, which modify the distribution of the surface temperature $\theta$ and surface velocity $v^2$ , is also significant, as shown in Fig. 4 for $R = 20$ and $a = 1$ , by comparing the bulk temperature and velocity vertical profiles for $\lambda_s = \mu_s = 0$ and $\lambda_s = \mu_s = 1$ .

For the latter case the surface temperature $\theta$ and the surface velocity $v^2$ along the interface are plotted in Fig. 5 and compared with the corresponding bulk temperature $(T)^-$ and velocity $(u^2)^-$ at the interface itself. The relative difference between the bulk and surface values give (see eqs. 17 , 18) a measure of the surface diffusive phenomena.

As expected, the vertical temperature gradients and the convective motions increase with the Reynolds number as shown in Fig. 6.

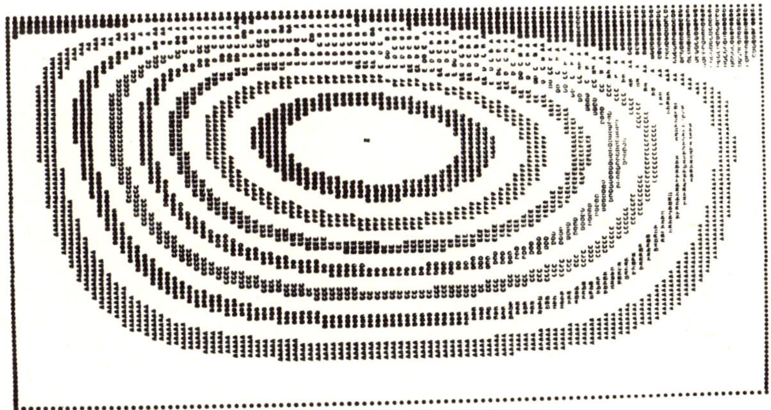

Fig. 2 -  Streamline computer plots for $R = 20$ , $a = .5$ for deformable diffusive interface.

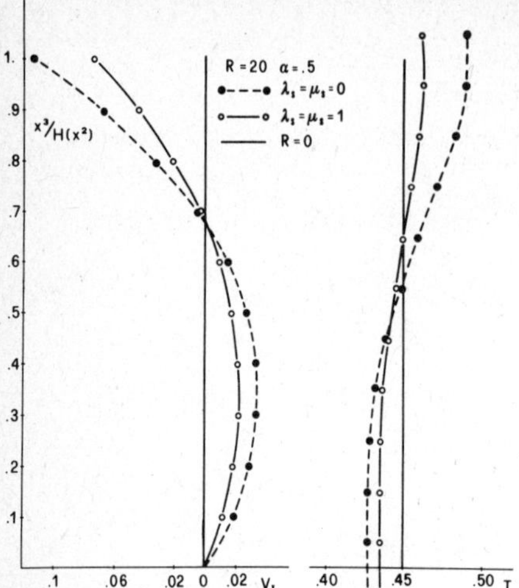

Fig. 3 - Temperature and velocity pro
files for cases (a) and (b).

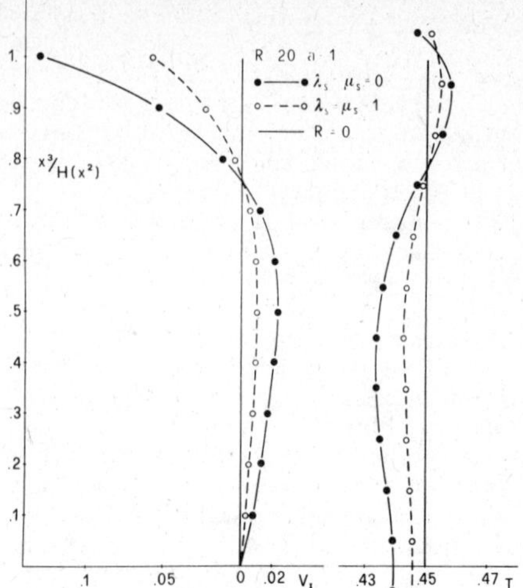

Fig. 4 - Temperature and velocity pro
files for $R = 20$ , $\lambda_s = \mu_s = \overline{1}$
and $\lambda_s = \mu_s = 0$ .

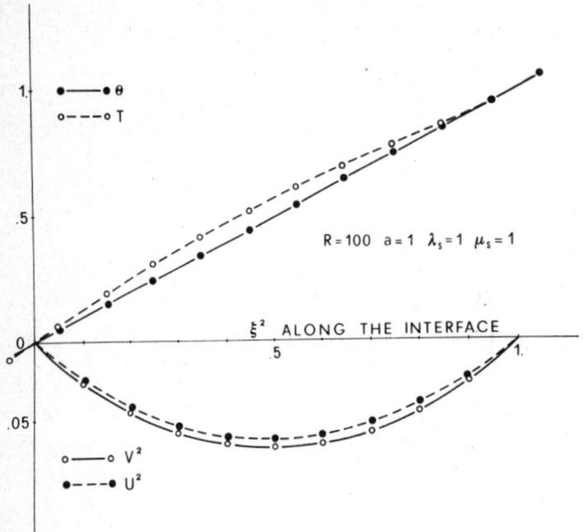

Fig. 5 - Surface and bulk temperatures
and velocities along the inter
faces.

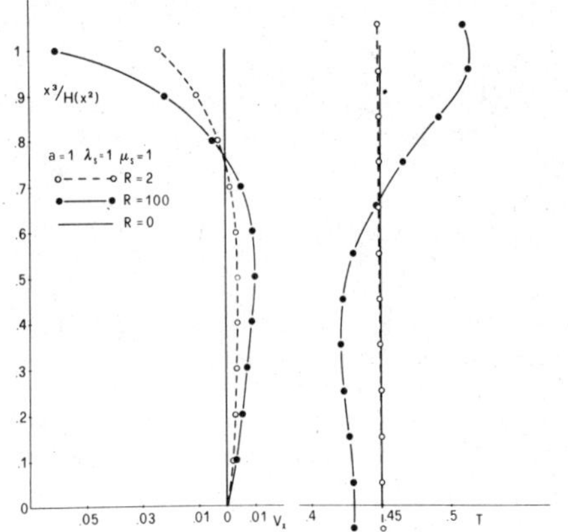

Fig. 6 - Temperature and velocity pro
files for $R = 2$ and $R = \overline{100}$.

References.


Bedeaux D., Albano A.M. and Mazur P., "Boundary Conditions and Non Equilibrium
Thermodynamics", Physica 82A, pp. 438-462, 1976.
Chang C.E. and Wilcox W.R., "Analysis of Surface Tension Driven Flow in Floating
Zone Melting", International Journal Heat & Mass Transfer, 19, pp. 355-366, 1976.
Daly B.J., "A Technique for Including Surface Tension Effects in Hydrodynamic
Calculations", J.C.P., 4, n° 1, June 1969.
Di Carlo A., Piva R. and Guj G., "Computational Schemes in General Curvilinear
Coordinates for Navier-Stokes Flows", Notes on Numerical Fluid Mechanics, vol. 2,
pp. 36-44, Viewag Verlag, Berlin, 1980.
Napolitano L.G., "Thermodynamics and Dynamics of Pure Interfaces", Acta Astronautica,
vol. 5, pp. 655-670, 1978.
Strani M. and Piva R., "A General Formulation of Boundary Conditions at Fluid
Interfaces", 1979, to be published.

FREE MOLECULAR FLOWS PAST A CONCAVE BODY

Dang Hong TIEM and Renée GATIGNOL
Laboratoire de Mécanique Théorique
Université Pierre et Marie Curie, Paris, France

In flows of a highly rarefied gas past bodies, the interactions of the molecules with the walls play a major role since the collisions between molecules are less important. In the limiting case of the free molecular flow, we can classify the problem into two cases : flows past a convex body and flows past concave bodies. In the first case there is no major difficulty ; aerodynamic forces and heat transfer have been calculated  [cf. the review paper of Schaaf and Chambe (1958)]. The second problem is more complex because the multiple reflections of the molecules on the walls (Fig. 1). Many research workers pay attention to internal flows namely flows through tubes and nozzles [Edwards (1977)]. Some external flows past concave bodies have been studied by Chahine (1961), Winberly (1968), Wuest (1975). Here we consider uniform flow past two perpendicular flat planes.

## I - FORMULATION OF THE PROBLEM

We consider uniform flow past two perpendicular flat planes that is a motionless body with a uniform temperature $T_w$ placed in an equilibrium infinite stream of velocity $U_\infty$ , density $n_\infty$ and temperature $T_\infty$. We assume that the molecules make diffuse reflection on the surface of the plates. We do not make any assumption on $U_\infty$. Body is made of two perpendicular planes of finite length and infinite width with or not an aperture at the vertex (Fig. 2).

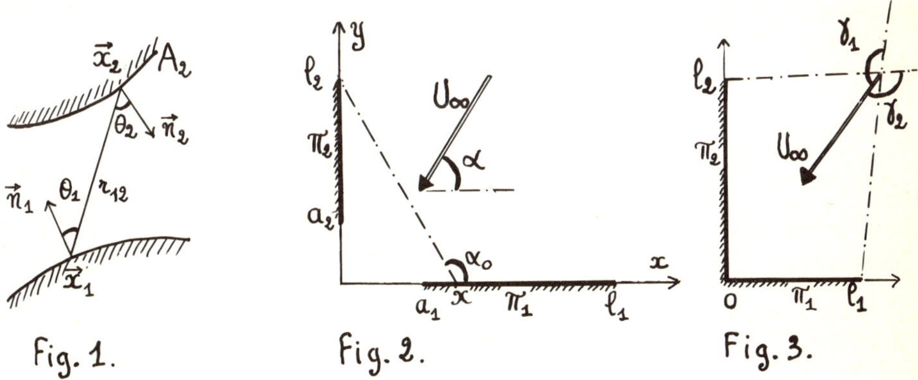

Fig. 1.     Fig. 2.     Fig. 3.

To determine the velocity distribution function of the gas molecules in the whole space, it is sufficient to know the quantities $N_1(x)$ and $N_2(y)$ which are the numbers of molecules impinging the first and the second plane respectively per unite time and per unite area. In the general case (Fig. 1) we write, with the notations of

Kogan (1969), the number of the impinging molecules is determined by the integral equation : $N(\vec{x}_1) = N(\vec{x}_1) + \int_{A_2} \dfrac{\cos_1 \cos_2}{\pi \, r_{12}^2} N(\vec{x}_2) \, dA_2$.

In our problem (Fig. 2) it is reduced to :

$$N_1(x) = N_{1\infty}(x) + \tfrac{1}{2} \int_{a_2}^{1_2} \frac{xy}{(x^2 + y^2)^{3/2}} N_2(y) \, dy$$

$$N_2(y) = N_{2\infty}(y) + \tfrac{1}{2}\int_{a_1}^{1_1} \frac{xy}{(x^2 + y^2)^{3/2}} N_1(x) \, dx \tag{2}$$

$N_{1\infty}(x)$ and $N_{2\infty}(y)$ are the numbers of incident molecules directly from infinity per unite of time and per unite area. Those are espressed by integrals and have been calculated explicitly by Tiem (1978). Here we give only the results for $a_1 = a_2 = 0$ :

$$\overline{N}_{1\infty} = v \sin \alpha \left[I(\alpha) + I(\beta)\right] + E(\alpha) - \cos \alpha_0 \, E(\beta) \tag{3}$$

where $\beta = \alpha_0 - \alpha$, $v = \sqrt{h_\infty} \, U_\infty$, $N = n_\infty (\pi h_\infty)^{-\frac{1}{2}} \overline{N}$,

$$I(\alpha) = \frac{2\alpha}{\sqrt{\pi}} + \sqrt{\pi} \; \mathrm{erf}(v \sin\alpha) + 2 \int_0^{v \sin\alpha} \mathrm{erf}(t \cot g\alpha) \, e^{-t^2} \, dt,$$

$$E(\alpha) = e^{-v^2 \sin^2 \alpha}\left[1 + \mathrm{erf}(v \cos\alpha)\right], \quad \mathrm{erf}\, u = \frac{2}{\sqrt{\pi}} \int_0^u e^{-t^2} \, dt.$$

We pay some attention to some special cases. First when $a_1$ and $a_2$ are zero and the lengths $1_1$ and $1_2$ infinite, we have a trivial exact solution. $N_{1\infty}$ and $N_{2\infty}$ are constant and $N_1 = 2/3 \, (2 N_{1\infty} + N_{2\infty})$ and $N_2 = 2/3 \, (N_{1\infty} + 2 N_{2\infty})$. Second, when the gas is at rest that is $U_\infty = 0$, the system (2) is reduced to :

$$\overline{N}_1(x) = 2 + \frac{x}{(x^2 + 1_2^2)^{\frac{1}{2}}} - \frac{x}{(x^2 + a_2^2)^{\frac{1}{2}}} + \tfrac{1}{2} \int_{a_2}^{1_2} \frac{xy}{(x^2 + y^2)^{3/2}} \, \overline{N}_2(y) \, dy \tag{4}$$

and a similar equation for $\overline{N}_2(y)$. It is easy to show that the solution is $\overline{N}_1 = \overline{N}_2 = 2$, which is equal to $N_{1\infty}$ or $N_{2\infty}$ in the case of one plane only. The gas is at rest everywhere. Tiem (1980) has made a complete analytic study for the case $a_1 = a_2 = 0$ and $1_1 = 1_2$. In this case, it is interesting to note that the two simultaneous integral equations is reduced to two uncoupled equations, one for $\overline{N}_1(x) + \overline{N}_2(x)$ and the other for $\overline{N}_1(x) - \overline{N}_2(x)$.

If $a_1$ or $a_2$ is not zero, then the kernels of the integral equations (2) have good properties (the kernels are bounded and their squares are summable), and existence and uniqueness of the solution can be proved by classical method. But when $a_1 = a_2 = 0$, the kernels have a singularity at the point $x = y = 0$. It is possible to prove rigorously that the number of incident molecules near the vertex is equal to that for $1_1$ and $1_2$ being infinite (Tiem (1978)). And also we prove the existence and uniqueness of the solutions by the aid of an iterative method : (5)

$$N_1(x) = N_{1\infty}(x) + \tfrac{1}{2}\int_0^{1_2} K(x, y) \, N_{2\infty}(y) \, dy + \frac{1}{2^2} \int_0^{1_2}\int_0^{1_2} K(x, y) \, K(y, u) \, N_{1\infty}(u) \, dy \, du + \dots$$

Thus :

$$N_1(x) \leqslant A + \tfrac{1}{2} A \int_0^\infty K(x, y) \, dy + \frac{1}{2^2} A \int_0^\infty \int_0^\infty K(x, y) K(y, u) \, dy \, du + \ldots$$

where $K(x, y) = x \, y \, (x^2 + y^2)^{-3/2}$ and $A$ is an upper bound for $N_{1\infty}(x)$ and $N_{2\infty}(y)$. The series (5) and the similar series for $N_2(y)$ are uniformly convergent and are the unique solution of the system.


## II - NUMERICAL METHOD

The computations have two purposes : the first to give the values of the quantities $N_{1\infty}$ and $N_{2\infty}$ and the second to solve the system (2) for the numbers of the incident molecules on the plates. When there is an aperture at the vertex, the system (2) is practically more complicate but theoretically simpler because the singularity at the origin is absent.

The approximate calculation of the integral $\mathfrak{J}(x) = \int_0^1 K(x, y) \, dy$ by the Simpson method gives a result with an error always equal to $10^{-2}$ for the value of $x$ equal to the abscisse of the division point nearest the origin. This error is reduced to $10^{-4}$ by an integration by parts that is :

$$2 \, \mathfrak{J}(x) = \frac{1}{x(x^2 + 1)^{\frac{1}{2}}} - \frac{1}{x} \int_0^1 \frac{y \, dy}{(x^2 + y^2)^{\frac{1}{2}}}$$

So we rewrite Eq. (2) in the form (with $a_1 = a_2 = 0$) :

$$N_1(x) = N_{1\infty}(x) + \frac{1_2^2 \, N_2(1_2)}{2x(x^2 + 1_2^2)^{\frac{1}{2}}} - \frac{1}{2 \, x} \int_0^{1_2} [N_2(y) + y \, N_2'(y)] \frac{y \, dy}{(x^2 + y^2)^{\frac{1}{2}}}$$

and we have a similar equation for $N_2(y)$. This system is solved by the method of iteration :

$$N_1^{\,0}(x) = N_{1\infty}(x)$$

$$N_1^{\,n}(x) = N_{1\infty}(x) + \frac{1_2^2 \, N_2^{n-1}(1_2)}{2x(x^2 + 1_2^2)^{\frac{1}{2}}} - \frac{1}{2 \, x} \int_0^{1_2} [\, N_2^{n-1}(y) + y \, N_2^{n-1}(y)] \, \frac{y \, dy}{(x^2 + y^2)^{\frac{1}{2}}} \qquad (7)$$

The derivatives of the right member are calculated by a Newton polynomial formula, which gives results with an error of $10^{-4}$ (even for $\sqrt{x}$) :

$$h \, f'(x) = - \frac{137}{60} f(x) + 5 \, f(x+h) - 5 \, f(x + 2h) + \frac{10}{3} f(x + 3h) - \frac{5}{4} f(x + 4h) + \frac{1}{5} f(x + 5h) + \frac{1}{6} h^6 f^{(6)}(\xi)$$

This method is well tested on the model system (4) where we know the exact solution $\overline{N}_1(x) = 2$.


## III - NUMERICAL RESULTS

The computations are made with various boundary conditions on the gas at infinity. The iterative process converges at about fifteenth iteration. In each case we obtain the physical quantities of the flow as density n, temperature T, flow velocity or stream function $\psi$. We have in the case $a_1 = a_2 = 0$ :

$$\bar{n} = I(\gamma_1) + I(\gamma_2) + \frac{1}{(\pi\,\tau)^{\frac{1}{2}}} \left\{ \int_0^{l_1} \frac{y\,N_1(t)\,dt}{(x-t)^2 + y^2} + \int_0^{l_2} \frac{x\,N_2(t)\,dt}{x^2 + (y-t)^2} \right\} \qquad (8)$$

$$\psi = [(x - l_1)^2 + y^2]^{\frac{1}{2}} [v \sin\gamma_1\, I(\gamma_1) + E(\gamma_1)] - [x^2 + (y - l_2)^2]^{\frac{1}{2}} [v \sin\gamma_2\, I(\gamma_2)$$
$$+ E(\gamma_2) + \frac{1}{2} \int_0^{l_1} \frac{(x - t)\,\bar{N}_1(t)\,dt}{[(x-t)^2 + y^2]^{\frac{1}{2}}} - \frac{1}{2} \int_0^{l_2} \frac{(y - t)\,N_2(t)\,dt}{[x^2 + (y - t)^2]^{\frac{1}{2}}} \qquad (9)$$

$$\bar{n}\,\bar{T} = \left(\frac{3}{2} + v^2\right) [I(\gamma_1) + I(\gamma_2)] + v \sin\gamma_1\, E(\gamma_1) + v \sin\gamma_2\, E(\gamma_2) - \bar{n}\,\bar{u}^2$$
$$+ \frac{3}{2} \left(\frac{\tau}{\pi}\right)^{\frac{1}{2}} \left\{ \int_0^{l_1} \frac{y\,\bar{N}_1(t)\,dt}{(x-t)^2 + y^2} + \int_0^{l_2} \frac{x\,\bar{N}_2(t)\,dt}{x^2 + (y-t)^2} \right\} \qquad (10)$$

when $n = \frac{n_\infty}{4\,(\pi)^{\frac{1}{2}}}\,\bar{n}$ , $T = \frac{2}{3}\,T_\infty\,\bar{T}$ , $\tau = T_w/T$ , and $\gamma_1$ and $\gamma_2$ are defined in Fig. 3.

The numerical results presented here concern the case $a_1 = a_2 = 0$ and $l_1 = l_2 = 1$, except in the figure 13. The behaviours of $\bar{N}_1$ as a function of $x$ are similar for different values of v. Countours of the density being constant are shown in Fig. 4. For all values of $\tau$, the gas density is the largest on the planes. Isothermal lines are drawn when $\alpha = 45°$, $v = 1$, $\tau = T_w/T = 1/10$ (Fig. 5) or $\tau = 10$ (Fig.6) The temperature of the gas is larger near the plates if $T_w > T$ or smaller if $T_w < T$ When the walls and the gas at infinity are at the same temperature ($\tau = 1$), we note the presence of a higher temperature zone (Fig. 7 and Fig. 8). The temperature distribution of the gas on the plane $\pi_1$ is shown in Fig. 9.

Streamlines do not depend on $\tau$ because $\tau$ does not come in the expression (9) Streamlines are drawn in Figs 10, 11 and 12. In each of them we note the presence of a boundary layer near the planes, in which the computations are in reality not correct. The hypersonic case (Fig. 12) and the case $v = 1$ (Fig. 11) are not very different. The hypersonic case is always with $v > 5$. Fig. 13 concerns the free molecular flow when there is an aperture at the vertex of the walls. For the symmetric flow similar to the one represented in Fig. 10, the most signifiant result is the disappearing of the stagnation point inside the gas.

We note, in conclusion, that the numerical method used for solving the integral equations (2) may be used to calculate the macroscopic quantities of free molecular flow past a body constructed with two non rectangular planes or, more generally, past any body constructed with several flat walls.

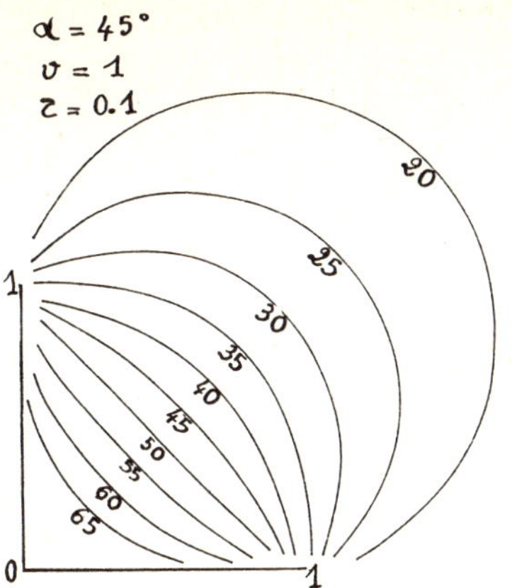


Fig. 4. Contours of density.

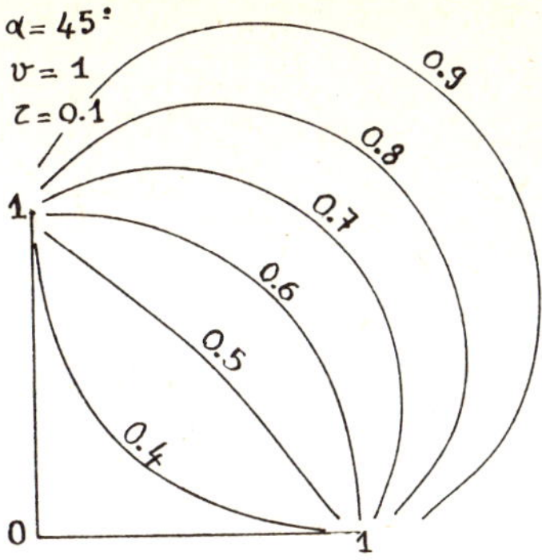


Fig. 5. Isothermal lines.

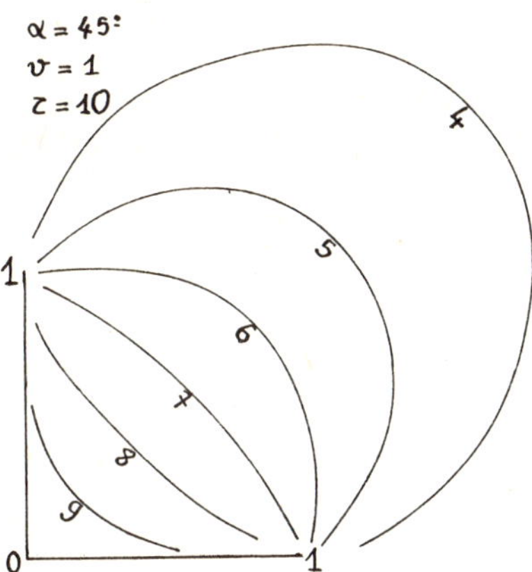


Fig. 6. Isothermal lines

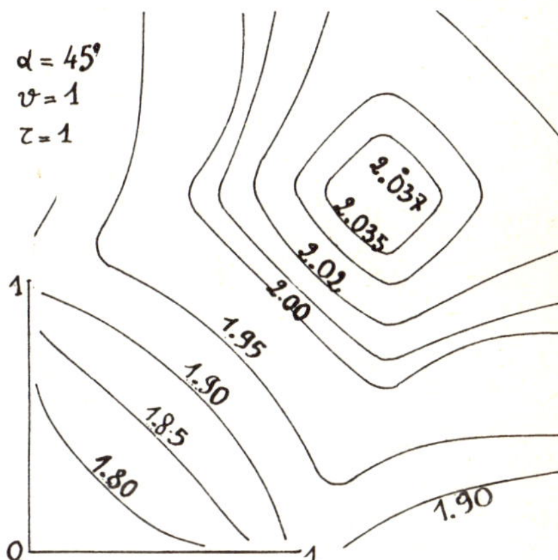


Fig. 7. Isothermal lines.

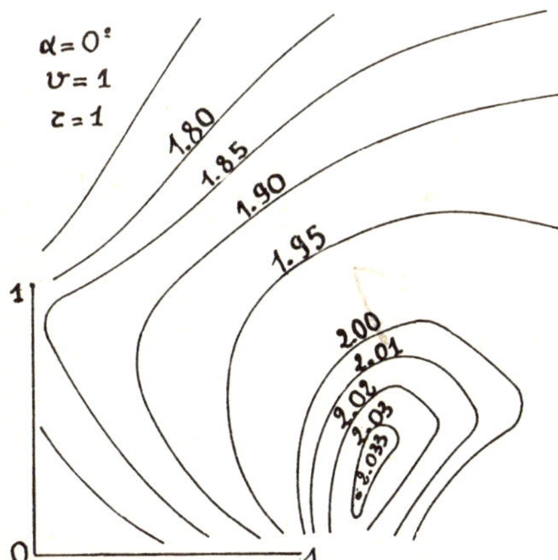


Fig. 8. Isothermal lines.

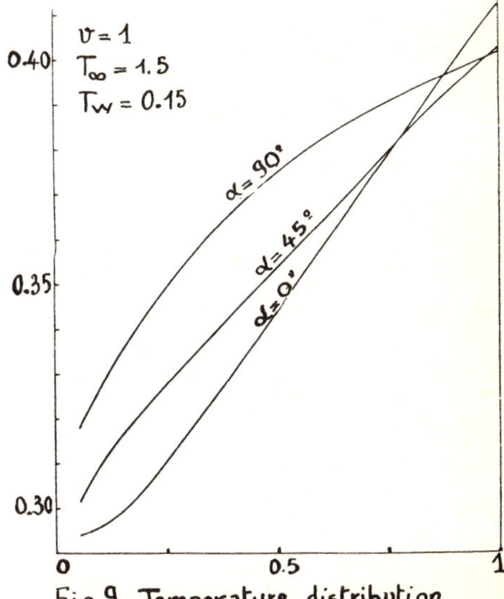


Fig. 9. Temperature distribution.

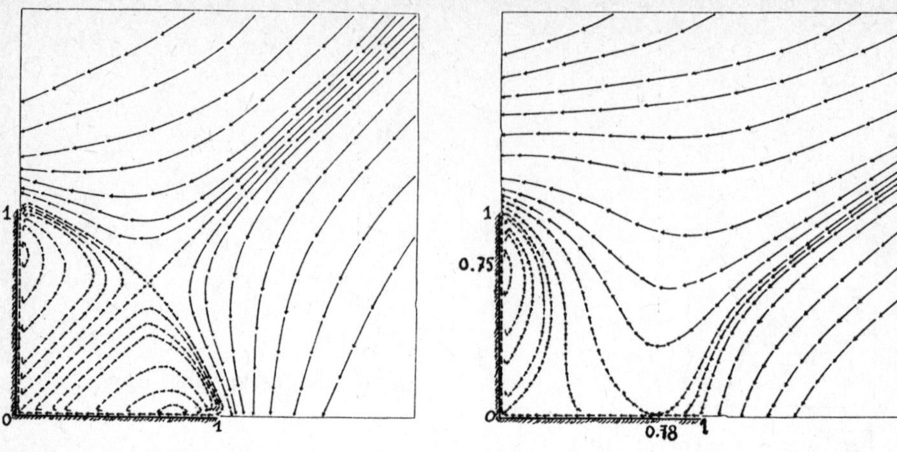

Fig. 10 _ Streamlines: $\alpha = 45°$, $\upsilon = 1$.

Fig. 11 _ Streamlines: $\alpha = 30°$, $\upsilon = 1$.

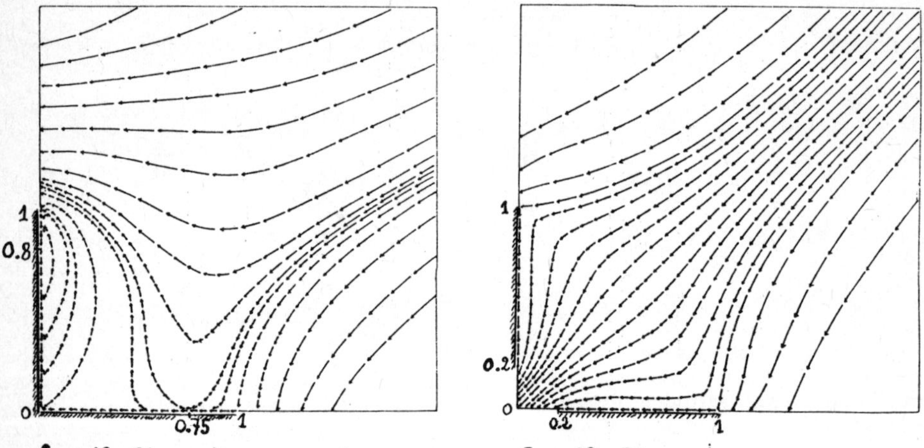

Fig. 12 _ Streamlines: $\alpha = 30°$, $\upsilon \gg 1$.

Fig. 13 _ Streamlines: $\alpha = 45°$, $\upsilon = 1$.


REFERENCES

CHAHINE, M. T., Rarefied Gas Dynamics, edited by Devienne Acad. Press, New York, 1961.

EDWARDS R. H., Rarefied Gas Dynamics, edited by J. L. Potter, Acad. Press, New York, 1977

KOGAN, M. N., Rarefied gas dynamics, Plenum Press, New York, 1969.

SCHAAF, S. A., CHAMBE , P. L., Fundamentals of Gas Dynamics, edited by H. W. Emmons, Princeton University Press, 1958.

TIEM, D. H., Thèse de 3ème cycle, Université P. et M. Curie, Paris, 1978.

TIEM, D. H., (To appear).

WINBERLY, C. R., AIAA Journal, 6, 1968, p. 734

WUEST, W., DLR - FB, Göttingen, 1975.

SUCCESSES AND SURPRISES
WITH COMPUTER-EXTENDED SERIES

Milton VAN DYKE

Stanford University, Stanford CA 94305

## Introduction

A promising alternative to purely numerical solution of flow problems is the semi-numerical technique that involves extending a perturbation series to high order by delegating the mounting arithmetic to a computer.  At Stanford we have in this way treated some 30 problems in various branches of fluid mechanics, and the technique is increasingly being used elsewhere.

Some of these solutions have been undisputed successes.  However, since the method is still under development, several erroneous conclusions have been published.  In other problems we have obtained results that we believe to be correct, but  which are surprising and hence controversial.

We first describe three clear successes of the method.  Second, as a warning we show how failure to exercise self-criticism has in two cases led to false conclusions. Third, we discuss two problems that yield surprising results not yet accepted by all our colleagues

## Some successes of the method

We illustrate the variety of flows that can be successfully treated with computer-extended series with problems in boundary-layer theory, water waves, and hypersonics. In the first case we indicate how the coefficients are analyzed and the series recast.

### The boundary layer on a parabola

Blasius (1908) showed how the laminar boundary layer can be expanded in powers of distance along the surface.  I computed (Van Dyke 1964) the six-term series for skin friction shown in Fig. 1 (where R is the Reynolds number based on nose radius).

This is a well behaved series.  The coefficients increase regularly in magnitude, showing that the radius of convergence is less than unity.  The signs alternate, indicating that the nearest singularity lies on the negative axis of $s^2$, so that it is not a physical singu-

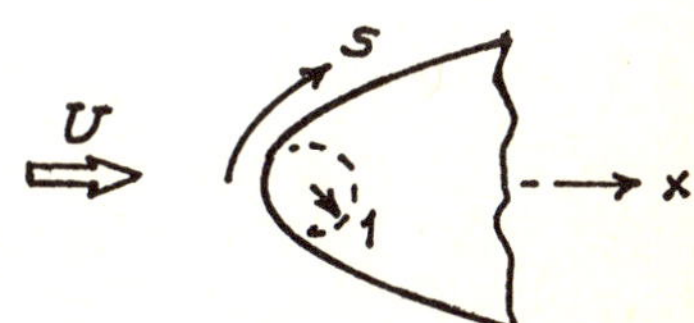


$$\tfrac{1}{2}\sqrt{R}\ c_f = 1.2326\,s - 1.9319\,s^3 + 3.1105\,s^5 - 5.0289\,s^7 + 8.1411\,s^9 - 13.187\,s^{11} + \ldots$$

FIG. 1. Blasius series for friction on parabola

larity. A graphical version of the D'Alembert ratio test, to be described presently, indicates that it has a simple pole at $s^2 = (\pi/4)^2$.

The range of convergence can be almost doubled by recasting the series in powers of the abscissa x rather than the curvilinear distance s. Then the ratio test shows a -3/2-power singularity at $x^2 = -1/2$. Finally, the convergence is extended to infinity by mapping that singularity away to infinity with an Euler transformation, recasting the series in powers of $z = 2x/(1+2x)$ as

$$\sqrt{R_x}\, c_f = 1.7431\, z - .69840\, z^2 - .15062\, z^3 - .06694\, z^4 - .03783\, z^5 - .0243\, z^6 - \ldots$$

1.0
.8
.6
.4
.2
0
√R_x c_f
1 term
2
6
numerical
0
.2
.4
.6
.8
1.0
2x/(1+2x)

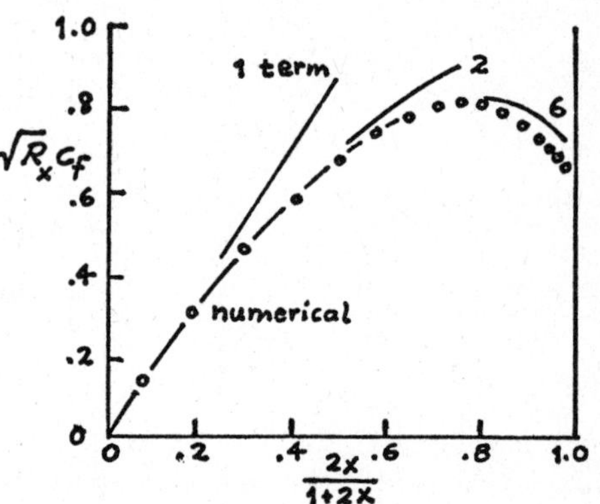

FIG. 2. Skin friction on parabola

Fig. 2 shows how the resulting distribution agrees with a numerical computation (circles) of Smith & Clutter (1963).

N. Curle has recently extended this series to 17 terms, and is trying to extract further details of the flow far downstream.

## Stokes water waves

Stokes (1880) computed the fifth approximation for plane periodic waves in deep water by expanding in powers of the first Fourier coefficient a, as indicated in Fig. 3. He conjectured that the series would converge to the highest wave, with a sharp 120-degree crest.

Schwartz (1974) extended the series to 117 terms by computer, and showed that it does not converge to the highest wave, but can be recast to do so. This solution has been completely accepted, and has inspired considerable further work.

y = a cos x + ½ a²(1+cos 2x) + ...
height = a + 3/2 a³ + 163/24 a⁵
+ 62 047/1440 a⁷ + 12 106 343/37 800 a⁹
+ ... + 5.868697 × 10³³ a⁷¹ + ...

FIG. 3. Stokes water waves

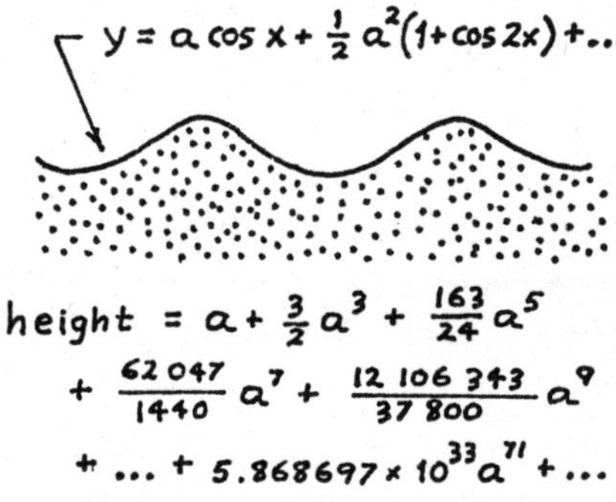

## Converging shock waves

Guderley (1942) showed that an imploding cylindrical or spherical shock wave approaches a power law as it collapses onto the center. For a spherical wave in a perfect diatomic gas, he calculated that the radius is locally proportional to the 0.717-power of time measured from the in-

stant of collapse.  However, Fujimoto & Mishkin (1978) have recently claimed that the problem can be solved analytically, and that the exponent is 0.707.

A. J. Guttmann and I have extracted Guderley's local solution from the global solution for a contracting spherical piston.  As indicated in Fig. 4, a sphere filled with cold quiescent gas begins at time $t = 0$ to contract with constant radial velocity, driving a converging spherical wave ahead of it.  We have computed 40 terms of an expansion in powers of time.

The series shown for the radius of the shock wave has fixed signs, corresponding to a singularity at the instant of collapse. From our 40 coefficients we estimate that time to the seven figures shown, and the exponent as 0.7172, in agreement with Guderley.

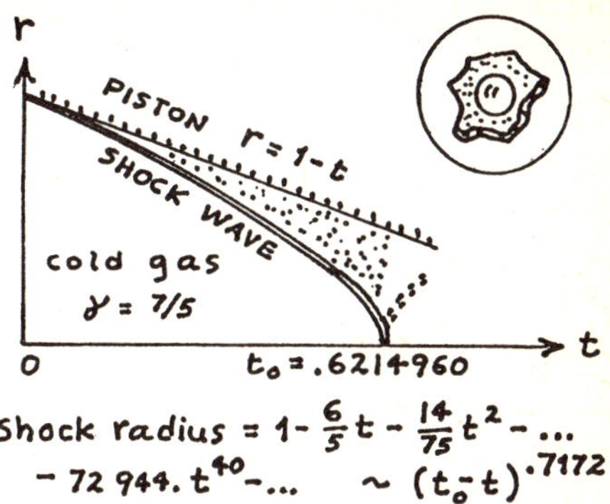


$$\text{Shock radius} = 1 - \frac{6}{5}t - \frac{14}{75}t^2 - \ldots$$
$$- 72\,944.\, t^{40} - \ldots \quad \sim (t_0^- t)^{.7172}$$

FIG. 4. Converging spherical piston

These estimates are based on the fact that if near its closest singularity $z = z_o$ a function known as a power series behaves like

$$f(z) = \Sigma \, c_n \, z^n \quad \sim \quad A(z_o - z)^\alpha \qquad \text{as} \quad z \to z_o$$

then the reciprocal D'Alembert ratios are eventually linear in $1/n$:

$$\frac{c_n}{c_{n-1}} \quad \sim \quad \frac{1}{z_o}\left(1 - \frac{1+\alpha}{n} + \ldots\right) \qquad \text{as} \quad n \to \infty$$

Domb & Sykes (1957) therefore suggest plotting $c_n/c_{n-1}$ versus $1/n$ and fitting a straight-line asymptote, to estimate simultaneously the radius of convergence and the exponent.  When the plot is smooth, as Fig. 5 shows it to be for our spherical piston, the fit can be refined by numerical means.

### Some false conclusions

In these three examples the series are well behaved, and our conclusions clearcut and convincing.  Often, however, the analysis of the coefficients is more delicate.  Then one must exercise self-criticism, and resist jumping to a hasty or beguiling conclusion.  Two examples serve as warnings.

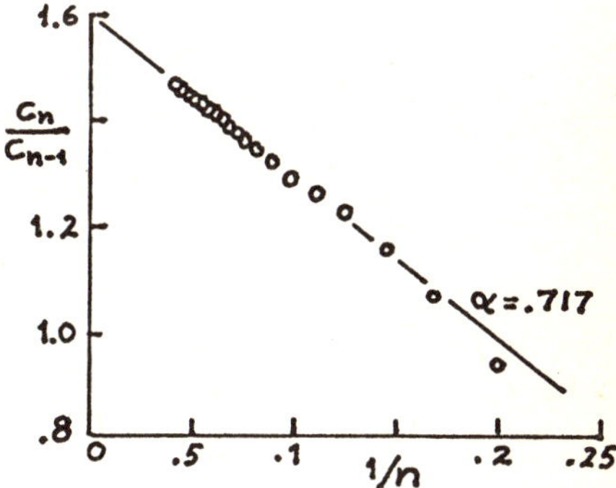


FIG. 5. Graphical ratio test of Domb & Sykes for converging shock wave

### Rotating fluid and disk

Weidman & Redekopp (1976) treated the problem indicated in Fig. 6: an infinite disk rotates steadily in a fluid that is rotating at a different rate. They perturbed von Kármán's solution for a non-rotating fluid to obtain nine terms of a series in the ratio of angular velocities of fluid and disk. The Domb-Sykes plots for various flow quantities are erratic, the smoothest being that shown in Fig. 7 for the azimuthal friction. Fitting a straight line by least squares, they deduced a 4/3-power singularity at $\varepsilon = -0.1433$.

Zandbergen & Dijkstra (1977) subsequently solved the problem using accurate numerical schemes. They find instead a square-root singularity at $\varepsilon = -0.16054$, corresponding to a turning back of the solution onto a second branch. They issue the "serious warning, when using the Domb-Sykes plot, to make oneself sure that the power of the singularity and its location have settled down."

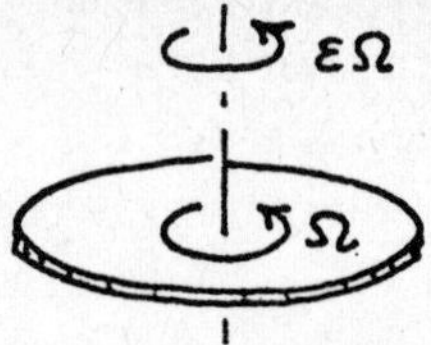



FIG. 6. Rotating fluid and disk

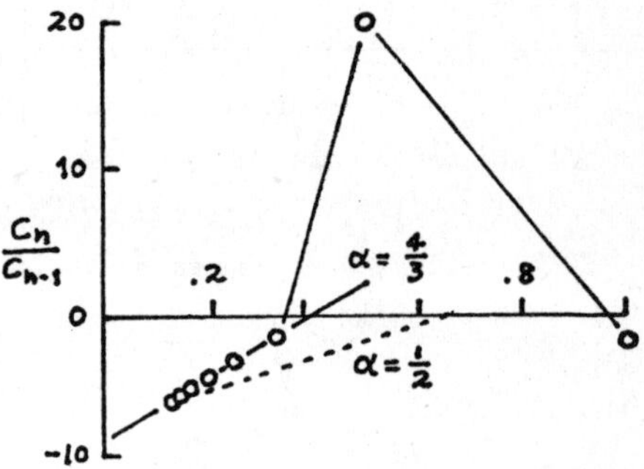

FIG. 7. Domb-Sykes plot for azimuthal friction

### The transonic controversy

For more than thirty years aerodynamicists have debated whether an airfoil can have a continuous range of shock-free flows above the critical Mach number. The majority believe not, but both experiments and refined finite-difference computations show no shock wave until an appreciable supersonic zone has developed.

Recently Guttmann and I extended the Janzen-Rayleigh or $M^2$ expansion for flow past a circle to 17 terms by computer (Van Dyke & Guttmann 1978). For the maximum surface speed in a perfect diatomic gas this gives

$$\frac{q_{max}}{U} = 2 + \frac{7}{6}M^2 + \frac{1547}{600}M^4 + \frac{405791}{54000}M^6 + \frac{1015683271}{39690000}M^8 + \ldots + 2162471459.08\,M^{32} + \ldots$$

We announced that "preliminary analysis of the coefficients ... suggests that the expansion converges up to values of M about four per cent greater than the critical value."

However, I have recently extended the series to 29 terms, including $M^{56}$. The Domb-Sykes plot continually curves upward, cutting into the supposed four-per-cent gap. Our present opinion (still somewhat tentative) is that we were mistaken in discerning a range of shock-free supersonic flow; the Janzen-Rayleigh expansion probably converges only up to the critical Mach number.

## Some surprising results

From two surprising conclusions that proved false, we turn to two others that are equally unexpected, and have been greeted by colleagues with disbelief, but which after some years of anxious reconsideration I believe to be correct.

### Laminar flow through a loosely coiled pipe

Dean (1928) discovered that although fully developed flow through a coiled pipe depends upon both the coiling ratio $a/L$ and the Reynolds number $Ua/\nu$ , those can to a good approximation for loose coiling be combined into the single similarity parameter shown in Fig. 8, which now bears his name. He computed four terms of an expansion in powers of Dean number as a perturbation of the Poiseuille flow through a straight pipe.

I extended Dean's series for the friction factor to 24 terms by computer (Van Dyke 1978). Analyzing the coefficients, I concluded that the friction eventually rises as the 1/4-power of Dean number based on mean flow speed, whereas all existing boundary-layer analyses and finite-difference computations show a square-root variation. Recent analysis of the series for local flow quantities suggests that the boundary-layer approximations are incorrect because they are based on the wrong structure and scaling. The discrepancy with finite-difference calculations will perhaps be resolved by the recent discovery of multiple solutions.

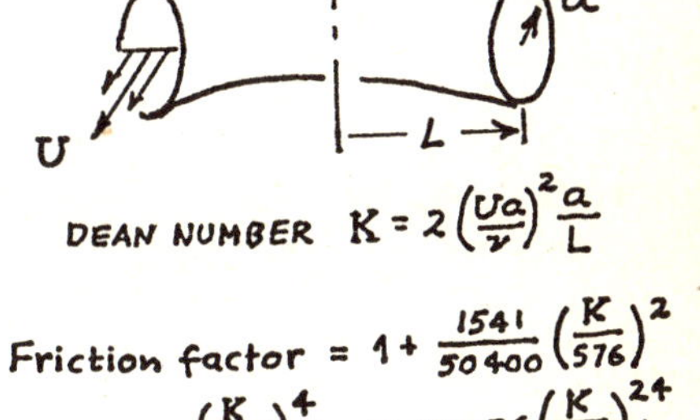



$$\text{DEAN NUMBER} \quad K = 2\left(\frac{Ua}{\nu}\right)^2 \frac{a}{L}$$

$$\text{Friction factor} = 1 + \frac{1541}{50400}\left(\frac{K}{576}\right)^2 -.01100\left(\frac{K}{576}\right)^4 + ... -.00056\left(\frac{K}{576}\right)^{24} + ...$$

FIG. 8. Flow through coiled pipe

### Shock wave from plane piston

According to the blast-wave analogy, the shock wave from a blunted plate in a hypersonic stream will spread as the 2/3-power of distance downstream. Likewise, moving a plane piston rapidly into quiescent gas and then stopping it would send off a plane shock wave to a distance that increases as the 2/3-power of time.

Reddall (1972) treated the smooth piston motion indicated in Fig. 9 by expanding the solution in powers of time. Analyzing 15

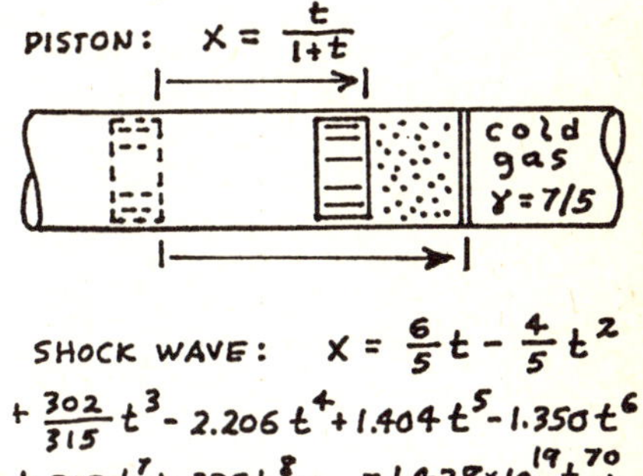



$$\text{PISTON:} \quad X = \frac{t}{1+t}$$

$$\text{SHOCK WAVE:} \quad X = \frac{6}{5}t - \frac{4}{5}t^2 + \frac{302}{315}t^3 - 2.206\,t^4 + 1.404\,t^5 - 1.350\,t^6 + .803\,t^7 + .335\,t^8 - ... - 1.428\times10^{19}\,t^{70} + ...$$

FIG. 9. Shock wave from plane piston

coefficients, he was forced to the conclusion that the blast-wave analogy is not exact; the shock-wave distance increases as a slightly smaller power of time, depending on the adiabatic exponent of the gas.  He has subsequently extended the series to 70 terms, and that conclusion remains unchanged: instead of 2/3, the exponent is about 0.636 for a monatomic gas and 0.626 for a diatomic gas

The recent work reported here was supported by the National Science Foundation under Grant ENG-7824412 and by the Office of Naval Research under Contract N00014-78-C-0373.

REFERENCES


BLASIUS, H.  1908  Grenzschichten in Flüssigkeiten mit kleiner Reibung.  Z. Math. Phys. 56:1-37; English transl., NACA TM 1256

DEAN, W. R.  1928  The stream-line motion of fluid in a curved pipe.  Phil. Mag. (7) 5:673-695

FUJIMOTO, Y. & MISHKIN, E. A.  1978  Analysis of spherically imploding shocks.  Phys. Fluids 21:1933-1938

GUDERLEY, G.  1942  Starke kugelige und zylindrische Verdichtungsstösse in der Nähe des Kugelmittelpunktes bzw. der Zylinderachse.  Luftfahrtforschung 19:302-312

REDDALL, W. F. 3rd  1972  The asymptotic trajectory of a strong planar shock wave arising from a non-self-similar piston motion.  Ph.D. diss., Stanford Univ.

SCHWARTZ, L. W.  1974  Computer extension and analytic continuation of Stokes' expansion for gravity waves.  J. Fluid Mech. 62:553-578

SMITH, A. M. O. & CLUTTER, D. W.  1963  Solution of the incompressible laminar boundary-layer equations; Solution of Prandtl's boundary-layer equations.  Douglas Aircraft Corp. Engg. Papers 1525, 1530

STOKES, G. G.  1880  Supplement to a paper on the theory of oscillatory waves.  Mathematical and Physical papers, 1:314-326.  Cambridge

VAN DYKE, M.  1964  Higher approximations in boundary-layer theory.  Part 3. Parabola in uniform stream.  J. Fluid Mech. 19:145-159

VAN DYKE, M. & GUTTMANN, A. J.  1978  Computer extension of the $M^2$ expansion for a circle.  Bull. Amer. Phys. Soc. 23:996, Abstract no. BD5

WEIDMAN, P. D. & REDEKOPP, L. G.  1976  On the motion of a rotating fluid in the presence of an infinite rotating disk.  Archives of Mechanics 28:1011-1024

ZANDBERGEN, P. J. & DIJKSTRA, D.  1977  Non-unique solutions of the Navier-Stokes equations for the Karman swirling flow.  J. Engg. Math. 11:167-188

# A FAST METHOD TO SOLVE INCOMPRESSIBLE
# BOUNDARY-LAYER INTERACTION PROBLEMS

A.E.P. Veldman          and        D. Dijkstra
National Aerospace                 Twente University of
Laboratory NLR,                    Technology
Amsterdam                          Enschede

The Netherlands



SUMMARY.

A new and fast method to solve incompressible boundary-layer interaction problems
will be presented. The method recognizes the important mutual influence of pressure
and displacement thickness by *simultaneously* updating *both* these quantities as the
boundary-layer is marched. The method can be overrelaxed and fast convergence is
obtained, typically within 10-15 iteration cycles.
The technique has been used with success for both a full boundary-layer formulation
(large, but finite Reynolds number) and a triple-deck approach (limit Re$\rightarrow\infty$). Results
obtained with these two formulations will be compared for a Carter/Wornom trough.



## 1. INTRODUCTION.

During the past decade many papers have been published on the subject of numerically
solving boundary-layer interaction problems, including triple-deck problems. For
supersonic flow the interaction is of a purely local nature and there exist satisfactory techniques for dealing with the interaction. In the incompressible case however the situation is much more difficult since the ellipticity of the outer flow
yields an interaction law which is global in character. Hence, an iteration technique
must be designed in order to solve the problem. It appears that the nature of this
iteration technique is *crucial* to its effectiveness. If the wrong iteration is used
then the procedure may collapse or it may require excessive underrelaxation in order
to obtain a convergent process.
In the present paper it is demonstrated that the equivalence (or lack of hierarchy)
between pressure and displacement thickness is a crucial property of the interaction
problem. All methods in the literature, as far as known to the present authors, use
some kind of hierarchy between pressure and displacement in the iteration, which in
some sense contradicts the nature of the problem. For the true interaction dictates
a lack of hierarchy in these variables, in agreement with the Stewartson-Messiter
triple-deck theory, see Veldman (1980 A).
The present method fully recognizes this equivalence between pressure and displacement by updating both these quantities simultaneously as the boundary-layer is swept.
The main ideas of the present method have been developed by the first author; the
roots of the method are already present in his thesis, Veldman (1976). The second
author has used the method for solving triple-deck flow structures over forward and
backward facing steps, including separation. An interesting standard geometry is the
Carter/Wornom trough (fig.1) at large Reynolds number. The first author solved that
problem with a full boundary-layer formulation, while the second author used a triple-
deck approach for the same geometry. A comparison of the results is made in section 5.

In order to stress the lack of hierarchy between pressure and displacement we will
concentrate in this paper on these two key quantities, ignoring the details of the
boundary-layer equations. A general survey of existing methods including the present
one is presented in sec. 2. A matrix formulation of these methods, which e.g. illu-
minates the need of underrelaxation when the inverse approach is used, will be presented by the first author in the very near future, Veldman (1980 B).

## 2. HIERARCHICAL AND NON-HIERARCHICAL WAYS OF SOLVING THE BOUNDARY-LAYER EQUATIONS.

In the theory of boundary-layer interaction problems the pressure and the displacement thickness appear to be fundamental quantities. As a consequence, the way in which a numerical method of solution incorporates the mutual influence of these variables, is crucial to its success. In our view this point has been overlooked in the majority of existing numerical techniques for solving incompressible interaction problems. In order to focus attention on this feature, we will describe the situation in terms of non-dimensional pressure P and some measure A for non-dimensional displacement thickness only.
A compact formulation of the problem may be given as follows

$$(2.1) \qquad \mathrm{Inv}\ (P, W + A) = 0 \ ; \quad \mathrm{Bl}\ (P, A) = 0,$$

where the first relation Inv denotes the relation furnished by the inviscid Euler equations between the pressure P, the wallshape W considered and the displacement A of the boundary-layer. The second relation Bl is the intimate connection between P and A as it follows from Prandtl's boundary-layer equations with boundary conditions. In order to elucidate our intentions regarding the system (2.1), we first consider the classical boundary-layer approach.

### 2.1 The classical boundary-layer approach.

In most cases the displacement A is small and the usual way to solve (2.1) at large Reynolds number is to neglect A in the relation Inv, in other words one starts to solve the inviscid equations for the pressure, symbolically

$$(2.2) \qquad \mathrm{Inv}\ (P, W) = 0 \ , \ W \text{ prescribed.}$$

Let $P_{inv}$ denote the result of this calculation. Then the next step is to solve the boundary-layer equations for the displacement:

$$(2.3) \qquad \mathrm{Bl}\ (P, A) = 0 \ , \ P = P_{inv} \text{ prescribed.}$$

With the resulting A one can correct the pressure via Inv and so on. It will be clear that there is a definite hierarchy in the outlined procedure, the hierarchy being dictated by the theory of Prandtl for large Reynolds number.
For attached flows good numerical results usually are obtained but the numerical method of solution fails at (or near) separation, where the Goldstein singularity enters into the solution of Bl (P,A) = 0 with P prescribed. On the basis of the paper by Catherall and Mangler (1966) we now know that this singularity can be avoided by prescribing displacement A rather than pressure at separation. In other words the appearance of the singularity can be explained by means of the implicit function theorem which states that Bl (P,A) = 0 can be solved for A provided that $\partial Bl/\partial A$ does not vanish. Although a rigorous proof of the vanishing of $\partial Bl/\partial A$ at separation does not seem to be available, there is ample numerical evidence for this to be correct. It is also clear from existing numerical solutions for separation bubbles that the pressure distribution $P_{inv}$ is completely inadequate over the range of the bubble. The reason is that a separation bubble gives rise to a significant change in the effective wallshape (W+A) as it is felt by the inviscid flow. It follows from the statements above that prescribing $P = P_{inv}$ is not the correct way of telling the boundary-layer that there is inviscid flow outside and hence some other mechanism to do so must be found. At the same time it will be clear that the hierarchy of the classical approach has to be discarded.

### 2.2 The interaction model.

A suitable interaction model may be derived by approximating the relation Inv in (2.1) by means of thin airfoil theory:

$$(2.4) \qquad P = P_{inv} - \frac{1}{\pi} \int_{x_b}^{x_e} \frac{A'(\xi)}{x-\xi} \, d\xi$$

where $[x_b, x_e]$ is the range over which the interaction is considered to be relevant.
Note that (2.4) reflects the ellipticity of the inviscid flow. Further note that triple-deck theory, Stewartson (1969) and Messiter (1970), also arrives at (2.4). The system (2.1) is written as follows

$$(2.5) \qquad \begin{array}{ll} P = I \ (A) & \text{(interaction law 2.4)} \\ P = B \ (A) & \text{(boundary-layer eqs.)} \end{array}$$

The interaction law I is elliptic and the boundary-layer equations are parabolic, so that (2.5) is an elliptic/parabolic problem which can be solved by iteration. The *organization* of the iteration technique appears to be crucial to its effectiveness: one should avoid any definite hierarchy between P and A in the iteration, since the interaction model dictates an equivalence between P and A, in agreement with triple-deck theory, see Veldman (1980 A). In the next section we describe what happens if this remark is ignored.

2.3 Hierarchical iteration.

Suppose that one tries to solve the system (2.5) with the classical hierarchy, viz.

$$(2.6) \qquad A^{(n)} = B^{-1} (P^{(n)}) \ ; \ P^{(n+1)} = I \ (A^{(n)}).$$

where n denotes the iteration index. Hence, the boundary-layer is marched with $P = P^{(n)}$ prescribed and the resulting displacement $A^{(n)}$ is substituted in the interaction law I to yield a new pressure. On the basis of our remarks in sec. 2.1 this procedure will fail at separation where $B^{-1}$ does not exist, so that we can reject this possibility.

Indeed, the majority of the existing techniques for solving (2.5) discards the classical hierarchy but they replace it by an inverse hierarchy, viz.

$$(2.7) \qquad P^{(n)} = B \ (A^{(n)}) \ ; \ A^{(n+1)} = I^{-1} \ (P^{(n)}).$$

In this case the boundary-layer is marched with prescribed displacement $A = A^{(n)}$ and the resulting pressure yields a new displacement via the inverse interaction law. The procedure does not fail at separation but now severe underrelaxation appears to be necessary in order to obtain a convergent procedure. Moreover the necessary underrelaxation factor tends to 0 if the Reynolds number grows or if the range $[x_b, x_e]$ in

(2.4) increases. Hence the iteration technique becomes time consuming since a large number of iteration cycles is required to obtain a solution with a given accuracy. It will be clear that neither the classical nor the inverse hierarchy are satisfactory to solve (2.5).

2.4 Semi-hierarchical iteration: a possibility.

A semi-hierarchical possibility which has appeared in the literature is the following

$$(2.8) \qquad P_1^{(n+1)} = I \ (A^{(n)}) \ ; \ P_2^{(n+1)} = B \ (A^{(n)}) \ ; \ A^{(n+1)} = \text{Relax} \ (P_1^{(n+1)}, P_2^{(n+1)}, A^{(n)})$$

Hence two pressure distributions are computed from a given distribution of the displacement A. Upon convergence the two pressure distributions should be the same which is achieved by a suitable relaxation represented by Relax in (2.8). The methods by Le Balleur (1978), Kwon and Pletcher (1979) and Carter (1979) have the form (2.8). Le Balleur still needs underrelaxation. The schemes designed by the other investigators can be overrelaxed but a simple design of the formula Relax in (2.8) is less straight forward and not so easy to extend to more general situations.
Characteristic  for all methods mentioned above is the fact that P and A are successively updated rather than simultaneously. Moreover, the process of updating is applied after completion of the boundary-layer sweep. In the present method P and A are updated simultaneously with the pass through the boundary-layer.

2.5 The present non-hierarchical iteration.

The present method of solving (2.5) reads

$$(2.9) \quad \begin{cases} P^{(n+1)} - I\,(A^{(n+1)}) = 0 \\ P^{(n+1)} - B\,(A^{(n+1)}) = 0 \end{cases}$$

Hence, at each boundary-layer station P and A are *simultaneously* updated from - essentially - a set of two equations in two unknowns. At each station the interaction is *instantaneously* incorporated in the results before proceeding to the next boundary-layer station. In order to implement (2.9) in a satisfactory way two ingredients are required:

(i) The boundary-layer code must be of a general nature in the sense that it expects a linear combination of pressure and displacement as input at each station. (Eventually, this can be avoided by using an inverse code and iterating at the current station).

(ii) The discrete approach to the elliptic interaction law P-I(A) = 0 should be as local as possible in the sense that the coefficient of the displacement A at the current station dominates over the coefficients of A at the other stations.

In the next section such a "localized" discrete representation is presented.


3. DISCRETE APPROACH TO THE INTERACTION LAW.

Two formulations of the interaction law (2.4) have been used, viz.

$$(3.1) \quad \pi\,P(x) = \pi\,P_{inv}(x) - \int_{x_b}^{x_e} \frac{A'(\xi)}{x-\xi}\,d\xi$$

$$(3.2) \quad \pi\,P(x) = \pi\,P_{inv}(x) - \int_{x_b}^{x_e} A''(\xi)\ln|x-\xi|\,d\xi.$$

For a treatment of (3.1) with the midpoint rule, see Veldman (1979). Here, we discuss the discretization of (3.2). We take the origin at the centre of $[x_b,x_e]$ and assume a uniform mesh over this interval with boundary-layer stations

$$(3.3) \quad \xi_i = i\,h \;,\; h = (x_b + x_e)/(2N) \;,\; i = -N(1)N$$

where stations $\pm N$ now correspond to $x_e$ and $x_b$. The interaction law is written as

$$(3.4) \quad \pi P_i = T_i - \sum_{j=-N}^{N} \int_{\alpha_j}^{\beta_j} A''(\xi)\ln|\xi_i-\xi|\,d\xi \;,\; \alpha_j = (j-\tfrac{1}{2})h \;,\; \beta_j = \alpha_j+h$$

where $T_i$ contains $P_{inv}$ together with effects arising from the endpoints of the range. On each subinterval we now set

$$(3.5) \quad A''(\xi) := A''_j := (A_{j+1} - 2A_j + A_{j-1})/h^2 \;,\; \alpha_j \le \xi \le \beta_j.$$

With the constant approximation (3.5) for A" the integration in (3.4) can be carried out and after some algebra the result becomes

$$(3.6) \quad \sum_{j=-N+1}^{N-1} C_{ij}A_j + h\pi\,P_i = (R.H.S.)_i \;,\; i = -N+1(1)N-1$$

where all quantities which are constant during the course of the iteration have been placed in the right hand side. The matrix $C = (c_{ij})$ has the following properties:

(i) C is translation invariant: $C_{ij} = g_k^j \;,\; k = |i-j|$

(ii) C is symmetric and diagonally dominant.

The translation invariance implies that the complete matrix C can be stored in a one-dimensional array with length 2N-1. The diagonal elements have the value 3ln3 and the first off diagonal entries are (5ln5-9ln3)/2. The ratio of these numbers is -3.58. Further away from the diagonal the elements $C_{ij}$ behave like $-1/k^2$ , $k = |i-j|$.

The desired "localized" representation of the interaction law is the Gauss-Seidel version of (3.6), viz.

$$(3.7) \qquad C_{ii}A_i^{(n+1)} + h\pi P_i^{(n+1)} = (R.H.S.)_i - \sum_{j=-N+1}^{i-1} C_{ij}A_j^{(n+1)} - \sum_{j=i+1}^{N-1} C_{ij}A_j^{(n)}$$

where $C_{ii}$ dominates over $\sum_{i \neq j} C_{ij}$ and n is the iteration index. The relation (3.7) is the linear combination of displacement and pressure which has been used to compute the boundary-layer solution at each station i.

## 4. DISCRETE APPROACH TO THE BOUNDARY-LAYER EQUATIONS.

Streamwise derivates have been discretized with a 3 point backward formula and lateral derivates have been replaced by central differences. A full Newton procedure was used to solve the set of non-linear algebraic equations at each station. Several modes have been run to cope with backflow:
(i)   Flare (set $u\partial u/\partial x = 0$ if $u < 0$)
(ii)  Ignore (simply march downstream).
(iii) Back (take backflow properly into account).

## 5. RESULTS AND CONCLUSIONS.

The present iteration technique (2.9) has been implemented by the first author in a full Boundary-Layer approach (BL), while the second author used a Triple-Deck formulation (TD). For the details see Veldman (1979) and Dijkstra (1978). Results have been calculated for the flat plate trailing edge interaction problem (BL), for backward and forward facing steps (TD) and for the Carter/Wornom trough (BL and TD), see fig. 1.
In all cases *fast convergence* was obtained; an example is the flat plate trailing edge problem - see Veldman (1979) - for which BL required 11 iteration cycles (using 1.5 overrelaxation) to satisfy

$$(5.1) \qquad \max_i \left| A_i^{(n+1)} - A_i^{(n)} \right| < 10^{-4} \ , \ A_i = \text{displacement at station i.}$$

According to TD calculations further acceleration of convergence is possible by making additional sweeps using (3.7) with fixed pressure. The CPU time for these additional sweeps (3.7) is negligibly small as compared to a boundary-layer cycle (2.9). In certain cases it was found that the number of boundary-layer cycles could be reduced by a factor 2 with this technique. Both BL and TD have been used to calculate the interaction solution over a trough in a flat plate at Re = 360000 (fig. 1). The BL program was run in the mode Ignore (sec.4), while TD used Flare. Both runs required 17 cycles to pass (5.1). In addition TD was run in the Back mode and the results are here compared with those produced by BL (Ignore). Although the plateau value of the pressure (fig. 2) shows a difference for the two approaches, the global agreement is good. In particular separation, re-attachment, dividing streamline and skin friction were found to be virtually identical (figures 3 and 4). Note that the absolute value of the skin friction in the bubble overtakes the Blasius value just prior to re-attachment where maximum backflow (10% of main stream) is encountered. Also note that Flare significantly underestimates the minimum of the skin friction in the bubble (fig. 4).
Inspection of our results obtained for various wall shapes revealed that small separation bubbles can be predicted by the present method within 10 iteration cycles: the change in the position of the bubble boundary is marginal if the iteration is continued. Originally the TD program was designed to resolve flow structures over wallshapes of O(1) when scaled to triple-deck demands, but -in order to match with fig. 1- the scaled depth of the trough became 27.5. Since this is far outside the original range, the comparison made above yields a strong argument for the wide scope and applicability of triple-deck theory.

REFERENCES.
Carter, J.E. (1979). AIAA paper 79-1450.
Catherall, D. and Mangler, K.W. (1966). JFM 26, 163-182.
Dijkstra, D. (1978). Proc.6th.Int.Conf.Num.Meth.Fl.Dyn., Tbilisi, U.S.S.R.
Kwon, O.K. and Pletcher, R.H. (1979). J.Fl.Eng.101, 466-472.
Le Balleur, J.C. (1978). La Rech.Aerospat. 183, 65-76.

Messiter, A.F. (1970). Siam J.Appl.Math. 18, 241-257.
Stewartson, K. (1969). Mathematika 16, 106-121.
Veldman, A.E.P. (1976). Ph.D.Thesis, Dep. of Math. Univ. of Groningen.
Veldman, A.E.P. (1979). NLR, TR 79023 U.
Veldman, A.E.P. (1980A). Proc.Bail I Conf., Dublin.
Veldman, A.E.P. (1980B). Proc.Agard Symp.Visc.-Inv.interactions, Colorado Springs.

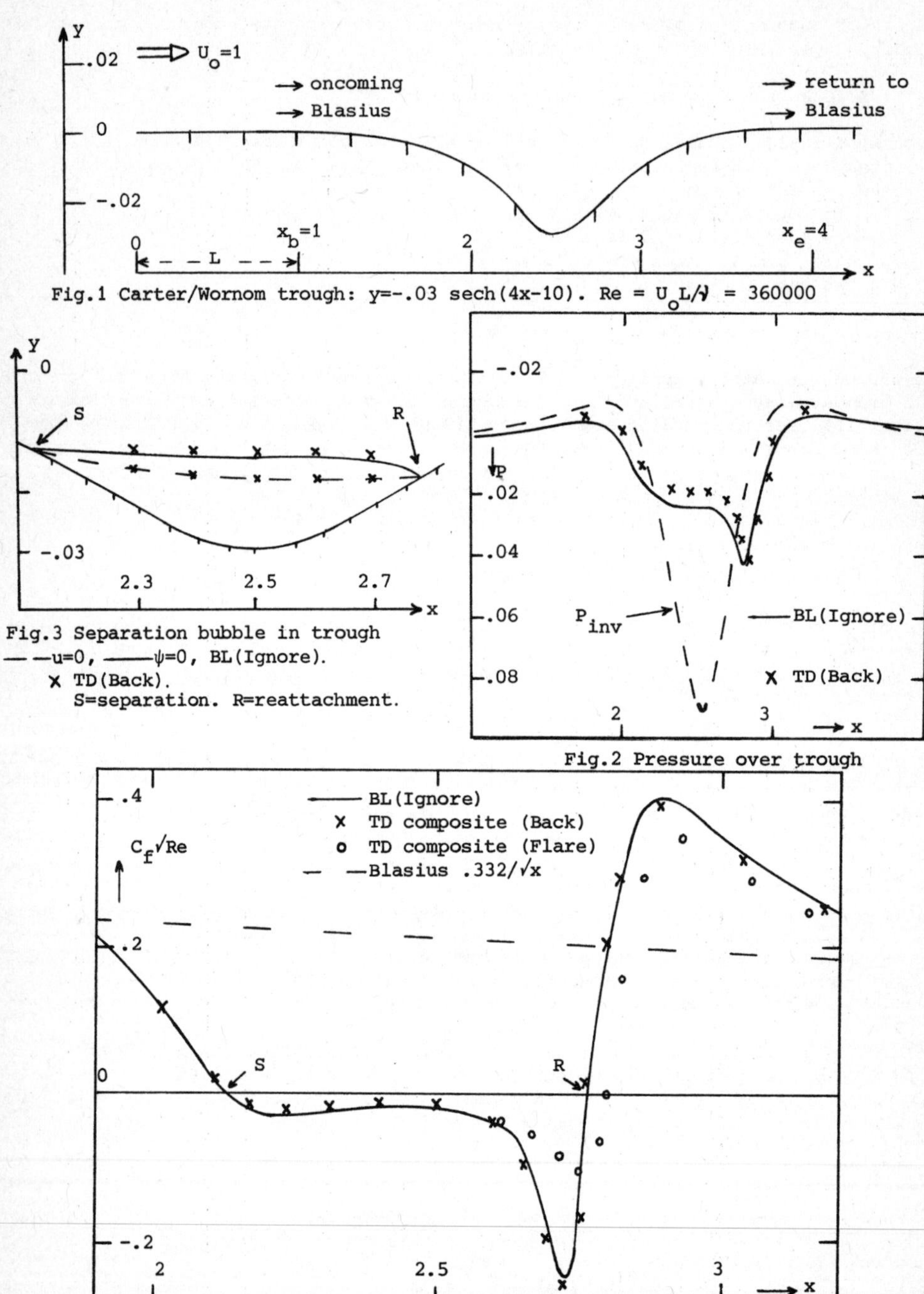


Fig.1 Carter/Wornom trough: y=-.03 sech(4x-10). Re = $U_o L / \nu$ = 360000



Fig.3 Separation bubble in trough
— —u=0, ———ψ=0, BL(Ignore).
 X TD(Back).
  S=separation. R=reattachment.



Fig.2 Pressure over trough



Fig.4 Skin friction.  S=separation , R=reattachment.

# VISCOUS TRANSONIC FLOW OVER AIRFOILS


by
J. C. Wai and H. Yoshihara
Boeing Military Airplane Company
Seattle, Washington


## SUMMARY


The TSFOIL small disturbance potential code is coupled to Green's lag entrainment method to compute transonic separated flows over airfoils. Green's equation serves as a viscous tangency condition for the TSFOIL code, eliminating an iterative procedure between the viscous and inviscid flows. The method is applied to the case of the NASA Supercritical Airfoil 12 and the 18% Circular Arc Airfoil, the latter with severe separation.


## I    PROBLEM

Computation of transonic turbulent viscous flows over airfoils including the effect of separation is considered.  The important viscous interactions expected are the shock-boundary layer interaction, camber modifications due to viscous displacement differences on the two sides of the airfoil, and of lesser importance the wake displacement effects.  These interactions can have a significant effect on the lift and moment on the airfoil.

Recent investigations on this problem include, for example, those of Le Balleur (Ref. 1), Melnik (Ref. 2), and Collyer and Lock (Ref. 3).  The present approach, in common with the above, uses an integral method for the boundary layer, but the manner of the coupling with the inviscid flow and the treatment of the shock-boundary layer interaction are different.

## II    APPROACH

The boundary layer is computed with the Green's lag-entrainment method (Ref. 4).  It is composed essentially of three ordinary first order differential equations.  Two of the equations are obtained by integrating the continuity and streamwise momentum equations across the boundary layer, while the third is derived from the Bradshaw-Ferriss turbulent energy equation (Ref. 5).  This set of equations is non-singular for cases of interest.

For the purpose of coupling the viscous flow to the inviscid flow, Green's equations are recast into the single equation derived by East, Smith, and Merryman (Ref. 6) given by

$$\phi_{xx} = F_2^{-1}(\delta_x^* - F_1) \tag{1}$$

where $\phi$ is the perturbation potential, $\delta^*$ is the displacement thickness, x the streamwise coordinate, $F_1$ and $F_2$ are functions of the boundary layer variables (see Ref. 6), and the subscripts denote differentiations.  The remaining boundary layer equations form the auxiliary equations from which the boundary layer development **is** determined.

For the inviscid flow, the transonic small disturbance code TSFOIL (Ref. 7) is used where the boundary condition at the airfoil is imposed in a quasi-planar fashion.  Thus with the viscous displacement layer added, the modified tangency condition at the airfoil becomes

$$\phi_y(x,0) = f_x(x) + \delta_x^*(x) \tag{2}$$

where $y = f(x)$ defines the airfoil.  The viscous tangency boundary condition for the inviscid flow is obtained from Eqs. (1) and (2); **that is,**

$$\phi_{xx} = F_2^{-1}(\phi_y - f_x - F_1).$$

(3)

Green's equations are also applied in the wake region where the wall skin friction is set equal to zero, and the dissipation length scale parameter halved to achieve the correct asymptotic far-field wake. The presence of the viscous wake generates a displacement afterbody downstream of the trailing edge across which a jump in the flow slope occurs. The equivalent condition to Eq. (3) is

$$[\phi_y] = [\delta_x^*] = [F_1 + F_2\phi_{xx}]$$

(4)

where the brackets denote the jump in the bracketed quantity across the wake which has been shrunk to the x-axis. In addition, the pressure across the wake (neglecting centrifugal effects) is assumed constant, and as in the inviscid condition

$$[\phi] = \text{constant.}$$

(5)

In the present approach, the iterative procedure between the inviscid and viscous flows used for example in Refs. 1-3, is not used. The outer inviscid and the boundary layer flows are integrated simultaneously, with the inviscid flow governed by the viscous tangency condition (3) and the wake contact jump conditions (4) and (5). In this way the need to pose the viscous problem in a direct or indirect mode to insure convergence (see Ref. 6) is avoided.

## III  SHOCK-BOUNDARY LAYER INTERACTION

Understandably there are fundamental and numerical difficulties associated with the modeling of the shock-boundary layer interaction. In the interaction region, the use of the boundary layer approximations fails and is complicated by the difficult matter of modeling the time-averaged turbulent transport process through the shock.

In the numerical procedure there is additionally the dilemma that the shock wave in the inviscid flow acquires a profile and hence a thickness that are unphysical being generated by truncation error viscosity. When such a pressure gradient is overlayed onto the boundary layer, the result is an unphysical mesh dependent interaction. This would then affect the post-shock state of the boundary layer and the subsequent prediction of the boundary layer and wake.

This dilemma can in principle be resolved phenomenologically by establishing the proper mesh spacing in the shock region to effect the desired test-theory agreement; that is, by requiring, for example, the post-shock pressure to assume a prescribed value as determined by wind tunnel measurements. For a given pre-shock pressure, a reasonable value is suggested in Reference 8 to be that midway between the sonic pressure and the shock detachment pressure on the shock polar.

## IV  SEPARATED FLOW

In separated flow an upstream influence is present via the reversed flow, thus making a downwind marching procedure, in general, ill-posed. In Green's procedure, a downwind marching is nevertheless employed, and it is rationalized by postulating the separated flow to be in near-equilibrium. That is, the velocity and shear profiles across the layer assume a near-invariance in the streamwise direction. This invariance then provides an a priori accounting of the upstream influence of the reversed flow, which then permits the downwind marching. The required definition of the equilibrium flow for the class of separated flows of interest must be obtained empirically, and in Green's method it is provided by Eqs. (A-28) to (A-33) and (A-18) of Reference 4.

In separated boundary layers, the absolute values of $F_1$ and $F_2$ of Eq. (3) or (1) assume very large values, so that at first glance it would appear that the pressure gradient $\phi_{xx}$ is independent of $\phi_y - f_x$ or $\delta_x^*$. That this is incorrect can be seen by inserting the equilibrium value of the entrainment coefficient $C_E$ into the expression for $F_1$ (Eqs. (A-29) into Eq. (A-6) of Ref. 6). The result is

$$\phi_{xx} \cong -\frac{\phi_y - f_x}{\bar{H}^2 \theta} \cong -\frac{\delta_x^*}{\bar{H}^2 \theta} \qquad (6)$$

where $\bar{H}$ is a form factor and $\theta$ is the momentum thickness.

## V  NUMERICAL ALGORITHM

The coupling of the inviscid and boundary layer codes as described above are implemented in the TSFOIL code in the following manner. The coupling procedure will differ for the attached and separated cases. In the case of attached flow, the viscous tangency condition (3) with the $\phi_y$ is used. The $\phi_{xx}$ and $\phi_y$ are differenced as in the inviscid flow and $F_1$ and $F_2$ are initially approximated by using the values from the prior iteration and then upgraded using the boundary layer variables at the new point obtained by an integration of the auxiliary boundary layer equations.

In the case of shock-induced separation, the above attached flow procedure is retained until the $\delta_x^*$ attains the shock polar angle corresponding to the postulated post-shock pressure. This point in general is located close to where incipient separation occurs. From this point to the post-shock point (the first subsonic point), the viscous tangency condition is switched to the form given in Eq. (1). In this interval, $\delta_x^*$ is set equal to the postulated wedge angle, and with a proper mesh this value will equal $\phi_y - f_x$ resulting from the inviscid calculations. This switching to the $\delta_x^*$ in the separated portion of the shock profile is necessitated by convergence considerations. It also provides a means to input the postulated wedge angle to insure the desired post-shock pressure and the proper shock location in the final result. Downstream of the shock to the trailing edge, the viscous tangency condition (3) is again used.

The wake is treated again following essentially the TSFOIL procedure, but now a jump in $\phi_y$ as given by Eq. (4) is additionally imposed across the wake slit. This is accomplished by introducing the analytic continuation across the slit, both from above and from below, and relating the potentials on the two sheets by the wake contact jump conditions.

## VI  EXAMPLES

The first case calculated is the NASA Supercritical Airfoil 12, an early Whitcomb airfoil, at M = 0.74, alpha = $2^o$, and the chord Re. No. = 7.7 X $10^6$ (63.5 cm chord). This case was selected since it is a challenging case to compute with the small disturbance code, and the viscous effects have an especially large effect. Furthermore, this airfoil is an extreme case for the small disturbance code because of the large blunt nose and the pronounced aft camber. In Figure 1 we first compare the inviscid results from the TSFOIL and the exact potential code, the latter developed by Holst (Ref. 9) to show the significant effects of the small disturbance assumption. Thus as a first step, following Reference 10, we determine shape changes which compensate for the small disturbance approximation such that the TSFOIL code yields the exact potential pressure distribution. Unfortunately, for the present example, only the corrective ramp for the lower surface could be determined since in the inviscid flow the aft upper surface pressures are supersonic while the corrective ramp must be determined in a subsonic environment occurring in the viscous flow. We can therefore expect some effects of the inviscid small disturbance errors in the aft upper surface pressure distribution comparisons of the viscous solution and experiments.

In the viscous calculations, for convenience, it was assumed that upstream of the shock on the upper surface and upstream of the adverse pressure gradient on the lower surface, the boundary layer displacement effects on the potential flow are negligible. This permitted the coupled treatment of the boundary layer and inviscid flow to be initiated downstream of the above points with the necessary initial values for the boundary layer furnished by a simple flat plate type theory.

The calculations converged without difficulty, and the resulting pressure distribution is shown in Figure 1 where reasonable agreement with experiments is seen. In particular, the significant effects of viscosity are evident here. In assessing this test-theory comparison, the effects of the unknown wind tunnel wall interference must be kept in mind. In Figure 2 the resulting boundary layer quantities are plotted. Corresponding experimental results were not measured.

The second example calculated is the 18% circular arc airfoil at $M = 0.78$, alpha $= 0^o$, and the chord Reynolds number $= 11 \times 10^6$ (20 cm chord). This case was tested at NASA-Ames (Ref. 11), and calculated both with the Reynolds equations (Ref. 11) and with the small disturbance code iterated semi-inversely with an integral boundary layer code (Ref. 1). The experiments indicated this to be an extreme case of shock-induced separation with reattachment occurring downstream of the trailing edge.

The resulting pressure distribution is compared both to experiments as well as to the calculated results of Refs. 1 and 11 in Figure 3. Improved agreement with experiments is seen here. The difference in shock location that still remains is in part intentional, mainly due to postulating a subsonic post-shock pressure rather than the slightly supersonic measured value which may be distorted by sidewall viscous interference effects.

Finally in Figure 4 we show the reasonably good agreement of the calculated and measured boundary layer variables, the latter obtained by laser velocimeter measurements reported in Ref. 11.

In both of the above examples, the standard TSFOIL mesh was used which in the region of the shock was 3% chord.

The addition of the viscous equations required little additional computing time, amounting to  20 percent additional time relative to the inviscid flow computation time of 40 seconds CPU time on the CYBER 175 computer.

VII  CONCLUDING REMARKS
In the present study the transonic small disturbance potential code (TSFOIL) was coupled directly with Green's lag entrainment boundary layer and wake equations. The latter served to define a viscous tangency condition at the airfoil and the wake afterbody. With this direct coupling, iterations between the inviscid and viscous flows were avoided, leading to a well-behaved numerical procedure.

The shock-boundary layer interaction required special treatment since the shock wave in the inviscid flow assumed an unphysical pressure-rise profile dependent upon the streamwise mesh size. With such an artificial pressure gradient superimposed on the boundary layer, a meaningless mesh dependent interaction would be obtained. To remedy this, a suitable mesh was used in the shock region to yield the "correct" post-shock pressure. Such a phenomenology would not, of course, insure a correct post-shock boundary layer; and for this, applicable boundary layer equations are required.

The results obtained suggest the present approach to be a viable base upon which to develop a transonic separated flow method. Further cases must be calculated and compared to experiments to develop and validate the procedure. For this purpose it would be desirable to replace the small disturbance inviscid code with an exact potential code to eliminate the disruptive influences of the inviscid small disturbance shortcomings.

VIII REFERENCES

1. Le Balleur, J., "Couplage Visqueux-Non Visqueux: Méthode Numérique et Applications Aux Écoulements Bidimensionnels Transsoniques et Supersoniques", Rech. Aerosp. Mar-Apr, 1978.

2. Melnik, R., "Recent Developments in a Boundary Layer Theory for Computing Viscous Flows over Airfoils", 4th US-FRG Meeting, BMVg-FBWT 70.31, 1979.

3. Collyer, M., and Lock, R., "Prediction of Viscous Effects in Steady Transonic Flows Past an Airfoil", Aero. Quart., Vol. 30, Part 3, 1979.

4. Green, J., Weeks, D., and Brooman, J., "Prediction of Turbulent Boundary Layers and Wakes in Incompressible Flow by a Lag-Entrainment Method", RAE Reports and Memoranda 3791, 1973.

5. Bradshaw, P., and Ferriss, D., "Calculation of Boundary Layer Development using Turbulent Energy Equation; Compressible Flow on Adiabatic Walls", J. Fluid Mech., Vol. 46, 1971.

6. East, L., Smith, P., and Merryman, P., "Prediction of the Development of Separated Turbulent Boundary Layers by the Lag-Entrainment Method", RAE Report 77046, 1977.

7. Murman, E., Bailey, R., and Johnson, M., "TSFOIL - A Computer Code for Two-Dimensional Transonic Calculations, Including Wind Tunnel Wall Effects and Wave-Drag Evaluation", NASA Report SP347, 1975.

8. Yoshihara, H., "Formulation of the 3D Transonic Unsteady Aerodynamic Problem", AFFDL TR-79-3030, 1979.

9. Holst, T., "A Fast Conservative Algorithm for Solving the Transonic Full-Potential Equation", AIAA Paper 79-1456, 1979.

10. Yoshihara, H., "Fixes to the 3D Transonic Small Disturbance Theory", Convair Report CASD-ERR-75-012, 1975.

11. McDevitt, J., Levy, L., and Deiwert, S., "Transonic Flows about a Thick Circular-Arc Airfoil", AIAA Journal, Vol. 14, No. 5, May 1976.

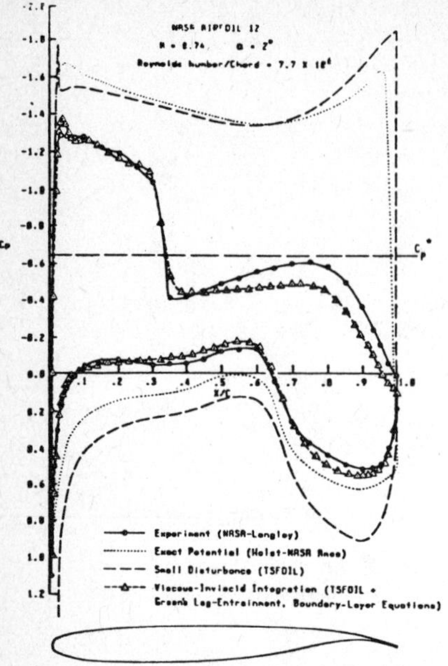

Figure 1. Theory/Experiment
Comparison-Pressure Distribution

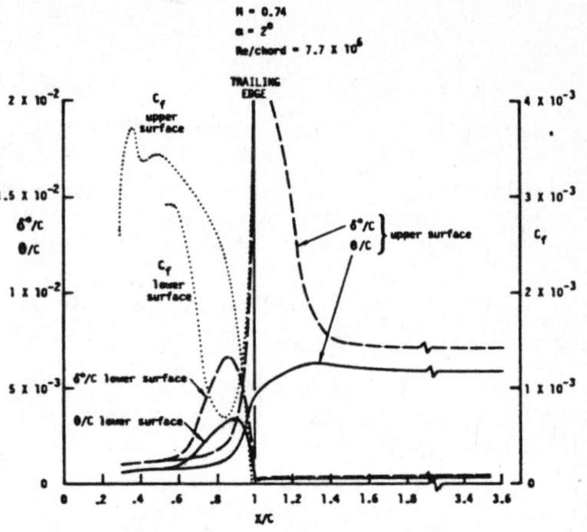

Figure 2. Boundary and Wake
Characteristics-NASA Airfoil 12.

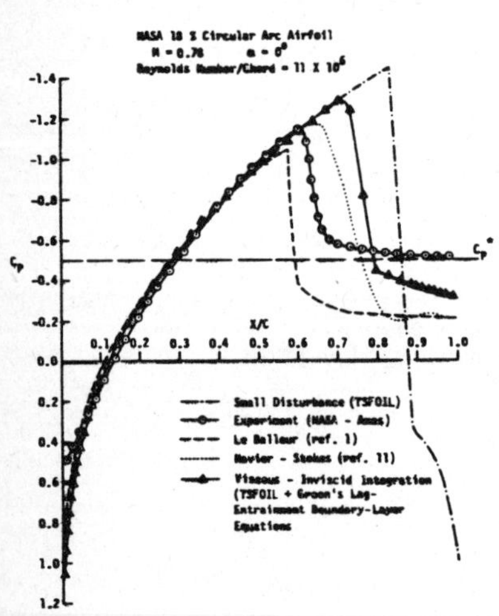

Figure 3. Theory/Experiment
Comparison-Pressure Distribution

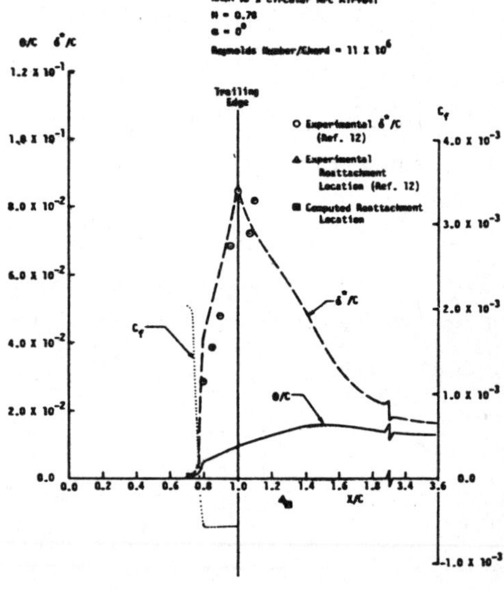

Figure 4.  Theory/Experiment Comparison-
Boundary Layer Characteristics.

# NUMERICAL METHODS FOR THE SOLUTION OF THE SIMPLIFIED NAVIER-STOKES EQUATIONS


R.Q.Wang  L.Q.Jiao  X.Z.Liu
Computing Center of Academia Sinica
Beijing, China


## 1.  INTRODUCTION

At present numerous works on the analysis and computation for the Simplified Navier-Stokes equations (SNSE) have appeared[1-6]. Among them, hypersonic viscous flow over sharp and blunt bodies is more interesting. However the troublesome thing is the numerical instability encountered in the computational process[1]. This phenomenon has been observed by us and has been investigated in detail. We conclude that the crucial reason for the phenomenon of numerical instability does not lie with the difference schemes, but with the ill-posedness of the system of PDE. The problem is ill-posed, if the initial-boundary value problem is solved by the marching technigue along the steamwise direction starting from the stagnation line.

## 2.  THE GOVERNING EQUATIONS

We have investigated the mathematical properties and the computational methods of the SNSE in the following nondimensional form:

$$(r\rho u)_x + (Hr\rho v)_y = 0 ,$$

$$\rho\left(\frac{u}{H}v_x + vv_y - k\frac{u^2}{H}\right) + P_y = 0 ,$$

$$(2.1) \quad \rho\left(\frac{u}{H}u_x + vu_y + \frac{kuv}{H}\right) + \frac{1}{H}P_x = \frac{1}{Re_\infty rH^2}\left[rH^2\mu\left(u_y - k\frac{u}{H}\right)\right],$$

$$\rho\left(\frac{u}{H}T_x + vT_y\right) - \left(\frac{u}{H}P_x + vP_y\right) = \frac{1}{Re_\infty rH}\left(\frac{\mu}{Pr}rHT_y\right)_y + \frac{\mu}{Re_\infty}\left(u_y - k\frac{u}{H}\right)^2,$$

$$p = \left[(\gamma-1)/\gamma\right]\rho T .$$

where $H = 1 + ky$, $k$ is curvature of a body, $Pr$ – Prandtl number, $Re_\infty$ – Reynolds number, L– Reference length, x is the coordinate tangent to the body, y is the coordinate normal to the surface of the body, u, v are components of a velocity, in the x and y directions respectively, P– pressure, $\rho$– density, T– temperature, $\mu$– dynamical viscosity coefficient, $\lambda$– themal conductivity coefficient, $C_p$– specific heat at constant pressure. All the dimensional quantities f are normalized in the following way:

$$X = \bar{x}/L , \quad y = \bar{y}/L , \quad u = \bar{u}/\bar{u}_\infty , \quad v = \bar{v}/\bar{u}_\infty ,$$

$$\rho = \bar{\rho}/\bar{\rho}_\infty , \quad p = \bar{p}/(\bar{\rho}_\infty\bar{u}_\infty^2) , \quad T = \bar{T}/(\bar{c}_p/\bar{u}_\infty) \quad and \quad \mu = \bar{\mu}/\bar{\mu}_\infty .$$

It is important to explain whether it is a well-posed or ill-posed initial-boundary value problem for the system(2.1).

## 3.  FORMULATION OF THE MATHEMATICAL PROBLEM

The classical system of the boundary layer equations is of the parabolic type and being an initial boundary value problem it is well-posed in the sence of Hadamard. But what about the SNSE? At present there are no theoretical proofs. Some authors treat it as a well-posed problem[5]. However, we found that as an initial boundary value problem it is ill-posed, when u<c (where u is the tangential

velocity and c is the sound speed), but it is reasonable in principle to pose a boundary value problem in the region $\Omega = \Omega 1 + \Omega 2 + \Omega 3$ (see Fig. 1.). Our analysis is as follows:

Consider a homogeneous system of equations with constant coefficients, corresponding to (2.1):

$$(3.1) \qquad A \frac{\partial X}{\partial x} + B \frac{\partial X}{\partial y} = C \frac{\partial^2 X}{\partial y^2} \ .$$

Where A,B,C are 4x4 matrices and $X$ is a column vector. Is the Cauchy problem for the system (3.1) well-posed, when $x \geqslant 0$ ? We represented the solutions in terms of the harmonic functions

$$(3.2) \qquad X(x,y) = \xi \, e^{i(\lambda x + \mu y)} \ .$$

Substituting (3.2) into the system (3.1), it is found that the necessary condition for (3.1) to be evolutional is that the roots of the characteristic equation

$$(3.3) \qquad det(\Lambda A - C) = 0$$

have non-negative real parts. All the eigenvalues are given below:

$$\Lambda_1 = \Lambda_2 = 0 \ ,$$

$$(3.4) \qquad \Lambda_{3,4} = \frac{m}{2R_{e\infty}} \left(\frac{H}{P_r \rho u}\right) \frac{1}{m^2-1} \left\{ \left[ (P_r + \gamma) m^2 - 1 \right] \pm \left[ ((P_r + \gamma) m^2 - 1)^2 - 4 P_r \gamma (m^2 - 1) \right]^{\frac{1}{2}} \right\}$$

where $m^2 = u^2 / c^2$, $\gamma = 1.4$.

It is clear that if $u \geqslant c$, then $\Lambda_3$, $\Lambda_4 \geqslant 0$, but if $u < c$, then $\Lambda_3 < 0$. Therefore, for the latter case the necessary condition for (3.1) to be evolutional is not satisfied and the Cauchy problem is ill-posed.

From the point of view of the method of integral relations(MIR) the conclusion obtained above can also be made. In this case the solution on the stagnation line can not be specified independently, it depends on the downstream flow[8].


## 4.  NUMERICAL METHODS

As mentioned above for our system of equations it is reasonable to pose the boundary value problem in a close domain including subsonic flow fields. However in order to save the computer time and storege significantly, the entire flow field is divided into four subregions $\Omega i$ (i=1,2,3,4). The principle of the division is that the mutual influence between two adjacent subregions be small. Then the solutions in $\Omega 1$ and $\Omega 3$ are obtained using methods for solving boundary value problems and the solutions in $\Omega 2$ and $\Omega 4$ are found by means of the marching technique. Up to now we have investigated numerical methods for the $\Omega 1$ and $\Omega 2$ only. In the computation, the position of the shock wave is determined automatically.

### Numerical Methods for the region $\Omega 1$

The method of lines was extended to viscous flow by the authors in 1970 and proved to be successfuly in the calculation of flow field $\Omega 1$ (Fig. 2.) Its basic idea is similar to [3], the differences being:

(A) Most of the subsonic region must be included in computational region $\Omega 1$;

(B) The resulting difference equations were rewritten in the form of a set of algebraic equations with dominant diagonal elements;

(C) The difference equations were solved using the method of circular iterations (MCI). The iterative process, in which for a fixed equation the ray lines were changed from one to another, was faster than the process, in which for a fixed ray line the equations were changed;

(D) The positions of the shock wave were determined simultaneous using the Newton's method.

In addition, the R.T.Davis' method[2] was applied also to the calculation of flow in $\Omega$1; as the first approximation the angle of the shock wave $\sigma$ and the normal velocity $V/V_S$ can be taken from the results of the corresponding inviscid flow for arbitrary Mach number. Hence, the thin shock layer approximation, as was used by R.T. Davis, was not necessary. Comparisions of the numerical results, obtained with the method of lines, with R.T.Davis' method, results of the inviscid flow [11] , and that of the full Navier-stokes equations (FNSE) [13] are given in Fig. 3.

### Numerical Methods for the region $\Omega$2

Now we present the results of our work during the period 1976-1978.

The SNSE can be also expressed in the form:

$$(4.1) \quad A_1 \frac{\partial X_1}{\partial x} + B_1 \frac{\partial X_1}{\partial y} = F_1 \left( x, y, X_1, X_2, \frac{\partial X_2}{\partial x}, \frac{\partial X_2}{\partial y} \right),$$

$$A_2 \frac{\partial X_2}{\partial x} + B_2 \frac{\partial X_2}{\partial y} = C_2 \frac{\partial^2 X_2}{\partial y^2} + F_2 \left( x, y, X_1, X_2, \frac{\partial X_1}{\partial x}, \frac{\partial X_1}{\partial y} \right).$$

where $X, Y \in D$ ( $0 \leqslant y \leqslant 1$, $0 \leqslant x \leqslant L$ ), $X_1 = (P, V)^T$, $X_2 = (u, T)^T$, $A_i$, $B_i$, $C_i$ are 2x2 matrices. In addition, the initial value and boundary value conditions on the body and the shock wave must be added to (4.1). The following implicit difference scheme is often used:

$$(4.2) \quad \begin{cases} A_{1, m+\frac{1}{2}}^{n+\frac{1}{2}} \left( \frac{\partial X_1}{\partial x} \right)_{m+\frac{1}{2}}^{n+s} + B_{1, m+\frac{1}{2}}^{n+\frac{1}{2}} \left( \frac{\partial X_1}{\partial y} \right)_{m+\frac{1}{2}}^{n+s} = F_{1, m+\frac{1}{2}}^{n+\frac{1}{2}}, \quad (m = 0, 1, \cdots, M-1) \\[2ex] A_{2, m}^{n+\frac{1}{2}} \left( \frac{\partial X_2}{\partial x} \right)_{m}^{n+s} + B_{2, m}^{n+\frac{1}{2}} \left( \frac{\partial X_2}{\partial y} \right)_{m}^{n+s} = C_{2, m}^{n+\frac{1}{2}} \left( \frac{\partial^2 X_2}{\partial y^2} \right)_{m}^{n+\frac{1}{2}} + F_{2, m}^{n+\frac{1}{2}}, \\[2ex] \quad (m = 1, 2, \cdots, M-1), \ (n = 0, 1, 2, \cdots). \end{cases}$$

$$\left( \frac{\partial X}{\partial x} \right)_{m+\frac{1}{2}}^{n+s} = \frac{(X_m^{n+1} + X_{m+1}^{n+1}) - (X_m^n + X_{m+1}^n)}{2\Delta x},$$

$$\left( \frac{\partial X}{\partial y} \right)_{m+\frac{1}{2}}^{n+s} = s \frac{X_{m+1}^{n+1} - X_m^{n+1}}{\Delta y} + (1-s) \frac{X_{m+1}^n - X_m^n}{\Delta y},$$

$$\left( \frac{\partial X}{\partial x} \right)_{m}^{n+s} = \frac{X_m^{n+1} - X_m^n}{\Delta x},$$

$$\left( \frac{\partial X}{\partial y} \right)_{m}^{n+s} = s \frac{X_{m+1}^{n+1} - X_{m-1}^{n+1}}{2\Delta y} + (1-s) \frac{X_{m+1}^n - X_{m-1}^n}{2\Delta y}.$$

For the solution of (4.2) the sweep method is applied for two coupled first order equations and uncoupled second order equations. In order to increase the stability of the sweep processes we suggest the following difference scheme:

$$(4.3) \quad \begin{cases} l_{i, m+\frac{1}{2}}^{n+\frac{1}{2}} \left[ \left( \frac{\partial X_1}{\partial x} \right)_{m+\frac{1}{2}}^{n+s} + \lambda_{i, m+\frac{1}{2}}^{n+\frac{1}{2}} \left( \frac{\partial X_1}{\partial y} \right)_{m+\frac{1}{2}}^{n+s} \right] = g_{i, m+\frac{1}{2}}^{n+\frac{1}{2}} \quad (i = 1, 2) \\[2ex] a_{j, m}^{n+\frac{1}{2}} \left( \frac{\partial f_i}{\partial x} \right)_{j, m}^{n+s} + b_{j, m}^{n+\frac{1}{2}} \left( \frac{\partial f_i}{\partial y} \right)_{m}^{n+s} = c_{j, m}^{n+\frac{1}{2}} \left( \frac{\partial^2 f_i}{\partial y^2} \right)_{m}^{n+s} + e_{j, m}^{n+\frac{1}{2}} \quad (j = 1, 2) \end{cases}$$

where $\lambda$ i are the eigenvalues of a system of two first order equations and $l$ i are the eigenvectors corresponding to the $\lambda$ i. A convection term $b(\partial f/\partial y)$ is treated in the following way:

$$\left(\frac{\partial f}{\partial y}\right)^{n+s}_{m} = \begin{cases} s\,\dfrac{f^{n+1}_{m} - f^{n+1}_{m-1}}{\Delta y} + (1-s)\,\dfrac{f^{n}_{m+1} - f^{n}_{m}}{\Delta y} \;, & \text{for } b \geqslant 0, \\[2ex] s\,\dfrac{f^{n+1}_{m+1} - f^{n+1}_{m}}{\Delta y} + (1-s)\,\dfrac{f^{n}_{m} - f^{n}_{m-1}}{\Delta y} \;, & \text{for } b < 0. \end{cases}$$

## 5.  NUMERICAL ANALYSIS OF THE MARCHING TECHNIQUE

In order to avoid the numerical instability encountered using the marching integration in $\Omega\,2$, two approaches can be used: A. change the structure of the coefficient matrix A of the system(3.1), B. control the increase of errors. In fact if the coefficient of any one of the x derivative term in (3.1) is specified, the roots of characteristic equation of PDE (3.1) becomes

$$(5.1) \qquad \Lambda_1 = \Lambda_2 = 0, \qquad \Lambda_3 = \frac{H\mu}{Re_\infty \rho u} \;, \qquad \Lambda_4 = \frac{\gamma}{P_r}\,\Lambda_3 \;.$$

This means that the necessary condition for the system of equations to be evolutional is satisfied. For example the marching technique will not encounter any difficulty by setting $P_x \equiv 0$ in the x-momentum equation [1].

We consider now a difference scheme with constant coefficient, similar to that by Helliwell-Lubard(1975):

$$(5.2) \qquad A\left(\frac{\partial X}{\partial x}\right)^{n+s}_{m} = C\left(\frac{\partial^2 X}{\partial y^2}\right)^{n+s}_{m} \;,$$

where

$$A = \begin{pmatrix} p/u & 0 & 1 & -p/T \\ 1 & 0 & 1/\rho u & 0 \\ 0 & 1 & 0 & 0 \\ 0 & 0 & 1/\rho & 1 \end{pmatrix} \;, \qquad C = \begin{pmatrix} 0 & 0 & 0 & 0 \\ E & 0 & 0 & 0 \\ 0 & 0 & 0 & 0 \\ 0 & 0 & 0 & E \end{pmatrix} \;, \qquad X = \begin{pmatrix} u \\ v \\ p \\ T \end{pmatrix} \;,$$

$$P_r \equiv 1 \;, \qquad E = \mu/(\rho u\,Re_\infty) \;.$$

Using the method of Fourier analysis it is not difficult to obtain all the eigenvalues of amplification matrix:

$$|\lambda_i| \leqslant 1 \qquad (i = 1,\,2,\,3)$$

$$|\lambda_4| \leqslant 1 \qquad \text{if} \quad u \geqslant c \;, \quad \tfrac{1}{2} \leqslant s \leqslant 1 \;,$$

$$|\lambda_4| > 1 \qquad \text{if} \quad u < c \;, \quad s = \tfrac{1}{2} \;\; (C\text{-}N \;\; scheme),$$

$$|\lambda_4| \leqslant 1 \qquad \text{if} \quad u < c \;, \quad \tfrac{1}{2} < s \leqslant 1 \;\; \text{and} \;\; \Delta x \geqslant \frac{N}{(2s-1)}\,\Delta y^2$$

where

$$N = (1/m^2 - 1) \,/\, \left(\frac{\mu\gamma\,(1-\cos k\Delta y)}{Re_\infty \rho u}\right) \;.$$

The latter inequality shows that the implicit treatment with an appropriate choice of step size is quite possible. This is because if the ill-posed problem is solved by difference schemes it is required that large step size and higher accuracy scheme in the marching integration be chosen. Our numerical tests show an approximate agreement with the above qualitative conclusion. We mention here that the authors have read paper [6] in which the results of the implicit treatment are also given for three-dimensional flow.

## 6.  COMPARISIONS OF THE VARIOUS TREATMENTS

For treating numerical instability during the marching integration explicit difference approximations has been used by different authors for $P_X$ or $U_X$ or $V_X$[1-4]. i. e.

$$(6.1) \quad \left(\frac{\partial \bar{f}}{\partial x}\right)^{n+s} = \frac{f^n - f^{n-1}}{\Delta x} + \frac{(2s+1)}{2}\Delta x \frac{\partial^2 f}{\partial x^2} + \cdots$$

It is first-order accurate only. So we suggested P-C procedure. i.e.

$$\left(\frac{\partial \bar{f}}{\partial x}\right)^{n+s} \simeq \frac{f^n - f^{n-1}}{\Delta x} \quad , \qquad \text{for the predictor step,}$$

$$\left(\frac{\partial f}{\partial x}\right)^{n+s} \simeq \frac{\bar{f}^{n+1} - f^n}{\Delta x} \quad , \qquad \text{for the corrector step.}$$

All the treatments were tested, in the begining of the marching integration all the different treatments for $(\partial f/\partial x)$ are very sensitive to the step size, but to the numerical results downstream the step size makes little difference. Finally, for illustration of the effects of the various schemes, some numerical results are presented in Fig. 3-7.

Now we are extending our algorithm to the case of three dimensional hypersonic flow over a sphere-cone.

## REFERENCES


1.  Lin,T.C. et al., Computers and Fluids, Vol.1, No. 1 (1973)
2.  Davis,R.T., AIAA J. Vol.8, No.5  (1970)
3.  Varonking,V.G. (Russian) Mechanics of liquid and gas, No.6 (1974)
4.  Helliwell,W.S. et al., Computers and Fluids, Vol.3,No.1(1975)
5.  Kaiser,J.E. et al., AD-669578 (1968)
6.  Murray,A.L. et al., AIAA J, Vol. 16, No. 12(1978)
7.  Gelfand,E.M., (Russian) Progress of Math. Sci., Vol. 14, No. 2(1959)
8.  Tolstykh,A.I. (Russian) J. Comp. Math. and Math. Phys. Vol.6, No. 1(1966)
9.  Waskiewicz,J.D. et al., AIAA J. Vol. 16, No.2(1978)
10.  Lees,L., Jet propulsion, Vol. 26, No.4 (1956)
11.  Chu,Y.L. et al., Difference methods for initial boundary value problems and flow around bodies, Science Press, Beijing, (1980)
12.  Yu,H.R., Private communication.
13.  Wu, H.M. et al., This proceeding.


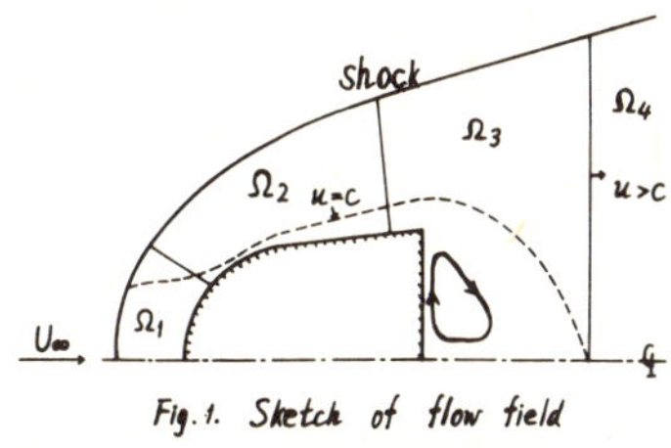


Fig.1. Sketch of flow field

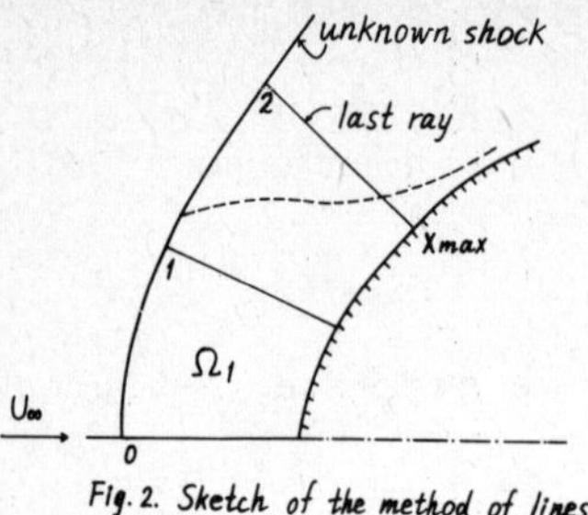

Fig. 2. Sketch of the method of lines

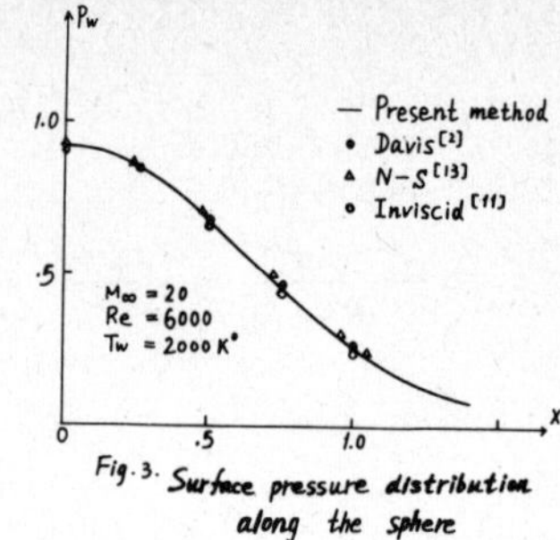

Fig. 3. Surface pressure distribution along the sphere

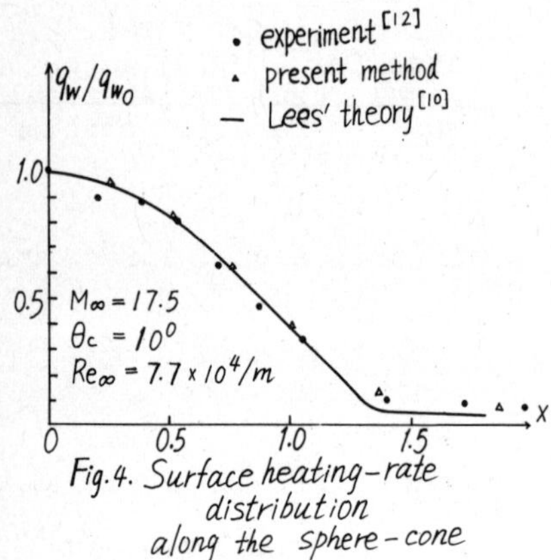

Fig. 4. Surface heating-rate distribution along the sphere-cone

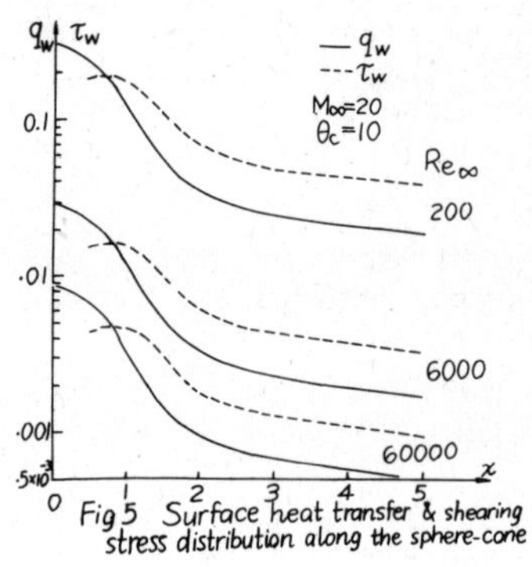

Fig 5  Surface heat transfer & shearing stress distribution along the sphere-cone

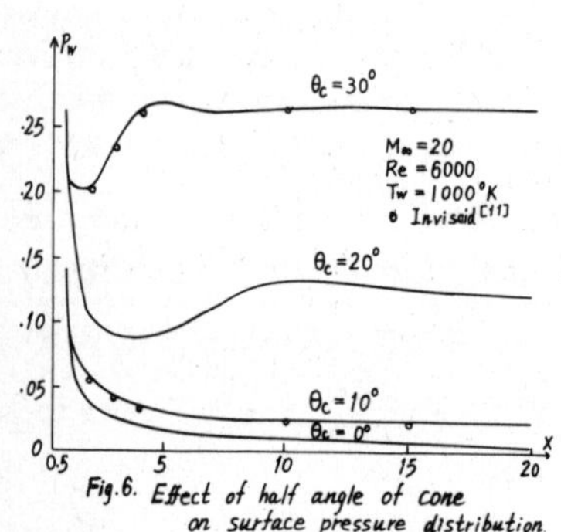

Fig. 6. Effect of half angle of cone on surface pressure distribution

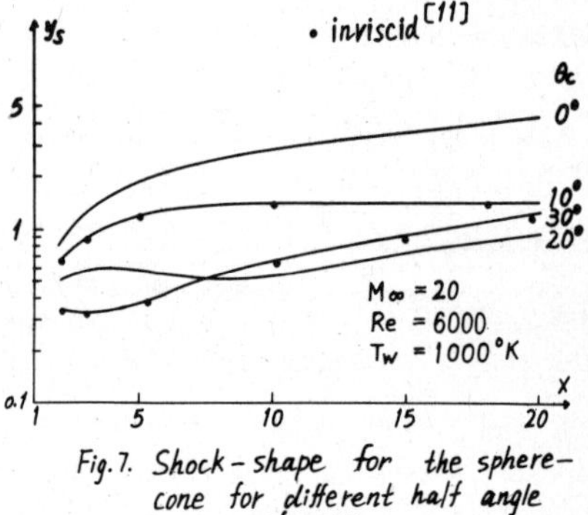

Fig. 7. Shock-shape for the sphere-cone for different half angle of cone

RECENT ADVANCES IN THE DEVELOPMENT OF IMPLICIT SCHEMES FOR THE
EQUATIONS OF FLUID DYNAMICS

R. F. Warming and Richard M. Beam
Ames Research Center, NASA, Moffett Field, California

## I.  INTRODUCTION

Physical problems described by the time-dependent gasdynamic equations are often
"stiff" by virtue of disparate characteristic speeds and/or length scales.  Examples
are readily found in astrophysics, meteorology, and aerodynamics.  In particular, the
unsteady compressible Navier-Stokes equations with turbulence models lead to an extremely stiff nonlinear system.  In the numerical solution of such problems, the
stability bound of an explicit algorithm forces a time step that can be orders of
magnitude smaller than required for accuracy and, as a consequence, leads to intoler-
ably long computation times.  This difficulty has led to the development of implicit
numerical techniques for stiff systems of partial differential equations (PDEs) (see,
e.g., Lindemuth & Killeen 1973; Briley & McDonald 1975; Berezin, Kovenya, & Yanenko
1975; Beam & Warming 1976,1978).

Consider a system of conservation laws of the general form

$$\frac{\partial U}{\partial t} = \frac{\partial}{\partial x}[-F(U) + V_1(U,U_x)] + \frac{\partial}{\partial y}[-G(U) + W_2(U,U_y)] + \frac{\partial}{\partial x}V_2(U,U_y) + \frac{\partial}{\partial y}W_1(U,U_x) \ , \qquad (1)$$

where  $U$  is the vector of conserved variables; $F$, $G$, $V$, and $W$  are flux vectors; and
$U_x = \partial U/\partial x$  and  $U_y = \partial U/\partial y$.  The appropriate time differencing method to be used on
(1) is determined by the eigenvalue spectrum associated with the operator on the
right-hand side.  At high Reynolds number, the operator has essentially pure imaginary
eigenvalues except for thin layers of the solution domain; in addition, the spectrum
is widely distributed, i.e., the physical problem is stiff.  Consequently, A-stable
linear multistep methods (Dahlquist 1963; Gear 1971) provide ideal time differencing
approximations.  The application of an A-stable method leads to an enormous banded
linear system of equations which can be efficiently solved by an alternating direction
procedure.

Classical alternating direction implicit (ADI) methods (Peaceman & Rachford 1955;
Douglas 1955; Douglas & Gunn 1964) were formulated for linear parabolic PDEs whose
operator spectrum is on the negative real axis, and the time differencing method was
generally the trapezoidal formula (Crank-Nicolson).  Since the eigenvalues associated
with (1) are predominantly imaginary and the trapezoidal formula is not usually an
effective method for stiff problems, it is clear that a generalization of the classical methods is needed.  (Classical ADI techniques were formally applied to PDE systems
by Lindemuth & Killeen 1973; Briley & McDonald 1975; Berezin, Kovenya, & Yanenko 1975;
and Beam & Warming 1976.)

In a recent paper (Warming & Beam 1979) we have shown that A-stable linear multistep methods (LMMs) combined with approximate factorization provide a natural framework for the construction of unconditionally stable ADI methods.  In a subsequent
paper (Beam & Warming 1979), the approach was extended to nonlinear PDEs.  The purpose
of this paper is to consider some new innovations and extensions. Before elaborating,
we briefly review some notation and theory for linear multistep methods.

## II.  PRELIMINARIES

A linear k-step method (Dahlquist 1963; Gear 1971) for a system of first-order
ordinary differential equations (ODEs)

$$\frac{du}{dt} = f(u,t) \ , \qquad u(0) = u_0 \qquad (2)$$

is defined by

$$\rho(E)u^n = \Delta t\sigma(E)f(u^n,t^n) \ , \qquad (3)$$

where $u^n$ is the numerical solution at $t = n\Delta t$, $\Delta t$ is the time step, $\rho$ and $\sigma$ are the generating polynomials

$$\rho(\zeta) = \sum_{j=0}^{k} \alpha_j \zeta^j , \qquad \sigma(\zeta) = \sum_{j=0}^{k} \beta_j \zeta^j , \tag{4}$$

and $E$ is the shift operator $Eu^n = u^{n+1}$. An LMM is said to be A-stable if its stability region contains all of the left half of the complex $\lambda\Delta t$ plane, including the imaginary axis (Dahlquist 1963; Gear 1971). (Here $\lambda$ is an eigenvalue of the Jacobian matrix $\partial f / \partial u$.) Dahlquist (1963) proved that the order of accuracy of an A-stable LMM cannot exceed two. A simple test for A-stability can be formulated in terms of positive real functions (Dahlquist 1978).

A class of methods closely related to LMMs is called one-leg methods (OLMs) and is defined by (Dahlquist 1976)

$$\rho(E)u^n = \Delta t f(\sigma(E)u^n, \sigma(E)t^n) . \tag{5}$$

The primary difference between (3) and (5) is that the OLM requires only a single function evaluation on the right-hand side.

The class of all two-step methods that are at least second-order accurate is given by

$$\rho(E) = (1 + \xi)E^2 - (1 + 2\xi)E + \xi , \tag{6a}$$

$$\sigma(E) = \theta E^2 + (\xi - 2\theta + 3/2)E - (\xi - \theta + 1/2) . \tag{6b}$$

The subclass of A-stable schemes is defined by the inequalities

$$\xi \leq 2\theta - 1 , \qquad \xi \geq -\frac{1}{2} .$$

The shaded region of Fig. 1 shows the domain of the parameters $(\theta, \xi)$ for which this class of methods is A-stable. Several well-known methods are special cases of the general two-step method (6) (see table 1 in Beam & Warming 1979).

A critical observation in constructing optimum unconditionally stable ADI methods for multilevel schemes is the choice of $\rho(E)u^n$ as the unknown variable rather than $u^{n+k}$ (Warming & Beam 1979). This choice ensures that an approximate factorization into a sequence of one-dimensional operators does not upset either the second-order temporal accuracy or the unconditional stability of the method.

## III. INNOVATIONS

In this paper we consider the following extensions of our earlier work (Warming & Beam 1979; Beam & Warming 1979).

A. _Implementation of one-leg methods._ The OLM (5) implemented with Newton's method and approximate factorization yields a computationally efficient ADI iterative method for solution of the PDE (1) and a noniterative ADI scheme if the Newton iteration is terminated after one-step. For brevity we present here only the noniterative scheme.

After a general coordinate transformation, the system of conservation laws (1) retains the same general form with $F(U) \leftarrow F(U,x,y,t), V_1(U,U_x) \leftarrow V_1(U,U_x,x,y,t)$, etc. We implement the one-leg (temporal) method (5) by using Newton's method. With the proper choice of unknown variable and initial value (Sec. IV of Beam & Warming 1979), the noniterative (one Newton step) scheme can, after approximate factorization, be implemented in the ADI form as

$$\left\{ I + \omega\Delta t \frac{\partial}{\partial x}[A(\bar{U},\hat{t}) - R(\tilde{U},\hat{t})\frac{\partial}{\partial x}]\right\}\rho(E)U^* = \text{RHS} , \tag{7a}$$

$$\left\{ I + \omega\Delta t \frac{\partial}{\partial y}[B(\bar{U},\hat{t}) - S(\tilde{U},\hat{t})\frac{\partial}{\partial y}]\right\}\rho(E)U^n = \rho(E)U^* , \tag{7b}$$

$$U^{n+k} = \frac{1}{\alpha_k}\left[\rho(E)U^n - \sum_{j=0}^{k-1} \alpha_j E^j U^n\right] \quad , \tag{7c}$$

where

$$RHS = \Delta t\left\{\frac{\partial}{\partial x}\left[-F(\bar{U},\hat{t}) + V_1(\tilde{U},\tilde{U}_x,\hat{t}) + V_2(\tilde{U},\tilde{U}_y,\hat{t})\right] + \frac{\partial}{\partial y}\left[-G(\bar{U},\hat{t}) + W_1(\tilde{U},\tilde{U}_x,\hat{t}) + W_2(\tilde{U},\tilde{U}_y,\hat{t})\right]\right\} \quad ,$$

$$\bar{U} = \left[\sigma(E) - \omega\rho(E)\right]U^n, \quad \tilde{U} = \sigma_e(E)U^n, \quad \hat{t} = \sigma(E)t^n, \quad \omega = \beta_k/\alpha_k, \quad A = \frac{\partial F}{\partial U}, \quad R = \frac{\partial V_1(U,U_x)}{\partial U_x}, \text{etc.}$$

The operator $\sigma_e(E)$ is an explicit operator, i.e., it can be computed from known data when advancing the numerical solution from $n + k - 1$ to $n + k$. The operator $[\sigma(E) - \omega\rho(E)]$ is also explicit by virtue of the choice of $\omega = \beta_k/\alpha_k$. The justification for explicit evaluation of the mixed derivative terms $\partial V_2/\partial x$ and $\partial W_1/\partial y$ is presented in the following section.

B. <u>ADI methods for equations with mixed derivatives.</u> The mixed spatial derivatives $\partial V_2/\partial x$ and $\partial W_1/\partial y$ on the right-hand side of Eq. (1) preclude the construction of an efficient implicit algorithm by a spatial factorization into a product of one-dimensional operators. This difficulty is avoided by applying an OLM to Eq. (1) where the right-hand side is split so that the mixed derivative terms are treated explicitly with second-order accuracy and the remaining terms treated implicitly (Beam & Warming 1980).

It is rather surprising that one can treat the mixed derivative explicitly with a second-order accurate scheme and still develop an unconditionally stable algorithm. Previously (Beam & Warming 1980) we determined the parameter space for which the resulting two-dimensional ADI schemes are second-order accurate and unconditionally stable. We have obtained similar stability results for the p-dimensional parabolic equation

$$\frac{\partial u}{\partial t} = \sum_{\ell=1}^{p} \sum_{m=1}^{p} c_{\ell m} \frac{\partial^2 u}{\partial x_\ell \partial x_m} \quad , \tag{8}$$

where $c_{\ell m} = c_{\ell m}(t,\vec{x})$ and $[c_{\ell m}]$ is a positive definite matrix at every point $(t,\vec{x})$. The p-dimensional ADI algorithm is

$$\prod_{m=1}^{p}\left(1 - \omega\Delta t c_{mm}\frac{\partial^2}{\partial x_m^2}\right)\rho(E)u^n = \Delta t \sum_{m=1}^{p} c_{mm}\frac{\partial^2 \bar{u}}{\partial x_m^2} + \Delta t\sum\sum_{m\neq\ell} c_{\ell m}\frac{\partial^2 \tilde{u}}{\partial x_\ell \partial x_m} \quad . \tag{9}$$

where $c_{\ell m} = c_{\ell m}(\hat{t},\vec{x})$ and $\bar{u}$, $\tilde{u}$ and $\hat{t}$ are defined following Eq. (7). The A-stable domains for the second-order two-step methods (6) when applied to (9) in two and three space dimensions are indicated in Fig. 2.

C. <u>Flux vector splitting.</u> The conservation-law form of the inviscid gasdynamic equations has the remarkable property that the nonlinear flux vectors are homogeneous functions of degree one in the conserved variables. Consequently, F(U) (see Eq. (1)) can be split into two parts (Steger & Warming 1979).

$$F = F^+ + F^- \quad , \tag{10}$$

where $F^+$ is a subvector associated with the positive eigenvalues of the Jacobian matrix $\partial F/\partial U$, and $F^-$ is associated with the negative eigenvalues. A similar splitting can be applied to $G(U)$. The principal advantage of flux vector splitting is that appropriate "upwind" differencing can be applied to the individual components $F^+$ and $F^-$ leading to robust implicit algorithms. In addition, the opposite phase error of symmetric and upwind schemes allows a considerable reduction of phase error if the two schemes are alternated on successive time steps. (This notion is the basis of Fromm's method of zero-average phase error (Fromm 1971).) These features are easily incorporated into the ADI-OLM of Sec. IIIA.

D.  <u>P-dimensional wave equation.</u>  As part of an extension of stability theory for ADI methods to p dimensions we consider the p-dimensional wave equation

$$\frac{\partial^2 u}{\partial t^2} = \sum_{\ell=1}^{p} \sum_{m=1}^{p} c_{\ell m} \frac{\partial^2 u}{\partial x_\ell \partial x_m} , \tag{11}$$

where  $c_{\ell m} = c_{\ell m}(t,\vec{x})$  and  $[c_{\ell m}]$  is a positive definite matrix.  The development follows that of (A) and (B) of this section for first-order equations.  A one-leg method for integrating the initial value problem

$$\frac{d^2 u}{dt^2} = f(u,t), \qquad u(0) = u_0, \qquad u'(0) = u_0', \tag{12}$$

is defined by

$$\rho(E)u^n = \Delta t^2 f(\sigma(E)u^n, \sigma(E)t^n) ,$$

where the generating polynomials  $\rho(\zeta)$  and  $\sigma(\zeta)$  have the same formal definitions as (4).  Using the class of unconditionally stable OLMs, one can construct the following ADI-OLM for the p-dimensional wave equation (11):

$$\prod_{m=1}^{p} \left( 1 - \gamma \Delta t^2 c_{mm} \frac{\partial^2}{\partial x_m^2} \right) \rho(E)u^n = \Delta t^2 \sum_{\ell=1}^{p} \sum_{m=1}^{p} c_{\ell m} \frac{\partial^2 \bar{u}}{\partial x_\ell \partial x_m} , \tag{13}$$

where

$$c_{\ell m} = c_{\ell m}(\hat{t},\vec{x}), \qquad \bar{u} \equiv [\sigma(E) - \gamma\rho(E)]u^n, \qquad \hat{t} = \sigma(E)t^n, \qquad \gamma = \beta_K/\alpha_K . \tag{14}$$

The method also has the property that the mixed derivatives are treated explicitly. The subclass of unconditionally stable two-step OLMs that are unconditionally stable for (13) is described elsewhere (Warming & Beam 1980).

E.  <u>Boundary conditions.</u>  A very difficult problem in constructing implicit algorithms for PDEs of the form (1) is the proper treatment of boundary conditions.  One must consider (1) boundary conditions for a well-posed PDE (see, e.g., Oliger & Sundstrom 1978); (2) extra boundary conditions often required for the finite-difference equations; (3) choice of variables (e.g., primitive or conservative); and (4) numerical stability.  The inclusion of the boundary conditions in the stability analysis presents a formidable mathematical problem and the numerical stability remains (in general) an open theoretical question.  In practice, the "unconditional stability" is generally lost when numerical oscillations appear near boundaries — either external or internal (shocks, slip surfaces, etc.).  Intuition and numerical experiments suggest that implicit treatment of the boundary conditions (Beam & Warming 1978) is necessary (but not sufficient) for unconditional stability; however, Gustafsson (1980) provides an example that shows that implicit treatment is not always a necessary condition.  Obviously, the improvement of boundary condition theory and implementation offers much promise for the improvement of implicit algorithms for the gasdynamic equations.

REFERENCES


Beam, R. M. & Warming, R. F. 1976 J. Comp. Phys. 22, 87–110.
Beam, R. M. & Warming, R. F. 1978 AIAA J. 16, 393–402.
Beam, R. M. & Warming, R. F. 1979 Paper 79-1446, Proc. AIAA 4th Comp. Fluid Dyn. Conf., Williamsburg, Va.
Beam, R. M. & Warming, R. F. 1980 SIAM J. Scient. & Stat. Comp., 1, 131–159.
Berezin, Y. A., Kovenya, V. M., & Yanenko, N. N. 1975 Proc. Fourth Int. Conf. on Num. Meth. in Fluid Dyn., Lect. Notes in Phys. 35, Springer-Verlag.
Briley, W. R. & McDonald, H. 1975 Proc. Fourth Int. Conf. on Num. Meth. in Fluid Dyn., Lect. Notes in Phys. 35, Springer-Verlag.

Dahlquist, G. 1963 BIT 3, 27–43.

Dahlquist, G. 1976 Lect. Notes in Math. 506 (eds. A. Dold and Eckmann) Springer-Verlag.

Dahlquist, G. 1978 Recent Advances in Numerical Analysis, de Boor, C. and Golub, G. eds., Academic Press, New York.

Douglas, J. 1955 J. Soc. Indust. Appl. Math. 3, 42–65.

Douglas, J. & Gunn, J. E. 1964 Numer. Math. 6, 428–453.

Fromm, J. E. 1971 Proc. of the Second Int. Conf. on Num. Meth. in Fluid Dyn., Lect. Notes in Phys. 8, Springer-Verlag.

Gear, C. W. 1971 Numerical Initial Value Problems in Ordinary Differential Equations, Prentice-Hall.

Gustafsson, B. 1980 Private communication.

Lindemuth, I. & Killeen, J. 1973 J. Comput. Phys. 13, 181–208.

Oliger, J. & Sundstrom, A. 1978 SIAM J. Appl. Math. 35, 419–446.

Peaceman, D. W. & Rachford, H. H. 1955 J. Soc. Indust. Appl. Math 3, 28–41.

Steger, J. L. & Warming, R. F. 1979 NASA TM 78605 (to appear in J. Comp. Phys.).

Warming, R. F. & Beam, R. M. 1979 BIT 19, 395–417.

Warming, R. F. & Beam, R. M. 1980 NASA TM (in preparation).

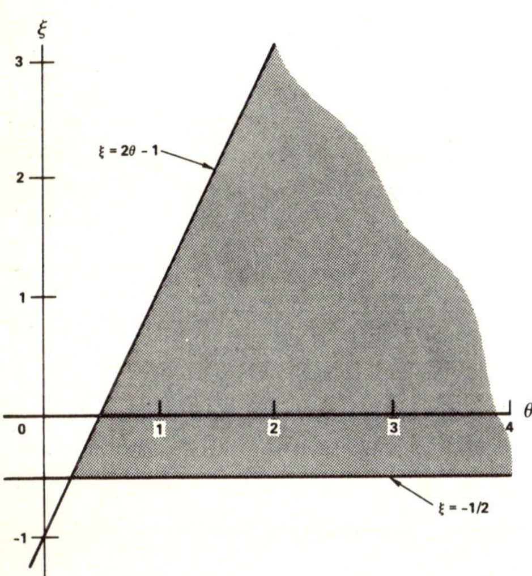


Fig. 1. A-stable domain of the parameters $(\theta, \xi)$ for the class of all second-order two-step methods.

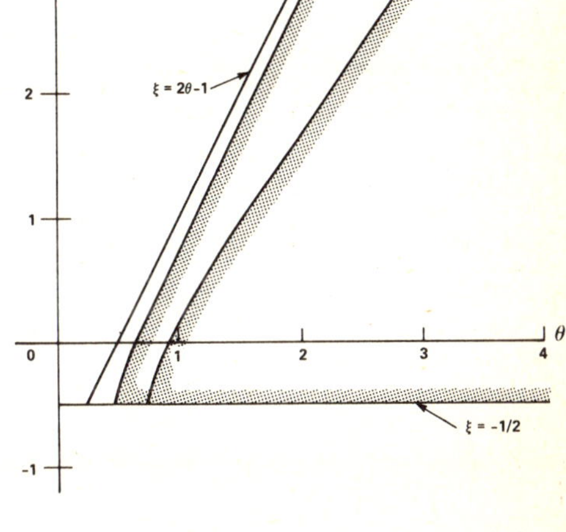


Fig. 2. A-stable domain of the parameters $(\theta, \xi)$ for the two-step method (6) applied to the ADI algorithm (9) in two and three dimensions.

High Resolution Difference Schemes
for Compressible Gas Dynamics


Paul Woodward and Phillip Colella*

E. O. Lawrence Livermore National Laboratory
Livermore, California  94550


The new schemes to be presented here have grown out of an extensive comparison
of a variety of difference methods.  The new schemes represent an attempt to
combine the advantages and avoid the disadvantages of the schemes which were
compared--namely, the von-Neumann-Richtmyer scheme [1], Godunov's scheme [2],
MUSCL [3,4], and Glimm's scheme [5,6,7].

We list the advantages of the various schemes first.  The principal advantage
of the von-Neumann-Richtmyer scheme is its use of a staggered grid.  Densities,
internal energies, and hence pressures also are prescribed at zone centers, while
velocities are prescribed at zone interfaces.  This grid structure is well suited
to the _Lagrangian_ equations of hydrodynamics, because it allows narrow-based
differences to be used to construct the necessary gradients.  The result is that
unusually high resolution of flow structure is obtained in Lagrangian problems.

The advantage of Godunov's scheme is the clear physical picture upon which it
is based.  Rather than replacing an infinite Taylor series by a truncated one,
this scheme replaces a physical system of complex structure by a simpler one
consisting of structureless zones.  This simpler system is evolved _exactly_ for a
time step, and then a similar replacement is made.  Naturally, the method rests on
the assumption that a Taylor series can be truncated, but the physical picture is
always clear.  To a physicist, this formulation has im- mense appeal.  By carrying
the accuracy of the physical representation one order higher than Godunov's
scheme, MUSCL combines the advantage of the clear physical picture with very high
resolution of the flow structure.

Glimm's scheme differs from Godunov's scheme in one essential way.  For the
variable values in its structureless zones it chooses those at some representative
point within each zone.  This point has the same location within each zone at a
given time step, and it follows some well-distributed, pseudo-random sequence from
one time step to another.  The most important effect of this procedure is to give
up exact conservation of mass, momentum, and energy in an effort to force all flow
discontinuities to zone boundaries, where they can be treated exactly by the

method.  Because errors arising in the improper treatment of discontinuities in
the flow can severely contaminate a computation with a standard difference method,
the treatment of discontinuities in Glimm's method gives that scheme unequaled
resolution of flow structure in one-dimensional problems.

When the above schemes are compared on a difficult two-dimensional flow
problem, their disadvantages are readily apparent.  The staggered grid of the
von-Neumann-Richtmyer scheme, which is so convenient in Lagrangian calculations,
is very badly suited to Eulerian calculations.  Particularly difficult to
formulate is the conservation of total energy.  An additional disadvantage is the
necessity to treat discontinuities as smooth flow regions with steep gradients.
This is done by adding in an artificial viscous pressure which smears out the
discontinuities over at least two zones.  The main disadvantage of Godunov's
scheme is its relatively poor resolution of flow structure.  MUSCL has the highest
resolution of these four schemes, but that resolution is limited by an
extrapolation procedure which is made at the beginning of each time step.  MUSCL
uses as data a zone-centered average value and first derivative of each variable.
From these, values of all variables at the zone interfaces must be constructed in
order to compute fluxes of conserved quantities during the time step.  The
extrapolation from the center of the zone to the zone interface is responsible for
most of the error in MUSCL.  Finally, the disadvantage of Glimm's scheme is that
its very special properties in one-dimensional problems are lost in two
dimensions, and the scheme must be abandoned in favor of a much lower resolution
method in the neighborhood of discontinuities (see [7]).  Because the principal
advantage of Glimm's scheme in 1-D flows was its treatment of discontinuities, the
hybridization of the scheme for 2-D problems results in a poorer scheme than
either MUSCL or the von-Neumann-Richtmyer scheme.

We have devised two new difference schemes which avoid all these disadvantages
and combine the advantages listed above.  The key ingredients are: (1) the
approach of Godunov's method in replacing a complicated physical system with a
simpler one of a standard form, (2) the translation of this assumed spatial
structure inside zones into temporal structure at the interfaces by solving
Riemann's problem as in Godunov's scheme and using the characteristic equations as
in MUSCL, and a new ingredient (3) the use of both zone-averaged values and
interface values of variables in order to define a distribution of each variable
at every point which is continuous except at true flow discontinuities and which
conserves mass, momentum, and energy exactly.  We have devised a second-order
method which uses a piecewise linear distribution for each variable with kinks at
zone centers and zone interfaces.  In addition we constructed a second-order
method which uses a piecewise parabolic distribution for each variable with kinks
only at the zone interfaces.  In 1-D test problems, both new schemes show at least

twice the resolution of MUSCL, the best of the four schemes discussed above.  Only the piecewise parabolic scheme has been run on 2-D problems.  It preserves the high resolution of its 1-D tests and is thus able to obtain a more accurate a flow description than MUSCL while using only half as many zones in each dimension.  The gains over the other schemes discussed are still larger.  The new algorithm is not yet optimized, but it presently requires only 30% more computer time per zone per time step than does MUSCL.  The gain in time consumed to achieve a given accuracy is thus a factor of 3 in 1-D.  In 2-D, a more complicated operator splitting algorithm doubles the time consumed, so that the gain is still only a factor of 3.  However, we expect that the new method can be speeded up considerably, and we hope to do so in the near future.

In Fig. 1 all the schemes discussed above are compared using the example of the flow of air (gamma is 1.4) through a duct containing a step.  Initially the flow is everywhere to the right at Mach 3, with $\rho = 1.4$, $p = 1$, $c = 1$.  The duct width is 1, its length is 3, and the step of height .2 is located a distance of .6 from the entrance.  All the results in Fig. 1 were obtained with a uniform Cartesian grid with $\Delta x = \Delta y = .05$.  At the exit a "flowout" boundary condition is applied, but this is unimportant because the flow to the right is always slightly supersonic there.  The system is shown at time 4, when a complicated system of shock reflections, rarefaction waves, and contact discontinuities is present.  This problem was used by Emery in 1969 [8] to compare the methods of Lax, Rusanov, and Lax and Wendroff (Emery used a very slightly different duct with a grid of nodal points with $\Delta x = \Delta y = 1/27$).  One of us also used this problem to demonstrate the MUSCL scheme described by van Leer in [3].  In that article the relatively structureless steady flow in the duct is displayed.  This steady flow is attained at about time 12.

Because of the lack of space, we show only the contours of density at time 4. These are the most difficult to compute correctly, because of the weak contact discontinuities which emanate from the two shock triple points associated with the two Mach reflections of the bow shock at time 4.  In Emery's article [8], only pressure contours are shown, and it is likely that the weak contact discontinuities in the flow were not resolved by any of the three methods he compared.  In Fig. 1a the results of Godunov's scheme are shown.  There is some indication of the Mach reflection at the upper wall.  In Fig. 1b, the Glimm-Godunov hybrid scheme shows only some improvement over Fig. 1a at the cost of introducing noise from the random choice feature of Glimm's scheme.  If the Mach reflection could be fully resolved (it is indeed resolved with 4 times as many zones), the weak contact discontinuity would be quite sharp.  After an initial smearing by Godunov's method, Glimm's method preserves the relatively narrow contact region.

A dramatic increase in resolution results from using a second-order accurate
scheme.  The results in Fig. 1c were obtained with the BBC code [9].  This code
uses a modified von-Neumann-Richtmyer scheme devised by DeBar [10] for its
Lagrangian step, and a MUSCL remap step on a staggered grid devised by Woodward.
To obtain the thin shocks shown here, the artifical viscosity was set to zero in
the Lagrangian step.  This has resulted in a mild oscillation behind each shock
which is most evident when the pressure is plotted.  When the artificial viscosity
is turned on, the shocks double in width and the flow resolution is significantly
degraded.  Especially to one who considers the staggered grid formulation both
confusing and inconvenient, these results are remarkably good.  The contact
discontinuity near the upper wall is spread over 2 to 3 zones, but it is clearly
visible.  Also the upper Mach stem is in the correct position directly above the
step, and it has the correct length (as proven by a run on a more refined mesh of
$\Delta x = \Delta y = .02$ which is shown in Fig. 2).  The MUSCL results shown in Fig. 1d
are of comparable quality.  They are superior in that the post-shock oscillations
of BBC are not present and the contact discontinuity is more sharply defined.
However, somewhat more entropy is artificially produced as the flow rounds the
corner of the step.  The result is a classic interaction of a shock with a
boundary layer, which produces the second, very weak reflected shock from the top
of the step at $x = 1.4$.  These two codes run at almost precisely the same
speed—2800 points per sec per cycle on a CDC 7600 and 20000 pts/sec/cy on a Cray
I.  They both make use of separate Lagrangian and remap steps in each 1-D pass.

In Fig. 1e we show the results of the new scheme which uses piecewise
parabolic interpolation.  In 1-D this scheme can be made third-order accurate, but
because of its use of 1-D passes it can only be second-order accurate in 2-D.  The
scheme used here is thus made only second-order accurate in 1-D, although several
gestures toward higher order are included.  The results of this new scheme are
comparable in quality to those of Fig. 2, which were obtained with BBC using a
much finer grid of $\Delta x = \Delta y = .02$.  The resolution of the weak contact
discontinuities from both Mach stems is particularly notable.  A "monotonicity
trick" has been used to constrain the interpolation parabolae so that the
post-shock oscillations usually associated with high-order schemes are completely
absent.  In Fig. 3 we show results of the new scheme using a grid with $\Delta x = \Delta y$
$= .1$.  Evidently, even on this coarse grid the new scheme correctly resolves all
the essential features of this complicated flow.

Finally, for comparison with the schemes discussed by Sod [11], we have shown
in Fig. 4 results of the new piecewise linear scheme on his shock tube problem.
The results in Fig. 4 use a grid of 50 zones rather than Sod's 100, and are more
accurate than the results of any of the 12 schemes he compared.  MUSCL results on
this problem have been given by van Leer [3].

This work was performed under the auspices of the U.S. Department of Energy by the E. O. Lawrence Livermore National Laboratory under contract number W-7405-ENG-48.

*Permanent address:  Lawrence Berkeley Laboratory, Berkeley, California.

References:

1.  von Neumann, J., and Richtmyer, R. D. 1950.  $\underline{\text{J. Appl. Phys.}}$, $\underline{21}$, 232.

2.  Godunov, S. K. 1959.  $\underline{\text{Mat. Sb.}}$, $\underline{47}$, 271.

3.  van Leer, B. 1979.  $\underline{\text{J. Comp. Phys.}}$, $\underline{32}$, 101.

4.  van Leer, B., and Woodward, P. 1979.  Proc. TICOM Conf., Austin, Texas.

5.  Glimm, J. 1965.  $\underline{\text{Comm. Pure Appl. Math.}}$, $\underline{18}$, 697.

6.  Chorin, A. J. 1976.  $\underline{\text{J. Comp. Phys.}}$, $\underline{22}$, 517.

7.  Colella, P. 1980.  preprint.

8.  Emery, A. F. 1968.  $\underline{\text{J. Comp. Phys.}}$, $\underline{2}$, 306.

9.  Sutcliffe, W. G. 1973.  "BBC Hydrodynamics," E. O. Lawrence Livermore National Laboratory Report UCID-17013.

10. DeBar, R. 1974.  "Fundamentals of the KRAKEN code," E. O. Lawrence Livermore National Laboratory Report UCIR-760.

11. Sod, G. A. 1978.  $\underline{\text{J. Comp. Phys.}}$, $\underline{27}$, 1.

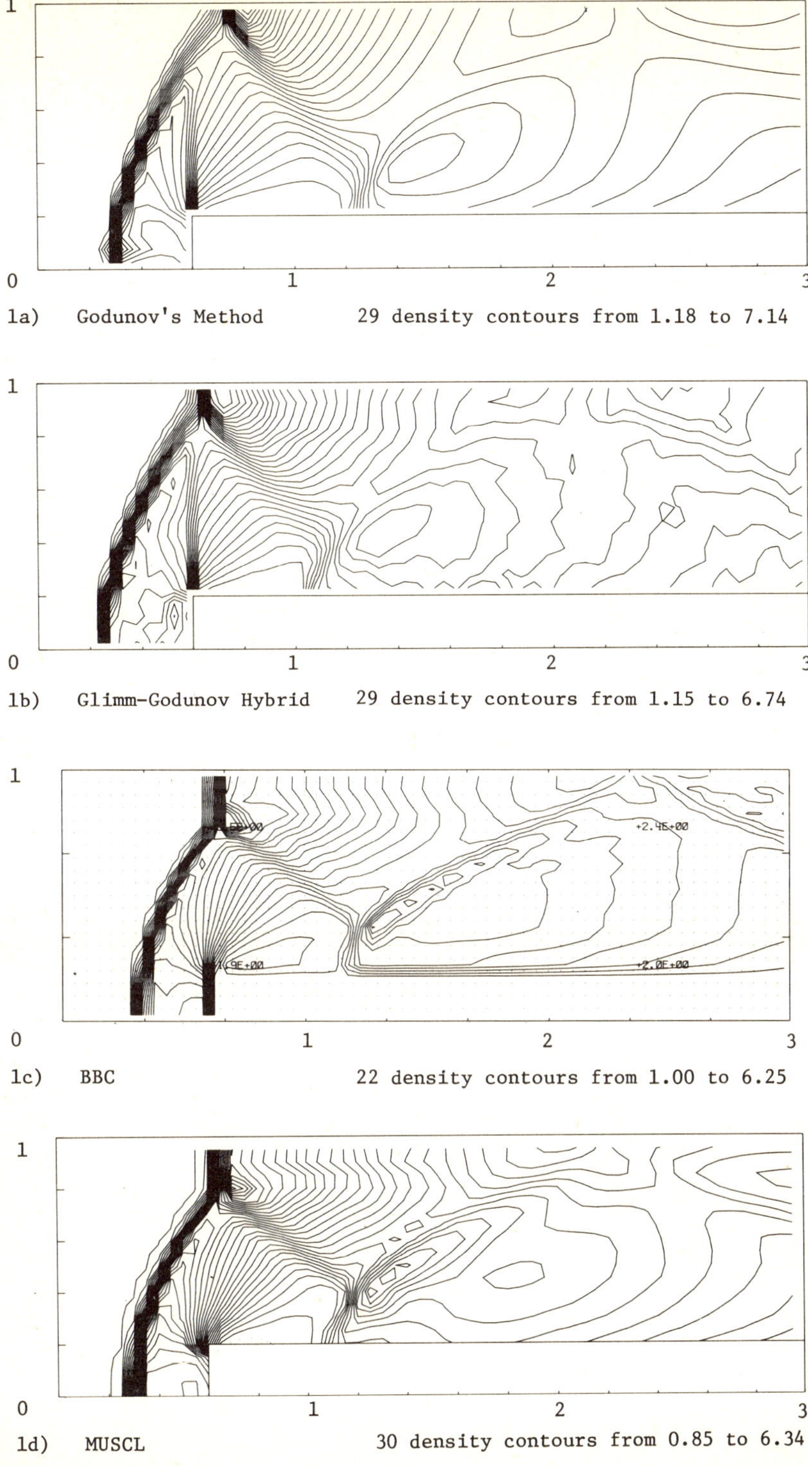

1a)    Godunov's Method        29 density contours from 1.18 to 7.14

1b)    Glimm-Godunov Hybrid    29 density contours from 1.15 to 6.74

1c)    BBC                     22 density contours from 1.00 to 6.25

1d)    MUSCL                   30 density contours from 0.85 to 6.34

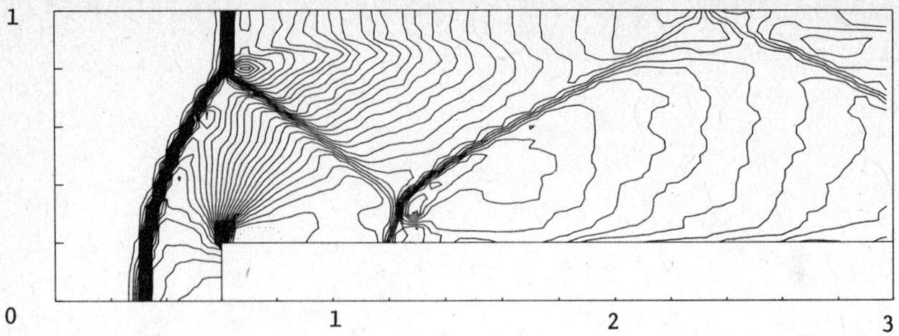

1e)    Piecewise Parabolic Method    30 density contours from 0.73 to
                                     6.31

Fig. 1.   Results of several difference schemes for the flow problem
          described in the text.  All schemes use a series of 1-D
          sweeps and Courant numbers of 0.8 or 0.9.  All use a uniform
          grid of 20 x 60 zones.  The methods are: (a) Godunov's method,
          (b) Glimm-Godunov hybrid, (c) BBC, (d) MUSCL, (e) the new piece-
          wise parabolic scheme.

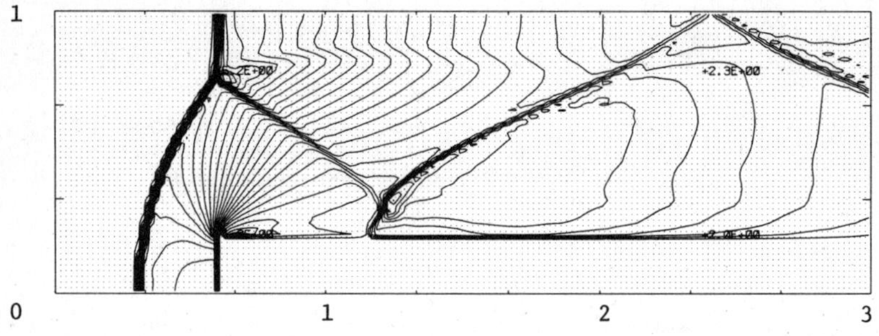

Fig. 2    BBC results for the same problem as in Fig. 1 but using
          a finer, uniform grid of 50 x 150 zones.

          23 density contours from 0.75 to 6.25 are shown as well
          as contours at densities 0.5 and 0.6.

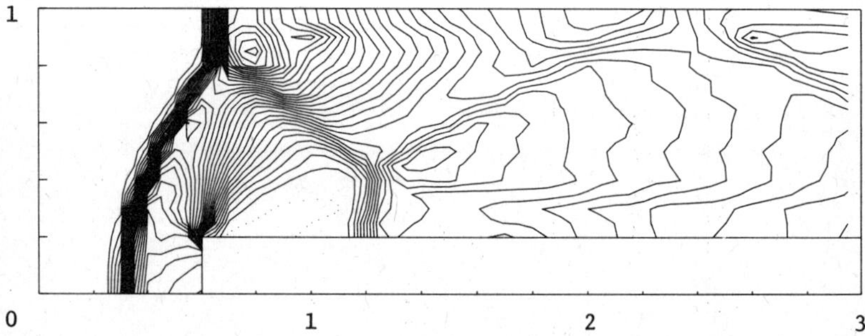

Fig. 3    Results of the new piecewise parabolic scheme for the same
          problem as in Fig. 1 but using a coarser, uniform grid of
          10 x 30 zones.

          30 density contours from 1.03 to 6.03 are shown.

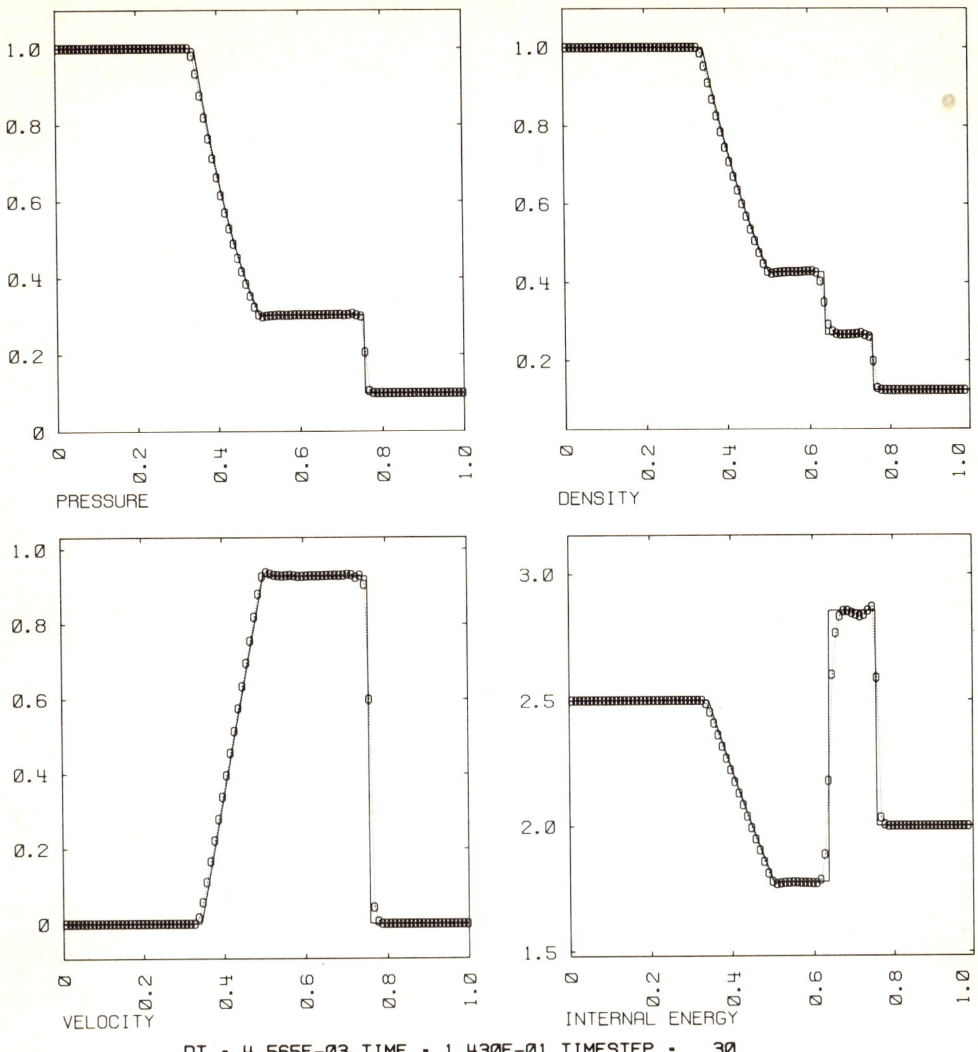



Fig. 4    Results of the new piecewise linear scheme for the shock tube
problem studied by Sod.  The solid line shows the exact
solution, and the zone average and interface values for the
grid of 50 zones are shown as circles.

# SOMS- A SECOND ORDER MONOTONE SCHEME FOR SHOCK CAPTURING AND ITS APPLICATION TO THE SOLUTIONS OF COMPRESSIBLE NAVIER-STOKES EQUATIONS


Hua-mo Wu   and Ming-liang Yang

Computing Centre of Academia Sinica, Beijing, China.


This work proposes the SOMS- a Second Order Monotone Scheme for hydrodynamic calculations. This scheme combines a first order and second order accurate schemes using a new switch function. The SOMS has been applied with success to the numerical solutions of compressible Navier- Stokes equations. The numerical tests show the advantage of this scheme in the monotone property of shock transition, the steepness of the shock profile and the high accuracy of the results in regions where the solution is smooth.

## 1.    INTRODUCTION.

As well known, for the numerical calculations in gas dynamics, the schemes with high order accuracy usually cause oscillations near the discontinuities, though they have a high order accuracy in the regions where the solution is smooth. On the other hand, the first order schemes are often monotonic, but have a lower accuracy for the smooth solutions. Therefore, it is natural to combine the merits of these two schemes.This idea is developing very rapidly today.  The first work in this direction, perhaps, was done by R. D. Fedorenko, later by Gol'den et al, van Leer, Harten and Zwas and so on ( see[1]-[4]). These authors constructed numerous switch functions for autonomous problems for inviscid flows. Some parameters in these switches have an empirical character without theoretical guarantee. Hence, it is difficult to choose these parameters for complicated calculations.

The switch proposed by van Leer is a very good one. It keeps the monotone property of the initial data, but it does not satisfy the maximum principle.

We had analysed the Fedorenko's switch[1] in 1963, and later suggested a new switch function, very similar to the van Leer's. We have proven that our hybrid scheme using our switch is monotonic and that it satisfies the maximum principle. We began to apply this scheme to the Navier-Stokes equations systematically since 1969. The calculation capturing the shock discontinuities ( i.e. without shock fitting ) was carried out for laminar blunt-body problems with bodies such as sphere, sphere-cone, leading edge, forward and backward steps, etc. The results of these calculations are satiafactory.

## 2. SOMS FOR THE CONVECTIVE EQUATION.

In general, suppose we have difference schemes $w^{n+1} = L_1 w^n$, and $w^{n+1} = L_2 w^n$ with first and second order accuracy for the differential equation $\partial w/\partial t = Mw$, then the combined scheme $w^{n+1} = (sL_1 + (1-s)L_2) w^n$ is called hybrid scheme and function $s$ is named as switch.

For the convection equation $\partial w/\partial t + a\partial w/\partial x = 0$, $a =$ const., we combine the one-sided scheme and the simple central scheme using a switch. The resulting scheme has the following forms:

SOMS: $\qquad w^0 = w_0 - a\Delta t/(2\Delta x)(w_1 - \bar{w}_{-1}) + a\Delta t/(2\Delta x)s_0(w_1 - 2w_0 + \bar{w}_{-1}),$ $\qquad$ (1)

SOMSc: $\qquad w^0 = w_0 - a\Delta t/(2\Delta x)(w_1 - \bar{w}_{-1}) + a\Delta t/(2\Delta x)(s_{\frac{1}{2}}(w_1 - w_0) - s_{-\frac{1}{2}}(w_0 - \bar{w}_{-1})),$

where $\qquad w^0 = w_j^{n+1} = w((n+1)\Delta t, j\Delta x),\ w_k = w_{j+k}^n = w(n\Delta t, (j+k)\Delta x),\ k = -1, 0, 1.$

$$\bar{w}_{-1} = \begin{cases} w^{-1} - \text{semi-implicit scheme for problems with} \\ \qquad\quad \text{steady solution,} \\ w_{-1} - \text{explicit scheme,} \end{cases}$$

and $s_0 = s_0(w)$ is a new switch function which we suggest to be

$$s_0 = \left( \frac{|w_1 - 2w_0 + \bar{w}_{-1}| + \varepsilon_1}{|w_1 - w_0| + |w_0 - \bar{w}_{-1}| + \varepsilon_2} \right)^k \qquad k \geq 0,\ \varepsilon_1 \approx 10^{-30},\ \varepsilon_2 \approx 10^{-10}, \qquad (2)$$

or

$$s_0 = \begin{cases} \left( \dfrac{|w_1 - 2w_0 + \bar{w}_{-1}|}{|w_1 - w_0| + |w_0 - \bar{w}_{-1}|} \right)^k \quad & \text{as } |w_1 - w_0| + |w_0 - \bar{w}_{-1}| \geq \varepsilon, \\[4pt] 0 & \text{as } |w_1 - w_0| + |w_0 - \bar{w}_{-1}| < \varepsilon, \end{cases}$$

and $\qquad s_{\pm\frac{1}{2}} = \max(s_0, s_{\pm 1}).$

Theorem: If $|a|\Delta t/\Delta x \leq 1/(1+k)$, then

i) the schemes SOMS and SOMSc are both monotonic, i.e. if $w_j - w_{j-1} \geq 0 (\leq 0)\ \forall j$, then $w^j - w^{j-1} \geq 0 (\leq 0)\ \forall j$;

ii) they satisfy the maximum principle, i.e.,

$$\min_j |w_j| \leq (\min_j |w^j|,\ \max_j |w^j|) \leq \max_j |w_j|. \qquad (3)$$

We note that van Leer proposed another switch, he took $||w_1 - w_0| - |w_0 - w_{-1}||$ instead of $|w_1 - 2w_0 + w_{-1}|$ in (2). The resulting scheme is monotonic too, but it dose not satisfy the maximum principle. It is easy to show that his switch can not wipe out the oscillations with equal amplitudes.

## 3.   SOMS FOR FLUID DYNAMIC EQUATIONS

The choice of the basic first order scheme and the switch function is the key problem in the construction of the hybrid schemes for the hydrodynamic calculations. One-sided scheme and Lax scheme are monotonic for inviscid flows and have been used as a basic first order scheme by many authors, but, as well known, for convection-diffusion equations Lax scheme will be  absolutely  unstable, so it can not be taken as a basic scheme for viscous calculations. One-sided scheme is conditional stable in this case, but its application in viscous gas dynamics will encounter some difficulties. The usual upwind schemes ( see[5,6]) and Rusanov scheme[7] are all non-monotonic in this case. These facts show the difficulties in the choice of the basic schemes in the construction of the hybrid scheme for Navier-Stokes equations, and it is probably one of the causes why people do not use the hybrid scheme for viscous calculations yet.

Consider the 1-D hydrodynamic system

$$\frac{\partial f}{\partial t}+\frac{\partial gu}{\partial x}+\frac{\partial b}{\partial x} = r\frac{\partial}{\partial x}\left(G\,\frac{\partial a}{\partial x}\right),\tag{4}$$

where

$$f=\begin{pmatrix}\rho\\ \rho u\\ \rho E\end{pmatrix},\quad g=\begin{pmatrix}\rho\\ \rho u\\ \rho E+p\end{pmatrix},\quad b=\begin{pmatrix}0\\ p\\ 0\end{pmatrix},\quad a=\begin{pmatrix}0\\ u\\ T\end{pmatrix},\quad G=\begin{pmatrix}0 & 0 & 0\\ 0 & 4/3\,\mu & 0\\ 0 & 4/3\,\mu u & \lambda\end{pmatrix},$$

$p,\rho,u,T,E,\mu,\lambda$ are pressure, density, velocity, temperature, total energy, coefficient of viscosity and heat conduction respectively. If $r=0$ we have inviscid flow, $r=1$ we have viscous flow.

Our hybrid scheme is as follows

$$f^O = f_0 - \frac{\Delta t}{\Delta x}\left(g_{\frac{1}{2}}u_{\frac{1}{2}} - \bar{g}_{-\frac{1}{2}}\bar{u}_{-\frac{1}{2}}\right)-\frac{\Delta t}{2\Delta x}\left(b_1-\bar{b}_{-1}\right)+\frac{\Delta t}{2\Delta x}F_0\tag{5}$$

$$+\frac{\Delta t}{\Delta x^2}r\left(G_{\frac{1}{2}}(a_1-a_0)-\bar{G}_{-\frac{1}{2}}(a_0-\bar{a}_{-1})\right),$$

where for SOMS

$$F_0 = s_0\left(u_{\frac{1}{2}}\,(g_1-g_0)- \bar{u}_{-\frac{1}{2}}\,(g_0-\bar{g}_{-1})\right),\tag{6}$$

and for SOMSc

$$F_0 = s_{\frac{1}{2}}\,u_{\frac{1}{2}}\,(g_1-g_0)-s_{-\frac{1}{2}}\,\bar{u}_{-\frac{1}{2}}\,(g_0-\bar{g}_{-1}),\tag{7}$$

$$s_{\pm\frac{1}{2}}=\max(s_0,\,s_{\pm 1}),\quad s_0 = s_0(\rho)\ \text{for}\ r=0,$$
$$s_0(p)\ \text{for}\ r=1,$$

$$h_{\frac{1}{2}} = \frac{1}{2}(h_1 + h_0), \quad h_{-\frac{1}{2}} = \frac{1}{2}(h_0 + \overline{h}_{-1}) \quad \text{for } h = g, u, G.$$

If $r=0$, $s_0=1$, the resulting scheme is a modification of the well known donor cell scheme[5,6], we call it MDC ( Modified Donor Cell) method. Usual donor cell method can be written in the form(5), if

$$g = \begin{pmatrix} \rho \\ \rho u \\ \rho E \end{pmatrix}, \quad b = \begin{pmatrix} o \\ p \\ pu \end{pmatrix}.$$

Our numerical tests show that the monotonic and the computational stability properties of MDC are much better than those of donor cell method and Rusanov's method. The MDC works well without any artificial viscosity to the pressure term. The tests show that it has better accuracy in comparision with other first order accurate schemes.

If we take $s_0=0$, then we have FTCS, which is of second order accuracy for steady state solutions.

Our SOMS combines the MDC and FTCS using the switch $s(\rho)$ in the inviscid case, but for the viscous calculations we prefer $s(p)$.


Remark 1. We may take other schemes such as Lax- Wendroff, Richtmyer, Mac Cormack as the second order scheme. In this case, for the convective equation our hybrid scheme will be

$$w^o = w_0 - a\Delta t/(2\Delta x)(w_1 - \overline{w}_{-1}) + \Delta t/(2\Delta x)F_0,$$

where for SOMS

$$F_0 = (|a| s_0 + a^2(1-s_0))(w_1 - 2w_0 + \overline{w}_{-1})$$

and for SOMSc

$$F_0 = (|a| s_{\frac{1}{2}} + a^2(1-s_{\frac{1}{2}}))(w_1 - w_0) - (|a| s_{-\frac{1}{2}} + a^2(1-s_{-\frac{1}{2}}))(w_0 - \overline{w}_{-1}).$$

We can prove the similar theorem with the only difference that the sufficient condition for monotone is $|a| \Delta t/\Delta x \leq 1/k$ ( $k=1,2,3,\ldots$).

Remark 2. SOMSc gives positions of shock better than SOMS, although the difference is very small.


## 4. <u>NUMERICAL TESTS AND CONCLUSIONS.</u>

The SOMS and SOMSc have been applied to the numerical solution of laminar flows with $M_\infty = 3- 20$, $Re_{\infty L} = 10^3 - 10^5$. We used always uniform mesh which contains about 30- 40 points in the direction normal to the body surface.

The first problem we have solved is the problem of solution of the Navier- Stokes equations in the near- stagnation region for blunt body We use a 40- grid point uniform mesh to solve this problem. Figure 1 shows the comparision of results using SOMS( $k=3$) with those using the Rankine- Hugoniot shock- fitting method[9]. We see that the agreement is very good and there are only two grid points in the shock transition region, and the error in the smooth region is less than 0.5%.

The typical 2-D laminar problems which we have solved are the problems of flows around bodies such as sphere , sphere - cone, leading edge and forward step and so on. Fig.2 to 4 show the distributions of surface pressure. We see that our results are always monotonic. Fig.4 show the comparision of our results with Carter's[10]. We see some strong oscillations in Carter's distribution. Fig.5 shows the distribution of surface pressure for the leading edge with M=14.1, and $Re_L$=1.04*10$^5$, the result is in good agreement with Holden's experiment [11], and  seems better than those given by Hung and Mac Cormack [12]. Fig.6 shows the heat transfer and the skin friction for the forward step. These distributions agree qualitatively with similar experiments with different M  and $Re_{\infty L}$. Fig.7 shows the distributions of pressure and temperature near the wall. We see some interesting oscillations in it. We are not able to explain this phenomena. It appears always in our tests.

Finally, we point out that both SOMS and SOMSc are efficient, accurate and adaptable to cases of high Mach number and Re number of the order from 10$^3$ to 10$^5$ and have the ability to capture well the strong shocks without oscillations. Besides they are easy to program.

## REFERENCES


1.   R.D.Fedorenko,   Soviet J.Comp.Math.Math.Phys.,2,6(1962).
2.   V.Y.Gol'den, et al, ibid, 5,6(1965).
3.   van Leer, J.Comp.Phys., 14, 361-370(1974).
4.   A.Harten, G.Zwas, J.Comp.Phys., 9,3(1972).
5.   A.Gentry, R.E.Martin,and B.J.Daly, J.Comp.Phys.,1,87-118(1966).
6.   C.Hirt, J.Comp.Phys., 2,339-355(1968).
7.   V.Rusanov, Soviet J.Comp.Math.Math.Phys., 1,2(1961).
8.   W.D.Hayes,R.F.Probstein, Hypersonic flow theory, Academia Press, New York and London, 1959.
9.   R.Q.Wang, et al, this Proceeding.
10.  J.E.Carter, NASA TR R-385, 1972.
11.  M.S.Holden, and J.R.Moselle, CAL Rep. No AF-2410-A-1 (1969).
12.  C.M.Hung, Mac Cormack, AIAA p 75-2 (1975).


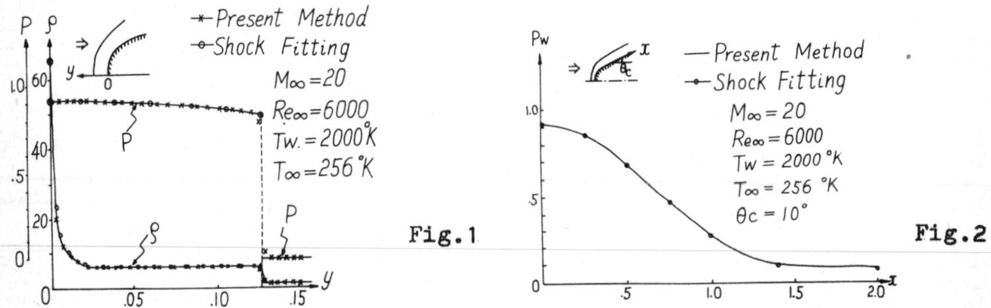

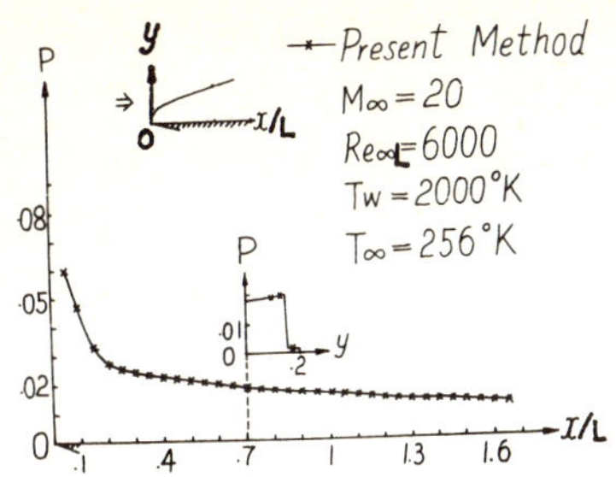


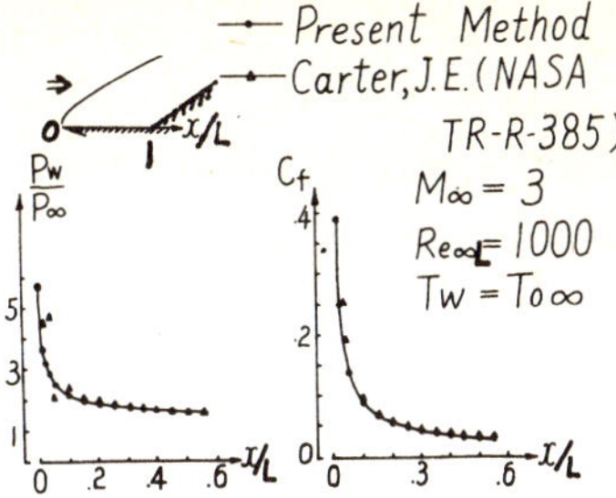


Fig. 3      Fig. 4

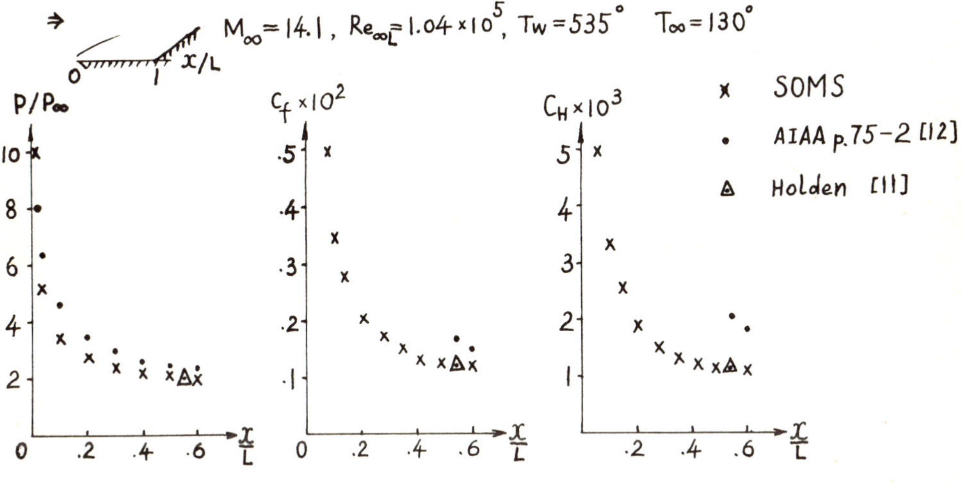


Fig.5

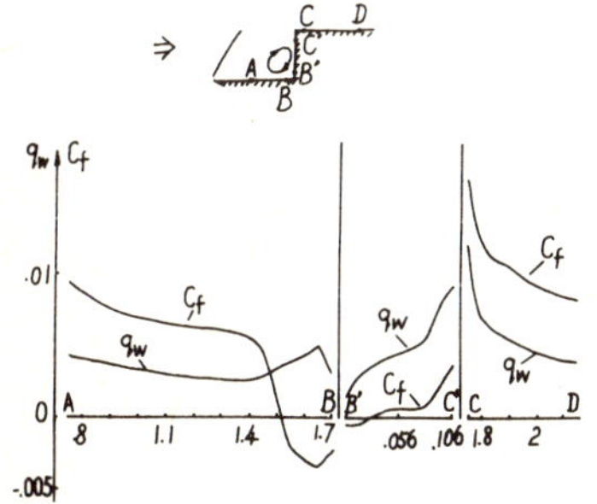


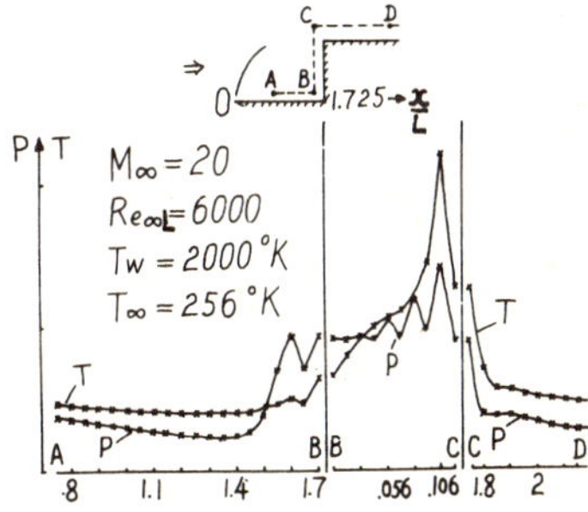


Fig. 6      Fig. 7

## ECONOMICAL METHODS FOR SOLVING THE PROBLEMS OF GAS DYNAMICS


N.N.Yanenko, V.M.Kovenya, G.A.Tarnavsky, S.G.Cherny
Institute of Theoretical & Applied Mechanics
USSR Academy of Sciences, Novosibirsk, 630090


Numerical simulation of multidimensional problems in the approximation of both gas dynamics and Navier-Stokes equations is associated with large expenditure of the machine time and computer storage needed. Therefore, the method of solution which should be economical, simple in operating and admit the generalization on the different physicomathematical models assumes a special actuality. The methods of splitting in terms of the physical processes and the spatial directions [1:4] permitting to reduce the solution of multidimensional problems to a set of their one-dimensional analogues satisfy the requirements mentioned.

The present paper is devoted to the construction of implicit difference schemes of splitting for the numerical solution of gas dynamics and Navier-Stokes equations of a compressible heat-conducting gas and simplified equations of a viscous gas. The application of splitting in terms of the physical processes and spatial directions enables one to construct the difference schemes operated in fractional steps by the scalar sweeps that renders them economical. The examples of calculations of the flow past the blunt and pointed bodies by the supersonic gas stream at the angle of attack are given. The algorithm of calculation of the subsonic zones on the basis of the method of regularization is proposed for the numerical solution of the simplified stationary equations of a viscous gas.

I. Let us consider the system of gas dynamics equations in the curvilinear orthogonal coordinates $x^\alpha$ ($\alpha = 1,2,3$):

$$\frac{\partial \vec{U}}{\partial t} = -\vec{W}, \quad \vec{W} = \frac{1}{d}\sum_{j=1}^{3} \frac{\partial}{\partial x^j}\left(d\frac{\vec{W_j}}{H_j}\right) + \vec{G}, \quad P = P(\rho, e), \quad (1)$$

where

$$\vec{U} = \begin{pmatrix} \rho \\ \rho v^1 \\ \rho v^2 \\ \rho v^3 \\ \rho E \end{pmatrix}, \quad \vec{W_j} = \begin{pmatrix} \rho v^j \\ \rho v^1 v^j + \delta_j^1 P \\ \rho v^2 v^j + \delta_j^2 P \\ \rho v^3 v^j + \delta_j^3 P \\ \rho v^j(E + P/\rho) \end{pmatrix}, \quad \vec{G} = \begin{pmatrix} 0 \\ v^2 d_{12} + v^3 d_{13} \\ v^1 d_{21} + v^3 d_{23} \\ v^1 d_{31} + v^2 d_{32} \\ 0 \end{pmatrix},$$

$$d = H_1 H_2 H_3, \quad d_{ij} = \frac{1}{H_j H_i}\left(v^i \frac{\partial H_i}{\partial x^j} - v^j \frac{\partial H_j}{\partial x^i}\right),$$

$\delta_j^i$ is the Kronecker symbol, $H_i$ are the Lamet coefficients, the rest of notations are conventional.

Introduce the non-degenerate transformation of coordinates

$$q^i = q^i(x^1, x^2, x^3) \tag{2}$$

transferring the computed region into a single cube $R\{0 \leqslant x^j \leqslant 1, j = 1,2,3\}$.
Represent the system of equations (I) in the variables in the form

$$\frac{\partial \vec{U}}{\partial t} = -\vec{W}, \quad \vec{W} = \sum_{i,j}^{3} z_j^i \frac{\partial \vec{W_j}}{\partial q^i} + \vec{G}, \quad z_j^i = \frac{\partial q^i}{\partial x^j} \tag{3}$$

or in the non-divergent form convinient for splitting

$$\frac{\partial \vec{f}}{\partial t} + \sum_{\ell=1}^{7} \Omega_\ell \vec{f} = 0$$

Gasdynamic functions or their combinations can be chosen as the components of the vector $\vec{f}$ sought for. The equations (4) take the simplest form in the variables $\vec{f} = (\rho, V^j, p)$ where $\Omega_\ell = \left(\sum_{j=1}^{3} z_j^\ell \frac{V^j}{H_j}\right) \times \frac{\partial}{\partial x^\ell} I$; $\ell = 1, 2, 3$; $I$ is the single matrix,

$$\Omega_{\ell+3} = \begin{pmatrix} 0 & 0 & 0 & 0 & 0 \\ 0 & 0 & 0 & 0 & S_1^\ell \\ 0 & 0 & 0 & 0 & S_2^\ell \\ 0 & 0 & 0 & 0 & S_3^\ell \\ 0 & z_1^\ell & z_2^\ell & z_3^\ell & 0 \end{pmatrix}, \quad \Omega_7 = \begin{pmatrix} 0 & 0 & 0 & 0 & 0 & 0 \\ 0 & 0 & 0 & d_{12} & d_{13} & 0 \\ 0 & d_{21} & 0 & d_{23} & 0 \\ 0 & d_{31} & d_{32} & 0 & 0 \\ 0 & 0 & 0 & 0 & 0 \end{pmatrix},$$

$$z_i^\ell = \frac{c^2 \rho}{d} \frac{\partial}{\partial x^\ell}\left(d \frac{V^i}{H_i}\right), \quad S_i^\ell = \frac{z_i^\ell}{H_i \rho} \frac{\partial}{\partial x^\ell}$$

The incorporation of the free terms arising as a result of coordinate curvilinearity into a separate operator $\Omega_7$ ($\Omega_7 \vec{f} = \vec{G}$) enables one to simplify the scheme operating in fractional steps. It is worthwhile noting that the operator $\Omega_7$ can be presented non-singlevaluedly.

The application of curvilinear coordinates and general transformation permits to consider within the framework of a single algorithm the different classes of problems choosing for each of them the most convenient system of coordinates.

2. Consider the difference scheme for the numerical solution of the system of equations (3)

$$C_h^n \frac{\vec{f}^{n+1} - \vec{f}^n}{\tau} = -(B\vec{W})_h^n, \tag{4}$$

where

$$C_h^n = \prod_{j=1}^{7} (I + \tau \alpha_j \Omega_j^n)_h, \quad B = \left(\frac{\partial \vec{U}}{\partial \vec{f}}\right)^{-1}$$

or the equivalent scheme in fractional steps

$$(I + \tau \alpha_1 \Omega_1^n)_h \vec{\xi}^{n+1/7} = -(B\vec{W})_h^n$$
$$(I + \tau \alpha_2 \Omega_2^n)_h \vec{\xi}^{n+2/7} = \vec{\xi}^{n+1/7}$$
$$\cdots \cdots \cdots \cdots \cdots \cdots \cdots \cdots \tag{5}$$
$$(I + \tau \alpha_7 \Omega_7^n)_h \vec{\xi}^{n+1} = \vec{\xi}^{n+6/7}$$
$$\vec{f}^{n+1} = \vec{f}^n + \tau \vec{\xi}^{n+1}$$

approximating the governing equations (I) with the order $O(\tau + h^\kappa)$ and in the case of obtaining the stationary solution - the stationary equations with the order $O(h^\kappa)$ . Here $\Omega_i^n$ are the difference operators approximating the corresponding differential operators with the order $O(h^\kappa)$ . As the analysis of stability for the linear equations shows, the difference scheme (4) is conditionally stable in the three-dimensional case. It is instructive to note that the difference schemes obtained on the basis of the method of approximate factorization (in more details, see [6]) possesses the proporty mentioned above.

The difference scheme (5) is operated in the first three fractional steps by either scalar sweeps or an implicit scheme of the running calculation for each component of the vector $\zeta^{n+\ell/7} (\ell = 1,2,3)$ as $\Omega_\ell$ are the diagonal operators. In the next fractional steps (4th--6th) the system of difference equations is reduced to the three--point difference equation for the component $\zeta_\rho^{n+(\ell+3)/7}$ by means of the elimination of components $\zeta_{v i}^{n+(\ell+3)/4}$ . The last fractional step of the scheme (5) is a system of linear algebraic equations with respect to the components of the vector $\zeta^{n+1}$ in one mesh point of the difference grid.

The difference scheme in the variables $\rho$ ,$v^i$, $e$ and its operating are considered in [7].

The scheme of the type (4) can be generalized for the case of Navier-Stokes equations. Thereat the viscous terms of the equation are divided into the two groups: the repeated derivatives are approximated implicitly in the operators $\Omega_\ell$ ($\ell$ =1,2,3) and the mixed derivatives - explicitly (see, [6]). The implicit approximation of the mixed derivatives leads to the difference schemes operated by matrix sweeps.

3. The calculations of the stationary flow past the bodies of a different shape by supersonic gas stream within a wide range of Mach numbers $M_\infty$ and angles of attack $\alpha$ are carried out with the help of the difference scheme proposed. The stationary solution was found by the relaxation method.

The systematic investigations of the convergence rate of the numerical solution to the stationary one were performed with the purpose of finding an optimum value of the iterative parameter. In Table I the results of the study on the convergence rate by the flow past the blunted in the sphere cone with the half-cone vertex angle equal to $20°$ at M=4, $\alpha$ =20° are presented. The solution was assumed stationary by satisfying the condition $|(\varrho^{n+1} - \varrho^n)/\tau \rho^n| \leqslant 10^{-3}$ throughout the whole flow field.

Table I

| $\tau$ | N | $K_o$ | $K_T$ |
|--------|-----|-------|-------|
| 0.03 | 394 | 3.54 | 1.16 |
| 0.04 | 260 | 4.71 | 2.23 |
| 0.05 | 215 | 5.89 | 2.78 |
| 0.054 | 206 | 6.36 | 3.00 |
| 0.055 | 173 | 6.48 | 3.06 |
| 0.056 | 199 | 6.60 | 3.12 |
| 0.07 | 340 | 8.25 | 3.90 |
| 0.08 | 472 | 9.43 | 4.45 |
| 0.10 | 700 | 11.78 | 5.57 |

N - is the number of iterations up to obtaining the stationary solution;

$K_o$ - is Courant number calculated with the help of parameters of the free-stream flow;

$K_T$ - is Courant number in the stagnation point.

As the initial data the undisturbed flow was defined and the bow

shock was separated in the process of calculation. We should note
the nonmonotonic dependence of the number of iterations up to the
convergence with the increase of numbers $K_0$ and $K_T$.

Several characteristic examples of spatial flows calculations are
given as an illustration of the method's possibilities.

a) The flow past a blunted in the sphere cone with the half-cone
vertex angle equal to 20° by the inviscid gas stream at $M_\infty=4$ at the
angle of attack $\alpha =40°$. A spherical system of coordinates with a
moving center is chosen as an initial one. The pressure isolines in
the problem symmetry plane are represented in Fig.I. The comparison
with the calculation of the flow past a body at $\alpha =0°$ with an appropriate angle turn of the coordinate system equal to the angle of attack (up to the values of $\alpha$ at which the sonic line is closed on a
spherical part of the body) was carried out to verify the computa-
tional accuracy. In Table 2 the values of Bernoulli integral $I = \dfrac{V^2}{2} + \gamma p/(\gamma-1)\rho$ in the winward semi-plane of symmetry are given.
Theoretical value of $I$ =0.6563.

Table 2

| Azimuth angle | 4.78° | 15.6° | 37,4° | 58.20° | 79.1° |
|---|---|---|---|---|---|
| Body Surface | 0.6555 | 0.6565 | 0.6560 | 0.6568 | 0.6559 |
| Relative Error % | 0.1219 | 0.0305 | 0.0457 | 0.0762 | 0.0609 |
| Shock Surface | 0.6557 | 0.6555 | 0.6573 | 0.6559 | 0.6552 |
| Relative Error % | 0.0914 | 0.1219 | 0.1523 | 0.0609 | 0.1676 |

A region of a strong rarefaction is formed on the lee side and the
regard for viscosity is necessary for obtaining a physical picture of
the flow. A separation zone stipulated by the approximate viscosity
of the scheme is formed within the framework of the ideal gas flow.

b) The flow past the axisymmetric body with a pointed tip at
$M_\infty=2$ and at the angle of attack $\alpha =7°$. In Fig.2 the pressure isolines
are given in the problem symmetry plane. The bow snock is detached.
The condensation of the mesh points to the pointed tip was carried
out to increase the computational accuracy. The accuracy in fulfill-
ing Bernoulli integral does not exceed 0.5%. The computational results are given in more details in papers [6,7].

4. The simplified equations of the viscous gas based on the estimate of the order of terms in a full system of Navier-Stokes equations (see, f.e. [5,8,9]) are used by studying the problems of stationary flow past a body by large and moderate Reynolds numbers.

The method of splitting in terms of the physical processes and the
spatial directions is generalized below for the numerical solution
of stationary simplified equations by marching method. For the simplicity of statement we confine ourselves to the consideration of iterative difference schemes for the two-dimensional equations of a viscous gas in Cartesian coordinates (schemes for the three-dimensional
equations are constructed similarly):

$$\sum_{i=1}^{2}\frac{\partial}{\partial x^i}\rho u^i = 0 , \tag{6}$$

$$\sum_{i=1}^{2}A_i\frac{\partial \vec{f}}{\partial x^i} = \frac{\partial}{\partial x^2}B\frac{\partial}{\partial x^2}D\vec{f} , \tag{7}$$

where

$$\vec{f} = \begin{pmatrix} u^1 \\ u^2 \\ p \end{pmatrix}, \quad A_i = \begin{pmatrix} \rho u^i & 0 & \delta_1^i \\ 0 & \rho u^i & \delta_2^i \\ \gamma p \delta_1^i & \gamma p \delta_2^i & u^i \end{pmatrix}, \quad B = \frac{1}{Re}\begin{pmatrix} \mu & 0 & 0 \\ 0 & \mu & 0 \\ 0 & 0 & \mu \end{pmatrix}, \quad D = \begin{pmatrix} 1 & 0 & 0 \\ 0 & 1 & 0 \\ 0 & 0 & 1/\rho \end{pmatrix}$$

$x^1$ is the coordinate directed along the flow direction (marching co-ordinate), $x^2$ is the coordinate directed orthogonally to the flow, are the corresponding velocity components.

Let us consider the iterative difference schemes for the numerical solution of the system of equations (6),(7)

$$C_{\varrho}^{\nu} \frac{\varrho^{\nu+1} - \varrho^{\nu}}{\tau} = -\Omega_{\varrho h}^{\nu} , \qquad (8)$$

$$C^{\nu} \frac{\vec{f}^{\nu+1} - \vec{f}^{\nu}}{\tau} = -\vec{\Omega}_h^{\nu} , \qquad (9)$$

where

$$C_{\varrho}^{\nu} = \left(1 + \frac{\tau}{h_1} u^{1*}\right)\left(1 + \alpha\tau\Lambda_2^{\kappa} u^{2*}\right) ,$$

$$C^{\nu} = \left(I + \frac{\tau}{h_1} A_1^*\right)\left(I + \tau\alpha A_2^{1*}\Lambda_2^{\kappa} + \tau\alpha\Lambda_2 B^*\Lambda_2 D^*\right)\left(I + \tau\alpha A_2^{2*}\Lambda_2^{\kappa}\right) .$$

Here

$$\Omega_{\varrho h}^{\nu} = \frac{(\varrho u^1)^{\nu} - \varrho u^1}{h_1} + \Lambda_2^{\kappa}(\varrho u^2)^* ,$$

$$\vec{\Omega}_h^{\nu} = A_1^* \frac{\vec{f}^{\nu} - \vec{f}}{h_1} + \left[A_2^*\Lambda_2^{\kappa} - \Lambda_2 B^*\Lambda_2 D^*\right]\vec{f}^* , \qquad \varphi^{\nu} = \left(\varphi_j^{n+1}\right)^{\nu} ,$$

$$\varphi^* = \alpha\varphi^{\nu} + (1-\alpha)\varphi , \qquad \varphi = \varphi_j^n$$

the difference approximation of the system of equations (6),(7) with the order $O(h_1^2 + h_2^{\kappa})$, r=2 at $\alpha$ =0.5, r=1 $\alpha \neq 0.5$, $\nu$ is the number of iteration, n, j are the mesh points in the direction $x^1$, $x^2$, res-pectively. The schemes in fractional steps of the type (5) are cons-tructed by analogy with (4) for the numerical operation of the itera-tive schemes (8),(9). The splitting of operator $A_2\Lambda_2^{\kappa}$ into the sum of operators $A_2^1\Lambda_2^{\kappa}$ and $A_2^2\Lambda_2^{\kappa}$ is carried out so that provide the convergence of the iterative process and scalar solvability of the scheme in fractional steps. A successive solution of the equations of motion and energy and continuity equations enables one to simplify the operator of iterative schemes and increase the convergence rate. It is well-known (see, [8]) that in the subsonic flow region the boundary value problem considered is incorrect. To eliminate the in-correctness by the numerical solution of the flow-past problem by the marching method, the regularizing functions are introduced into the momentum equation in the marching direction

$$(1+\varphi)u^1 \frac{\partial u^1}{\partial x^1} + u^2 \frac{\partial u^2}{\partial x^2} + \frac{1+\psi}{\varrho} \frac{\partial p}{\partial x^1} = \frac{1}{Re\varrho} \frac{\partial}{\partial x^2}\int^{\mu} \frac{\partial u^1}{\partial x^2} , \qquad (10)$$

where

$$\varphi = \delta F^{\alpha} , \psi = (\delta F/(1+F))^{\beta} , \delta = O(1/\sqrt{Re}) , F = \left|\frac{\partial\varrho}{\partial x^1}\right| + \left|\frac{\partial u^1}{\partial x^1}\right| + \left|\frac{\partial u^2}{\partial x^1}\right| + \left|\frac{\partial p}{\partial x^1}\right|$$

The regularizing functions $\varphi$ and $\psi$ are chosen so that provide the correctness of the problem in the subsonic region and in the superso-nic flow region their influence was $\approx O(1/\sqrt{Re})$, i.e. did not exceed the error of the model chosen (see, [10]).

Calculations of the flow past a blunt cylinder are carried out at $M_{\infty}$ =6, Re=$10^7$ with the help of the scheme considered. In Fig.3 the distribution of the longitudinal velocity component $u^1$ (dashed line) at $x^1$ =0.63 is represented and the comparison with calculations car-ried out by the model of Navier-Stokes equations (solid line) and simplified models ( O -[8], $\Delta$ - [9]) is given.

# References


1. N.N.Yanenko. The Method of Fractional Steps, Springer Verlag, Berlin, 1971.
2. V.M.Kovenya, N.N.Yanenko, Zhurnal Vychislitel'noi Matematiki i Matematicheskoi Fiziki, t.19, No.1, 1979.
3. R.M.Beam, R.F.Warming, J. of Comp. Physics, 22, Sept.
4. W.R.Briley, H.McDonald. J. of Comp. Physics, v.24, (4), 1977.
5. N.N.Yanenko, V.M.Kovenya, S.G.Cherny. DAN SSSR, t.245, No.6, 1979.
6. V.M.Kovenya, G.A.Tarnavsky, N.N.Yanenko. Zhurnal Vychislitel'noi Matematiki i Matematicheskoi Fiziki (in press).
7. V.M.Kovenya, G.A.Tarnavsky. In: Chislennyje Metody Mekhaniki Sploshnoi Sredy, Novosibirsk, t.10, (6), 1979 and t.9,(1), 1978
8. T.S.Lin, S.G.Rubin, Computers and Fluids, 1 (1973).
9. Baum, Denison, AIAA, v.5, No.7 (1975).
10. V.M.Kovenya, S.G.Cherny. In: Chislennyje Metody Mekhaniki Sploshnoi Sredy, Novosibirsk, 10(I), 1979.


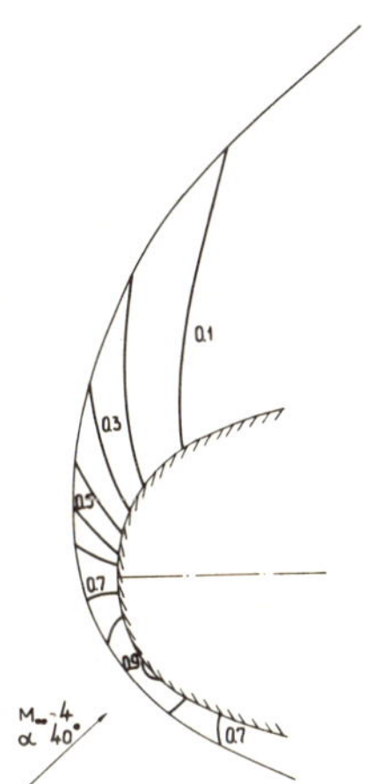


Fig. 1

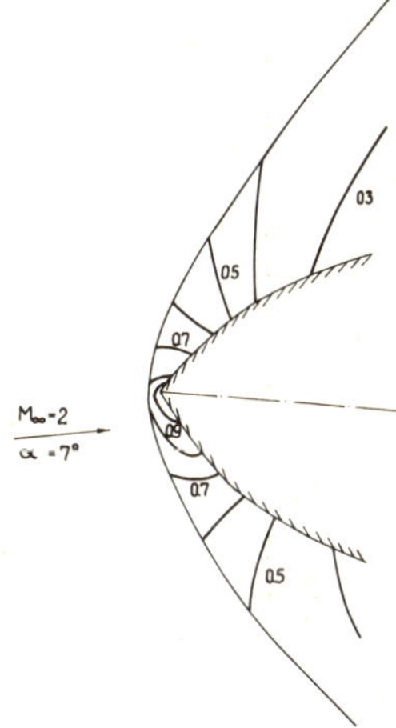


Fig. 2

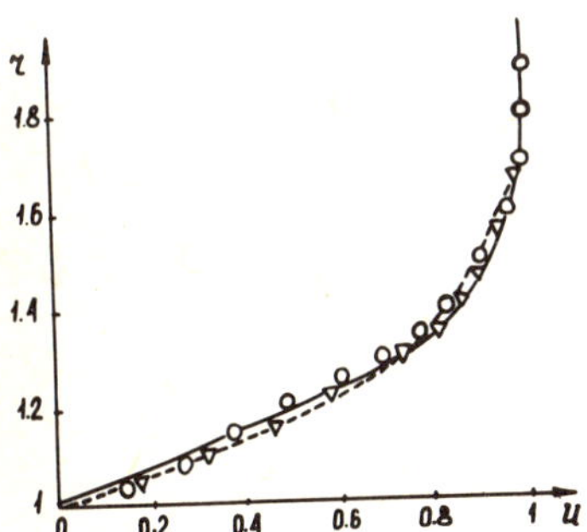


Fig. 3

# NUMERICAL SIMULATION OF THE RAREFIED GAS FLOWS


N.N.Yanenko, Yu.N.Grigoryev, M.S.Ivanov
Institute of Theoretical & Applied Mechanics
USSR Academy of Sciences, Novosibirsk, 630090


For description of rarefied gas flows in transient regime, when a characteristic dimension of the flow $L$ is commensurable with molecular free path $\lambda$ $(Kn = \lambda/L \sim 1)$, the kinetic approach, i.e. the formulation of appropriate boundary problem for nonlinear kinetic equation of the theory of gases appears to be natural.

A complicated mathematical structure of the nonlinear kinetic equations impedes the practical application of regular numerical methods because of extremely high requirements of the storage capacity and speed of computers. Therefore, the Monte Carlo methods find a wide application nowadays in practice of numerical calculations of the rarefied gas dynamics problems.

The lack of a strict mathematical connection of numerical algorithms with original boundary value problem for the kinetic equation has proved to be an essential shortcoming of the Monte Carlo methods applied in the rarefied gas dynamics.

I. A main goal of the present paper consisted in constructing the numerical Monte Carlo algorithms for stationary problems of the rarefied gas dynamics proceeding directly from the formulation of appropriate boundary value problem for Boltzmann equation and the class of model kinetic equations both. Thus, in the case of the external flow past a body the boundary value problem for the model statistical equation has the form

$$\bar{V} \cdot grad\, f(\bar{z}, \bar{V}) = \nu([M_\nu])\big(f_+(\bar{z}, \bar{V}, [M]) - f(\bar{z}, \bar{V})\big) \tag{1}$$

$$f(\bar{z}_s, \bar{V}) = f_0(\bar{z}_s, \bar{V}); \quad z_s \in \Gamma; \quad (\bar{V} \cdot n_\Gamma) > 0 \tag{2}$$

$$|\bar{V} \cdot \bar{n}_\gamma| \cdot f(\bar{z}_s, \bar{V}) = \int |\bar{V}' \cdot \bar{n}| f(\bar{z}_s, \bar{V}') \tilde{T}(\bar{V}, \bar{V}'; \bar{z}_s) d\bar{V}' \tag{3}$$

$$(\bar{V}' \cdot \bar{n}) < 0$$

Here the flow region $G$ is bounded by an external surface $\Gamma$; $\gamma$ is the surface of the body flowed past, $f$ is the distribution function for molecules having a sense of the local density in the neighbourhood of the point $(\bar{r}, \bar{V})$ of the sixdimensional phase space of coordinates and velocities. $\nu$, $f_+$ are some nonlinear functionals of $f$ depending on the finite sets $[M]_\nu$, $[M]_+$ of moments of a function taken with respect to the velocity space which determine the specific model equation.

The condition (2) determines the distribution function of molecules entering into the region and the condition (3) determines the equality of the flows of the incident and reflected molecules on the body surface flowed past. $\tilde{T}(\bar{V},\bar{V}';\tau_\delta)$ is the boundary shock transformant (Vallander, 1960). It is assumed that the law of conservation of particles number holds for (I)

$$\int f(\bar{z},\bar{V})\,d\bar{V} = \int f_+(\bar{z},\bar{V},[M]_+)\,d\bar{V} = n(\bar{z}) \tag{4}$$

The iterative process of the form

$$\bar{V}\cdot\text{grad}\,f^{(K)}(\bar{z},\bar{V}) + \nu([M]_\nu^{(K-1)})f^{(K)}(\bar{z},\bar{V}) = \int_V f^{(K)}(\bar{z},\bar{V}') \times$$

$$\times f_+(\bar{z},\bar{V},[M]_+^{(K-1)})/n^{(K-1)}(\bar{z})\,\nu([M]_\nu^{(K-1)})\,d\bar{V}' \tag{1'}$$

$$f^{(K)}(\bar{z}_\delta,\bar{V}) = f_o(\bar{z}_\delta,\bar{V}) \tag{2'}$$

$$|\bar{V}\cdot\bar{n}_\gamma|\,f^{(K)}(\bar{z}_\delta,\bar{V}) = \int_{(V_i'n)<0} |\bar{V}'\cdot\bar{n}_\gamma|\cdot f^{(K)}(\bar{z}_\delta,\bar{V}')\,T(\bar{V},\bar{V}';\bar{z}_\delta)\,d\bar{V}' \tag{3'}$$

is introduced for (I)–(3).

The boundary value problem for the equation (3) in k-th iteration is transformed by means of inversion of the left-hand side operator and application of the procedure analogue of Peierls' equation derivation in the transport theory of neutrons to the integral equation of Fredholm type of the second kind

$$\varphi^{(K)}(x) = \int_X K^{(K-1)}(x',x)\,\varphi^{(K)}(x')\,dx' + \varphi_1^{(K)}(x) \tag{5}$$

where $\varphi^{(K)} = f^{(K)}(\bar{r},\bar{V})\cdot\nu([M]_\nu^{(K-1)})$ is the density of collisions; $X = R\times V$ is the phase space of the problem. Here the additional variables of integration are compensated by introducing the $\delta$-functions.

The probabilistic treatment of the kernel and free term (5) enables one to construct the process of direct simulation, i.e. determine probability density of initial states $r_o(x)$, transient density $r(x',x)$ and breakdown probability $p(x)$ of Markov chain.

Thereat

$$z_o(x) = \varphi_1(x)\Big/\int_X \varphi_1(x)\,dx\,;$$

$$z(x',x) = K^{(K-1)}(x',x)\Big/q(x')\,; \quad q(x') = \int_X K^{(K-1)}(x',x)\,dx < 1\,;$$

$$p(x') = 1 - q(x').$$

The kernel of equation (5) has the form

$$K(\bar{z}',\bar{V};\bar{z},\bar{V}) = \frac{f_+(\bar{z}',\bar{V},[M]_+^{(K-1)})}{n^{(K-1)}(\bar{z}')} \times \frac{\nu([M]_\nu^{(K-1)})}{|\bar{V}|} \times$$

$$\times \exp\left\{-\int_0^{|\bar{z}-\bar{z}'|} \nu(\bar{z}'+s\cdot\bar{V}/|\bar{V}|,[M]_\nu^{(K-1)})/|\bar{V}|\,ds\right\}\,\delta\left(\frac{\bar{z}-\bar{z}'}{|\bar{z}-\bar{z}'|}-\frac{\bar{V}}{|\bar{V}|}\right)\Big/|\bar{z}-\bar{z}'|^2\,,$$

and the free term $\varphi_1(x)$ is determined only by the given distribution function on the boundary of the flow region.

Thus, the probability densities determining the Markov chain are determined $[M]_\nu^{(k-1)}$, $[M]_+^{(k-1)}$ and hence by realizing this algorithm on a computer unlike Haviland (1965) there is no necessity to store the array of $f^{(k-1)}(\bar{r},\bar{v})$ values, it is sufficient to store only the set of its moments.

For statistical estimate of functionals of the solution of eqn.(6) in each iteration it is necessary to simulate the trajectories of appropriate Markov chain by means of realizing on a computer random walks of the probe particles according to the following scheme:
a) The simulation of initial points of trajectories of the probe particles.
At first a point $\bar{z}_s$ on the boundary $\Gamma$ is simulated with the use of the probability density

$$p(\bar{z}_s)=\int_{(\bar{v}\cdot\bar{n})>0} |\bar{v}\cdot n_\Gamma|\cdot f(\bar{z}_s,\bar{v})\,d\bar{v}\Big/\int_\Gamma \int_{(\bar{v}\cdot\bar{n})>0} |\bar{v}\cdot\bar{n}_\Gamma|f(\bar{z}_s,\bar{v})\,d\bar{v}\,d\Gamma$$

and then the probe molecule velocity is determined by the probability density

$$p(\bar{v}/\bar{z}_s)=|\bar{v}\cdot\bar{n}|\cdot f(\bar{z}_s,\bar{v})\Big/\int_{(\bar{v}\cdot\bar{n})>0} |\bar{v}\cdot\bar{n}|\,f(\bar{z}_s,\bar{v})\,d\bar{v};$$

b) The molecular free path of a probe particle in the direction $\bar{\Omega}=\bar{v}/|\bar{v}|$ is simulated by the probability density

$$p(\ell)=\nu(\bar{z}',[M]_\nu^{(k-1)})\Big/|\bar{v}|\times\exp\left\{-\int_0^{|\bar{z}-\bar{z}'|} \nu(\bar{z}+s\bar{\Omega}),[M]_\nu^{(k-1)}\Big/|\bar{v}|\cdot ds\right\};$$

c) After collision of a probe molecule with a field one a new velocity of the probe molecule is simulated by the distribution:

$$p(\bar{v}')=f_+(\bar{z}',\bar{v}',[M]_+^{(k-1)})\Big/n^{(k-1)}(\bar{z}')$$

d) By intersection of the probe molecule trajectory of the body flowed past the velocity of the reflected molecule is simulated in accordance with (3).
The form of the distributions mentioned above depends on the form of intermolecular forces, boundary conditions etc.
The iterative process (1-3) possesses the property of conservativeness in the mass flux in each iteration. Really,

$$\int_V \left\{n^{(k)}(\bar{z})f_+(\bar{z},\bar{v},[M]_+^{(k-1)})\Big/n^{(k-1)}(\bar{z})-f^{(k)}(\bar{z},\bar{v})\right\}d\bar{v}=0$$

in virtue if the normalization condition (4). This condition contributes to the acceleration of the iterative process convergence (Ivanov, 1979).
Earlier a similar development was carried out by Grigoryev et al. (1971) for the nonlinear Boltzmann equation of the theory of gases that allowed to give a strict substantiation of the well-known Haviland algorithm.
Moreover, the algorithm stated was generalized by Grigoryev, Ivanov (1977b) for the case of the binar mixture of gases described by the system of statistical kinetic equations.
II. The numerical algorithm of the Monte Carlo method was used for the solution of a number of one-dimensional, plane and axisymmetrical problems of the rarefied gas dynamics in a transient regime( Grigoryev, Ivanov, 1972, 1974, 1978; Ivanov, 1977.)
The first test of the algorithm approbation carried out on the example of a classical Couette flow of the rarefied gas (Grigoryev et al. 1971) has shown a good compatibility of results with other numerical methods including the regular ones.
The considerable theoretical and applied interest present the supersonic flows of rarefied gases with large gradients of macroparame-

ters: density, velocity, temperature. As the equations of BGK type
are in a certain sense the approximation of Boltzmann equation, it turned out to be necessary to study the applicability of different kinetic model equations in this field.

Such a study was carried out on the basis of the problem on the
shock wave structure in which a series of characteristic peculiarities
is displayed. The comparison of computational and experimental density profiles (Schmidt, 1969) has shown that in a wide range of M numbers (2.8÷10) the BGK model (Grigoryev, Ivanov, 1972) gives a good
fit to the experiment (Fig.I)

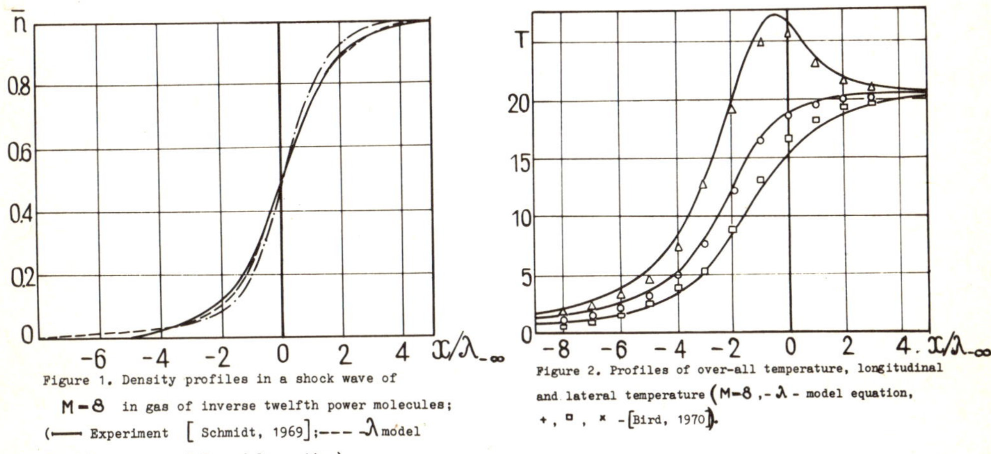

Figure 1. Density profiles in a shock wave of
$M=8$ in gas of inverse twelfth power molecules;
(—— Experiment [ Schmidt, 1969];−−− $\lambda$ model
Equation;—·—·— E.S. model equation).

Figure 2. Profiles of over-all temperature, longitudinal
and lateral temperature ($M=8$ ,- $\lambda$ - model equation,
+ , □ , × -[Bird, 1970]).

Since temperature was not measured in the experiment, the obtained
computational temperature profiles were compared (Fig.2) with data of
the direct statistical simulation (Bird, 1971) which can be considered as a result of the approximate solution of Boltzmann equation.

Systematic studies of the supersonic flow past the axisymmetric
blunt bodies of a simple aerodynamic form: sphere and segmental-conical body were carried out. In computations omitting the distribution
function the gasdynamic parameters of the flow - density, velocity,
temperature, heat flux on the body and its aerodynamic drag were directly calculated.

The free molecular solution was taken usually for initial approximation. In some cases the iterations started from the undisturbed flow
for the control. The quantity of the trajectories modelled in one iteration made up $(2÷5)\cdot10^4$. The quantity of iterations varied from 5 to
20. The iterations in Knudsen number were used to accelerate the convergence. Calculations were carried out in a wide range of numbers
$Kn = 2\cdot10^{-2} ÷ 10^1$ and $M = 3.8 ÷ 17$.

The determination of the flow parameters in a near wake which affect directly the aerodynamics is of great importance when computing
the flow past the blunt bodies. At the same time a relative small number of probe particles reaches this zone in the process of simulation
that leads to a large dispersion of statistical estimates of the flow
parameters. In calculations the special methods for decreasing the
dispersion - essential sampling and splitting of trajectories - were
used to increase the accuracy in the neighbourhood of the body flowed
past.

In Fig.3 the computational results of the sphere drag within a range of numbers $Kn$ from $10^{-2}$ up to $10^1$ are shown. The results are well
compared with the experiment (Phillips, Kuklthau, 1971)

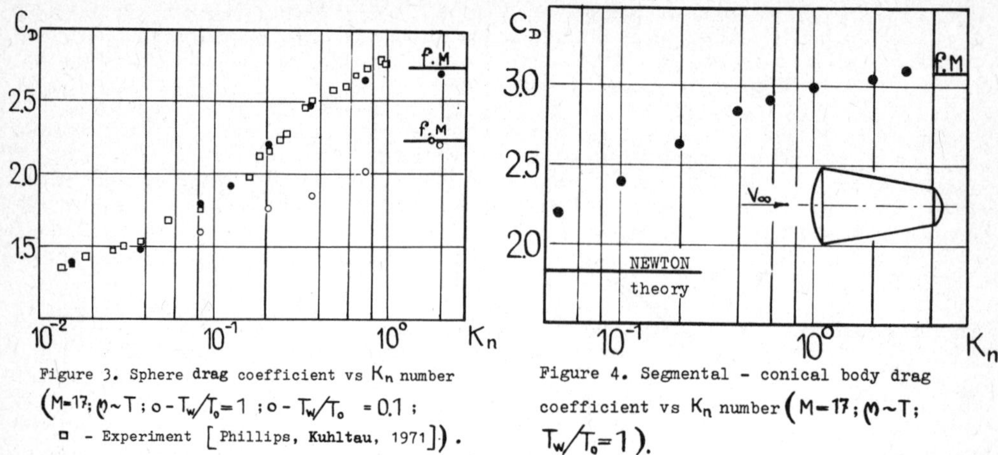



Figure 3. Sphere drag coefficient vs $K_n$ number $\left(M=17; \mu \sim T; o - T_w/T_0 = 1; o - T_w/T_0 = 0.1; \square - \text{Experiment} \left[\text{Phillips, Kuhltau, 1971}\right]\right)$.

Figure 4. Segmental - conical body drag coefficient vs $K_n$ number $\left(M=17; \mu \sim T; T_w/T_0 = 1\right)$.

In a number of papers, in particular, experimental ones a possibility of exceeding the sphere $C_x$ at the finite numbers over the free-molecular value $C_x$ was proposed, especially in the case of a cold $(T_w/T_0 \ll 1)$ surface. It was explained by the increase of the momentum flow on the body owing to the collisions of the free-stream flow molecules with a large mass of "cold" molecules being formed before the body.

Accurate computations of the supersonic flow past the cold spheres were carried out for the purpose of veryfying this hypothesis. A characteristic result for the sphere is given in Fig.3 at $T_w/T_0 = 0.1$ (light points). It is seen that there is no such exceeding throughout the whole range of $K_n$ numbers computed.

The comparison of Fig.3 (squares) and Fig.4 shows that at the same cross-section area the segmental-conical body $C_x$ is more then $C_x$ of the sphere.

The calculations of the flow past the axisymmetric blunt bodies were carried out up to the regimes near to the flow of continuum medium described by Navier-Stokes equations. The comparison of computational density profiles along the sphere stagnation stream line with the experiment (Russel, 1968) is given in Fig.5. The results of the flow past the sphere on the basis of Navier-Stokes equation are given in the same figure. Some discrepancy of results is explained by the fact that no-slip conditions were taken for Navier-Stokes equations. The estimates show that the use of the slipping conditions leads to a satisfactory coincidence of results.

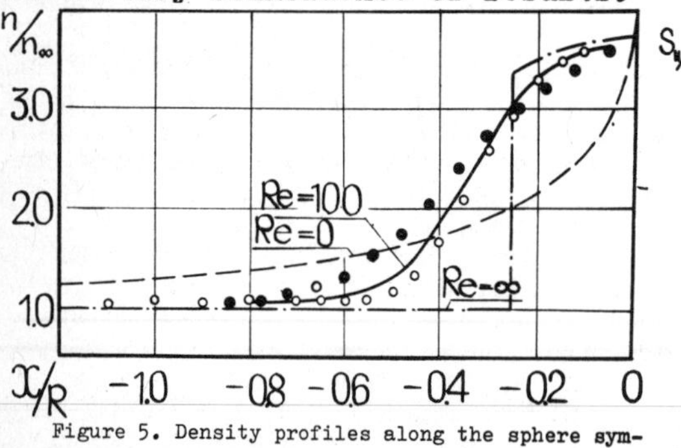



Figure 5. Density profiles along the sphere symmetry axis vs $X/R$ (M =3.8; $\mu \sim T$; $T_w/T_0 = 1$; — -Experiment [Russel, 1968], o – $\lambda$-model equation, ● - Navier-Stokes equations).

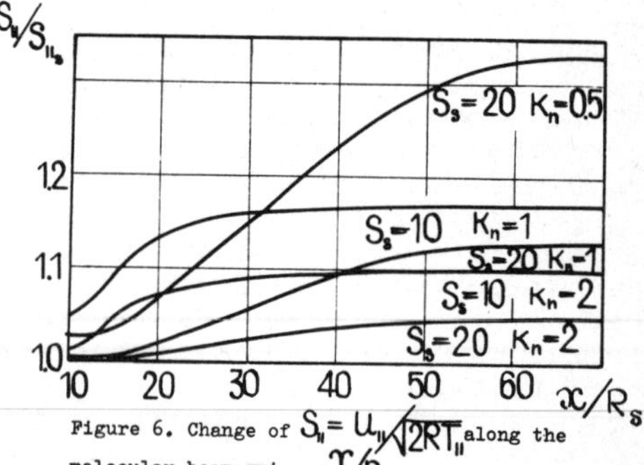



Figure 6. Change of $S_‖ = U_‖/\sqrt{2RT_‖}$ along the molecular beam axis vs $X/R_s$.

The interesting results are obtained by the solution of the problem on the influence of collisions in the flow between the scimmer and detector on the parameters of rarefied gas jets measured by the time--flighting method. This method finds a wide application in the up-to--date experimental studies of the rarefied gas flow structure. The problem has attracted our attention in connection with the assumption made by Singh et al.(1977) that the collisions inside the molecular beam cut out by scimmer can lead to distortion of the parameters measured. Calculations have confirmed this assumption. In Fig.6 the change of the longitudinal number $S_{\parallel} = U_{\parallel}/\sqrt{2RT_{\parallel}}$ is given which characterizes the distortion of the measured distribution function of the longitudinal molecular velocities.

It is well-known (Kogan,1967) that the aerodynamic characteristics of the thin bodies in a region of finite Knudsen numbers change unmonotonically. The regard for collisions between the molecules in the case of the flow past the thin bodies by the rarefied gas gives a principal change of the behaviour of the aerodynamic characteristics whereas in the case of the flow past the blunt bodies the regard for collisions leads only to some corrections of the free-molecular values of aerodynamic characteristics.

A thin plane wedge has been chosen for studying the features peculiar to the flow past the thin bodies in transient regime. In these calculations an essentially inhomogeneous mesh was used as for the thin bodies it was necessary to study in details the flow region in the neighbourhood of the body flowed past.

An example of dependence $C_x$ of the wedge on the Knudsen number at two angles of attack is shown in Fig.7. At a large angle of attack ($\alpha = 40°$) the free-molecular value of the drag coefficient is maximum and with decreasing Knudsen number becomes smaller. However, at $\alpha = 5°$

of the wedge at finite numbers $Kn$ the drag coefficient exceeds essentially its free-molecular value. Even at $Kn = 10^{-2}$ the drag coefficient of the wedge is somewhat more than $C_{xf.m}$. The lift coefficient increases in both cases independently of the angle of attack.

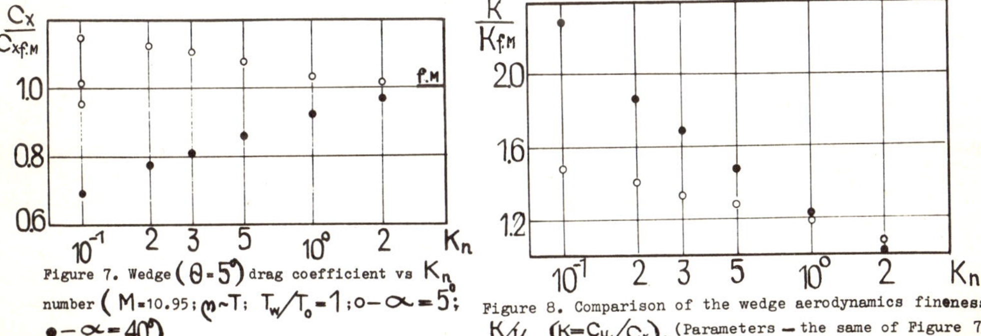

Figure 7. Wedge $(\theta = 5°)$ drag coefficient vs $Kn$ number $(M = 10.95; m \sim T; T_w/T_0 = 1; o - \alpha = 5°; \bullet - \alpha = 40°)$.

Figure 8. Comparison of the wedge aerodynamics fineness $K/K_{f.m}$ $(K = C_y/C_x)$. (Parameters — the same of Figure 7).

The comparison of aerodynamic lift-drag ratio (l.d.r.) of the wedge at $\alpha = 5°$ and $\alpha = 40$ given in Fig.8 shows that the relative l.d.r. at $\alpha = 40°$ increases essentially faster than at $\alpha = 5°$.

Thus, the aerodynamic l.d.r. of a thin wedge increases independently of the angle of attack by decreasing Kn although the drag behaves differently depending on the regard for attack.

III. Proposed by the authors Monte Carlo algorithm for the first time in practice of the statistical methods in the rarefied gas dynamics has a strictly formalized connection with original boundary value problem for the nonlinear kinetic equation. The reformulation of the boundary value problem in the form of the integral equation of the second kind with a substochastic kernel gives a possibility for the systematic improvement of the algorithm on the basis of the general theory of Monte Carlo methods.

An important physical consequence of the systematic numerical investigation of the various rarefied gas dynamics problems carried out is that fact the model kinetic equations can be applied for obtaining the integral characteristics of the rarefied gas flows throughout the whole transient region up to the regimes described by Navier-Stokes equations.

Bird G.A. Phys. Fluids, 1970, vol.13, No.   , pp.1172-1177.
Grigoryev Yu.N., Ivanov M.S., Kcharitonova N.I. Chisl. met. mech.
spl. sr. AN SSSR Sib. otd. V.C., Nov. 1971a, 2, N 4, s.101-107.
Grigoryev Yu.N., Ivanov M.S., Kcharitonova N.I. In: Ver. met. resh.
zad. matem. phys. 1971b, V.C. SO AN SSSR, Nov., s.71-90.
Grigoryev Yu.N., Ivanov M.S. Izv. SO AN SSSR, ser. tech. nauk, 1972,
13, N 3, s.33-38.
Grigoryev Yu.N., Ivanov M.S. Chisl. met. mech. spl. sr., AN SSSR, Sib.
otd. V.C., Nov., 1974, 5, N 1, s.152-156.
Grigoryev Yu.N., Ivanov M.S. In: Pricl. aerod. cosm. app., Kiev, Nau-
kova Dumka, 1977a, s.27-31.
Grigoryev Yu.N., Ivanov M.S. Chisl. met. mech. spl. sr. AN SSSR, Sib.
otd. V.C. Nov., 1977b, 8, N 6, s.30-37.
Haviland J.K. In: Methods in Computational Physics, vol.4, New York,
1965.
Ivanov M.S. In: Tr. IV Vses. conf. dyn. razr. gas. Moskva, 1977,
s.385-391.
Ivanov M.S. Diss. ITPM SO AN SSSR, Nov., 1979.
Kogan M.N. Dyn. razr. gas. M., Nauka, 1967.
Phillips W.H., Kuhlthau A.R. AIAA J., 1971, vol.9, No.7, pp.277-278.
Russell B.A. Phys. Fluids, 1968, vol.11, No.8, pp.1679-1685.
Sohmidt B. J. Fluid Mech. 1969, vol.39, pt.2, pp.361-373.
Singh R., Hurlbut F.G., Robben F. In:Rarefied Gas Dynamics, DFVLR -
- Press, Göttingen, 1977, vol.11.
Vallander S.V. Dokl. AN SSSR, 131, N 1, 1960.

MIXED SPECTRAL/FINITE DIFFERENCE

APPROXIMATIONS FOR SLIGHTLY VISCOUS FLOWS


Thomas A. Zang

College of William & Mary, Williamsburg, VA

M. Y. Hussaini[*]

Institute for Computer Applications in Science and Engineering


## I.  Introduction

In high Reynolds number flows the numerical dissipation resulting from the truncation error of the advection terms may overwhelm the physical dissipation from the viscous terms, as noted, for example, by Roache (1975).  For incompressible flows, pseudo-spectral methods have proven their ability to reduce greatly the advection errors compared to a standard finite difference calculation with the same number of grid points [Orszag & Israeli (1974)].  The viscous errors are reduced as well, but for slightly viscous flows, these are of much less concern.  For compressible flows, however, pseudo-spectral methods are unproven.  Shocks present an obvious challenge to a global pseudo-spectral method.  Moreover, as Table 1 demonstrates for the two-dimensional Navier-Stokes equations, the pseudo-spectral evaluation of the viscous terms is much more expensive for compressible flows than for incompressible ones.  A numerical scheme which employs a standard finite difference approximation for the viscous terms but which resorts to pseudo-spectral methods for the advection terms is discussed here.  This method offers the prospect of greatly reducing the major source of numerical error without dramatically increasing the computational cost.  The details of the method will be described below in connection with three model problems.

## II.  Results

### A.  Burgers' Equation

The first model problem is the familiar Burgers' equation

$$u_t + uu_x = \nu u_{xx}$$

on the interval $[0,2\pi]$ with periodic boundary conditions.  The initial condition is based on the exact solution for a diffused sawtooth wave [Whitham (1975)]:

$$u(x,t) = -2\nu \frac{\phi_x(x-4t,t+1)}{\phi(x-4t,t+1)} \quad , \quad \text{where} \quad \phi(x,t) = \sum_{n=-\infty}^{\infty} e^{-(x-2\pi n)^2/4\nu t} \quad .$$

This solution is a wave with mean height $4$ which travels to the right with speed $4$ and gradually diffuses toward a uniform state.


* Research supported by NASA Contracts No. NAS1-14472 and NAS1-15810 while in residence at ICASE, NASA Langley Research Center, Hampton, VA  23666.

**Table 1**

NUMBER OF FAST FOURIER TRANSFORMS

FOR PSEUDO-SPECTRAL SOLUTION OF TWO-DIMENSIONAL NAVIER-STOKES EQUATIONS

| FORM OF EQUATIONS | ADVECTION TERMS | DIFFUSION TERMS | TOTAL |
|---|---|---|---|
| INCOMPRESSIBLE (NON-CONSERVATIVE) | 9 | 2 | 11 |
| COMPRESSIBLE (CONSERVATIVE) | 8 | 27 | 35 |

**Table 2**

SPECTRAL/FINITE DIFFERENCE METHOD FOR BURGERS' EQUATION

$$u_J^{N+1} = u_J^N - 1.5\,\Delta T\,(F_X)_J^N + 0.5\,\Delta T\,(F_X)_J^{N-1}$$

$$+\,0.5\,\frac{\nu\Delta T}{(\Delta x)^2}\,(u_{J+1}^{N+1} - 2u_J^{N+1} + u_{J-1}^{N+1} + u_{J+1}^N - 2u_J^N + u_{J-1}^N)$$

$$F = u^2/2 \qquad\qquad u_J^N = u(J\Delta x, N\Delta T)$$

$(F_X)_J$ IS EVALUATED BY (FOURIER) COLLOCATION:

$$A_K = \frac{1}{N}\sum_{J=0}^{N-1} F(x_J)\,E^{-IKX_J}$$

$$(F_X)_J = \sum_{K=-N/2}^{N/2-1} IK\,A_K\,E^{IKX_J}$$

The interval $[0,2\pi]$ is divided into $N$ equal sub-intervals with grid points at the locations $x_j = \pi j/N$, $j = 0, 1, \ldots, N-1$. The spectral/finite difference (SFD) method uses a Fourier pseudo-spectral technique to evaluate the advection term and second-order central differences to evaluate the diffusion term. Specific formulae are provided in Table 2 for a time-differencing scheme employing an explicit Adams-Bashforth method on the advection term and an implicit Crank-Nicolson method on the diffusion term. This problem was also solved by a fully finite difference (FFD) method — using second-order central differences on both terms — and by a fully pseudo-spectral (PSS) method. To facilitate comparison of the three methods, a fixed time-step was used. It was specified by the CFL number at $t = 0$ (when the maximum velocity is approximately $4 + \pi$). In all the cases cited here this method is linearly stable; the implicit diffusion treatment counter-balances the weakly unstable explicit advection treatment for small time-steps. The FFD method, however, is stable for larger CFL numbers than the other two.

**Table 3**

$L_2$ ERRORS FOR BURGERS' EQUATION WITH $\nu = 0.25$

| N | FFD | SFD | PSS |
|---|---|---|---|
| 16 | .312 | .0710 | .0244 |
| 32 | .0797 | .0143 | .00631 |
| 64 | .0193 | .00337 | .00157 |
| 128 | .00477 | .000831 | .000392 |

**Table 4**

$L_2$ ERRORS FOR BURGERS' EQUATION WITH $\nu = 0.0625$

| N | FFD | SFD | PSS |
|---|---|---|---|
| 16 | .390 | .636 | .404 |
| 32 | .183 | .0582 | .0381 |
| 64 | .0512 | .00866 | .00603 |
| 128 | .00648 | .00162 | .000506 |

The numerical results are presented for $t \simeq 1.56$ , the time for which the wave has travelled a distance of $2\pi$ . The errors measured in the discrete $L_2$-norm are shown in Table 3 for the case $\nu = 0.25$ . The FFD errors actually worsen as the CFL number is reduced, presumably because of increased phase errors. The SFD results undergo only a modest improvement as the time-step is reduced. The errors of these two methods are dominated by the spatial discretization. However, the SFD errors are noticeably smaller. On the other hand, the PSS error can be reduced substantially by using smaller time-steps. In fact, for $N = 32$ , the time-differencing errors dominate until the initial CFL number reaches $0.05$ . Bear in mind that in a realistic computation, the PSS method is much costlier than the others.

The FFD and SFD solutions for $\nu = 0.0625$ are shown in Figure 1. The PSS solution is graphically indistinguishable from the SFD one in this example. All three methods are subject to oscillations for this sharp shock situation. The FFD oscillations were reduced by means of a Shuman filter. The SFD and PSS results were plotted (and $L_2$ errors computed) only after Lanczos averaging was applied. For this problem, the computed solution itself was not altered by the averaging. The physical diffusion here gradually reduces the oscillations. Figure 2 shows some SFD results before and after Lanczos averaging. The $L_2$ errors for this value of viscosity are given in Table 4. To prevent the $L_2$ errors from being overwhelmed by the relatively large errors near the shock, a region of width $2\pi/16$ centered on the shock was ignored in the $L_2$ error determination of this second case. The results all undergo negligible improvement for smaller CFL numbers. The SFD results are again clearly better than the FFD ones, not only in the $L_2$ errors but also in the resolution of the shock. Now, however, the PSS method provides a smaller improvement over the SFD method.

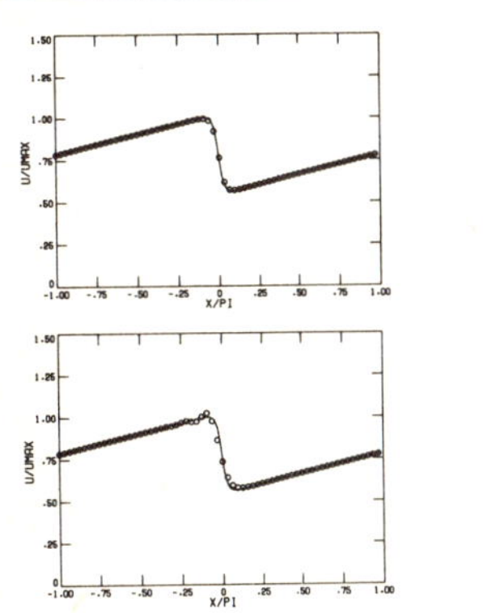

Figure 1. SFD (top) and FFD (bottom) solutions of Burgers' equation with N=64 for $\nu = 0.0625$. The solid curve represents the exact solution. The open circles denote the computed solution.

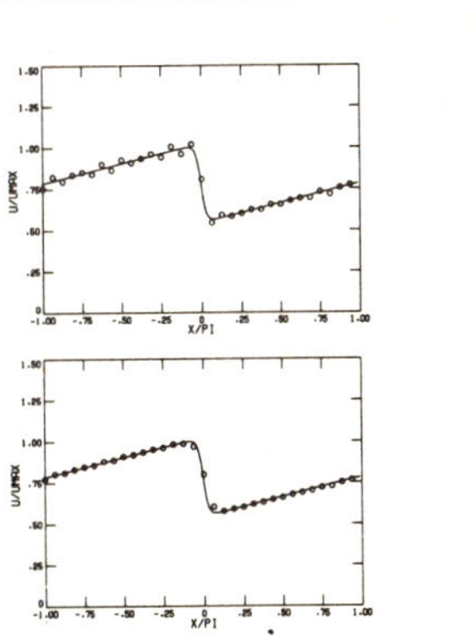

Figure 2. SFD solution of Burgers' equation with n = 32 for $\nu = 0.0625$. The open circles denote the computed solution before (top) and after (bottom) Lanczos averaging.

### B.  Quasi-One-Dimensional Nozzle Flow

A second application of the SFD method was made to the quasi-one-dimensional
flow in a nozzle of slowly-varying cross section.  The equations are given in
Table 5.  The spatial derivatives on the left-hand-side of the equals sign were
evaluated by a Chebyshev pseudo-spectral method.  The derivatives on the right-hand-
side were handled by central differences.  The time-differencing was again Adams-
Bashforth for advection and Crank-Nicolson for diffusion.  At the inflow boundary
a zero pressure gradient was enforced, along with prescribed values of density and
velocity.  At outflow only the pressure was fixed; the density and velocity were
evaluated by zero-order extrapolation.  The initial flow was the symmetric critical
flow having a sonic point at the throat of the nozzle.

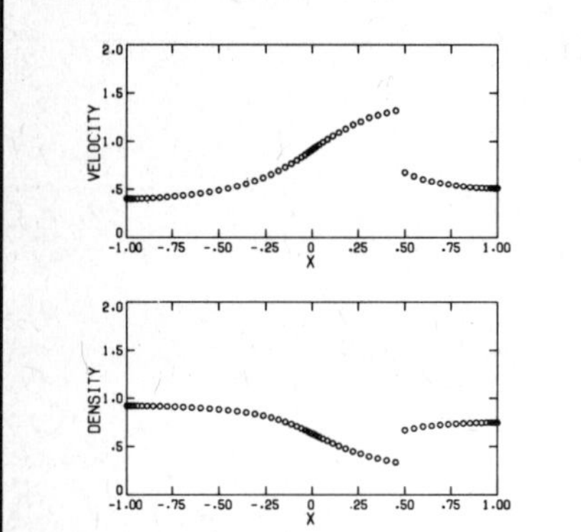

Figure 3.  SFD solution of quasi-one-
dimensional nozzle flow using two
patched Chebyshev grids totalling 65
points.  The variables are scaled to
their upstream stagnation values.

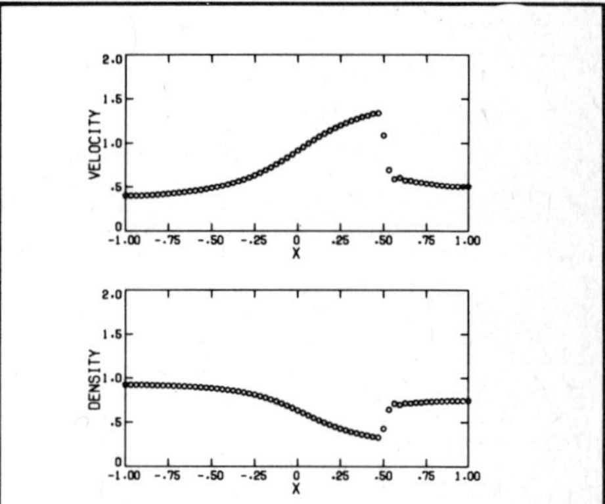

Figure 4.  FFD solution of quasi-one-
dimensional nozzle flow using a uniform grid with 65 points.  The variables are scaled to their upstream
stagnation values.

The steady-state results from one SFD calculation are shown in Figure 3.  The
Reynolds number (based on stagnation conditions upstream) was  3458 .  To improve
the resolution at the sonic point, as well as to ease the CFL restriction, two
Chebyshev grids were patched together at the throat.  The grid Reynolds number
ranges from  7  near the boundaries to over  200  near the shock.  Strong filtering
was required in order to obtain this result, for the unfiltered SFD calculation
develops severe oscillations.  The filtering used here is the extension to a non-
uniform grid of the averaging procedure

$$u_j \leftarrow (u_{j-1} + 2u_j + u_{j+1})/4 \ .$$

This type of filtering — also known as a von Hann window [Hamming (1977)] — has
been used by Orszag & Gottlieb (1979) in inviscid pseudo-spectral calculations.
One-sided averaging was applied at the shock.  Unlike the purely cosmetic filtering
applied to the Burgers' equation results, the filtering here actually altered the

computed solution.  In the calculation shown in Figure 3 this averaging was applied every 100 time-steps.  This filtering can be interpreted as the addition of an extra dissipative term.  Over the course of 100 time-steps, the effective viscosity of this filtering is an order of magnitude smaller than the physical viscosity, except near the shock.

An FFD calculation using MacCormack's method [Holst (1976)] is shown in Figure 4. The FFD resolution near the shock is better, since the shock happens to occur at the coarsest location on the Chebyshev grid.  Nevertheless, the FFD and SFD calculations differ by only a few tenths of a percent away from the shock.  The small grid spacing of the SFD method near the boundaries and at the throat lead to a more severe CFL restriction than occurs for the uniform grid of the FFD method.  Consequently, it took the SFD method about 7 times as many iterations to reach the steady-state.  The SFD method can probably be made more competitive, for example, by using a semi-implicit advection treatment.  But this example does illustrate that the SFD method can indeed handle a strong shock.

<table>
<tr><td>

**Table 5**

QUASI-ONE-DIMENSIONAL NOZZLE FLOW

$$(A\rho)_T + (A\rho u)_X = 0$$

$$(A\rho u)_T + [A(\rho u^2 + P)]_X - PA_X = [\tfrac{4}{3} A\mu u_X]_X$$

$$(AE)_T + [A(E + P)u]_X = [\tfrac{AK}{C_V} (E/\rho)_X]_X + [(\tfrac{2}{3}A\mu - \tfrac{AK}{2C_V})uu_X]_X$$

BOUNDARY CONDITIONS

    INFLOW:  ALL VARIABLES PRESCRIBED

    OUTFLOW:  PRESSURE PRESCRIBED
           OTHER VARIABLES EXTRAPOLATED

INITIAL CONDITIONS:  CRITICAL FLOW

</td><td>

**Table 6**

ASTROPHYSICAL PROBLEM

$$\rho_T + (\rho u)_\phi = 0$$

$$(\rho u)_T + [\rho(u^2 + c^2)]_\phi = 2\Omega(v-V)\rho + \rho A \sin\phi$$

$$(\rho v)_T + (\rho uv)_\phi = -(\kappa^2/2\Omega) (u-U)\rho$$

BOUNDARY CONDITIONS:  PERIODIC

INITIAL CONDITIONS:  STEADY-STATE SOLUTION

</td></tr>
</table>

C.  Astrophysical Problem

The last example is a set of inviscid fluid equations used by Woodward (1975) to model periodic gas flow driven by the gravitational field in a flat galaxy.  The equations are given in Table 6.  The steady-state solution exhibits a strong shock for which the advection steepening is balanced by Coriolis rather than by pressure forces.  After adding a small, explicit viscous term to the equations, the SFD method was applied, with the results shown in Figure 5.  The same strong filtering used in the previous example was also employed here.  However, nearly the same results are achieved if the filter is applied only cosmetically.  An FFD computation using MacCormack's method is shown in Figure 6.  With the use of the one-sided filter near the shock the SFD method resolves the discontinuity quite well.  Away from the shock, however, the FFD method appears to do a slightly better job, as indeed is reflected by an $L_2$ error computation.

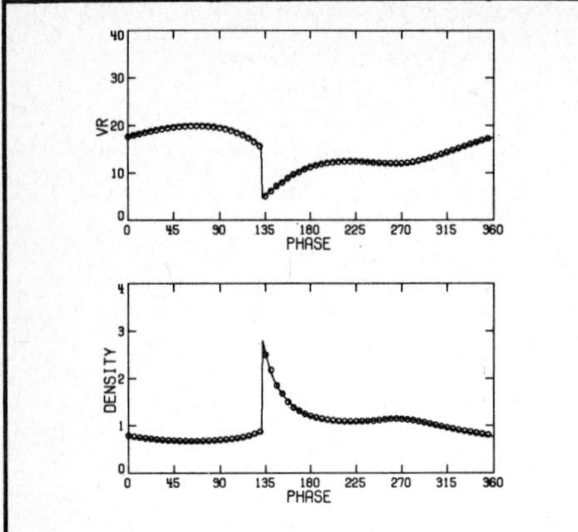

Figure 5. SFD solution of astrophysical problem on a uniform grid with 64 points. The solid curve represents the exact, steady-state solution. The open circles denote the computed solution after 1200 time-steps with a CFL number of 0.25.

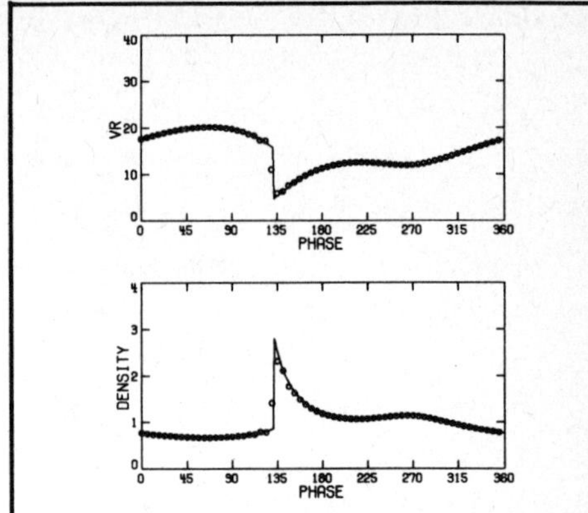

Figure 6. FFD solution of astrophysical problem on a uniform grid with 64 points. The solid curve represents the exact, steady-state solution. The open circles denote the computed solution after 600 time-steps with a CFL number of 0.50.

## III.  Summary

The SFD method has exhibited the ability to handle strong shocks and to outperform the FFD method at moderate viscosity.  On the more difficult problems, the SFD results are not yet as good as those of the best finite difference methods.  Possibly the development of special filtering techniques especially suited for the global oscillations occurring in spectral methods will change the balance.

## Acknowledgement

M.Y.H. is happy to acknowledge helpful discussions with Steven Orszag.

## References


Hamming, R. W. 1977, Digital Filters, Prentice-Hall.

Holst, T. L. 1976, "The Relative Merits of Numerical Techniques for Solving Compressible Navier-Stokes Equations," NASA CP2001, 1467-1481.

Orszag, S. A. & Gottlieb, D. 1979, "High Resolution Spectral Calculations of Inviscid Compressible Flows," preprint.

Orszag, S. A. & Israeli, M. 1974, "Numerical Simulation of Viscous Incompressible Flows," Ann. Rev. Fluid Mech., 49, 75-112.

Roache, P. J. 1975, "A Review of Numerical Techniques," Proceedings of the First International Conference on Numerical Ship Hydrodynamics.

Whitham, G. B. 1974, Linear and Nonlinear Waves, Wiley-Interscience.

Woodward, P. R. 1975, "On the Nonlinear Time Development of Gas Flow in Spiral Density Waves," Astrophys. J., 195, 61-75.

AN ACCURATE METHOD FOR CALCULATING
THE INTERACTIONS BETWEEN DISCONTINUITIES IN
THREE DIMENSIONAL FLOW

Y.-l. Zhu          B.-m. Chen

(The Computing Center, Academia Sinica, Beijing, China)

## I.  Introduction

In steady supersonic flow around bodies with complicated shapes, there appear interactions between two shocks and between a shock and a contact discontinuity.  It is difficult to calculate this problem accurately.  The reason is that you can obtain satisfactory results for this problem only when you have a method which can accurately calculate shocks, contact discontinuities, central waves, and their locations.

We have accurately calculated the interactions between discontinuities in two dimensional steady flow (Zhu et al., 1977).  In three dimensional steady flows, this problem is more complicated.  In this paper we briefly introduce a complete numerical method with high accuracy for calculating the interactions between discontinuities in three dimensional steady flow, and present some typical results.

## II.  The "Singularity-Separating" Difference Method

The problem we want to solve is the flow around bodies with complicated shapes (see Figures 1,4).  There are several shocks in these flow fields.  Moreover, because of the interactions between shocks, there are also several contact discontinuities and central waves.

In order to solve this problem accurately, we have developed a method named "Singularity-Separating" Difference Method.  "Singularity-Separating" means as follows.

1)  The singular surfaces -- the shocks, contact discontinuities, and boundaries of central waves -- are always coordinate surfaces in the calculating processes, and the singular lines --  the lines of intersection between discontinuities, and the edges of bodies -- are always coordinate lines.

2)  The jump conditions are used on the discontinuities, the continuity conditions are used on the weak discontinuities, and the jump conditions and the central wave differential equations, which are the generalization of the Prandtl-Mayer flow equations in three-dimensional problems, are used at the singular lines;

3)  The appearance of differences over discontinuities or stronger weak-discontinuities is not allowed in the difference equations.

In the following, we briefly describe our method. From that, you can see how the three points above are realized.

As is well known, the fluid dynamics equations for the inviscid, non-heat conducting flow in cylindrical coordinates $\{z,r,\phi\}$ are

$$
\begin{cases}
u \dfrac{\partial u}{\partial z} + v \dfrac{\partial u}{\partial r} + \dfrac{w}{r} \dfrac{\partial u}{\partial \phi} + \dfrac{1}{\rho} \dfrac{\partial p}{\partial z} = 0 \;, \\[2ex]
u \dfrac{\partial v}{\partial z} + v \dfrac{\partial v}{\partial r} + \dfrac{w}{r} \dfrac{\partial v}{\partial \phi} + \dfrac{1}{\rho} \dfrac{\partial p}{\partial r} - \dfrac{w^2}{r} = 0 \;, \\[2ex]
u \dfrac{\partial w}{\partial z} + v \dfrac{\partial w}{\partial r} + \dfrac{w}{r} \dfrac{\partial w}{\partial \phi} + \dfrac{1}{r\rho} \dfrac{\partial p}{\partial \phi} + \dfrac{vw}{r} = 0 \;, \\[2ex]
\rho \left( \dfrac{\partial u}{\partial z} + \dfrac{\partial v}{\partial r} + \dfrac{1}{r} \dfrac{\partial w}{\partial \phi} \right) + u \dfrac{\partial \rho}{\partial z} + v \dfrac{\partial \rho}{\partial r} + \dfrac{w}{r} \dfrac{\partial \rho}{\partial \phi} + \dfrac{\rho v}{r} = 0 \;, \\[2ex]
u \dfrac{\partial p}{\partial z} + v \dfrac{\partial p}{\partial r} + \dfrac{w}{r} \dfrac{\partial p}{\partial \phi} - a^2 \left( u \dfrac{\partial p}{\partial z} + v \dfrac{\partial \rho}{\partial r} + \dfrac{w}{r} \dfrac{\partial \rho}{\partial \phi} \right) = 0 \;,
\end{cases}
\tag{1}
$$

where $u,v,w$ are components of velocity corresponding to the $z$, $r$ $\phi$-directions respectively; $p$ is pressure; $\rho$ is density; $a$ is sonic speed. In our processes of calculation, instead of $z$, $r$, we introduce two new independent variables $\xi$, $\eta$ such that all the shocks, contact discontinuities et al. are coordinate surfaces. At this case, the problem we want to solve is reduced to the following problem: on an initial plane $\eta =$ const., the initial values are given; on several planes $\xi =$ const., the boundary conditions or internal boundary conditions, for example, the jump conditions are given; on the regions between these planes, a hyperbolic system is given; the flow field, including the locations of boundaries, needs to be determined. Therefore, in order to solve this problem, the only thing we need to do is to construct a difference scheme for general initial-boundary-value problems for hyperbolic systems.

We have presented a difference scheme for initial-boundary-value problems and applied it to calculating flow around bodies (Zhu et al., 1977, 1979b, 1980b). In our scheme, every difference equation approximates to the following equation of characteristic form

$$
G_n^* \left( \frac{\partial u}{\partial \eta} + \lambda_n \frac{\partial U}{\partial \xi} \right) + C_n^* \frac{\partial U}{\partial \phi} = f_n,
\tag{2}
$$

where $G_n^*$, $C_n^*$, $U$ are vectors, and $\lambda_n$, $f_n$ are scalars. Moreover, the differences over "singularities" are not allowed to appear in difference equations. Our scheme is a mixed one: when $\lambda_n$ is not too large, (2) is approximated by a second-order explicit scheme; and when $\lambda_n$ is large, (2) is approximated by a second-order implicit scheme. In order to determine the location of the boundaries, besides the difference equations approximating (2) and the boundary conditions, we also use some difference equations approximating

$$r_\ell = \int \frac{\partial r_\ell}{\partial \eta} \, d\eta, \quad \frac{\partial r_\ell}{\partial \phi} = \int \frac{\partial^2 r_\ell}{\partial \eta \partial \phi} \, d\eta \,,$$

where $r = r_\ell(\eta, \phi)$ are equations of boundary-locations. If the explicit scheme is adopted everywhere, it is simple to determine the values of functions. However, if the implicit scheme is adopted in some places, in order to determine the values of functions, we must solve a system of equations which consists of many linear equations and a few nonlinear equations. We solve this system by using a block-double-sweep method presented by us.

In our procedure of calculation, at some certain cases, we have to do some other things. For example, when two discontinuities intersect, a three-dimensional Riemann problem should be solved before proceeding with further calculation. When an embedded shock will appear, we must determine where the embedded shock will appear. Our method to determine the locations of the embedded shocks is based on the characteristic theory in three independent variables.

In order to be sure that our method will give good and reasonable results, we have carefully studied the theoretical soundness of our method. We have proved that our initial-boundary-value difference scheme is unconditionally stable with respect to initial and boundary values (Zhu et al., 1978a, 1979a), and that the approximate solutions converge to the real solutions with a second-order rate of convergence (Zhu, 1980c). We have also proved that the block-double-sweep method presented by us and used in our calculation is stable (Zhu, 1978b).

### III. Some Results

Because (i) we separate all singularities, (ii) our scheme is a second-order scheme and (iii) it has theoretical soundness, we have obtained a series of very accurate results by using grids with a small number of mesh-points, and have not confronted with any difficulties in our calculations. In this paper we present two typical results.

In Figures 1-3, some results about the flow around the axisymmetric combined body-I are given, where the angle of attack $\alpha = 2^\circ$, the freestream Mach number $M_\infty = 4$, the gas is perfect and the ratio of specific heats $\gamma = 1.4$. Because of the complicated shape of this body, there appear the interactions between two shocks and between a shock and a contact discontinuity in this flow field. Therefore, the structure of this flow field is very complicated. In Figure 1, the shapes of shocks, contact discontinuities and boundaries of central waves on the windward and leeward sides are given. In Figures 2-3, the pressure curves and the density curves in the $\xi$-direction on the section $z = 34$ are drawn. As you see, there is no oscillation on our curves of results.

In Figure 4, the structure of the flow field around the axisymmetric combined body-II is given, where $M_\infty = 15$, $\alpha = 6^0$ and the gas is perfect with $\gamma = 1.4$. Because this body has a surface which presses continuously on the flow, there is an embedded shock in this field. Moreover, the embedded shock intersects the main shock in this field. Therefore, the structure of the flow field is also complicated.

According to these results, we think that we have successfully calculated those problems with interactions between discontinuities and with automatically formed shocks in three dimensional steady flow.

In our monograph (Zhu et al., 1980a), we systematically describe our method, give the complete proofs about those theoretical results, and present a series of computed results. For the details of our work, readers are referred to this monograph.

## References


Zhu, You-lan et al., <u>1977</u>, A numerical method for initial-boundary-value problems of hyperbolic systems and applications, Acta Mathematicae Applogatae Sinica, 1977. No. 3, 12-27.

Zhu, You-lan et al., <u>1978a</u>, Difference schemes for initial-boundary-value problems of hyperbolic systems and examples of application, Scientia Sinica, 1978, No. 2, 125-138, Chinese edition, or 1979, Special Issue (II), 261-280, English edition.

Zhu, You-lan, <u>1978b</u>, A block-double-sweep method for "noncomplete" linear algebraic systems and its stability, Mathematicae Numericae Sinica, 1978, No. 3, 1-27.

Zhu, You-lan, <u>1979a</u>, Difference schemes for initial-boundary-value problems of first order hyperbolic systems and their stability, Mathematicae Numericae Sinica, 1979, No. 1, 1-30.

Zhu, You-lan et al. <u>1979b</u>, The numerical calculation of the supersonic flow around combined bodies, Acta Mechanica Sinica, 1979, No. 3, 209-218.

Zhu, You-lan et al., <u>1980a</u>, Difference methods for initial-boundary-value problems and flow around bodies, Science Press, Beijing, China, 1980.

Zhu, You-lan and Chen, Bing-mu, <u>1980b</u>, A numerical method with high accuracy for calculating the interactions between discontinuities in three independent variables, Scientia Sinica (in press).

Zhu, You-lan, <u>1980c</u>, Stability and convergence of difference schemes for linear initial-boundary--value problems. Mathematicae Numericae Sinica (to appear).

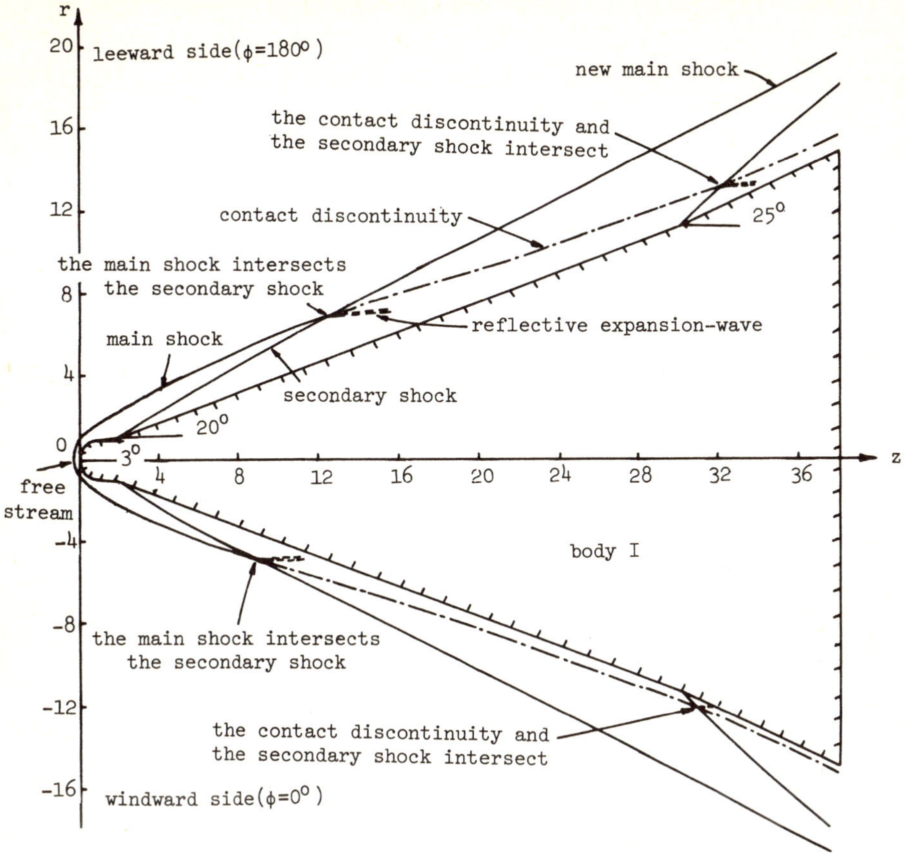

Figure 1    Construction of Flow on the Windard and Leeward Sides
( $M_\infty=4$    $\alpha=2^\circ$    perfect gas    $\gamma=1.4$ )

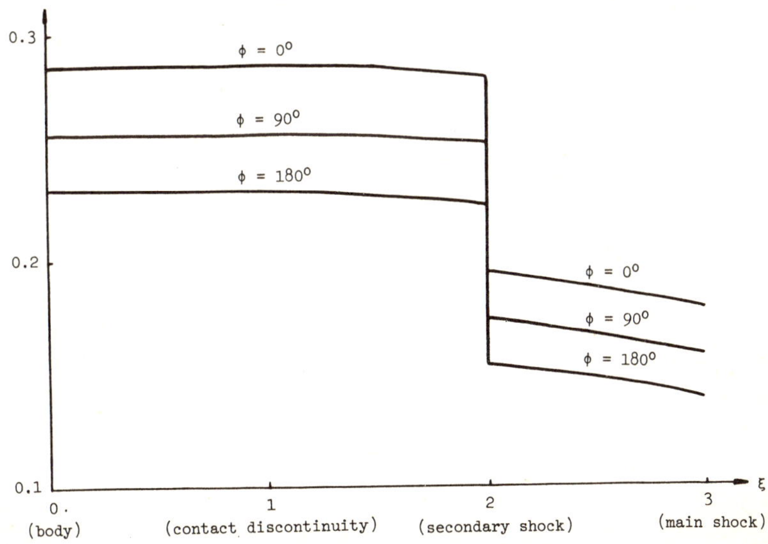


Figure 2.  Pressure Curves in the $\xi$-Direction

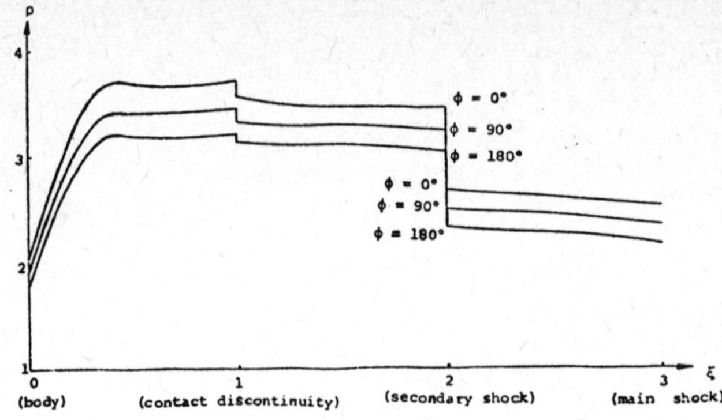



Figure 3.  Density Curves in the ξ-Direction

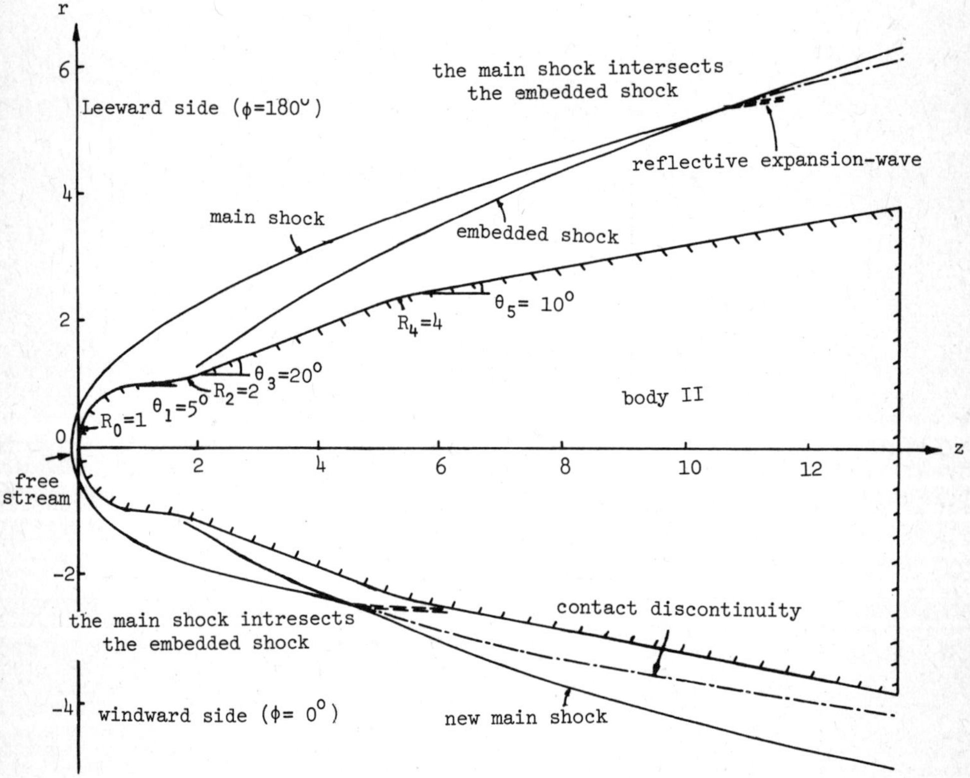

Figure 4    Construction of Flow on the Windward and Leeward Sides
($M_\infty$ =15    α=6°    perfect gas    γ=1.4 )

Mr. Michael Abbett
Mgr. Comp. Fluid Dynamics
Acurex Corp.
485 Clyde Avenue
Mountain View, CA 94042

Prof. Robert C. Ackerberg
Polytechnic Institute of New York
Route 110
Farmingdale, N. Y. 11735

Mr. Robert Altman
NASA-Ames Research Center
M. A. 223-6
Moffett Field, CA 94035

Dr. Bert Arlinger
Aerospace Division
SAAB-SCANIA AB S-58188
Linköping, SWEDEN

Mr. Wilbur Armstrong
ARO, Inc.
OGM/CSB
Arnold AFS, Tenn. 37389

Dr. William T. Ashurst
Gas Dynamics Div. 8354
Sandia Labs
Livermore, CA 94550

Dr. Essam Atta
Lockheed Georgia Co.
Department 72/74, Zone 403
Marietta, GA 30063

Mr. Frank Bailey
M. S. 233-1
NASA-Ames Research Center
Moffett Field, CA 94035

Mr. Harry Bailey
M. S. 202A-1
NASA-Ames Research Center
Moffett Field, CA 94035

Dr. Gregory Baker
Mail Code 2-337
Dept. of Mathematics
Mass. Inst. of Technology
Cambridge, MA 02139

Dr. Fritz Bartels
David W. Taylor
   Naval Ship R&D Center
Bethesda, MD 20084

Mr. Mark Barnett
M. S. 229-1
NASA-Ames Research Center
Moffett Field, CA 94035

Dr. John Barton
M. S. 202A-1
NASA-Ames Research Center
Moffett Field, CA 94035

Dr. Howard Baum
Center for Fire Research
National Bureau of Standards
Washington, D. C. 20234

Mr. Richard Beam
M. S. 202A-1
NASA-Ames Research Center
Moffett Field, CA 94035

Dr. Matania Ben-Artzi
18 Golomb St.
Giryat-Tivon, ISRAEL

Mr. Jean-Pierre Benque
6, Quai Watier
78400 Chatou, FRANCE

Mr. Michael H. Berger
Union Carbide Corp., Nuclear Div.
P. O. Box P, K-1004D, MS-282
Oak Ridge, TN 37830

Ms. Muriel Bergmann
M. S. 229-1
NASA-Ames Research Center
Moffett Field, CA 94035

Prof. Yuriy Berezin
Inst. of Theor. & Appl. Mechanics
Novosibirsk, 630090, USSR

Mr. David Black
M. S. 245-1
NASA-Ames Research Center
Moffett Field, CA 94035

Dr. Frederick Blottner
Div. 5511, P. O. Box 5800
Sandia National Laboratories
Albuquerque, NBM 87185

Dr. David L. Book
Code 4042
Naval Research Lab
Washington, D. C. 20375

Dr. Jay Boris
Naval Research Laboratory
Washington, D. C.  20375

Mr. A. Bossavit
1 Avenue General DeGaulle
Clamart, FRANCE  92141

Mr. John Bridgeman
M. S. 202A-14
NASA-Ames Research Center
Moffett Field, CA  94035

Dr. W. Roger Briley
Scientific Research Associates
P. O. Box 498
Glastonbury, CT  06033

Mr. David Brown
NASA-Informatics, Inc.
Caltech  217-50
Pasadena, CA   91125

Mr. James Brown
M. S. 229-1
NASA-Ames Research Center
Moffett Field, CA  94035

Mr. O. Buneman
Electrical Engr. Dept.,  ERL 314
Stanford University
Stanford, CA  94305

Mr. Pieter Buning
M. S. 202A-1
NASA-Ames Research Center
Moffett Field, CA  94035

Prof. Odus Burggraf
Aero-Astro Engineering
Ohio State University
2036 Neil Avenue Mall
Columbus, OH  43210

Prof. Brian Cantwell
371 Durand
Stanford University
Stanford, CA  94305

Dr. Enrique Caponi
TRW DSSG
One Space Park
Redondo Beach, CA  90278

Mr. Frank Caradonna
M. S. 215-1
NASA-Ames Research Center
Moffett Field, CA  94035

Mr. Ralph Carmichael
M. S. 227-2
NASA-Ames Research Center
Moffett Field, CA  94035

Dr. Sukumar Chakravarthy
Rockwell Int'l. Science Center
P. O. Box 1085
Thousand Oaks, CA  91360

Dr. Frank W. K. Chan
JAYCOR
626 39th Ave.
San Francisco, CA  94121

Dr. Dean Chapman
Aero-Astro Dept.
Stanford University
Stanford, CA  94305

Dr. Jean-Jacques Chattot
ONERA
29 Avenue Dir. Leclerc
92320 - Chatillon, FRANCE

Dr. Denny Chaussee
Flow Simulations, Inc.
298 S. Sunnyvale Ave., Suite 204
Sunnyvale, CA  94086

Mr. Hai-chow Chen
The Boeing Co.
P. O. Box 3999,  M. S. 3N-19
Seattle, WA  98124

Mr. Lee-tzung Chen
McDonnell Douglas Corp.
P. O. Box 516
St. Louis, MO  63166

Dr. Donald R. Chenoweth
Org. 8124, Appl. Mech. Dept.
Sandia Laboratory
P. O. Box 969
Livermore, CA  94550

Mr. Richard Chipman
Grumman Aerospace Corp.
9 Lone Oak Court
Centerport, N. Y.  11721

Mr. Leslie Chow
M. S. 202A-14
NASA-Ames Research Center
Moffett Field, CA  94035

Dr. Melvyn Ciment
Mathematical Analysis Division
National Bureau of Standards
Washington, D.C. 20234

Dr. Michael Cline
Lox Alamos Scientific Laboratory
M. S. 216
Los Alamos, NM  87545

Dr. Lawrence Cloutman
M.S. 216
Los Alamos Scientific Laboratory
Los Alamos, NM  87545

Mr. Thomas Coakley
M. S. 229-1
NASA-Ames Research Center
Moffett Field, CA  94035

Mr. Phillip Colella
50A-2129 CSAM
Lawrence-Berkeley Laboratory
UC-Berkeley, CA  94720

Dr. Raul J. Conti
Dept. 52-33, Bldg. 201
Lockheed Palo Alto Research Laboratory
3251 Hanover St.
Palo Alto, CA  94306

Mr. Morton Cooper
Office of Naval Research
800 N. Quincy St.
Arlington, VA  22217

Dr. Benoit Couët
Inst. for Plasma Research
ERL 312
Stanford University
Stanford, CA  94305

Dr. Anthony J. Crisalli
Westinghouse Marine Division
Advanced Concepts Engineering
Hendy Ave.
Sunnyvale, CA  94088

Dr. Cornelis Cuvelier
Mathematical Department
Delft University of Technology
Julianalaam 132
Delft, THE NETHERLANDS

Dr. Ronald Davis
Fluid Engrg. Div., 771, FMI05
National Bureau of Standards
Washington, D. C.  20234

Mr. William Davy
M. S. 229-4
NASA-Ames Research Center
Moffett Field, CA  94035

Dr. Gerrit J. de Bruin
Room WB/W250
Tech. Univ. Twente
Postbus 217, 7500 AE Enshede
THE NETHERLANDS

Mr. Herman Deconinck
Dept. of Fluid Mechanics
Vrije Universiteit
Pleinlaan 2
1050 Brussel, BELGIUM

Mr. George Deiwert
M. S. 202A-1
NASA-Ames Research Center
Moffett Field, CA  94035

Prof. Stanley C. R. Dennis
Dept. of Applied Mathematics
University of Western Ontario
London, Ontario, N6A 5B9, CANADA

Mr. A. Dervieux
88 rue de Petit Bois
Plaisir, FRANCE  78370

Mr. F. Carroll Dougherty
M. S. 202A-14
NASA-Ames Research Center
Moffett Field, CA  94035

Mr. John Philip Drummond
M. S. 168
NASA-Langley Research Center
Hampton, VA  23665

Dr. John Dukowicz
Los Alamos Scientific Laboratory
Los Alamos, NM  87544

Dr. Djordje S. Dulikravich
M. S. 5-9
NASA-Lewis Research Center
21000 Brookpark Road
Cleveland, Ohio  44125

Dr. Douglas L. Dwoyer
M. S. 360,
NASA-Langley Research Center
Hampton, VA  23665

Mr. Harry A. Dwyer
Dept. of Mechanical Engrg.
UC-Davis
Davis, CA  95616

Dr. Douwe Dykstra
Twente Univ. of Technology
T.H.T. Dept. TW,  POB 217
7500 AE Enschede, THE NETHERLANDS

Mr. Yutaka Ebi
Ricoh Company, Ltd.
2071 Concourse Drive
San Jose, CA  95131

Dr. Joseph Eisenhuth
Applied Research Laboratory
P. O. Box 30
State College, PA  16801

Dr. Peter Eiseman
M. S. 132C
NASSA-Langley Research Center
Hampton, VA  23665

Prof. Nabil Esmail
Dept. of Chem. & Chem. Engrg.
University of Saskatchewan
Saskatoon, Saskatchewan,
Canada S7N OWO

Dr. Hermann F. Fasel
Dept. of Mech. & Aerospace Engrg.
Engineering Quadrangle
Princeton University
Princeton, N. J.  08544

Dr. William Feiereisen
Department TX-2
Brown Boveri & Cie
CH-5401  Baden, Switzerland

Prof. Joel Ferziger
Dept. of Mechanical Engrg.
Stanford University
Stanford, CA  94305

Dr. Clive A. J. Fletcher
Dept. of Mech. Engrg.
University of Sydney
Sidney, N.S.W. 2006, AUSTRALIA

Mr. Clifford Forester
1920 Blaine
Seattle, WA  98112

Mr. Jacob Fromm
IBM Research  K31/281
5600 Cottle Road
San Jose, CA  95193

Dr. Kozo Fujii
Inst. of Space & Aero Science
University of Tokyo
4-6-1 Komaba, Meguroku
Tokyo, Japan  153

Mr. John H. Gardner
Naval Research Laboratory
Washington, D. C.  20375

Mr. G. Roger Gathers
Lawrence Livermore Nat'l. Lab.
L-355
P. O. Box 808
Livermore, CA  94550

Dr. Nima Geffen
Mathematics Dept.
Tel Aviv Univesity
Ramat Aviv, ISRAEL

Mr. Peter Goorjian
M. S. 202A-14
NASA-Ames Research Center
Moffett Field, CA  94035

Dr. Randolph A. Graves
Code RTF-6
NASA Headquarters
Washington, D. C.  20546

Mr. Michael Green
M. S. 229-4
NASA-Ames Research Center
Moffett Field, CA  94035

Dr. Taeyoung Han
Dept. of Mech. Engrg.
6124-C  Etcheverry Hall
University of California
Berkeley, CA  94720

Mr. C. Frederick Hansen
M. S. 223-6
NASA-Ames Research Center
Moffett Field, CA  94035

Prof. Amiram Harten
Tel-Aviv University
Ramat Aviv, ISRAEL

Dr. Thorwald Herbert
Inst. Aerodynamik
Universität Stuttgart
Pfaffenwaldring 21
D-7000 Stuttgart 80, GERMANY

Ms. Kristin Hessenius
M. S. 202A-14
NASA-Ames Research Center
Moffett Field, CA  94035

Dr. Naoki Hirose
2nd Aerodynamics Div.
National Aerospace Laboratory
1880 Jindaiji
Chofu-shi, Tokyo 182, JAPAN

Dr. Richard S. Hirsh
JAYCOR
P. O.  Box 370
Del Mar, CA  92014

Dr. Gilbert H. Hoffman
Applied Research Laboratory
P. O. Box 30
State College, PA  16801

Mr. Terry Holst
M. S. 202A-14
NASA-Ames Research Center
Moffett Field, CA  94035

Mr. Kiyosi Horiuti
Faculty of Engineering
The University of Tokyo
Hongo-7-3-1
Bunkuo-ku, Tokyo 113, JAPAN

Mr. Dun Huang
Dept. of Mathematics
Beijing University
Beijing, People's Republic of China

Mr. Ching-mao Hung
M. S. 202A-1
NASA-Ames Research Center
Moffett Field, CA  94035

Dr. Mohammed Hussaini
ICASE
NASA-Langley Research Center
Hampton, VA  23665

Mr. James M. Hyman
T-7, MS233
Los Alamos Scientific Laboratory
Los Alamos, NM  87545

Mr. Mamoru Inouye
M. S. 202A-1
NASA-Ames Research Center
Moffett Field, CA  94035

Dr. Osamu Inoue
Inst. of Space & Aero Science
Komaba 4-6-1
Meguro-ku, Tokyo 153, JAPAN

Prof. Moshe Israeli
Computer Science Dept.
Technion City
Haifa, ISRAEL  32000

Dr. Raad Issa
Mech. Engrg. Dept.
Imperial College
London S.W.7, ENGLAND

Mr. Adrien Jami
ENSTR Chemin de la Huniere
43 rue Maurice Berbeaux
91120 Palaiseau, FRANCE

Dr. Dennis Jespersen
Mathematics Dept.
Oregon State University
Corvallis, OR  97331

Mr. Forrester T. Johnson
The Boeing Company
P. O. Box 3999,  M.S. 3N-19
Seattle, Washington  98124

Dr. Gary Johnson
M. S. 5-9
NASA-Lewis Research Center
Cleveland, Ohio  44135

Mr. Wm. Pritchard Jones
M. S. 233-10
Moffett Field, CA  94035

Dr. Wen-Huei Jou
Flow research Co.
21414 - 68th Ave. S.
Kent, WA  98031

Prof. Krishna Karamcheti
Aeronautics & Astronautics
Stanford University
Stanford, CA  94305

Dr. Michael Kascic
Control Data Corp.
4201 Lexington Avenue, North
Arden Hills, MN  55112

Dr. Theodore Katsanis
NASA-Lewis Research Center
21000 Brookpark Rd.
Cleveland, Ohio  44135

Prof. Czeslaw P. Kentzer
Aero/Astro,  Grissom Hall
Purdue University
W. Lafayette, IN  47907

Prof. Prem Khosla
Dept. of Aero. Engrg. & Appl. Mech
University of Cincinnati
Cincinnati, Ohio  45221

Prof. John Kim
M. S. 202A-1
NASA-Ames Research Center
Moffett Field, CA  94035

Dr. Goetz Klopfer
Nielsen Engineering
510 Clyde Avenue
Mountain View, CA  94043

Mr. Karl Kneile
Computational Systems Branch,
ARO, Inc.
Arnold AFS, Tenn.  37389

Mr. Mark Koenig
M. S. 303A-1
NASA-Ames Research Center
Moffett Field, CA  94035

Dr. Cornelis Korving
Delft University of Technology
Julianalaan 132
Delft, THE NETHERLANDS

Dr. Edward M. Kraft
ETF/TAB
Sverdrup/ARO, Inc.
Arnold AFS, Tenn.  37389

Prof. Egon Krause
Director, Aerodynamisches Institut
Wullnerstr. ZW 5UF
57 Aachen, WEST GERMANY

Dr. Allen L. Kuhl
R&D Associates
P. O. Box 9695
Marina Del Rey, CA  90291

Mr. Paul Kutler
M. S. 202A-14
NASA-Ames Research Center
Moffett Field, CA  94035

Dr. Kunio Kuwahara
Dept. of Applied Physics
University of Tokyo
Hongo, Bunkyo-ku
Tokyo, JAPAN

Dr. Dochan Kwak
M.S. 202A-14
NASA-Ames Research Center
Moffett Field, CA 94035

Mr. Jean-Pierre La Hargue
Compagnie Internationale
    de Services en Informatique
BP.n°24, 91190
GIF-s/-Yvette, FRANCE

Dr. B. Lakshminarayana
Dept. of Aerospace Engrg.
The Pennsylvania State University
233 Hammond Buildings
University Park, PA  16802

Mr. Jacques Laminie
12 Avenue de Paris
78000 Versailles, FRANCE

Mr. Martin Lanfranco
Informatics, Inc.
1121 San Antonio Road
Palo Alto, CA  94303

Mr. William E. Langlois
IBM Research Laboratory
5600 Cottle Road
San Jose, CA  95193

Prof. Patrick Lascaux
Dept. de Mathematique Informatique
Conserv. National de Arts et Métiers
292 Rue St. Martin
75141 Paris, Cedex 03, FRANCE

Mr. Thomas Lasinski
M. S. 202A-14
NASA-Ames Research Center
Moffett Field, CA  94035

Dr. John A. Laurmann
Applied Mechanics, Durand Bldg.
Stanford University
Stanford, CA  94305

Prof. Peter Lax
Courant Inst. NYU
251 Mercer Street
New York, N. Y.  10012

Mr. Ki D. Lee
The Boeing Co.
P. O. Box 3999, M/S 3N-19
Seattle, WA  98124

Dr. Shen Lee
Dept. of Mechanical Engrg.
University of Missouri-Rolla
Rolla, MO  65401

Mr. Noel Legrand
C.E.A.  B.P. 27
94190 Villeneuve St. Georges, FRANCE

Mr. Patrick LeQuere
Laboraidire d'Energetique Sulaire
40 Avenue de Recteur Pineau
86000 Poitiers, France

Mr. Oleg Lesovoy
Presidium of Siberian Branch
USSR Science Academy
17, Prospect Nauki
Novosibirsk, USSR

Dr. Ron Levine
Advanced Computation Laboratory
3990B Freedom Circle
Santa Clara, CA.  95051

Mr. Lionel Levy
M. S. 229-1
NASA-Ames Research Center
Moffett Field, CA  94035

Mr. Markku Lindroos
Institute of Mathematics
Helsinki University of Technology
SF-02150  Espoo 15, FINLAND

Mr. Meng-Sing Lion
McDonald Douglas
P. O. Box 516
St. Louis, MO  63166

Mr. Dominique Litaise
Centre d'Etudes de Gramat
46500 Gramat, FRANCE

Mr. Eli Livne
Dept. of Theoretical Physics (Rakavy)
Hebrew University
Jerusalem, ISRAEL

Dr. Harvard Lomax
M. S. 202A-1
NASA-Ames Research Center
Moffett Field, CA  94035

Dr. Charles Lombard
Pac. Engineering Design Analysis Co.
1150 Fife Street
Palo Alto, CA  94301

Professor M. S. Longuet-Higgins
University of Cambridge
Dept. of Appl. Math. & Theor. Physics
Silver Street
Cambridge CB3 9EW,  ENGLAND

Mr. Ciro W. Lucchi
Messerschmitt-Boelkow-Blohm, FE 122
Postfach 801160
D-8000  München,  80, W. GERMANY

Mr. Robert MacCormack
M. S. 202A-1
NASA-Ames Research Center
Moffett Field, CA  94035

Dr. Nagi N. Mansour
General Motors Research Lab.
Fluid Dynamics Department
Warren, MI  48090

Mr. Bernard Marquié
Commissariat a l'Energie Atomique
B. P. n° 7
93270 Sevran, FRANCE

Mr. Dale Martin
M. S. 202A-1
NASA-Ames Research Center
Moffett Field, CA  94035

Mr. Joseph Marvin
M. S. 229-1
NASA-Ames Research Center
Moffett Field, CA  94035

Dr. Ken Marx
Division 8354
Sandia Laboratories
Livermore, CA  94550

Mr. Mats Mattsson
Heat Transfer & Fluid
ASEA AB, Mechanical Systems Office
Vaesteras, SWEDEN

Mr. Unmeel Mehta
M. S. 202A-1
NASA-Ames Research Center
Moffett Field, CA  94035

Mr. Daniel Meiron
Dept. of Mathemtics
Mass. Institute of Technology
77 Massachusetts Ave.
Cambridge, Mass.  02139

Dr. James Meng
Fluid Dynamics Div.
Science Applications, Inc.
P. O. Box 2351
La Jolla, CA  92038

Mr. Marshal Merriam
M. S. 202A-1
NASA-Ames Research Center
Moffett Field, CA  94035

Dr. Ralph Metcalfe
Flow Research Co.
21414 - 68th Ave. S.
Kent, WA  98031

Mr. Gérard Meurant
C.E.A. - Centre de Limeil
B. P. 27
94190 Villeneuve-St-Georges, FRANCE

Mr. Richard Miller
M. S. 245-3
NASA-Ames Research Center
Moffett Field, CA  94035

Prof. Morton Mitchner
Dept. of Mech. Engrg.
Stanford University
Stanford, CA  94305

Mr. Richard Miller
M. S. 245-3
NASA-Ames Research Center
Moffett Field, CA  94035

Dr. Parviz Moin
Dept. of Mech. Engrg.
Stanford University
Stanford, CA  94305

Dr. Wim Mol
Mathematical Centre, Room 46
2e Boerhaavestraat 49
1091 AL Amsterdam, THE NETHERLANDS

Mrs. Elizabeth Moore
FM105
National Bureau of Standards
Washington, D. C.  20234

Mr. John Murphy
M. S. 227-8
NASA-Ames Research Center
Moffett Field, CA  94035

Dr. Brian McCay
Cabot Corporation  2-336
Mass. Institute of Technology
Cambridge, MA  02139

Mr. W. J. McCroskey
M. S. 215-1
NASA-Ames Research Center
Moffett Field, CA  94035

Mr. H. McDonald
Scientific Research Associates
P. O. Box 498
Glastonbury, CT  06033

Mr. James D. McLean
The Boeing Co., M/S 3N-19
P. O. Box 3999
Seattle, WA  98124

Dr. John McLean
Fluid Mechanics Dept.
TRW,  One Space Park
Redondo Beach, CA  90278

Major D. Scott McRae
M. S. 206-3
NASA-Ames Research Center
Moffett Field, CA  94035

Mr. Jean Claude Nedelec
Ecole Polytechnique
3 Place de Bretagne
91430 Igny, FRANCE

Mr. Billy Nichols
Los Alamos Scientific Laboratory
P. O. Box 1663
Los Alamos, NM  87545

Prof. Helmer L. Nielsen
1980 Laver Court
Los Altos, CA  94022

Dr. David Nixon
Nielsen Engineering & Research, Inc.
510 Clyde Ave.
Mountain View, CA  94043

Mr. Grove Nooney
Informatics, Inc.
1121 San Antonio Road
Palo Alto, CA  94303

Dr. Robert Numrich
4201 N. Lexington Ave.
St. Paul, MN  55112

Dr. Ed Oliver
AFOSR/NM
Bolling AFB, D.C.  20332

Dr. Kiyoaki Ono
Dept. Mech. Engrg.
Hihon University
1-8 Kanda-Surugadai
Chiyoda-ku, Tokyo, 101, JAPAN

Dr. Elaine Oran
Code 4040
Naval Research Laboratory
Washington, D. C.  20375

Prof. Paolo Orlandi
Istuto de Aerodinamica
Universita di Roma
Via Eudossiana 16
00184 Roma, ITALY

Prof. Steven A. Orszag
Mathematics Dept.  2-347
Mass. Inst. of Technology
Cambridge, MA  02138

Prof. Koichi Oshima
Inst. Space Aero Sci.
University of Tokyo
Komaba, Meguro-ku
Tokyo 153,  JAPAN

Prof. Maurizio Pandolfi
Institut de Thermique Appliquée
Ecole Polytechnique Federale
Lausanne, SWITZERLAND

Dr. James E. Park
Nuclear Division
Union Carbide Corp.
Box P, MS 53
Oak Ridge, TN  37830

Prof. Vithal Patel
Applied Mechanics
Stanford University
Stanford, CA  94305

Mr. Anthony Patera
Applied Math Dept. - Rm. 2-347
Mass. Institute of Technology
77 Massachusetts Ave.
Cambridge, MA  02139

Dr. Alvin J. Paullay
25 Berry Ct.
Tappan, N. Y.  10983

Mr. Richard Pelz
5 Linden Blvd.
Hicksville, N. Y.  11801

Mr. Victor Peterson
M. S. 229-3
NASA-Ames Research Center
Moffett Field, CA  94035

Mr. William J. Phares
ARO, Inc.
Arnold AFS, Tenn.  37389

Dr. Thomas Phillips
Nat'l. Center for Atmospheric Research
P. O. Box 3000
Boulder, CO  80307

Dr. Gershon Pinchuk
Ministry of Defense
P. O. Box 2250
Haifa, ISRAEL

Dr. Renzo Piva
Istituto di Aerodinamica
University of Rome
Via Eudossiana
18 Roma, ITALY

Dr. Hartley Pond
Code 401
Naval Underwater Systems Center
New London Laboratory
New London, CT  06320

Mr. Thomas Pulliam
M. S. 202A-1
NASA-Ames Research Center
Moffett Field, CA  94035

Mr. John Rakich
M. S. 229-1
NASA-Ames Research Center
Moffett Field, CA  94035

Mr. Alan Ratliff
Lockheed Missiles & Space Co.
4800 Bradford Blvd.
Huntsville, Alabama  35807

Mr. K. C. Reddy
Univ. of Tennessee Space Inst.
Tullahoma, TN  37388

Mr. Walter Reinhardt
M. S. 202A-14
NASA-Ames Research Center
Moffett Field, CA  94035

Dr. T. Reyhner
M. S. 73-07
Boeing Commercial Airplane Co.
P. O. Box 3707
Seattle, WA  98124

Dr. William C. Reynolds, Chairman
Mechanical Engineering Dept.
Stanford University
Stanford, CA  94305

Dr. James Riley
Flow Research Co.
21414 - 68th Ave. South
Kent, WA  98031

Dr. Arthur Rizzi
FFA, Aeronautical Res. Inst. of Sweden
P. O. Box 11021
S-16111  Bromma, SWEDEN

Mr. Leonard Roberts
M. S. 200-3
NASA-Ames Research Center
Moffett Field, CA  94035

Mr. Philip L. Roe
3 Ft. Tunnel
Royal Aircraft Establishment
Bedford, UNITED KINGDOM

Prof. Karl G. Rosener
Inst. f. Mechanik
Techn. Hochschule
Hochashul St., D-6100
Darmstadt, WEST GERMANY

Dr. Paul E. Rubbert
M. S. 3N-19
The Boeing Co.
P. O. Box 3999
Seattle, WA  98124

Mr. Morris Rubesin
M. S. 202A-1
NASA-Ames Research Center
Moffett Field, CA  94035

Prof. Stanley G. Rubin
Dept. of Aerospace Engrg. & Appl.
Mech.
University of Cincinnati
Cincinnati, Ohio  45221

Prof. Viktor Rusanov
Keldysh Inst. of Applied Math.
Miusskaya Pl. 4
Moscow A-47, USSR

Dr. Oleg Ryzhov
Computing Center
Academy of Sciences of the USSR
ulitsa Vavilova, 40
Moscow V-333, USSR

Mr. Shinichiro Sakamoto
48B Escondido Village
Stanford, CA  94305

Dr. Billy R. Sanders
Division 8354
Sandia Laboratories
Livermore, CA  94550

Dr. Lakshmi Sankar
Lockheed Georgia Co.
36 South Cobb Drive
Marietta, GA  30063

Dr. Bruno Scheurer
C.E.A. - Centre de Limeil
B. P. 27
94190 Villeneuve-St-Georges,  FRANCE

Mr. Lewis Schiff
M. S. 227-8
NASA-Ames Research Center
Moffett Field, CA  94035

Prof. Robert Schreiber
Dept. of Computer Science
Stanford University
Stanford, CA  94305

Dr. Samuel P. Shanks
Mail Zone 2882
General Dynamics Corp.
P. O. Box 748
Fort Worth, Texas  76101

Prof. F. Sherman
261 Grizzley Peak Blve.
Kensington, CA  94708

Dr. Vsevolod Shidlovskiy
Computing Center
Academiy of Sciences of the USSR
ulitsa Vavilova 30
Moscow  V-333, USSR

Dr. C. C. Shir
IBM Research Laboratory
K31/282
5600 Cottle Road
San Jose, CA  95193

Mr. Bruce Smith
M. S. 245-3
NASA-Ames Research Center
Moffett Field, CA  94035

Dr. Helmut Sobieczky
DFVLR Göttingen
Görlitzer Str. 62
D 3400 Göttingen, W. GERMANY

Dr. Peter Sockol
M. S. 5-9
NASA-Lewis Research Center
21000 Brookpark Road
Cleveland, Ohio  44135

Prof. Gary Sod
Dept. of Mech. & Aero Engrg.
Princeton University
Princeton, NJ  08540

Mr. Reese Sorenson
M. S. 202A-14
NASA-Ames Research Center
Moffett Field, CA  94035

Dr. Soubbaramayer
C.E.N./SACLAY
Départment de Génie Isotopique
B.P. n°2
91190 Gif sur Yvette, FRANCE

Prof. John R. Spreiter
Applied Mech., Mech. Engrg. Dept.
Stanford University
Stanford, CA  94305

Dr. Joseph Steger
Flow Simulations, Inc.
Suite 204
298 S. Sunnyvale Ave.
Sunnyvale, CA  94086

Dr. John Steinhoff
Research Cept.
Grumman Aerospace Corp.
Bethpage, N. Y.  11714

Prof. Massimo Strani
Istituto di Macchine
Universita' di Roma
via Eudossiana
Roma, ITALY

Mr. Gordon Strate
M. S. 202A-1
NASA-Ames Research Center
Moffett Field, CA  94035

Dr. Robert M. Stubbs
NASA-Lewis Research Center
21000 Brookpark Road
Cleveland, Ohio  44135

Dr. Myron Sussman
5026 Belmont Ave.
Bethel Park, PA  15102

Mr. Blair K. Swartz
T-7, MS233
Los Alamos Scientific Laboratory
Los Alamos, NM  87545

Prof. John C. Tannehill
Dept. of Aerospace Engrg.
Iowa State University
Ames, Iowa  50011

Mr. Yehuda Tassa
Dept. 72-74
Lockheed Georgia Co.
86 South Cobb Dr.
Marietta, GA  30063

Prof. Roger Temam
Laboratoire d'Analyse Numerique
Universite Paris Sud, Bat. 425
91405-Orsay, FRANCE

Mr. Chen-Huan Teng
Mathematics Dept.
University of California
Berkeley, CA  94720

Dr. Frank Thiele
Technische Univesitaet Berlin
Strasse d. 17 Juni 135
D-1000 Berlin 12, WEST GERMANY

Prof. James Thomas
Mathematics Dept.
Cole Stte University
Ft. Collins, Colo.  80523

Mr. P. Denis Thomas
Dept. 52-33, Bldg. 201
Lockheed Palo Alto Research Laboratory
3251 Hanover St.
Palo Alto, CA.  94304

Mr. Scott Thomas
M. S. N202A-14
NASA-Ames Research Center
Moffett Field, CA  94035

Prof. William T. Thompkins, Jr.
M. I. T., 37-375
77 Massachusetts Ave.
Cambridge, MA  02139

Mr. Jeffry Tollinger
Advanced Computation Laboratory
Technology Development Corp.
1095 E. Duane
Sunnyvale, CA  94086

Mr. Nelson Tunstall
Nuclear Computer Sciences Div.
Union Carbide
P. O. Box P, MS-53
Oak Ridge, Tennessee  37830

Mr. Marius Ungarish
Computer Science Dept.
Technion
Haifa, ISRAEL

Mr. William Van Dalsem
M. S. 202A-14
NASA-Ames Research Center
Moffett Field, CA  94035

Prof. Adrian I. van de Vooren
Mathematisch Instituut
University of Groningen
P. O. Box 800
9700 AV Groningen, THE NETHERLANDS

Prof. Milton Van Dyke
Applied Mechanics Div.
Stanford University
Stanford, CA  94305

Dr. Bram van Leer
ICASE, M. S. 132C
NASA-Langley Research Center
Hampton, VA  23665

Mr. E. Venkatapathy
M. S. 229-1
NASA-Ames Research Center
Moffett Field, CA  94035

Mr. John Viegas
M. S. 229-1
NASA-Ames Research Center
Moffett Field, CA  94035

Mr. Alain Viault
CISI
BP.n°24
91190 GIF-s/-Yvette, FRANCE

Mr. Peter Vinella
6395 Colby St.
Oakland, CA  94618

Dr. Marcel Vinokur
919 Channing Avenue
Palo Alto, CA  94301

Dr. Henry Viviand
Head, Theor. Aerodynamics Div.
ONERA
92320 Chatillon, France

Dr. Giuseppe Volpe
Grumman Aerospace Corp.
Bethpage, N. Y.  11714

Prof. Ru-quan Wang
Computing Center of Academia Sinica
Beijing, CHINA

Mr. Sang Wang
JAYCOR
1401 Camino Del Mar
Del Mar, CA  92014

Mr. Robert Warming
M. S. 202A-1
NASA-Ames Research Center
Moffett Field, CA  94035

Mr. Val Watson
M. S. 202A-1
NASA-Ames Research Center
Moffett Field, CA  94035

Dr. John Weiss
3638 Pocahontas Ct., Apt. A
San Diego, CA  92117

Prof. Pieter Wesseling
Dept. of Mathematics
Delft University of Technology
Julianalaan 132
2628 BL  Delft, THE NETHERLANDS

Dr. Robert Whitehead
Office of Naval Research
800 N. Quincy St.
Arlington, VA  22217

Dr. George F. Widhopf
Aerospace Corp.
P. O. Box 92957
Los Angeles, CA  90009

Dr. Graham Wilks
Dept. of Mathematics
University of Strathclyde
Glasgo SCOTLAND

Dr. Karl Heinz Winkler
MPI für  Astrophysik
D-8046 Garching 6
München, WEST GERMANY

Mr. Calvin Wolf
Acurex Corp.
485 Clyde Ave.
Mountain View, CA  94042

Mr. Paul R. Woodward
Lawrence Livermore National Lab
P. O. Box 808
Livermore, CA  94550

Mr. Alan Wray
M. S. 202A-1
NASA-Ames Research Center
Moffett Field, CA  94035

Mr. Nickolay Yanenko
Inst. of Theo. & Appl. Mechanics
USSR Academia of Sciences
Novosibirsk  630090,  USSR

Ms. Helen Yee
M. S. 202A-1
NASA-Ames Research Center
Moffett Field, CA  94035

Prof. S. M. Yen
Aero. & Astro. Engrg. Dept.
University of Illinois
Urbana, Ill.  61801

Dr. Woon-Shing Yeung
Department of Mechanical Engrg.
University of Lowell
1 University Avenue
Lowell, MA  01854

Mr. Kenneth Yoshikawa
M. S. 229-4
NASA-Ames Research Center
Moffett Field, CA  94035

Dr. Chi Yuan
SST: 245-3
NASA-Ames Research Center
Moffett Field, CA  94035

Mr. Steven T. Zalesak
Code 4780
Naval Research Laboratory
Washington, D. C.  20375

Prof. Pieter F. Zandburgen
Technische Hogeschool Twente
P. O. Box 217
Enschede, THE NETHERLANDS

Prof. Thomas Zang
Dept. of Mathematics & Comp. Sci.
College of William & Mary
Williamsburg, VA  23185

Prof. Luca Zannetti
Politecnico di Torino-Italy
C.so Duca Delli Abruzzi 24
Torino, ITALY

Prof. Yu-lan Zhu
Computing Center of Academia Sinica
Zhongguancun, Beijing
People's Republic of China

# Turbulent Shear Flows I

Selected Papers from the First International Symposium on Turbulent Shear Flows, The Pennsylvania State University, University Park, Pennsylvania, USA, April 18–20, 1977

Editors: F. Durst, B. E. Launder, F. W. Schmidt, J. H. Whitelaw

1979. 256 figures, 4 tables. VI, 415 pages
ISBN 3-540-09041-X

**Contents:** Free Flows. – Wall Flows. – Recirculating Flows. – Developments in Reynolds Stress Closures. – New Directions in Modeling.

# Turbulent Shear Flows II

Selected Papers from the Second International Symposium on Turbulent Shear Flows, Imperial College London, July 2–4, 1979

Editors: L. J. S. Bradbury, F. Durst, B. E. Launder, F. W. Schmidt, J. H. Whitelaw

1980. 310 figures, 12 tables. IX, 391 pages
ISBN 3-540-10067-9

**Contents:** Turbulence Models. – Wall Flows. – Complex Flows. – Coherent Structures. – Environmental Flows. – Index of Contributors.

Springer-Verlag
Berlin
Heidelberg
New York

# Lecture Notes in Physics